Soil Liquefaction

A Critical State Approach

Second Edition

Applied Geotechnics Series

Series Editor: William Powrie,
University of Southampton, United Kingdom

PUBLISHED TITLES

David Muir Wood, *Geotechnical Modelling*
Hardback ISBN 978-0-415-34304-6 • Paperback ISBN 978-0-419-23730-3

Alun Thomas, *Sprayed Concrete Lined Tunnels*
Hardback • ISBN 978-0-415-36864-3

David Chapman *et al.*, *Introduction to Tunnel Construction*
Hardback ISBN 978-0-415-46841-1 • Paperback ISBN 978-0-415-46842-8

Catherine O'Sullivan, *Particulate Discrete Element Modelling*
Hardback • ISBN 978-0-415-49036-8

Steve Hencher, *Practical Engineering Geology*
Hardback ISBN 978-0-415-46908-1 • Paperback ISBN 978-0-415-46909-8

Martin Preene *et al.*, *Groundwater Lowering in Construction*
Hardback • ISBN 978-0-415-66837-8

Steve Hencher, *Practical Rock Mechanics*
Paperback • ISBN 978-1-4822-1726-1

Mike Jefferies *et al.*, *Soil Liquefaction, 2nd ed*
Hardback • ISBN 978-1-4822-1368-3

Paul F. McCombie *et al.*, *Drystone Retaining Walls: Design, Construction and Assessment.*
Hardback • ISBN 978-1-4822-5088-6

FORTHCOMING

Zixin Zhang *et al.*, *Fundamentals of Shield Tunnelling*
Hardback • ISBN 978-0-415-53597-7

Christoph Gaudin *et al.*, *Centrifuge Modelling in Geotechnics*
Hardback • ISBN 978-0-415-52224-3

Kevin Stone *et al.*, *Weak Rock Engineering Geology and Geotechnics*
Hardback • ISBN 978-0-415-56071-9

Soil Liquefaction

A Critical State Approach

Second Edition

Mike Jefferies
Golder Associates (UK), Canada

Ken Been
Golder Associates, Houston, USA

CRC Press
Taylor & Francis Group
Boca Raton London New York

CRC Press is an imprint of the
Taylor & Francis Group, an **informa** business
A SPON PRESS BOOK

CRC Press
Taylor & Francis Group
6000 Broken Sound Parkway NW, Suite 300
Boca Raton, FL 33487-2742

First issued in paperback 2019

ISBN-13: 978-1-4822-1368-3 (hbk)
ISBN-13: 978-0-367-87340-0 (pbk)

Visit the Taylor & Francis Web site at
http://www.taylorandfrancis.com

and the CRC Press Web site at
http://www.crcpress.com

Contents

Preface

Soil liquefaction is a phenomenon in which soil loses much of its strength or stiffness for a generally short time but nevertheless long enough for liquefaction to be the cause of many failures, with both deaths and some very large financial losses. Unsurprisingly, there is a vast literature on liquefaction. Apart from a sustained set of contributions to the usual journals over the past 30 years or so, there have been publications by university departments specializing in the subject, specialty conferences, theme sessions at geomechanics conferences, state-of-the-art papers, two Rankine lectures, the Canadian Liquefaction Experiment (Canlex), a book by the National Research Council (United States) and research competitions sponsored by the National Science Foundation to test theories and modelling techniques (VELACS, in particular).

So where does this book fit in? The first edition was far from a balanced view of the literature 10 years ago, the first edition taking the view that liquefaction was simply a particular aspect within a wider spectrum of soil behaviour. Significantly, it was a computable aspect with details of stress–strain behaviour being readily computed in a spreadsheet using conventional soil properties. Necessarily, dealing with liquefaction as a computable behaviour required critical state soil mechanics and a generalized version of CSSM, which is not as difficult as it might seem. The first edition was well received, quickly garnering extensive citations and shifting the discussion away from the *engineering geology* view that so underpinned North American liquefaction practice at the time. This edition extends and builds on the first edition in three main ways.

The first enhancement is the treatment of cyclic mobility. There are now more data available, both laboratory data and with most of the case history record becoming public domain rather than proprietary.

The second enhancement is computational. While the first edition was based on constitutive modelling, it was mostly at the *element* level of laboratory tests. Given our acceptance of the doctrine *if you cannot compute, then you have got nothing*, this second edition now extends from laboratory tests to finite element analyses with a new chapter (Chapter 8) and a new appendix (Appendix D). This new material is usable in engineering practice, at least for those willing to roll up their sleeves and learn how, as it is based on an extension of public-domain finite element software. It is not as convenient as a user-defined model in a geotechnical modelling platform, but vastly better for research students who are our target readers for this new material. Much better analyses of the case history record are needed, as there is still a mismatch between the laboratory and full-scale experience. There are reasonable grounds for thinking that the causes of the mismatch come down to local pore water migration and stochastic variation in void ratio, but fairly sophisticated finite element analyses are needed to investigate these effects and the code offered is a starting point for such work.

The third enhancement is application in practice. We have given a number of talks based on the first edition, and it has become apparent that while the first edition was a 'pretty good text on constitutive modelling' (quote from an international conference), mathematical rigour was less useful to consulting engineers. So we have added a new chapter (Chapter 9) to assist with bringing the critical state approach into engineering practice.

Of course, there are other additions such as tabulation of more laboratory tests on silt materials, CPT calibration chamber data and a new appendix (Appendix H) to convince you that we are continuing a 125-year thread of understanding soil behaviour (and with Original Cam Clay being a special case of this more general view).

We have also removed some material from the first edition, mostly where issues have been clarified and a 'on one hand this, on the other that' is no longer needed. There is now a settled view.

As the aim of this book is to help and not to intimidate, the first edition made the files used to create plots downloadable from the Internet, including the supporting raw data. We have gone further for this second edition, both with better cross-referencing to data/routines in the text and a lot more being made downloadable. There are now more programs, and these programs are better documented with comments back to the source equations in this book. How to download is presented after the Acknowledgements.

We also draw your attention to the appendices, which contain detailed derivations that you will not find in published papers (and which textbooks do not usually explain either). We trust that there is no equation in this book that appears without an appropriate derivation. The appendices also document testing procedures, give the entire database of cone chamber calibration tests and have details on interesting case histories as well as records of difficult-to-find information. These appendices are a substantive part of the book.

Mike Jefferies, PEng
Consulting Engineer

Ken Been, PEng
Principal
Golder Associates Ltd.
Burnaby, British Columbia, Canada

Acknowledgements

Those of you who have read some of our papers will recognize that we owe a great deal to the Canadian offshore oil industry and Golder Associates. Much of what is presented in this book was developed on construction projects for Gulf Canada Resources, Esso Resources Canada and Dome Petroleum. We thank in particular John Hnatiuk, Bill Livingstone, Brian Wright, Brian Rogers, Chris Graham, David James and Sanjay Shinde. Without the support from these gentlemen, much of what you will find in this book would not exist.

We also owe an enormous debt to our colleagues in Golder Associates who have both encouraged and humoured us. Our co-authors are acknowledged by reference to the various joint papers, and we thank in particular Jack Crooks, Brian Conlin, John Cunning, Dave Horsfield and Joe Hachey.

A feature of this second edition is that we have coerced some people into more active contributions. Chapter 8 and Appendix D were authored by Dawn Shuttle who developed the software discussed. Roberto Olivera expanded the original Appendix B into the new version here with its far greater detail on testing methods.

We also owe a debt to Robb Moss, who shared the case history data files developed as part of his research at Berkeley. Putting this information into the public domain was a great contribution to geotechnical engineering and sharing the results with us in a ready-to-use form greatly assisted with Chapter 7.

Many friends and acquaintances not already mentioned also provided useful case history and laboratory data over the years. At the risk of missing a few names, we acknowledge data provided by Mike Jamiolkowski (calibration chamber data), Ramon Verdugo (Toyoura sand test data), Mike Hicks and Radu Popescu (probabilistic analyses), Gonazalo Castro (Lower San Fernando data), Scott Olson (various CPT data), Professor Yoshimine (Jamuna Bridge data) and Howard Plewes (Sullivan Dam data). Crown copyright material is reproduced with the permission of the Controller of HMSO and the Queen's Printer for Scotland.

It is also appropriate that we acknowledge intellectual guidance over the years. Professor Bob Gibson first introduced one of us (MJ) to theoretical soil mechanics and has helped over the years. Other guidance and encouragement came from Professors Dan Drucker, Andrew Palmer and Peter Wroth. We are very appreciative of their time and interest. Hopefully, this book will not leave any of them disappointed.

Finally, both of us must thank our families for their forbearance. This book and the papers that led to it have taken many hours of our lives. Our families have humoured us by not complaining as we disappeared into our studies for many nights over the best part of the 30 years, but we expect to be summarily shot if we ever suggest or agree to a third edition.

Downloading soil liquefaction files

Downloadable files are an important feature of this book. They proved popular with the first edition, and we have enhanced and expanded what can be downloaded for this second edition. The background to these downloads is as follows.

A premise of this book is that soil liquefaction is simply a computable aspect of soil stress–strain behaviour requiring no new soil properties or concepts. All fine, but the methods to do this are nearly all numerical. And while we suggest that at least the simpler stuff (e.g. Cam Clay) be programmed in a spreadsheet by every reader (as this will show how simple the framework truly is), it gets a bit tedious to program things like cyclic simple shear from scratch. Thus, all the programs underlying this book are provided as commented and documented open-source code. Most programs run in the VBA environment lying behind the Excel spreadsheet, with the source code being visible (and modifiable) from within Excel. All you need to do is download the relevant *xls* file and the source code comes along with it. Other source code is Fortran90 and C++, both of which require a little more effort with compilers.

A further reason to download the soil modelling *xls* files is seeing the simulations come alive on a screen, which is way more interesting than looking at the plots only on paper, and it allows easy exploration of the effect of changing properties and state on soil response. Running simulations gives a feel for things. Plus, plot scales can be changed and so forth if you are interested in some detail that is not apparent at the normal scale.

A second premise of this book is that we want all readers to follow the Russian proverb 'trust is wonderful, but distrust is better'. Flippancy aside, this principle is becoming embedded in Codes of Practice, which place an obligation on engineers to validate their design data and methods (e.g. see EN1997-1 Clause 1.5.2.2 and Clause 2.1-7). Validation is much more than citing a particular paper for some aspect, as peer review and publication is in itself not a guarantee of adequacy (evident from instances of competing, different views being published concurrently across many areas of science, including geotechnical engineering). Validation requires that engineers dig into the data/information behind their methods to satisfy themselves that the data are relevant, and the method(s) appropriate, for the situation being considered. Accordingly, we have made our data public domain, with the various figures in this book being traceable back to the source data. The data also provide an archive source for those who wish to go further.

This sharing of knowledge is a *Golder Value* from the founding of the company, reflecting, in part, the older value of learned societies that dominated the progress in engineering from the start of the industrial revolution in the United Kingdom. Downloads at the time of writing are hosted by Golder Associates at www.golder.com/liq, but this is anticipated to change in the future. Golder has established a foundation for *Furthering Knowledge and Learning*; see www.golderfoundation.org (Golder Associates, 2015). It is expected that the downloads for this book will be transferred to the Golder Foundation as that sharing of knowledge is our intent. Do check the Golder Foundation website.

Given the speed at which the first edition sold out, and the citations in the literature to it, we expect this second edition to be in print long after we have retired. So we have established a second source of downloads via the publisher available at the CRC Press website: http://www.crcpress.com/product/isbn/9781482213683.

When downloading, please check the *readme* file as that provides the structure to what is where, as well as notes about the various files. Because much of the interesting data on liquefaction are hidden in hard-to-access files at universities, consultants and research organizations, we have also made available all sorts of source data that we have found useful.

All of the downloads are subject to Terms and Conditions which add up to *use at your own risk*. Hopefully, this will not be an issue for data. In the case of programs and source code, this is all released under the *GNU General Public License Version 2* published by the Free Software Foundation. This means you may freely copy, use and distribute the downloaded source code provided that it remains open source with its origin acknowledged.

Contributors

Robb Moss
Department of Civil and Environmental Engineering
California Polytechnic State University, San Luis Obispo
San Luis Obispo, California

Roberto Olivera
Golder Associates Ltd.
Burnaby, British Columbia, Canada

Dawn Shuttle
Consulting Engineer
Lincoln, United Kingdom

Chapter 1

Introduction

1.1 WHAT IS THIS BOOK ABOUT?

Soil liquefaction is a phenomenon in which soil loses much of its strength or stiffness for a generally short time but nevertheless long enough for liquefaction to be the cause of many failures, deaths and major financial losses. For example, the 1964 Niigata (Japan) earthquake caused damage for more than $1 billion and most of this damage was related to soil liquefaction. The Aberfan (Wales) colliery spoil slide was caused by liquefaction and killed 144 people (116 of whom were children) when it inundated a school. Liquefaction was involved in the abandonment of the Nerlerk (Canada) artificial island after more than $100 million had been spent on its construction. Liquefaction at Lower San Fernando Dam (California) required the immediate evacuation of 80,000 people living in its downstream. Liquefaction is an aspect of soil behaviour that occurs worldwide and is of considerable importance from both public safety and financial standpoints.

In terms of age of the subject, Ishihara in his Rankine Lecture (Ishihara, 1993) suggests that the term spontaneous liquefaction was coined by Terzaghi and Peck (1948). The subject is much older than that, however. Dutch engineers have been engineering against liquefaction for centuries in their efforts to protect their country from the sea. Koppejan et al. (1948) brought the problem of coastal flowslides to the soil mechanics fraternity at the Second International Conference in Rotterdam. In the last paragraph of their much-cited paper, they mention flowslides in the approach to a railway bridge near Weesp in 1918 triggered by vibrations from a passing train. They claim that this accident, with heavy casualties, was the immediate cause of the start of practical soil mechanics in the Netherlands.

At about the same time, Hazen (1918) reporting on the Calaveras Dam failure clearly recognized the phenomenon of liquefaction and the importance of pore pressures and effective stresses. If the files of the U.S. Corp of Engineers are consulted, one finds that Colonel Lyman densified fill for the Franklin Falls Dam (part of the Merrimack Valley Flood Control scheme) in the late 1930s specifically to ensure stability of the dam from liquefaction based on the concept of critical void ratio given in Casagrande (1936). Reading these files and reports is an enlightening experience as Casagrande discussed many topics relevant to the subject today, and Lyman's report (Lyman, 1938) of the Corps' engineering at Franklin Falls is a delight that historians of the subject will enjoy.

We will not attempt our own definition of liquefaction, or adopt anyone else's, beyond the first sentence of this book. Once it is accepted that liquefaction is a constitutive behaviour subject to the laws of physics, it becomes necessary to describe the mechanics mathematically and the polemic is irrelevant.

Liquefaction evaluation is only considered worthwhile if it may change an engineering decision. Testing and analysis of liquefaction potential is only undertaken in practice in the context of a particular project. As will be seen later, liquefaction is an intrinsically brittle process and the observational method must not be used. If liquefaction is a potential problem, it must be engineered away. Engineering to avoid liquefaction involves processes with often relatively high mobilization (start up) costs, but a low marginal cost of additional treatment. So if liquefaction may be a problem, then the construction costs to avoid it may show considerable independence from the potential degree of liquefaction or desired safety factor against liquefaction. Many practical problems simply become go/no-go decisions for near fixed price ground modification. In such situations, there is nothing to be gained from elaborate testing and analysis if it will not change the decision. As it turns out, many projects with possible liquefaction issues fall into this class.

One might mistakenly take such a decision-driven approach to be pragmatic and anti-theory. This is not the case and theory has a crucial role. More particularly, relevant theory must be much more than some vague liquefaction 'concept' or definition. A full constitutive model is used to predict not just when liquefaction occurs but also the evolution of pore pressures and strains. It is self-evident that liquefaction, in all its forms, is simply another facet of the constitutive behaviour of soil and as such can only be properly understood within a background of constitutive theory. Why such an elaborate demand for theory? There are two reasons.

Firstly, the literature abounds with liquefaction concepts developed on the basis of incomplete information or poorly formed theory. How are these many concepts to be distinguished? Which concepts are erroneous if applied to other situations because they missed crucial factors? Which concepts assume one form of behaviour which, while a reasonable approximation within any given limited experience, wrongly predicts outside its experience base? Which concepts are reasonable representations of the micro-mechanical processes actually happening between the soil particles? Demanding a full constitutive model which works for all testable stress paths, while not guaranteeing future adequacy (and there is nothing that can give such assurance), does at least ensure as good a job as possible and eliminates ideas that are intrinsically wrong.

Secondly, it is a fact of life that civil engineering is an activity with little opportunity for full-scale testing. Correspondingly, much experience is based on a few (relative to the total construction market) failures for which there is often limited and uncertain information. This necessity to deal with what might be called rare events distinguishes civil engineering from the many other branches of engineering in which it is feasible to test prototypes. Further, in geotechnical experience there are generally enough free (unknown) parameters to fit any theory to any failure case history. Correspondingly, there is enormous potential for misleading theories to be perceived as credible by the incautious. Only by using theories whose adequacy is established outside the case history can the profession limit its potential for being misled. It is not an overstatement that the profession can be misled. One of the empirical graphs in common usage is a plot of residual shear strength after liquefaction against SPT blow count (N value), first proposed by Seed (1987). This graph has been reproduced many times with additional data since then but the graph is fundamentally wrong – it ignores the initial stress level on one axis but uses a pressure normalized penetration resistance on the other axis. The implied relationship for post-liquefaction strength is dimensionally inconsistent, and the chart cannot possibly be useful in a predictive sense.

In effect, the approach in this book demands that the profession's view of liquefaction should pass what could be called a variation of the Turing test: 'if it does not compute, then you have nothing'. Turing, 1912–1954, a mathematician and logician, pioneered in computer theory and logical analyses of computer processes. Turing computability is the property of being calculable on a Turing machine, a theoretical computer that is not subject to malfunction or storage space limits, that is, the ultimate PC. There should be no interest in non-computable 'concepts' and liquefaction dogma. The approach to liquefaction presented in this book meets this criterion of computability.

Surely, one might ask, if mechanics is so good then why not rely on it? The answer is that there are factors such as time, scale effects, pore water migration, strain localization and soil variability that are routinely neglected in most theories, testing and design methods but these factors are real and important. Emphasizing practical experience, properly set in a plasticity framework, develops an understanding of what is formally known as model uncertainty.

So what this book provides is a mathematically and physically consistent view of an important subject with a strong bias to real soils and decisions that must be faced in engineering practice. Because a lot of the material may seem intimidating, derivations are included in detail so that the origin of the equations is apparent. Samples of source code are available from a website so that the reader can see how complex-looking differentials actually have pretty simple form. The source data are provided as downloadable files, so that this book can be used as a tutorial. The book is also quite a bit more than a compendium of papers and many ideas have evolved since first publication.

1.2 A CRITICAL STATE APPROACH

This book is sub-titled *A Critical State Approach* for a particular reason. Density affects the behaviour of all soils – crudely, dense soils are strong and dilatant, loose soils weak and compressible. Now, as any particular soil can exist across a wide range of densities it is unreasonable to treat any particular density as having its own properties. Rather, a framework is needed that explains why a particular density behaves in a particular way. The aim is to separate the description of soil into true properties that are invariant with density (e.g. critical friction angle) and measures of the soils state (e.g. current void ratio or density). Soil behaviour should then follow as a function of these properties and state.

The first theory offered that captured this ethos was what became known as critical state soil mechanics, popularized by Schofield and Wroth (1968) with the Cam Clay idealized theoretical model of soil. The term critical state derives from anchoring the theory to a particular condition of the soil, called the critical void ratio by Casagrande in 1936. The definition of the critical state will come later but for now just note that the critical state is the end state if the soils is deformed (sheared) continuously. The neat thing about the critical state, at least mathematically and philosophically, is that if the end state is known it then becomes simple to construct well-behaved models. You always know where you are going.

The need to have a model comes back to the earlier statement paraphrasing Turing – if you cannot compute, you have nothing. Computing needs a model. These days there are a number of appropriate models to choose, but the choice is more a matter of detail than fundamental. Given the philosophical view that the model should explain the effect of density on soil behaviour, it turns out that (to date) only models incorporating critical state concepts are available. So, one way or another, things get anchored to a critical state view once the requirement is invoked for computable behaviour with density-independent properties.

Before going much further, it is appropriate to acknowledge a related school of thinking. At about the same time as critical state soil mechanics was developing in England, workers in the United States, in particular Castro (1969) with guidance from Casagrande at Harvard, put forward the view that the critical state during rapid shearing was the end point and knowledge of this end point allowed the solution of most liquefaction problems. The critical state after rapid shearing was termed the *steady state*. On the face of it, this approach allowed exceedingly simple analysis of a complex problem – a post-liquefaction strength (the steady state) allows engineering of stability using straightforward undrained analysis.

Mathematically there is no difference between the definitions of steady and critical states and they are usually taken to be the same. So, does this book belong to the Steady State School? The answer is an emphatic no. The Steady State School does not provide a computable model or theory. In contrast, this book offers a constitutive model in accordance with established plasticity theory that computes the details of strains and pore pressures during liquefaction. Critical state theory, being formulated under the framework of theoretical plasticity, insists on consistent physics and mathematics. However, it is also true that many of the ideas incorporated in critical state soil mechanics owe as much to Cambridge, Massachusetts, as Cambridge, England. The similarities and differences will be discussed in Chapters 2 and 3, as these aspects are interesting both historically and intellectually.

The basic approach is to anchor everything to the state parameter, ψ, defined in Figure 1.1. The state parameter is simply the void ratio difference between the current state of the soil and the critical state at the same mean stress. The critical state void ratio varies with mean effective stress and is usually referred to as the critical state locus (CSL). Dense soils have negative ψ, and loose contractive soils have positive ψ. Soil constitutive behaviour is related to ψ, and liquefaction behaviour is no different from other aspects of stress–strain response.

In summary, a critical state approach and associated generalized constitutive model (*NorSand*) provide a simple computable model that captures the salient aspects of liquefaction in all its forms. This critical state approach is easy to understand, is characterized by a simple state parameter (ψ) with a few material properties (which can be determined on reconstituted samples) and lends itself to all soils.

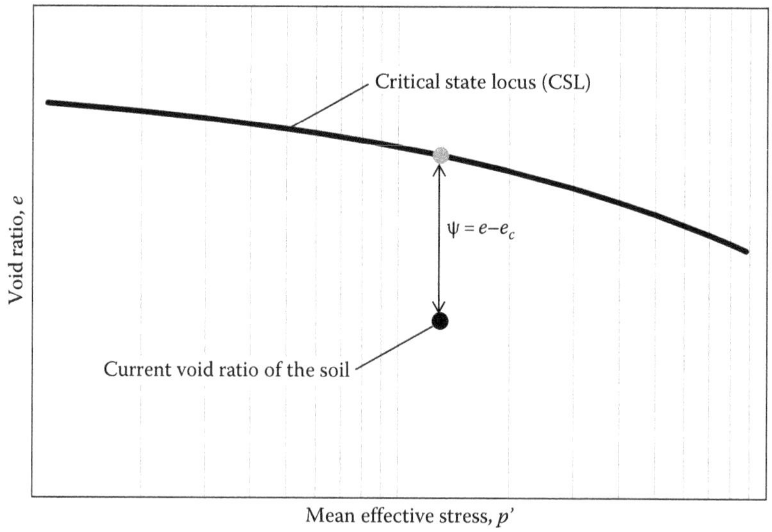

Figure 1.1 Definition of state parameter ψ.

1.3 EXPERIENCE OF LIQUEFACTION

In writing about a subject like liquefaction there are always questions as to how much background and experience to present before introducing the framework in which this information is to be assessed. On the one hand, giving all the theory first does not lead to the easiest book to read as there is no context for the theory. On the other hand, all information without any framework leads to confusion. However, given the premise that only full-scale field experience can provide verification of models, a good place to start is to describe a variety of liquefaction experiences. That comprises the remainder of this chapter. What liquefaction is, places it occurred and under what circumstances are all illustrated by examples. This also gives an appreciation for the history and economic consequences. However, this chapter largely stays away from numbers – it is intended to give a feel for the subject, no more.

1.3.1 Static liquefaction of sands: (1) Fort Peck Dam

Fort Peck Dam is a classic example of a static liquefaction failure. Dam construction was started in 1934 on the Missouri River in Montana, about 70 miles south of the Canadian border. The hydraulic fill method was used with four electrically operated dredges assembled at the site. River sands and fine grained alluvial soils were pumped and discharged from pipelines along the outside edges of the fill, thus forming beaches sloping towards the central core pool. The resulting gradation of deposited material was from the coarsest on the outer edge of the fill to the finest in the core pool. The foundation consisted of alluvial sands, gravels and clays, underlain by Bearpaw shales containing bentonitic layers.

A large slide occurred in the upstream shell of the dam near the end of construction in 1938. At the time of failure, the dam was about 60 m high with an average slope of 4H:1V. The failure occurred over a 500 m section and was preceded by bulging over at least 12 h prior to the failure. At some time after these initial strains, a flowslide developed, with very large displacements (up to 450 m) and very flat (20H:1V) final slopes. About 7.5 million m^3 of material was involved in the failure and eight men lost their lives. The post-failure appearance was that of intact blocks in a mass of thoroughly disturbed material. There were zones between islands of intact material that appeared to be in a quick condition with sand boils evident. Figure 1.2 shows an aerial view of the Fort Peck Dam failure illustrating the nature of the slide and the great distance moved by the failed mass.

A history of Fort Peck Dam and its construction can be found at http://www.fortpeckdam.com (Sigmundstad, 2015), giving many engineering details and photographs of practical aspects and the characters involved. This website also records observations by the workmen on the dam leading up to the failure and it is well worth the time to visit. The Fort Peck story is fascinating as a history of a large civil engineering project from the depression era, as well as a humbling example of the practical consequences of liquefaction.

The Fort Peck Dam slide was investigated by a nine-man review board, whose members held diverse views about the cause of failure. A majority of the board concluded that the slide was caused by shear failure of the shale foundation and that 'the extent to which the slide progressed upstream may have been due, in some degree, to a partial liquefaction of the material in the slide'. The minority (including Casagrande) view was that 'liquefaction was triggered by shear failure in the shale, and that the great magnitude of the failure was principally due to liquefaction'. Interestingly, Casagrande (1975) reports that he was also forced to the conclusion that sand located below the critical void ratio line, as he had defined it in 1936, can also liquefy. (This aspect will be discussed later.) Studies by the Army Corp of Engineers both soon after the slide and during a re-evaluation of the stability of the dam

Figure 1.2 Aerial view of Fort Peck failure. (From U.S. Army Corp of Engineers, Report on the slide of a portion of the upstream face at Fort Dam. US Government Printing Office, Washington, DC, 1939.)

in 1976 (Marcuson and Krinitzsky, 1976) indicate that the relative density of the sand was probably about 45%–50%. This is not especially loose.

The Fort Peck Dam failure is important as it appears to have effectively put an end to the practice of hydraulic fill construction of water retention dams in the United States. After Fort Peck it became normal practice to compact sand fills in dams. Failure of the Calaveras Dam, which was also constructed by hydraulic fill, had been reported as early as 1918 by Hazen and been attributed to liquefaction (Hazen, 1918, 1920) so Fort Peck Dam was not a 'one-off' event.

1.3.2 Static liquefaction of sands: (2) Nerlerk berm

As if to emphasize that Fort Peck was not a one-off, a very similar failure arose through a similar basal extrusion mechanism nearly 50 years later (once the lessons of Fort Peck had been forgotten?). Oil exploration of the Canadian Beaufort Sea Shelf is constrained by the area being covered by ice for nine months of the year. This ice can move, and moving can cause large horizontal loads on structures. As the ice crushes, the loading has the nature of fluctuating and periodic force. The technology that developed for oil exploration in this region was to use caisson retained islands, in which a caisson is combined with sand fill. Ice forces are resisted by the weight of the sand fill, while the caisson minimizes the volume of sand and allows construction in the limited period of the summer open water season. Sandfill is typically hydraulically placed and usually undensified. Achieved density depends on the details of the hydraulic placement, but is usually a little to significantly denser than the critical state when clean (<3% silt) sands are used. Jefferies et al. (1988a) give the background to the development of the technology and details of some islands.

Nerlerk B-67 was to be an exploration well drilled in 45 m of water in the Canadian Beaufort Sea in the winter during 1983/1984. The platform was to be Dome Petroleum's

SSDC structure, founded on a 36 m high sand berm constructed on the seabed. Foundation conditions consisted of a 1–2 m thick veneer of soft Holocene clay underlain by dense sand. Construction of the berm started in 1982, using dredged sand fill from the distant Ukalerk borrow source brought to site in hopper dredges. This sand was bottom dumped in the central area of the berm, in an attempt to promote displacement of the clay layer outwards. Because of the large fill volumes required for an island of this height, the local seabed sand was also exploited by dredging and pumping it to site through a floating pipeline. This local borrow was a finer sand and was placed on the outer parts of the berm through various designs of discharge nozzle to maximize the side slopes obtained. About 3 million m^3 of material was placed in 1982 before the end of the offshore construction season.

Construction commenced again in July 1983, using only the local Nerlerk borrow and pipeline placement of fill through a new 'umbrella' discharge nozzle. Care was taken about placing the fill around the slopes of the island to maintain side slopes of 5H:1V. A week after restart of construction, on 20 July 1983, bathymetric surveys revealed that a significant part of the Nerlerk berm had disappeared with the berm still 10 m below its design level. Construction continued and more failures occurred. A total of six large mass failures of the Nerlerk berm were reported. Actually, the first failure occurred at the end of the 1982 construction season and was seemingly unrecognized at the time. Slide 5 in 1983 was triggered as part of an experiment to obtain a better understanding of the problem. The volume of sand fill involved in each failure is enormous.

A typical cross-section through the Nerlerk berm at the time of the first failure in 1983 is shown in Figure 1.3 (Been et al., 1987a), while Figure 1.4 shows a plan sketch of the failures reported by Sladen et al. (1985a) as well as the pre- and post-failure profiles at the location of Slide 3. The post-failure slopes were very low at the toe (about 1V:30H), but also quite steep in the back-scarp zone (about 1V:7H). The run out of the failed slopes appeared considerable, but not very well defined by the bathymetric surveys which were at the limit of their resolution. It appears that the Nerlerk berm liquefied under static loading conditions, and that it involved mainly the local Nerlerk sand placed through a pipeline with the denser bottom dumped Ukalerk sand in the centre of the berm largely unaffected.

Nerlerk failures sparked considerable interest and discussion amongst the Canadian operating companies regarding the rational design and safe performance of hydraulic fills and

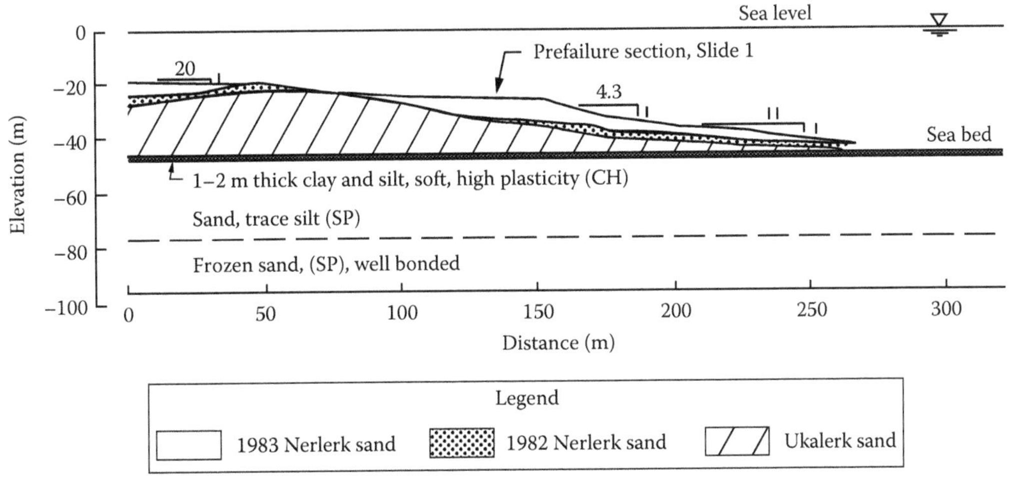

Figure 1.3 Nerlerk B-67 berm and foundation cross-section. (From Been, K. et al., *Can. Geotech. J.*, 24(1), 170, 1987a. With permission from the NRC, Ottawa, Ontario, Canada.)

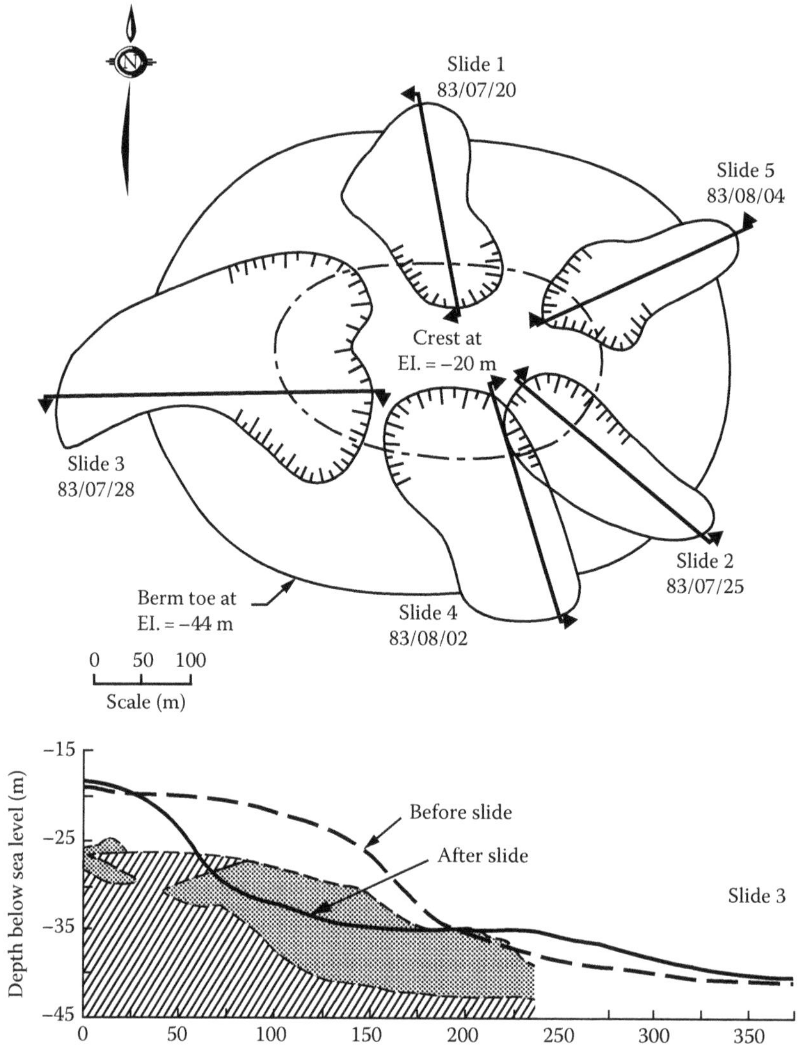

Figure 1.4 Plan of failures that occurred at Nerlerk B-67 and cross-section through Slide 3. (From Sladen, J.A. et al., *Can. Geotech. J.*, 22(4), 579, 1985a. With permission from the NRC, Ottawa, Ontario, Canada.)

caisson islands in the Beaufort Sea. Besides the estimated $100+ million cost of the Nerlerk failure, there were issues related to the feasibility of exploration from sand islands in deep water, pipeline placement of dredged fill, and whether compaction of fills would be necessary.

The undrained strength of the clay layer underlying the Nerlerk berm was estimated to be about 7 kPa at the time of the first failure in 1982. This strength is based on samples of the clay taken adjacent to the site in 1988, because no data were available prior to construction. It is assumed that no strength gain had occurred by the end of 1982 construction season. The strength at the time of failures in 1983 would likely have been greater as a result of consolidation under the loads imposed by the 1982 fill, although there is evidence that confined shear strains substantially delay consolidation and can even lead to rising pore pressures for a year or more (Becker et al., 1984). Assuming that consolidation was not delayed,

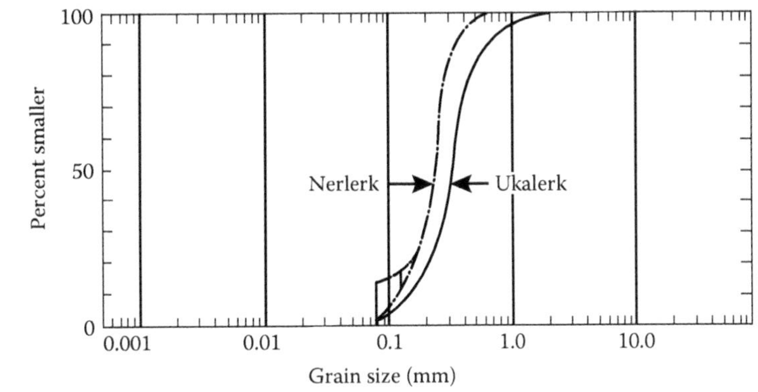

Figure 1.5 Grain-size distribution information for Nerlerk B-67 materials. (After Sladen, J.A. et al., *Can. Geotech. J.*, 22(4), 579, 1985a. With permission from the NRC, Ottawa, Ontario, Canada.)

undrained strengths were probably 20–25 kPa under the centre of the island and 12–15 kPa on average under the side slopes.

Typical grain-size distributions of Nerlerk and Ukalerk sand (Sladen et al., 1985a) are given in Figure 1.5. However, the sand was not uniform with the median grain size generally lying between 0.260 and 0.290 mm. The silt content was less than 2% in most cases. CPTs were routinely carried out in the Nerlerk berm as a method of QA during construction. Figure 1.6 shows a example CPT sounding, including a clearly identifiable clay layer at the

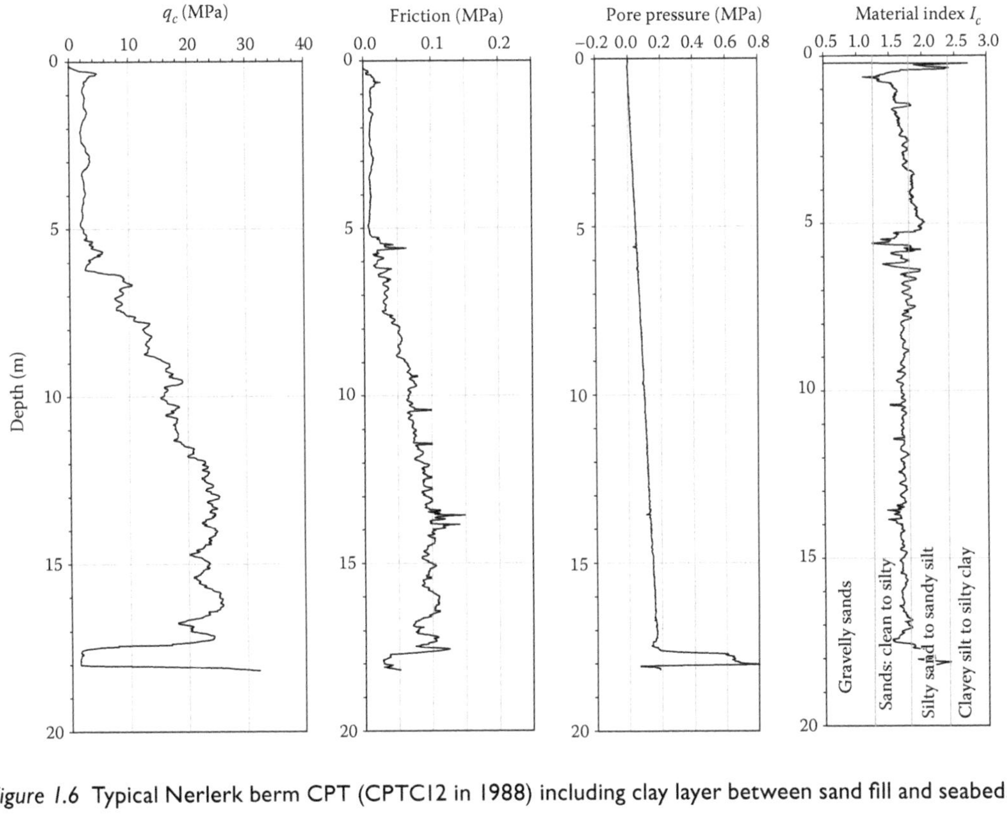

Figure 1.6 Typical Nerlerk berm CPT (CPTC12 in 1988) including clay layer between sand fill and seabed.

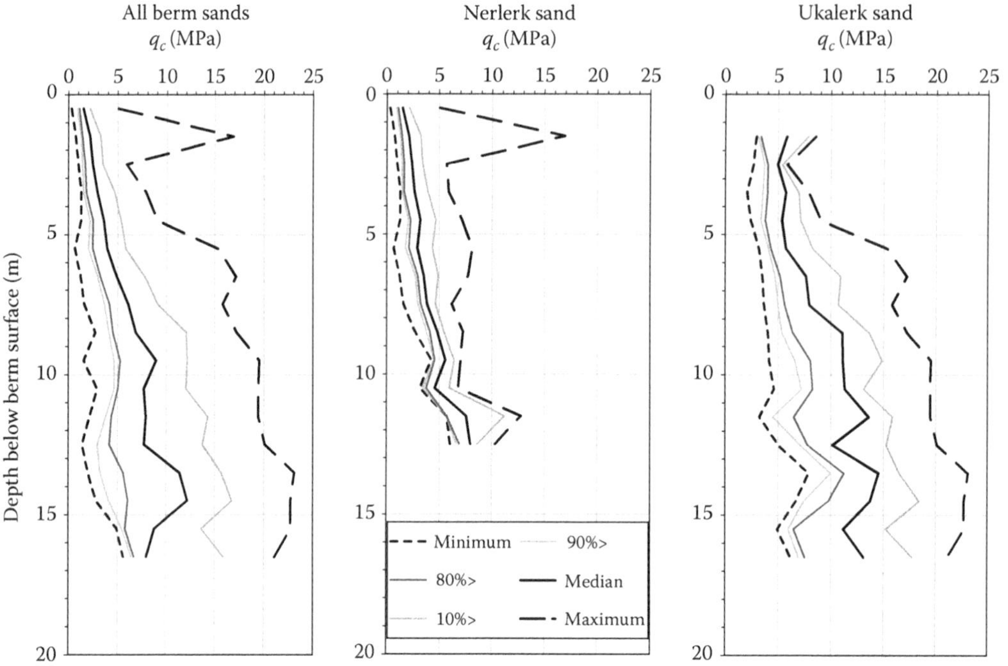

Figure 1.7 Summary of CPT distributions in Nerlerk B-67 berm, in Nerlerk sand and Ukalerk sand.

base of the sand fill. Figure 1.7 is a statistical summary of CPTs in the Nerlerk berm. The first graph shows the range, median and selected percentile values of the CPT tip resistance (q_c) for all tests in the Nerlerk berm based on values in depth intervals of 1 m below surface of the berm. The 80% and 90% greater-than lines are close together, which is significant when characteristic values are discussed in Chapter 5. It is also interesting that the maximum value is much higher than the 10% greater than line. Based on the construction records (regular bathymetric surveys were undertaken during fill placement), it is possible to identify which CPTs are in Nerlerk sand only, in mixed Nerlerk and Ukalerk sand, or in Ukalerk sand only. Quite clearly the Nerlerk sand is looser than the berm sand in general as shown in Figure 1.7.

Given these data, there was much discussion on the exact nature of the Nerlerk berm failures (in particular, Been et al., 1987a; Sladen et al., 1987). In summary, the Nerlerk engineers, as reported by Sladen et al. (1985a), worked from the morphology of the slides as defined by the bathymetry, and back-analysed the failures assuming the failures were static liquefaction and only in the Nerlerk sand. They concluded that the slides were caused by liquefaction of the fill which was in a very loose state, triggered by essentially static loading (i.e. placement of additional fill). Their view, which was a consequence of the adopted scenario, was that the sand fill had to be extremely loose ($\psi \sim +0.1$). This requirement of very loose in-situ sand state was largely dominated by taking the toe slope as characteristic of the residual stability angle and neglecting any influence of the underlying soft clay. The view requires abandoning CPT calibration chamber test data as misleading, and raises quite a dichotomy in understanding soil strength from the CPT.

The contrasting view was that, based on the then best CPT interpretation in Beaufort Sea sands, most of the sand appeared to be at a state parameter of $\psi < -0.03$, which is marginally dilatant and would not give flowslide-like behaviour under normal compression load paths in a triaxial test. Been et al. (1987a) considered other failure modes in addition to static liquefaction, including failure through the underlying clay, instability due to construction pore pressures in the sedimenting sand fill and combinations of these mechanisms. Failure mechanisms involving the soft clay were always more likely than those involving the Nerlerk sand alone, whatever properties were taken for the Nerlerk sand.

Notice the parallel of Nerlerk with Fort Peck. There was a similar situation of an apparently dense-enough fill liquefying, the presence of a basal weak soil leading to lateral strains in the fill, and a dispute as to what happened. Nerlerk is nevertheless interesting as it has much more measured data than Fort Peck which allows for more thorough analysis. Nerlerk, however, was under water so there are no photographs and even the morphology of the slides is speculative as it is an interpretation of imprecise bathymetric surveys. Nerlerk continues to be studied (e.g. Hicks and Onisiphorou, 2005) and will be revisited in detail later, but for now the following key issues should be kept in mind:

- CPT interpretation in terms of state parameter in 1983/1984 was missing the key ingredient of elastic shear modulus, which resulted in an apparent stress level bias in the interpretation.
- For various reasons, there was reluctance to bring the underlying clay layer into the equation and the effect that strains in the foundation may have on the berm fill.
- The dangerous nature of declining mean-stress paths in terms of liquefaction behaviour, caused by basal extrusion, was not understood in 1983.
- Variability of density of fills, and its impact on behaviour, was not recognized.

These issues are explored in Chapter 4 (CPT interpretation), Chapter 5 (variability), and Chapter 6 (static liquefaction physics). Suffice it to say that the Nerlerk berm failures would be avoided today if the influence of the underlying soft clay and variability in density of sand fills were taken into account in the design. The Nerlerk berm could be built today (although not as designed in 1983) but in 1983 the project was doomed as evidenced by the fact that the slopes failed when there was still another 10 m of fill and the drilling structure to be added. After several slope failures, and with abundant evidence that minor slope flattening was not going to solve the situation, the project was abandoned.

1.3.3 Liquefaction in Niigata earthquake

Niigata is important because it was largely that earthquake, and the major earthquake in Alaska the same year, which raised the awareness amongst geotechnical engineers of earthquake-induced liquefactions and its catastrophic consequences. There have been many more examples since then: San Fernando Valley (1971), Haicheng (1975), Tangshan (1976), Imperial Valley (1979), Armenia (1988), Loma Prieto (1989), Turkey (1999), to name but a few. The National Information Service for Earthquake Engineering at the University of California, Berkeley, has an excellent website (http://nisee.berkeley.edu, University of California Berkeley, accessed 15 March 2015) with information on most major earthquakes.

The Niigata earthquake, on 16 June 1964, inflicted major damage on the city of Niigata on the west coast of Japan. The epicentre was about 35 miles north of the city (offshore) and the recorded magnitude was 7.3 on the Richter Scale.

Niigata lies on the banks of the Shinano River where it enters the sea. The city is underlain by about 30 m of fine alluvial sand. Damage due to the earthquake resulted mainly from liquefaction of the loose sand deposits in low-lying areas. An apartment building founded on the sands tilted about 80° because of bearing capacity failure in the liquefied ground and is now a frequently used illustration of the results of liquefaction, Figure 1.8.

Figure 1.8 Apartment building at Kawagishi-cho that rotated and settled because of foundation liquefaction in 1964 Niigata earthquake. (From Karl V. Steinbrugge Collection, Earthquake Engineering Research Center, University of California, Berkeley, CA.)

Underground structures such as septic tanks, storage tanks, sewage conduits and manholes floated upwards out of the ground. Sand flows and mud volcanoes ejected water shortly after the earthquake and were reported to continue for as much as 20 min after shaking had stopped. Sand deposits 20–30 cm thick covered much of the city, and five simply supported girders of the Showa Bridge across the river fell when pier foundation piles deflected because of lateral support loss from liquefaction.

It is interesting to look at the Niigata case record and examine two sites: one where liquefaction occurred and one where no liquefaction was observed. Ishihara and Koga (1981) provide just such a study, and much of what follows is their work. The Kawagishi-cho site, situated about 500 m north of the Shinano river, is representative of the area where ground damage due to liquefaction was most severe during the 1964 earthquake. The apartment block shown in Figure 1.8 was at this site. At a second test site, simply called the South Bank site, practically no damage occurred. These two sites are within 2 km of each other and lay in the same general area of severe damage (Figure 1.9), but exhibited sharply contrasting behaviour.

A summary of the soil profile, SPT N value and CPT resistance at the Kawagishi-cho site, is shown in Figure 1.10 (from Ishihara and Koga, 1981). Similar data were also obtained for the South Bank site as shown in Figure 1.11. At the Kawagishi-cho site, the CPT resistance is typically around 5 MPa down to a depth of 12.5 m (and the corresponding N value is around 10). In the fine sand at the South Bank site, the CPT resistance is between 15 and 20 MPa while the N value is generally 27 or greater. Clearly, the sand at the South Bank site is very much denser than at the Kawagishi-cho site.

Much of the traditional empirical approach to liquefaction evaluation (Seed et al., 1983) is based on plotting the soil strength characterized by the penetration resistance against the

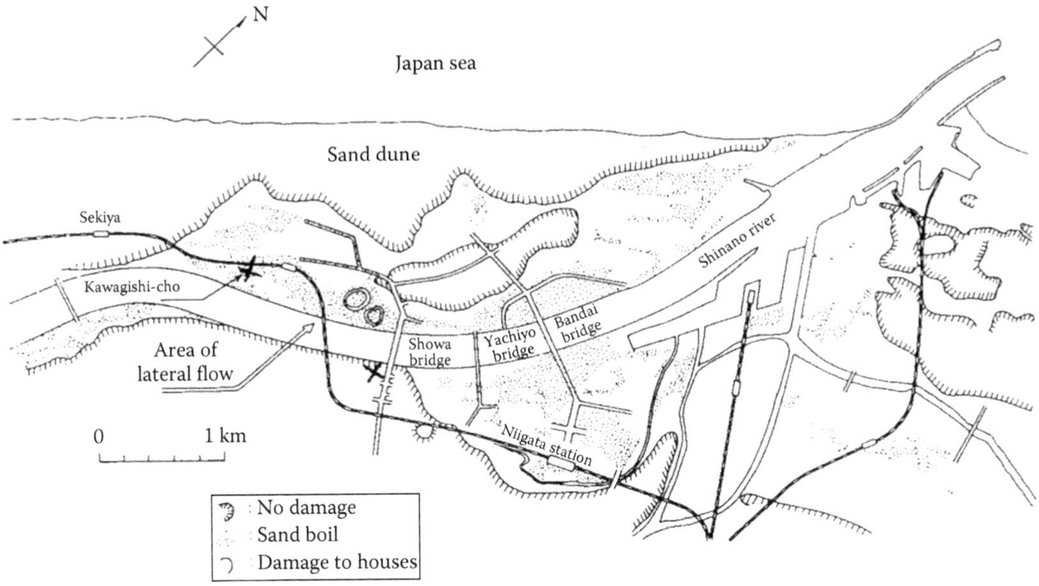

Figure 1.9 Sketch plan of Niigata, showing main area of damage. Kawagishi-cho and South Bank sites marked with X. (After Ishihara, K., *Géotechnique*, 43(3), 349, 1993. With permission from the Institution of Civil Engineers, London, U.K.)

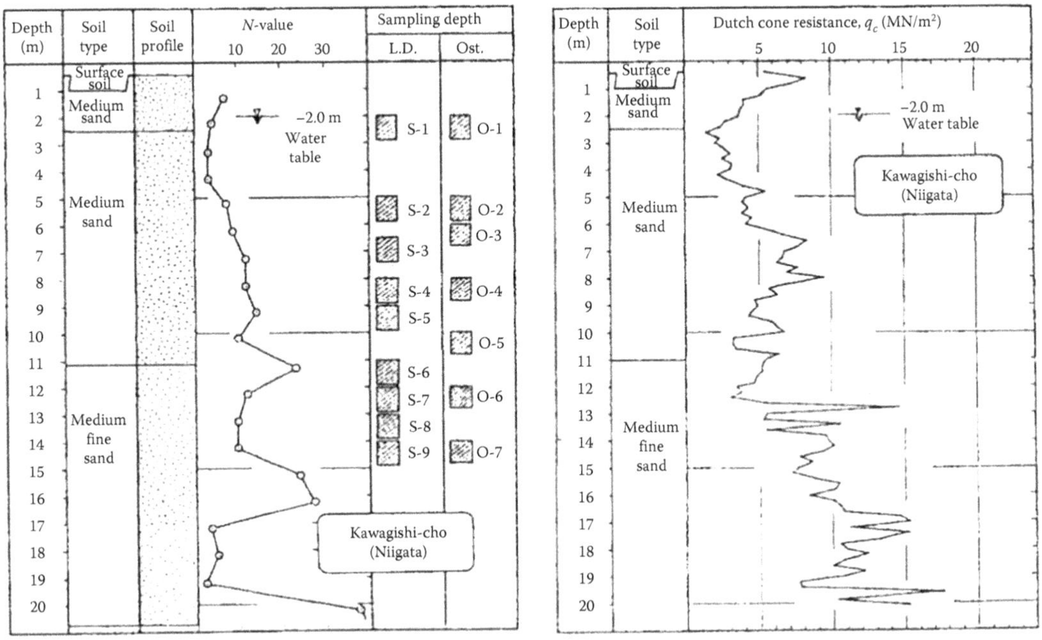

Figure 1.10 Soil profile and CPT resistance at Kawagishi-cho site. (From Ishihara, K. and Koga, Y., *Soils and Foundations*, 21(3), 35, 1981. With permission from the Japanese Geotechnical Society, Tokyo, Japan.)

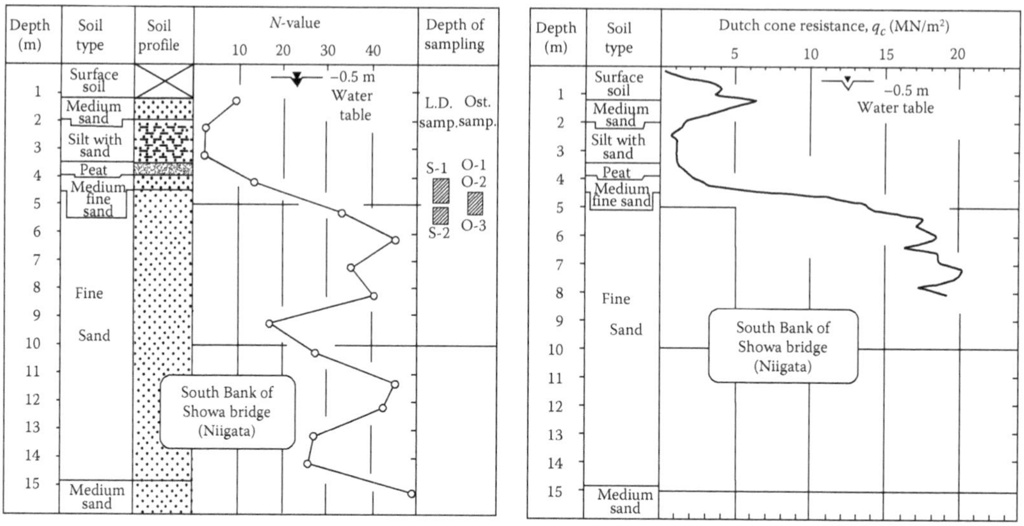

Figure 1.11 Soil profile and CPT resistance at South Bank site. (From Ishihara, K. and Koga, Y., *Soils and Foundations*, 21(3), 35, 1981. With permission from the Japanese Geotechnical Society, Tokyo, Japan.)

applied earthquake loading (characterized as a cyclic stress ratio, CSR = τ_{cyc}/σ'_{vo}). Those sites where liquefaction occurred are then distinguished from those where no liquefaction was observed. Figure 1.12 shows the Seed liquefaction assessment chart which uses the SPT penetration resistance, adjusted for stress level and energy level, $(N_1)_{60}$, as the preferred measure of penetration resistance. The peak acceleration during the earthquake in the basement of the apartment building No 2 at Kawagishi was around 0.16 g, which corresponds to a cyclic stress ratio of about 0.19. At this CSR, an $(N_1)_{60}$ of about 18 would separate liquefaction from non-liquefaction behaviour. A similar procedure with the CPT resistance, after Stark and Olson (1995), puts the dividing line at q_{c1} ~ 11 MPa.

Without getting into the detail of 'adjustments' to the measured N values and q_c (dealt with in detail in Chapter 4), it is easy to see how Seed's chart was developed and that liquefaction is likely at Kawagishi-cho and unlikely at the South Bank site.

Seed's liquefaction assessment diagram has been added to and modified by many people since 1983, but largely has not changed in nature or location of the dividing line between liquefaction and non-liquefaction for clean sands. The additions have largely been more data and looking at soils other than clean sands. Seed's approach is in essence a geological classification scheme, taking minimal account of soil properties and treating the silt-sized fraction of the soil as a key index to anticipated behaviour for a given penetration resistance.

1.3.4 Post-earthquake liquefaction: Lower San Fernando Dam

By the late 1960s, a debate was developing in North America about the term liquefaction. In essence, two types of liquefaction were recognized: static liquefaction as occurred at Fort Peck Dam and liquefaction that was observed during earthquakes. Various terms such as cyclic mobility, cyclic liquefaction, cyclic softening and initial liquefaction were proposed for the latter phenomenon, and actual liquefaction or flow liquefaction for the static version. Needless to say, the subject becomes confusing and there is a third variation on the theme.

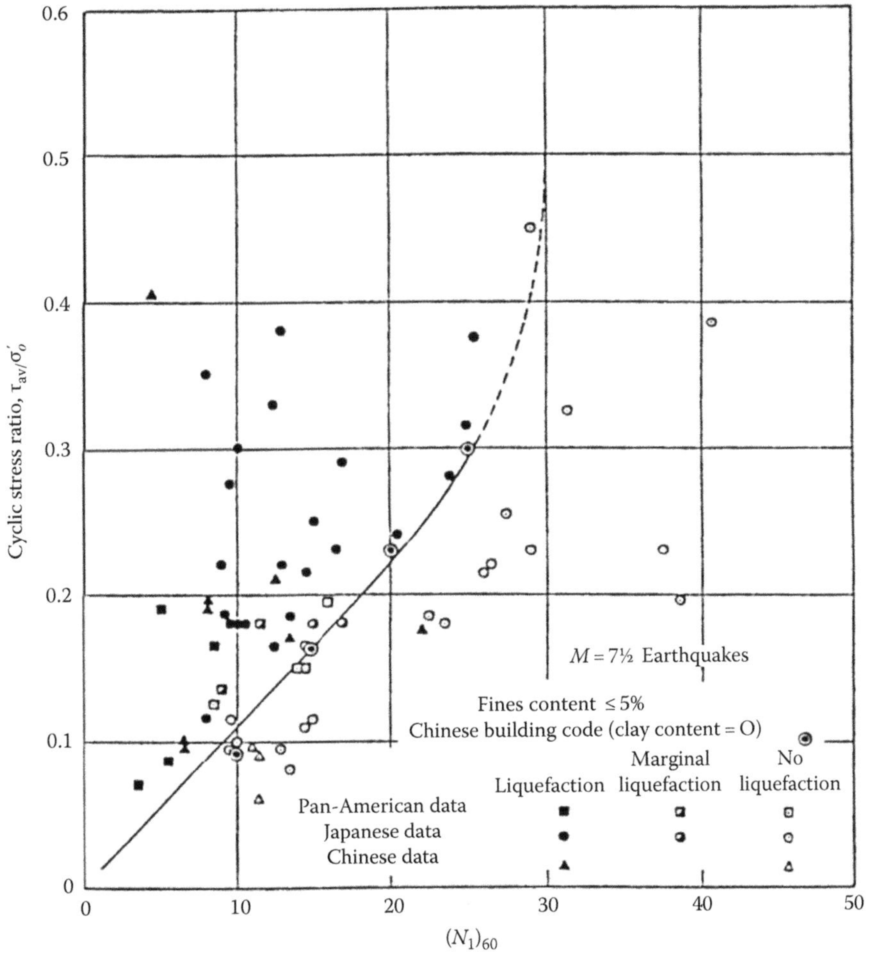

Figure 1.12 Seed liquefaction assessment chart. (From Seed, H.B. et al., *J. Geotech. Eng. Division*, 109(GT3), 458, 1983. With permission from the ASCE, Reston, VA.)

In 1971 the upstream slope of the Lower San Fernando Dam in California failed about a minute after the end of an earthquake. Here was a static failure as in Fort Peck Dam, but it is a result of pore pressures generated during an earthquake and without any earthquake-related inertial forces during the actual failure. The Lower San Fernando Dam failure is an excellent case history, well studied and researched, with much information readily accessible in the public domain.

The reservoir retained by the 43 m high Lower San Fernando Dam was the terminus of the main aqueduct system for Los Angeles, which supplied 80% of the city's water. Shaking during the 1971 San Fernando earthquake caused a slide of the top 30 ft of the dam, illustrated in two photographs in Figure 1.13. About 80,000 people lived in a 6 mile long area down the valley from the dam and were threatened by the very real possibility that the dam would fail completely, inundating the area by a catastrophic flood wave. Disaster was narrowly averted by drawing down the reservoir before the remnant of the crest gave way.

One significance of Lower Sand Fernando is the scale of the potential disaster and the public policy changes that then ensued. This 1971 event started a widespread evaluation of

(a)

(b)

Figure 1.13 Liquefaction failure of Lower San Fernando Dam after the 1971 earthquake. (a) Situation at the end of slide before reservoir drawdown and (b) details of failed slope revealed after reservoir drawn down. (Photographs from Karl V. Steinbrugge Collection, Earthquake Engineering Research Center, University of California, Berkeley, CA.) Note paved crest of dam descending into water in top photograph.

earth dam vulnerability to earthquakes within North America and many remedial works on dams then followed. Technically it is interesting, in that a salient feature of the dam's failure was essentially ignored, that is, that it was caused by migration of excess pore water pressure. It is only in 2003 that an analysis of this dam failure including pore pressure migration was published. This will be discussed at some length in Chapter 6.

1.3.5 Mine waste liquefaction: (I) Aberfan

So far liquefaction has been considered in its most common context of loose sands to silty sands. In such materials, with hard mainly quartzite grains, the residual state after liquefaction or during sliding can be identified with the critical state. In particular, continued shearing does not cause further changes in the shear stress and void ratio. Mine waste can be rather different, and mine waste is an important area for liquefaction. Two case histories illustrate some of the issues.

The first example is a liquefaction-induced flowslide of coal waste onto the village of Aberfan, South Wales, in 1966 in which 144 people lost their lives. Aspects of people and government-related issues about this disaster can be found at the website http://www.nuff.ox.ac.uk/politics/aberfan/home.htm (Johnes and Maclean, 2008), although this material mainly concentrates on the social aspects and neglects the uncertainty in liquefaction knowledge of the time.

The Aberfan colliery waste was tipped from rail trucks over the face of the tip (Tip 7) located on the side of a hill. The material was loose, but the triggering mechanism lay in the hydrogeology of the site which after heavy rain set up artesian pore pressure in the sandstone beneath the less permeable glacial deposits at the toe of the slope. Figure 1.14 shows the possible failure mechanism (after Bishop, 1973).

Tip 7 was about 67 m from toe to crest when the big slide occurred on 21 October 1966. When the first work crew reached the top of the tip at about 0730 h, the crest had dropped about 3 m over a distance of 10–13 m behind the edge. An hour later this settlement had increased to about 6 m and at about 0910 h the toe was observed to start moving forward. This movement continued for a few minutes before the rapid flow of material down the hillside began. The slide moved down the 12.5° slope for a distance of about 500 m to the junior school in Aberfan, which it largely destroyed, and continued for a further 100 m or so covering the road in the terminal area to a depth of 9 m. Of the 144 killed in the Aberfan slide, 116 were young children in the junior school at that time. The photograph in Figure 1.15 gives an idea of the nature of the failure. The Aberfan slide involved only about 107,000 m^3 of material but the destruction was caused by this material moving at between 10 and 20 miles per hour.

Density measurements in the unfailed part of the tip indicated a few very low values (1.5–1.7 kN/m^3 at depths of 13–28 m). This material was clearly capable of a very large decrease in volume as a result of shear displacements, typical of liquefaction and pre-requisite for large undrained brittleness (Bishop, 1973). Material from the waste tip contained on

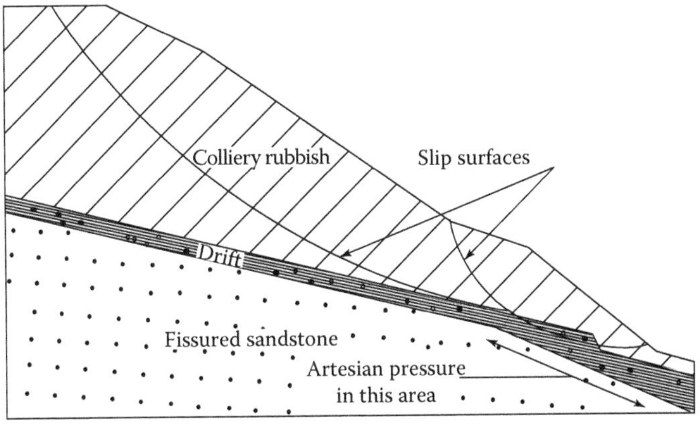

Figure 1.14 Possible failure mechanism for Aberfan Tip No 7. (From Bishop, A.W., *Quart. J. Eng. Geol.*, 6, 335, 1973. With permission from the Geological Society, London, U.K.)

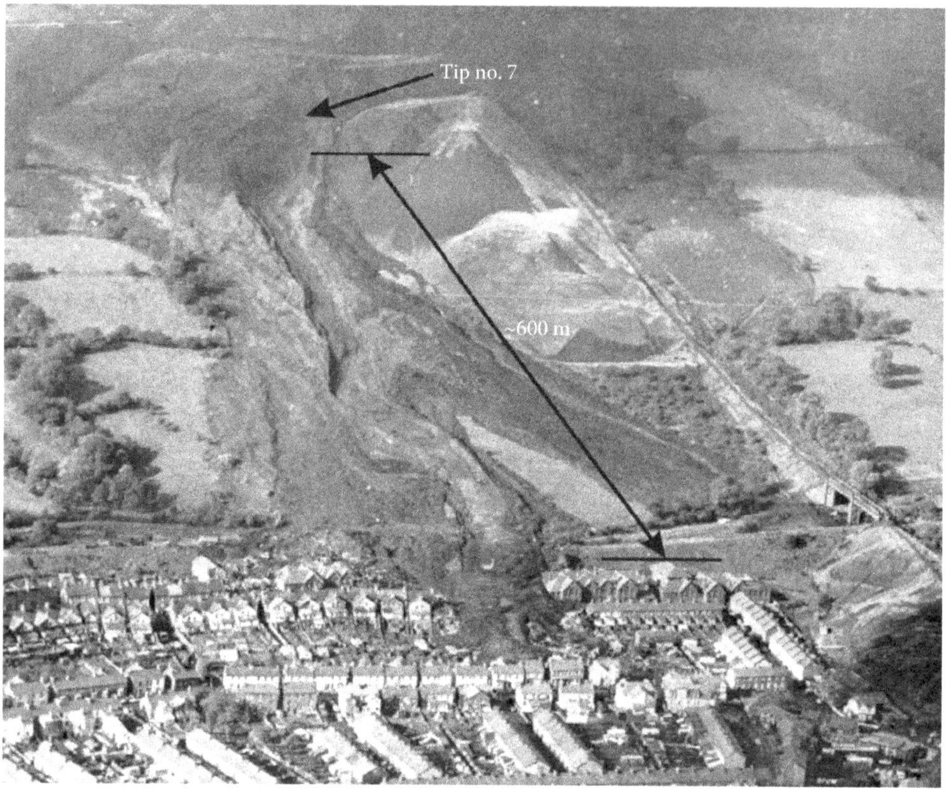

Figure 1.15 Aberfan flowslide shortly after the failure. (Reproduced from Welsh Office, A selection of technical reports submitted to the Aberfan Tribunal, Her Majesty's Stationary Office, London, U.K. With the permission of Controller of HMSO and the Queen's Printer for Scotland; Flowslide distance added by authors.)

average only 10% passing the 200 sieve giving a peak friction angle of $\varphi' = 39.5°$ in drained triaxial tests. Bishop (1973) reports that the flowslide material may also have degraded to a cohesive material, with a plasticity index of 16 and a residual value of $\varphi' \approx 18°$ on the slip surface where a displacement of at least 21 m was estimated.

1.3.6 Mine waste liquefaction: (2) Merriespruit tailings dam failure

The Merriespruit gold tailings dam failure in 1994 is an interesting case history in that much the same method of gold mine tailings disposal has been used in South Africa for decades, but Merriespruit was the first catastrophic flowslide occurrence. Fourie et al. (2001) argue that the tailings were in a very loose state in-situ and that overtopping and erosion of the impoundment wall exposed this material, resulting in static liquefaction of the tailings and the consequent flow failure.

The failure of the Merriespruit tailings dam occurred a few hours after a thunderstorm during which about 50 mm rainfall. About 600,000 m^3 of tailings flowed from the dam through the village and came to rest 3 km downstream of the dam. Figure 1.16 shows the destruction caused. The primary cause of the failure was overtopping, which resulted in large-scale removal of tailings from the slope face by water flowing over the crest for more than an hour. Removal of tailings from the outer slope would have exposed tailings inside

Figure 1.16 Aerial view of the Merriespruit tailings dam failure showing the path of the mudflow that occurred. (From Fourie, A.B. et al., *Can. Geotech. J.*, 38(4), 707, 2001. With permission from the NRC, Ottawa, Ontario, Canada.)

the dam that had previously been confined. Conventional wisdom in South Africa, however, would be that gold tailings are strongly dilatant and should not have moved a significant distance. The complex sequence of retrogressive failures postulated by Wagener et al. (1998) is shown in Figure 1.17.

Fourie et al. (2001) try to address the question of the nature and state of the impounded tailings that liquefied and flowed. They examined samples of the unfailed tailings material taken in Shelby tubes or block samples adjacent to the failure scarp. Firstly, it was apparent that there was no single particle size distribution, but a broad range with fines contents ranging from 40% to 100%. About 60% of the samples had fines content greater than 80%. Fourie and Papageorgiou (2001) tested a selection of samples to determine the critical state line (or steady-state line in their terminology) with results shown in Figure 1.18.

The distribution of in-situ void ratios is shown in Figure 1.19, but it is not straightforward to move from a knowledge of void ratios and CSL to the state parameter ψ as each sample for which the void ratio was measured does not usually match the grain-size distribution of one of the critical state lines. Although various interpolation schemes can be derived to map the CSL in terms of fines contents when there is a range of data available as at Merriespruit, it is usually simpler to use the CPT and infer ψ from such data rather than to directly measure void ratio and then use laboratory testing to estimate the CSL. Chapter 4 is all about using the CPT (and other in-situ tests) to determine ψ. Returning to Merriespruit, a simple comparison of Figures 1.18 and 1.19 shows that much of the tailings was in a loose state well above the critical state line, conditions conducive to brittle stress-stain behaviour and liquefaction.

Fourie et al. continue their discussion to make the point that liquefaction and flow of the tailings at Merriespruit are only part of the equation. A trigger mechanism had to develop before the tailings would liquefy. If the tailings impoundment had been operated according

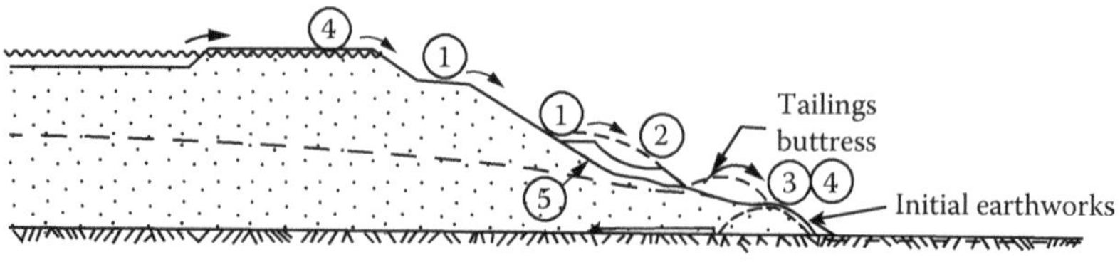

1. Berms overtop after thunderstorm
2. Loose tailings infill to earlier failures on lower slope erodes
3. Tailings buttress starts to fail
4. Pool commences overtopping and erodes slopes and buttress
5. Unstable lower slope fails and failed material is washed away

(a)

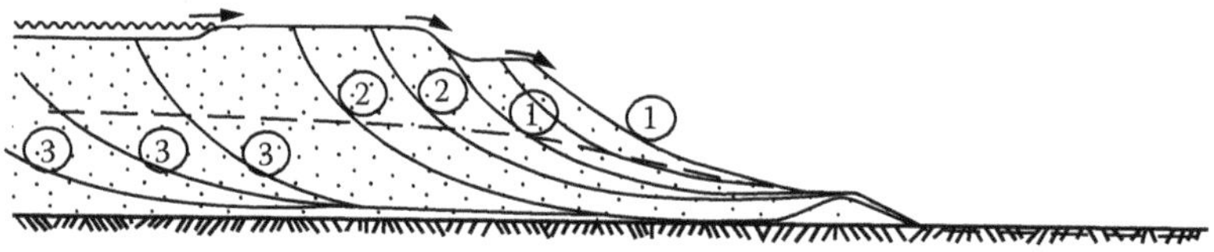

1. Lower slopes fail and are washed away
2. Domino effect of local slope failures which are washed or flow away
3. Major slope failures with massive flow of liquid tailings

(b)

Figure 1.17 Sequence of retrogressive failures of Merriespruit containment postulated by Wagener et al. (1998). (a) Critical section of north wall during early stages of failure and (b) critical section of north wall during failure. (From Fourie, A.B. et al., *Can. Geotech. J.*, 38(4), 707, 2001. With permission from the NRC, Ottawa, Ontario, Canada.)

to the regulations and the water levels properly managed, the impoundment would not have been breached and the liquefaction failure would not have occurred.

A further lesson from this failure is the dominant influence of water. If there is available water, then once a liquefaction flowslide occurs it can pick up the water and the resulting slurry can travel for, literally, miles.

1.3.7 High cycle loading

The case histories discussed so far are what might be regarded as conventional when looking at the liquefaction literature, in that they are well-known earthquake and static cases. Liquefaction can also occur in cyclic loading at much smaller cyclic stress levels if that loading goes on long enough. One situation like this is storm loading to offshore platforms with thousands of cycles in contrast to the ten or so dominant in most earthquakes. These long periods of cyclic loading can be viewed as high-cycle cases. An example of such a situation arose in another Canadian Beaufort Sea project, the Amauligak I-65 island of Gulf Canada Resources Ltd.

Gulf's caisson was a re-deployable bottom-founded caisson–type drilling unit called the Molikpaq (Figure 1.20). It was deployed at a site known as Amauligak I-65 in 1985, about 50 miles offshore in 30 m water depth. The core was filled using hydraulic placement of Erksak sand to an achieved state of better than $\psi < -0.05$. This sand was extensively tested, both in-situ and in the laboratory, and will feature a lot in the data presented later.

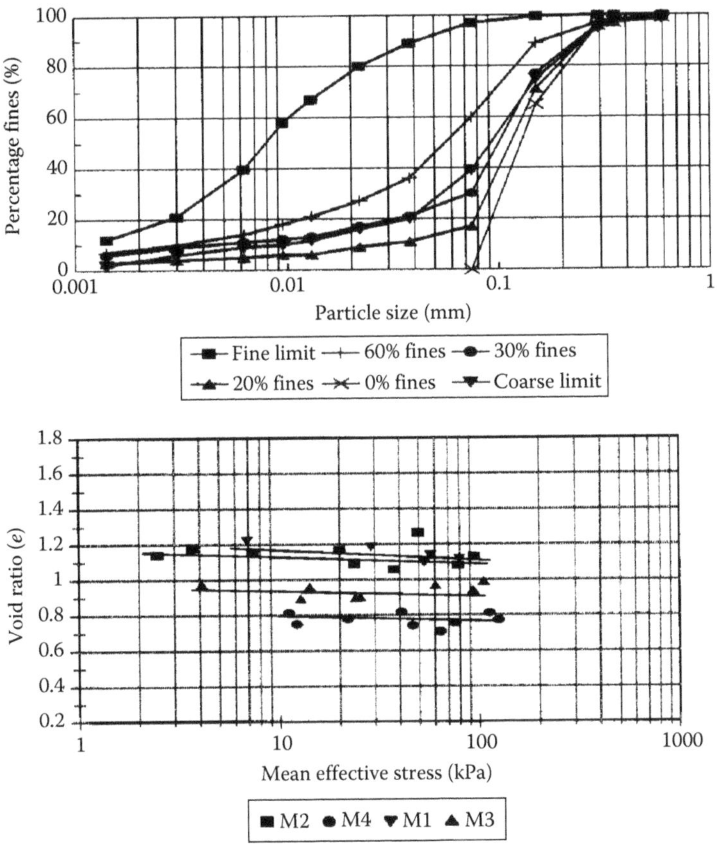

Figure 1.18 Grain-size distribution and critical state line of Merriespruit tailings materials. M1 is 0%, M2 is 20%, M3 is 30% and M4 is 60% fines content. (From Fourie, A.B. and Papageorgiou, G., *Can. Geotech. J.*, 38(4), 695, 2001. With permission from the NRC, Ottawa, Ontario, Canada.)

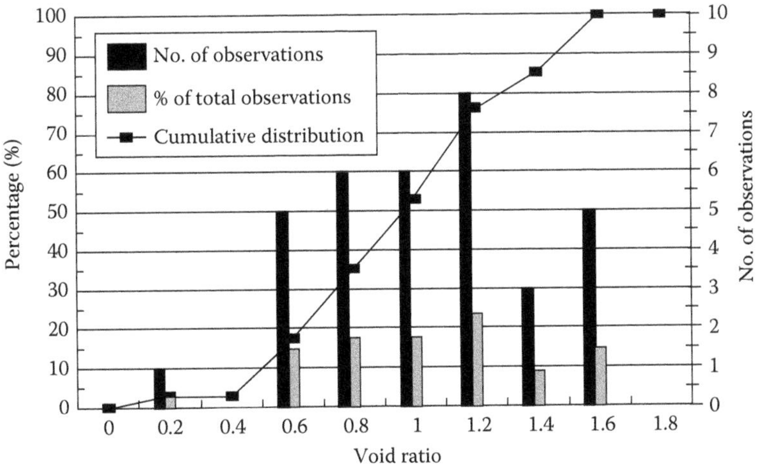

Figure 1.19 Distribution of in-situ void ratios obtained during post-failure investigation of Merriespruit tailings dam. (From Fourie, A.B. et al., *Can. Geotech. J.*, 38(4), 707, 2001. With permission from the NRC, Ottawa, Ontario, Canada.)

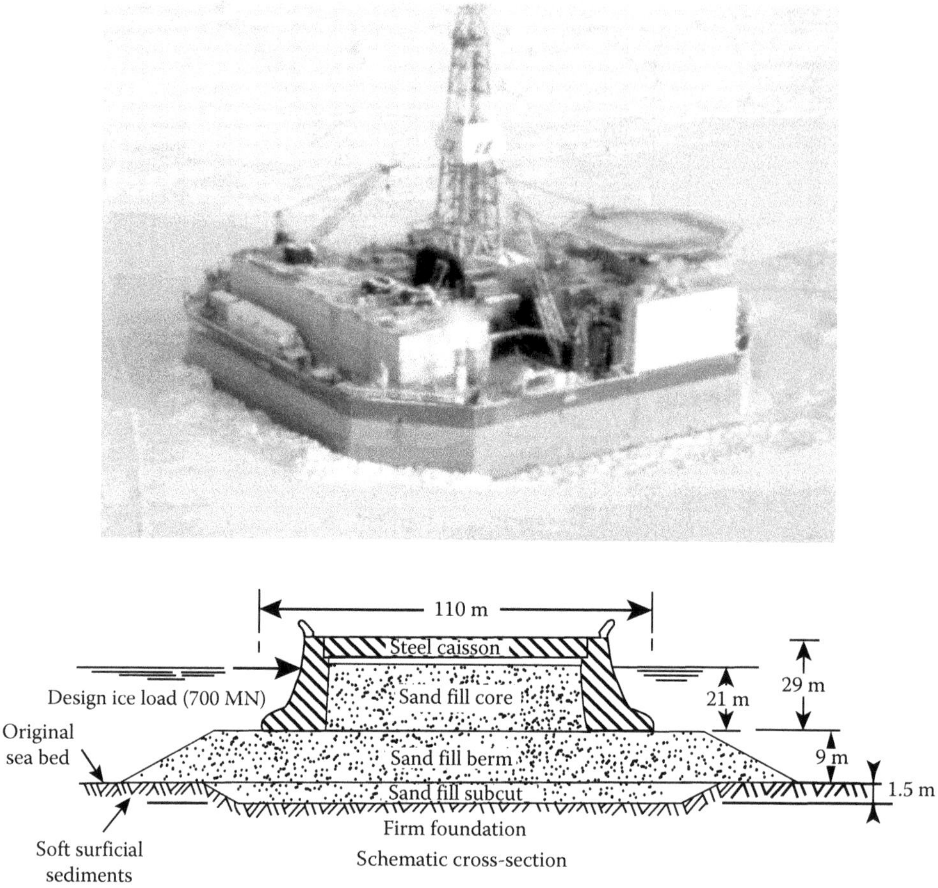

Figure 1.20 Gulf Canada's Molikpaq structure in the Beaufort Sea.

Regarding liquefaction, the performance of the Molikpaq was completely satisfactory up to the 12 April 1986 ice event. It had been used a year earlier for the Tarsuit P-45 exploration well, and had withstood several ice loadings with all evidence pointing to better than anticipated behaviour. At Amauligak, the Molikpaq withstood substantial loads from second-year and multi-year ice flows impacting the structure during 5–8 March 1986 and all evidence confirmed what was seen at Tarsuit P-45.

The situation changed with strong offshore winds during April 10 and 11, and on the morning of 12 April 1986 the ice was moving past the Molikpaq at several knots. At approximately 0830 h a relatively small hummock of multi-year ice came in contact with the east face of the Molikpaq. The 75 m long and approximately 12 m thick ice ridge caused extreme vibrations on the structure, measured to be in excess of 0.11 g at deck level. The hummock was contained in a much larger ice sheet (about 1 × 2 km) which had sufficient driving force to cause complete crushing of the hummock against the face of the Molikpaq. The horizontal loads from ice crushing were slightly in excess of the design load of 500 MN, but this was not of particular concern given the observed better than expected performance in somewhat thinner ice. However, the 12 April ice event caused large cyclic loads, in the frequency range 0.5–2 Hz, which lasted for about fourteen minutes. This amounted to about 900 cycles of similar magnitude loading.

Later in the day, it was discovered that the sand surface within the core on the eastern side of the Molikpaq had settled up to 1.5 m. This led to a rapid and general check of the instrumentation records. (The Molikpaq was equipped with some 600 sensors because of the novelty of deploying large offshore platforms in moving ice.) The instrumentation data revealed that part of the core had liquefied during the ice loading, although dilation and load transfer to the non-liquefied portions kept the resulting displacements small.

The nature by which the cyclic ice loads lead to liquefaction of the Molikpaq core is illustrated in Figure 1.21, which shows 12 cycles of ice loading on the structure and the corresponding response at a piezometer in the centre of the locally liquefying zone in the sand core. For any single load increase, a positive pore pressure is induced (in part from

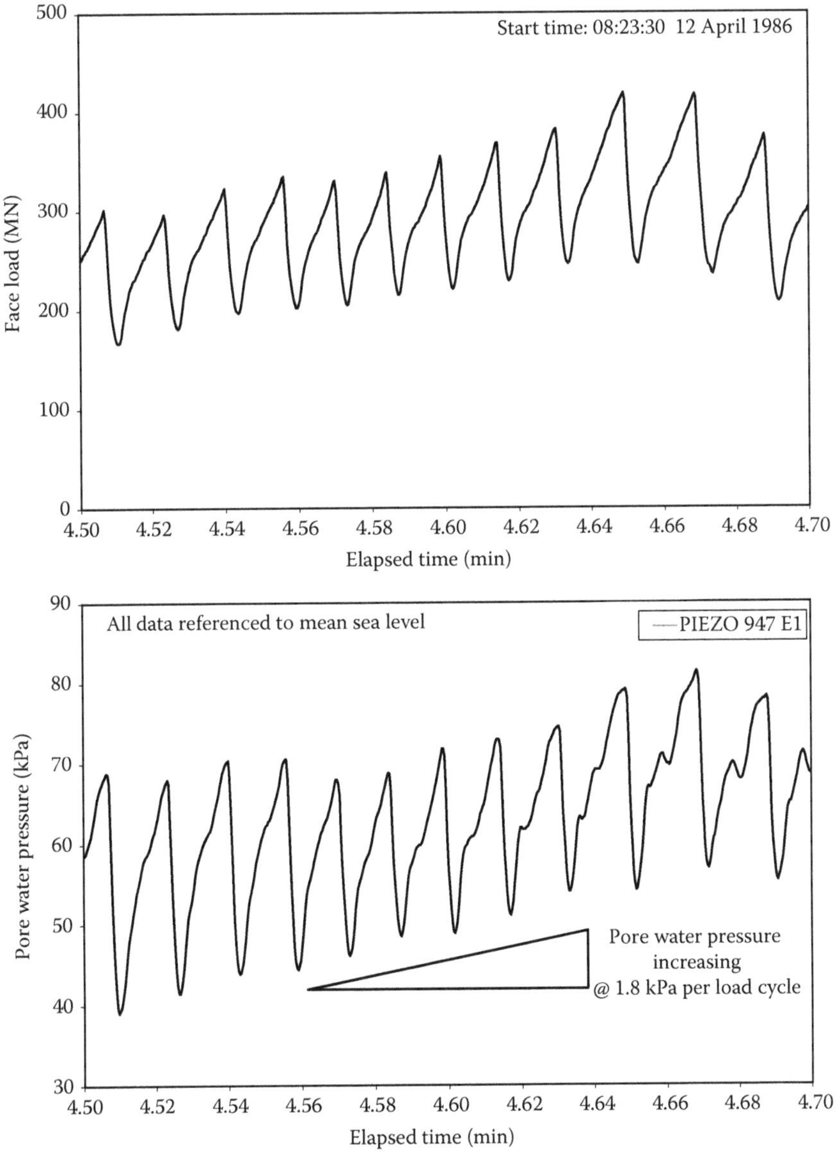

Figure 1.21 Details of cyclic ice loading and excess pore pressure 12 April, 1986.

the increase in mean stress as the load comes onto the sand) but the situation is stable in the sense that there is no run-away generation of pore pressure, but the induced pore pressure is not completely recovered during the unloading part of the cycle. The effect of continuous load cycling is therefore to ramp the average pore pressure upwards. The full pore pressure response in Figure 1.22 shows that ramping continued at an average rate of accumulation of excess pore pressure of 0.8 kPa/cycle. The process terminated when a zero-effective stress condition was reached. Thereafter the pore pressure increased about 15 kPa caused by settlement of the piezometer by 1.5 m. Fluidization of the sand core was progressive, initiating at mid-height on the loaded side and propagating downwards and into the core.

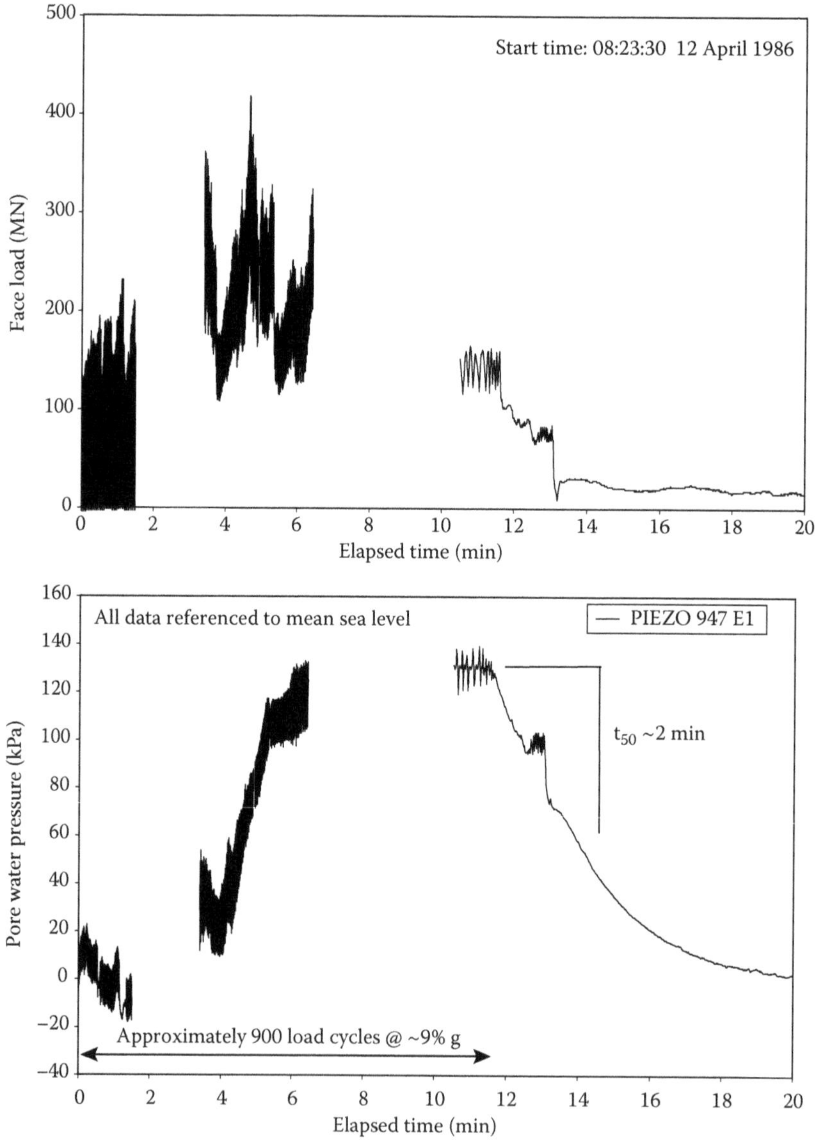

Figure 1.22 Cyclic loading and piezometric response showing accumulating excess pore pressure to liquefaction (piezometer E1, mid depth in centre of loaded side). The thick bandwidths of left-hand parts of plots are caused by cyclic variations; see Figure 1.21.

The following observations can be made about the behaviour of the sand in cyclic loading from the Molikpaq data:

- Lightly dilatant sands ($\psi \sim -0.05$) have substantial reserves of cyclic strength in compression loading.
- Excess pore pressures continue to be generated as long as cyclic loading occurs.
- With sands, redistribution of pore pressure can occur on a similar timescale to cyclic loading.

These observations show that liquefaction under cyclic loading cannot reliably be treated by total stress concepts (e.g. the steady-state concept of a minimum assured undrained strength or Seed's simplified liquefaction approach). For important structures, the full mechanics of sand behaviour and boundary conditions need to be considered. This Molikpaq case history actually has a wealth of information and detail. In particular, an unusual aspect is that the Erksak sand used for filling the Molikpaq was also used in CPT calibration chamber tests. Consequently, the inference of ψ from the CPT is much more precise than for other case histories. The in-situ data also comprise a large number of CPTs, self-bored pressuremeter tests and geophysical tests. It is also one of the few full-scale case histories with a detailed instrumentation record of excess pore pressure generation and dissipation.

Although the Molikpaq history is interesting from a liquefaction mechanics view, there are also two very important and disturbing engineering aspects about the experience: the usefulness of centrifuge tests and the relevance of the observational method.

Regarding the value of centrifuge tests, the possibility of cyclic-induced liquefaction with undensified hydraulic sandfills was recognized during Molikpaq design in 1981. At that time, computational approaches to cyclic mobility did not exist and therefore centrifuge tests were carried out, by one of the leading researchers in the field, to simulate the cyclic loading of the Molikpaq with an undensified core. Although details of the appropriate design ice loading scenarios were uncertain, as was the actual sand to be used in construction, model testing used the best engineering estimates. These turned out to be conservative choices with regard to pore pressure dissipation time factors. The model tests showed the potential for modest cyclic ratcheting of the structure to accumulate 500 mm or so of horizontal displacement in a cyclic loading event. The ratcheting was always drained and there was no tendency to liquefaction in any of the centrifuge tests. This was regarded as uncontroversial, as at that time there was much credence in the literature as to how 'static bias' reduced liquefaction potential and the Molikpaq had a lot of static bias. Therefore the centrifuge results were relied on as substantiating the use of undensified hydraulic fill to resist cyclic ice loading.

In hindsight, the 12 April 1986 event was remarkably similar in many aspects to one of the centrifuge runs, yet the actual sand behaviour was utterly different. What is unclear, even today, is what would have happened had cyclic loading continued for 30 minutes rather than 14 minutes. Nevertheless, this was a 'Class A' test of the adequacy of the centrifuge and it turned out that the centrifuge results mislead about the liquefaction potential.

Although the centrifuge was relied on as substantiating the use of undensified hydraulic fill, there was a conscious appreciation that factors might have been missed or over-idealized in the modelling. An observational approach was adopted to deal with these issues, the Molikpaq being equipped with some 600 sensors and a state-of-the-technology (for 1986) data acquisition system with rolling buffered scanning to capture snatches of dynamic response as well as a continuous archive of average and peak sensor values. In addition, there were many ice loading events over the winter operating seasons. Detailed analysis of measured response to the events prior to 12 April 1986 indicated that the structure and its undensified hydraulic fill was behaving as well, or even better, as predicted by the centrifuge. This was misplaced confidence in the observational method.

Liquefaction is a brittle mechanism and as such is poorly handled with the observational method. For the observational method to be reliable, it needs a continuous increase in loading to produce a continuing increase in displacements and/or pore pressures. This allows the opportunity to evaluate divergence from predicted behaviour well in advance of critical situations. Where behaviour can snap-through to an undrained liquefaction, the observational method is simply inapplicable. Rather, uncertainties must be formally addressed by detailed engineering ahead of time.

1.3.8 Liquefaction induced by machine vibrations

So far the examples of liquefaction have been rather large in extent. An interesting case at the other end of the size spectrum, which also illustrates the effect of high cycles, was the failure of a road embankment that crossed a lake in Michigan. The failure and its circumstances were described by Hryciw et al. (1990) and their description of the event was as follows.

The road embankment allowed Michigan Highway 94 to cross Ackerman Lake. The embankment fill was a clean medium to fine sand. The below water portion was placed by end dumping after removing peat and soft sediments from the lake bottom. Above the lake level, the fill was compacted. The road surface varied from about 2 to 4 m above water level, being graded from one side of the lake to the other. Side slopes were 2H:1V on one side and 4H:1V on the other.

On 24 June 1987, the embankment was traversed by a train of six vibroseis trucks which were carrying out geophysical surveys for oil exploration. A vibroseis is a vibrating plate that is pressed against the ground and excited using an eccentric weight vibrator under computer control. Typically frequency is changed linearly during the excitation, in this case from 8 to 58 Hz over a time of 8 s. This gave 264 cycles of uniform amplitude in any particular seismic shot. This particular survey used six trucks in a train with the vibroseis units linked electronically to keep them in phase, the train being spread out over some 74 m length from bumper to bumper. Figure 1.23a shows a similar train of vibroseis trucks.

Figure 1.23b shows the failure caused by the trucks when the vibroseis units were activated on top of the road embankment. Notice that two trucks are almost submerged at the toe of the failed slope. The driver of the last truck in the train saw the failure develop in front of him and was able to reverse; the second, third and fifth trucks slid into the lake as the embankment liquefied. The drivers fortunately escaped through the doors or windows as the trucks sank. The forth truck remained upright on a failed section of road. The drivers reported feeling as if the ground had completely disappeared beneath them and free-falling rather than sliding into the lake. The failure was sufficiently rapid to cause a 4.5 m high wave that crossed the lake and destroyed a boat dock.

1.3.9 Instrumented liquefaction at Wildlife Site

The Wildlife Site in Imperial Valley, California, is perhaps unique worldwide in that pore pressures and ground response were recorded when the site liquefied during an earthquake. This was not a coincidence. The U.S. Geological Survey (USGS) recognized the need for an instrumented liquefaction site in 1982 and the Wildlife Site was selected for this based on its location in a highly active seismic zone and the fact that the site was susceptible to liquefaction (which had been confirmed when liquefaction was observed during the Westmoreland earthquake in 1981). After only a 5-year wait, on 24 November 1987, the USGS investigators were rewarded by the M6.6 Superstition Hills earthquake. Youd and Holtzer (1994) describe

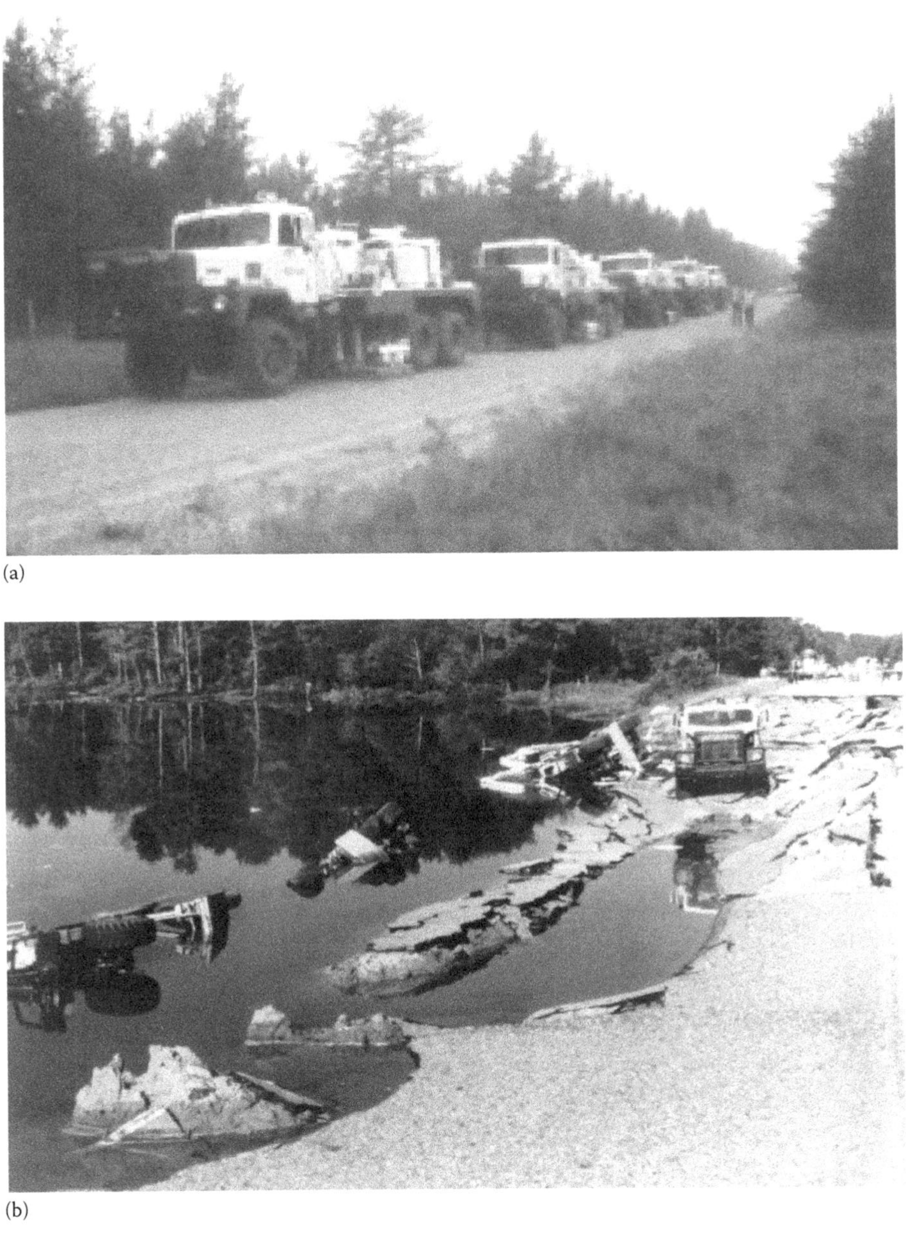

(a)

(b)

Figure 1.23 Failure of embankment on Ackermann Lake triggered by vibroseis trucks. (a) Train of vibroseis trucks (the vibroseis is the plate tamper beneath the center of the trucks) and (b) liquefaction failure induced by vibroseis excitation on loose saturated sand.

the instrumentation and its performance in detail, while Zeghal and Elgamal (1994) provide an analysis of the liquefaction event based on the records.

Figure 1.24 shows a plan and cross-section of the instrumentation. The liquefiable material was a silty sand from about 2.8 to 6.5 m below ground surface. Instrumentation included two triaxial force-balance accelerometers and six electronic piezometers. The accelerometers were located at ground surface and in the silty clay unit underlying the

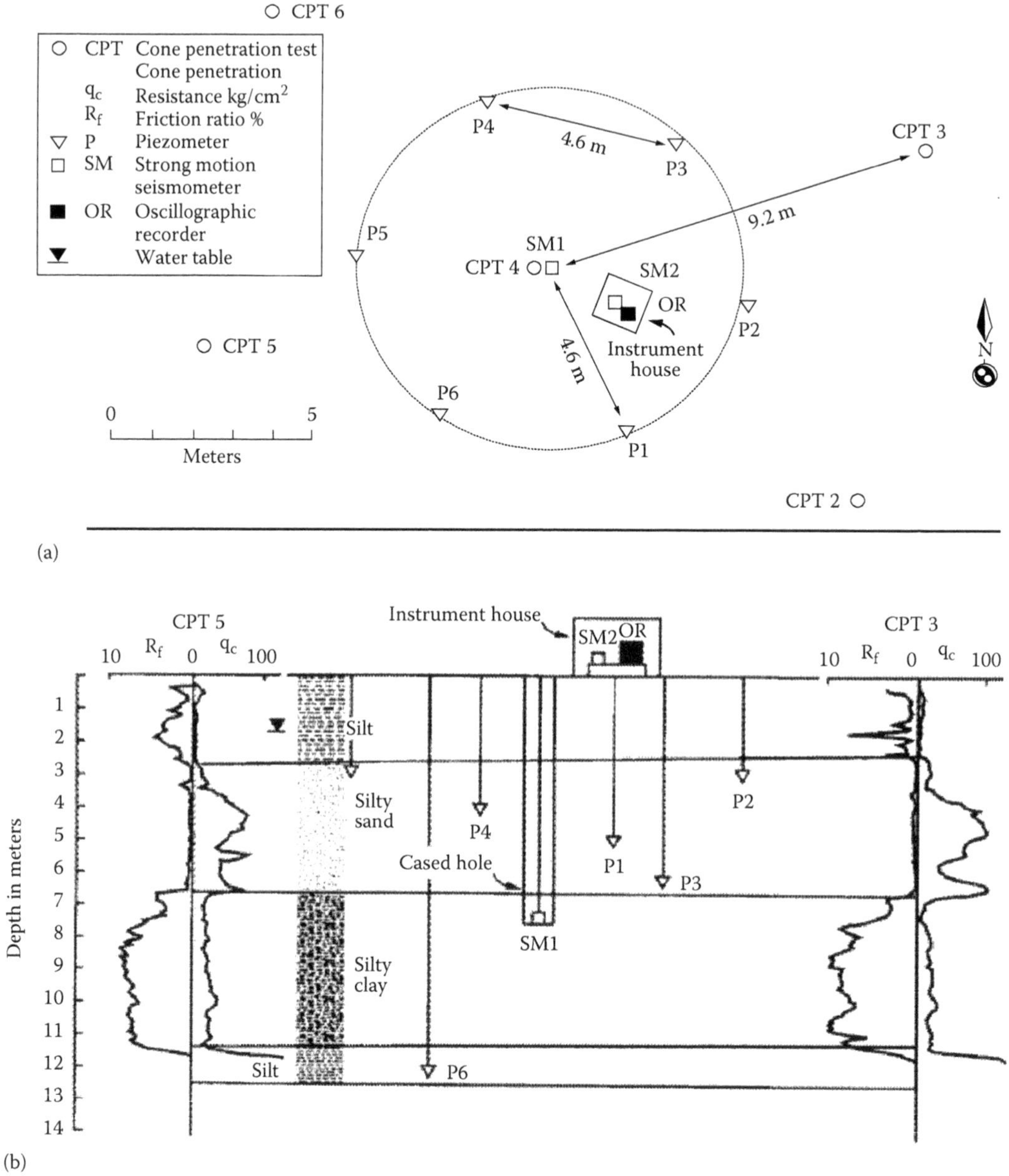

Figure 1.24 Plan (a) and cross-section (b) of the Wildlife instrumentation array. (From Youd, T.L. and Holzer, T.L., *J. Geotech. Eng.*, 120(GT6), 975, 1994 based on Bennett, M.J. et al., U.S. Geological Survey Open File Report, 84, 1984. With permission from the ASCE, Reston, VA.)

loose silty sand at a depth of 7.5 m. Five piezometers were installed at different depths in the loose silty sand while the sixth was embedded in a deeper dense silt. The instruments were designed to be triggered by an acceleration pulse of 0.01 g in the surface accelerometer.

The responses of the North–South surface accelerometer and Piezometer P5 at a depth of 2.9 m are shown in Figure 1.25. Significant excess pore pressure was not generated until the strongest acceleration pulse jolted the site 13.6 s after triggering of the instruments. That pulse

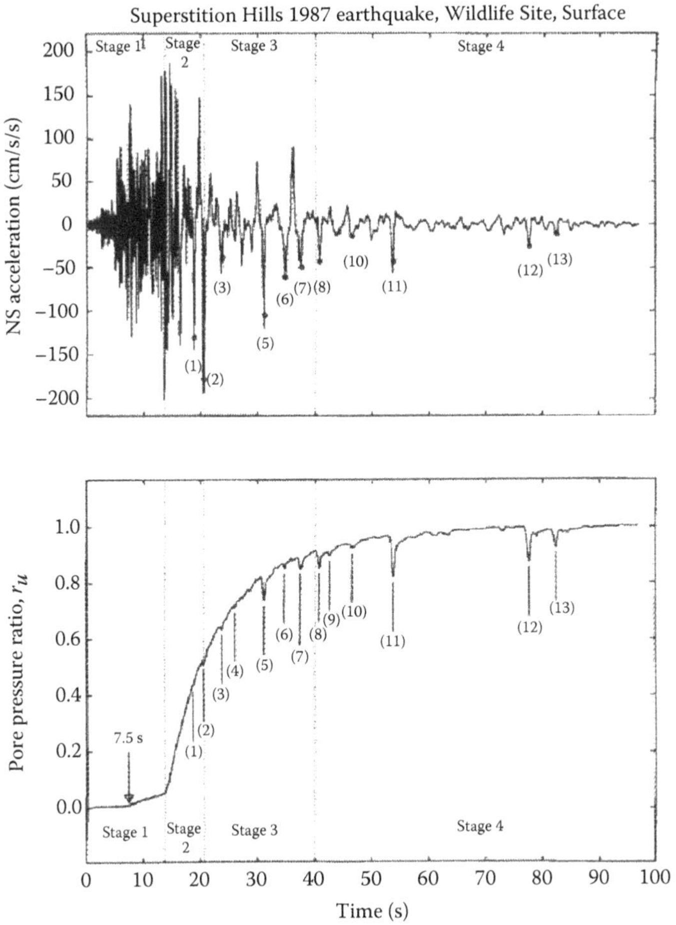

Figure 1.25 Surface accelerometer (N–S) and piezometer P5 (2.9 m) at Wildlife Site during Superstition Hills 1987 earthquake. (From Youd, T.L. and Holzer, T.L., *J. Geotech. Eng.*, 120(GT6), 975, 1994. With permission from the ASCE, Reston, VA.)

immediately generated a rise of excess pore pressure as monitored by each of the four functioning piezometers in the liquefying layer. Pore pressures continued to rise until the recorded pressures approached the initial overburden pressure 60–90 s after instrument triggering. At the end of strong acceleration pulses (about 26.5 s after triggering), the monitored pore pressures had only risen to about 50% of the initial overburden pressure in the lower part of the layer and to 70% in the upper part. The pore pressures continued to rise after the cessation of strong accelerations, which at first sight is surprising. Zeghal and Elgemal integrated the acceleration histories and showed that the strain history at the elevation of piezometer P5 gives a somewhat different picture (Figure 1.26) with significant strains occurring after 26.5 s and for most of the 90 or so seconds of the record. Figure 1.26 also shows their interpretation of the shear stress history based on the measured accelerations. Based on this information, the continued rise of pore pressures after 26.5 s is easily explained.

How liquefaction developed is also illustrated by the interpreted shear stress–shear strain response shown for six specific loading cycles in Figure 1.27 (after Zeghal and Elgamal but reported by Youd and Holtzer). Also shown in each of Figures 1.27 is the pore pressure

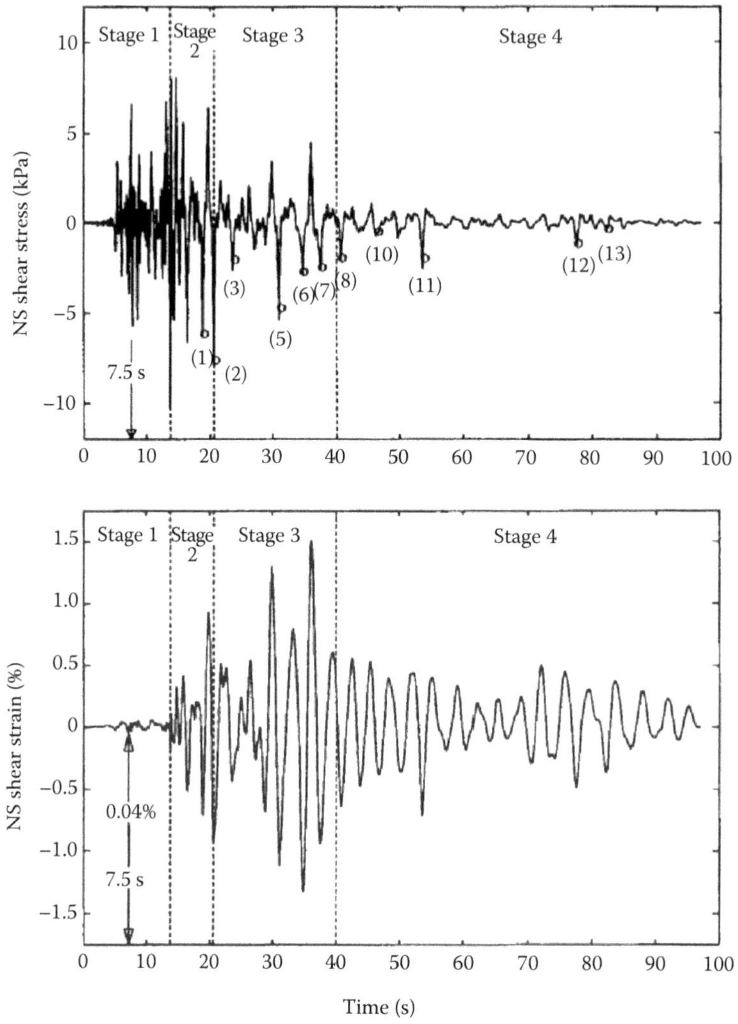

Figure 1.26 Shear stress and shear strain history at depth of piezometer P5 at Wildlife Site, interpreted from accelerometers by Zeghal and Elgemal. (Reproduced from Zeghal, M. and Elgamal, A.-W., *J. Geotech. Eng.*, 120(6), 996. With permission from the ASCE, Reston, VA.)

ratio and the time of the loading cycle which corresponds to an acceleration spike in Figure 1.25. Figure 1.27 clearly shows the progressive softening of the ground as liquefaction develops. Initially, when $r_u = 0.07$, the stress–strain curve is steep, but the stiffness rapidly reduces until very little stress can be sustained when $r_u > 0.97$ as indicated. Another interesting facet of the Wildlife record is the correspondence of the downward spikes in the P5 piezometer record with specific acceleration spikes. Youd and Holtzer suggest that these are caused by shear dilation and occur only during 'downward' acceleration spikes because greater movement (and therefore dilation) was possible towards the incised Alamo River Valley than in the opposite direction.

The Wildlife records show nicely how liquefaction developed during the Superstition Hills earthquake, and also how nothing occurred during the slightly less severe M6.2 Elmore Ranch earthquake the previous day.

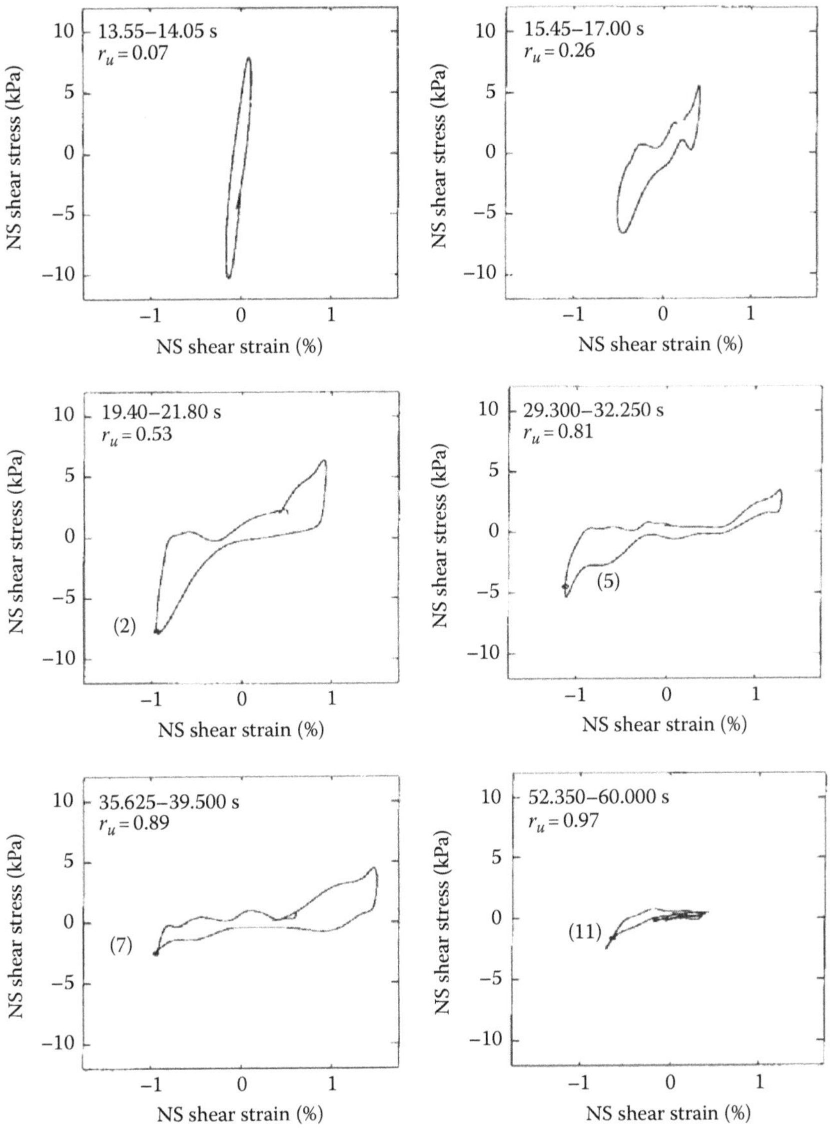

Figure 1.27 Average stress–average strain graphs for selected time increments interpreted from NS accelerometers at Wildlife Site. (After Youd, T.L. and Holzer, T.L., *J. Geotech. Eng.*, 120(GT6), 975, 1994. With permission from the ASCE, Reston, VA.)

1.3.10 Summary of lessons from liquefaction experiences

The case histories of liquefaction presented earlier allow some inferences to be drawn about liquefaction which will then guide how to approach liquefaction engineering. Key observations include the following:

- Liquefaction is a soil behaviour associated with excess pore water pressure, but it is not necessarily undrained and the movement of excess pore pressures throughout the soil over time may be crucial (e.g. Lower San Fernando Dam).

- Excess pore pressures can arise from cyclic loading of soil, either by earthquakes (e.g. Niigata) or by external forces (e.g. Molikpaq).
- Excess pore pressures can arise through static loading if the soils are loose enough (e.g. Fort Peck, Nerlerk). Even though straining may be evident for days before the failure, the transition to high excess pore pressures is normally very rapid. Any attempt at an observational approach is likely futile and certainly dangerous.
- Reducing mean effective stress because of water seepage can trigger liquefaction (e.g. Aberfan).
- Liquefaction involves increasing strains and may become a flowslide if the soil is loose enough. Even if not a flowslide, strains can be large enough to cause functional failure of structures (e.g. the buildings in Niigata).
- Soil strata have naturally variable density, and the distribution and structure of these natural variations can play a crucial role (e.g. at Nerlerk).

1.4 OUTLINE OF THE DEVELOPMENT OF IDEAS

The overview of liquefaction given by the few case histories just discussed provides the context to develop the theoretical aspects needed to properly understand liquefaction. Chapter 2 introduces the critical state and how it may be measured. The state parameter ψ is used to show how many soil behaviours (e.g. peak friction angle) are unified regardless of soil type, fines content, etc. Aspects such as fabric and over-consolidation and how they affect behaviour are also considered.

At this point, the framework for liquefaction is defined, but not computable. So Chapter 3 introduces *NorSand*, a generalized critical state model based on the state parameter ψ. *NorSand* is presented in three steps (starting with triaxial monotonic loading, followed by the generalization to 3-D stress space and finally the general model for cyclic loading) as this makes things a lot easier to understand and explain. Chapter 3 gives the first two steps. Calibrations are presented against laboratory data, for triaxial and plane strain drained tests. Undrained conditions are illustrated in Chapter 6, while cyclic loading aspects are left to Chapter 7.

Having introduced the state parameter and what it can do, from normalization of soil behaviour through to the computable *NorSand* model, the next logical step is to measure the state parameter in-situ, which is the subject of Chapter 4. The usual difficulty that undisturbed samples are impossible to obtain, at least practically and on a routine basis, means that penetration tests are the basis for most work. Determining ψ from penetration tests is covered in considerable detail as the method relies on the CPT. Other needed parameters (such as shear modulus) and alternatives to the CPT are also considered.

Chapter 5 moves into the realm of real soils, as opposed to the largely clean quartz sands used in research testing and, in particular, addresses the issue of how to select a characteristic state for design. Calculations are normally done using a single value for state or strength throughout a domain. The domain might be broken into a few layers or zones, but that is usually the limit of design idealizations. The reality is that soil state and properties vary in-situ even in the same geologic stratum – there is a distribution of properties. Correspondingly, one of the issues in design is the value to be chosen as representative (or characteristic in limit states jargon) for design calculations or analysis. Often guidance would be sought from a code of practice on this choice, but there is precious little such guidance for foundation design, never mind liquefaction problems. However, the situation with characteristic values for liquefaction is not totally grim as there are three interesting and important studies that give some clues on selecting characteristic values. These studies are summarized and discussed in Chapter 5, which is a bit of a leap as these results depend on advanced simulation methods which are not covered in this book. The reader is asked to accept this leap in order to appreciate the effect of variability of state to understand the limitations of the various case histories used to underpin design methods for liquefaction.

Chapter 6 presents liquefaction and large-scale deformations under essentially static loads, as these fall within the same theoretical framework. The triggering mechanism may be static, cyclic or hydraulic. This chapter builds on the triaxial theory based on laboratory experience of drained tests to the undrained conditions usually encountered in liquefaction. Key case histories are included. The outcome of this chapter is both an understanding of flaws in some existing approaches and a calibrated methodology for going forward with liquefaction engineering.

Chapter 7 gets to cyclic loading and liquefaction, or cyclic mobility as it is more usually called. First, laboratory test data are used to examine trends in the data from different kinds of cyclic tests. Current design practice is dominated by correlations to case histories; so this practice is presented before showing how empirical factors are predicted from theory developed in the earlier chapters and, importantly, where theory suggests existing experience is being wrongly extrapolated (mainly the effects of depth and soil compressibility). *NorSand* is then extended from monotonic to cyclic loading and in particular the effect of rotation of principal stress directions is introduced. This sets the theoretical framework for cyclic liquefaction.

Although we stated earlier that this book is not about methods of analysis, we did assert that the critical state theory was 'computable'. Chapter 8 shows implementation into a finite element code and validates that the numerical solution is indeed performing as it should. We have used the publically available source code of Smith and Griffiths (1998) as a platform for this work. After testing the solutions, we look at the computed behaviour of a simple slope consisting of loose sand subject to additional loading at the top of the slope, or unloading at the toe of the slope. These are, of course, the generally accepted triggers of static liquefaction failures.

We have tried to pull together the practical implementation of the theory in some of the downloadable material that comes towards the end of this book in Chapter 9. This chapter walks through using laboratory test data to determine the critical state line and estimate the stress–dilatancy parameters that are needed to use this book in practice. Similarly, it shows how to use CPT data to arrive at estimates of the in-situ state parameter. With these in place, the engineer has the parameters they need for analysis.

Finally, in Chapter 10 we bring together some of the threads within the book, including a summary of the material presented and a look to the future of how the subject may develop.

To keep the book more readable, some details are presented in appendices. One thing needed to work with the material in this book is a consistent set of stress and strain measures and Appendix A defines those. Laboratory testing procedures are important to obtain accurate, consistent critical state data; so Appendix B details our experience in this regard. Appendices C and D are concerned respectively with the theoretical derivations of *NorSand* (in particular the full 3D version) and the numerical implementation of the *NorSand*. Appendix E contains CPT calibration chamber test results as these are not readily found in the published literature. Appendices F and G document many of the case records needed to evaluate residual strength after static liquefaction and the earthquake case records upon which much of the empirical work on liquefaction is based. A question we have been asked frequently is how *NorSand* relates to the very much more familiar Cam Clay theoretical model of critical state soil mechanics. So in this edition we provide a detailed theoretical comparison in Appendix H, to show that Cam Clay, or rather Original Cam Clay, fits perfectly within the *NorSand* framework. All that is needed is to set the conditions in *NorSand* to correspond to the restrictive assumptions within Modified Cam Clay to obtained practically identical results.

Chapter 2

Dilatancy and the state parameter

This chapter covers the experimental evidence and historical basis for state parameter and stress–dilatancy, without straying into too much theory. As such, this is needed background information to understand the material presented in Chapters 6 and 7 about static and cyclic liquefaction behaviour, although Sections 2.7 and 2.8 could be skipped if less interested in generalized (non-triaxial) stress conditions and the finer details of stress–dilatancy.

Understanding soil behaviour needs familiarity with stress invariants, which are used in preference to mobilized friction angles, and these stress invariants require work conjugate strain invariants – Appendix A works through these ideas and the various definitions. Appendix A may best be read before this chapter if unfamiliar with work conjugate invariants. These work conjugate invariants are the basis on which this chapter looks at soil behaviour and are easy to appreciate (they make things very much clearer) despite being unfamiliar.

2.1 FRAMEWORK FOR SOIL BEHAVIOUR

2.1.1 Dilatancy

Volume change behaviour largely distinguishes soils (or, more generally, particulate media) from other engineering materials. When soils are sheared, they increase in volume if they are initially dense or contract if they are initially loose. The tendency of soils to change volume while shearing is called dilatancy, which being a fundamental aspect of soil behaviour has been known for more than a century since the work of Reynolds (1885). Indeed, there was a view in the late 1800s that particulate materials were a fourth state of matter to the familiar gas, liquid and solid. We do not need that fourth state view today, but grasping that soils have a range of states (void ratios, densities, stresses) and that those states change with deformation is crucial to a rational approach to soil mechanics.

Care is needed in discussing dilatancy because the common usage is for positive dilatancy to be regarded as volume increase during shear, which is opposite to the compression positive convention widely used in geotechnical engineering. There is also an issue in the two definitions of dilatancy which are already in use, often without people seemingly being aware of the difference. The alternatives will be called the absolute and rate definitions in this book:

- The absolute definition: Dilation is the change in volumetric strain incurred since the initial condition.
- The rate definition: Dilation is the ratio of rate (or increment) of volume change with rate (or increment) of shear strain.

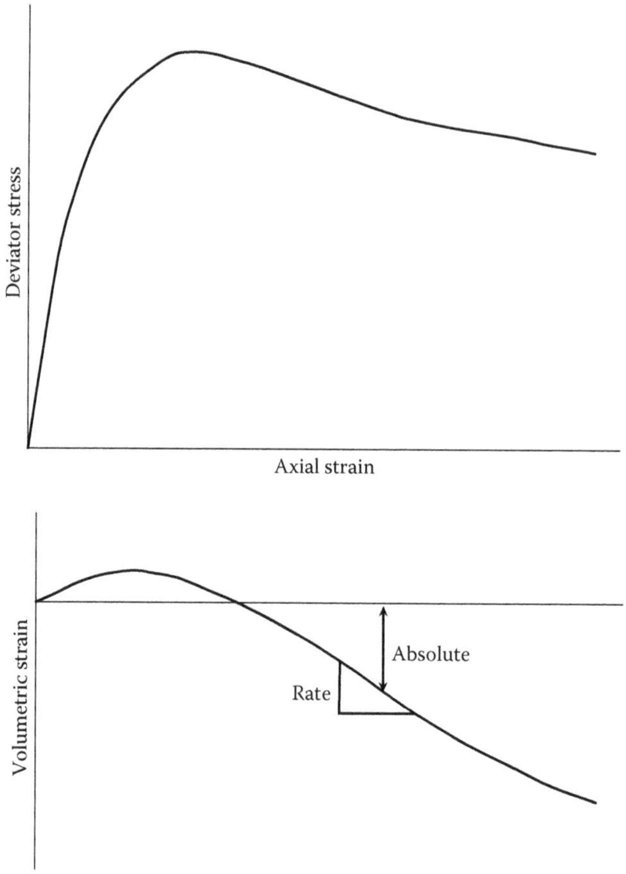

Figure 2.1 Difference between rate and absolute definitions of dilatancy.

These two definitions of dilatancy are interrelated, with the first definition simply being the integral of the second over the particular stress path imposed on the soil. In many cases, the volume change may be contractive after a particular shear strain even though the material is dilating. Figure 2.1 illustrates the two different concepts of dilation.

The rate definition of dilation traces back to the idea that soil has a true friction that can only be understood or determined by accounting for the work done as the soil dilates. This true friction concept was suggested by Taylor (1948) and formalized by Bishop (1950), when considering soil strength. The understanding of the role of dilation was greatly extended by Rowe (1962) who showed that dilation operated throughout the stress–strain behaviour of particulate materials and was not something associated only with peak strength. The rate definition of dilation is intrinsically linked to understanding soil behaviour through applied mechanics.

The absolute definition of dilation seems to trace to the Netherlands where there is long-standing experience of liquefaction problems (going back at least a century). A subtlety in the Dutch use of the volume change is that only the volume change due to shearing is considered (e.g. Lindenberg and Koning, 1981). Thus, volumetric strains as a result of mean stress changes need to be subtracted from the total volumetric strains to obtain the volume change due to shear-induced dilation. The absolute definition becomes useful here because, if the soil dilates according to the absolute definition, then the undrained strength will be greater than the drained, and hence, liquefaction becomes impossible (as long as the situation was originally statically stable).

As well as this issue of two definitions, the geotechnical literature uses the terms dilation and dilatancy interchangeably; they are commonly also expressed as an angle analogous to the friction angle. This leads to a further niggle that maximum dilation angles refer to the greatest rate of increase in the soil's void ratio, but this is actually a negative or minimum strain rate ratio because of the compression positive convention of soil mechanics.

This book uses the *rate* definition of dilation. The rate definition is preferred because it is an expression of the work flow in the soil, which is fundamental to plasticity-based constitutive models. We will come to this aspect in the next chapter. Where the absolute definition of dilation might be relevant we refer to the actual positive or negative volumetric strain, so reserving the words dilation and dilatancy for the applied mechanics view.

2.1.2 Critical state

Given that dense soil increases in volume during shear while loose soil contracts, it is natural to wonder how the two behaviours are related. Casagrande, in 1936, explored this issue. Using shear box tests, it was found that loose sands contracted and dense sands dilated until approximately the same void ratio was attained at large strains as shown in Figure 2.2. This large strain void ratio distinguished which mode of behaviour the soil exhibited. Looser sand was contractive under either definition of dilation. Casagrande termed the void ratio that demarked the volumetric strain behaviour as the critical void ratio. Interestingly, Casagrande's work was not blue sky research but came out of the need to engineer a hydraulic fill dam that would not liquefy – Franklin Falls Dam in New Hampshire. Arguably, this project was the real start of modern soil mechanics.

The critical void ratio is affected by mean effective stress, becoming smaller as the stress level increases – a behaviour first reported in Taylor (1948). The relationship between critical void ratio and mean effective stress is called the critical state locus (or CSL).

Traditional geotechnical practice has taken account of the density, which affects whether a soil will dilate or contract, rather simply by assigning different properties to the soil according to whether it is dense or loose. For example, the same geological material may be assigned a friction angle $\phi' = 32°$ in a loose in-situ state, but given $\phi' = 36°$ for design after densification or compaction. No relationship is offered between density and behaviour, with each density of the soil in effect treated as a different material whose properties must be established by testing. This is certainly inconvenient, but it is also an intellectual failure since intrinsic properties are not a function of soil density. Soil is a material that exists across a range of states, with the state determining how the true or intrinsic properties are transformed into engineering behaviour such as strength and stiffness.

The first theoretical development that captured the density of soils as a state variable, rather than a soil property, and thereby accounted for volume changes during shearing, was the framework that became known as critical state soil mechanics (Schofield and Wroth, 1968). The name critical state derives from anchoring the theory to Casagrande's critical void ratio. The critical state is taken to be the ultimate state the soil reaches if we keep deforming (shearing) the soil. Critical state soil models are formulated from theoretical plasticity modified to take into account volume changes during loading. Critical state theory, or more precisely a generalization of the framework put forward by Schofield and Wroth, is going to form the basic approach to liquefaction.

However, before getting into critical state frameworks, a few definitions are needed to avoid any potential confusion with terminology and concepts surrounding critical states, steady states, dilation and volume changes. The critical state was defined by Roscoe et al.

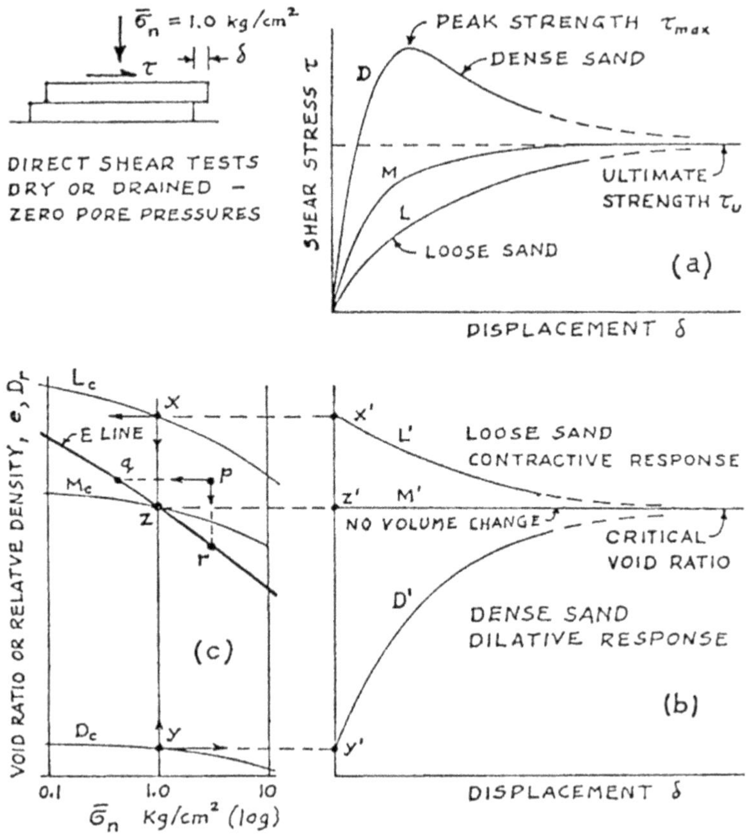

Figure 2.2 Early hypothesis of critical void ratio from direct shear tests. (a) Shear stress vs. displacement, (b) void ratio vs. displacement and (c) void ratio vs. normal stress. (From Casagrande, A., Liquefaction and cyclic deformation of sands: A critical review, *Proceedings of Fifth Pan-American Conference on Soil Mechanics and Foundation Engineering*, Buenos Aires, Argentina, Vol. 5, pp. 79–133, 1975.)

(1958) as the state at which a *soil continues to deform at constant stress and constant void ratio* – essentially a formalization of Casagrande's idea. Note that there are two conditions in the definition: (1) the soil is at constant void ratio and (2) it has no propensity to change from this constant void ratio condition. Much confusion, past and present, arises if condition (2) is ignored.

Reliable methods to determine the CSL have been remarkably elusive ever since Casagrande first suggested the presence of a critical void ratio in 1936. However, in the mid-1960s, one of Casagrande's students, Gonzalo Castro, undertook a series of stress-controlled triaxial tests in an attempt to reproduce field loading conditions which Casagrande surmised were stress controlled. These tests on loose samples systematically resulted in liquefaction failures leading to a well-defined *steady state* at the end of the tests.

Castro achieved a steady state in his tests by starting with loose samples, and using a load-controlled loading device. A hanger was placed on the triaxial loading piston and dead weights added to the hanger at a rate of about 1 every 30 s. Eventually, the sample would reach its peak deviator stress condition and start strain softening. With the weights still on the hanger, the strain rate would rapidly increase and the test would be over in a fraction of a second. The weights would hit the stop plate with a big thump.

Poulos (1981) formalized the definition of the steady state as follows: 'The steady state of deformation for any mass of particles is that state in which the mass is continuously deforming at constant volume, constant normal effective stress, constant shear stress and constant velocity'. The locus of steady-state void ratios with mean effective stress is often referred to as the *steady-state locus* (SSL).

Initially, it was thought that this high strain rate, load-controlled testing was an essential part of achieving the steady state. However, this supposition has not been borne out by subsequent studies in which it has been shown that strain-rate-controlled tests result in the same steady-state condition. Strain-controlled testing is actually preferable, as it requires less in the way of transducer response time and data acquisition rates, avoids inertial corrections to measured loads and provides more detailed data on the post-peak behaviour. Appendix B provides details of the laboratory procedures that can be used to obtain reliable measurements of the critical state; these procedures are not difficult or costly but they are a further step beyond standard laboratory practice and so need attention.

An interesting topic of discussion in the 1970s and 1980s was whether the critical and steady-state lines are in fact the same (Casagrande, 1975; Poulos, 1981; Sladen et al., 1985b; Alarcon-Guzman et al., 1988). Intellectually, there appears to be little distinction between the critical and steady state except for the method of measurement. Critical state researchers have generally relied on drained strain-rate-controlled tests on dilatant samples to determine the critical state. In undrained tests, the steady state is usually measured on loose, contractive samples. This led Casagrande (1975) to define the S line based on drained tests (the slow, or critical state line [CSL]) and the F line based on Castro-type undrained stress-controlled tests (the fast, or steady-state line). For the moment, we will assume that the two are equivalent, as Been et al. (1991) examined the question in some detail and concluded that, for practical purposes, equivalence could be assumed. In previous studies where large differences were found, it was as much the interpretation of the steady or critical state data from individual tests, as any other factor, that gave rise to an apparent difference. Further, it will be shown in Chapter 6 that the S and F lines reported can be computed from the proposition of a unique CSL by introducing the strain limits of the triaxial equipment.

Comparison of the definitions of the critical state and steady state shows that the steady-state definition has a velocity term. This velocity is never specified and could in principle be vanishingly small, at which point the two definitions become identical.

It is common to treat the CSL as semi-logarithmic for all soils (at least as an engineering approximation):

$$e_c = \Gamma - \lambda \ln(p'_c) \tag{2.1}$$

where Γ and λ are intrinsic soil properties, that is properties that are not affected by fabric, stress history, void ratio, etc. The subscript 'c' denotes critical state conditions. Caution is needed when looking at quoted values of λ as both log base 10 and natural logarithms are used. Natural logarithms are more convenient for constitutive modelling, whereas base 10 logarithms arise when plotting experimental data; we use the notation λ (or λ_e where emphasis is needed) and λ_{10} (= 2.303 λ), respectively. The parameter Γ also has an associated stress level, which is $p' = 1$ kPa by convention. More sophisticated variants on (2.1) exist, but the validity of the CSL as a frame of reference does not depend on a semi-log approximation – it is only a modelling detail (we show an alternative idealization of the CSL later).

The discussion so far has largely been about void ratio. The critical state is more than this, however, as a shear stress is required to keep the soil deforming. Commonly, this shear stress is expressed in terms of the constant volume friction angle ϕ_{cv}, the critical friction angle ϕ_c or the critical shear stress ratio M. This shear stress at the critical state is far less controversial than the critical void ratio, and the two notations ϕ_c and ϕ_{cv} are conceptually identical (although it might be argued that ϕ_c links to a particular theory while ϕ_{cv} is general).

2.1.3 Stress–dilatancy

A basic and excellent framework for understanding soil is its stress–dilatancy behaviour. That dense sands dilate and are markedly stronger than loose sands led to interest in energy corrections in the early years of soil mechanics (e.g. Taylor, 1948). The idea was that the friction angle should be broken down into a dilational component and a frictional component, as illustrated in Figure 2.3 (from Bishop, 1950), with the frictional component being perceived to be a true fundamental value. Of course this offers little insight without understanding dilatancy. The seminal paper by Rowe (1962) related the mobilized stress ratio to the plastic strain rates, in what has become known as *stress–dilatancy* theory. Rowe, who was also at the University of Manchester where Reynolds had done the original experiments demonstrating dilatancy almost a century earlier, considered the mechanics of an assembly of spheres/rods. Rowe showed that there was an intimate relationship between plastic strain rates and the mobilized stress ratio. Most importantly, this relationship applied to the whole strain history, not just peak strength values. Rowe's original proposition can be stated as (for a compression positive convention):

$$\frac{\sigma_1'}{\sigma_3'} = K\left(1 - \frac{\dot{\varepsilon}_v}{\dot{\varepsilon}_1}\right) \tag{2.2}$$

When stress–dilatancy theory was introduced, it was thought that K might be a constant and related to the soil mineral–mineral friction (which was perceived at that time as the fundamental property controlling the behaviour). This is now known to be not the entire story with a sense of disappointment apparent in Rowe's paper as even the earliest experimental data showed that K evolved with strain. Nevertheless, Equation 2.2 recognizes dilation as a work transfer mechanism between the principal stress directions. This provides an enormously insightful way of plotting soil test data so as to understand the underlying physics.

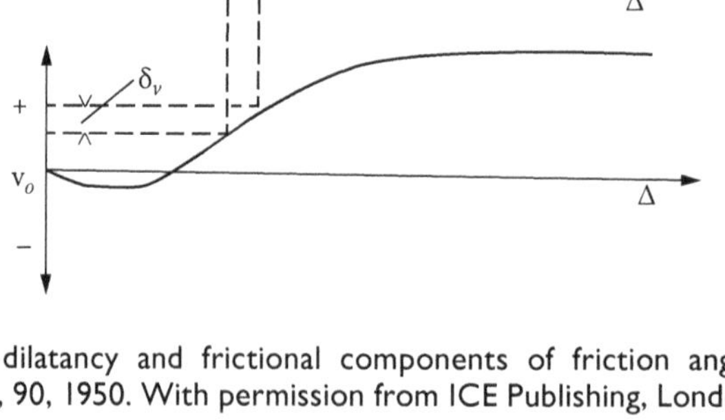

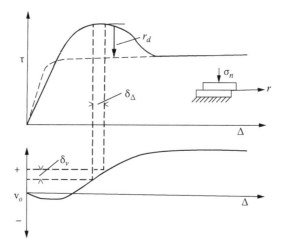

Figure 2.3 Illustration of dilatancy and frictional components of friction angle. (From Bishop, A.W., *Géotechnique*, 2, 90, 1950. With permission from ICE Publishing, London, UK.)

Having introduced some concepts, things need to be brought on to a more formal basis before proceeding. It is helpful to get away from friction angles in favour of stress invariants and, because soils are controlled primarily by the relative amounts of shear stress to mean stress, to use the stress ratio η defined as

$$\eta = \frac{q}{p'} \tag{2.3}$$

Dilatancy must be defined so that it can be used quantitatively. Dilatancy D is defined as being the ratio between the two work conjugate strain increment invariants:

$$D = \frac{\dot{\varepsilon}_v}{\dot{\varepsilon}_q} \tag{2.4}$$

Following on from Rowe's stress–dilatancy approach, it is anticipated that soil behaviour would follow a function of the form

$$\eta = f(M_f, D^p) \tag{2.5}$$

where
the function f () is not yet defined
the superscript 'p' has been introduced to indicate that it is the plastic component of the strain rates that is relevant

Practically, it is often useful when first looking at test data to neglect elasticity and to use total strain (i.e. D rather than D^p) because elastic property data may not be available and working out D is a trivial transform of drained triaxial test data after importing it into a spreadsheet. M_f is the equivalent of Rowe's mobilized critical friction ϕ_f, which varies a little with strain and rather more with state.

Equation 2.5 suggests that reducing test data to η and D values should give insight into that data, and it is used for the test data in this chapter. Notice that this approach is automatically dimensionless and stress-level independent – one of the goals of a proper model falls into your lap when a stress–dilatancy view is taken. Further understanding then follows by relating D to density (void ratio), which is accomplished using the state parameter.

2.2 STATE PARAMETER APPROACH

2.2.1 Definition

As soil is a material which exists in a range of states, the first requirement is a measure of that state. The concept of relative density is exactly this. The maximum and minimum densities define reference conditions, and relative density is a measure of the sand state relative to these reference conditions. Relative density can be improved upon very significantly, however, as a measure of sand state.

The kernel concept for the measurement of sand state is that the critical state defines a reference state and the distance of the sand from the reference state in void ratio – stress space is a first-order measure of that sand's structure. Casagrande's observations of sand behaviour (Figure 2.2) were that sands dilate or contract when they are sheared until they reach the critical state. The further away from the final critical state, the faster dilation or contraction happens. The state parameter, ψ, is simply defined as a measure of that deviation (see Figure 1.1):

$$\psi = e - e_c \tag{2.6}$$

where

e is the current void ratio of the soil
e_c is the void ratio of the critical state at the current mean stress

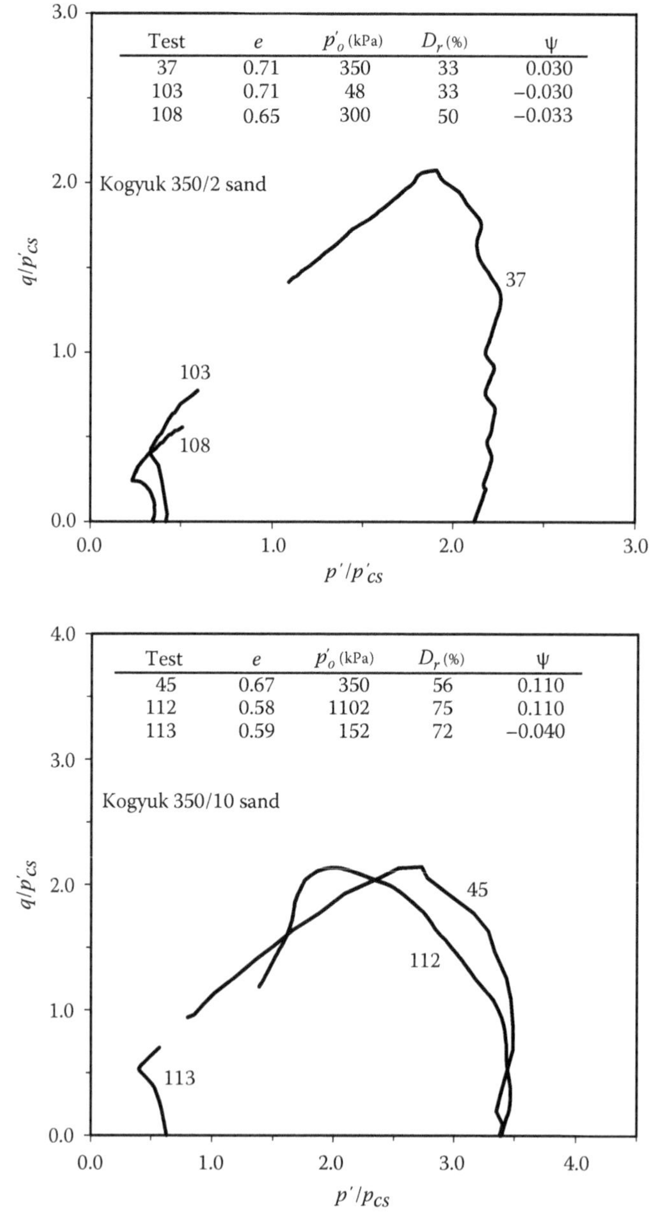

Figure 2.4 Comparison of behaviour of sand as a function of relative density and state parameter for Kogyuk 350/2 and Kogyuk 350/10 sands.

Using e_c in the definition captures the effects of changes in the reference soil structure, while using e captures the current soil density.

Why use ψ rather than void ratio or relative density directly? Because high confining stress levels tend to suppress dilatancy, the definition of state must take into account the stress level. It is the magnitude of dilation that determines strength, not the void ratio or density at which dilation occurs. This is exactly Rowe's stress–dilatancy concept in operation. Figure 2.4 presents some undrained tests on Kogyuk sand as experimental evidence of these effects. Samples at the same relative density (e.g. 37 and 103, or 112 and 113) show widely differing stress paths because of changed stress levels. In contrast, samples at the same ψ but different densities and stress levels (e.g. 108 and 103, or 45 and 112) show similar behaviour.

The value of the state parameter approach, to a practical engineer, is that many soil properties and behaviours are simple functions of the state parameter. There is more to the state parameter, however, than the utility of powerful normalizations. It turns out that the state parameter is fundamental to constitutive models of soil which have properties that are invariant with soil density and stress level, and one such model is presented in the next chapter.

2.2.2 Theoretical basis

When we first used ψ it provided a very useful normalization of test data (Been and Jefferies, 1985) which was seemingly independent of sand gradation, silt content, mineralogy, etc. However, there is a sound theoretical basis as to why ψ should be the first-order expectation of an appropriate normalizing variable, a basis that relates directly to the dilation rate.

Consider an element of soil at a void ratio e. Soils not at the critical state must change volume as they are sheared, since it is a basic postulate that the critical state represents the ultimate condition that will be attained after sufficient shear. As interest is in distortion effects, the argument is kept simple by assuming that shear is under the condition of constant mean stress. An expected state path for the shear of the soil can be approximated as a straight line, Figure 2.5. In reality, if starting from isotropic conditions, there will be an initial contraction before dilation sets in, but this is a detail on the basic state path vector.

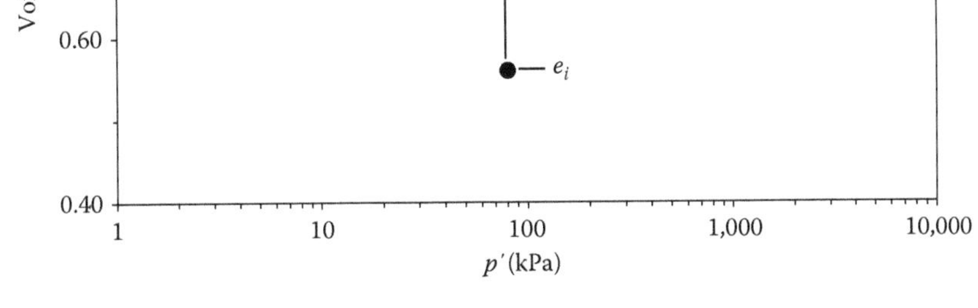

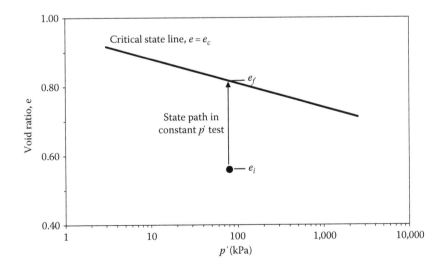

Figure 2.5 Idealized state path to illustrate the relationship of dilatancy to state parameter.

From the definition of volumetric strain in terms of void ratio, for the test illustrated in Figure 2.5, the volume change is given by

$$\varepsilon_v = -\frac{\Delta e}{1+e_i} = -\frac{e_f - e_i}{1+e_i} \tag{2.7}$$

where the subscripts i and f refer to initial and final conditions. For the idealized path, the final state is on the CSL (i.e. $e_f = e_c$). Introducing the definition of state parameter $\psi = e_i - e_c$ in (2.7) gives

$$\varepsilon_v = \frac{\psi}{1+e_i} \tag{2.8}$$

Equation 2.8 states the total volume change from the initial condition to the critical state during shearing. As the critical state is an equilibrium condition in which as many soil particles are moving into void space as are being displaced out of it by shear, it seems reasonable that this kind of dynamic balance will require displacements of at least two to three particle diameters to mobilize the condition in a shear zone. If shear zones are about 10 particles thick, this purely geometric consideration of critical state, ignoring formal strain measures, would also give $\varepsilon_q = 0.2 - 0.3$ as a first approximation. The average dilation on the approximate state path is then

$$D_{ave} = \frac{\varepsilon_v}{\varepsilon_q} \approx 4\frac{\psi}{1+e} \tag{2.9}$$

The exact coefficient in (2.9) will come from experimental data – 4 is only a plausible estimate of likely magnitude. The important point is that dilation will be approximately linearly related to the state parameter. Also notice that (2.9) is independent of stress level. Average dilatancy depends strongly on the state parameter and with a slight contribution from void ratio. This thought experiment was the basis of the proposition by Been and Jefferies (1985) that the state parameter ψ is a normalizing index for soil behaviour irrespective of stress level or material type.

As is evident from stress–dilatancy theory, actual (rather than average) D values depend on η as well as other factors. Because of this, the approach is to relate the maximum dilation (which is actually $D_{\min}$ because of the compression positive convention) to ψ. An alternative that follows from stress–dilatancy is the relationship of $(\phi_{\max} - \phi_c)$ to ψ.

The aforementioned argument is small strain which gives simplicity. This seems sufficient for a normalizing approach. A large strain variant of ψ is also readily defined. In passing it can be noted that Bolton's relative dilatancy index (Bolton, 1986) has the same intent as ψ, but ψ is both more convenient and fundamental.

2.2.3 Using initial versus current void ratio

It is convenient to use conditions at the start of a test when reducing laboratory data to develop soil properties, as that requires the least effort and is readily done from standard test report sheets. However, using initial void ratio when reducing data can lead to some not so obvious offsets from what you may have expected – for example, a dilation trend with state parameter that does not go through zero dilation for zero-state parameter (an effect caused by initial sample densification when sheared). Conversely, using the current value of the state parameter when considering a soil behaviour (say peak strength) requires the test data to be processed to update the void ratio from its initial value to the value at the soil behaviour being considered (peak strength in this example, where both void ratio and mean stress have changed from their initial values).

In our early publications using the state parameter (up to about 1990), everything was done in terms of initial conditions, that is, nice and simple standard engineering practice. However, working in terms of initial conditions is not smart when moving from practical engineering to doing the math and putting things in a formal framework – the math is far simpler if expressed in terms of current values. Since each approach has its own merits and uses, here we use the notations

- ψ_o = state parameter as measured at start of the loading path using initial void ratio and critical void ratio at initial mean effective stress: $\psi_o = e_o - (e_c \text{ at } p_o)$
- ψ = state parameter measured using current void ratio and critical void ratio at the current mean effective stress: $\psi = e - e_c$

Broadly, the ψ_o approach may be the most convenient when assessing test data and we follow that approach here. But when getting involved with constitutive modelling, soil properties will be best expressed in terms of ψ which is easy enough to do provided that the testing laboratory delivers data that can be imported into a spreadsheet. We will introduce the soil property χ later in this chapter using ψ, not ψ_o.

2.2.4 Experimental evidence for approach

So much for the thought experiment, what about data? There is, in current geotechnical literature, a substantial body of test data on the drained triaxial compression behaviour of sand. Mostly, this is for laboratory standard sands (e.g. Monterey, Ottawa, Toyoura and Ticino) that have been used in various academic studies, for example how strength is related to cone penetration resistance or how elasticity depends on void ratio and stress. Because most of this work was done outside a critical state context, the CSL was not usually determined. With the advent of interest in liquefaction, and the realization that liquefaction was most appropriately dealt with in a critical state context, Golder Associates was retained by the Canadian oil industry in the 1980s to determine the CSL of these standard sands. This means that the database from the literature can be used to assist in evaluating liquefaction potential.

Real sands found in-situ are not like laboratory sands in that they have at least a few (and often much more) fines in them, and they also tend to have a wider distribution of sand size particles. There are also mine tailings to consider, the sand- and silt-sized particles created when ore is crushed to extract metals. Over the past 30 years, Golder Associates has tested many of these in-situ sands and tailings, and most of these testing programmes have included the determination of the CSL. In addition, test data from the literature or universities have been obtained and digitized where possible to expand the database.

In total, both drained and undrained data are available for some 35 sands to sandy silts, and this database is summarized in Table 2.1, along with CSL parameters for several other sands that have been published. The determination of the critical state parameters for these sands is discussed shortly. The triaxial test data for most sands in Table 2.1 are provided as Excel spreadsheet files that can be downloaded from the web as detailed in the Preface. These files contain, for each test, the material type, sample preparation method, laboratory that carried out the test and the initial conditions (void ratio, confining stress and state parameter). For drained tests, the peak friction angle, dilation rate at peak and volumetric strain at peak are also tabulated, while for undrained tests, the undrained shear strength

Table 2.1 Critical state properties for some soils

	D_{50} (micron)	Fines (%)	e_{max}	e_{min}	G_s	Γ_I	λ_{10}	M_{tc}	χ_{tc}	Source reference
a) Laboratory standard sands										
Brasted sand	260	2	0.792	0.475		0.912	0.05	1.3	2.8	Cornforth (1964)
Castro S and B	150	0.12	0.84	0.5		0.791	0.041	1.22		Castro (1969)
Castro S and C	280	0.12	0.99	0.66		0.988	0.038	1.37	2.8	Castro (1969)
Hokksund	390	0.12	0.91	0.55		0.934	0.054	1.29	3.4	Golder project files
Leighton Buzzard	120	5	1.023	0.665		0.972	0.054	1.24	3.1	Golder project files
Leighton Buzzard	500	0.12	0.79	0.515		0.69	0.04			Hird and Hassona (1990)
Leighton Buzzard: 10% Mica	500	0.12	1.07	0.591		0.99	0.145			Hird and Hassona (1990)
Leighton Buzzard: 17% Mica	470	0.12	1.32	0.615		1.11	0.16			Hird and Hassona (1990)
Leighton Buzzard: 30% Mica	450	0.12	1.789	0.823		1.61	0.385			Hird and Hassona (1990)
Monterey	370	0.12	0.82	0.54		0.878	0.029	1.29		Golder project files
Nevada	150	7.5	0.887	0.511		0.91	0.045	1.2		Velacs project
Ottawa	530	0.12	0.79	0.49		0.754	0.028	1.13	4.8	Golder project files
Reid Bedford	240	0.12	0.87	0.55		1.014	0.065	1.29	3.8	Golder project files
Ticino-4	530	0.12	0.89	0.6	2.67	0.986	0.056	1.24	2.9	Golder project files
Ticino-8	530	0.12				0.943	0.031			Golder project files
Ticino-9	530	0.12				0.97	0.05			Golder project files
Toyoura	210	0.12	0.873	0.656		1	0.039	1.24	2.9	Golder project files
Toyoura	160	0.12	0.981	0.608	2.65	1.043	0.085		5.1	Golder project files
b) Natural sands										
Amauligak F-24	140	10			2.67	0.946	0.083	1.37		Golder project files
Amauligak F-24	144	21			2.69	0.966	0.124	1.33		Golder project files
Amauligak I-65	80	48			2.65	1.634	0.358	1.29		Golder project files
Amauligak I-65	310	9			2.67	1.018	0.153	1.42		Golder project files
Amauligak I-65	290	3			2.65	1.023	0.095	1.31		Golder project files
Erksak	320	1	0.808	0.614		0.875	0.043	1.27	3.4	Golder project files
Erksak	355	3	0.963	0.525	2.67	0.848	0.054	1.18	3.5	Golder project files
Erksak	330	0.7	0.747	0.521	2.66	0.816	0.031	1.27	3.5	Golder project files
Isserk	210	2	0.76	0.52	2.67	0.833	0.043	1.22	4.2	Golder project files

(Continued)

Table 2.1 (Continued) Critical state properties for some soils

	D_{50} (micron)	Fines (%)	e_{max}	e_{min}	G_s	Γ_1	λ_{10}	M_{tc}	χ_{tc}	Source reference
Isserk	210	5	0.83	0.55		0.879	0.089	1.24		Golder project files
Isserk	210	10	0.86	0.44		0.933	0.123	1.24		Golder project files
Kogyuk	350	2	0.83	0.47		0.844	0.064	1.31	4.5	Golder project files
Kogyuk	350	5	0.87	0.49		0.924	0.104	1.31		Golder project files
Kogyuk	350	10	0.93	0.46		1.095	0.205	1.24		Golder project files
Kogyuk	280	5	0.87	0.56		0.902	0.062	1.2	3.7	Golder project files
Nerlerk	270	1.9	0.812	0.536	2.66	0.849	0.049	1.29	5.2	Golder project files
Nerlerk	280	2	0.94	0.62		0.88	0.04	1.2		Sladen et al (1985)
Nerlerk	280	12	0.96	0.43		0.8	0.07	1.24		Sladen et al (1985)
Alaskan Beaufort	140	5	0.856	0.565	2.7	0.91	0.037	1.22	3.6	Golder project files
Alaskan Beaufort	140	10	0.837	0.53	2.7	0.92	0.053	1.20	3.6	Golder project files
West Kowloon sand	730	0.5	0.685	0.443	2.65	0.71	0.08			Golder project files
Chek Lap Kok	1000	0.5	0.682	0.411	2.65	0.905	0.13	1.3	4.0	Golder project files
Fraser River (Massey)	200	<5	1.1	0.7	2.68	1.071	0.038			Robertson et al (2000)
Duncan Dam	200	6.5	1.15	0.76	2.77	1.17	0.0854			Robertson et al (2000)
San Fernando 3	290	11			2.69	0.869	0.093			Seed et al (1988)
San Fernando 7	75	50			2.69	0.815	0.106			Seed et al (1988)
Bennett silty sand (a)	270	34	0.678	0.178	2.7	0.457	0.041	1.4		Golder project files
Bennett silty sand (b)	370	26	0.524	0.332	2.7	0.435	0.05	1.43		Golder project files
Bennett silty sand (c)	410	20	0.509	0.337	2.7	0.43	0.034	1.43		Golder project files
Estuarine sand	170	2	0.887	0.58	2.67	0.915	0.04			Golder project files
Northwest Brook Pit	720	2	0.682	0.392	2.74	0.665	0.036	1.53	5.3	Golder project files
c) Tailings sands and silts										
Hilton Mines	200	2.5	1.05	0.62		1.315	0.17	1.42	2.7	Golder project files
Highland Valley copper	200	8	1.055	0.544	2.66	0.98	0.068			Robertson et al (2000)
Faro lead-zinc	100	30	0.99	0.556	4.48	0.921	0.082	1.19		Golder project files
Faro lead-zinc	50	65	2.017	0.837	3.97	1.076	0.159	1.2		Golder project files
Sudbury (nickel) - A	115	35	1.032	0.537	3.03	0.938	0.112	1.45		Golder project files

(Continued)

Table 2.1 (Continued) Critical state properties for some soils

	D_{50} *(micron)*	*Fines (%)*	e_{max}	e_{min}	G_s	Γ_1	λ_{10}	M_{tc}	χ_{tc}	*Source reference*
Sudbury (nickel) - B	50	65			2.98	0.868	0.108	1.45		Golder project files
Syncrude oil sand tailings	207	3.5	0.898	0.544	2.64	0.86	0.065	1.33	5.3	Golder project files
Syncrude (Mildred Lake)	160	10	0.958	0.522	2.66	0.919	0.035			Robertson et al (2000)
Yatesville silty sand	100	43			2.67	0.653	0.164	1.33		Brandon et al (1991)
Merriespruit gold tailings	140	0	1.221	0.738		1.24	0.07			Fourie & Papageorgiou (2001)
Merriespruit gold tailings	130	20	1.326	0.696		1.18	0.05			
Merriespruit gold tailings	110	30	1.331	0.577		0.96	0.035			
Merriespruit gold tailings	60	60	1.827	0.655		0.8	0.02			
Endako silt	5.5	99.2			2.69	2.063	0.541	1.37	0.74	Golder project files
Nevada copper tailings	60	53	1.056	0.586	3.03	0.858	0.111	1.57		Golder project files
Argentina mixed tailings	60	53				0.74	0.08	1.56		Golder project files
Conga dry tailings	25	80			2.75	0.89	0.128	1.56		Golder project files
Tailings beach	75	51	1.015	0.685	2.68	0.713	0.086	1.44		Golder project files
Tailings sand	170	22	1.065	0.512	2.68	0.914	0.115	1.45		Golder project files
Oxide tailings	43	75			2.78	0.763	0.084	1.45		Golder project files
Toromocho	60	58			3.14	1.023	0.145	1.56	5.9	Golder project files
Skouries	55	65			2.77	0.736	0.069	1.4		Golder project files
d) Other materials										
Dogs Bay sand (carbonate)	280	1	2.2	1.5	2.71	3.35	0.77	1.65		Coop (1990), index data are for pretest (uncrushed) sand
TVA Kingston coal ash	37	72			2.37	1.08	0.12	1.19		AECOM (2009)

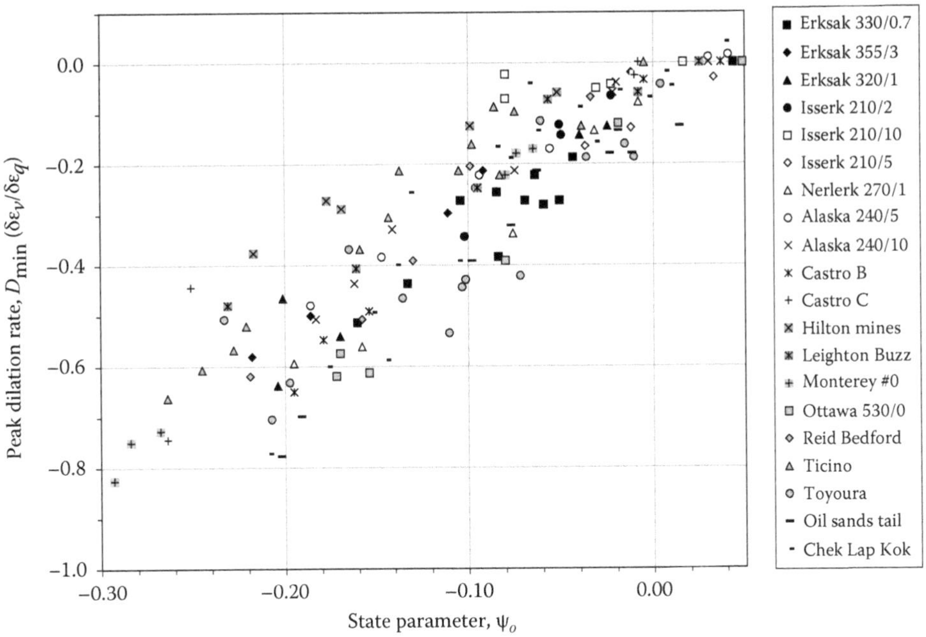

Figure 2.6 Maximum dilatancy (D_{min}) of 20 soils in standard drained triaxial compression.

and associated pore pressure parameter A_f (for loose samples) or the mobilized friction angle and rate of change in pore pressure (for dense samples) are provided. A shorthand notation for the grain size of sands has been used throughout: Erksak 360/3 sand denotes Erksak sand with a D_{50} of 360 μm (0.360 mm) and a fines content of 3%. Where the full grain size curves are available, these are also given as downloadable files.

Figure 2.6 shows dilation versus the state parameter for 20 of the soils listed in Table 2.1, presented in the form of D_{min} against ψ_o as explained earlier. This figure is an extended and re-plotted version of the figure published some 30 years ago (Been and Jefferies, 1985, 1986), using the work conjugate strain invariant definition of dilatancy (Equation 2.4). The data lie within about ±0.2 of $D_{min} = 3\ \psi_o$, which compares nicely to Equation 2.9. The data plotted on this figure range from clean quartz sands through to silty sands and the mean effective confining stress from 19 to 1200 kPa. Hopefully, these data are sufficiently convincing as to ψ being an effective and universal normalizing parameter that works exactly as expected.

When the utility of ψ was proposed, interest centred on the peak friction angle rather than dilatancy, for which the relationship is shown in Figure 2.7. There is arguably no more scatter in the peak friction angle than with the relationship to D_{min}. Given the understanding from stress–dilatancy theory, it is appropriate also to plot $\phi_{tc} - \phi_{cv}$ versus ψ_o. In effect, this highlights the component of peak strength that can be attributed to stress–dilatancy behaviour rather than the intrinsic friction angle ϕ_c. Such a plot is presented in Figure 2.8, with no apparent difference in scatter in the data. Generally, $\phi_c \approx 31°$ for sub-rounded hard quartz sands, but care is needed as ϕ_c can change markedly with mineralogy and grain shape. How to measure ϕ_c is covered shortly.

Volumetric strain at peak strength is of interest for liquefaction evaluation, because if the volumetric strain is compressive in drained shear, then in an undrained case the pore pressures would be positive and the undrained strength of the sand would be less than the

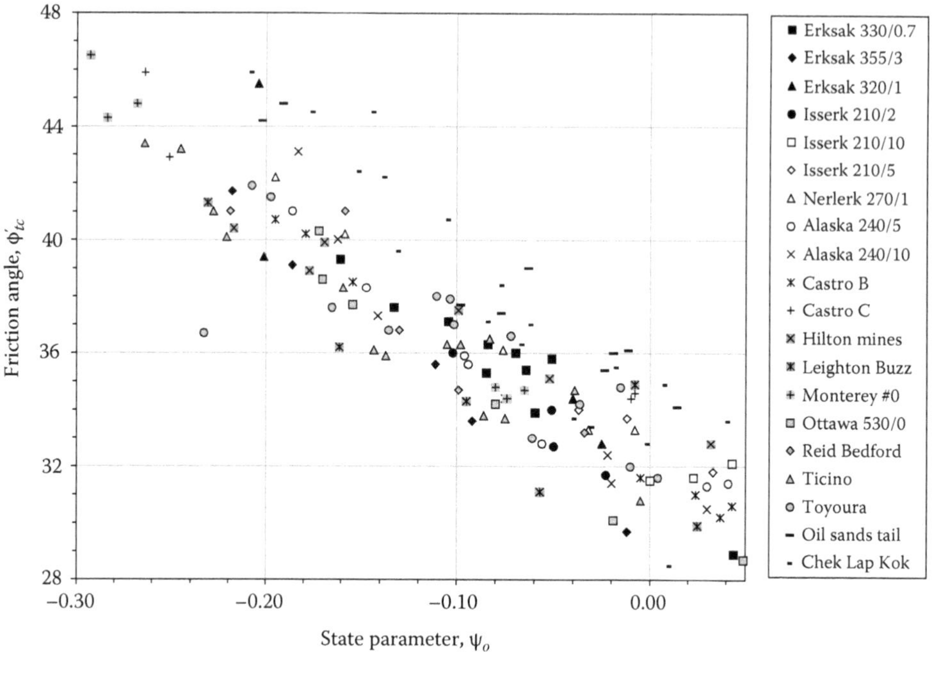

Figure 2.7 Peak friction angle in standard drained triaxial compression.

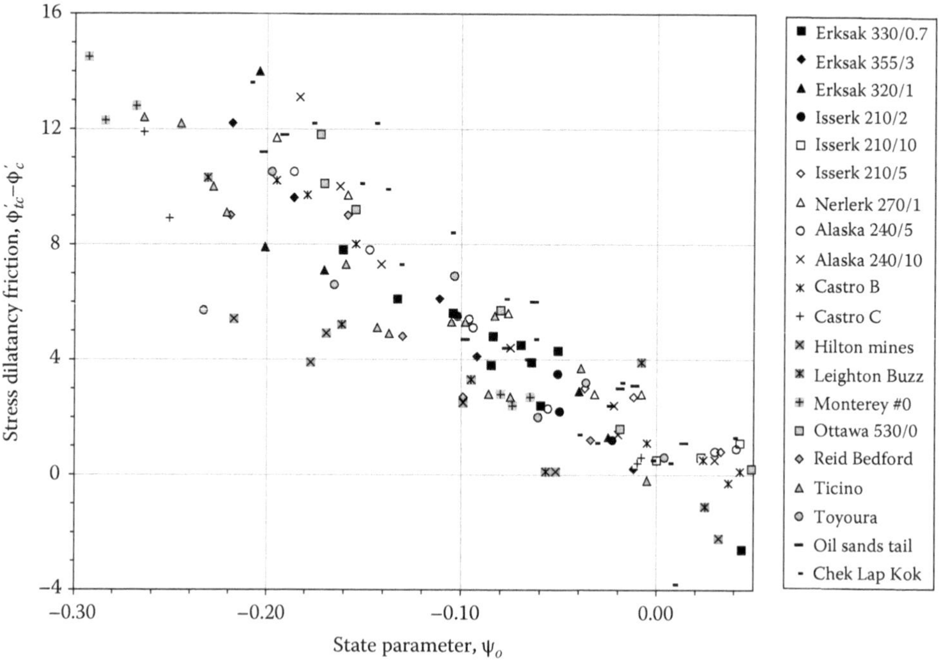

Figure 2.8 Stress–dilatancy component of peak strength of 20 soils in standard drained triaxial compression.

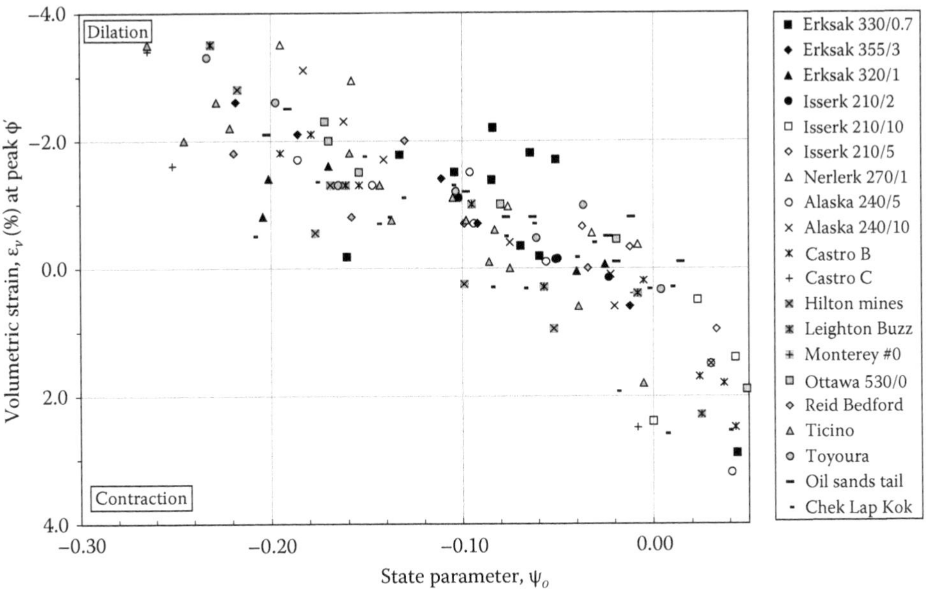

Figure 2.9 Volumetric strain at peak stress for drained triaxial compression tests on 20 sands.

drained strength. Positive pore pressures at peak strength may indicate the potential for large strains and rapid failures. In contrast, if the total volumetric strain is dilatant, pore pressures would be negative in the undrained case, and therefore, the undrained strength would be greater than the drained strength. Rapid failures or flow slides would be unlikely. Figure 2.9 therefore shows the total volumetric strain at peak strength for standard drained isotropic compression triaxial tests on sands. There is a clear trend of larger overall volumetric dilation (more negative ε_v) as ψ_o becomes more negative.

Recall the earlier discussion on how a dilatant sand by the rate definition would not necessarily show dilatant volumetric strain at peak strength. Figure 2.9 shows that an initial state of something like $\psi_o = -0.06$ is necessary on average to ensure that a net dilation occurs at peak ϕ. (Note that the stress path influences the volumetric strain at peak ϕ. In standard drained triaxial compression tests, the mean stress increases at the same time as the deviator stress. The corresponding volume change is a combination of volumetric compression due to the mean stress increase, and a dilational component [positive or negative] due to shearing. Constant mean stress or extension tests would be expected to show volumetric strains slightly different from Figure 2.9.)

In particular, notice from the data in Figure 2.9 that $\psi_o = 0$ does not distinguish tests that are overall contractive at peak strength from those that are not. This is a direct consequence of ψ operating in relation to the rate definition of dilation. For a soil to show an undrained strength that is greater than its drained strength, which is the basic proposition of the Steady State School (Chapter 1), there must be a net volume increase by peak strength and something denser than $\psi_o = 0$ is needed for this. How much denser depends on the in-situ geostatic stress, the relative amounts of increased shear to increased mean stress and some soil properties. As noted earlier, about $\psi_o < -0.06$ is required before there can be reasonable confidence that the undrained strength exceeds the drained. Simply being denser than the CSL (SSL) is not going to provide adequate engineering performance for post-liquefaction stability, as is now widely known.

2.2.5 Normalized and other variants of the state parameter

Despite the theoretical basis for ψ, somewhat curiously, people have sought to improve on it by introducing normalizations. The desire for a normalized state parameter seemingly developed because of the scatter in the ϕ_{max} versus ψ relationship. One aspect of ψ is that it is not the sole parameter controlling soil behaviour, and the role of soil fabric and overconsolidation ratio will be discussed shortly. Before that, however, any notions of improving the state parameter by normalization need to be buried. The normalizations considered here include the use of index void ratios, λ, and fines content corrections.

The first normalized state parameter was suggested by Hird and Hassona (1986), who introduced both maximum and minimum void ratio into the grouping $\psi/(e_{max} - e_{min})$. This suggestion was subsequently reiterated by Konrad (1990a). There are three difficulties with this normalization. First, it does not improve the correlation between ϕ and ψ with substantial scatter continuing to exist, Figure 2.10. Second, there is no role for $(e_{max} - e_{min})$ in the theoretical framework. Finally, e_{max} is a dubious parameter in its own right, as it is unclear that there is a maximum void ratio under an isotropic stress state. Soil will become progressively more susceptible to the influence of shear stress as it becomes looser, but this does not mean that very high void ratios are not possible (in principle) under isotropic stress. Also, as is well known, e_{max} is difficult to measure with any degree of repeatability between different testing laboratories.

A more interesting proposal is to use the parameter group ψ/λ_e. This parameter group arises for the particular case of undrained loading in which constant void ratio is enforced and provided that the CSL can be approximated as a straight line in semi-log space. Given this situation, the ratio of mean stresses $p'/p'_c = \exp(-\psi/\lambda_e)$ from (2.1). However, for the case of drained shear under constant mean stress, there is theoretically no role for λ. Because undrained loading is the imposition of a boundary condition (i.e. no drainage) and is not a fundamental and

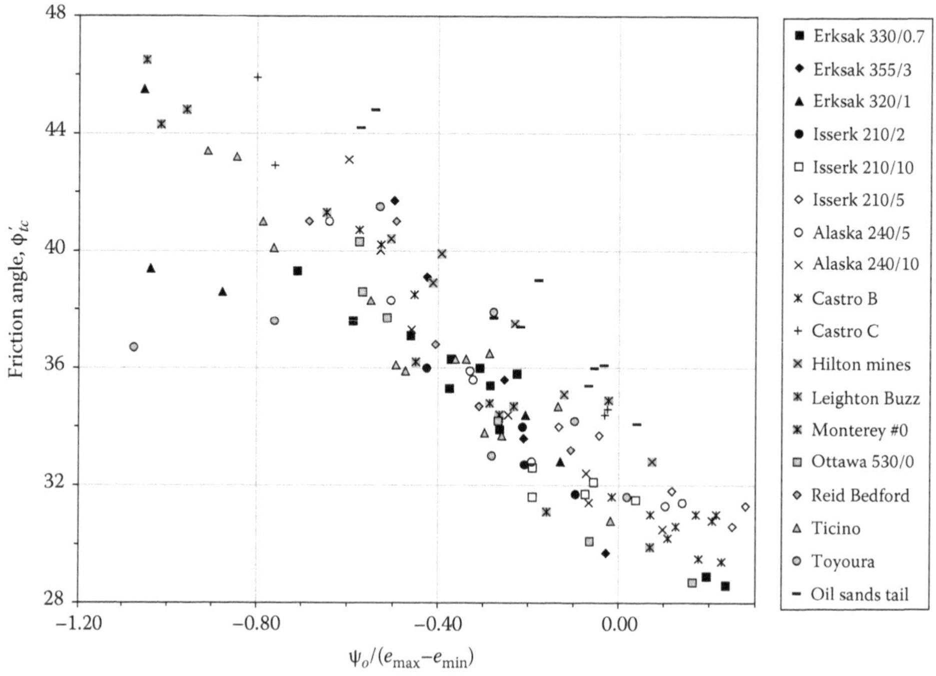

Figure 2.10 Maximum dilatancy (D_{min}) versus state parameter normalized by range of accessible void ratios ($e_{max} - e_{min}$). Compare the lack of improvement over Figure 2.6.

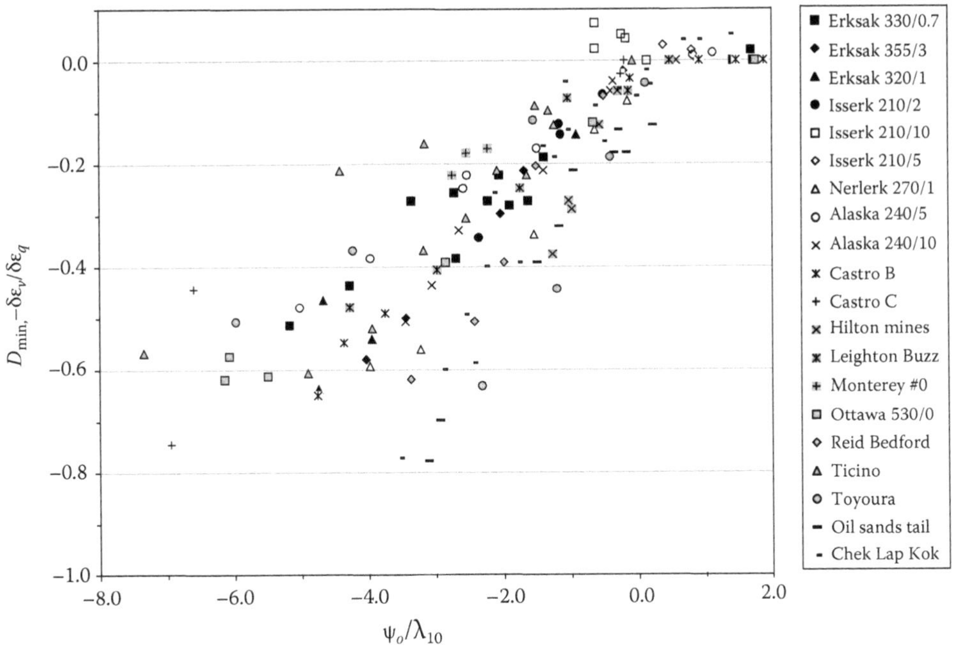

Figure 2.11 Maximum dilatancy (D_{min}) as a function of ψ_o/λ_{10}. There is no improvement in the correlation compared to ψ_o alone (compare with Figure 2.6).

intrinsic behaviour of the soil skeleton, using a grouping that arises only in undrained loading cannot be correct. This can be checked using the database by plotting D_{min} against ψ_o/λ, Figure 2.11. The scatter in values of ψ_o/λ close to zero is similar to that for the basic state parameter plot (Figure 2.6), and the scatter increases markedly for denser states. Adopting ψ_o/λ is a backward step.

Some improvement in precision might be expected if ψ was normalized by a $(1+e)$ term, based on the considerations that lead to Equation 2.9. This normalization does indeed reduce the scatter in predicting D_{min} from ψ_o as illustrated in Figure 2.12. However, the improvement is small and practical engineering seems best served using the simple and original definition of ψ_o.

An interesting suggestion is that the state parameter should only be computed on the sand-sized fraction, with at least some of the fines being viewed as inert particles filling void space but not transferring any forces between the sand particles. Taking fines (f_c) as material passing the 75 μm sieve, an equivalent granular void ratio e^* can be defined (e.g. Thevanayagam et al., 2002; Yang et al., 2006) as

$$e^* = \frac{e + (1-b)f_c}{1-(1-b)f_c} \tag{2.10}$$

where b is the fraction of fines that are active in transferring forces between soil grains and which is unmeasurable from first principles. In practice, (2.10) is applied by regressing data to coalesce a family of CSLs in e-log(p) space into a single relationship in e^*-log(p) space. The coalesced line is known as *equivalent granular steady-state line* (EG-SSL). For example, Chu and Leong (2002) reported a single EG-SSL for a marine sand with variable fines to a maximum of 10%. What can be the objection to this? There are two objections, one fundamental

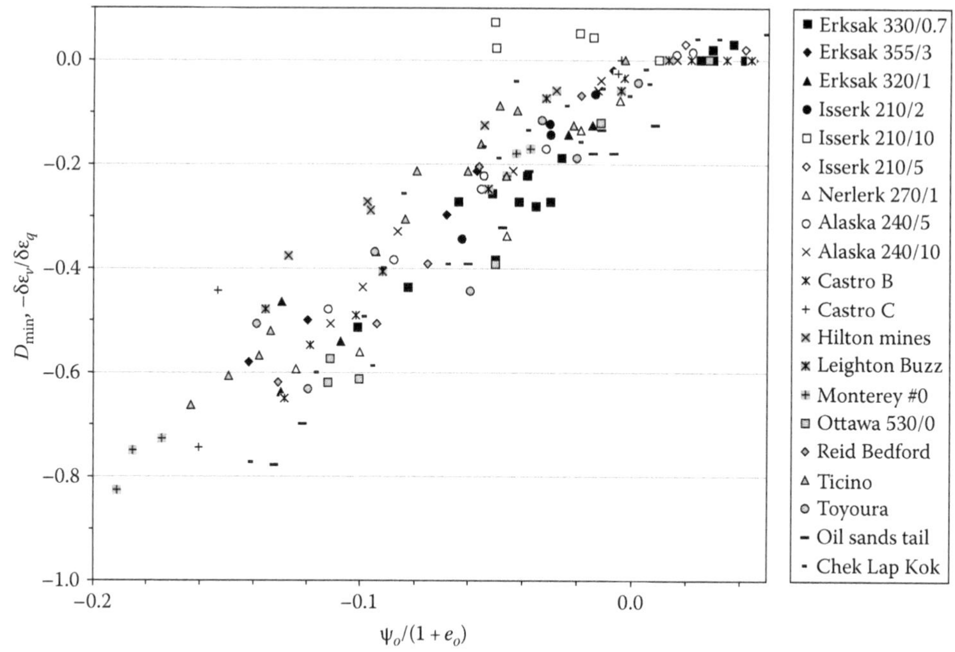

Figure 2.12 Maximum dilatancy (D_{min}) as a function of ψ_o normalized by ($1 + e_o$). There is a small improvement compared to state parameter alone (compare with Figure 2.6).

and one experimental. From the fundamental standpoint, there is nothing in the definition of ψ that limits the framework to sands. The mathematics actually has no intrinsic relation to geology, with the CSL simply being a representation of the end state of a particulate material. It could be applied to, say, granulated sugar and we will show data on pure silts later (i.e. for soils with $f_c = 100\%$) where all the particles are fines and where it is self-evident that one must work with the void ratio of those fines. From the experimental standpoint, the idea behind the EG-SSL really comes down to the concept that fines only affect the soil property Γ. This concept of fines only affecting Γ is incorrect in general, although it works in some instances, as can be seen from the earliest data we published (Been and Jefferies, 1985) where fines changed λ as well as Γ in the sands tested for construction of artificial islands.

Finally, there is the *relative state parameter index* (ξ_R) proposed by Boulanger (2003) that uses relative density instead of void ratio as the state variable. This is ill conceived for two reasons. First, you need to determine e_{max} and e_{min} which, as discussed earlier, have no place in a critical state framework and are problematic to measure. Second, as proposed by Boulanger, the CSL is not measured for each sand, but is computed from Bolton's (1986) dilatancy index. This is equivalent to assuming a single, unmeasured, CSL for all sands in terms of relative density – an idealization that does not fit the data except for a narrow range of single-sized quartz research sands. The relative dilatancy index is an unnecessary backward step offering no additional simplicity for the loss in generality.

2.2.6 State–dilatancy (soil property χ)

Section 2.2.2 presented a thought experiment from which it was deduced that dilation should be related to the state parameter. This idea was behind the original Been and Jefferies (1985) paper which went one step further and suggested that the trend between D_{min} and ψ_o was unique based on testing a range of sands with variable fines contents (see Figure 2.6). This was an interesting idea for its time, but it has been overtaken as the state parameter approach has moved from sands to till-like soils and to pure silts. Broadly, well-graded soils have less free void space and correspondingly show a greater sensitivity to the effect of volumetric strain while silts do the opposite. This leads to a choice: (1) continue with the original idea while accepting it only gives approximate trends or (2) accept that there is a soil property involved in relating state parameter to maximum dilatancy. The second approach broadens the utility of critical state soil mechanics and, accordingly, a *state–dilatancy* law is defined (analogous to stress–dilatancy) as

$$D_{min} = \chi_{tc}\psi \tag{2.11}$$

where χ_{tc} is a soil property defined under drained triaxial compression. Importantly, note that ψ is defined as its current, not initial, value and D_{min} generally occurs at the peak stress ratio. D_{min} is preferred to strength (i.e. η_{max}) to quantify the effect of state as D_{min} is related to the change in void ratio and ψ has void ratio as its input – essentially, the same quantity is used on both sides of Equation 2.11. It also harks back to Reynolds (1885) who showed that dilation is a kinematic consequence for deformation of particulate materials. It is all a matter of particle geometry and the ability of particles to move relative to one another. Stress change is the consequence of strains, not the input. Using the current value of ψ means that there is no offset (or constant) in (2.11) and the condition $\psi = 0$ naturally gives $D_{min} = 0$, which is the critical state of course. Figure 2.13 shows several examples of D_{min} versus ψ from drained triaxial tests from which the values of χ_{tc} in Table 2.1 were determined. If we then replot the data from Figure 2.6 using D_{min}/χ_{tc}, there is a notable reduction in scatter of the data set as shown in Figure 2.14. Accepting that a soil property is involved greatly improves the available precision from the state parameter approach, although we have added a little bit of complexity in processing laboratory test data to gain a simple and elegant framework. This is not quite as onerous as it might seem, because, as D_{min} is a limiting condition, there are no elastic strain rates at peak strength so you do not have to bother with elasticity (i.e. at peak strength $D_{min} = D^p_{min}$). In practice all you need to do is account for the change in void ratio from the start of the test to peak strength and to use the mean stress at peak strength when computing the state parameter – the mechanics of this data processing are presented in Chapter 9.

2.2.7 Influence of fabric

While ψ is the major controlling influence on the behaviour of sand, it does not provide a complete description of sand behaviour. This is well illustrated by Figure 2.15, which shows test results from two samples of Kogyuk sand. These samples were prepared by two different methods, moist tamping and wet pluviation, to essentially the same density and subjected to equal confining pressures. Thus, the state parameter and all other testing conditions were the same, except for the preparation method. The only difference between the samples was the soil structure or fabric induced by sample preparation.

For these two samples, neither the deviator stress curves nor the volumetric strain curves are similar. In particular, the volumetric strain behaviour is substantially different. Despite this, both the peak and ultimate shear stresses are very similar, as is the maximum dilation rate. Consequently, ϕ' values are also similar, while volume changes (and pore water pressures if the tests were undrained) are very different. The question remains: what is the influence of fabric on ϕ' for sands?

Arthur and Menzies (1972) showed the importance of initial fabric to sand behaviour, with differences of over 200% in axial strain to reach a given stress ratio for a sand at the same void ratio and stress level. Further, they found a typical variation of about 2° in ϕ as a

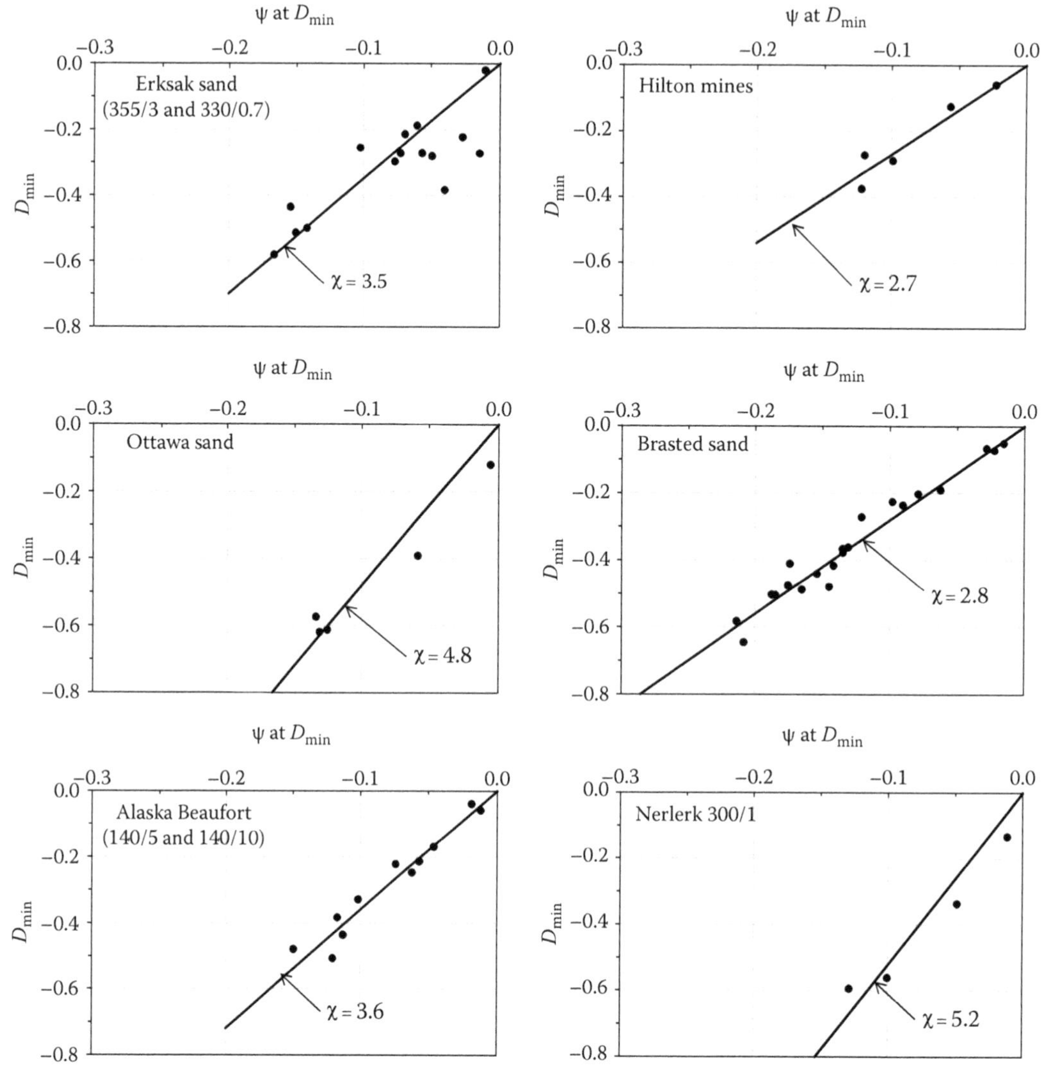

Figure 2.13 Derivation of χ parameter for selected sands (χ is a material property relating maximum dilatancy to the state parameter for each sand).

function of fabric. Similar direct effects of fabric have been reported by Oda (1972a,b), Oda et al. (1980) and Muira et al. (1984). Figure 2.16 shows data from Tatsuoka (1987) where the friction angle and dilatancy in triaxial tests vary quite significantly as the direction of loading relative to deposition changes. Reported effects of initial fabric on static and cyclic sand behaviour in laboratory tests indicate that initial fabric generally appears to have an influence of ±2° on the friction angle. The reported effects of initial fabric by Oda (1972a) and Tatsuoka (1987) are superimposed on the ϕ versus ψ_o relationship in Figure 2.17. It is readily apparent that the variation due to initial fabric is comparable to the overall scatter in the data.

For cyclic loading, which is discussed in more detail in Chapter 7, the aforementioned conclusion – that ψ is the most important variable for the behaviour of sand and that fabric

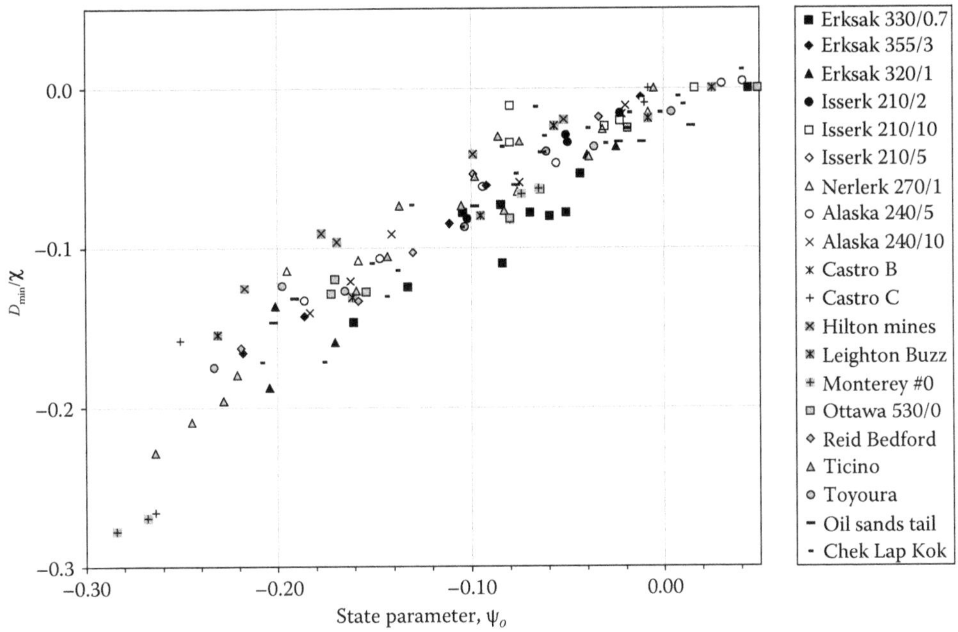

Figure 2.14 Maximum dilatancy (D_{min}/χ) versus ψ in triaxial compression. There is a notable improvement normalizing D_{min} by χ compared with Figure 2.6.

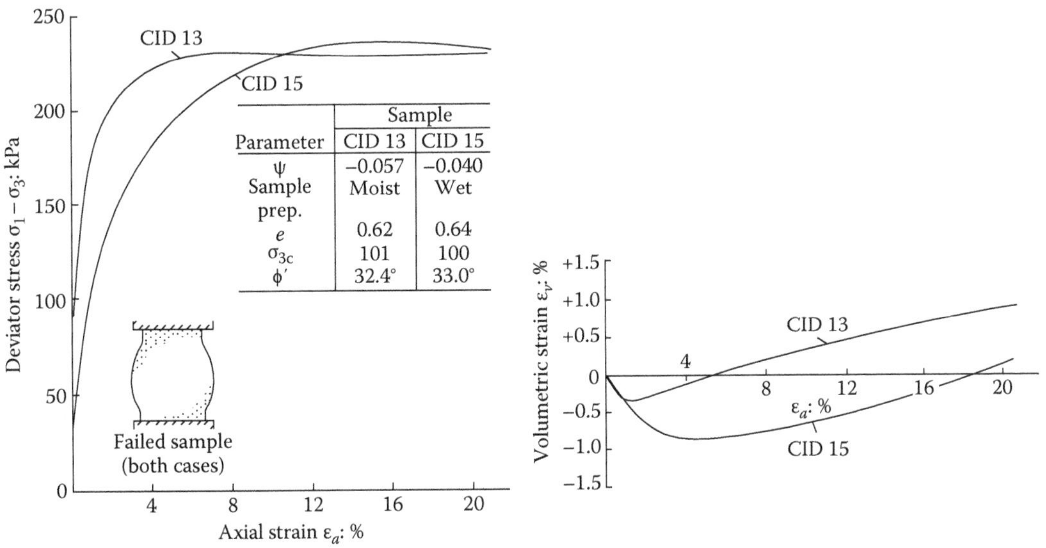

Figure 2.15 Effect of sample preparation on the behaviour of Kogyuk sand. (From Been, K. and Jefferies, M.G., *Géotechnique*, 35, 2, 99, 1985. With permission from the Institution of Civil Engineers, London, U.K.)

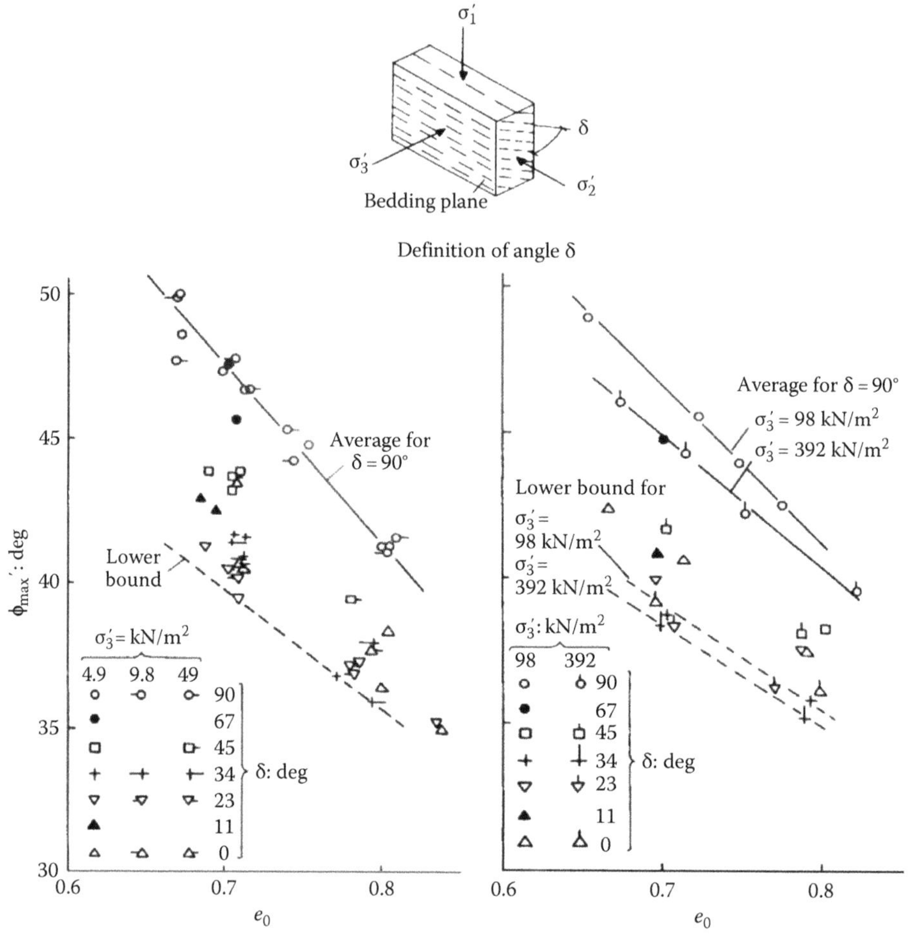

Figure 2.16 Effect of loading direction to soil structure on the strength of Toyoura sand. (After Tatsuoka, F., *Géotechnique*, 37, 2, 219, 1987. With permission from the Institution of Civil Engineers, London, U.K.)

is a second-order effect – is not necessarily appropriate. Figure 2.18 illustrates data from Nemat-Nasser and Tobita (1982) that show the independent influence of fabric and void ratio on the cyclic strength of sand. For a state parameter change of 0.07 (a void ratio change from 0.65 to 0.72), the shear stress for liquefaction in 10 cycles reduces by 32%. The shear stress to liquefaction in 10 cycles increases by a similar amount if pluviation is used as a sample preparation method, rather than moist tamping. This demonstrates that fabric is of equal importance for the cyclic behaviour of sands – a little inconvenient given that there is no easy and standard method for measuring fabric.

One of the more promising approaches to deal with fabric is to invoke particle mechanics, where the interparticle contact orientations and forces are explicitly included in the constitutive behaviour (Rothenburg and Bathurst, 1989, 1992). Within the context of these particulate models, the coordination number, that is the average number of physical contacts per particle, represents the equivalent of state parameter as a scalar description of packing density. Fabric, or anisotropy, is captured through parameters representing the

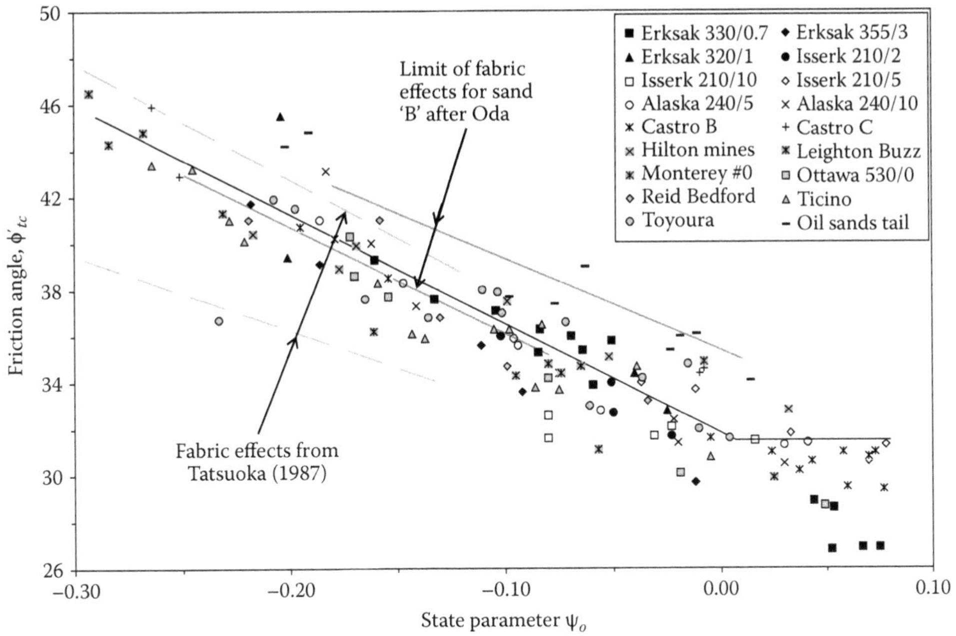

Figure 2.17 Effect of fabric on friction angle of sands reported by Tatsuoka and Oda, compared to general correlation of friction angle to state parameter.

distribution of particle contact directions, normal force directions and tangential force directions.

The practical difficulty is in measuring the parameters of fabric anisotropy, either in the field or in laboratory samples. This is still an area of active research. For the moment, the best that can be done is to recognize that fabric is an important influence and to be conservative in how the effects are included in engineering design and assessment.

2.2.8 Influence of OCR

Overconsolidation ratio (OCR) is a logical and frequently used state parameter for clays. A limited number of tests were carried out on Erksak sand to determine the effect of OCR in sand, and the measured influence of OCR on friction angle and dilatancy is illustrated in Figure 2.19. These data show that when OCR = 4 (based on isotropic compression loading), if the state parameter is calculated on the basis of the state at the maximum past mean stress (i.e. the yield point), then the influence of OCR on ϕ is negligible.

Importantly, this also confirms that ψ is associated with the current yield surface, since overconsolidation is an unloading from a state of plastic yield into an elastic domain. When a sample is unloaded to increase its OCR, the yield surface remains unchanged (actually, this is not entirely true, as we will see in Chapter 3, but it is a sufficient idealization here). Hence, the state parameter should be associated with the condition prior to unloading.

Figure 2.19 also shows that the effect of calculating the state parameter using the void ratio at the start of shearing, again for OCR = 4, is about a 1° drop in ϕ, erroneously suggesting that the effect of OCR in sand is to make the soil weaker. In reality, the effect of the OCR is to change both the state and fabric of the sand, usually resulting in an increase in ϕ as seen later in Chapter 6.

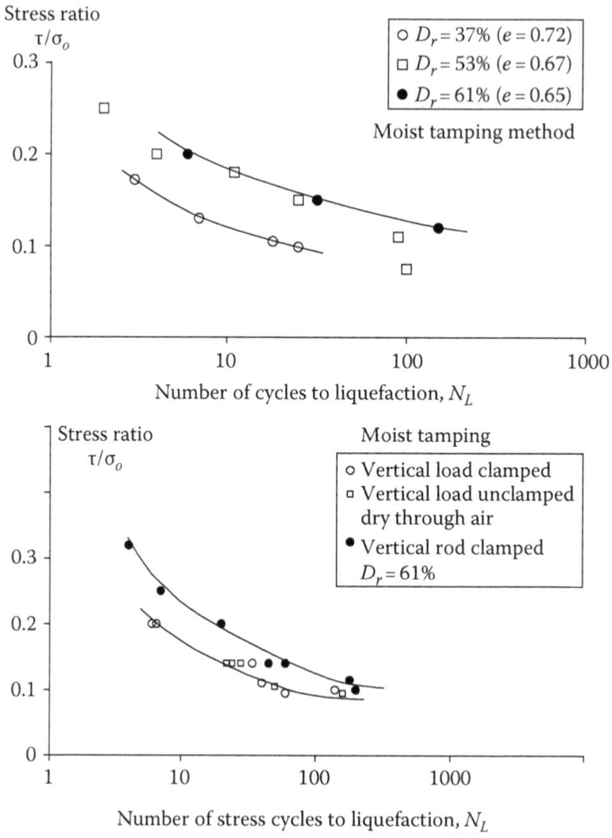

Figure 2.18 Comparison of the effect of void ratio and sample preparation method on the cyclic strength of two sands in simple shear. (After *Mech. Materials*, 43, Nemat-Nasser, S. and Tobita, Y., The influence of fabric on liquefaction and densification potential of cohesionless sand, 43–62, Copyright (1982), with permission from Elsevier.)

2.2.9 Effect of sample size

It is well known that localized shear bands form in dense sand samples sheared under drained conditions. This effect is seldom considered when assessing sand behaviour, although there are theories which predict and describe this phenomenon (bifurcation).

Given that shear band thickness is expected to be a function of grain size, rather than sample size, there might be an influence of sample size on the measured behaviour of sand samples. The importance of sample scale, or sample size, on behaviour is best appreciated when considering that our present understanding of sand behaviour is almost entirely based on laboratory tests carried out on samples a few centimetres in size, whereas actual construction involves soil typically at a scale of metres (for footings) or hundreds of metres (land reclamation, tailings dams, etc.). The scale factors in terms of volume tested are enormous.

Jefferies et al. (1990) report some testing on Ticino sand to investigate the effect of scale on behaviour. A series of eight tests were carried out, with all factors except the sample size kept the same. The samples were all prepared by dry pluviation to a void ratio of 0.59 ± 0.02,

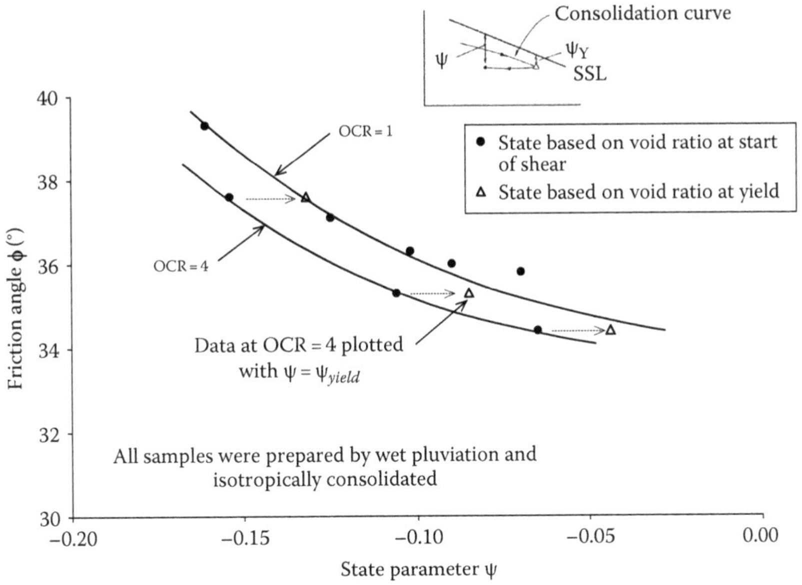

Figure 2.19 Influence of overconsolidation ratio on the friction angle of Erksak 330/0.7 sand.

or a relative density close to 100%, and consolidated to an isotropic confining stress of 100 kPa. Two tests each were carried out on samples of 36, 75, 154 and 289 mm in diameter, with a height to diameter ratio of approximately two.

The measured behaviour is presented in Figure 2.20, as deviator stress and volumetric strain plotted against nominal axial strain. For clarity, only one test is shown at each sample size; duplicate tests showed similar behaviour at each sample size. It is immediately apparent that the sample behaviour has changed as a consequence of a change in sample size and that an intrinsic scale effect exists.

It is helpful to consider first what has not changed as a result of sample size. The peak dilation rate is the same for all tests. Although there is some change in peak deviator stress, there is no systematic bias with sample size, and the variation (±5%) is within normal test repeatability. Scott (1987) reported similar conclusions on the effect of scale on sand behaviour.

Three clear effects of scale are apparent in Figure 2.20 – initial modulus, volumetric strain at a given axial strain and post peak brittleness. The effect of sample size on the slope of the stress–strain curve in initial loading (apparent modulus) was also evident in the data reported by Scott (1987). However, Scott's data showed an increase in stiffness, whereas our data show a reduction in stiffness with sample size. Although peak dilatancy is identical in all samples, there is a progressive change with sample size in the overall volumetric strain from first loading. The total volumetric strain at an axial strain of 6%, which corresponds to the flat portion of the stress–strain curve before the onset of softening, varies from +4.5% to only 3% as sample size is changed by an order of magnitude. This change in strain might seem small, but it is nevertheless a 30% reduction in the gross volume change and will clearly influence pore pressures in an undrained soil mass.

The post peak brittleness is an anticipated result if one considers shear band formation. The smaller sample shows a small decline in strength with strain after peak, whereas

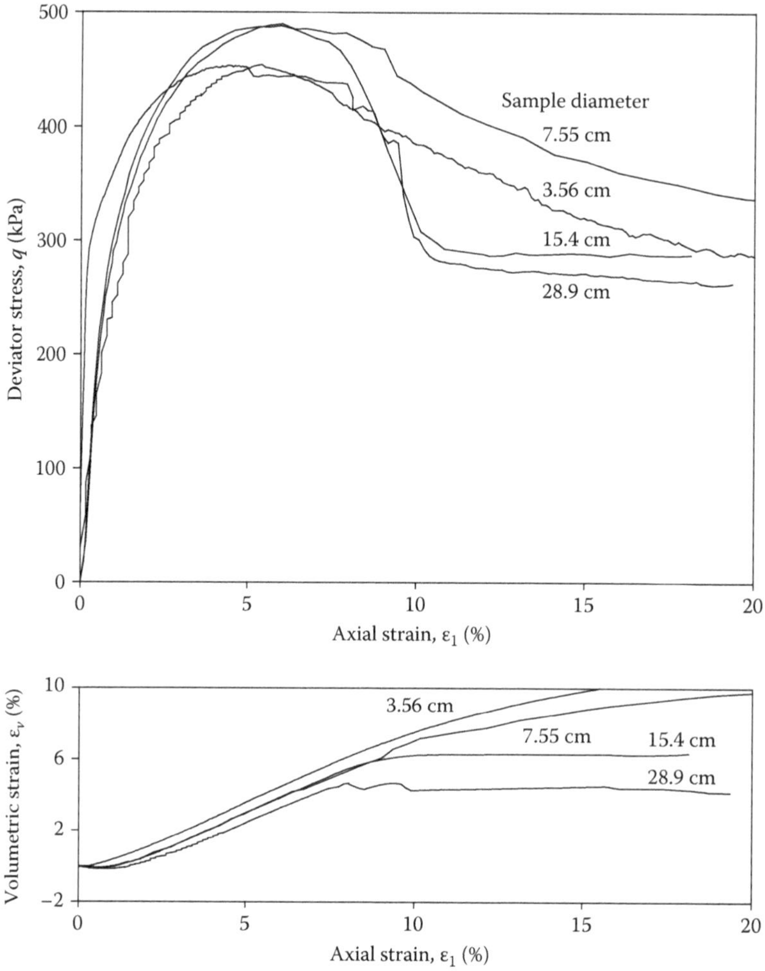

Figure 2.20 Effect of sample size on the behaviour of dense Ticino sand.

the larger samples show a more rapid drop to a stable residual strength. The nature of shear zonation encountered is illustrated in Figure 2.21, which presents a photograph of the shear bands observed on the large (289 mm diameter) sample. The occurrence of multiple shear bands was a surprising result, as generally only a single shear band is found in the triaxial compression failure of conventional sized samples.

There are many interesting studies of the formation of shear bands and void ratio evolution in shear zones (e.g. Desrues et al., 1996). At this stage, an awareness of the phenomenon and how it may affect field and laboratory behaviour of sands is needed.

2.3 EVALUATING SOIL BEHAVIOUR WITH THE STATE PARAMETER

Given the premise of the state parameter approach, that the behaviour of sand under shear loading is primarily a function of its state and a few properties and that the preferred measure of in-situ state is ψ, the question now is how do you set about using this approach? The short answer is that both laboratory and in-situ tests are required, but first some clarification

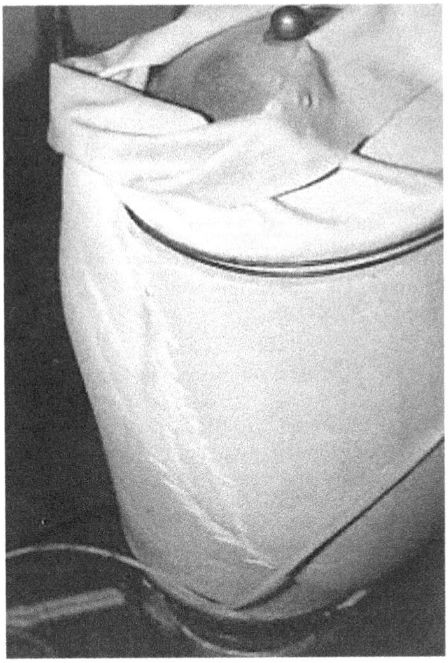

Figure 2.21 Multiple shear bands evident through membrane in large (300 mm diameter) sample after drained shearing.

is needed on the important differences between the properties of a soil, the state of a soil, boundary conditions and a soil's behaviour. These terms are often confused or used loosely, which is unhelpful. Table 2.2 describes the terms and gives some examples.

The techniques developed in geotechnical engineering to measure the behaviour of sands (and silts for that matter) are based on the premise that there is no such thing as an undisturbed sample of sand. This contrasts with the engineering approach to clays, which tends to be dominated by laboratory element tests on high-quality samples.

At the first-order level, the behaviour of sand under loads is determined by

- The density aspect, which is expressed in terms of the state parameter ψ
- The fabric, or structural arrangement of the particles (which is sometimes also termed the anisotropy or inherent anisotropy of the sand)

Even the density of sand is not accurately measured from samples. Sampling results in dilation (in dense sands) or contraction (in loose sands). The best we can do with samples is to reconstitute them to measure properties that do not depend on density.

Preserving the fabric of sand samples is even more problematic than measuring its density because minute strains can significantly alter the relative position of sand grains and how the interparticle forces are distributed in the sand mass. It has to be assumed that fabric is not preserved during the sampling process, even when going to the extreme of freezing the soil in-situ and then coring the frozen ground. (We cannot demonstrate that this assumption is true or false, because there is no accepted method of quantifying the fabric of a sand in the laboratory or in-situ. However, the assumption is reasonable given the little confidence in preservation of density during sampling.)

Table 2.2 Clarification of terminology for describing soils

Term	*Meaning*	*Examples*
Intrinsic properties	Properties that do not vary as a function of state or boundary conditions. They should be unambiguously measurable.	Grain size distribution Grain shape Mineralogy Interparticle friction Critical state locus
State	The conditions under which a soil exists, in particular the void ratio, the stress conditions and the arrangement of the particles (fabric).	ψ for all soils Overconsolidation ratio for all soils Relative density for sands Fabric when technology develops Stresses (K_o)
Boundary conditions	Conditions that are imposed from the outside on the soil mass. Boundary conditions do not change the properties of a soil but they may affect its behaviour.	Drained or undrained in the lab Distance to free surface or drainage zone
Behaviour	Response of a soil to the applied boundary conditions. This will depend on a soil's properties, its state and the imposed boundary conditions.	Undrained strength ϕ' Dilation rate Stress–strain curve SPT *N* value Cone penetration resistance
Parameters	Quantified properties that describe the behaviour of soils for engineering design. Parameters relate characteristics of the soil to the engineering model or framework. Some parameters, like those describing the critical state locus or normalized small strain shear modulus, can be considered properties because they are independent of state or boundary conditions, while others are behaviours.	Property parameters Γ, λ, M_{tc} Poisson's ratio Plastic hardening modulus Elastic rigidity I_r Behaviour parameters ϕ', s_u, K_o

The approach to the measurement of sand properties and behaviour must therefore rely on both laboratory and in-situ tests, as follows (Figure 2.22):

- Laboratory tests to determine intrinsic properties of sand and behaviour as a function of ψ
- In-situ tests to determine the in-situ state
- In-situ tests to determine properties not measurable in the laboratory

Why not simply determine design parameters directly from in-situ tests, such as the CPT? Interpretation of in-situ tests is an inverse problem (or more correctly an inverse boundary value problem). In-situ tests do not measure any particular parameter directly. Rather, the response of the soil to an applied displacement is measured with the soil properties evaluated from this response. In general, some form of constitutive model is assumed in this property evaluation. It is, therefore, necessary to know certain properties of the sand before a rational interpretation can be made, and this always requires some laboratory testing. Everything cannot be entirely anchored to in-situ tests and iteration between laboratory and in-situ testing is necessary. Chapter 4 is devoted to in-situ tests, especially the CPT, that determine the in-situ ψ. It turns out that the elastic shear modulus (generally referred to as

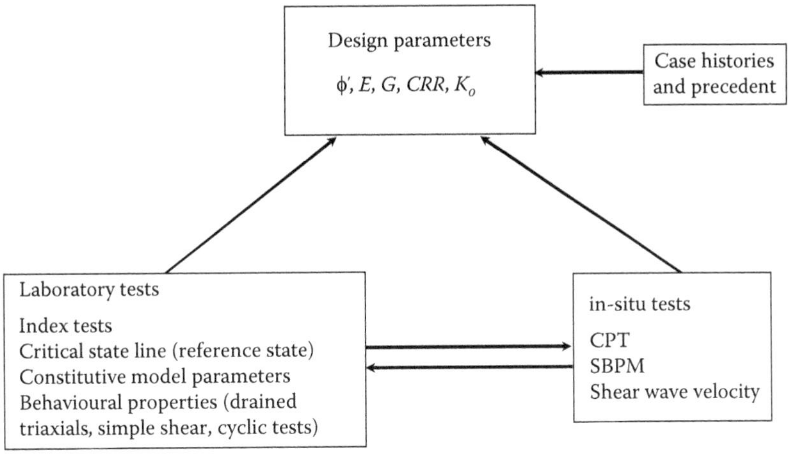

Figure 2.22 Schematic illustration of relationship between parameters and testing methods.

G_{max} in the literature) must also be determined in-situ, as G depends a lot on fabric as well as the more familiar void ratio and stress level. Fabric, as we noted previously, is lost in the sampling process.

Intrinsic properties are unrelated to sample state or fabric and may be determined by testing reconstituted samples in the laboratory. Intrinsic properties determined in the laboratory include the critical state parameters M_{tc}, Γ and λ that describe the CSL. Critical state parameters need to be determined for two reasons. First, when adding laboratory data on new soils to the existing state parameter framework, the new data trends should be assessed as well as the nature of any departure from the already established trends. Second, the in-situ test data will need to be evaluated and both M_{tc} and λ are required to do this.

The remainder of this chapter is mainly about determining soil properties for sands to silts. The other design parameters that need to be determined will largely depend on the specific project or problem at hand. A few of the more common laboratory tests and their use for liquefaction evaluation and design with sands are listed in Table 2.3. Testing should also include the standard index tests to document the soil type.

Table 2.3 Laboratory tests for design parameters in sands and silts

Symbol	*Description*	*How measured*	*Applications*
Γ, λ	CSL parameters	Drained and undrained triaxial tests	All static design calculations (bearing capacity, earth pressures, etc.)
M_{tc}	Critical friction ratio		
H, χ	Plastic modulus, stress–dilatancy	Triaxial tests; possibly self-bored pressuremeter	Displacement calculations, finite element models
G	Shear modulus	Seismic shear wave velocity, resonant column test, self-bored pressuremeter	Dynamic analyses, displacement calculations
D_r	Damping ratio	Resonant column test	Dynamic analysis
CRR	Cyclic strength	Cyclic triaxial, cyclic simple shear tests	Assessment of cyclic liquefaction
H_r	Plastic hardening degradation with principal stress rotation	Hollow cylinder test; possibly cyclic simple shear	Cyclic liquefaction, especially low loads for large numbers of cycles (cyclic mobility)

Liquefaction analyses will very likely require knowledge of the plastic hardening or modulus of the soil at hand. This is the parallel property to the elastic shear modulus G, and controls the relative magnitude of the plastic strains. The plastic modulus could be expected to be at least in part dependent on soil fabric. However, to date, no techniques have emerged for assessing plastic modulus in-situ. Accordingly, drained triaxial tests on reconstituted samples consolidated to estimated in-situ void ratio and stress conditions are used. This appears to be the general practice for estimating plastic strains, although the tacit assumption that the fabric in the laboratory is the same as the fabric in-situ tends to be glossed over.

The alternative to assessing plastic behaviour is the back-analysis of case histories to establish properties exhibited at full scale in similar materials at similar states. This methodology has substantial application in engineering practice and has largely dominated liquefaction assessments for the past three decades. The critical state framework turns out to fit rather well with the case history record.

2.4 DETERMINING THE CRITICAL STATE

2.4.1 Triaxial testing procedure

The critical state comprises two aspects, a locus or line in void ratio–mean stress space and the ratio between the stresses at the critical state. The void ratio aspect is the most difficult to measure, and so, that is dealt with first.

Experience indicates that the preferred method of determining the CSL is a series of triaxial compression tests on loose samples, generally markedly looser than the critical state. Loose samples do not form shear planes and do not have the tendency to localization that is normal in dense (dilatant) sands. Originally, the standard protocol followed Castro and concentrated on undrained tests. Undrained tests should always be the starting point for the practical reason that the strains required to reach the critical state are well within the limits of triaxial equipment for loose samples. Small strains result in large pore pressure changes, and therefore, undrained samples can change state (i.e. move to the critical state) relatively quickly. However, it turns out that it is quite difficult to obtain data on the aforementioned CSL at about $p'=200$ kPa with undrained tests, as it is necessary to consolidate the sample to $p'>2$ MPa prior to shear. Such high pressures are inconvenient for most commercial triaxial equipment as well as often involving grain crushing effects. Drained tests are therefore used as well as undrained. In drained tests on loose samples, the sample moves to the critical state at a much slower rate and displacements to the limits of the triaxial equipment are required. The goal in the testing is to determine both the void ratio and the mean stress at the critical state accurately over a range of critical state conditions. The undrained test data are presented in $q-\varepsilon_1$ and $\Delta u-\varepsilon_1$ plots and a $q-p'$ stress path for review and picking the critical conditions. Drained test data are presented as $q-\varepsilon_1$ and $\varepsilon_v-\varepsilon_1$ plots. Often the sample may have to be taken to 20% axial strain, and lubricated end platens are essential.

In developing and testing the critical state approach, we were fortunate enough to be able to carry out an extensive series of tests on Erksak 330/0.7 sand, taken from the core of the Molikpaq structure (Section 1.3.7). This series of tests is listed in Table 2.4 and forms the basis of much of what follows here and the theoretical developments in Chapter 3. It is the same testing documented in Been et al. (1991), and test data are downloadable as described in the Preface so that you can play with the information.

Table 2.4 Triaxial tests on Erksak 330/0.7 sand to determine CSL and other soil properties parameters

	Initial conditions				*Test conditions*			*End of test*						
Test #	*Sample prep.*	*Void ratio*	p' *(kPa)*	ψ_0	*Drainage*	*Strain rate (%/h)*	*Stress path*	*Steady state*	σ'_3 *(kPa)*	p' *(kPa)*	q *(kPa)*	e	ϕ_c *(deg.)*	*Remarks*
L-601	MC	0.757	499	0.025	U	L	C	Yes	64	100	108	0.754	27.2	Load-controlled test series
L-602	MC	0.712	500	−0.020	U	L	C	Dil	285	460	500	0.711		
L-603	MC	0.787	300	0.048	U	L	C	Yes	4	6	5	0.780	22.6	
L-604	MC	0.772	699	0.044	U	L	C	Yes	55	72	50	0.768	18.2	
L-605	MC	0.771	500	0.039	U	L	C	Yes	16	19	8	0.766		
L-606	MC	0.763	701	0.035	U	L	C	Yes	52	74	65	0.759	22.6	
L-607	MC	0.751	701	0.023	U	L	C	Yes	170	260	270	0.748	26.3	
C-641a	PL	0.666	140	−0.084	U	4	C	Cav	560	1060	1500	0.668		Test series with strain control
C-609	MC	0.800	500	0.068	U	54	C	Yes	8	10	6	0.793	15.8	
C-610	MC	0.754	1200	0.033	U	53	C	Yes	220	350	400	0.751	28.4	
C-611	MC	0.738	1450	0.020	U	51	C	Mdil	660	1107	1340	0.737		
C-612	MC	0.773	500	0.041	U	51	C	Yes	40	47	22	0.769	12.5	
C-613	MC	0.740	1300	0.020	U	49	C	Mdil	340	540	601	0.737		
C-614	MC	0.740	1200	0.020	U	51	C	Yes	323	520	595	0.738	28.6	
C-616	MC	0.703	1200	−0.017	U	48	C	Dil	970	1664	203	0.703		
C-620	MC	0.616	8000	−0.079	U	49	C	Yes	1500	2350	3000	0.613		Test series in high-pressure apparatus
C-621	MC	0.618	8000	−0.077	U	47	C	Yes	1850	3250	4500	0.616		
C-622	MC	0.659	8000	−0.036	U	47	C	Yes	1300	2000	2400	0.656		
C-623	MC	0.662	8100	−0.033	U	49	C	Yes	1146	2207	3184	0.659		
C-624	MC	0.745	3000	0.037	U	47	C	Yes	256	460	613	0.741	33.0	
C-625	MC	0.703	7000	0.006	U	48	C	Yes	910	1570	1960	0.700	31.2	
C-626	MC	0.751	4000	0.047	U	48	C	Yes	401	708	920	0.748	32.3	

(Continued)

Table 2.4 (Continued) Triaxial tests on Erksak 330/0.7 sand to determine CSL and other soil properties parameters

	Initial conditions				Test conditions			End of test						
Test #	Sample prep.	Void ratio	p' (kPa)	ψ_0	Drainage	Strain rate (%/h)	Stress path	Steady state	σ'_3 (kPa)	p' (kPa)	Q (kPa)	e	ϕ_c (°)	Remarks
C-631	MC	0.694	200	−0.051	U	9	C	Dil	424	826	1205	0.695		Examination of sample preparation effects and dense states
C-632	PL	0.652	200	−0.092	U	10	C	Yes	860	1650	3022	0.655		
C-633	MC	0.655	200	−0.089	U	10	C	Dil	387	737	1064	0.656		
C-634	PL	0.667	200	−0.077	U	10	C	Yes	1240	2137	2692	0.670		
C-635	MC	0.588	200	−0.156	U	10	C	Yes	1048	2351	3900	0.591		
C-636	PL	0.618	200	−0.126	U	9	C	Yes	1529	2767	3714	0.621		
C-637	MC	0.580	50	−0.183	U	10	C	Yes	895	2232	4011	0.584		
C-639	MC	0.596	800	−0.130	U	10	C	Yes	1323	2768	4335	0.597		
C-641b	PL	0.687	200	−0.058	U	10	C	Yes	808	1476	2010	0.689		
C-642	PL	0.566	800	−0.160	U	10	C	Yes	1500	2799	3897	0.567		
E-641c	MC	0.732	500	0.000	U	10	E	Mdil	39	90	75	0.728	29.4	Extension test series (unloading)
E-642	MC	0.767	500	0.035	U	14	E	Mdil	55	100	70	0.764	22.9	
E-643	MC	0.747	500	0.015	U	10	E	Mdil	111	207	144	0.745	23.2	
E-644	MC	0.783	500	0.051	U	10	E	Yes	11.5	13	4.5	0.777		
E-645	MC	0.766	500	0.034	U	9	E	Dil	112	234	183	0.764		
E-646	MC	0.750	500	0.018	U	9	E	Yes	50	100	78	0.746	26.0	
E-647	MC	0.776	500	0.044	U	9	E	Mdil	24	51	41	0.771	27.4	
E-648	MC	0.702	500	−0.030	U	10	E	Dil	167	349	273	0.700		
D-661	PL	0.680	140	−0.069	D	5	C	Mdil	140	240	301	0.735	31.2	Drained tests on dense sands
D-662	PL	0.677	60	−0.084	D	6	C	Yes	60	104	131	0.752	31.5	
D-663	PL	0.675	300	−0.064	D	6	C	Dil	300	565	790	0.714		
D-664	PL	0.635	300	−0.104	D	6	C	Maybe	300	473	520	0.691	27.7	
D-665	PL	0.691	130	−0.060	D	5	C	Dil	130	240	323	0.737		
D-666	PL	0.710	60	−0.051	D	5	C	Dil	60	114	151	0.778		
D-667	PL	0.590	130	−0.161	D	5	C	Dil	130	253	355	0.702		

(Continued)

Table 2.4 (Continued) Triaxial tests on Erksak 330/0.7 sand to determine CSL and other soil properties parameters

	Initial conditions				*Test conditions*			*End of test*						
Test #	*Sample prep.*	*Void ratio*	*p′ (kPa)*	ψ_0	*Drainage*	*Strain rate (%/h)*	*Stress path*	*Steady state*	σ'_3 *(kPa)*	*p′ (kPa)*	*Q (kPa)*	e	ϕ_c (°)	*Remarks*
DR-668	PL	0.680	140	−0.069	D		C/R	Dil	180	504	486	0.680		Radial loading
D-760	PL	0.698	250	−0.044	D	5	C	Dil	250	356	620	0.742		OCR = 4
D-761	PL	0.657	250	−0.085	D	5	C	Dil	250	356	620	0.716		OCR = 4
D-762	PL	0.609	250	−0.133	D	5	C	Dil	250	360	630	0.708		OCR = 4
D-874	MC	0.798	200	0.053	D	5	C	Con	200	323	370	0.753	28.7	OCR = 8.8
D-681	MC	0.775	1000	0.052	D	10	C	Yes	1000	1542	1613	0.703	26.5	Drained tests on loose sands
D-682	MC	0.776	500	0.044	D	8	C	Yes	500	812	938	0.725	28.9	
D-684	MC	0.820	200	0.075	D	10	C	Yes	200	308	321	0.775	26.4	
D-685	MC	0.812	200	0.067	D	10	C	Con	200	283	273	0.749	23.9	

Source: Been, K. et al., *Géotechnique*, 41(3), 365, 1991.

MC, moist compaction; PL, wet pluviation; U, undrained; D, drained; L, load control; C, triaxial compression; E, triaxial extension; C/R, triaxial with increasing radial stress; Dil, dilation; Mdil, mild dilation; Con, contracting; Cav, pore fluid cavitation at end of test.

Successful CSL testing is dependent on getting certain details of the triaxial testing correct:

- Uniform samples must be prepared to a loose and predetermined void ratio (the operator must be able to achieve a desired void ratio).
- Samples must be fully saturated.
- The void ratio must be known accurately (to within about ±0.003).
- The measurement system must be capable of measuring low stresses as well as pore pressures at a high rate with very little system compliance (a liquefied sample may be at a mean effective stress of ≈5 kPa, derived as the difference between a measured total stress of 300 kPa and a pore pressure of 295 kPa).

The required procedures to deal with each of these aspects are covered in Appendix B.

By far the greatest aid to critical state testing of sands is a good computer-controlled testing system. Computer data acquisition is now a generally accepted tool, but the inclusion of feedback and control of the test by computer is not widely practiced. Computer control, such as that provided in the GDS testing systems (Horsfield and Been, 1987; Menzies, 1988), provides the flexibility to test along any desired stress path, under stress or strain rate control (as distinct from load or displacement rate control). Measured data also need correcting for membrane penetration and cross-sectional area changes during the test, and these corrections are easily done with computer-controlled work. True constant volume tests compensating for membrane errors can be achieved.

Regardless of whether computer control is used, it is essential to use computerized data acquisition. At some point, models are going to be fitted to the data to evaluate design parameters, and possibly calibrate properties for numerical analysis. Doing this on a few data points or on data scaled off hard copies is tedious and there are chances of information being lost. Without computer-based data acquisition, testing is simply not to modern standards.

2.4.2 Determining CSL from test results (soil properties Γ, λ)

Regardless of whether testing is drained or undrained, there is judgement involved in identifying critical conditions from the test data. Interpretation of the critical state from triaxial tests is conceptually straightforward, but it is also surprising how much confusion and disagreement there has been on the existence, or otherwise, of a critical state. Much of this disagreement is based simply on different interpretations of the test. It is appropriate first to repeat the definition of the critical state as the state at which the soil 'continues to deform at constant stress and constant void ratio' (Roscoe et al., 1958). Implicit in the definition is the expectation that the sample will continue to deform in the same way with further strain, so that a temporary condition where void ratio and effective stresses are constant does not represent the critical state. The critical state is defined very simply in terms of the dilatancy: both dilatancy and rate of change of dilatancy must be zero, that is $D = 0$ and $\dot{D} = 0$. It is this second condition that assures that the true critical state has been reached and not a transient condition. Figures 2.23 through 2.25 show the results of a series of undrained tests on Erksak sand (Been et al., 1991) to illustrate some of the details of test interpretation.

The stress–strain and pore pressure behaviour of a loose sample are shown in Figure 2.23. A clearly defined steady state is reached after about 8% axial strain, and the soil deforms at this constant state to 20% strain, at which time the test was terminated. In this test, the critical state stresses can be determined unambiguously, but corrections to the void ratio for the effects of membrane penetration are important for the accuracy of the final critical state point (Appendix B).

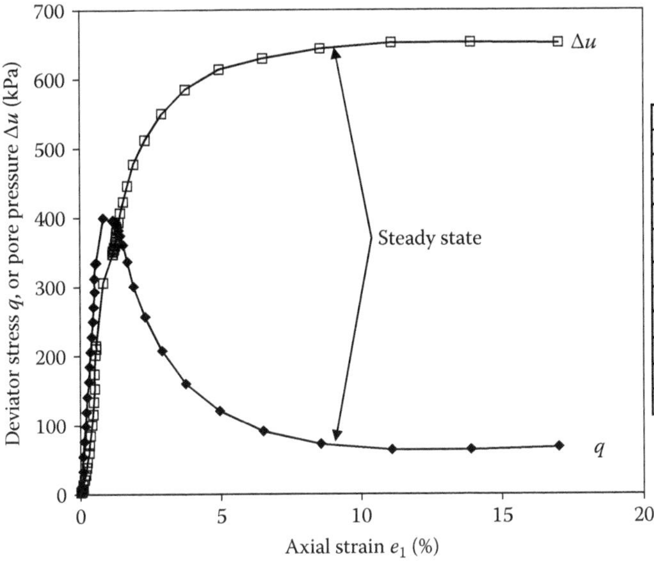

Test number	Liq-G606
Test type	CIU
Load method	Load control
Material tested	Erksak 330/0.7
Preparation method	Moist tamped
G_s	2.655
Consolidation	
Void ratio	0.763
σ' axial (kPa)	701
σ' radial (kPa)	701
Back pressure (kPa)	258
ψ	0.035

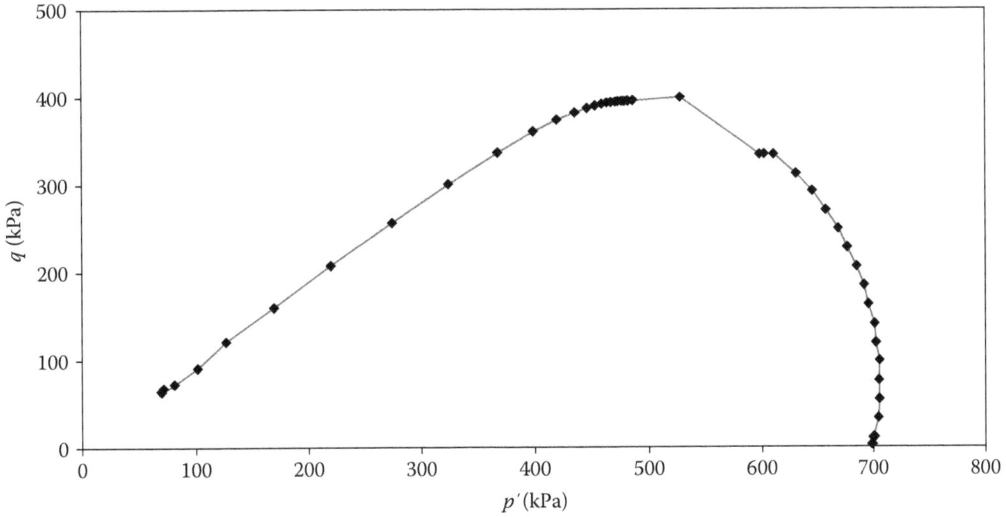

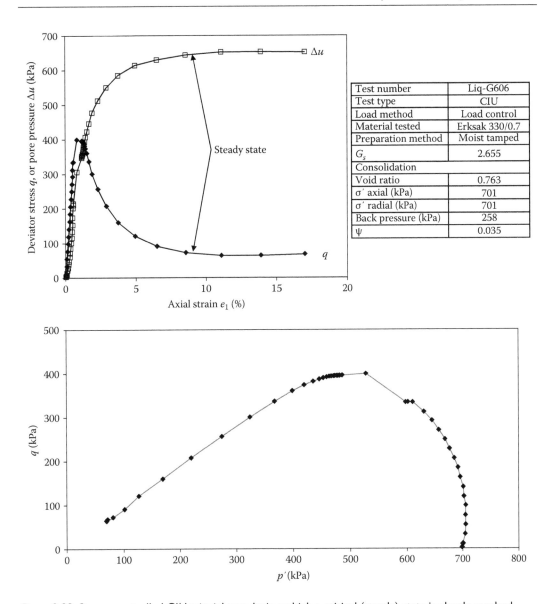

Figure 2.23 **Stress-controlled CIU triaxial test during which a critical (steady) state is clearly reached.**

In some cases, samples appear to reach a steady state at about 8% strain, but then start to dilate at higher strains. Figure 2.24 shows the results of such a test. The *quasi-steady state* in this test must not be interpreted as a critical state. It is a temporary condition, with the sample moving from a contractive to a dilative behaviour as is readily seen in the test data. This temporary condition has been termed the phase transformation by Ishihara and various co-workers (e.g. Ishihara et al., 1990), the quasi-steady state by Alarcon-Guzman et al. (1988) and the lower limit of steady-state strength by Konrad (1990b).

The quasi-steady state is very much influenced by the test conditions and sample fabric. For a test result such as that in Figure 2.24, it is more meaningful to plot the condition at the end of the test on a state diagram to determine the CSL. For such tests, the notation of an arrow on the plotted point is used to indicate that the sample was still evolving to the

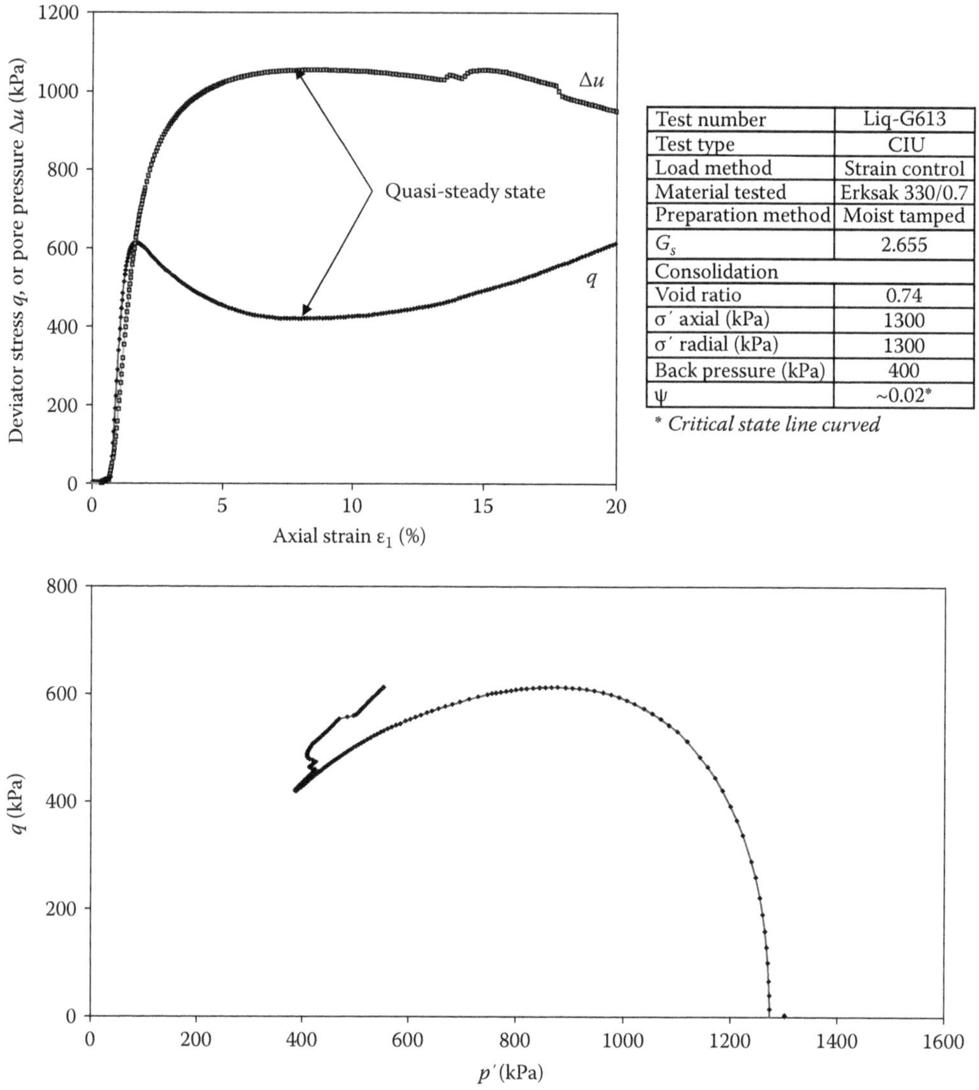

Test number	Liq-G613
Test type	CIU
Load method	Strain control
Material tested	Erksak 330/0.7
Preparation method	Moist tamped
G_s	2.655
Consolidation	
Void ratio	0.74
σ' axial (kPa)	1300
σ' radial (kPa)	1300
Back pressure (kPa)	400
ψ	~0.02*

* *Critical state line curved*

Figure 2.24 CIU triaxial test showing dilation at large strains. The quasi-steady state must not be interpreted as the critical state.

critical state in the direction shown. Treating the quasi-steady state as the true critical state is a surprisingly common error, and is the principal reason various workers have reported non-unique CSL that vary with stress path, strain rate and sample preparation.

Figure 2.25 shows a selection of further tests that were used to determine the CSL in Figure 2.26. Notice that real test data do not always match the idealized descriptions in technical papers, and so a degree of judgement and knowledge of the test equipment and procedures is needed in picking the steady state from the graphs and also in estimating the void ratio accurately. With care and consistency, the CSL can be determined with an accuracy of about ±0.01 in terms of void ratio.

Figure 2.26 plots the CSL for Erksak sand (from Been et al., 1991). The critical state obtained using load control cannot be distinguished from the critical state using displacement

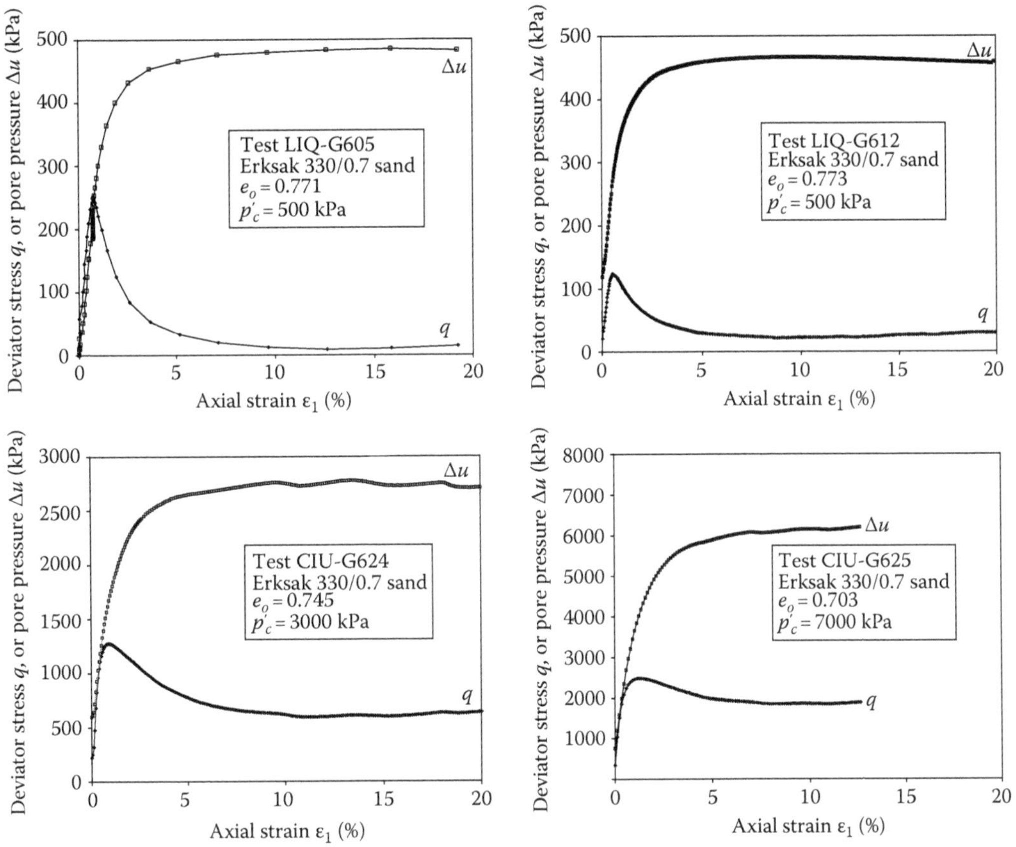

Figure 2.25 Selection of undrained tests used to give critical state line in Figure 2.26.

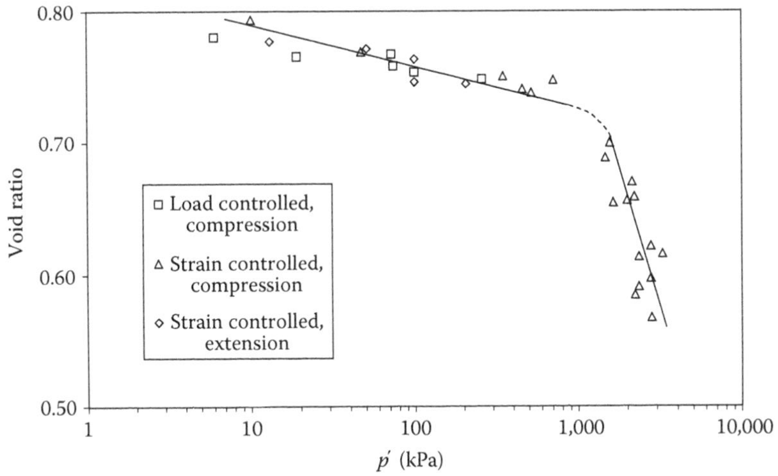

Figure 2.26 Critical state line for Erksak 330/0.7 sand from undrained tests that reached a distinct critical (steady) state.

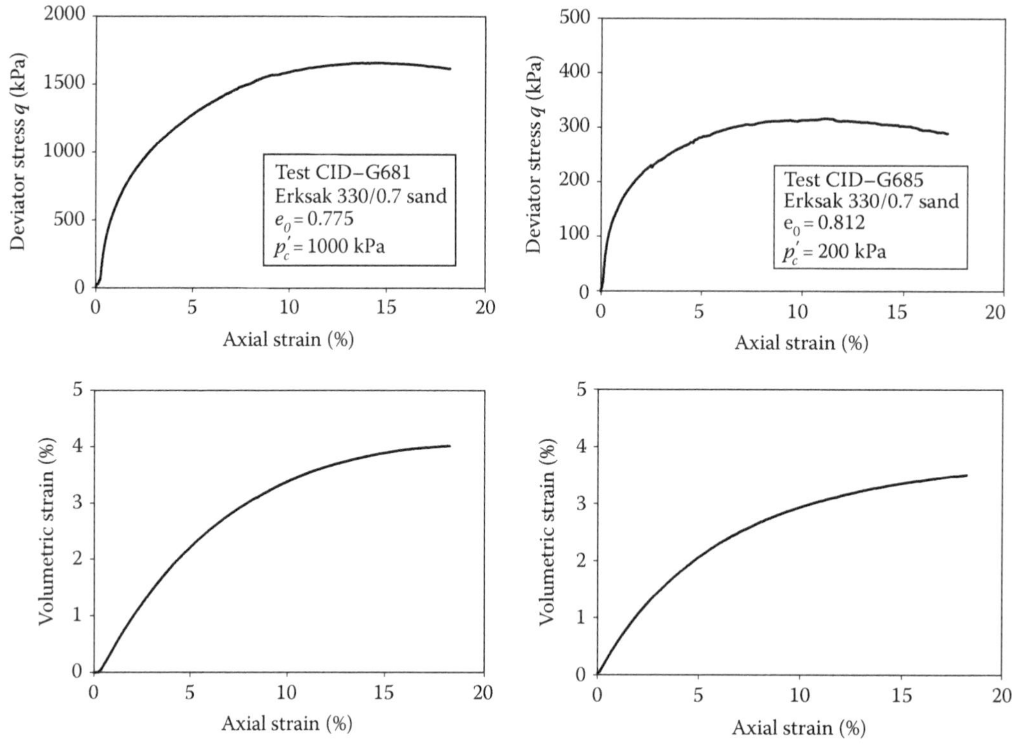

Figure 2.27 Examples of drained triaxial tests on loose samples reaching critical state.

control. Extension tests also give similar critical states. As long as the pseudo–steady state is not used, rather good results in defining the CSL can be obtained with diligent and unbiased processing of the data.

So far only loose undrained tests have been used. Drained tests on loose samples can also be useful. Figure 2.27 shows an example of such drained data. Even though it takes more than 20% strain before critical conditions become established, the soil does get there and it is very easy to identify the critical state. (In both tests, but particularly CID-G685, the deviator stress starts dropping at high strains while the sample is still contracting as indicated by the volumetric strains. This is probably due to bulging of the sample so that the area correction used to calculate deviator stress is likely inaccurate.)

The usefulness of the drained tests becomes apparent when plotting up the end points to estimate the CSL. Figure 2.28 shows the results of three undrained tests on very loose samples of a sandy silt and three drained tests on loose samples. This soil comprised about 65% silt-sized particles. The end points at which the critical state was achieved are circled for each test (and this is not at the end of the state path for tests that showed S-shaped stress paths). The loosest drained test has a critical state very close to that of an undrained test of comparable critical void ratio. The denser drained test nicely extends the data defining the CSL to $p' > 1500$ kPa. As can be seen from the state paths of the undrained tests, to achieve such a critical mean stress with an undrained sample by relying on a brittle stress drop would have required an initial confining stress in excess of 8000 kPa. This is beyond the equipment in most testing laboratories, whereas the drained test used routine equipment.

Based on the earlier behaviours, and the experience accumulated in the various Golder Associates laboratories doing this testing, determination of reliable CSLs in the laboratory

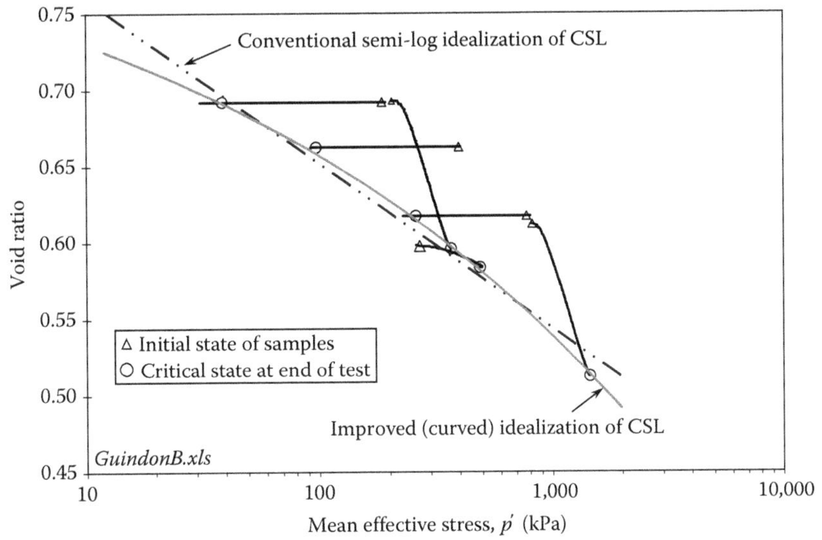

Figure 2.28 Critical state line for Guindon Tailings B (67% fines) showing use of drained tests on loose samples to define critical state at higher stresses.

has evolved to about eight triaxial tests per soil of a given gradation. Testing to determine the critical state should be undertaken in two phases:

Phase 1: Three undrained tests at initial void ratios equivalent to a relative density (density index) of 10%, 20% and 30%. Initial confining stress should be about 350 kPa. These are followed by two drained tests on samples at initial void ratios equivalent to a relative density (density index) of 20%, one with an initial confining stress of 200 kPa and one with an initial confining stress of 800 kPa. The lower stress drained sample should give a critical state in similar stress range to the undrained tests and is used as a check. The higher stress drained test is there to extend the CSL to stresses of about 1000 kPa, the upper range of usual practical interest.

Phase 2: A further three or so tests at initial void ratios and confining stresses are selected on the basis of the initial estimate of the CSL. The aim is to provide a uniform distribution of data points to define the CSL and to avoid a single test result unduly influencing the slope or position of the CSL. By the time these tests are carried out there should be a good feel for how a particular density is obtained during sample preparation.

Some abortive tests should be allowed for in planning testing to determine a CSL. About 1 in 10 samples is typically discarded as a result of membrane leaks, failure to maintain effective stresses during saturation, inconsistencies in void ratio measurements, etc. In addition, the desired void ratios are sometimes difficult to achieve, with repeated attempts at sample preparation being needed (there is uncertainty in the volume changes that will occur during saturation and consolidation of any particular sand until a few tests have been carried out). This is the primary reason why phased testing is needed.

Provided reasonable diligence is used, and in particular care taken in void ratio determination, the CSL should be defined to within a precision of about ±0.01 in terms of void ratio in good quality commercial testing. The CSL of Figure 2.28 has a slight curvature and although

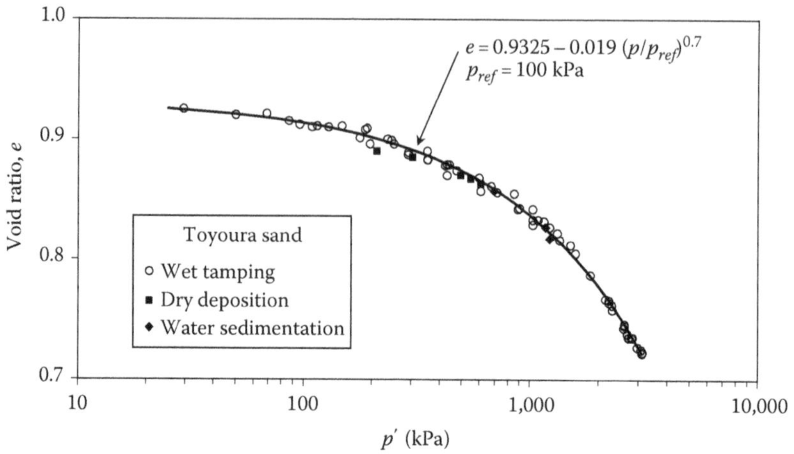

Figure 2.29 Critical state locus for Toyoura sand. (Data from Verdugo, R., *Géotechnique*, 42(4), 655, 1992.)

a semi-log trend line of the form of Equation 2.1 can be fitted to the data, this approximation gives up a precision of about 0.02 in void ratio in modelling the critical state. Much better can be achieved in a research environment, as illustrated by Figure 2.29 that shows results obtained on Toyoura sand (a Japanese standard research soil) by Verdugo (1992). There is a much more marked curvature to the CSL for Toyoura sand in Figure 2.29. Although in many instances the familiar semi-log form of (2.1) is an acceptable engineering approximation for the CSL, there are higher-order idealizations that capture CSL curvature. A power law equation is popular, and the data shown in Figure 2.29 are nicely fitted with

$$e = 0.9325 - 0.19\left(\frac{p'}{p'_{ref}}\right)^{0.7} \tag{2.12}$$

There is no particular reason to choose the form of (2.12) over (2.1) from a mathematical or physical standpoint. It is simply a matter of which best fits the data to hand, although there is much published work that uses the form for the CSL of (2.1) and simply accepts it as an engineering approximation. The definition of the state parameter is utterly independent of the particular equation used to represent the CSL, and the state parameter approach does not depend on the semi-log idealization for the CSL. A curved CSL was used when introducing ψ in Chapter 1 to emphasize this point.

2.4.3 Critical friction ratio (soil property M_{tc})

Although discussion about the critical state tends to concentrate on the CSL or void ratio aspects, the critical state is also associated with a particular stress ratio. The stresses at the critical state can be expressed simply as $q_c/p'_c = M$. With quartz sands, M is essentially independent of pressure level to at least initial confining stresses in excess of 2.5 MPa (Vaid and Sasitharan, 1992).

Using the parameter M is the preferred way of representing the critical state as it is a dimensionless ratio of stress invariants. However, much of the literature, and especially that from decades ago, uses a friction angle notation and also associates the idea of constant volume rather than explicitly critical conditions. Thus, it is common to encounter ϕ_{cv} (the friction angle at constant volume) although this is of course the same idea as the critical state. It is preferable to use the notation ϕ_c instead of ϕ_{cv} as this makes it clear that it is the critical state value and not something related to a quasi condition.

For triaxial compression conditions (denoted by the subscript *tc*), the soil property M_{tc} is directly related to the corresponding critical friction angle, $\phi_{c,tc}$, by:

$$M_{tc} = \frac{6\sin\phi_c}{3-\sin\phi_c} \quad \textit{in triaxial compression} \tag{2.13}$$

Triaxial compression conditions are specified because M varies with the magnitude of the intermediate principal stress relative to the other principal stress. Broadly, in friction angle terms, ϕ_c increases by a couple of degrees for plane strain conditions compared to triaxial compression and then falls to something less when the soil is in triaxial extension. Because of this variation it is usual to take triaxial compression as the reference condition defining the soil property. The variation of M with proportion of intermediate stress (i.e. the Lode angle) is discussed later in this chapter as a behaviour around the value of the soil property.

Three ways of determining M_{tc} (or $\phi_{c,tc}$) are found in the literature: (a) multiple drained triaxial compression tests on samples of varying density; (b) one or more triaxial compression tests on loose samples and (c) ring shear, despite this test being self-evidently not the reference triaxial compression condition.

Method (a) was first suggested by Bishop (1966) and is arguably still the standard method. It involves carrying out a series of drained triaxial tests on samples of varying density. Each test is reduced to a value of peak measured dilatancy D_{min} at peak strength η_{max} (peak dilatancy should occur at the same point in the test as peak strength). Plotting the results of several tests and projecting the result to zero dilation indicates M. Figure 2.30 shows data

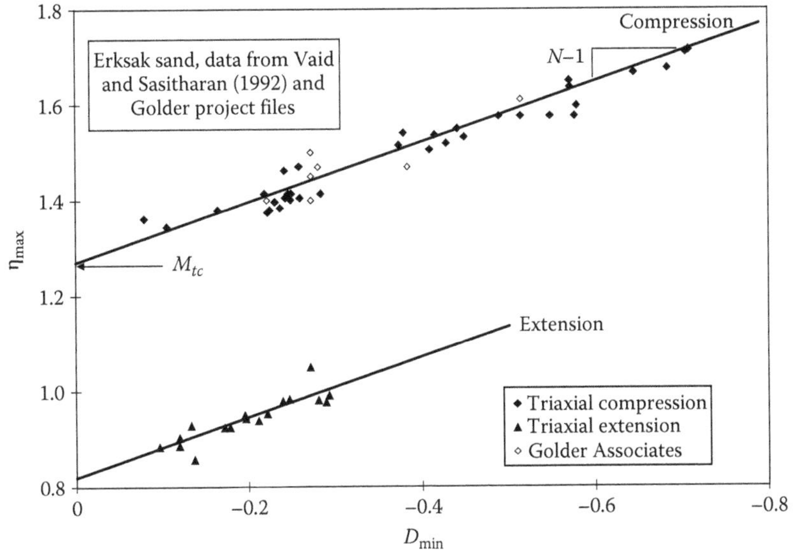

Figure 2.30 Experimental data for relation between peak strength and peak dilatancy for Erksak sand in triaxial compression and extension.

on Erksak sand from two laboratories and illustrates the scatter within the overall trend as well as inter-laboratory repeatability of the trend. A clear trend is evident, with no difference between the two laboratories, giving $M_{tc}=1.26$. The only negative factor to this method is the number of tests needed, at least five evenly distributed over a range of initial states (relative density range from 40% to 100%). Research testing often has many more tests, as can be seen in Figure 2.30.

Method (b) to determine M_{tc} makes use of the testing carried out to determine the CSL. In principle, each of these tests (drained or undrained) is strained until it reaches the critical state. Therefore, plotting the stress ratio η against strain and simply picking up the maximum (which should also be the terminal) value of η when critical conditions have been reached gives M_{tc} directly. With this method, it is crucial that strain-controlled tests with computer-based data acquisition be used, otherwise the time lag in measured pore pressure, or the lack of simultaneous measurements, can give a misleading calculation. It can also be subject to transducer inaccuracy for very loose samples. In the case of loose drained samples, some inaccuracy may arise because of the effect of the area correction in calculating axial stress. However, these loose tests will have been done in any event to determine the e-log p' line, so look at the data but bear in mind possible limitations in the measurements.

Method (c) is to shear the soil to large displacement in a ring shear device and measure the limiting stress conditions. Negussey et al. (1988) describe a series of experiments they performed with a ring shear device on Ottawa sand (a quartz sand), two tailings sands, granular copper, lead shot and glass beads. These tests showed how the large strain ϕ_c for each material is invariant with normal stress. The great difficulty with ring shear, however, is that the complete stress conditions are not known and only a friction angle can be determined (i.e. p' is simply unknown in ring shear). The method also relies on the tacit assumption that ϕ_c is invariant with the proportion of intermediate principal stress. This tacit assumption is known to be incorrect.

The best approach to determining M_{tc} is to use a mixture of method (b) and method (a). Testing to determine the CSL will provide data that should be reduced as per method (b), as it is in effect free information. Then a few drained triaxial compression tests should be carried out on at least one dense (say $\psi < -0.2$) and one compact ($\psi \approx -0.1$) sample. These tests allow a plot as per method (a), albeit with limited data points. For large projects or research programs, it is best to carry out the detailed testing of method (a).

2.5 UNIQUENESS OF THE CSL

The definition of critical state and its application to liquefaction evaluation requires that the CSL is unique and independent of test conditions or the stress path followed to reach the critical state (although a strain rate effect is permissible, as discussed in Chapter 3). Uniqueness is normally taken to mean that there is only one critical void ratio for one mean effective stress. A more general requirement for uniqueness is that the CSL is a single-valued function of effective stress (potentially allowing a role for the Lode angle θ as well as p'), the view of which is taken here.

The reason that uniqueness is important is that uniqueness leads to physical simplicity and thus easily understandable idealizations of soil behaviour. Put simply, if the CSL is unique you always know where the soil is going when it is sheared. Studies suggesting non-uniqueness in the CSL logically imply a range of possible end points all of which depend on stress or strain history and which become questionably computable and questionably usable in engineering practice. Uniqueness, or lack of it, must be taken very seriously.

Published studies on the existence and uniqueness of the CSL include a range of conclusions. Key views can be summarized as follows:

- There is a unique CSL, but care is needed in testing techniques and interpretation to establish the CSL (Poulos et al., 1988; Been et al., 1991; Ishihara, 1993).
- A band of states represents steady-state conditions, depending on the initial density and stress level, so that the CSL becomes a zone rather than a line (Konrad, 1993).
- Sample preparation methods result in different anisotropies, giving different stress paths and different critical states (Kuerbis and Vaid, 1988; Vaid et al., 1990a).
- Extension and compression tests will result in very different stress paths and critical states (Vaid et al., 1990a; Negussey and Islam, 1994; Vaid and Thomas, 1995).
- There is an S line from drained tests and an F line from undrained tests that differ as a result of the collapse potential of the soil (Alarcon-Guzman et al., 1988).
- Strain rate affects the CSL (Hird and Hassona, 1990).

Konrad (1993) apparently obtained a band of steady states for Hostun RF sand and indicates that his data imply non-uniqueness of the CSL. However, Konrad's ideas have grown from an initial attempt (Konrad, 1990a,b) to determine the minimum undrained shear strength of the sand, rather than the ultimate, critical state. Examination of his subsequent work indicates that the accepted CSL is in fact identical to his UF line, which is used as a reference condition for CPT interpretation (Konrad, 1997). Clarity in terminology and interpretation is needed, and this is a classic example of confusing the transient pseudo–state state with the critical state.

The suggestion that the CSL might depend on sample preparation method (i.e. fabric) is interesting. Figure 2.15 illustrated how sample preparation changes the stress–strain response of Kogyuk sand in drained compression, but it is Vaid and co-workers who have been the most active in asserting that sample preparation affects the CSL (Kuerbis and Vaid, 1988; Vaid et al., 1990a). The effect of sample preparation method on the CSL was examined in detail for Erksak 330/0.7 sand. Tests were carried out on paired samples of compact to dense samples prepared using two methods, moist tamping (MT) and below water pluviation (PL). Figure 2.31a and b shows the different stress–strain and stress path behaviour encountered with these paired sets of MT and PL samples, which were in other respects as similar as experimentally possible apart from the preparation method. Both sets of paired samples were slightly denser than critical, and the initial conditions of these samples are summarized in each figure. A comparison of wholly contractive undrained behaviour was impossible because samples could not be prepared looser than the CSL using pluviation.

Comparison of Figure 2.31a and b illustrates that moist tamping consistently produces stiffer soil behaviour at small strains than pluviation, but less dilatant behaviour after the yielding becomes established. In the case of Figure 2.31b, the pluviated sample was initially far softer, but by 5% axial strain had become markedly more dilatant than the paired moist tamped sample. It is clear from Figure 2.31 that sample preparation or fabric also affects the undrained behaviour of sand, but the question is whether the critical state is also affected by these differences. Figure 2.32 summarizes the data on the critical state of Erksak 330/0.7 sand in terms of tests using moist tamped and pluviated samples. Within the experimental precision, the CSL is independent of sample preparation method.

Casagrande (1975) and Alarcon-Guzman et al. (1988) identified separately S and F lines from drained triaxial tests and consolidated undrained tests, respectively. The implication is that the CSL is not unique, in that different tests result in different ultimate states. This is incorrect and is an artefact of test procedures and interpretation. The main reason the data are misinterpreted in this way is that drained and undrained triaxial tests on dense

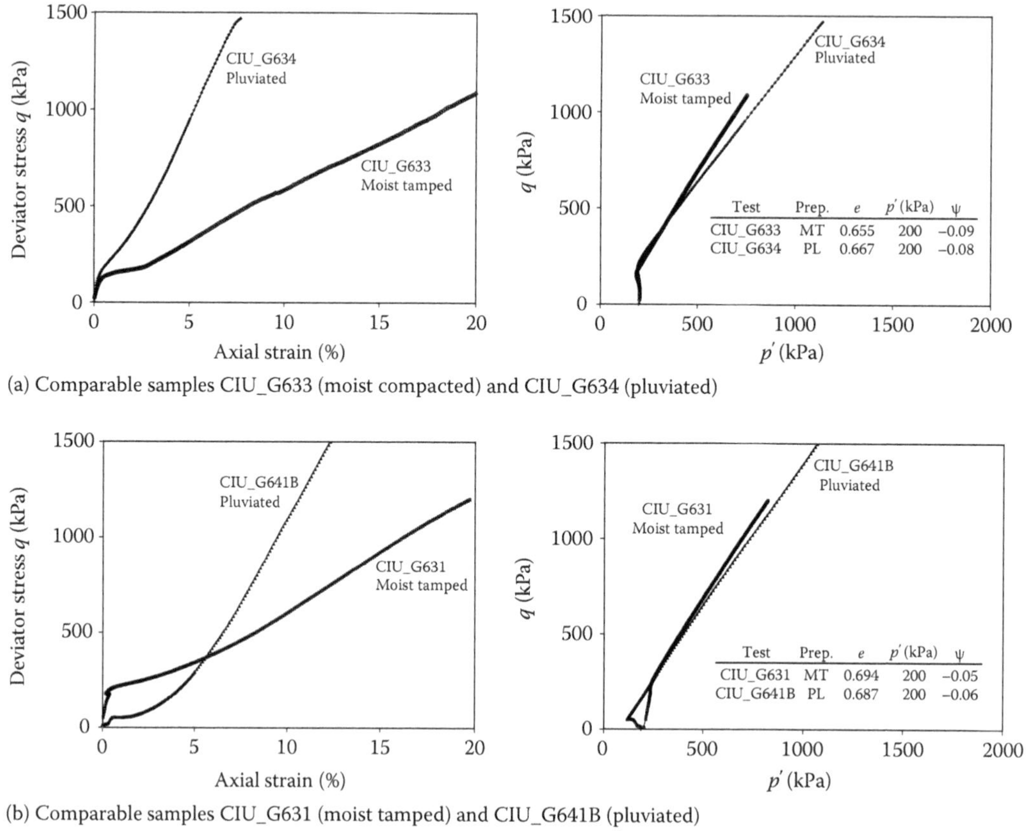

Figure 2.31 Effect of sample preparation on undrained behaviour of Erksak 330/0.7 sand.

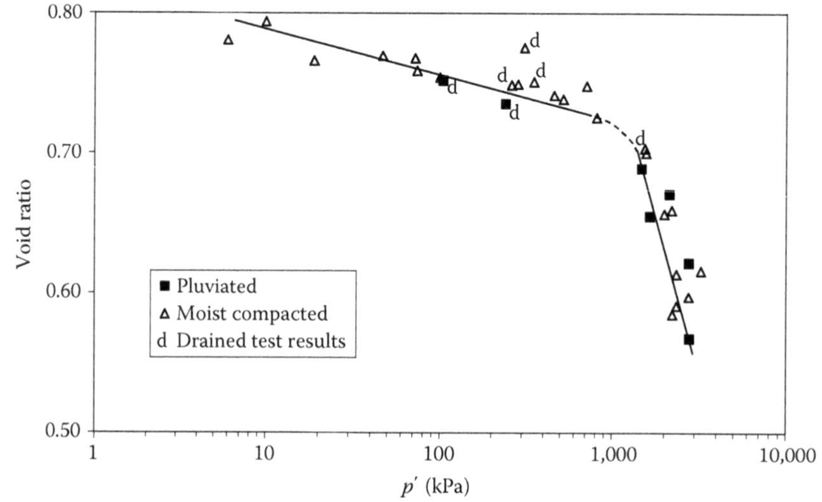

Figure 2.32 Comparison of critical states from pluviated and moist compacted samples of Erksak 330/0.7 sand. (Data from Been, K. et al., *Géotechnique*, 41(3), 365, 1991.) Note that pluviated samples cannot be prepared at high void ratios.

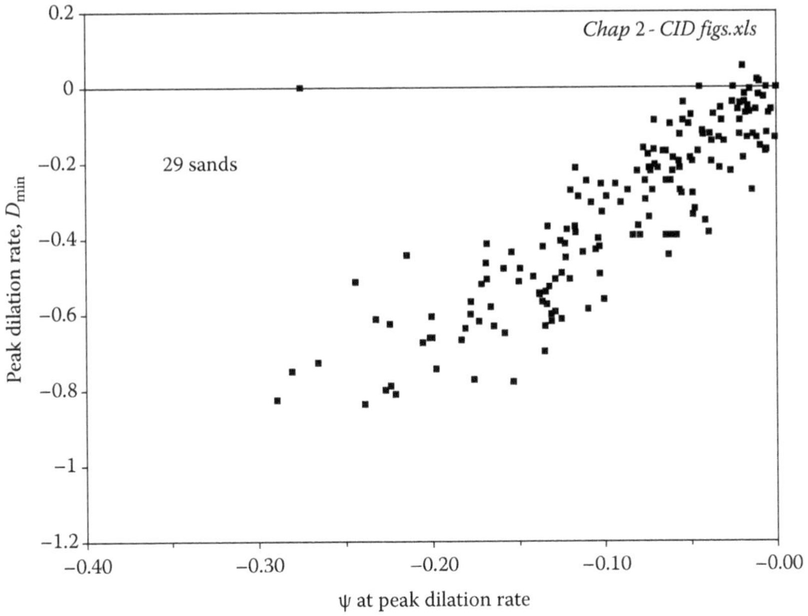

Figure 2.33 Peak dilation rate in drained triaxial compression tests as a function of distance from critical state line determined from undrained tests. The trend line passes close to zero, indicating that drained and undrained behaviour relate to the same CSL.

sands seldom reach the critical state. That drained tests on samples tend to reach the critical state (determined from undrained tests) can be shown by plotting the rate of volume change (or triaxial dilation rate) not against ψ_o but against ψ or by calculating the distance the sample state is from critical state at any time during a test (usually at peak stress conditions and before localization effects occur in the sample). Figure 2.33 shows a set of such data for 29 sands (Been et al., 1992). The dilatancy is the slope of the volumetric versus shear strain (Equation 2.4) from conventional drained triaxial tests, while the x-axis is the difference between the void ratio of the sample at the time the dilatancy is determined and the CSL (determined from undrained tests) at the same stress level. It is clear from the data in Figure 2.33 that the rate of volume change is proportional to distance from the CSL. Similar methods were used by Parry (1958) to support the existence of a CSL for London and Weald clays. If the critical state from drained tests were different from that determined from undrained tests, a mean line through the data in Figure 2.33 should not pass through the origin.

The remainder of this important discussion on uniqueness of the critical state appears in Chapter 6, as that allows the introduction of a constitutive model (Chapter 3) to give insight into the confusing observations of non-uniqueness. In particular, it will be shown that different behaviours as a result of stress path (extension or compression), fabric and drainage can be predicted perfectly well using a model that assumes uniqueness of the CSL.

2.6 SOIL PROPERTIES

2.6.1 Summary of properties for a mechanics-based framework

Soil properties have been introduced throughout this chapter as various aspects of soil behaviour were presented and discussed. We now collect these various strands together to

Table 2.5 Soil properties for a critical state framework

Property	*Typical range*	*Remark*
Γ	0.9–1.4	Altitude of CSL, defined at 1 kPa by convention
λ_{10}	0.02–0.07: uniform quartz sands 0.10–0.25: uniform sandy silts to silts 0.04–0.07: well-graded sandy silts	Slope of CSL when approximated by straight line in e-log(p') space; defined on base 10 logarithms
M_{tc}	1.20–1.35: quartz sands 1.15–1.25: soft sands 1.30–1.60: tailings sands or silts	Critical friction ratio, the limiting large strain ratio q/p'; triaxial compression as reference condition
χ_{tc}	>4: well-graded soils ~4: uniform quartz sands ~3: uniform soft sands	Relates minimum dilatancy to ψ; triaxial compression as reference condition

summarize the basic properties needed to characterize the behaviour of particulate soils. None of these properties are associated with a particular constitutive model, the properties being quite general in themselves and all are dimensionless.

Accepting the semi-log approximation of the CSL as a reasonable engineering approximation, Table 2.5 summarizes the various properties and which aspect of soil behaviour they capture, while values for particular soils are provided in Table 2.1. All properties are defined and measured under triaxial conditions, with Section 2.7 and Chapter 3 discussing how these triaxial properties extrapolate into general stress conditions (and plane strain in particular). Elasticity is omitted from Table 2.5 as it needs to be measured in-situ (covered in Chapter 4).

2.6.2 Example properties of several sands and silts

Quite a few soils have been tested to define their critical state, and Figure 2.34 illustrates some of the CSL that have been determined. Although the CSL is not necessarily linear on the semi-log plots used, in particular at stresses in the range of 1000 kPa or more, the range of interest for most engineering purposes is about 20–500 kPa. Over this range, it is possible to treat the CSL as being linear in most cases, but do not blindly put a regression line through all the data (as you easily could do in a spreadsheet) as that will mislead. Use engineering judgement to best fit the semi-log CSL that honours the test data over the stress range relevant to the situation being assessed.

Turning to the wider issue of soil properties associated with a critical state approach, Table 2.1 lists the critical state soil properties (Γ, λ, M_{tc}, χ_{tc}) for a number of soils. Index properties are also given in this table for reference. Figure 2.35 illustrates some of the particle size distribution curves for the soils in Table 2.1 to define the range of soil types over which this framework has been found to be valid – broadly clean coarse sands to pure silts.

It would be convenient if the CSL parameters for sands were correlated to index properties. In general, the slope of the CSL (λ) does appear to be related to the fines content for a given sand. Been and Jefferies (1985) show how λ increases for Kogyuk sand with higher silt content, and Hird and Hassona show a similar trend for increasing mica content in Leighton Buzzard sand. Similarly, the parameter Γ, which describes how high the CSL is located on the void ratio axis, appears to be somewhat a function of the maximum void ratio.

Figure 2.36 shows the relationship between fines content and λ_{10} for the sands in Table 2.1, while Figure 2.37 illustrates how Γ is related to e_{max}. (In fact it turns out that Γ_{10}, i.e. the CSL intercept at $p' = 10$ kPa, is more closely related to e_{max}.) While Γ, λ do appear to be loosely

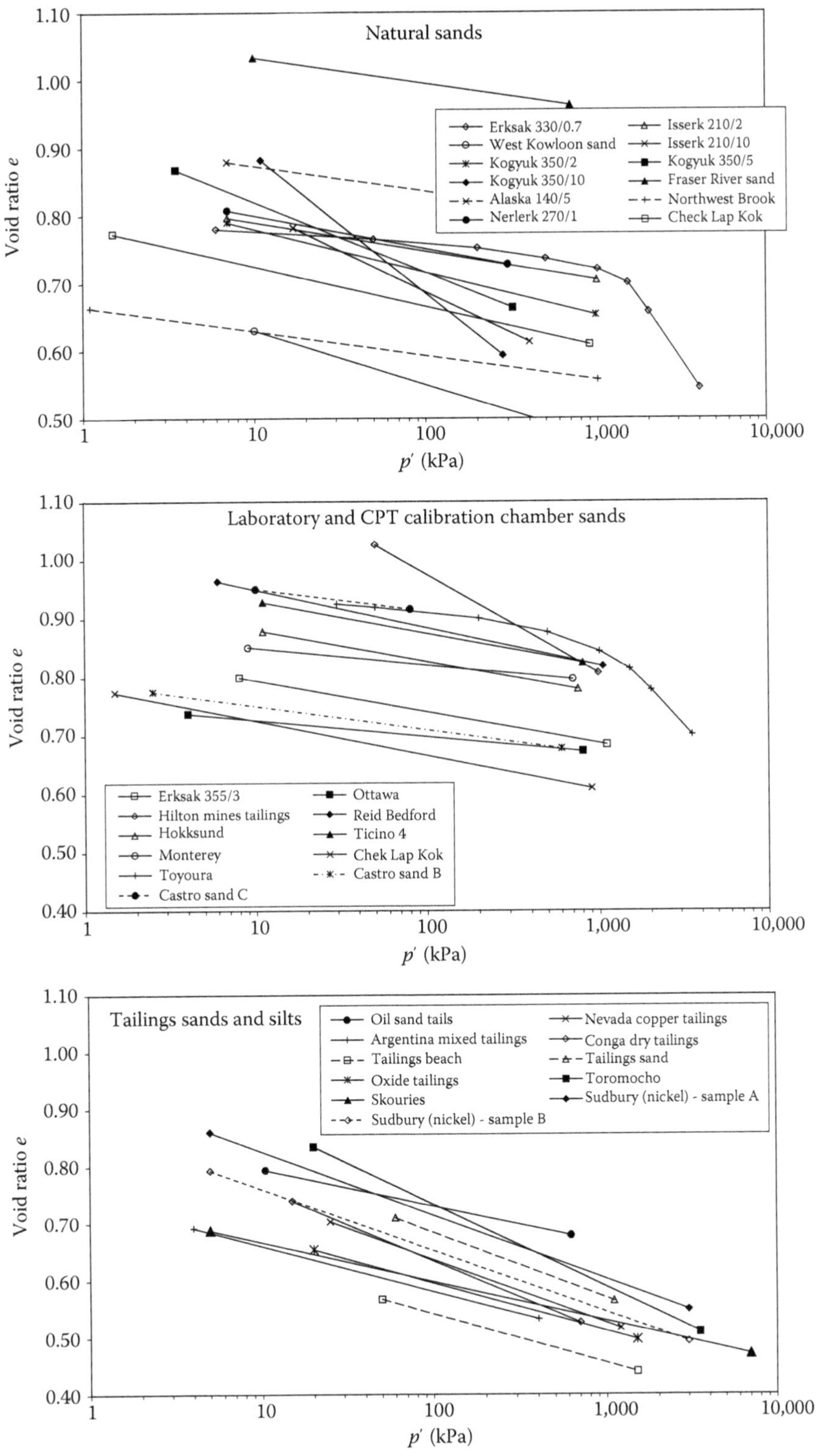

Figure 2.34 Critical state loci for several sands whose properties are given in Table 2.1.

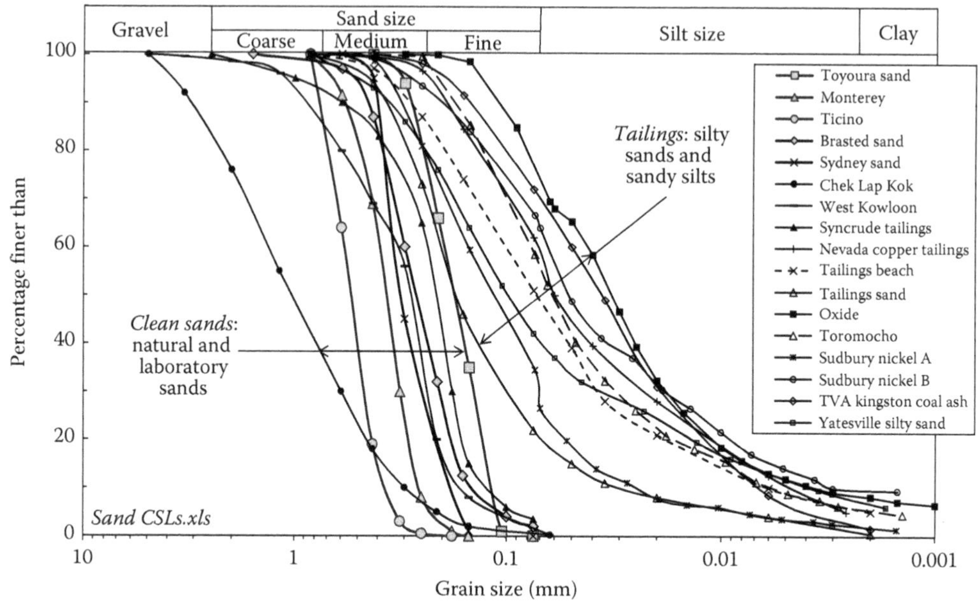

Figure 2.35 Grain size distribution for selected sands and silts whose properties are given in Table 2.1.

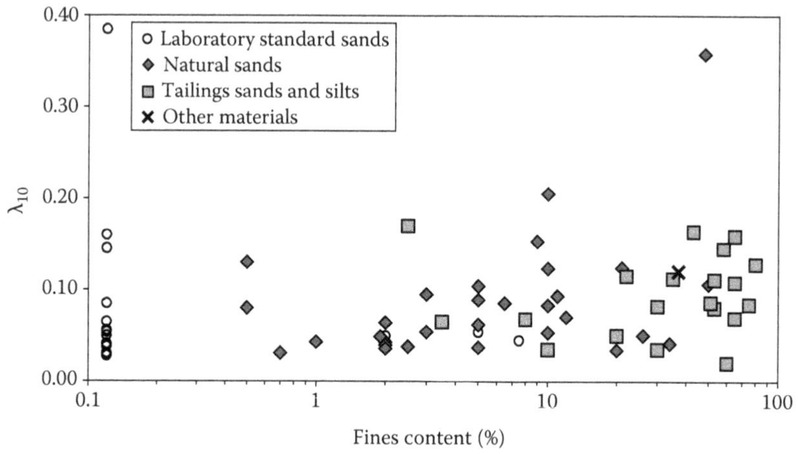

Figure 2.36 Relationship between slope of the critical state line and fines content.

correlated to fines content and maximum void ratio, there is really too much scatter in the relationships for the correlations to be of any great practical use. Minor variations in intrinsic properties of sand have a major influence on the CSL including grain shape, mineralogy, grain size distribution and surface roughness of grains. In general, clean sands with hard, rounded quartzitic grains, such as Ottawa sand, will have a low value of λ_{10} (about 0.03), while angular, crushable silty sands will have greater values of λ_{10} (in the range 0.15–0.2, which are approaching the compressibility associated with low plasticity clays). Use Figures 2.36 and 2.37 only for preliminary estimates of critical state parameters.

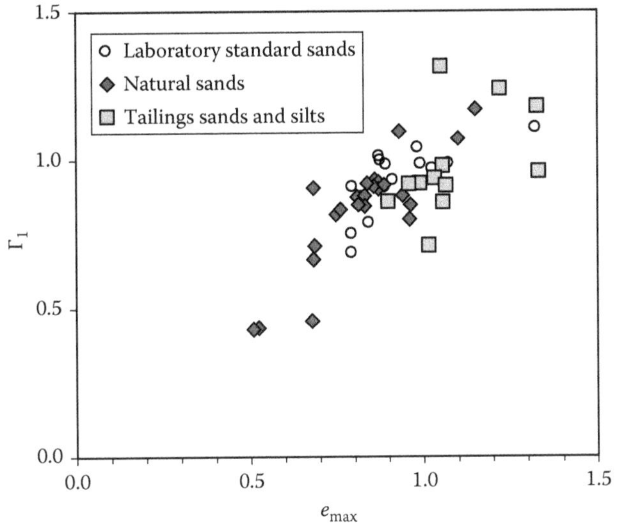

Figure 2.37 Relationship between altitude of critical state line at $p' = 1$ kPa (Γ_1) and maximum void ratio (e_{max}).

The data shown in Figures 2.36 and 2.37 are for rather uniformly graded soils. Well-graded soils are different. Figure 2.38 shows CSL for a well-graded silty sand (the core of Bennett dam) and a uniform silty sand (foundation of the Guindon dam in Sudbury); both soils had about 34% fines. In the case of the well-graded silty sand, λ_{10} approaches the value expected for uniform quartz sands despite the high fines content. Fines fraction on its own is not a good predictor of the corresponding CSL. What appears to happen is that as fines increase from zero, the fines initially facilitate grain separation and subsequent movement during shear. However, there comes a point when the fines fully occupy the interstitial space between the sand grains and this then forces a return to far less compressible behaviour.

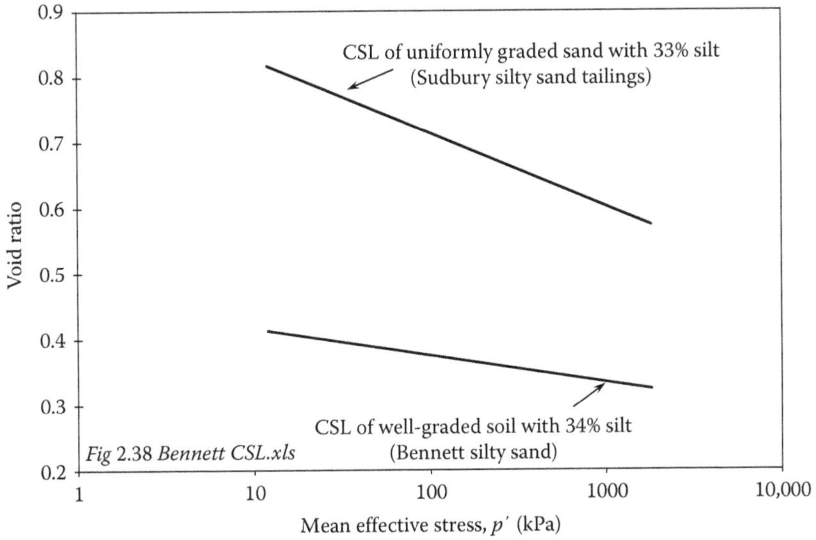

Figure 2.38 Comparison of critical state lines for uniformly graded and well-graded silty sands.

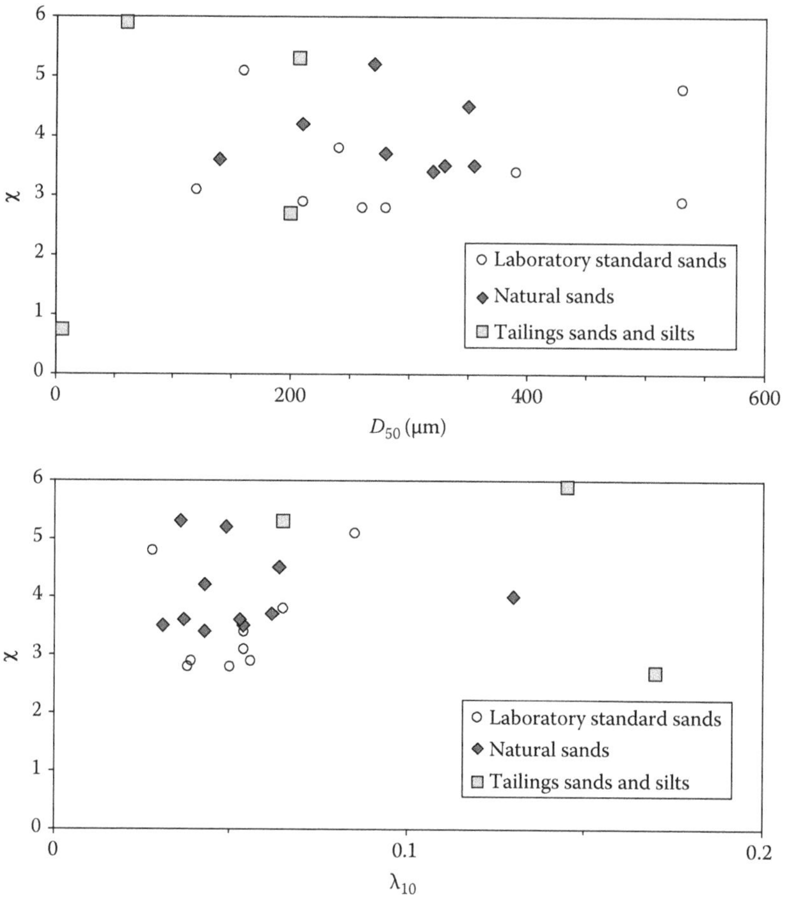

Figure 2.39 Absence of relationship between dilatancy parameter χ, D_{50} and λ_{10}.

The critical friction ratio M_{tc} and the state–dilatancy property χ_{tc} are equally difficult to pin down without soil specific tests. M is generally considered to increase with increasing angularity of soil grains, and we have seen values as large as $M_{tc} \sim 1.7$ for tailings materials and carbonate sands (compared to 1.2–1.3 for subrounded silica sands). The minimum dilatancy plausibly depends on factors such as the crushability of the sand and grain size distribution, with uniformity of particle sizes and crushing both tending to reduce dilatancy. Well-graded, non-crushable sand grains should increase dilatancy. However, with the available database (a good set of both dense drained and loose undrained tests is needed) we have not been able to establish any such relationship with index properties. For example, Figure 2.39 shows the value of χ_{tc} against D_{50}, as the most basic index property of the sands we have tested, and λ_{10}. Further work would be needed to better understand all of the factors that influence how minimum dilatancy varies with state parameter for each sand type.

2.6.3 Soil property measurement

Overall, although judgements can be made about critical state properties of a soil (Γ, λ, M_{tc}, χ_{tc}) given information on particle size distribution and mineralogy, these properties are very sensitive to geologic factors. For any particular project or soil type, dedicated triaxial testing is needed. Assumptions should never be made when dealing with mine tailings or carbonate sands – almost invariably they will be atypical, and must be tested. The required testing is not necessarily expensive (commonly about $15,000 at the time of writing for a comprehensive suite of testing), and it is mainly a question of thoroughness. If the engineering project is more than minor building and carries any seismic risk to life, sound practice demands that laboratory testing is carried out. Table 2.5 is a guide for developing engineering judgement, it is not an excuse to avoid appropriate testing.

2.7 PLANE STRAIN TESTS FOR SOIL BEHAVIOUR

This section considers information from other than triaxial tests to move understanding of soil behaviour from the confines of the laboratory element tests to situations of engineering practice. We do this by considering two types of plane strain test, which better approximate conditions encountered in civil engineering works than can be achieved with the triaxial test. The subsequent Section 2.8 then shows how the information from such plane strain tests affects geotechnical understanding of the general behaviour of soils.

2.7.1 Simple shear

So far the discussion has been in terms of triaxial tests because the stress and strain conditions in a triaxial test are largely uniform and measurable – reasons that have made the triaxial test the basis on which soil behaviour has been understood for at least 50 years. But triaxial tests have a special symmetry and fixed principal stress directions that do not relate to much civil engineering, and these limitations have produced interest in the simple shear test.

Simple shear is the plane strain condition in which a shear stress is imposed, typically horizontally, with an imposed condition of no horizontal extension. Figure 2.40 shows this schematically as well as the stress conditions in terms of Mohr's circles. A feature of the basic simple shear test is that the principal stresses are not measured, with the data comprising the vertical effective stress, the imposed (horizontal) shear stress, the vertical strain and the shear strain.

Knowledge of the stresses on the top and bottom of the sample is not enough to define the stress state in the sample. Only one point on the Mohr circle of stress can be defined from these measurements and the lateral stresses on both horizontal axes are needed to complete the picture. However, even if these measurements are made, the complementary shear stresses are largely absent from the edges of the sample so that only the middle third of the sample has approximately uniform stresses (Muir Wood et al., 1979). Simple shear is some way from a good element test for assessing soil behaviour even though it is an attractive analogue for soil loading in many situations.

It is these rather basic limitations of the simple shear test that gave the widespread preference for the triaxial test for measuring soil behaviour during the past 50 years. However, some workers have tried to remove the deficiencies of the basic simple shear test. The Cambridge University simple shear device (Roscoe, 1953) uses rigid metallic walls lined with a rubber membrane to contain the sample. Pressure transducers in the walls measure normal and shear stresses at transducer locations. Bjerrum and Landva (1966) describe the development of a simple shear device at the Norwegian Geotechnical Institute that used a steel reinforced rubber membrane, developed from the ideas of Kjellman (1951), to constrain a cylindrical sample. Stresses on the lateral boundaries cannot be measured, but because of its simplicity this device is popular, especially when implemented within a modern computer-controlled test. Today, the computer-controlled simple shear test has moved

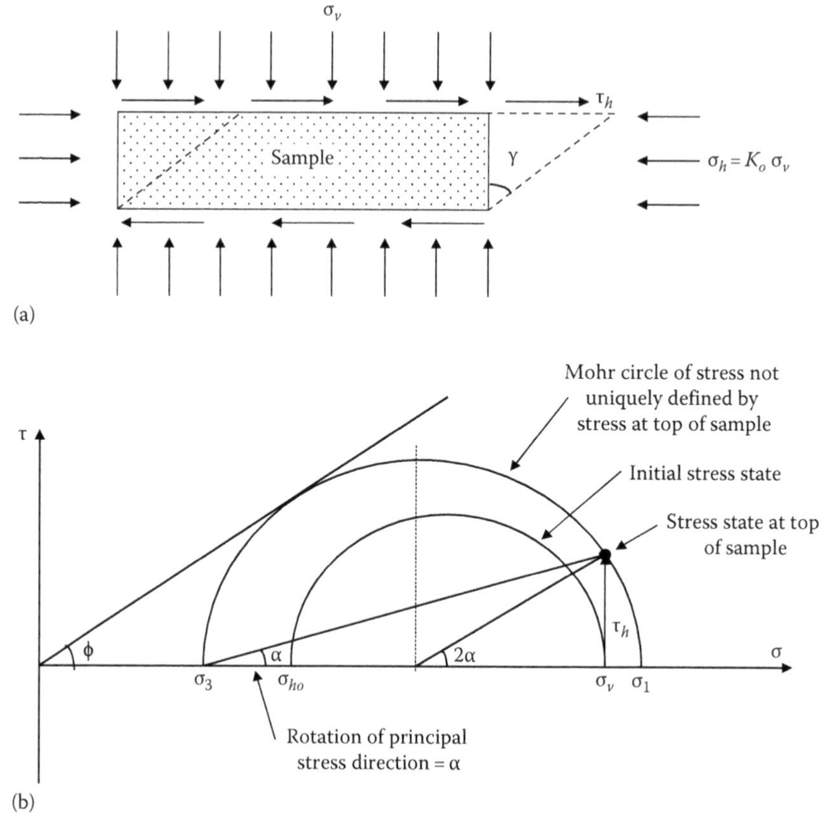

Figure 2.40 Stress conditions in the simple shear test. (a) Stresses applied to sample and (b) Mohr's circle for simple shear.

to mainstream consulting practice, especially for cyclic loading of soils to mimic an earthquake (i.e. liquefaction studies).

Despite the popularity of the simple shear test, there are very few published liquefaction studies using it, particularly tests on sands with initial states looser than critical. An exception is the study of Fraser River Sand by Vaid and Sivathayalan (1996a). The Fraser River flows through the lower mainland of British Columbia, and has deposited sands over recent geologic time. The sand sample used in the study was a uniformly graded medium-grained sand with $D_{50}=300$ μm and a fines content of about 1%. A careful series of simple shear tests were carried out (using a Geonor-type device with a reinforced rubber membrane) on samples prepared as loosely as possible by water pluviation and then consolidated to the desired vertical stress level. Figure 2.41 shows some of Vaid and Sivathayalan's results as confining stress and void ratio were varied. Consistent trends are observed.

The interesting point, however, is that Vaid and Thomas (1995) carried out triaxial compression and extension tests on this same sand, also prepared by water pluviation. Compression tests were loaded in the direction of pluviation while for extension tests the major principal stress direction was perpendicular to the direction of pluviation. In simple shear, the principal stress direction varies but lies between the triaxial testing bounds. As shown in Figure 2.42, the behaviour in compression, extension and simple shear of samples consolidated to the same void ratio and stress level is very different.

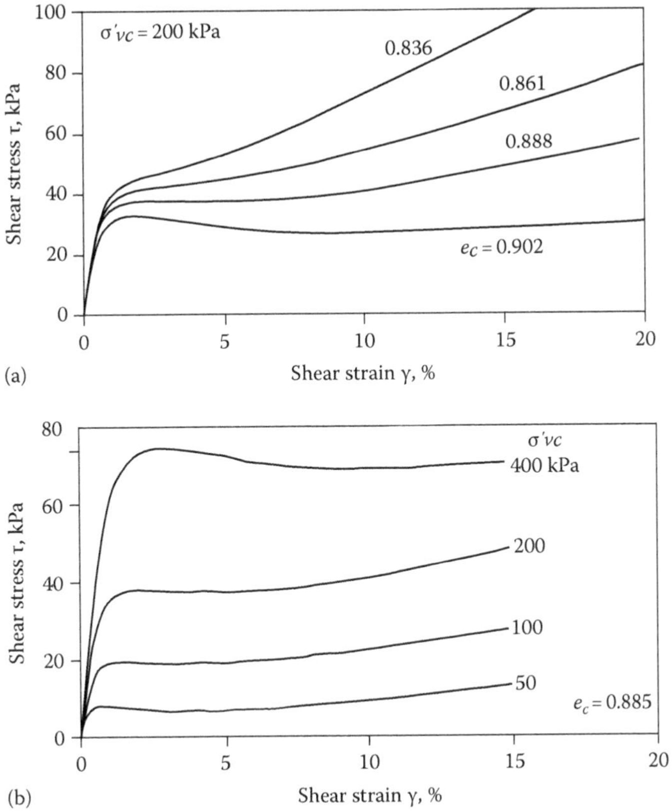

Figure 2.41 Undrained simple shear tests on Fraser River sand. (a) Void ratio varied and (b) vertical consolidation stress varied. (From Vaid, Y.P. and Sivathalayan, S., *Can. Geotech. J.*, 33(2), 281, 1996a. With permission from the NRC, Ottawa, Ontario, Canada.)

This comparison cannot be pushed too far as the triaxial samples were isotropically consolidated to 100 kPa, while the vertical stress in simple shear was 100 kPa. The initial stress conditions are therefore not exactly the same between the two test types. Note also that the maximum shear stress is assumed to occur on the horizontal plane in simple shear which is an approximation.

The triaxial compression sample, loaded in the direction of pluviation, is stiffer and more dilatant at small strains than the simple shear and extension samples. The simple shear and extension tests are remarkably similar in terms of peak shear stress, but the initial stiffness and large strain behaviour are clearly different. Figure 2.42 suggests that the three samples consolidated to approximately the same initial state will end at very different ultimate (critical) states. The challenge for any critical state-based constitutive model is to explain this observed behaviour.

2.7.2 Imperial College and Nanyang Technical University plane strain test

An alternative to the simple shear test to investigate plane strain was developed at Imperial College 50 years ago by Wood (1958) and was, in effect, a variation on the triaxial test.

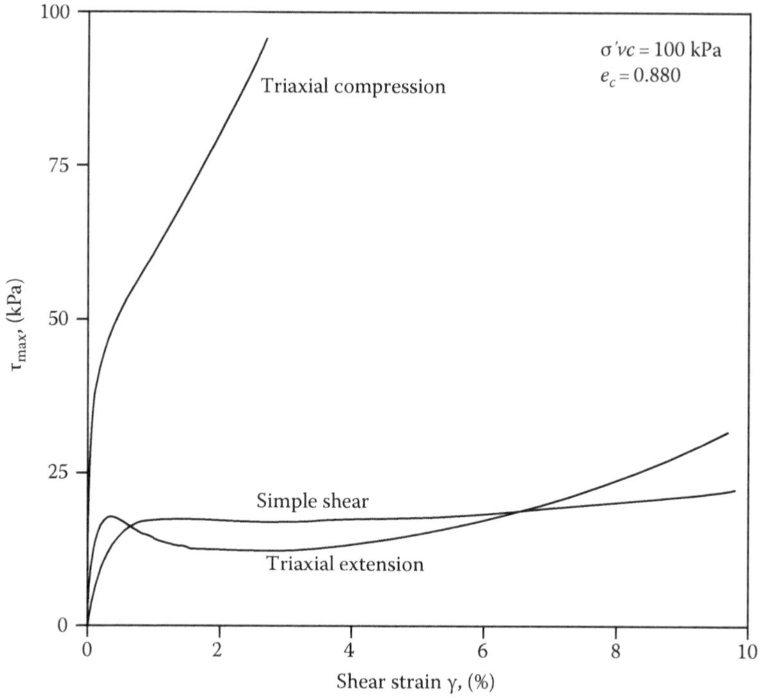

Figure 2.42 Comparison of triaxial compression, extension and simple shear behaviour of Fraser River sand. (From Vaid, Y.P. and Sivathalayan, S., *Can. Geotech. J.*, 33(2), 281, 1996a. With permission from the NRC, Ottawa, Ontario, Canada.)

Plane strain was enforced by end platens, with the intermediate principal stress being measured by the infinitely stiff null method. Deviator load was applied vertically by two loading rams using zero axial friction rotating bushings, bearing on a rigid platen at the quarter length points. Cell pressure was applied in the same way as a triaxial test. Lubricated end platens were used. Although this plane strain apparatus does not have the convenience of modern transducers and data acquisition systems (and a large number of data points), it does provide accurate results for slow drained tests. Figure 2.43 shows a picture of a failed sand sample using this equipment.

A modern variation on the Imperial College approach to plane strain was developed at Nanyang Technological University by Wanatowski and Chu (2006). The plane strain condition was imposed by two metal vertical platens, fixed in position by two pairs of horizontal tie rods. The lateral stress in the $\varepsilon_2 = 0$ direction (i.e. intermediate principal stress, σ_2) was measured by four submersible pressure cells with two on each vertical platen. Figure 2.44 shows a photograph and details of this development. The plane strain testing system was fully automated.

2.8 GENERAL SOIL BEHAVIOUR FROM TRIAXIAL PROPERTIES

2.8.1 Critical friction ratio in 3D stress states: $M(\theta)$

Taking M as a material constant was a dominant view when critical state models were first being developed. Constant M is acceptable in triaxial compression but is rather poor in a

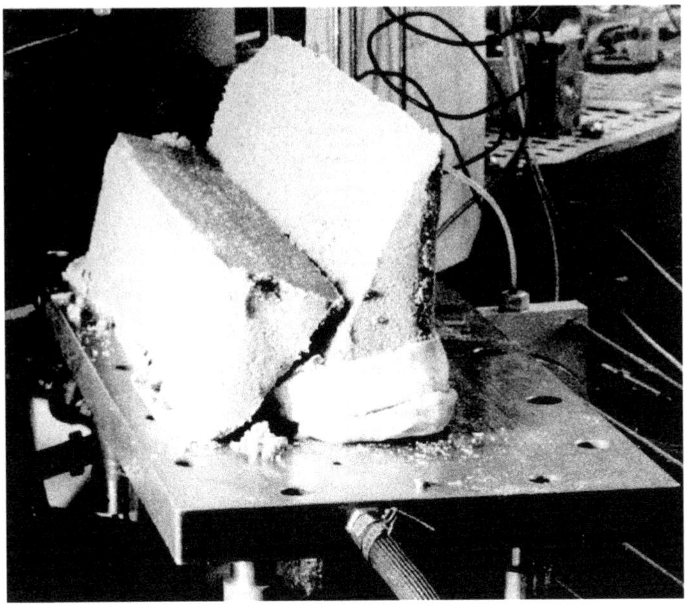

Figure 2.43 Dense sand after failure in Imperial College plane strain apparatus, tested by Cornforth, 1964. (From Cornforth, D.H.: *Landslides in Practice: Investigation, Analysis and Remedial/Preventative Options in Soils*. 2005. Copyright Wiley-VCH Verlag GmbH & Co. KGaA. Reproduced with permission.)

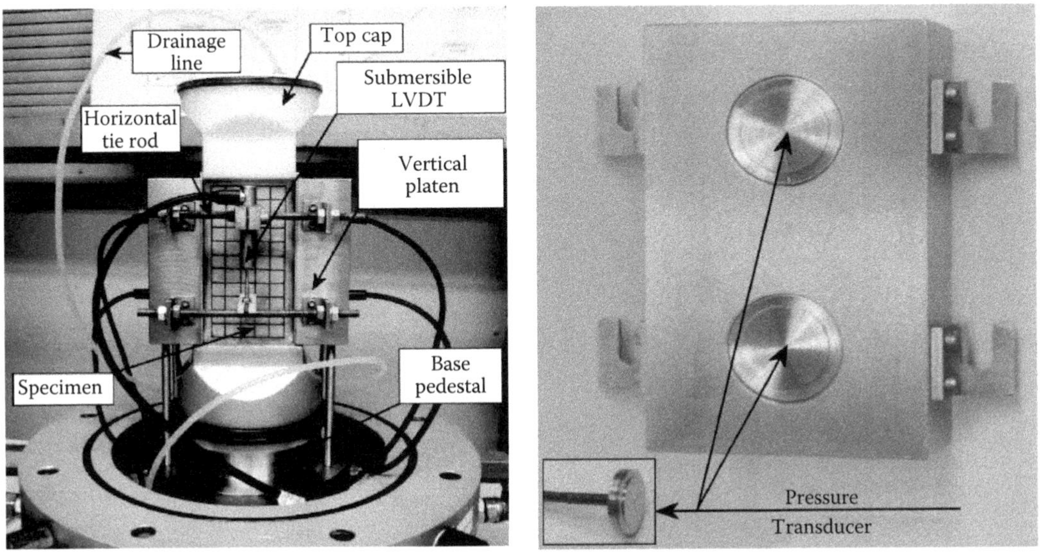

Figure 2.44 Fully automated plane strain device at Nanyang Technological Institute described by Wanatowski and Chu (2006). (Courtesy of Prof. D. Wanatowski.)

more general context (Bishop, 1966). What is needed is to assess the effect of the varying proportion of intermediate principal stress on M. This proportion is given by the Lode angle θ, a third stress invariant (see Appendix A) that ranges from $\theta = +30°$ in triaxial compression ($\sigma_2 = \sigma_3$) to $\theta = -30°$ in triaxial extension ($\sigma_2 = \sigma_1$); all other stress states lie between these limits. Thus, interest moves to $M(\theta)$ as a general concept for the critical friction ratio and where M_{tc} is used as the soil property to scale $M(\theta)$.

The effect of intermediate principal stress on the failure criteria of sand was actively researched throughout the 1960s and early 1970s (e.g. Cornforth, 1964; Bishop, 1966; Green and Bishop, 1969; Green, 1971; Reades, 1971; Lade and Duncan, 1974). This interest covered a wide range of sand densities, but was directed at peak strength with substantial dilatancy. However, it is essential to define the zero dilation rate critical friction.

Cornforth (1961, 1964) combined conventional triaxial testing with the newly developed Imperial College plane strain test to examine the behaviour of Brasted sand. Cornforth's data have the attraction of providing grouped tests at three densities and two stress levels for each of triaxial compression, triaxial extension and several stress paths in plane strain; this is exactly what is needed to evaluate a critical state approach, since critical state theory claims independence of the material properties from density. It is difficult to overstate the importance of Cornforth's work as it underlies most understanding of the effect of intermediate principal stress on sand behaviour, and it remains relevant 60 years after the work was done.

Figure 2.45 shows Cornforth's data plotted in the stress–dilatancy form of η_{max} versus D_{min}. A linear trend fits the Brasted triaxial compression data and gives $M_{tc} = 1.27$ (equivalent to $\phi_c = 31.6°$). For the triaxial extension data, the trend line suggests that $M_{te} = 0.81$ (equivalent to $\phi_c = 27.9°$). In plane strain, the best-fit trend line is still linear and indicates that $M_{ps} = 1.08$ where the subscript ps denotes plane strain conditions; the plane strain data indicate $\phi_c = 33.4°$ for the average Lode angle at the critical state in plane strain of about $\theta = 15°$ found in Cornforth's tests, an increase in frictional strength of about 2° over the triaxial compression conditions.

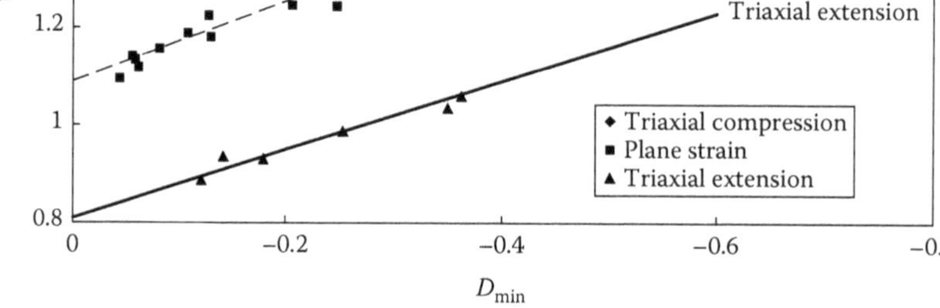

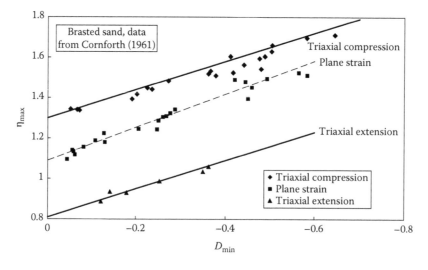

Figure 2.45 Brasted sand stress–dilatancy in plane strain and triaxial conditions. (After Jefferies, M.G. and Shuttle, D.A., *Géotechnique*, 52(9), 625, 2002. With permission from the Institution of Civil Engineers, London, U.K.)

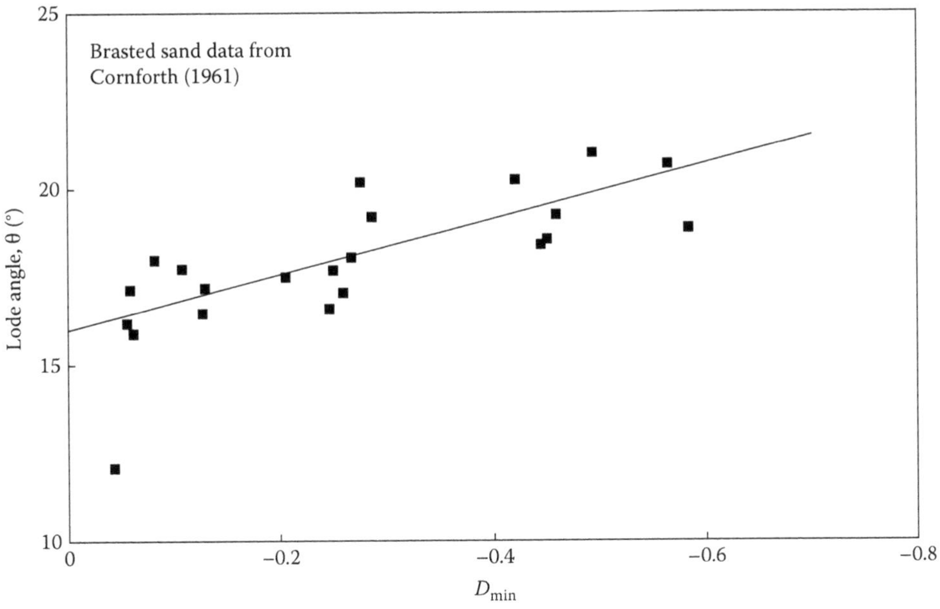

Figure 2.46 Lode angle at peak strength in plane strain. (After Jefferies, M.G. and Shuttle, D.A., *Géotechnique*, 52(9), 625, 2002. With permission from the Institution of Civil Engineers, London, U.K.)

The trends found in Cornforth's results differ from the prevailing view, which is that ϕ_c is invariant with Lode angle (e.g. Bolton, 1986). Green (1971) also suggested invariance of ϕ_c based on Ham River sand, although allowing for the possibility that ϕ_c might increase by perhaps 2° in plane strain over triaxial compression because of experimental uncertainties. More recently, Schanz and Vermeer (1996) concluded that ϕ_c was sensibly invariant with strain conditions based on triaxial compression and plane strain tests on Hostun sand. The clear reduction of ϕ_c in triaxial extension seen in both the Erksak data (Figure 2.30) and Brasted data (Figure 2.45) is apparently presently unrecognized.

One common error, which arises from plots like Figure 2.45, is to treat plane strain as an alternative condition to (say) triaxial compression. The problem in doing this is that plane strain is not a condition in which the stress state is similar from one plane strain state to another. Rather, as its name suggests, in plane strain the stress state develops to accommodate the imposed strain condition. Figure 2.46 shows resulting variation in the proportion of intermediate principal stress (which is the out-of-plane stress) plotted in terms of the Lode angle at peak strength versus the dilation rate for that peak strength. While there is no difficulty for the soil actually at the critical state, M in plane strain cannot be assessed from Figure 2.46 without further processing. That processing is discussed in Appendix C, and it is useful to show the results here to aid insight.

Figure 2.47 shows Cornforth's plane strain data after it has been transformed into an operating critical friction ratio and plotted versus Lode angle in the test. The data for M for the range of Lode angles developing in plain strain cluster around a trend which is shown as a dotted line on the figure. What is this trend line? It is the average M from two failure criteria, the familiar Mohr–Coulomb representation and the less familiar Matsuoka–Nakai (1974) failure criteria. The Mohr–Coulomb criterion with a constant critical friction angle matches the prevailing view for the critical state as noted earlier, but consistently gives weak strengths based on a triaxial calibration. The Matsuoka–Nakai criterion is based on the

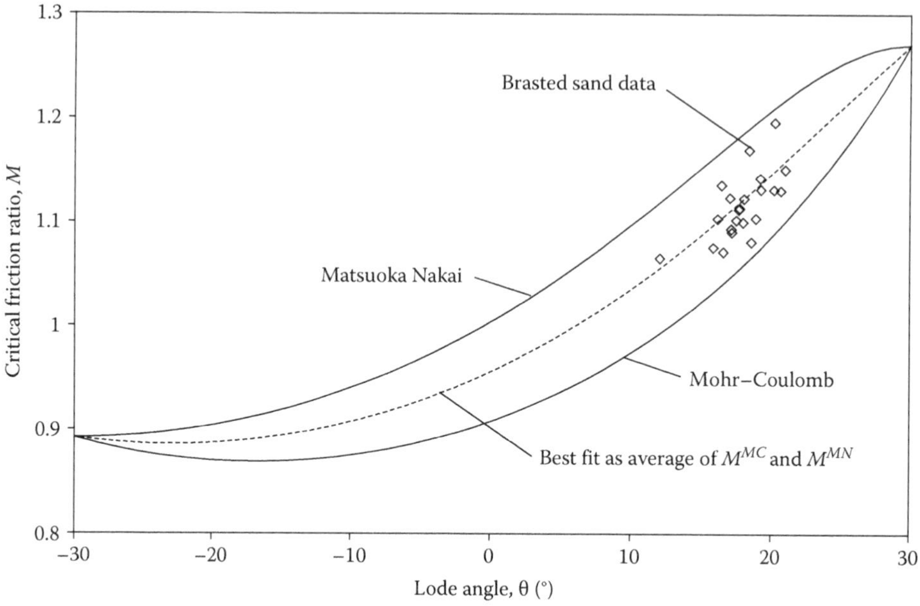

Figure 2.47 Comparison of functions for $M(\theta)$ with Brasted sand data. (After Jefferies, M.G. and Shuttle, D.A., *Géotechnique*, 52(9), 625, 2002. With permission from the Institution of Civil Engineers, London, U.K.)

physically appealing concept of spatially mobilized planes, but shows too great strengths in general. The suggestion that an average of Mohr–Coulomb and Matsuoka–Nakai criteria was a reasonable representation of Brasted sand behaviour (Jefferies and Shuttle, 2002) was further supported by tests on Changi sand that showed much the same results (Wanatowski and Chu, 2007, see their Figure 11).

For completeness, we also note that there is a general absence of information on the trends in M (or ϕ_c for that matter) within geotechnical engineering. There are sets of M values for triaxial compression, plane strain and triaxial extension with all workers drawing smooth curves based on their physical idealizations to fill in the data gap between $\theta = -30°$ and $+15°$.

2.8.2 Operational friction ratio in stress–dilatancy: M_i

Rowe's stress–dilatancy relationship is given as Equation 2.2 where the parameter K is related to the mineral to mineral friction ϕ_u. Rowe suggested that the operating sliding contact friction angle in (2.2) was ϕ_f, not ϕ_u, and was such that $\phi_u \le \phi_f \le \phi_c$ (where ϕ_c is the critical state angle). Rowe's suggestion has not been disputed and appears in current geotechnical textbooks (e.g. Muir Wood, 1990). But minimal guidance is provided in the literature on how ϕ_f evolves and K is generally taken as constant. In Chapter 3, we discuss the proposition that M captures the aspect of soil that dissipates plastic work (as heat and following the laws of thermodynamics). In such a context, Rowe's idea translates into a variable work dissipation mechanism that might depend on void ratio, soil particle arrangement (aka fabric), etc.; there is also the possibility of inelastic energy storage to offset work dissipation. Let us look to some data.

Information on stress–dilatancy can be obtained from drained triaxial tests if the data are logged digitally using a high scan rate. This allows subsequent numerical differentiation of the measured results into a stress–dilatancy form. Thus, rather than plotting q versus ε_1 (say), the data are transformed into η versus D^P. Figure 2.48 shows the results of such numerical differentiation of the stress–dilatancy behaviour of Erksak sand in η – D^P space. Dense and loose sand data are shown separately for clarity in Figure 2.48a and b, and together in Figure 2.48c.

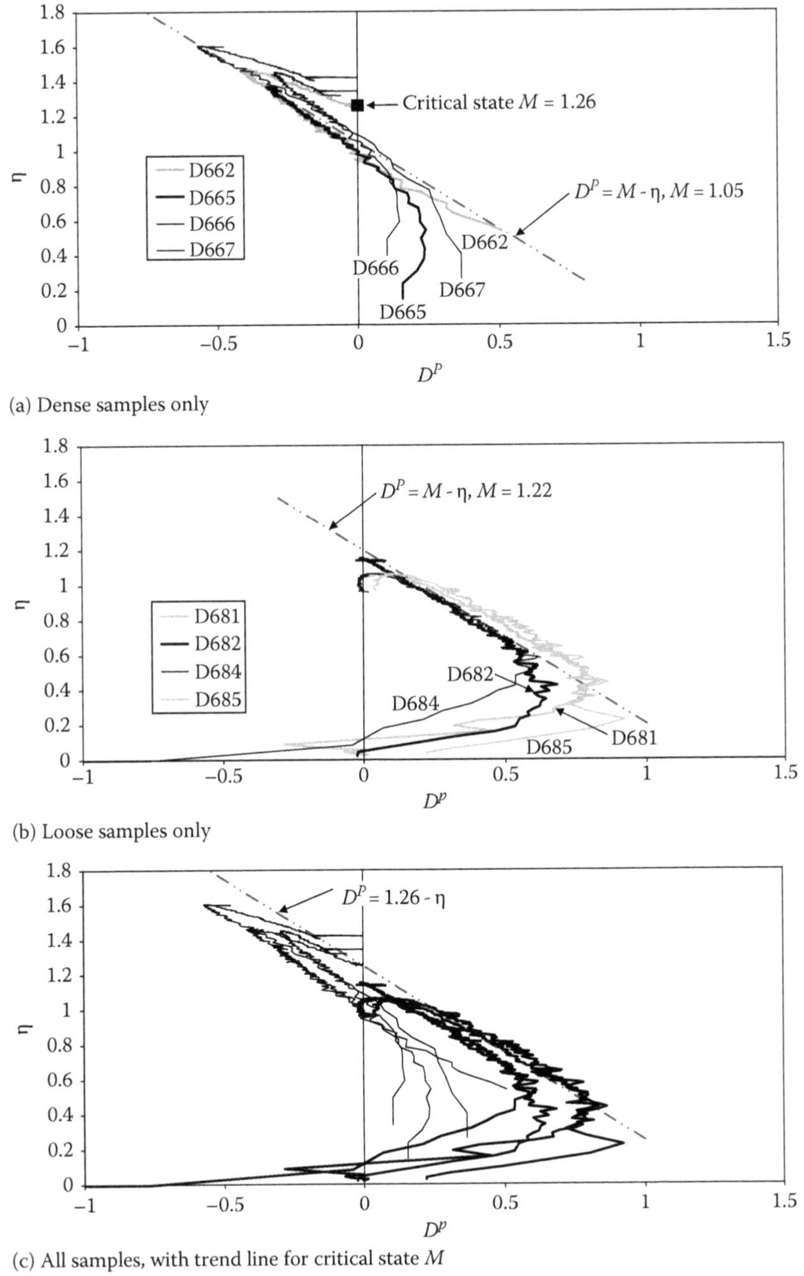

Figure 2.48 Drained triaxial data for Erksak sand reduced to stress–dilatancy form. (After Been, K. and Jefferies, M.G., *Canadian Geotechnical Journal*, 41(1), 972, 2004. With permission from the NRC, Ottawa, Ontario, Canada.)

Stress–dilatancy, for both loose and dense samples, is approximately linear in the $\eta - D^P$ space between the low stress ratio initial part of the curve and the peak stress ratio. In the case of the dense sand, the stress–dilatancy plots naturally reverse once the stress ratio reaches a peak, giving a hook in the curve as it drops to the critical state. (The initial non-linear part of the curve at low stress ratio is not considered representative of stress–dilatancy and is attributed to the effects of initial fabric and apparent overconsolidation.)

For dense sands, Figure 2.48a, the test data establish an initial crossing of the $D^P = 0$ axis less than M, which is exactly the behaviour identified by Rowe (referred to earlier as the parameter M_f), reaches a peak and then forms a second branch. A best estimate is about $M_f \approx 1.05$ based on the pre-peak linear part of the stress–dilatancy plot for $\psi < 0$, where M_f corresponds to where the straight line crosses the $D^P = 0$ axis. This is obviously not the critical state because dilation continues into negative dilatancy space. Extrapolation of the declining η branch in negative D^P space to the $D^P = 0$ axis gives M_{tc}. In the case of the test data on Erksak sand shown in Figure 2.48a, the procedure indicates $M_{tc} = 1.26$. This is the same estimate of M_{tc} obtained from many more tests using Bishop's method (Figure 2.30).

The best-fit linear trends through the loose sand data (Figure 2.48b) project to the $D^P = 0$ axis at a value of η that suggests $M \approx 1.22$ for $\psi > 0$. This projection is based on the trends seen in the data in the range $0.1 < D^P < 0.6$ so as to avoid the effects of initial elasticity and inaccuracy at large displacements. The projected value is M (rather than M_f), as the $D^P = 0$ situation is the critical state for the loose samples. The fit of one straight line to the data suggests that $M_f \approx M$ for most of the stress path.

The transient condition where $D^P = 0$ in the dense samples, that is where volumetric strain rate changes from contraction to dilation, is called the *image condition*. Ishihara et al. (1975) called this transient condition the phase transformation while others have called it the pseudo–steady state. The term image condition is preferred as the state is a projection of the critical stress ratio on the $q - p'$ plane, whereas it is neither a phase change nor a steady state nor a pseudo condition. Accordingly, we prefer to switch from Rowe's notation and use M_i rather than M_f, where the 'i' subscript denotes the current image condition.

Overall, there are good reasons to expect M_i to evolve with strain, with the general requirement only being that $M_i \Rightarrow M$ as $\varepsilon_q \Rightarrow \infty$. But strain is in itself not an admissible input to M_i because, in any piece of ground in-situ, it is not possible to know the current strains. M_i needs to be expressed in terms of state variables, all of which can be measured (at least in principle). The general requirement on M_i is most elegantly, and correctly, expressed in the alternative state parameter form, suggested by Dafalias and co-workers: $M_i \Rightarrow M$ as $\psi \Rightarrow 0$. Within this general framework, several alternative ideas have been suggested which are summarized in Table 2.6 and these proposals are also plotted in Figure 2.49 to compare with the Erksak data.

Taking the case of dense sand, the relation of Manzari and Dafalias (1997) seems close to a lower bound. The slightly more complex form for M_i suggested by Li and Dafalias (2000) is similar for dense sand, but predicts $M_i > M$ for loose sands. This is at variance with test data, and it is difficult to understand how loose sand could dissipate plastic work at

Table 2.6 Summary of proposed relationships for M_i

Originator	*Relationship*	*Comments*
Manzari and Dafalias (1997)	$M_i = M + m\psi$	$m \approx 4$ for Toyoura sand.
Li and Dafalias (2000)	$M_i = M \exp(m\psi)$	$m \approx 4$ for Toyoura sand.
Jefferies and Shuttle (2002)	$M_i = M(1 - \lvert\psi_i\rvert / M_{tc})$	Matches Nova's rule on average at D_{min} for all sands; see Appendix C.

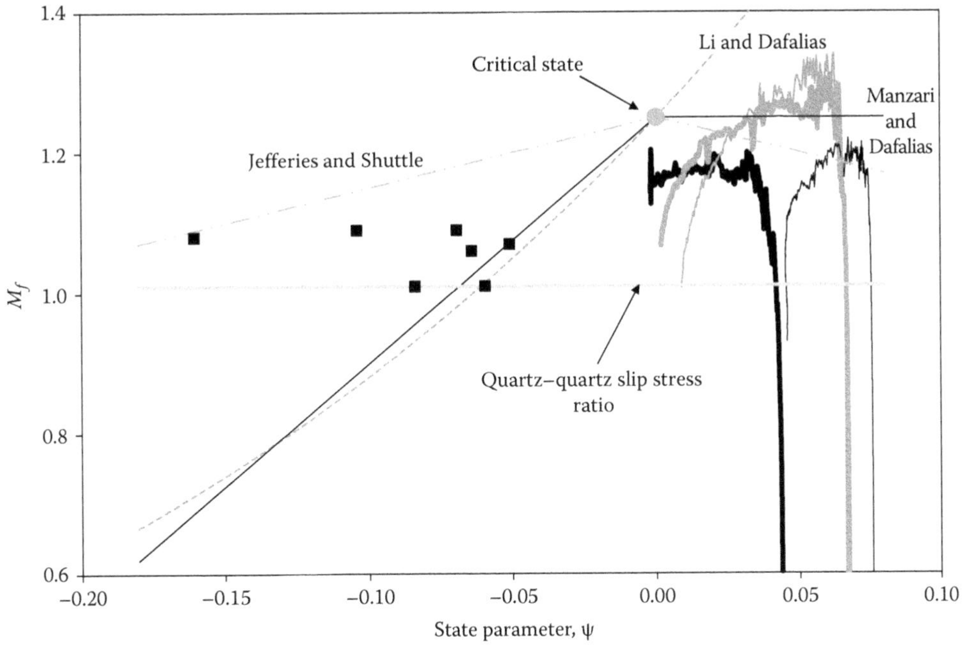

Figure 2.49 Relationship of mobilized friction ratio M_i to ψ for Erksak data. (After Been, K. and Jefferies, M.G., *Can. Geotech. J.*, 41(1), 972, 2004. With permission from the NRC, Ottawa, Ontario, Canada.) Dense sand data at initial $D^p = 0$ shown as filled squares; loose sand data shown as traces for complete strain path. Also shown are several proposed constitutive model relationships.

a greater rate than in the critical state. The Jefferies and Shuttle (2002) relationship is an upper bound to the dense sand data, with symmetry around the critical state. The relationship developed in Appendix C is an enhanced version of Jefferies and Shuttle (2002) and which now uses the standard soil properties N, χ_{tc}, discussed earlier in this chapter as inputs to the computed M_i.

Looking at the data in Figure 2.49, it is unclear that ψ alone is a sufficient choice for the state variable controlling M_i. The fit of these alternative suggestions to the Erksak data suggests something is missing, with a fabric tensor needed in addition to ψ to define the state of sand. A product of these two state measures can ensure that $M_i \Rightarrow M$ as $\psi \Rightarrow 0$ in accordance with the idealization of the critical state while allowing for soil fabric detail.

Finally, we note that the effect of state and/or fabric applies on top of the effect of Lode angle. Although this may seem complicated, it actually has rather simple operational form and is easily implemented as a user-defined function in spreadsheets; the relationship for M_i developed in Appendix C can be found as the function $M_{psi}\_v_3(M_{tc}, N, \chi_{tc}, \theta, \psi)$ in the various downloadable spreadsheets.

2.8.3 General state–dilatancy

The influence of soil state on maximum dilatancy (= D_{min}) was discussed earlier in the context of triaxial tests, with the soil property χ being introduced to capture this aspect of soil behaviour. How does this generalize to other stress states? The immediate difficulty in answering the question is lack of data. Many workers have looked at peak strength in plane strain but, as noted earlier, there has been an absence of a stress–dilatancy framework in

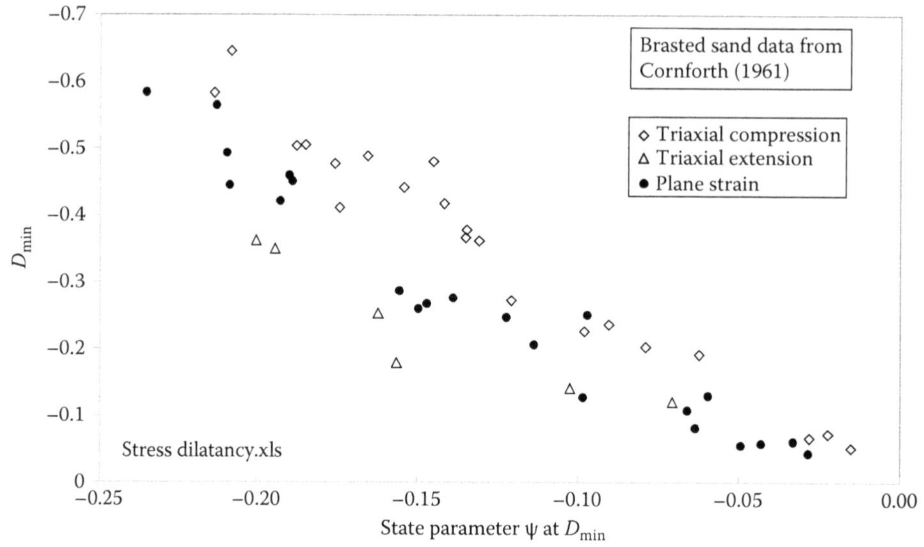

Figure 2.50 Maximum dilatancy D_{min} versus ψ at D_{min} for Brasted sand in triaxial compression, triaxial extension and plane strain. (Adapted from Jefferies, M.G. and Shuttle, D.A., *Géotechnique*, 52(9), 625, 2002, based on data in Cornforth, 1961.)

the way the testing was carried out. The exception is Cornforth's testing that has been relied on to infer the nature of the critical friction ratio M. Using Cornforth's data, Figure 2.50 compares trends in the maximum dilatancy (i.e. D_{min}) versus state parameter ψ at D_{min} of Brasted sand in triaxial compression, triaxial extension and plane strain. The slope of the D_{min} – ψ line is the soil property χ which appears to be different depending on the loading condition. Figure 2.51 takes the same data and scales D_{min} by the ratio M_{tc}/M, where M

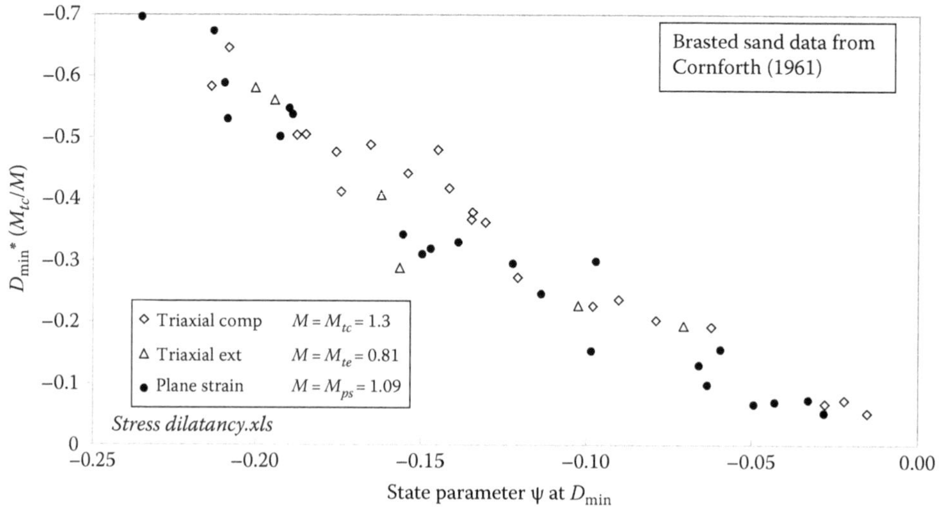

Figure 2.51 Scaled maximum dilatancy versus ψ at D_{min} for Brasted sand (D_{min} is scaled by ratio of M in compression, extension and plane strain).

is the appropriate value for compression, extension or plane strain. This has the effect of unifying the data trends. Thus, χ is properly defined under triaxial compression and scales to other loading conditions, exactly as with M_{tc}. The theoretical reason why this scaling works, and why it is necessary, comes from constitutive modelling which we now turn to in the next chapter.

Chapter 3

Constitutive modelling for liquefaction

3.1 INTRODUCTION

3.1.1 Why model?

This chapter is about modelling soil stress–strain behaviour, which will lead into modelling liquefaction. Before doing so, an obvious question needs to be addressed: why model? There are several reasons and more to modelling than just a precursor to stress analysis using finite elements.

The understanding of liquefaction has been plagued by dubious 'concepts', many of which run counter to a sound appreciation of soil mechanics. This has led to mutually exclusive propositions and the notion that grasping the subject of liquefaction requires great wisdom and many years of experience. Constitutive modelling, using an appropriate model (and what is appropriate follows later in this chapter), shows that liquefaction is simply another soil behaviour that is relatively easily understood. Given this understanding, which is accessible to anybody given a little diligence, how to engineer liquefaction resistant works becomes far less contentious. Hypotheses ('explanations') such as 'metastable particle arrangement' are tossed out in favour of conventional soil properties and proper geomechanics for engineering designs.

Modelling is also important because geotechnical engineers depend to a large extent on in-situ tests to determine sand or silt properties, but in-situ tests do not really measure soil properties: rather, in-situ tests measure a response to a loading process. Obtaining soil properties from in-situ tests involves solving an inverse boundary value problem, and a model is required for this.

Modelling is also an excellent way to capture full-scale experience. Because civil engineering must rely largely on case histories from failures rather than testing of prototypes, a great weight is placed on such experience and properly so. But full-scale experience needs to be understood using a sound framework and this framework necessarily comes from mechanics. Mechanics, in turn, is based upon understanding soil constitutive behaviour. Wroth (1984) made this point years ago, but it is still not always fully appreciated.

3.1.2 Why critical state theory?

A basic premise is that a proper constitutive model for soil must explain the changes in soil behaviour caused by changes in density. Despite the obvious nature of this premise, void ratio (or any related variable such as relative density) is rarely included as a variable in constitutive models for soil, as can be ascertained from the proceedings of a workshop (Saada, 1987) where some 30 different models for sand were represented. The exceptions are models based upon critical state theory and this naturally sets up critical state theory as the preferred starting point.

There is more to critical state theory than independence of the properties relating to density. Constitutive models for soil cover a philosophical range from descriptive to idealized. Descriptive models are intrinsically curve fitting and anchored to test data – they can be very suitable for computing if the stress paths in the problem of interest are similar to the test conditions. However, the accuracy of descriptive models in representing a particular situation is too often offset by an absence of insight into the underlying physical processes. Idealized models, on the other hand, start from postulated mechanisms from which behaviours are then derived. Idealized models trade possibly reduced accuracy in a particular situation for a consistent (and known) physics. Critical state theory is very much an idealized framework but one that traces to the second law of thermodynamics – and that makes it fundamental.

3.1.3 Key simplifications and idealization

There are two key simplifications in what follows: isotropy and small strain theory. Isotropy is familiar to everyone exposed to engineering science, but there is also a litany that soils are intrinsically anisotropic. Isotropy will nevertheless be assumed because there is little point in getting involved in the complexity of anisotropic behaviour if the isotropic version is not functional – anisotropy is further detail, not a fundamental premise that will make or break the model. Further, anisotropy can be approximated in an isotropic model, as will be shown, if there is a supporting evidence. This then leads to the point that most practical engineering has enough difficulty in obtaining characteristic parameters for a simple isotropic model and that anisotropy is therefore (at least presently) a distraction from more important issues.

Small strain theory is a far more important point. Almost all degree courses in engineering science teach small strain theory in which higher than first-order terms are dropped in moving from displacement gradients to strain. This small strain approximation (small meaning major principal strain of the order of 0.1% or so) is very reasonable within the context of elasticity. However, soil behaviour may involve strains to failure of as much as 50%. The standard elasticity-based small strain theory taught in engineering courses is not genuinely adequate for geomechanics and even routine work should invoke large strain theory of one sort or another.

To save layering additional information and complexity, what follows will be presented within the familiar small strain context with two exceptions. First, incremental volumetric strains will be integrated to obtain void ratio change; tracking void ratio this way is a large strain approach and assures that the correct end state will be reached. Second, large strain analysis will be used formally in evaluating in-situ tests. In doing this, there is an implied assumption that the properties determined in calibrating with small strain definitions are sufficiently representative for reasonable subsequent use in large strain analysis. There is as well, of course, the ever-present problem with laboratory test equipment and observations where there is a limit to how accurately the strain can be determined and the accuracy of measurements at larger strains.

3.1.4 Overview of this chapter

Although this book is about liquefaction and it has just been suggested that liquefaction is to be modelled, that modelling will not be found in this chapter. Instead, the principle of effective stress will be invoked and the desired framework will be developed in a drained approach. This is done deliberately as undrained behaviour will never be properly simulated unless the drained response is properly understood and captured. Undrained behaviour arises because of boundary conditions, and is not in itself fundamental soil behaviour. Readers unable to restrain their enthusiasm, or wishing to verify that the *NorSand* model does describe liquefaction well, can flip to Chapters 6 and 7 to look at the plots.

In presenting the material, a fair bit of mathematics is inevitable. However, the underlying ideas are rather easy to understand (most can be visualized geometrically). So, to avoid unnecessary confusion, most of the formal derivations have been bundled in Appendix C. This way the ideas can be presented more simply and the derivations left until the whole picture has been obtained.

Critical state ideas are indeed straightforward but they do have the feature that there is no practically useful closed-form solution (i.e. an equation directly relating stress and strain) for the models that follow. This is the case even for something as straightforward as a drained triaxial compression test in which the whole sample is at the same stress state and a known stress path is applied. However, the incremental form of critical state models relating changes in stress to strain increments is delightfully simple, and these relations can be numerically integrated easily in a spreadsheet to provide the desired results. Spreadsheets have been provided on the website that implement critical state models for triaxial compression, plane strain and cyclic loading. The spreadsheets are set up as providing an 'interface' with input properties and graphical output with the code that implements the model written in the VBA programming environment that lies behind Excel (VBA is standard to Excel and does not require an enhanced version of the program). The open-source and commented code can be found under the Visual Basic Editor (use the 'Alt' + 'F11' keys) when in Excel; nothing is hidden and we are not asking you to believe in a 'black box'. Each routine is discussed in this book and the code follows the variable use and equations derived in Appendix C.

A feature of critical state theory is that the ideas were developed in the context of the triaxial test. This test has axial symmetry of strains and fixed principal stress directions, both substantial simplifications for a model to be used in real engineering. However, the first step is to understand and appreciate the generalization of the state parameter framework with *NorSand* and how this computes real soil behaviour. Triaxial conditions are sufficient for this; the relevant downloadable file is *NorSandM.xls*. On the other hand, much of real engineering involves plane strain. The chapter, therefore, concludes with a validation of the 3D version of *NorSand* under plane strain conditions (*NorSandPS.xls*). The full derivation of this 'industrial strength' *NorSand* is given in Appendix C. This chapter, then, provides enough guidance to look at in-situ tests and how the state parameter is determined in-situ (Chapter 4) and gives the context for the ideas. The application of *NorSand* to the full-scale experience of static liquefaction follows in Chapter 6, with cyclic-induced liquefaction presented in Chapter 7.

3.2 HISTORICAL BACKGROUND

As always, the history of developments helps to understand the formalism that has developed around a subject and the reasons for such formalism. The history of critical state theory goes back a surprisingly long way and centres as much upon Cambridge, Massachusetts (Harvard and MIT), as upon Cambridge, England (Cambridge University). There were also particular contributions and insights from Manchester (Victoria University of Manchester), London (Imperial College) and Providence, Rhode Island (Brown University). Perhaps even more surprisingly, the U.S. Army Corps of Engineers had a guiding role in the early days. The history is much wider than you might believe from associating the subject with the Schofield and Wroth (1968) book. Let us trace developments.

Soil has two behaviours that are apparent to the casual observer: plasticity and density dependence. Plasticity is apparent because deformations imposed on soil are largely irrecoverable. Density dependence is obvious because soil can exist over a range of densities at constant stress, and dense soil behaves quite differently from loose soil. A slightly less obvious behaviour that is a corollary to density dependence is that dilatancy is an intrinsic kinematic consequence of soil deformation (Reynolds, 1885). A most interesting, and fundamental, aspect of geotechnics is how these behaviours may be represented within a single, complete constitutive model.

Most soil models are based on plasticity, which is in itself a macro-scale abstraction of the underlying micromechanical reality of grain realignments and movements. Plasticity theory is the dominant methodology for constitutive modelling of geomaterials, as it reasonably captures their behaviour in a computable way. Although purists might argue the necessity for a micromechanical approach, the reality is that micromechanical models are complicated and generally unusable. Much as engineering with metals uses plasticity, even though it is dislocation movements that matter to the metal behaviour, so too can soil mechanics use plasticity without worrying about the internal mechanisms of the soil skeleton. In fact, plasticity theory can be given a fundamental slant through thermodynamics (Drucker, 1951). Plasticity theory will, therefore, be used for soil without further discussion of its relevance.

Plasticity is the idea that some (and usually most) strains are not recovered when a body is unloaded, an idea that dates back 150 years. Tresca (1864) first proposed a yield condition that distinguished between those stress combinations that cause yield (or irrecoverable strains) from those that do not. During yielding, strains are viewed as comprising two mechanisms, one elastic (denoted by the superscript 'e') and one plastic (denoted by the superscript 'p'), with the *strain decomposition*

$$\varepsilon = \varepsilon^e + \varepsilon^p \tag{3.1}$$

One of the differences between plasticity and elasticity is the treatment of strains. In elasticity, principal strain increments are in the same direction as principal stress increments. This relationship is intrinsic to the way the theory is developed. In plasticity, however, theoretical development first concentrates on identifying the stress conditions under which plastic strain occurs to define the yield surface. The magnitude and direction of the plastic strains require further thought.

A simple thought experiment to show the direction of plastic strains is shown in Figure 3.1. An ice hockey puck is sitting on the ice and is about to slide under the action of two forces, both of which are applied by strings acting at an angle. Sliding starts when the force on one string is increased slightly. One can immediately appreciate that the puck starts moving in the direction of the force resultant, not the force increment that initiated sliding. This is the simplest explanation of what is termed the normality principal (alternatively called associated flow). Plastic strain increments are directed normal to the stresses defining the yield surface, not to the stress increment that initiates the yielding. In a slightly more sophisticated explanation, Calladine (1969) shows that normality is a way for a material to maximize the energy absorbed during yielding.

With normality, the principal stresses and principal strain increment directions are aligned, and the net strain increment vector is normal to the yield surface at the stress state corresponding to the present yielding. This is illustrated in Figure 3.2, which plots principal stress and principal strain on the same axes. Normality is an important principle of plasticity, and it was given further impetus by Drucker (1951), who showed that it was necessary for unique solutions to boundary value problems with plastic flow.

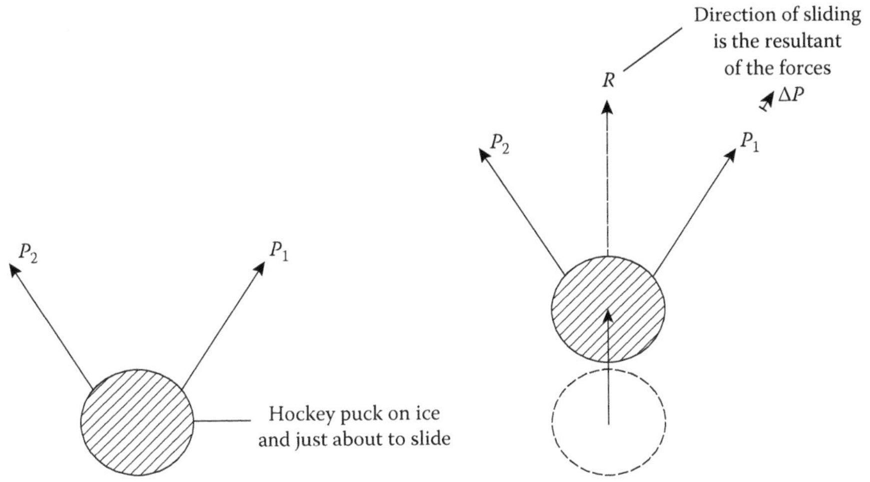

Figure 3.1 Illustration of normality through hockey puck analogy.

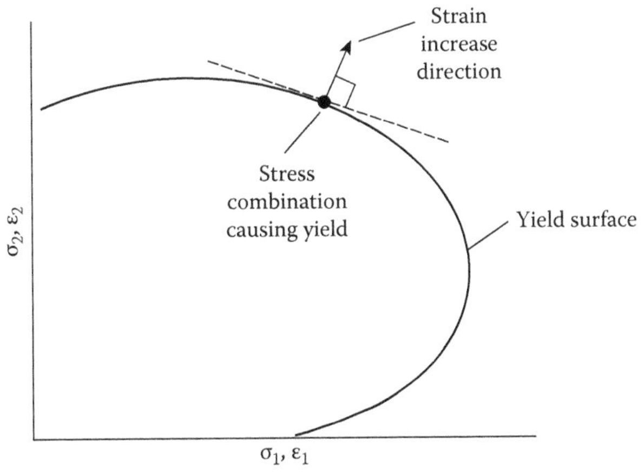

Figure 3.2 Definition of normality (associated plastic flow).

Soil failure had long been represented by the Mohr–Coulomb criterion, which builds on the idea of friction as controlling soil behaviour. When plasticity theory was first applied to soils, the Mohr–Coulomb criterion was viewed as a yield surface (since it was a limiting stress ratio) and normality applied. Figure 3.3 illustrates the consequence: a dilation angle equal to friction angle is implied with frictional yielding. This dilatancy is many times greater than what is measured in real soils and is grossly unrealistic as a basis of a soil model.

Drucker et al. (1957) showed that the correct way to apply plasticity to soil behaviour was to recognize that the Mohr–Coulomb limiting stress ratio was not a yield surface at all. Rather, the yield surface must intersect the normal compression locus (NCL), since normal compression produced irrecoverable strains. Hence, the spectrum of soil density states conventionally classified as consolidation theory was coupled to all aspects of soil constitutive behaviour, because the yield surface size was coupled to the stress-causing yield in isotropic compression. This isotropic yield stress would conventionally be recognized as the

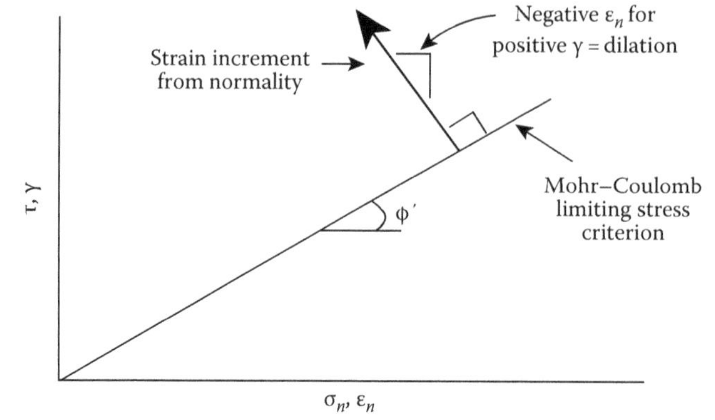

Figure 3.3 Dilation implied by normality to Mohr–Coulomb surface.

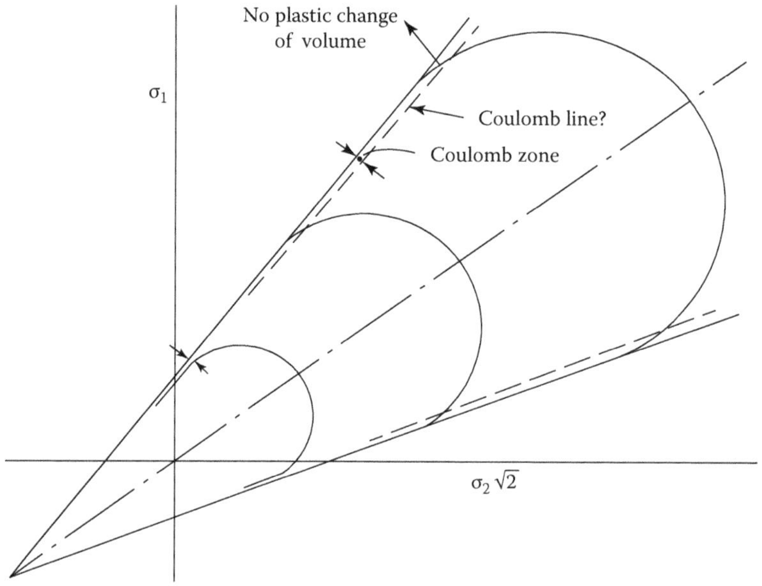

Figure 3.4 Correct association of yield surface with soil strength. Note explicit linking of the Mohr–Coulomb line with zero dilation rate. (Reproduced from Drucker, D.C. et al., *Trans. Am. Soc. Civil Eng.*, 122, 338, 1957. With permission from the ASCE, Reston, VA.)

preconsolidation pressure, and there is considerable existing experience as to how preconsolidation varies with the void ratio of soils. Figure 3.4 shows the Drucker et al. idealization. Theories developed directly from this idealization are sometimes referred to as cap models.

Dilatancy of soils was known from early shear box experiments and had been recognized as a work transfer mechanism rather than an intrinsic strength. A school of thought then developed that the true soil behaviour should be understood in terms of friction angles corrected for the work transferred by dilation. Taylor (1948) was the original proponent of this view, although its first formal statement as an equation appears to be Bishop (1950). The subsequent framework of stress–dilatancy theory (Rowe, 1962) generalized the concept of work transfer by dilatancy to the entire stress–strain behaviour.

Casagrande's (1936) critical void ratio, which was based on testing carried out at Harvard (also in Cambridge, MA) to engineer a liquefaction resistant dam (Franklin Falls, NH) for the Corps of Engineers, is an obvious alternative to the NCL as the basis of the hardening law in a Drucker et al. (1957)-type model. The critical state locus (CSL) also provides a relation between void ratio and stress and is a more natural basis for the hardening law, as the critical state involves shear strain and, by definition, a zero dilation rate. In effect, the critical state is the same thing as the Taylor/Bishop energy corrected friction angle concept.

These ideas of critical friction and correctly associated yield surfaces were pulled together by various people at Cambridge University (Roscoe et al., 1963; Roscoe and Burland, 1968; Schofield and Wroth, 1968) to produce a predictive constitutive framework known as critical state soil mechanics, or CSSM. Key features of CSSM include stress invariants to express behaviour in terms of the ratio of distortional to mean stress, yield surfaces based on idealized mechanisms for plastic work, normality and use of the CSL to relate yield surface size to void ratio.

The theory of CSSM put forward by Schofield and Wroth (in 1968) is based on two idealized and related models for triaxial conditions: Granta Gravel (rigid plastic) and Cam Clay (elastic plastic). Both models adopt a postulated mechanism for dissipation of plastic work, which then allows the development of a full constitutive model. The advantage of this approach is that many behaviours can be developed from a few axioms and postulates by invoking self-consistency and energy conservation. This provides a complete, self-contained and thorough model. Whether this model is useful for the task at hand depends on the degree to which its predictions match the behaviour of real soils.

CSSM explicitly recognizes that any particular soil can exist over a spectrum of densities, and CSSM quantifies the effect of void ratio (density) on soil behaviours. This predictive power develops because in standard CSSM models the yield surface always intersects the CSL in the e–p' plane. This CSL then becomes the hardening law for all stress paths and adds the effect of density and pressure changes to the models. Despite these attractions, the well-known variants of Original Cam Clay (Schofield and Worth, 1968) and Modified Cam Clay (Roscoe and Burland, 1968) are avoided in modelling most real soils, including sands and silts, because of their inability to dilate and yield in a manner approximating real sand behaviour (Mroz and Norris, 1982). The issue is simple. In the various variants of Cam Clay, soils markedly denser than the CSL are treated as overconsolidated and this treatment of density as an effective overconsolidation generally causes unrealistic stiffness with absurd strengths. This is a bit surprising from a train of formal theoretical developments and raises the question: what is wrong with 'the math'?

It turns out the issue is not 'the math' but an unnecessary assumption: that all yield surfaces intersect the critical state. The deficiencies of CSSM are removed by returning to the original premise of Drucker et al. (1957) regarding the association between yield surfaces and void ratio and by further recognizing that soils (not just sands) exist in a spectrum of states. There is an infinity of normal compression loci in the e–p' plane, depending on the initial void ratio at deposition, which is in general arbitrary. Figure 3.5a compares the standard idealized view of a single NCL and its relation to the CSL – the two are parallel, offset by a 'spacing ratio', and there is only one isotropic NCL. Real soils are rather different, with many NCL that are not parallel to the CSL, as sketched in Figure 3.5b. There is actually an infinite number of NCL. Which NCL the soil is on depends on its formation void ratio as the particles coalesce to form soil from a suspension of particles.

The concept of an infinity of NCL was first suggested by Ishihara et al. (1975) based on laboratory experiments, and a detailed experimental verification is presented later in this chapter. An infinity of NCL is not an artefact of the laboratory, with the same behaviour encountered in large-scale hydraulic fills, which is also discussed shortly. Given this pattern of many NCL, it becomes necessary to use two parameters to characterize the state of

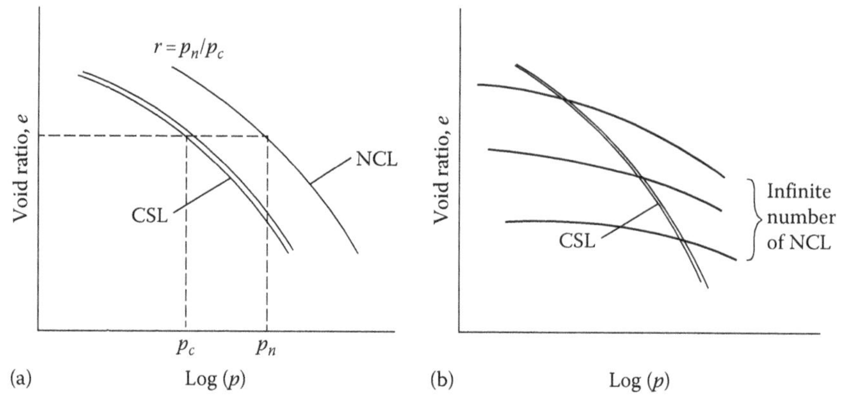

Figure 3.5 Comparision of isotropic compression idealization. (a) Usual Cam Clay idealization and (b) real soil behaviour (and *NorSand*).

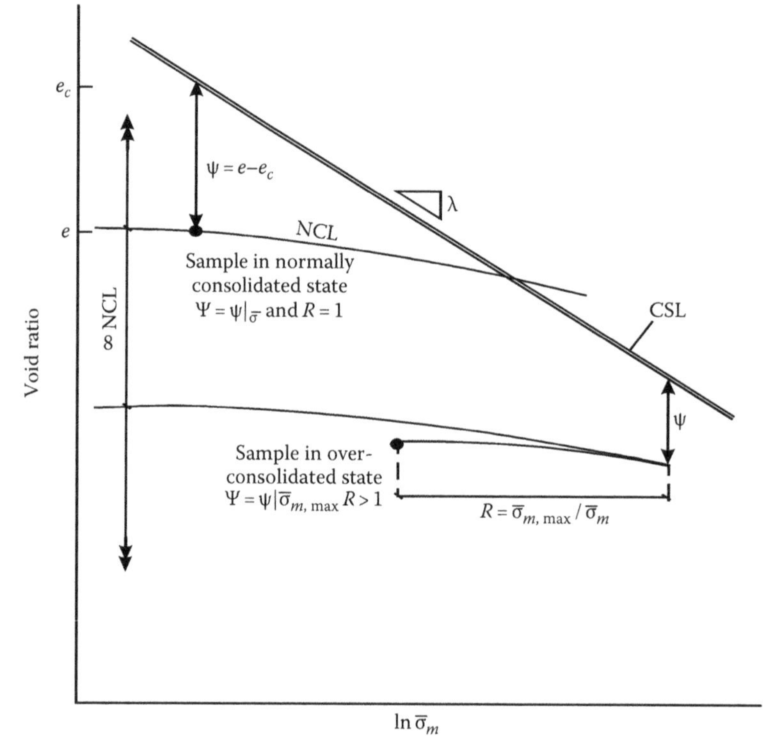

Figure 3.6 Separation of state parameter from overconsolidation ratio. (From Jefferies, M.G. and Shuttle, D.A., *Géotechnique*, 52(9), 625, 2002. With permission from the Institution of Civil Engineers, London, U.K.)

a soil: ψ and R. The state parameter ψ is a measure of the location of an individual NCL in e–p' state space and allows realistic models to be developed around the conceptual framework of Drucker et al. (1957). The overconsolidation ratio R represents the proximity of a state point to its yield surface when measured along the mean effective stress axis. Figure 3.6 illustrates the difference between ψ and R. Their respective definitions are

$$\psi = e - e_c \tag{3.2}$$

$$R = \frac{p'_{\max}}{p'} \tag{3.3}$$

NorSand (Jefferies, 1993) was the first state parameter-based model and generalized critical state theory. Original Cam Clay (OCC) model is a special case of *NorSand* (Appendix H examines and explains this statement in more detail). Subsequently, other authors introduced the state parameter into bounding surface models (Manzari and Dafalias, 1997; Li et al., 1999), a simple hyperbolic plastic stiffness model (Gajo and Muir Wood, 1999), a critical state-like model using Rowe's stress–dilatancy (Wan and Guo, 1998) and a unified clay and sand model (Yu, 1998).

There is now widespread recognition that the state parameter is a fundamental characterizing parameter for particulate materials, and thus far more than just a useful way of normalizing laboratory data. It has resulted in the extensive citations of Been and Jefferies (1985) and, in essence, state parameter is the key to using the powerful unifying ideas of CSSM as a basis of civil engineering (indeed, that is how the state parameter came about.) We also hope this historical overview of ideas has shown that the state parameter is built on theoretical understanding that developed over the past 125 years with cross-fertilization from both sides of the Atlantic. ψ did not appear from 'out of the blue'.

3.3 REPRESENTING THE CRITICAL STATE

The critical state needs to be formalized before being used as a basis of models and then its representation developed. Representing the critical state is usually broken down into two parts: the relationship between the critical void ratio and the mean effective stress and the relationship between the stresses in the critical state. Of course, the soil must meet both sets of criteria when shearing at the critical state.

3.3.1 Existence and definition of the CSL

Critical state models are based on the existence of a unique CSL, formally expressed as an axiom (the First Axiom of CSSM):

$$\exists C(e,q,p')\big|_{\dot{p}'=0} \ni \dot{\varepsilon}_v \equiv 0 \wedge \ddot{\varepsilon}_v \equiv 0 \quad \forall \varepsilon_q \tag{3.4}$$

where $C()$ is the function defining the CSL. There are two conditions: first (3.4), the volumetric strain rate must be zero; second (3.5), the rate of change of this strain rate must also be zero. This can equivalently be stated as the requirement on dilatancy D:

$$\exists C(e,q,p')\big|_{\dot{p}'=0} \ni D \equiv 0 \wedge \dot{D} \equiv 0 \quad \forall \varepsilon_q \tag{3.5}$$

Both dilatancy and rate of change of dilatancy must be zero during shearing at the critical state, and in some ways (3.5) is the best way of thinking about the critical state given that it is stress–dilatancy that controls soil behaviour. Regardless of whether a void ratio or dilatancy view is taken, there are no strain rate terms in $C()$, making the CSL identical to the steady state of Poulos (1981). Constant mean stress is invoked in (3.3) to avoid a less easily understood definition for the situation in which mean stress is increased while the soil is continuously sheared at the critical state.

Uniqueness of the CSL simply refers to C() being a single-valued function of void ratio and effective stress. For any given set of effective stresses, there is only one value of e_c. This critical void ratio e_c is independent of the strain conditions and direction from which the critical state is approached.

Why start with an existence axiom? The short answer is to avoid getting bogged down in arguments over the interpretation of experimental data on the CSL. In developing a mathematical framework, an axiom is invoked as the starting point for the theory and the relevance of the theory is justified when it is seen how well it matches the stress–strain data of soils. Only the relevance of the theory may be disputed, as starting from an axiom means that the framework is always correct provided that the mathematics and physics are consistent. Starting from axioms also allows behaviours to be predicted, as aspects of the theory become necessary for self-consistency with the axioms. This can provide enormous insight into soil behaviour, one such example being plastic yield in unloading (which is pervasive with soils, but not an obvious consequence of plasticity theory).

Critical state frameworks can be developed with strain rate dependency. Having a rate effect does not negate the theory put forward here. Of course, the way in which the CSL varies with shear–strain rate will need to be defined and this will involve additional parameters. To date, there seems to be little data on what a strain rate effect might look like (other than studies for confinement of underground nuclear explosions, but these are not usual engineering material velocities), and the experience in standard soil mechanics tests on sands is that there is no measurable rate effect. Also note that the flow structure at the steady state postulated by Poulos (1981) is inadmissible mechanics – this flow structure is undefined and, apparently, not measurable.

3.3.2 Critical state in void ratio space

Conventionally, the critical state is represented in e–p' space using the semi-log form:

$$e_c = \Gamma - \lambda \ln(p'_c) \tag{3.6}$$

There is abundant data to show that the CSL is more complex than (2.1) when viewed over a wide range of mean stress. Figure 2.26, for example, shows that the critical state line of Erksak sand has a distinct 'knee' or crushing point (for want of a better phrase) at a mean stress of about 1000 kPa, while Figure 2.28 shows that a smooth curve provides a better fit for Guindon tailings. Li and Wang (1998) suggested plotting e against $(p')^\alpha$ and that $\alpha = 0.7$ will generally linearize the relationship for sands as is done for Toyoura sand in Figure 2.29. Verdugo (1992) points out that the location of the curvature of the critical state line can depend on whether one views the data on logarithmic or arithmetic scales and that a bi-linear relationship usually suffices. The semi-log representation is convenient for engineering of soils as it arises naturally when the stiffness is proportional to the mean stress. The knee in the Erksak sand CSL occurs when this is no longer the case and the stiffness becomes approximately constant.

Despite the additional complexity that can be captured in representing the CSL, for most practical engineering, Equation 2.1 is a sufficient approximation for commonly encountered stress levels in engineering of about $10 \text{ kPa} < p' < 500 \text{ kPa}$. More elaborate representations of the critical state, such as those shown in Figure 2.29, are relatively easily adopted and are merely additional details to the basic framework – they do not affect the reasonableness of the approach.

To illustrate that the form of the CSL does not matter, all the downloadable spreadsheets invoke *Function e_crit(p)*; this function returns the critical void ratio for the current stress state. If a curved CSL is desired, or the specification of the stress state is desired more generally (using σ_1, σ_2, σ_3), put the preferred equation in this function and *NorSand* will then look after the maths automatically.

3.3.3 Critical stress ratio *M*(θ)

The critical state friction angle ϕ_c or ϕ_{cv} is not especially helpful for constitutive modelling, as models are cast in terms of stress invariants, and the relationship between these invariants needs to be expressed for the critical state. The convention is to introduce a critical stress ratio, M, so that, at the critical state,

$$q = Mp' \tag{3.7}$$

The parameter M (and it is deliberately not called a property) was initially viewed as a constant. However, it became obvious quite early (Bishop, 1966) that constant M implied a soil with tensile strength, and, patently, this does not fit the idealization of soil as a collection of particles in sliding contact with each other. M has to be treated as a function of the intermediate principal stress, represented by the Lode angle θ. In doing this, triaxial compression conditions are taken as the reference case in which soil properties are determined. Thus, M_{tc} becomes the soil property (where the subscript 'tc' denotes triaxial compression), and $M(\theta)$ is developed in terms of this property. A range of alternative views exist for $M(\theta)$ but these are in the nature of modelling detail. What the constitutive law needs is that any particular combination of void ratio and stress state returns a single value for M.

Chapter 2 considered the experimental data and suggested that a best-fit approach to M was an average of the Mohr–Coulomb criterion and the Matsuoka–Nakai criterion (see Figure 2.47). Although appealing, this average is computationally inefficient because the Matsuoka–Nakai criterion is implicit. Jefferies and Shuttle (2011) suggested it would be better to use an operationally similar but far more efficient expression for finite element simulations, this function being (for θ measured in radians)

$$M(\theta) = M_{tc} - \frac{M_{tc}^2}{3 + M_{tc}} \cos\left(\frac{3\theta}{2} + \frac{\pi}{4}\right) \tag{3.8}$$

This idealization is programmed as *Function Mpsi_v3()* in the downloadable spreadsheets. The previous idealization of an average between Mohr–Coulomb and Matsuoka–Nakai continues to exist in the VBA code as *Function Mpsi_v1()*. Choose which you prefer, although that choice only matters when looking at simple shear or plane strain conditions.

The critical state has been fully defined. Attention now focuses on stress–strain theories developed using the critical state as a kernel idea for the soil model.

3.4 CAMBRIDGE VIEW

Before continuing, apologies are due to the University of Florida and Imperial College, since it will be appreciated from the historical development sketched out earlier that the paper by Drucker, Gibson and Henkel in 1957 was key to clarification of the critical state model. Drucker was based at the University of Florida, while both Gibson and Henkel were at Imperial College. It is, however, appropriate to call the final model a Cambridge view as it was pulled together in Cambridge (England) while also using crucial insights from Cambridge (Massachusetts). Perhaps not coincidentally, Drucker had been on sabbatical at Cambridge University immediately prior to the theory coalescing.

In presenting the Cambridge view, the use of triaxial conditions under which the theory was developed is followed, and the familiar q, p' notation used to reinforce the point that it was developed for these conditions. The effective stress subscript is not necessary, however, as this work deals only with effective stresses and the 'dash' notation is dropped for the moment. Unloading is also skipped and only plastic straining is considered. The aim is to give the essence of the Cambridge view before going on to the general framework incorporating the state parameter (which is only a modest extension of the original and widely accepted ideas).

3.4.1 Idealized dissipation of plastic work

The starting point is to return to the Taylor/Bishop energy correction and think about the work done on an element of soil by the stresses acting on it as the soil undergoes a strain increment. The rate of working on the soil skeleton by the external loads per unit volume is

$$\dot{W} = q\dot{\varepsilon}_q + p\dot{\varepsilon}_v \tag{3.9}$$

where

q, p are the usual triaxial stress invariants
$\dot{\varepsilon}_q, \dot{\varepsilon}_v$ are the corresponding work conjugate strain increments

Notice that the work on the soil element is being broken down into that part caused by shear and that part caused by mean stress, the latter being associated with volumetric strain, and hence the equivalent of Taylor/Bishop energy 'correction'. It is the plastic work absorbed by the soil skeleton that is of interest, as the elastic strains are recoverable. So invoking the strain decomposition (3.1),

$$\dot{W}^p = \dot{W} - \dot{W}^e = q\dot{\varepsilon}_q^p + p\dot{\varepsilon}_v^p \tag{3.10}$$

where the superscript 'p' denotes plastic. Dividing this plastic work rate first by the mean effective stress (to make it dimensionless) and then by the plastic shear strain increment (so that it becomes a normalized rate of working per unit plastic distortion of the soil element),

$$\frac{\dot{W}^p}{p\dot{\varepsilon}_q^p} = \frac{\dot{\varepsilon}_v^p}{\dot{\varepsilon}_q^p} + \frac{q}{p} = D^p + \eta \tag{3.11}$$

Nothing is assumed in (3.11) about how the work is stored (elastically) or dissipated (plastically) by the soil skeleton. No constitutive model is involved. Yet, it is seen that the plastic work done to the soil skeleton, in dimensionless terms, is just the sum of the stress ratio η and the plastic dilation rate D^p. Soil mechanics is as simple and as fundamental as that.

The assumption introduced to develop the Cambridge models, Cam Clay and Granta Gravel, was that the dimensionless plastic energy dissipation rate was constant throughout plastic shearing:

$$\frac{\dot{W}^p}{p\dot{\varepsilon}_q^p} = M \tag{3.12}$$

The idealization of constant M does not fit detailed soil data, but consider constant M for the moment as interest here centres on the Cambridge view and that was their idealization. Equating (3.11) and (3.12) gives a stress–dilatancy relationship (also called a flow rule):

$$D^p = M - \eta \tag{3.13}$$

Notice that in the critical state, for which dilation $D^p = 0$, $\eta = M$ follows from (3.13). Dilatancy acts as a work transfer mechanism between the principal stresses acting on the soil element, and it is the critical friction ratio M that dissipates the plastic work.

3.4.2 Original Cam Clay and Granta Gravel

Although (3.13) gives the relationship between plastic volumetric and shear strains during yielding, it does not explain when plastic yielding occurs. A yield surface is needed to do this, the yield surface being the locus of stress states leading to plastic strains. Within the yield surface, everything is elastic (or rigid). The derivation of the yield surface depends on just two assumptions: normality in the q–p plane and the stress–dilatancy relationship of Equation 3.13.

From the definition of the stress ratio $q = \eta p$, the differential is taken to express the change in shear stress as

$$\dot{q} = p\dot{\eta} + \eta\dot{p} \tag{3.14}$$

Assuming that soil can be treated as a work hardening plastic material, and following Drucker (1951), from normality,

$$\frac{\dot{q}}{\dot{p}} = \frac{-1}{\left(\dot{\varepsilon}_q^p / \dot{\varepsilon}_v^p\right)} = -D^p \Rightarrow \dot{q} = -D^p \dot{p} \tag{3.15}$$

Substituting (3.14) in (3.15) gives

$$\frac{\dot{p}}{p} + \frac{\dot{\eta}}{D^p + \eta} = 0 \tag{3.16}$$

This equation is an identity of the normality condition and is true regardless of the soil's internal dissipation mechanisms, so long as work hardening or perfectly plastic conditions prevail. Substitute (3.13) in (3.16) and integrate (it is a separated equation) with the integration coefficient chosen as the value of $\ln(p)$ when $\eta = M$, which in the Cambridge models is of course the critical state. So, (3.16) becomes

$$\frac{\eta}{M} = 1 - \ln\left(\frac{p}{p_c}\right) \tag{3.17}$$

This is the equation of the Granta Gravel and Original Cam Clay yield surface. Its linkage to the void ratio of the soil becomes obvious when we introduce the CSL. Cambridge models use the standard idealized CSL:

$$e_c = \Gamma - \lambda \ln(p'_c) \tag{3.6}$$

which can be substituted in (3.17) to make the relationship between yield stresses and void ratio explicit:

$$\frac{\eta}{M} = 1 - \ln(p) + \frac{\Gamma - e_c}{\lambda} \tag{3.18}$$

where e_c is the current critical void ratio and is best viewed as a hardening parameter. The manner in which e_c evolves controls how the size of the yield surface evolves, and hence the evolution of stresses during shearing because those stresses remain on the yield surface. The question then becomes: what is the relation of e_c to the soils' current void ratio?

In the Granta Gravel model, it is assumed that there are no elastic strains of any kind. The model is rigid plastic. Under this idealization, $e_c = e$. Equation 3.18 can now be used directly:

$$\text{Granta Gravel:} \quad \frac{\eta}{M} = 1 - \ln(p) + \frac{\Gamma - e}{\lambda} \tag{3.19}$$

In Cam Clay, volumetric elasticity is added. The aim is to avoid locking up under undrained conditions, since constant void ratio implies constant e_c for Granta Gravel, and hence no hardening of the yield surface and no strains. Cam Clay adopts a clay-like idealization of volumetric elasticity with a semi-log relationship between void ratio and confining stress during elastic mean stress changes, using a new soil property conventionally called κ:

$$e = A - \kappa \ln(p) \textit{ during unloading or reloading} \tag{3.20}$$

with A a constant whose value depends on the soil's void ratio at the start of unloading. Although (3.20) is conventional, clarification follows from expressing this equation in differential form as an elastic bulk modulus, K:

$$\frac{K}{p} = \frac{1+e}{\kappa} \tag{3.21}$$

It can be seen from (3.21) that Cam Clay is invoking almost (almost because there is a small effect of void ratio) a constant bulk rigidity (rigidity equals modulus/stress) to go with the assumption of infinite shear rigidity.

Because there is volumetric elasticity, in Cam Clay e_c is no longer equivalent to e. Rather there is an elastic void ratio change as the stress changes on, or within, the yield surface. The κ model for elasticity (3.20) leads to the simple expression for elastic void ratio change from current to critical conditions Δe^e:

$$\Delta e^e = -\kappa \ln\left(\frac{p}{p_c}\right) \tag{3.22}$$

The minus sign arises in (3.22) because an increase in mean stress causes a decrease in void ratio. The relevant critical void ratio for Cam Clay is then related to the current void ratio by

$$e_c = e + \Delta e^e \tag{3.23}$$

On combining (3.23) with (3.22) and (3.17), the equation of the Original Cam Clay yield surface is obtained:

$$\text{Original Cam Clay:} \quad \frac{\eta}{M} = 1 - \ln(p) + \frac{\Gamma - e - \kappa \ln(p)}{\lambda - \kappa} \tag{3.24}$$

Notice how (3.24) neatly resolves model lock-up of Granta Gravel under undrained conditions. Putting constant e in the right-hand side of (3.24) does not stop plastic hardening because the current mean effective stress is also present in the hardening term, and this mean stress varies with plastic strain.

Nothing so far actually provides a stress–strain curve. A stress–dilatancy relationship (3.13) and a yield surface (3.24) have been presented, but these do not provide what is needed. The missing step is the consistency condition.

3.4.3 Numerical integration and the consistency condition

Work hardening (and softening) plastic models change the size of their yield surface with plastic strain. The consistency condition is simply that the stress state must remain on the yield surface during plastic strain, so that the stress state evolves on a one-to-one basis with the evolution of yield surface size. The consistency condition is illustrated in Figure 3.7, which shows an initial yield surface that has hardened after an increment of plastic strain. The question then is how the stress state has evolved. Conventionally, this is done incrementally rather than by solving simultaneous equations. The equation of the yield surface, in our case (3.17), is differentiated (see Appendix C for details) to give

$$\dot{\eta} = M\left(\frac{\dot{p}_c}{p_c} - \frac{\dot{p}}{p}\right) \tag{3.25}$$

There are two terms in the right-hand side of (3.25). The first term is the change in yield surface hardening parameter (the current critical state mean stress for the Cambridge models). The second term is the change in the current stress state.

In the case of Cambridge models, only volumetric strain matters, and this controls the yield surface size by changing the critical void ratio. This is best expressed in a conventional work hardening framework. So, differentiating the idealized CSL, (3.6), and writing in terms of plastic rather than total strain gives (see Appendix H)

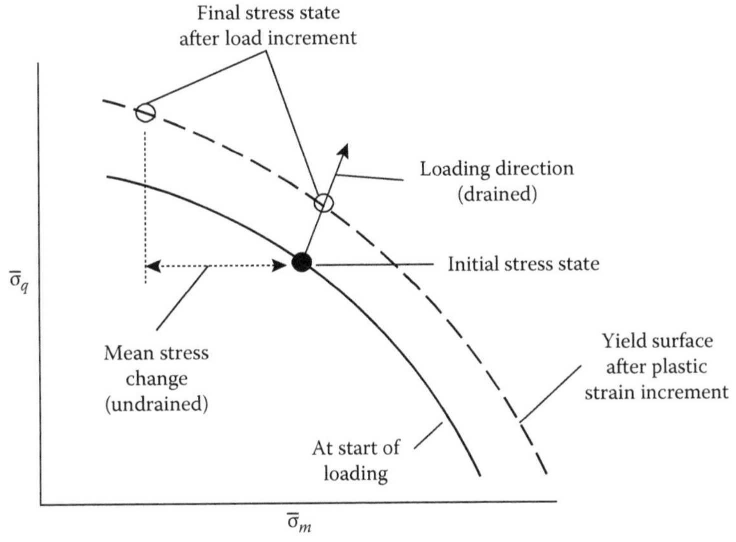

Figure 3.7 Illustration of the consistency condition.

$$\frac{\dot{p}_c}{p_c} = \frac{(1+e)}{\lambda - \kappa}\dot{\varepsilon}_v^p \tag{3.26}$$

Notice the clarity (3.26) adds to the understanding of Cam Clay: (1) the evolution of the yield surface is controlled by plastic volumetric strain, not void ratio per se; (2) no plastic strain increment means no change in the yield surface and (3) the parameter group $(1+e)/(\lambda - \kappa)$ is a conventional and dimensionless plastic hardening modulus.

Arriving at the final stress–strain model now depends on the second term of the right-hand side of (3.25). Undrained loading is particularly simple as the mean stress changes directly with the volumetric strain using the relation

$$\dot{\varepsilon}_v = 0 \Rightarrow \dot{\varepsilon}_v^p + \dot{\varepsilon}_v^e = 0 \Rightarrow \dot{p} = \dot{\varepsilon}_v^e K \Rightarrow \dot{p} = -\dot{\varepsilon}_v^p K \tag{3.27}$$

where K is the elastic bulk modulus. In other words, the elastic and plastic volumetric strain components must be equal and opposite to satisfy the constant volume loading condition. Imposing a plastic strain increment gives the plastic volumetric strain increment through stress–dilatancy, and this immediately gives the change in mean effective stress by elastic rebound to maintain the undrained (constant void ratio) condition. For drained loading it is a little more complicated, as finding the new stress state from an increment of plastic strain involves working out a relationship between the change in mean stress and the change in η. This relationship depends on the stress path. In the case of laboratory experiments, the stress path depends on the equipment and the test conditions chosen (e.g. triaxial or simple shear). For standard triaxial compression, for example, the mean stress (p) increment is one-third of the deviator stress (q) increment. Appendix C derives relationships for the common laboratory stress paths.

It will have been noticed by now that everything is worked out incrementally and that the stress–strain curve is developed by integration. There is no closed-form solution in general and numerical methods are used. This comprises a simple four-step loop of Euler integration:

- Impose a plastic strain increment
- Calculate the plastic yield surface hardening
- Derive the new stress state using the consistency condition
- Add in the elastic strain increments

Further details on numerical integration are in Appendix C, and numerical integration is used throughout the routines provided in the *NorSand**.xls* files on the website. For interest, Cam Clay is also implemented numerically in *OrigCamClay.xls* and a verification given using the closed-form solution for constant p from Schofield and Wroth (1968), which is one of the very few closed-form solutions for critical state models. *OrigCamClay.xls* is a good place to start in getting comfortable with numerical integration.

3.5 STATE PARAMETER VIEW

3.5.1 Trouble with Cam Clay

So far the Cambridge view has been presented, in particular the Original Cam Clay model that follows from it. Having made the case for introducing void ratio into the representation of soil behaviour, and having shown how Cam Clay simply and elegantly implements this, one might wonder why critical state models are not in widespread use in civil engineering practice.

Early objections to the Cambridge models (e.g. Bishop, 1966) centred on the idealization of M as a soil constant, and this was touched on earlier. These objections are about model detail, however, and are not fundamental. It is entirely possible to develop M as a function of Lode angle, $M(\theta)$, such that it nicely fits soil behaviour. Further, because $M(\theta)$ is not a function of η or p', dependence of M on θ has no effect on the validity of the integration in going from (3.13) to (3.18).

More seriously, Cam Clay does not dilate realistically for denser than critical soils, a defect that has been known since very shortly after Cam Clay became popular (e.g. Mroz and Norris, 1982). Although this appears to be an objection about the accuracy of the model, it is actually far more fundamental, and something one can easily appreciate. Consider a void ratio denser than the CSL and lying on its κ line, as illustrated in Figure 3.8. Rewriting the relation for the elastic void ratio change (3.22) in terms of the state parameter gives the ratio of critical to current mean stress:

$$\frac{p_c}{p} = \exp\left(\frac{-\psi}{\lambda - \kappa}\right) \tag{3.28}$$

The relationship between the critical state stress and the stress on the isotropic NCL (i.e. p_{nc}) is only a function of the shape of the yield surface. Putting $\eta = 0$ in (3.17) gives

$$\frac{p_{nc}}{p_c} = 2.718 \tag{3.29}$$

Combining (3.28) and (3.29) allows us to write an equation for the overconsolidation ratio (R) implied by Cam Clay for a given state parameter:

$$R = 2.718 \exp\left(\frac{-\psi}{\lambda - \kappa}\right) \tag{3.30}$$

If one now looks at a typical state parameter plot for sands, Figure 2.7, for example it is apparent that it is very easy to prepare samples to states of at least as dense as $\psi < -0.2$.

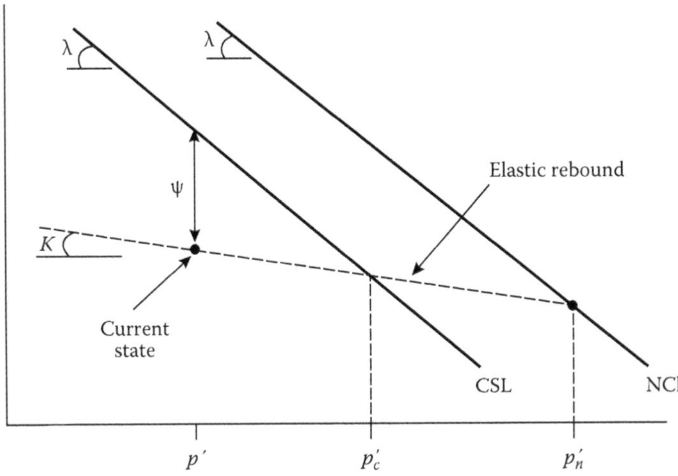

Figure 3.8 Implied overconsolidation for a given state ψ in Cambridge models.

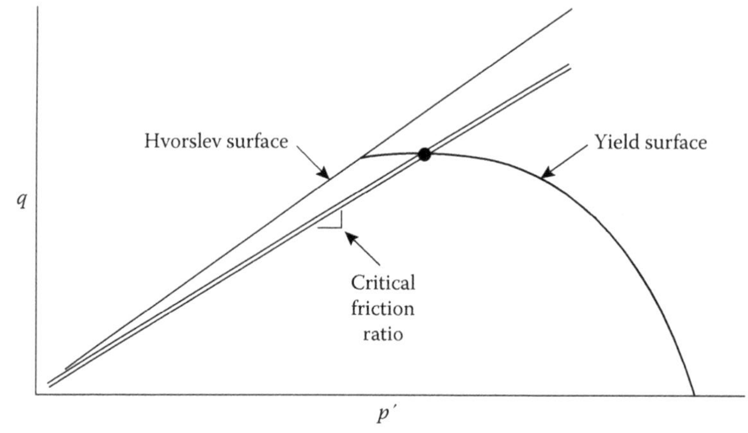

Figure 3.9 Illustration of the Hvorslev surface idealization.

Typical λ values for such soils can be as little as 0.02 (recall $\lambda \approx \lambda_{10}/2.3$), see Table 2.1. The κ model is not a particularly good idealization for sand, but representing elastic volumetric stiffness with the κ model gives $0.1\lambda < k < 0.4\lambda$ (approximately). Inserting these numerical values into Equation 3.30, one finds a very large overconsolidation ratio, $R > 10^5$, for an entirely reasonable $\psi = -0.2$. This implies huge strengths and essentially only elastic behaviour for the stress ratios that can be imposed in triaxial compression. Of course, the reality is that the sample yields and dilates from the start of loading.

This problem of predicting near elastic behaviour for real soil samples that yield plastically has been long recognized, and so a second yield surface was introduced (often called the Hvorslev surface). Figure 3.9 illustrates the approach. This second yield surface abandons the proper association of plasticity with soil behaviour, as the strain increments are most certainly not normal to the Hvorslev surface. Further, the Hvorslev surface is not related to void ratio through the CSL. The model is no longer based on simple postulates. In short, the insight on the effect of density that was the original motivation to develop the Cambridge approach has been lost.

An artefact like the Hvorslev surface is unnecessary. The difficulty with Cam Clay stems from the assumption that yield surfaces go through the CSL. Recall that the original proposition of Drucker et al. was to associate hardening with the NCL (Figure 3.4), not the CSL. So, the next step is to look at NCLs of soils and how they relate to void ratio.

3.5.2 Infinity of NCL

Earlier, the point was made that soils exist in a spectrum of states with an infinite number of NCL, each of which depends on the formation density and subsequent strain history of the soil. Figure 3.5 illustrates the idea. The importance of the infinity of NCL follows immediately by going back to the concept of Drucker et al. (1957), which is illustrated in Figure 3.4. Each NCL can be viewed as a hardening law for an associated yield surface. An infinity of NCL means a multiple infinity of yield surfaces because any NCL can be viewed as the trace of a set of yield surfaces as they harden. Correspondingly, soil in any part of the $e - p'$ domain can plastically strain. There is no 'elastic wall' confining plastic behaviour. This is how real soils behave. It is now useful to look at the history of how the idea of an infinity of NCL developed, and some experimental evidence for this, before further mathematics.

Poorooshasb et al. (1966, 1967) had used triaxial tests to define yield surfaces in dense sand, which motivated Tatsuoka and Ishihara (1974) to use similar testing to define yield surfaces in loose sand and, more generally, to investigate the effect of density on the yield surface. Tatsuoka and Ishihara noted in their conclusion 'It was recognized that yield loci change to some extent depending on the density of the sample'. This was an understatement, to put it mildly.

At a completely different scale, the perception of a multiple infinity of yield surfaces for sand was independently found in the result of CPT soundings in an underwater sand berm for the Molikpaq in the Canadian Arctic (Chapter 1). During 1982, Gulf Canada Resources constructed an underwater berm at the Kogyuk N-67 wellsite. This berm involved the underwater and slurried placement of some 600,000 m^3 of clean medium fine sand to form a structure rising 9 m above the original seabed at a depth of 28 m. Figure 3.10 shows the dredge depositing 5000 m^3 of sand in about ten minutes by discharging the sand through valves in the base of the ship. The sand pluviates through the water column to accrete and form the berm – a perfect example of the geological concept of 'normal consolidation'. Extensive cone penetration test (CPT) soundings after construction gave the relative density distributions shown in Figure 3.11 (Stewart et al., 1983). A wide distribution of void ratios existed at any stress level and the distribution evolved smoothly from one stress level to the next with no tendency for void ratio to coalesce into a single relationship: there was 'an infinity' of normal compression loci (NCL).

So far, this idea of an infinity of NCL amounted to passing observations of a research triaxial program and inference from CPT data in large-scale hydraulic fills. These observations were then further investigated by comparing direct isotropic compression of samples prepared at different densities with an independently determined CSL (Jefferies and Been, 2000). Four samples of Erksak 330/0.7 sand were tested using multiple unload–reload stages carried out during the isotropic compression of the sample to separate the elastic and plastic components of behaviour. These four samples ranged in void ratio from 0.84 to 0.60,

(a) (b)

Figure 3.10 Hydraulic deposition of sand into berm (see Figure 1.20). About 5000 m^3 sand was deposited by bottom-valve discharge through water in 10 min between photo on left and photo on right. (a) With hoppers full and (b) after discharging load.

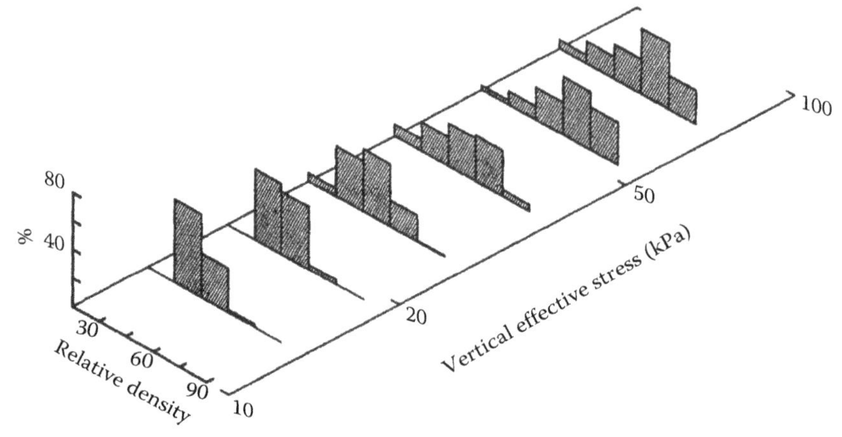

Figure 3.11 Distribution of fill density in normally consolidated hydraulic sand fill. (From Stewart, H.R. et al., Berm construction for the Gulf Canada Mobile Arctic Caisson, *Proceedings of the 15th Offshore Technology Conference*, Paper OTC 4552, 1983. With permission from OTC.)

a range that nearly spans the accessible range of void ratios for Erksak sand. In addition to the four samples tested in detail, all Erksak sand samples tested in triaxial compression for any purpose had their consolidation behaviour measured. All the isotropic compression data are presented in Figure 3.12a and compared to the reference CSL for Erksak 330/0.7 sand (Figure 2.26). The behaviour measured in the four load–unload samples is shown in Figure 3.12b through e at an expanded scale to distinguish loading, unloading and reloading (note the expanded plots are at different scales to maximize the clarity of each).

The compression behaviour in the e–log(p) space of Figure 3.12a shows the isotropic loading line for each of the sand samples. Each line is regarded as a true NCL because of two factors. First, the samples were prepared under low stresses (some by gentle moist tamping, some by pluviation) and never overconsolidated. Second, the unload–reload loops define a classic elastic–plastic form, as illustrated in the expanded views, even when the samples are denser than the CSL. These data support the model of multiple or non-unique NCL for sands.

A possible alternative view to an infinity of NCL is that there is actually a unique NCL, but sands only approach this at high stress. This alternative view was apparently first suggested by Atkinson and Bransby (1978) and subsequently elaborated as the *Limiting Compression Locus* by Pestana and Whittle (1995). To examine this unique NCL as an alternative explanation, a dotted line is drawn parallel to the CSL in Figure 3.12a and denoted as the PNCL, where P stands for pseudo. The PNCL is drawn parallel to the CSL from Cam Clay theory and separated from it by the Original Cam Clay spacing ratio ($p_{nc}/p_c = 2.718$). If the proposition that true normal compression is only apparent at high stresses is correct, then we should see no samples above this PNCL. In reality, it is quite straightforward to prepare samples looser than any credible PNCL, and they are not unstable when exposed to increasing mean stress. As can be seen by inspection of Figure 3.12a, there is a smooth spectrum of behaviour from loose to dense states.

The existence of an infinity of NCL is the best explanation for the test data of Figure 3.12. Since there is nothing special about Erksak sand and there are data showing the same behaviour of Ticino sand, for example, this means that any soil can exist in a whole spectrum of normally consolidated states. A requirement to measure the state of a soil immediately follows. This is analogous to using temperature as a measure of a gas state (recall $pV = nRT$). The kernel concept then becomes: the critical state defines a reference state and the distance

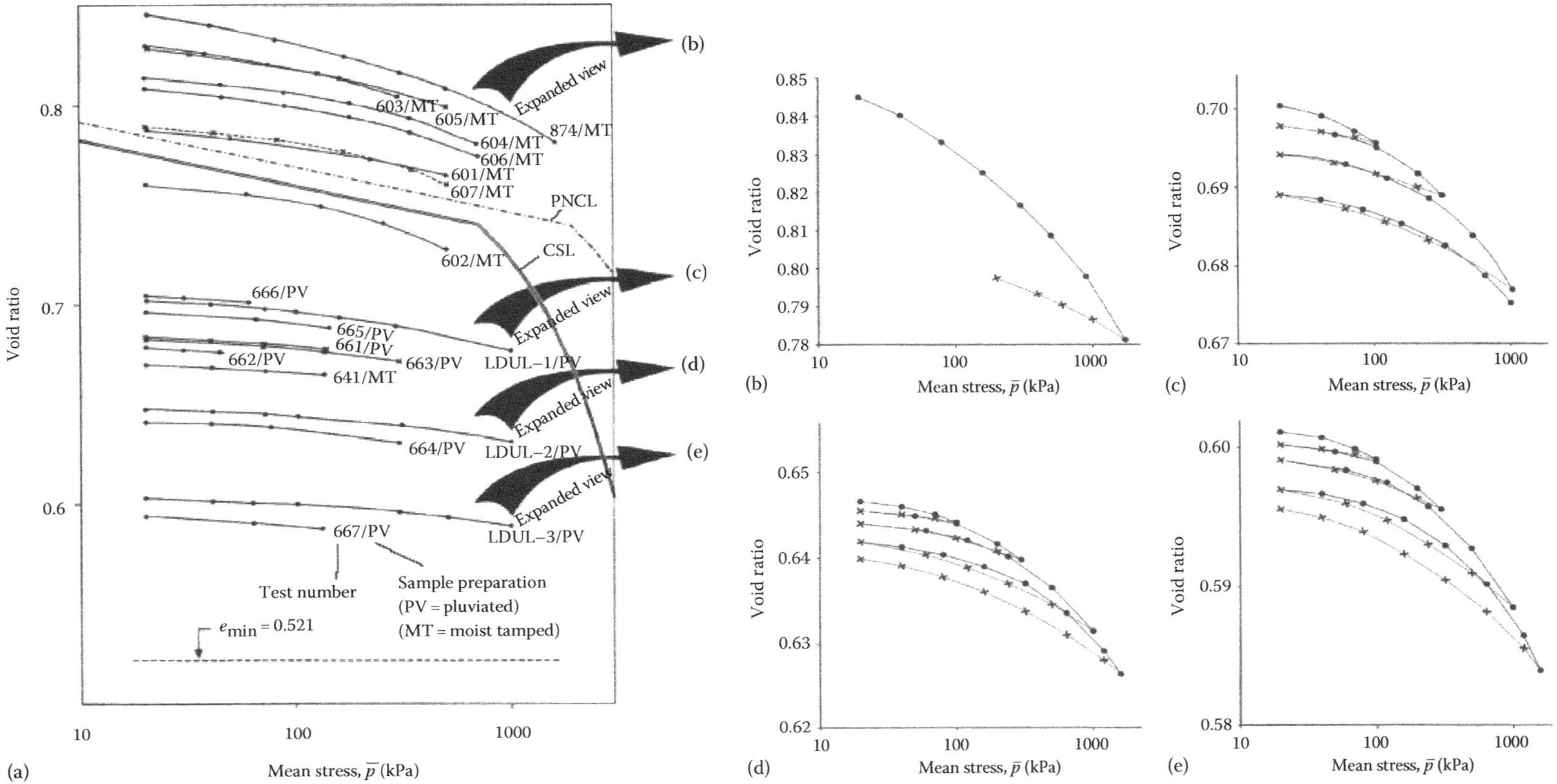

Figure 3.12 Experimental evidence for an infinity of NCL: isotropic consolidation of Erksak 330/0.7 sand. (a) All tests at same scale, (b) test 874/MT at expanded scale, (c) test LDUL-1/PV at expanded scale, (d) test LDUL-2/PV at expanded scale and (e) test LDUL-4/PV at expanded scale. (After Jefferies, M.G. and Been, K., Géotechnique, 50(4), 419, 2000. With permission from the Institution of Civil Engineers, London, U.K.)

of the sand from the reference state in void ratio – stress space is a first-order measure of that sand's structure. This is the underpinning of the state parameter. It amounts to a step back from a basic proposition of CSSM (see Schofield and Wroth, 1968, p.20, third paragraph) in the sense that, whereas CSSM requires that all yield surfaces intersect the CSL, which enforces a single plastic modulus, the state parameter is simply a rate variable much like in radioactive decay or other classic processes in physics. The further the current state is away from the final or equilibrium state, the faster the changes in state occur.

Importantly, notice from Figure 3.12 how overconsolidation exists alongside the state parameter. The four samples taken through unloading–reloading paths had both a finite overconsolidation ratio and a state parameter during the unloading–reloading stages. More generally, each of the NCL is associated with the geological concept of normal consolidation and hence has an overconsolidation ratio (OCR) of $R=1$.

3.5.3 State as an initial index versus state as an internal variable

The state parameter as used earlier relates various soil behaviours (e.g. peak friction angle) to the state parameter at the start of shearing. This uses the state parameter ψ as an initial index of soil behaviour. Although this is a useful way to unify test data and carry out basic engineering analyses, the state parameter is best used as an internal state variable in a complete constitutive model. Using the state parameter as an internal variable does not require a different definition – the basic equation defining the state parameter, $\psi = e - e_c$, remains the same. But now the initial conditions become explicitly recognized by the subscript 'o', as in ψ_o, and the state parameter becomes the basis for properly generalizing critical state ideas.

There are several constitutive models that incorporate the state parameter as the method of making models independent of void ratio and confining stress. *NorSand* (Jefferies, 1993) will be pursued here because it explicitly fits within CSSM. And it turns out that Cam Clay is just a particular parameter set in *NorSand* (Appendix H).

3.6 *NorSand*

3.6.1 Triaxial compression version

The approach followed in presenting *NorSand* is to go through the triaxial compression case for drained loading and compare this to the more familiar Cam Clay. In presenting the framework of ideas, this chapter also skips over the detailed derivations. This way, the simplicity of *NorSand* can be seen as well as how *NorSand* fits within the historical framework of ideas. The meaning of the various model parameters will also be examined. Once the basic framework is grasped, the detailed derivations will be most readily understood and these are given at length in Appendix C. If one needs to use *NorSand* in finite element formulations, then the 'industrial strength' version will be needed which incorporates unloading, full 3D stress conditions and principal stress rotation. These aspects of *NorSand* are also given in Appendix C, and feature later in this chapter as well as in Chapters 6 and 7.

The starting point for *NorSand*, like Cam Clay, is stress–dilatancy. When *NorSand* was first published, it incorporated Nova's rule:

$$D^p = \frac{M_{tc} - \eta}{1 - N} \tag{3.31}$$

Some five years later, Dafalias and co-workers returned attention to Rowe's observation that the apparent material property in the stress–dilatancy relationship, here M_{tc}, was not itself a property but evolved with strain. This topic was explored in detail in Chapter 2. *NorSand* was modified in the light of these new ideas (it actually gives a simpler and more accurate model than the first published versions) and now adopts a Cam Clay like stress–dilatancy relationship:

$$D^p = M_i - \eta \tag{3.32}$$

Equation 3.32 differs from Original Cam Clay (Equation 3.11) in using image stress ratio M_i as an evolving parameter that tends to the critical state M with shear strain (Section 2.8.2). Perhaps the reason that Rowe's initial observations lapsed into obscurity was that Rowe did not suggest a relationship for how the coefficient in stress–dilatancy evolved. The advance produced by Dafalias and co-workers was to make Rowe's observation a function of the state parameter. *NorSand* follows closely on the ideas suggested by Dafalias and adopts

$$M_i = M\left(1 - \frac{\chi_i N |\psi_i|}{M_{itc}}\right) \tag{3.33}$$

where N is the volumetric coupling coefficient from Nova's flowrule at peak strength (see Figure 2.30) and now controls the evolution of M_i.

Determination of the yield surface follows exactly as presented for Cam Clay earlier. Putting the revised stress–dilatancy rule in the normality condition (3.16) leads to the following yield surface equation:

$$\eta = M_i\left[1 - \ln\left(\frac{p}{p_i}\right)\right] \tag{3.34}$$

So where is the difference between the *NorSand* yield surface (Equation 3.34) and that of Cam Clay (Equation 3.17)? *NorSand* does not use the mean stress at the critical state p_c or the critical stress ratio M. Instead, it uses another condition, that of the *image state*, denoted by the subscript i. Recall that there are two conditions for the critical state, $D = 0$ and $\dot{D} = 0$. The condition where $D = 0$ alone exists is referred to as an *image* of the critical state, since only one of the two conditions for criticality is met.

The mean stress at the image state, p_i, describes the size of the yield surface exactly, as p_c was used in Cam Clay to link the yield surface to the CSL. Figure 3.13 illustrates the *NorSand* yield surface for very loose sand and very dense sand. The main difference between Figure 3.13a and b is the location of the critical state relative to the image condition; for loose sand the image condition is at a higher effective stress and the yield surface will shrink to the critical state, while for dense sand the yield surface must harden (expand) to reach the critical state.

Further, $\eta \equiv M_i \Leftrightarrow p \equiv p_i$ but the question then arises as to how the image stress is related to the soil's void ratio. There is no direct relationship. Rather, *NorSand* uses a differential form based on the state parameter and invokes the *Second Axiom* of critical state theory. The Second Axiom of critical state theory is simply that the soil state moves to the critical state with increasing shear strain. This is most naturally stated as

$$\psi \rightarrow 0 \text{ as } \varepsilon_q \rightarrow \infty \tag{3.35}$$

The Second Axiom never arose in Cam Clay explicitly because it is enforced tacitly through the assumption that the yield surface intersected the CSL. This is why Cam Clay is viewed as a degenerate case of *NorSand*. Cam Clay has assumed the end condition as its starting point,

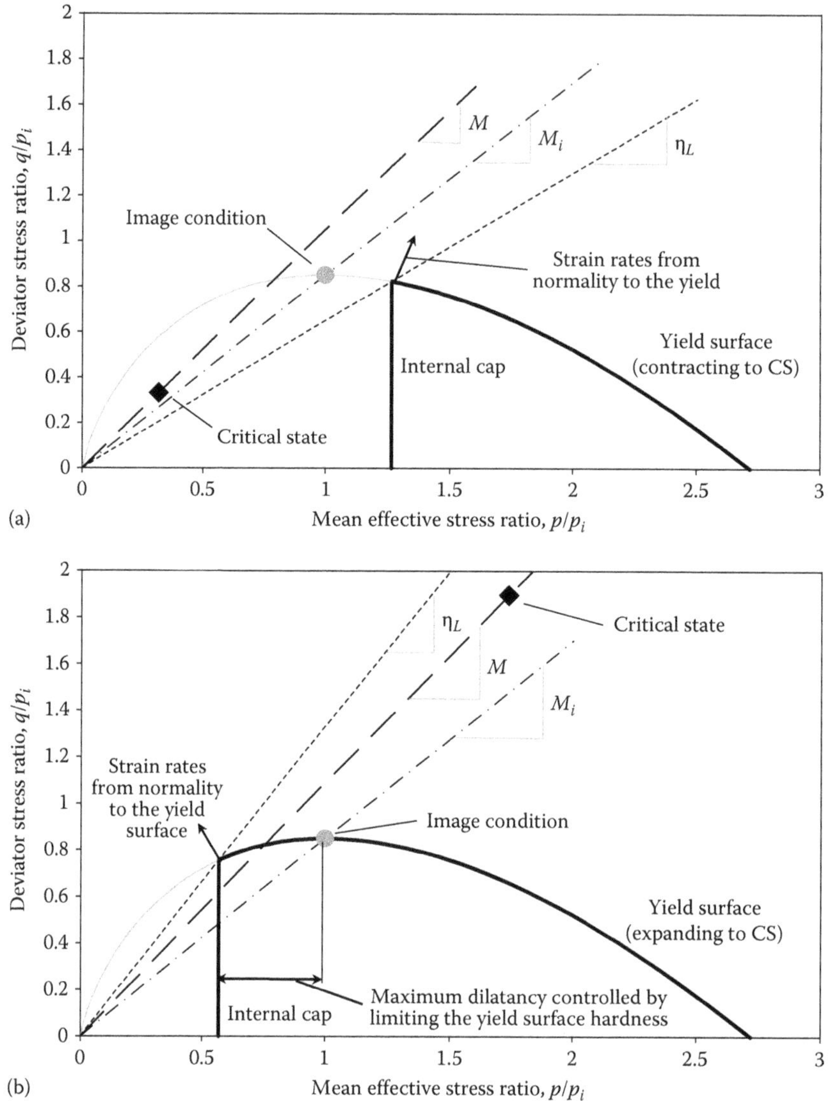

Figure 3.13 Illustration of *NorSand* yield surface, limiting stress ratios and image condition. (a) Very loose soil and (b) very dense soil.

but that cannot be done with an infinity of NCL, and so the Second Axiom becomes the basic hardening law. Also notice that the Second Axiom is stated in terms of shear strain because the critical state is a condition of shear. There are conditions of loading at low stress ratios where soil states will move away from the critical state but these involve small and finite shear strains.

A key feature of dense soils, whether sands or clays, is that dilatancy is limited to a maximum value for that specific soil state. Experimentally, there is the very strong relationship between D^p_{min} and ψ, shown in Figure 2.33. A little care is needed in transforming this very strong, and expected, relationship to a plasticity model. The difficulty is that in 'neutral' loading where a yield surface has constant size, ψ varies around a yield surface as p changes. But there is only one D_{min} for that yield surface. Accordingly, we anchor yet another aspect to the image condition and define

$$D^p_{\min} = \chi_i \psi_i \tag{3.36}$$

which has no theoretical difficulties since ψ_i is unique for any particular yield surface. The practical problem then arises that testing engineers and laboratory staff work with the soil property χ_{tc} as that property is both simple to calculate from a set of triaxial tests and independent of any constitutive model. We need to relate the *NorSand* parameter χ_i to the soil property χ_{tc}. Appendix C derives the relationship and for practical purposes leads to

$$\chi_i = \frac{\chi_{tc}}{(1 - \chi_{tc}\lambda / M_{tc})} \tag{3.37}$$

Or, for typical soil properties, a small shift in value with $1.05\chi_{tc} < \chi_i < 1.10\chi_{tc}$. Equation 3.37 is built into the downloadable spreadsheets, so that the soil property χ_{tc} is the input in the worksheet with VBA then transferring the appropriate model parameter to the *NorSand* routines.

Conventionally, dilation is limited by invoking a non-associated flow rule with an appropriate choice of dilation angle. That approach is not acceptable within a strict critical state framework because normality is used to derive the yield surface. Instead, realistic maximum dilatancy is controlled through the hardening parameter, p_i. The basic concept is illustrated in Figure 3.13, which shows a section through the yield surface in the q–p plane. The evolution of p_i can be specified such that the value of p_i is limited with respect to the current stresses, and this then controls the dilatancy through normality. The limit on hardening is given by

$$\left(\frac{p_i}{p}\right)_{\max} = \exp\left(-\frac{\chi_i \psi_i}{M_{itc}}\right) \tag{3.38}$$

Equation 3.38 gives an internal limit or planar cap to the yield surface, as is illustrated in Figure 3.13. Triaxial compression is used as the reference condition for the soil properties, with these properties then transformed into the two internal model parameters χ_i and M_{itc} (the current value of M_i for a triaxial compression stress state). Equation 3.38 is derived for triaxial compression but is actually general in 3D because *NorSand* is isotropic and the limiting condition is expressed as a ratio of mean stresses.

The natural form for a hardening law that complies with the Second Axiom, while respecting the constraint on dilatancy, is a simple difference equation:

$$\dot{p}_i = H(p_{i,\max} - p_i)\dot{\varepsilon}_q \tag{3.39}$$

Because of the form of the hardening limit (3.38), the hardening law is better expressed in dimensionless form by dividing using the current mean stress, and noting that $p_{i,\max}$ is not the true maximum of p_i, but rather its maximum for the current state. Two additional features also prove convenient. First, because the consistency condition is expressed in terms of the ratio $\dot{p}_i / p_i$, it is helpful if hardening is in the same terms as needed in the consistency condition. Second, it turns out that a better fit is obtained to experimental data if the hardening is given a dependence on the shear stress level. After some algebra, the hardening law becomes (see Appendix C)

$$\frac{\dot{p}_i}{p_i} = H\left(\frac{p}{p_i}\right)^2 \left[\exp\left(\frac{-\chi_i \psi_i}{M_{itc}}\right) - \frac{p_i}{p}\right]\dot{\varepsilon}_q \tag{3.40}$$

The hardening parameter H is a model soil property, necessary because the decoupling of the yield surface in *NorSand* from the CSL means that λ can no longer serve as a plastic compliance. H is ideally a constant, but in principle could also be a function of ψ. It is to be determined by calibration of the model to experimental data.

Those are the basics of the *NorSand* model. It has a related yield surface and stress–dilatancy to Original Cam Clay, but fundamentally decouples the hardening from the CSL. It is this decoupling that gives the model its versatility and relevance to real soil behaviour. The decoupling is nevertheless directly related to the CSL using the state parameter and the associated dilatancy relationship (3.36).

3.6.2 Elasticity in *NorSand*

Isotropic elasticity is adopted, despite ample evidence that soils tend to be cross-anisotropic, on the grounds that isotropic plastic theory is used. More to the point, it is unhelpful to leap into a model with many parameters, as are necessarily involved with anisotropic models, when there is not a proper understanding of the basic soil behaviour within the simpler isotropic framework. From a practical engineering viewpoint, it is difficult enough to determine parameters for an isotropic model; anisotropy remains highly academic and theoretical for now and is likely to continue as such for many years.

The isotropic elastic κ-model of Cam Clay neglects elastic shear. Elastic shear strains exist in soil, ubiquitously, and are also necessary for finite element formulations. A model that is rigid in shear is actually unattractive. Further, soil elasticity seems very dependent on the local grain arrangement, and is not simply related to void ratio and stress, so that a case can be made for treating the elastic modulus as a further state variable rather than a property. However, the κ-model has the nice feature that it is approximately related to a dimensionless approach, Equation 3.21, and it is desirable to keep elasticity dimensionless (as are all the other properties). Based on these considerations, the basic form of elasticity adopted is a shear rigidity I_r and a constant Poisson's ratio. These are related to other elastic properties as follows:

$$I_r = \frac{G}{p} \tag{3.41}$$

$$\frac{K}{p} = \frac{1+e}{\kappa} = I_r \frac{2(1+\nu)}{3(1-2\nu)} \tag{3.42}$$

The elastic shear modulus in (3.41) is commonly called 'G_{max}' in engineering practice. It is discussed at some length in Chapter 4 as G_{max} is often measured during site investigations. Poisson's ratio is rarely measured with the 'not unreasonable' range $0.15<\nu<0.25$ being adopted without testing (Poisson's ratio estimate can be refined when fitting undrained triaxial tests with *NorSand*).

3.6.3 *NorSand* summary and parameters

Table 3.1 summarizes the *NorSand* equations for fixed principal stress direction. *NorSand* is a sparse model, with the variant presented here requiring seven parameters to span the entire behaviour over the range of accessible void ratios. Table 3.2 summarizes the soil properties used in the model. Of these seven properties, two are used to define the reference CSL. Three properties define the plastic behaviour. Two properties define the elastic behaviour. Table 3.2 also indicates the range of values commonly encountered with sands. Triaxial compression is used as the reference condition in Table 3.2 because, although only

Table 3.1 Summary of *NorSand*

Aspect of NorSand	*Equations*
Internal model parameters	$\psi_i = \psi - \lambda \ln(\bar{\sigma}_{m,i}/\bar{\sigma}_m)$ where $\psi = e - e_c$ $\chi_i = \chi_{tc}/(1 - \chi_{tc}\lambda/M_{tc})$ $M_i = M(1 - \chi_i N\lvert\psi_i\rvert/M_{tc})$
Critical state	$e_c = \Gamma - \lambda \ln(\bar{\sigma}_m)$ and $\eta_c = M$ where... $M = M_{tc} - (M_{tc}^2/(3 + M_{tc}))\cos(3\theta/2 + \pi/4)$
Yield surface and internal cap	$\dfrac{\eta}{M_i} = 1 - \ln\left(\dfrac{\bar{\sigma}_m}{\bar{\sigma}_{m,i}}\right)$ with $\left(\dfrac{\bar{\sigma}_{m,i}}{\bar{\sigma}_m}\right)_{max} = \exp(-\chi_i\psi_i/M_{itc})$
Hardening rule	$\dfrac{\dot{\bar{\sigma}}_{m,i}}{\bar{\sigma}_{m,i}} = H\dfrac{M_i}{M_{itc}}\left(\dfrac{\bar{\sigma}_m}{\bar{\sigma}_{m,i}}\right)^2\left[\exp\left(\dfrac{-\chi_i\psi_i}{M_{tc}}\right) - \dfrac{\bar{\sigma}_{m,i}}{\bar{\sigma}_m}\right]\dot{\varepsilon}_q^p$
Stress–dilatancy	$D^p = M_i - \eta$
Elasticity	$I_r = \dfrac{G}{\bar{\sigma}_m}$ and ν = constant

Note that these equations are in general stress notation with $\bar{\sigma}_m = p'$.

Table 3.2 *NorSand* parameters and typical values for sands

Parameter	*Typical range*	*Remark*
CSL		
Γ	0.9–1.4	'Altitude' of CSL, defined at 1 kPa
λ	0.01–0.07	Slope of CSL, defined on natural logarithm
Plasticity		
M_{tc}	1.2–1.5	Critical friction ratio, triaxial compression as the reference condition
N	0.2–0.5	Volumetric coupling coefficient for inelastic stored energy
H	25–500	Plastic hardening modulus for loading, often $f(\psi)$; as a first estimate for refinement, use $H = 4/\lambda$
χ_{tc}	2–5	Relates maximum dilatancy to ψ. Triaxial compression as the reference condition
Elasticity		
I_r	100–600	Dimensionless shear rigidity (G_{max}/p')
ν	0.1–0.3	Poisson's ratio

triaxial compression has been considered so far, these model parameters will turn out to be functions of the intermediate principal stress in the more general formulation. Triaxial compression is also a common test and easy to carry out.

3.6.4 Numerical integration of *NorSand*

NorSand is a differential model in which there is a relationship between stress and strain increments. Obtaining the stress–strain behaviour requires integration and this is necessarily numerical, as there are no analytical solutions for *NorSand*. *NorSand* is readily implemented in finite element programs, the topic of Chapter 8. However, for known stress paths, such as for laboratory tests, the *NorSand* equations can be integrated directly using the consistency condition.

Direct integration is most useful for the present, as the goal here is to understand liquefaction. Much of this understanding comes from laboratory tests and the framework that follows from these tests is then used to look at full-scale data which is based on case histories (and which also needs to be viewed in the context of simple shear). So, the starting point is direct integration, which is easily done within a spreadsheet.

Direct integration requires that the stress path be known and the details of the integration vary with the laboratory test. Appendix C presents the derivations of the necessary equations and these are implemented in various *NorSand**.xls* spreadsheets, which can be found on the website. For ease of use, these spreadsheets work for triaxial compression or simple shear separately (otherwise you get too many plots to page through), although nearly all the VBA is common between them. Test data for comparison with the model are opened using Excel and copied onto a worksheet in *NorSand**.xls*. The appropriate data can then be plotted in the relevant plot windows provided. The *NorSand* behaviour plots presented in this book, as well as the comparisons with test data, were all generated using these spreadsheets. The intent is that these spreadsheets be used alongside this text, by loading in the test data (all supplied on the website) to look at the effect of model parameters, calibrations, etc.

3.7 COMPARISON OF *NorSand* TO EXPERIMENTAL DATA

3.7.1 Determination of parameters from drained triaxial tests

It is interesting now to compare how well *NorSand* fits sand behaviour, as that is a basic 'validation' that the model is useful. The theory is all very nice, but how well does it fit reality?

The test data on Erksak 330/0.7 sand are used for this validation. As discussed in Chapter 2, Erksak sand was used for a comprehensive investigation into the uniqueness of the CSL. One by-product of this investigation is that a data set exists with a thoroughly defined CSL and with a selection of drained (as well as undrained) tests covering a spectrum of densities and initial confining stresses.

The data from the various tests define the CSL for Erksak sand (see Figure 2.26). For the stress range of $p < 1000$ kPa, a semi-log idealization of the CSL gives the best-fit properties $\Gamma = 0.816$, $\lambda_{10} = 0.031$. This fixes two of the seven *NorSand* parameters for Erksak sand. The initial state parameter of each test is immediately known from the measured void ratio and stress level (reported in Table 2.4, and embedded in the downloadable data).

The critical friction ratio is determined from a stress–dilatancy plot, presented earlier in Figure 2.30. This figure is based on testing done at the University of British Columbia (UBC), with sand supplied by Golder Associates from their stock of Erksak sand. It is used here because a wider range of stresses were explored at UBC than in the Golder drained testing. This plot gives $M_{tc} = 1.26$, M_{tc} being the intersection of the trend line with the zero dilation intercept.

The state–dilatancy parameter χ is obtained by plotting maximum dilation versus the state parameter at maximum dilation, presented earlier in Figure 2.13. This gives $\chi_{tc} = 3.5$. There are now only two parameters to be determined, H and I_r. Both are dimensionless shear moduli, one plastic and one elastic.

Ideally, I_r will be determined using bender elements located on triaxial samples or from unload–reload cycles with local strain measurements. None of this was done for the Erksak samples. However, the bulk modulus of Erksak sand was carefully measured in a series of isotropic compression tests on Erksak sand (Figure 3.12). These measurements were on samples prepared in exactly the same way in the same apparatus as used for the triaxial

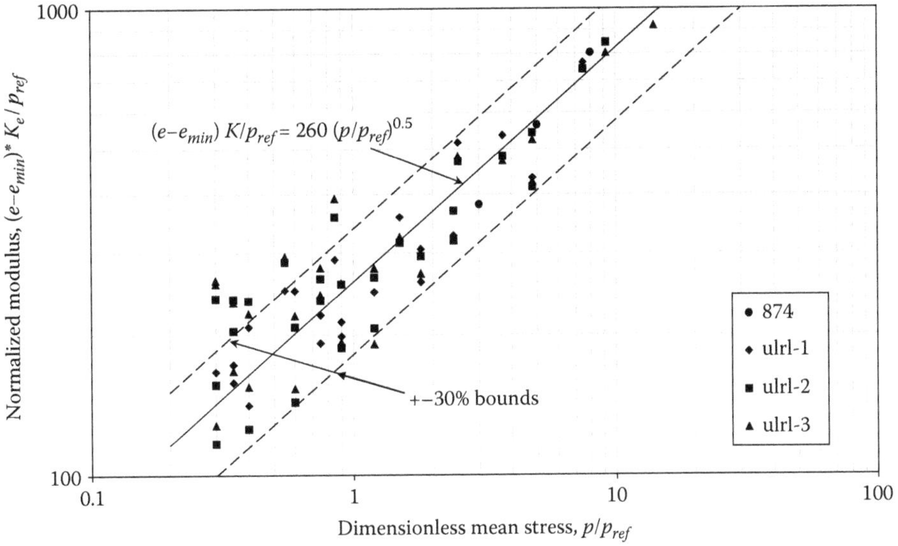

Figure 3.14 Measured bulk modulus of Erksak sand in isotropic unload–reload tests. (All stresses as effective, from Jefferies, M.G. and Been, K., *Géotechnique*, 50(4), 419, 2000. With permission from the Institution of Civil Engineers, London, U.K.)

compression tests. Elastic bulk modulus was determined during the unloading portion of isotropic unloading and reloading tests. Figure 3.14 shows the elastic bulk modulus data. The bulk modulus is a function of void ratio and stress level, given by

$$\frac{K^e}{\sigma_{ref}} = \frac{C}{e - e_{\min}} \left(\frac{p'}{\sigma_{ref}} \right)^b \tag{3.43}$$

where the reference mean stress is $\sigma_{ref} = 100$ kPa as is conventional. The properties for Erksak sand are $C = 260$, $b = 0.5$ and $e_{\min} = 0.355$. Adopting a constant Poisson's ratio $\nu = 0.2$, as Poisson's ratio does not vary greatly from one soil to another, gives the result that elastic shear modulus $G = 3/4\ K$ using the usual relationships of isotropic elasticity. This gives the dimensionless shear rigidity as

$$I_r = \frac{G}{p'} = \frac{3}{4}\frac{K^e}{p'} = \frac{3}{4}\frac{C}{(e - e_{\min})} \left(\frac{p'}{\sigma_{ref}} \right)^{b-1} \tag{3.44}$$

The remaining parameter H is determined by iterative forward modelling (IFM) of drained triaxial compression tests. In IFM, parameters are estimated and behaviour computed using the model. Comparing the computed behaviour with that measured then leads to revised parameter estimates, the process being repeated until a good match of model with data is obtained. IFM is preferred to parameter isolation (in which a particular aspect of a test, such as initial stiffness, is used to identify a parameter) because all models are idealizations of behaviour, and IFM optimizes the idealization over the whole range of behaviour to be represented. IFM was carried out using the *NorSandM.xls* spreadsheet (on the website for downloading).

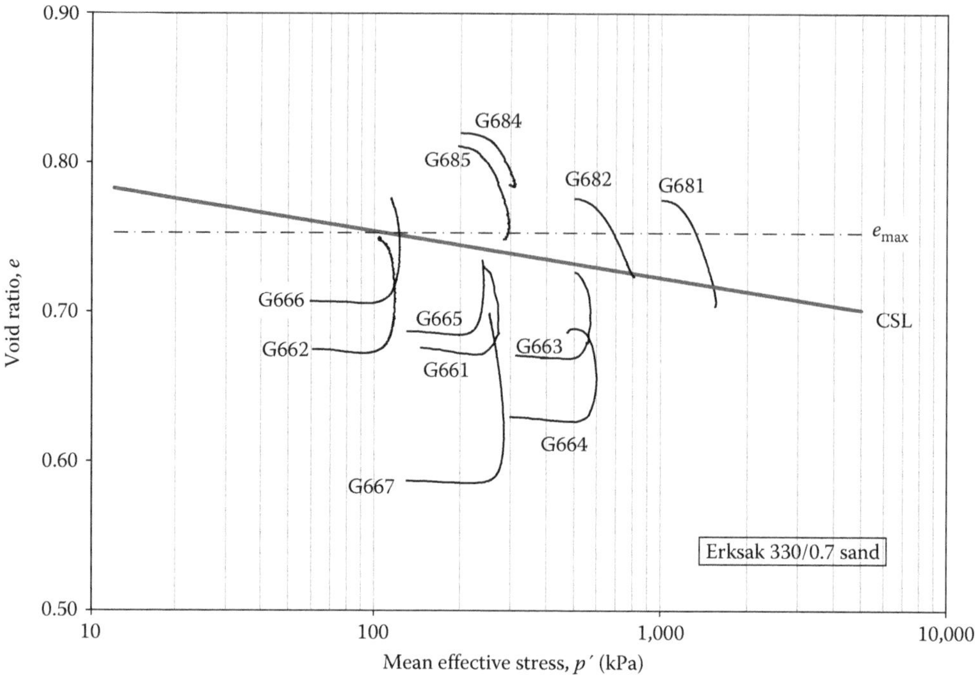

Figure 3.15 State diagram for drained tests on Erksak 330/0.7 sand.

Figure 3.15 shows the state diagram for the drained compression tests used to determine *H*, together with the idealized linear CSL. The range of test conditions is from very loose to quite dense, and the confining stress ranges from a low of 60 kPa to a high of 1000 kPa. Some of these tests were carried out early in the test program for Erksak sand and in particular before the adoption of sample freezing for accurate void ratio determination; Tests 684 and 666 appear denser than the reported void ratios. Test 681 is at the start of CSL curvature.

Four examples of the calibrated fit of *NorSand* to the various tests are shown in Figure 3.16. The parameter values used to achieve these fits are given in Table 3.3. All the Erksak triaxial data are downloadable and these fits can be verified, and other fits can be examined.

Excellent calibration of model to the data is evident across the spectrum of void ratios and confining stress. In fitting the model to data, the post-peak behaviour has been neglected. This is because once the peak strength has been attained, the strains localize, and the average of strains over the whole sample no longer represents what is happening in the zone of shearing.

Fitting model to data makes it evident that *H* is a function of ψ with Erksak sand. Figure 3.17 shows the relationship that has the simple form

$$H = a + b\psi_o \tag{3.45}$$

where for Erksak sand $a = 103$ and $b = -980$. The functional dependence of *H* on ψ_o is admissible from a theoretical standpoint and appears to be common.

The calibration values shown in Table 3.3 are for treating the soil properties as 'freedoms' to best fit the model to data. Of course, the essence of models like *NorSand* is that

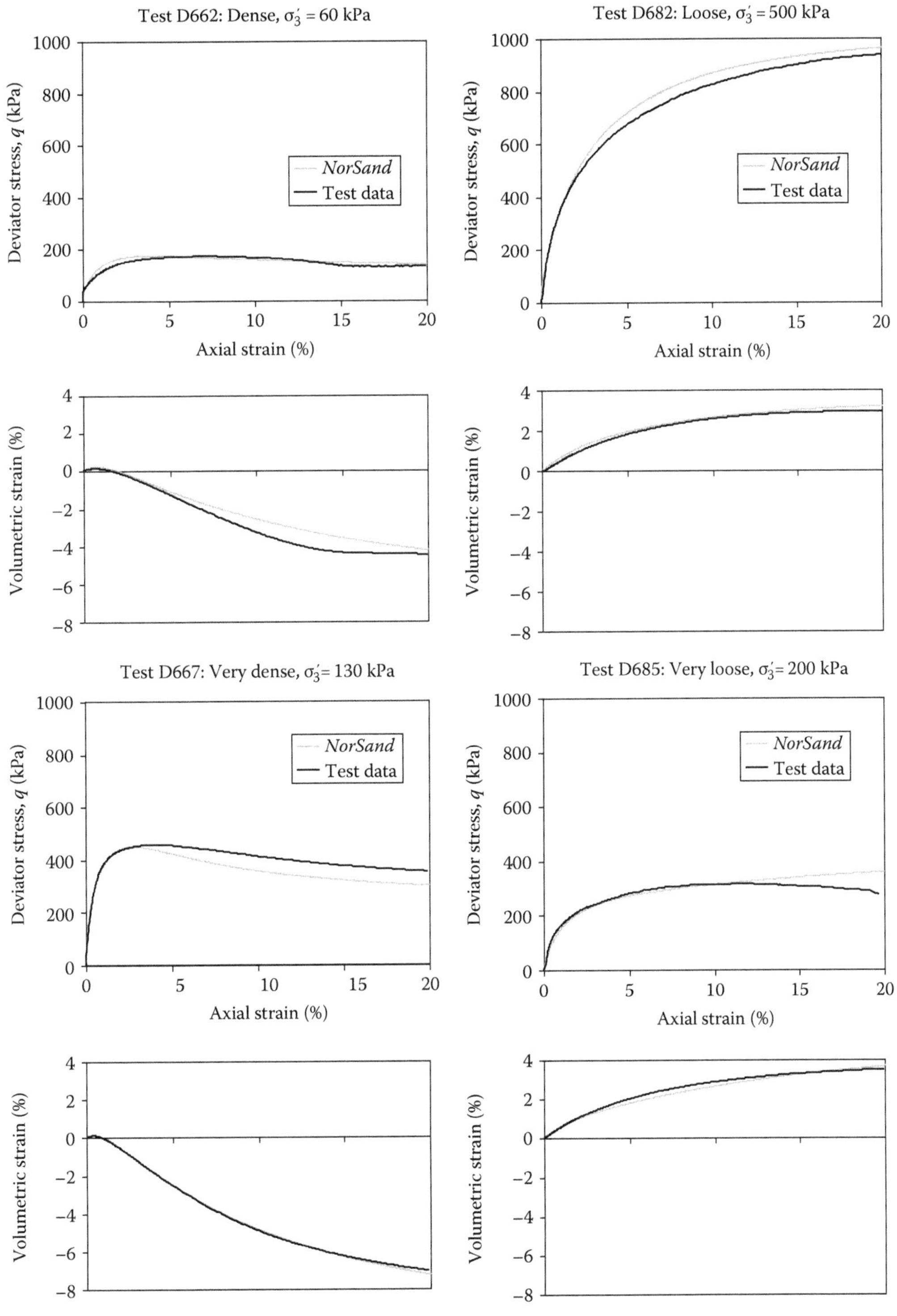

Figure 3.16 Examples of calibrated fit of *NorSand* to Erksak 330/0.7 sand in drained triaxial compression.

Table 3.3 *NorSand* Erksak 330/0.7 drained triaxial calibration

	Test values		*Best-fit parameters for test simulation*					
Test	p_o *(kPa)*	ψ_o	ψ_o	*M*	*N*	χ_{tc}	*H*	I_r
D661	140	−0.069	−0.080	1.26	0.35	3.7	150	600
D662	60	−0.084	−0.095	1.26	0.35	4	200	1000
D663	300	−0.064	−0.080	1.26	0.35	4	160	400
D664	300	−0.104	−0.110	1.26	0.35	3.9	200	400
D665	130	−0.059	−0.070	1.26	0.35	4.5	130	700
D666	60	−0.051	−0.080	1.26	0.35	4.5	170	1000
D667	130	−0.160	−0.171	1.3	0.35	3.8	300	600
D681	1000	+0.052	+0.070	1.18	0.35	4	50	150
D682	500	+0.044	+0.045	1.18	0.35	3.7	45	250
D684	200	+0.075	+0.040	1.18	0.35	3.7	70	400
D685	200	+0.067	+0.067	1.18	0.35	4.5	80	400

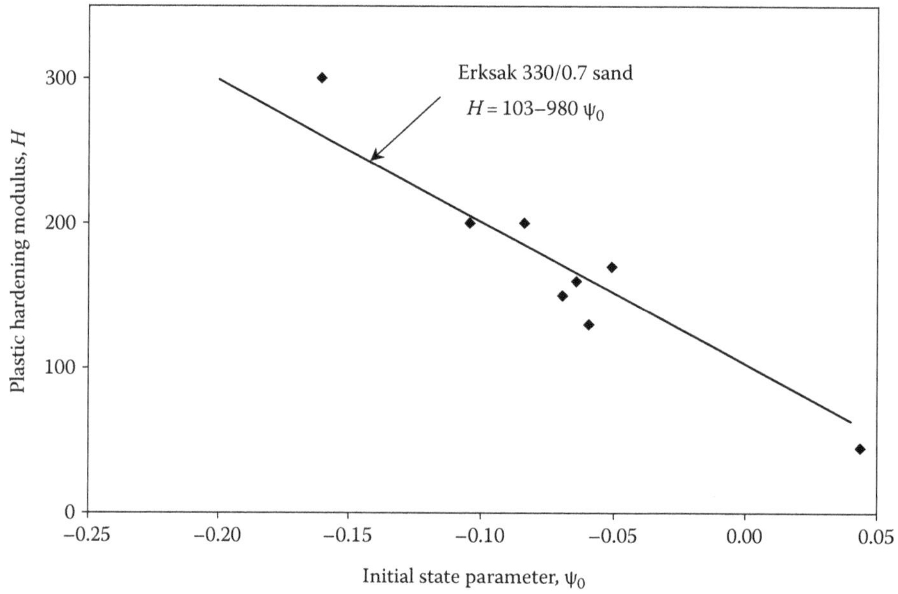

Figure 3.17 Plastic hardening modulus versus state parameter ψ_o for Erksak sand.

the properties do not change from test to test. The difference between the best-fit and trend values reflects unrepresented aspects of soil in the model, that is 'fabric'. In this context,

- M_{tc} does not vary from expectation ($M_{tc} = 1.26$) across the dense tests, but the best fit for loose is systematically less at $M_{tc} = 1.18$. This aspect was commented on in Been et al. (1991) and shows up clearly when modelling the loose tests in detail.
- N was not varied from one test to another and left at the trend value ($N = 0.35$).
- χ_{tc} did vary from test to test for best fit, which was anticipated to be an effect of fabric. The average of the best-fit values corresponds to the trend fit ($\chi_{tc} = 4.0$), with test to test variation being ±10% of the trend value.

- H values lie within ±10% of the overall trend (considering the first-order dependence on ψ), also thought to be a fabric effect.
- ψ_o was varied from test to test, but the shifts were within the accuracy of void ratio measurement in the laboratory and thus of no great significance.

3.7.2 Influence of *NorSand* properties on modelled soil behaviour

Having seen the fit of *NorSand* to test data, it is now helpful to look systematically at the effect of the model parameters on the computed behaviour. Because *NorSand* is properly dimensionless, it is sufficient to present the results normalized by the initial effective confining pressure. Figure 3.18a through f shows the effects of the model parameters. In all cases, the base parameter set is a parameter set broadly representative of quartz sands and is as follows: $\Gamma=0.8$, $\lambda_{10}=0.05$, M_{tc}=1.25, N=0.3, , χ_{tc}=3.5, H=250, I_r=600, ν=0.25.

The effect of the initial state ψ_o on the peak stress ratio and dilatancy is large. Initial state also affects the stiffness, with a change from $\psi_o=+0.04$ to $\psi_o=-0.2$ approximately tripling the initial stiffness. The peak shear stress ratio should be directly related to the critical state stress ratio M and this is indeed the case. Turning to the plastic hardening H, this affects the initial stiffness as expected but it also affects the volumetric strain and in particular the volumetric strain at peak stress ratio. This arises because shear and volumetric strains are directly coupled through D^P.

The effect of the intrinsic soil compressibility represented by the slope of the CSL λ is to change the dilatancy. It does this through the definition of the image state, with the most compressible soils (large λ) showing the least dilatancy. The soil property χ_{tc} that limits maximum dilatancy has a similar effect to λ on the post-peak dilatancy. Finally, it is apparent

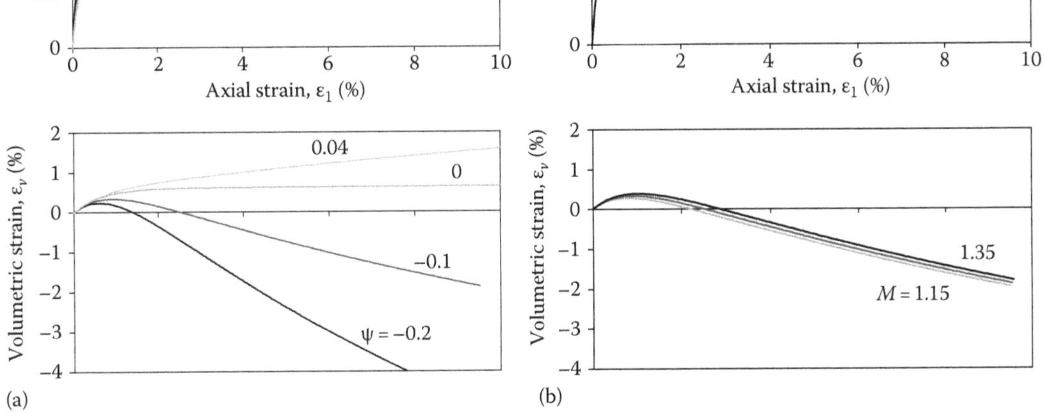

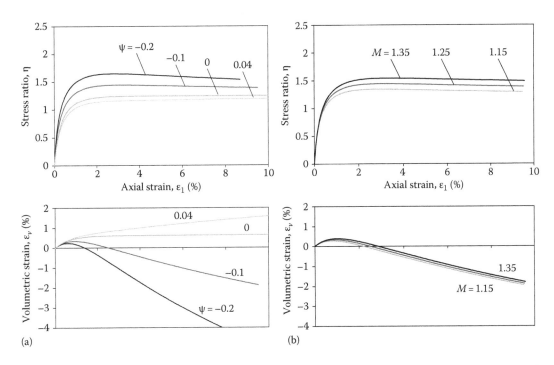

Figure 3.18 Effect of *NorSand* model parameters on drained triaxial compression behaviour. (a) Effect of initial state ψ_o and (b) effect of critical stress ratio M. (*Continued*)

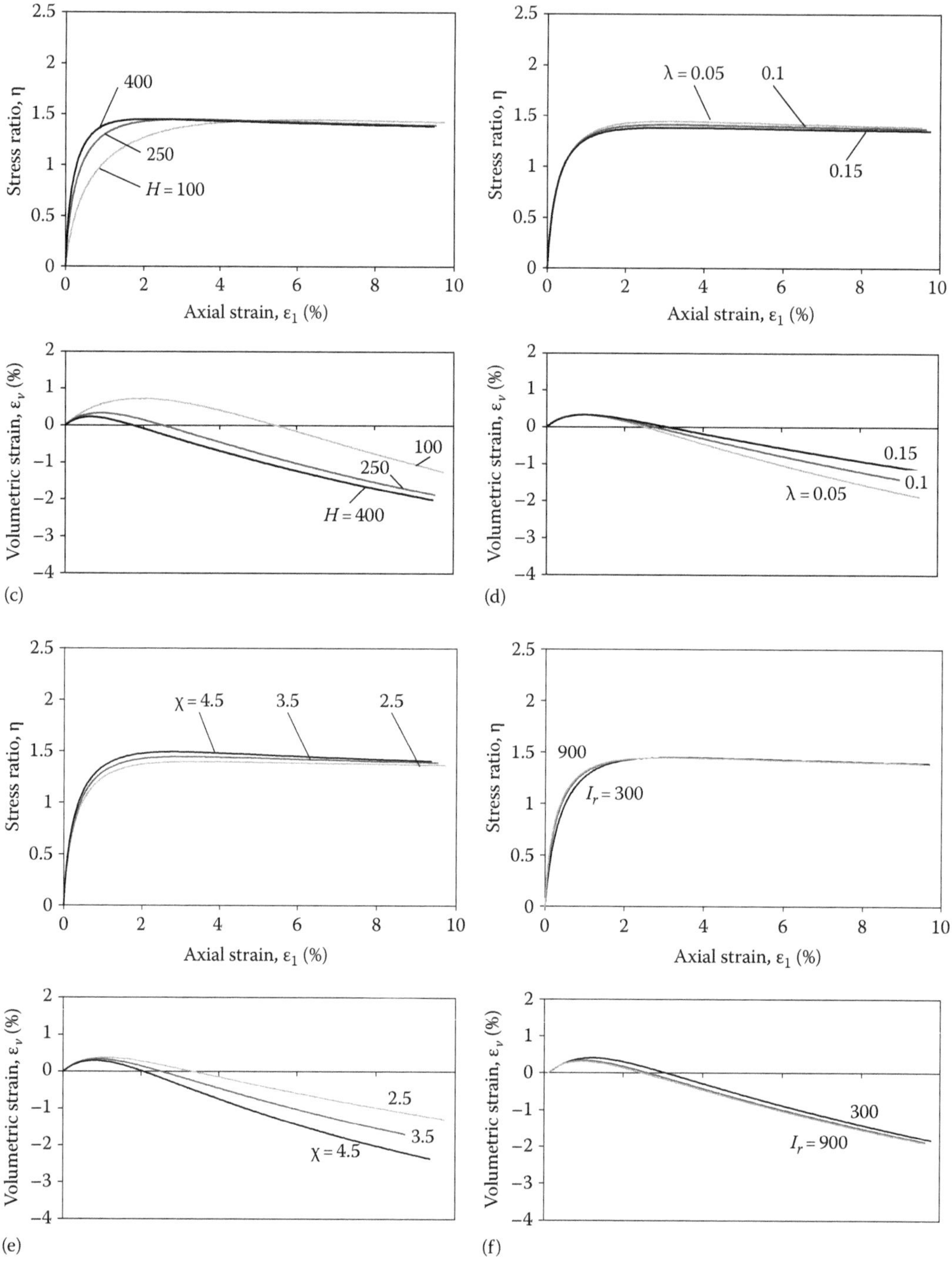

Figure 3.18 (Continued) Effect of *NorSand* model parameters on drained triaxial compression behaviour. (c) effect of plastic hardening H, (d) effect of critical state line slope λ_{10}, (e) effect of dilatancy scaling χ and (f) effect of elastic rigidity I_r.

that the elastic shear stiffness, represented by I_r, has a relatively minor influence because plastic shear strains dominate the response beyond the initial loading. Poisson's ratio is not shown as its effect is minor.

3.8 COMMENTARY ON *NorSand*

The presentation of *NorSand* has gone through the model quickly, primarily focusing on the prior Cam Clay model as the reference for discussion. However, there are aspects of the model that need further comment now that the fit to test data has been seen.

3.8.1 Yield surface shape

Those with some background in constitutive modelling of sands may have questioned the *NorSand* yield surfaces presented in Figure 3.13 on the basis that sand yields predominantly in shear. This is not the case and the evidence is the yield apparent in isotropic compression, as shown in Figure 3.12. A closed (capped) yield surface is required for sands, exactly as suggested in the original work on soil plasticity by Drucker et al. (1957). Evidence for *NorSand*-like yield surfaces is also found in the literature.

Yield surfaces were determined experimentally for Fuji River sand in triaxial compression by Tatsuoka and Ishihara (1974) and Ishihara et al. (1975). Using stress probing to determine the onset of plastic deformations, families of yield surfaces were mapped. These families of yield surfaces depended on sample density and mean pressure, and examples are shown in Figure 3.19. Corresponding yield surfaces from *NorSand* are presented in Figure 3.20, the parameters having been chosen to match the Fuji River data. The yield surfaces in Figure 3.20 are shown for a pattern of progressively increasing hardening, with each yield surface being associated with a particular value of p'_i. These yield surfaces were developed by selecting the properties and the hardness to illustrate the basic shapes obtainable, rather than a detailed simulation of experimental stress paths. The similarity with the experimental results is striking.

Tatsuoka and Ishihara (1974) considered a yield surface similar to that proposed here and concluded that it poorly fits their data. The difference in producing Figure 3.20 is threefold. First, *NorSand* decouples the image stress from the critical state, and thus, the surface can harden far more than expected with Granta Gravel or Cam Clay. This flattens the yield surfaces in Figure 3.20 as hardening increases. Second, limiting hardening has been introduced, with the yield surfaces terminating on this locus (conventionally called the Hvorslev or failure surface). This aspect is controlled through Equation 3.38 and one can see from this equation that the Hvorslev surface is not a surface at all, as it varies with the state parameter. For any mean effective stress, there are an infinite number of surfaces because there is an infinite range of accessible state parameters. Third, M_i as a function of ψ has been introduced into the flow rule.

Of course, a further departure from conventional expectations of yield surfaces in sands is the use of an internal cap, which comes from the concept of limited hardening. This limited hardening in *NorSand* is actually no more than a strict implementation of the concept suggested by Drucker et al. (1957) and it gives realistic dilatancy with an associated flow rule (normality). Non-associated models are unnecessary.

While thinking about the hardening limit, recall that *NorSand* does not fit the post-peak behaviour observed in tests because of non-uniform strain within the samples. As peak strength is approached, hardening of the yield surface continues even though dense soil will

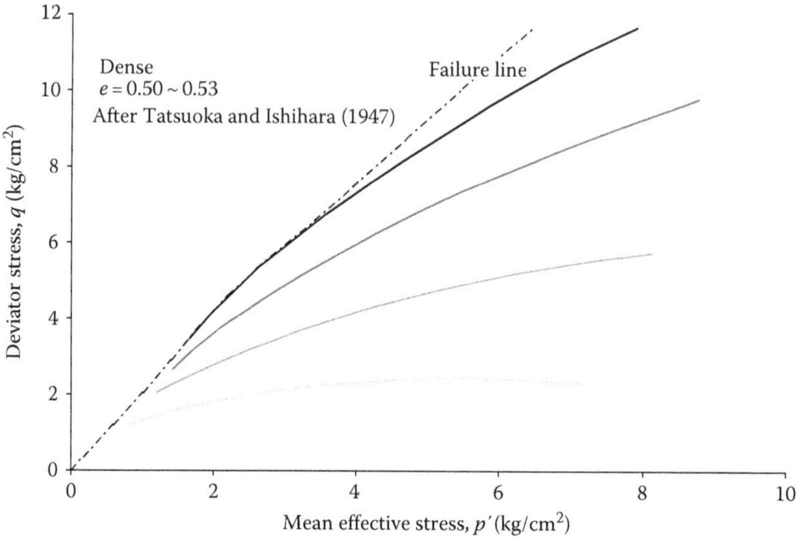

Figure 3.19 Example of experimentally determined yield surfaces in Fuji River sand. (From Tatsuoka, F. and Ishihara, K., *Soils Found.*, 14, 63, 1974. With permission from the Japanese Geotechnical Society, Tokyo, Japan.)

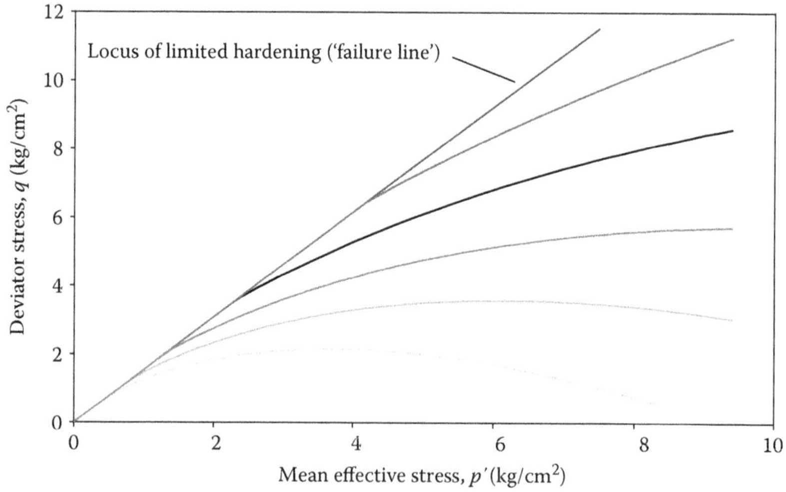

Figure 3.20 *NorSand* yield surfaces for comparison with experimental results on Fuji River sand.

be dilating and smooth behaviour is expected. However, post-peak when the hardening limit is imposed, the model becomes strain softening and the yield surface shrinks towards critical conditions. At this point, Drucker's stability postulate is violated. The effect on the soil is that less work needs to be done if the shear strain concentrates in a band (localizes), with the tendency to critical conditions being accelerated in the shear band. Although *NorSand* captures both aspects, the test can no longer be treated as uniform stress and strain, and a far more sophisticated integration using finite elements and additional idealizations about the nature of the localization is needed. This is presently an area of intense research.

3.8.2 Effect of elastic volumetric strain on ψ

A criticism of the state parameter (and *NorSand*) is that volumetric elasticity has been neglected in the definition of the state parameter. Recall that in Section 3.4.2, the change from Granta Gravel to Cam Clay was the addition of an elastic term Δe^e to account for the elastic volumetric changes caused by the change in mean stress. Conceptually, this term means that the yield surface is unchanged by elastic strains – which is consistent for a plastic model. An equivalent term is readily added to the definition of the image state parameter for *NorSand*:

$$\psi_i = e - e_{c,i} - \kappa \ln\left(\frac{p}{p_i}\right) \tag{3.46}$$

The void ratio correction term has not been used for several reasons, despite its theoretical attraction. The elastic void ratio change is second (third?) order compared to the state parameter and it is not essential to include it in the definition of state – undrained loading is computed very nicely by *NorSand* (Chapter 6). The apparent sophistication of the elastic void ratio change is just that, apparent not real, at least as far as practical application of *NorSand* is concerned and the simplest definition of ψ is preferred. There is also the issue that κ is not a particularly good representation of elasticity for particulate materials.

3.8.3 Volumetric versus shear hardening and isotropic compression

The Modified Cam Clay yield surface has not been used, despite the preference for it in many finite element programs. This is not an oversight. The problem with Modified Cam Clay is that it is volumetric hardening, and this means that if $D^p = 0$ then no hardening takes place with continued shear. $D^p = 0$ is exactly the image condition and Second Axiom hardening cannot work using volumetric plastic strain. Now consider the opposite case of shear strain hardening, as used in *NorSand*, which elegantly copes with the image condition. In the case of a Modified Cam Clay yield surface, there is no plastic shear strain on the isotropic axis for plastic volumetric strain. Because of this, shear hardening has no control over the NCL, and this most fundamental aspect of soil behaviour is then lost. Yield surfaces of the Modified Cam Clay style cannot be used with a single plastic hardening modulus within a state parameter framework. At least two plastic moduli are needed, and this implies two yield surfaces. Double yield surface models exist (Vermeer, 1978; Molenkamp, 1981), but to date these have not incorporated critical state concepts.

Discussion of the NCL brings up an interesting aspect of Second Axiom hardening. The premise of the state parameter approach is that soil exists across a range of states in e–p space. Self-consistency means that these states must be accessible and develop through isotropic compression, since the depositional void ratio is arbitrary. How can this be reconciled with the requirement that soil moves to critical conditions with shear strain, given the assertion that a vertex is needed on the yield surface at the NCL, such that the plastic strain can be controlled by a plastic shear modulus? In fact, no inconsistency arises because the stress–dilatancy relation allows contractive strains for $\eta < M$. Figure 3.21a is a familiar e–log(p) plot showing the isotropic NCL paths because of plastic strains, obtained by direct integration of the hardening rule (3.38) for the case $\eta = 0$.

Constant properties have been used, and the computed NCL are shown for various initial void ratios. Notice the similarity to the experimental data of Figure 3.12 (there is not a direct comparison as Figure 3.21 does not include elastic strains). Figure 3.21b shows computed NCL for one initial void ratio and three different values of H. Decreasing H models more compressible sand.

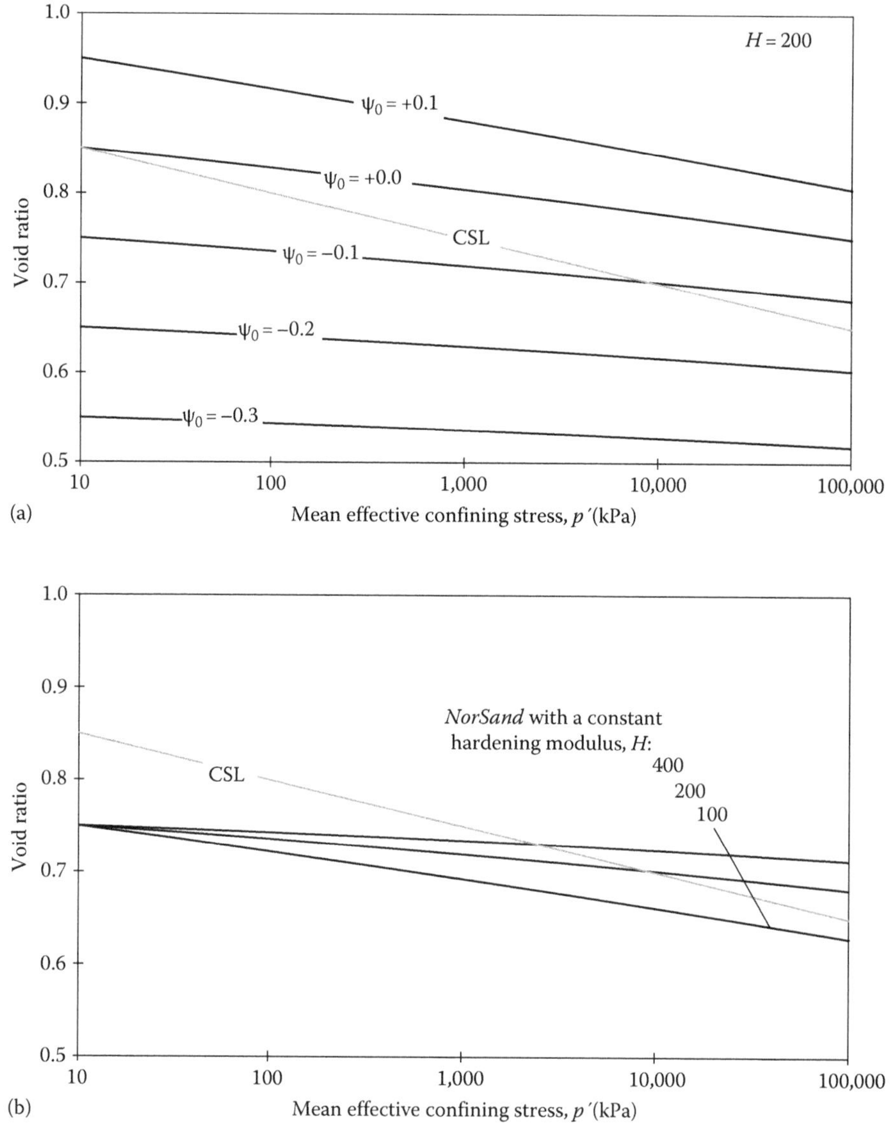

Figure 3.21 Isotropic plastic compression behaviour of *NorSand*. (a) For constant plastic hardening, $H=200$, and various initial states and (b) for constant initial state parameter, $\psi_o = -0.1$, and various plastic hardening moduli.

3.8.4 Limit on hardening modulus

Interestingly, although the concept of an infinity of NCL has led to a decoupling of isotropic compression from the slope of the CSL, λ, self-consistency within the conceptual framework requires that isotropic compression paths should extend to the loose side of the CSL. This aspect is derived in Appendix C, from which a constraint on H is obtained:

$$H > \frac{1}{\lambda - \kappa} \quad \forall\, \psi_i < 0 \tag{3.47}$$

No sands encountered to date have had the *NorSand* calibration limited by (3.47) and, as a rule of thumb, a first estimate is that $H \approx 4/\lambda_{10}$.

3.8.5 Plane strain and other non-triaxial compression loadings

In the presentation of *NorSand* so far, the aim has been to follow along the historical route, pointing out where these past ideas fit within the broader framework of geomechanics. However, many practical situations involve plane strain and one might reasonably ask for validation under such plane strain conditions. The generalization of *NorSand* from the triaxial compression model to arbitrary stress states/paths and with rotating principal stress is given in detail in Appendix C. This is the form of *NorSand* used in finite element calculations. Broadly, the generalizations involve some changes in notation and a few additional features, but the basic framework remains that of the triaxial model and the additional ideas are straightforward.

Given sufficient parameters, it is always possible to fit a model to data. The challenge comes when conditions are changed and the model has to predict the consequence of these changes. This chapter, therefore, also presents a validation of 3D *NorSand* for the case of plane strain. The validation takes the form of predicting plane strain behaviour based on triaxial calibration. This verifies both that the 3D model computes as advertised and that the computed behaviour matches reality (the validation) within reasonable accuracy (although reasonably depends somewhat on the task at hand). Plane strain was chosen as the validation case because it is both practically important and very different from triaxial conditions (triaxial extension is an inadequate validation because it still invokes the symmetry of the triaxial test).

Cornforth (1961, 1964) examined the behaviour of Brasted sand in drained loading under triaxial compression, triaxial extension and plane strain conditions. The triaxial equipment used by Cornforth was the Imperial College apparatus described in Bishop and Henkel (1957) and is comparable to the triaxial equipment used for the other tests discussed. The plane strain apparatus was developed at Imperial College by Wood (1958) and is described in Section 2.7. Figure 2.43 shows a picture of a failed sample in this apparatus.

Cornforth's testing encompassed a wide range of initial void ratios and a somewhat more restricted range of initial stresses. The resultant data have been presented earlier (Figure 2.45) as summary values in terms of stress–dilatancy, but interest here centres on the comparative constitutive behaviour between pairs of triaxial compression and plane strain tests at similar void ratio and stress level. Table 3.4 summarizes the initial conditions for triaxial/plane strain pairs or loose, compact and dense sands.

The CSL is best determined through several undrained triaxial compression tests on loose to compact samples, but no such tests are available for Brasted sand. The CSL parameters were therefore estimated as follows. Plotting the maximum dilatancy versus the initial void ratio for the triaxial compression tests at an initial confining stress of 276 kPa (40 psi) showed that zero dilatancy corresponded to $e_c \approx 0.77$ at this stress. The data at an initial confining stress of 414 kPa (60 psi) are more scattered and do not allow for sensible determination of the critical void ratio for that stress. Therefore, $\lambda = 0.02$ was adopted as not unreasonable based on other sands, then giving $\Gamma = 0.902$ from $e_c \approx 0.77$ at 276 kPa. The critical friction ratio M_{tc} is defined by the stress–dilatancy plot (Figure 2.45), giving $M_{tc} = 1.27$.

The elasticity of Brasted sand is uncertain because there are no load–unload stages in the test data to identify the elastic component, nor was shear wave velocity measured directly. For calibration, $I_r = 500$ and $\nu = 0.2$ were adopted as not unreasonable for sand at the density

Table 3.4 Paired tests on Brasted sand

Reference[a]	*Test*	e_o	ψ_o	K_o	$\bar{\sigma}_{m,0}$ *(kPa)*
Loose					
A10-31	Triaxial: C21	0.754	−0.02	0.447	390
A10-14	Plane strain: P20	0.721	−0.05	0.444	391
Compact					
A10-34	Triaxial: C25	0.664	−0.11	0.448	389
A10-12	Plane strain: P18	0.650	−0.12	0.435	395
Dense					
A10-41	Triaxial: C34	0.570	−0.20	0.379	426
A10-17	Plane strain: P23	0.572	−0.20	0.381	425

Source: Data from Cornforth, D.H., Plane strain failure characteristics of a saturated sand, PhD thesis, University of London, London, England, 1961.

[a] The reference is to the figure number in Cornforth (1961).

and confining stress of Cornforth's experiments. No effect of density was used. The power law exponent was taken as $b=0.5$.

The dilatancy parameter χ_{tc} was determined by plotting $D_{\min}$ in triaxial compression versus ψ (Figure 2.50). The remaining plastic parameter is the hardening modulus, H_{tc}, which was determined through IFM of the triaxial compression data. Figure 3.22 shows the achieved fits for the three triaxial tests listed in Table 3.4. In achieving these fits, it turns out that H_{tc} varies with the state parameter: for the loose test, $H_{tc}=75$; for the compact test, $H_{tc}=150$ and for the dense test, $H_{tc}=275$. The dependency of H_{tc} on ψ_o for Brasted sand is similar to Erksak sand (Figure 3.17) and is as follows:

$$H = 50 - 1125\psi_o \tag{3.48}$$

The comparison of measured and predicted constitutive behaviour in plane strain is presented in Figure 3.23. These plane strain simulations used the sand properties obtained in triaxial calibration, described earlier, without any modification. The three tests plotted in Figure 3.23 are the plane strain tests paired with the triaxial tests (Table 3.4 and Figure 3.22).

The volume change behaviour (or dilatancy), maximum dilation rate, peak strength and the increased stiffness in plane strain are well predicted in all three plane strain tests using the parameters obtained by calibration to triaxial tests. The intermediate principal stress is also well predicted in two tests, but diverges a little from the data for the dense sample.

Post-peak behaviour shows an interesting divergence of theory with data in plane strain. The theory shows a relatively slow reduction in peak strength with axial strain, much like under triaxial conditions, but the experimental results show a rapid reduction to what appears to be the critical state condition.

All in all, the 3D *NorSand* validates well in plane strain based on triaxial calibration. In addition, *NorSand* is based entirely on ψ to drive the computed soil behaviour. So, the next chapters are about determining ψ in-situ and developing characteristic values for design in real soils.

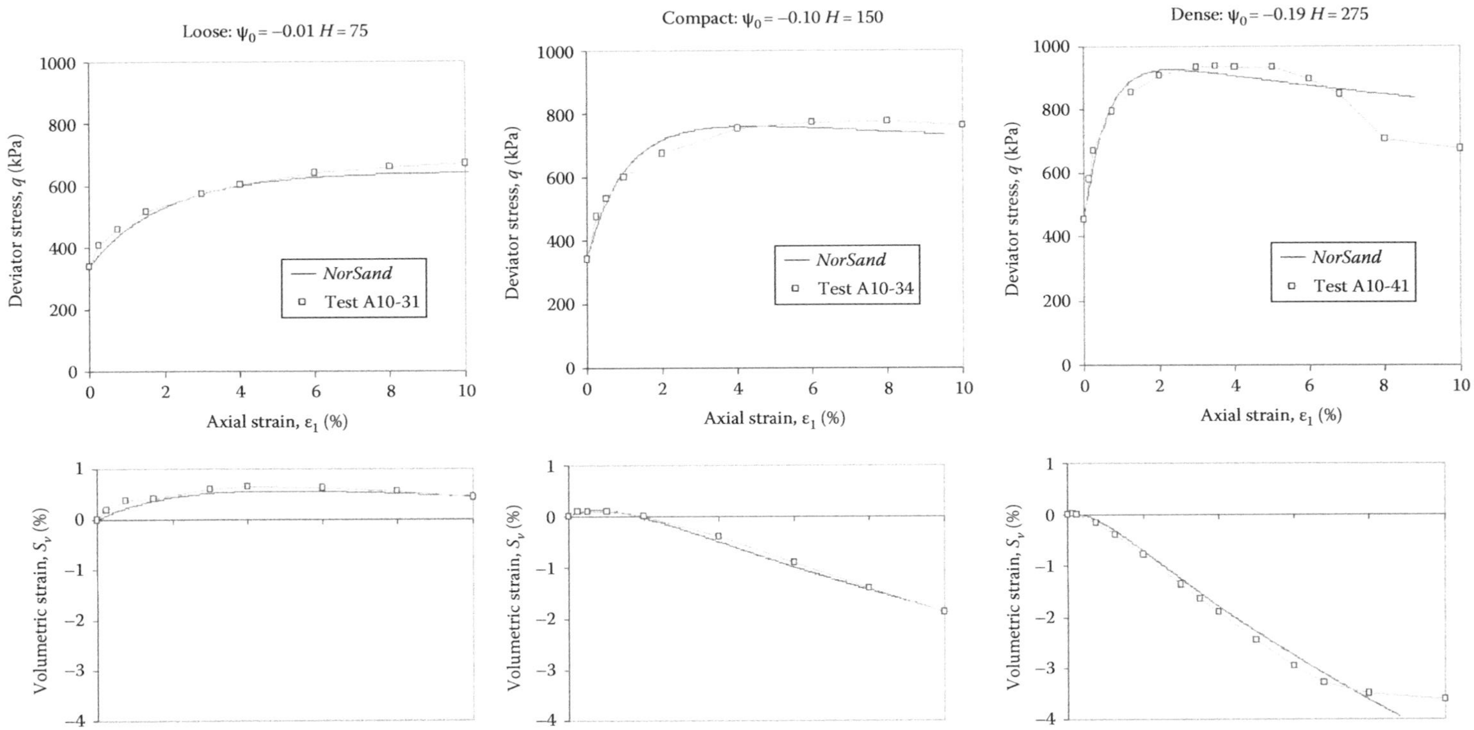

Figure 3.22 Calibration of *NorSand* to Brasted sand in triaxial compression. (Updated from Jefferies, M.G. and Shuttle, D.A., *Géotechnique*, 52(9), 625, 2002. With permission from the Institution of Civil Engineers, London, U.K.)

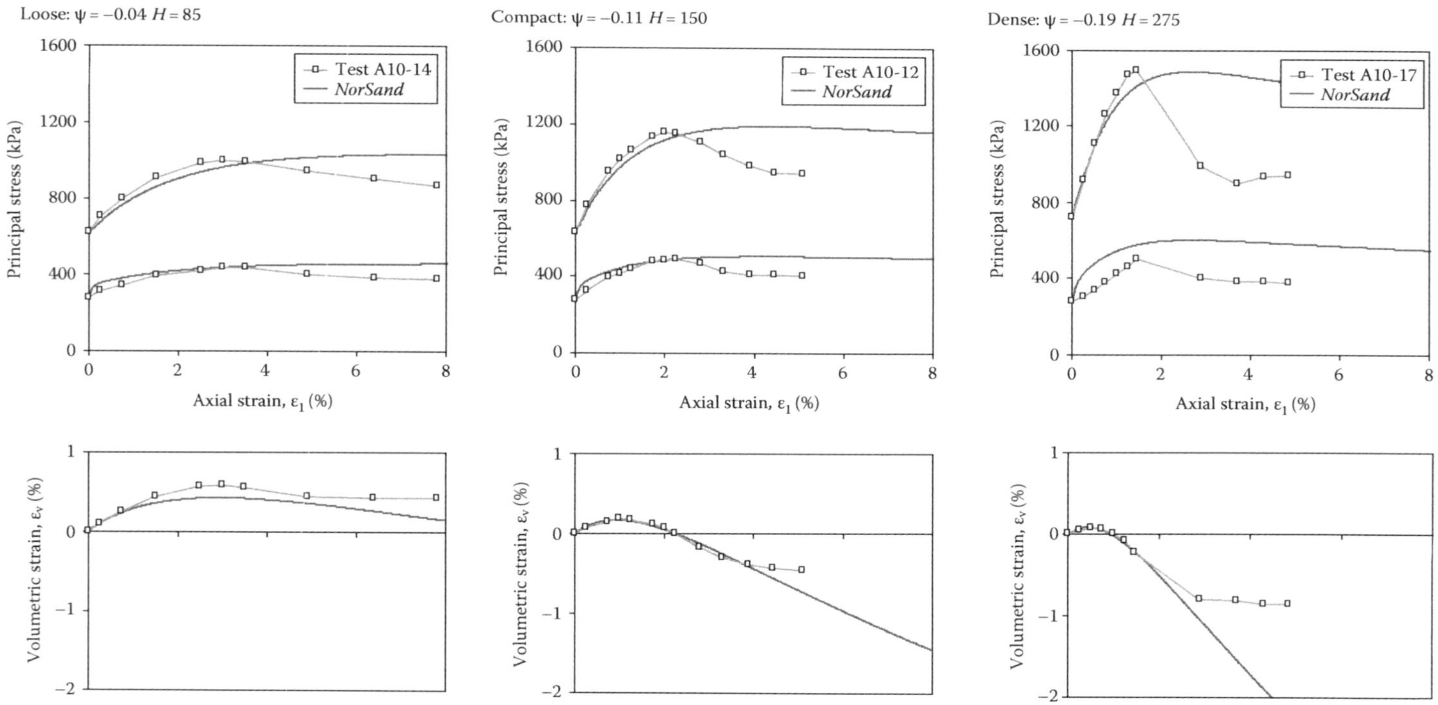

Figure 3.23 Verification of *NorSand* in plane strain by comparison of predictions versus data for Brasted sand. (Updated from Jefferies, M.G. and Shuttle, D.A., *Géotechnique*, 52(9), 625, 2002, using revised M_i function of Table 3.1.)

Chapter 4

Determining state parameter in-situ

4.1 INTRODUCTION

Because of the difficulty in sampling cohesionless soils in anything like an undisturbed condition, engineering of sands and silts depends largely on penetration tests. Penetration tests are also simple and inexpensive, attributes that allow large numbers of tests to characterize the variability of ground properties in the strata of interest. This is not to say that there are no alternatives to penetration testing; there are, and these will be discussed later. But alternatives to penetration tests are difficult to execute, usually offer less accuracy, and do not necessarily provide the information that is needed directly. Estimating soil state from penetration tests is the backbone of liquefaction assessments.

Penetration testing for many years almost always meant the standard penetration test (SPT), a test in which a weight is dropped from a fixed height as many times as it takes to drive a standard sampler a distance of 300 mm (1 ft) into the ground. During the last 35 years, the SPT has been progressively replaced by the cone penetration test (CPT). Today, CPT is most often conducted using an electronic piezocone.

Regardless of the type of penetrometer used, a basic issue arises with the fact that the measured data need to be *interpreted* as the process is typically described. The word *interpreted*, however, is slightly misleading. What happens in penetration testing is that the soil response (resistance) to an enforced displacement is measured, but it is the properties/states of the soil that are sought. This is an inverse boundary value problem. It is a boundary value problem in that we have a domain with boundaries. It is inverse in that we know the answer (load or response) but need to identify the properties or state (which would usually be the input). Inversion of penetration test data is what is sought, not interpretation. Much of this chapter is devoted to inversion.

Penetration test resistance is very dependent on the stress level, all other factors being equal. This makes the in-situ stress one of the key considerations in inversion. Two alternative approaches to allow for the stress level have developed. Within North American practice, it has become common to 'correct' the measured data to a reference stress level. Correction is a misleading word, as there is nothing wrong with the original data. What happens in the reference stress–level approach is that the measured data are mapped to what would have been measured at the reference stress level if nothing else (e.g. void ratio) were changed.

The alternative approach to the reference stress level is to work in a conventional framework of applied mechanics and use dimensionless parameters. A dimensionless approach allows scaling through the laws of mechanics and avoids adjustments or corrections. Since it is so common, the reference stress approach will be examined, but in doing this the errors and simplifications will become obvious. Dimensionless inversions for CPT data will therefore be pursued as the core method.

Table 4.1 Dimensionless parameter groupings for CPT interpretation

Dimensionless parameter group	*Description*
$Q_p = (q_t - p_o)/p'_o$	Tip resistance normalized by mean effective stress
$Q = (q_t - \sigma_{vo})/\sigma'_{vo}$	Tip resistance normalized by vertical effective stress
$F = f_s/(q_t - \sigma_{vo})$	Normalized friction or friction ratio (usually expressed in %)
$B_q = (u - u_o)/(q_t - \sigma_{vo})$	Normalized excess pore pressure
$Q(1 - B_q) + 1 = (q_t - u)/\sigma'_{vo}$	Suggested by Houlsby (1988). It is an effective stress form of tip resistance, useful in silts and when considering classification of soil types with the CPT. It allows greater differentiation between silty clays and clayey silts at the low q_t undrained end of the spectrum.
$Q_p = 3Q/(1 + 2K_o)$	Relationship between normalized resistances Q_p and Q

Note: q_t is the end area corrected tip resistance, and q_c is the raw measured value. For sands and silts, the pore water pressure is small compared to q_c and the values of q_t and q_c are essentially identical. We use q_t and q_c somewhat interchangeably in this book because before the mid 1980s only q_c was reported (because pore pressure measurements were uncommon).

There has also been a dominant tendency to rely on empirical correlations to determine in-situ properties. This reliance on correlations is unfortunate, as correlations can often be found where there is no basis for a causal relationship, or certain factors have been missed. Wroth (1988) addressed this issue and emphasized that correlations between the results of in-situ tests and soil properties should be: (1) based on physical insight; (2) set against a theoretical background; and (3) expressed in dimensionless form.

The CPT will be used as the principal basis for assessing the in-situ state. The relevant dimensionless parameters for the CPT are presented in Table 4.1. These dimensionless parameters follow Wroth's strictures, most being developed from cavity expansion as the analogue to the CPT.

The reason to introduce both Q and Q_p is that, while Q_p is the fundamental dimensionless variable of interest, it does require knowing the geostatic stress ratio (K_o). This requirement to know K_o is inconvenient, and so some people have tried to avoid the issue by working in terms of only vertical stress or Q. There are useful soil classification charts using Q, and we need to retain it for that purpose. However, inversions with greatest precision require Q_p. Practically, with sufficient accuracy, there is a very simple relationship between the two dimensionless variables which is shown in the final row of Table 4.1.

4.2 SPT VERSUS CPT

The SPT is a test that has been in use for more than 75 years. Originating in the United States, it is now used worldwide. The advantages of the SPT are that the equipment is rugged, the procedure is easy to carry out, a sample of the soil tested is usually recovered and most soil types can be tested. The word 'standard' is misleading though, as there are quite large variations in the test procedure. Attempts have been made in recent years to properly standardize the SPT, with an international reference test procedure being proposed in 1988. But the principal difficulty with the SPT is its lack of repeatability, even with the same equipment and borings directly adjacent to each other. The poor repeatability of the SPT arises because of two factors: the variability in the input energy and the dynamic nature of the test.

The original form of the SPT used a spinning cylinder (called a cathead) driven off the power unit of the drilling rig, with a slipping rope around it as the means by which the weight was raised. The driller pulled on the rope, which stopped the rope slipping on the cathead, which raised the weight until the driller thought that it had reached the standard height, whereupon the rope was released again. Although it sounds like something out of Monty Python, this is in fact the way things were done. Not surprisingly, there was a lot of variation in the input energy both from the variable height of each blow and from the differing drag of the rope on the cathead. It has been argued that this procedure should continue, as it represents the standard on which the database of liquefaction case histories is based. Aside from the technical arguments, the cathead system should be avoided on health and safety grounds – drillers have had fingers and hands mangled in rotating catheads.

Improvements to the SPT initially concentrated on mechanizing the hammer system, so that the blows were truly delivered in free fall and from the same height each time. The goal was to achieve a repeatable 60% of the free fall energy delivered to the SPT rods (selected as an average target for equivalence with cathead systems) and hence the nomenclature for the blow counts as N_{60}. This improvement does not, however, address a basic problem. The energy transfer through the anvil to the penetrometer depends on various impedances including those of the ground and the inertia of the drill rods. To circumvent variable energy transfer, people began using transducers and electronic measuring systems to measure the actual energy transferred to the soil. Robertson et al. (1992) give an interesting account of these developments and an example of one such system. ASTM published standard D4633 for energy measurement during the SPT. Of course, once the energy is measured electronically, the simplicity of the SPT is lost with the attempt to apply sophisticated electronics and data acquisition systems to a poor test. Once computers are involved, the CPT offers so much more with so much less effort.

To illustrate the poor repeatability of the SPT even with energy correction, Figure 4.1a shows the results of two borings with SPTs in alluvial sands of the Fraser Delta, near Vancouver, Canada. These borings were only 2 m apart. The SPTs were conducted by the same drillers with the same mechanized hammer equipment, with energy measurement. Nevertheless, the N_{60} values fluctuate more than ±25% about the mean trend. This extremely poor test repeatability is a principal objection to the SPT. It is pointless to attempt to base any engineering on a test that is inaccurate to this extent.

The CPT has also been in use for a long time, nearly as long as the SPT in fact. Originating in the Netherlands (hence its early name of 'Dutch cone') as a way of probing soft clay to find the bearing stratum for a piled foundation, the CPT developed into a more widely used test for pile capacity throughout the Low Countries. With the growth of the offshore oil industry, the CPT developed very rapidly and became the reference test for that industry. A considerable CPT business has developed worldwide, with many specialist contractors testing many thousands of meters of soundings per year. Since the mid-1970s, there has been an explosion of knowledge and technology associated with the CPT. Key developments include the following:

- Standardization on right cylindrical geometry with 60° conical point
- Addition of piezometric measurements
- Addition of inclinometers to track CPT deviation during sounding
- Addition of geophones for concurrent vertical seismic profiling
- Microprocessors in the penetrometer for transducer stabilization
- Analogue to digital conversion of data in the penetrometer
- Digital data transmission to the surface
- Recording of data in digital form using microcomputers

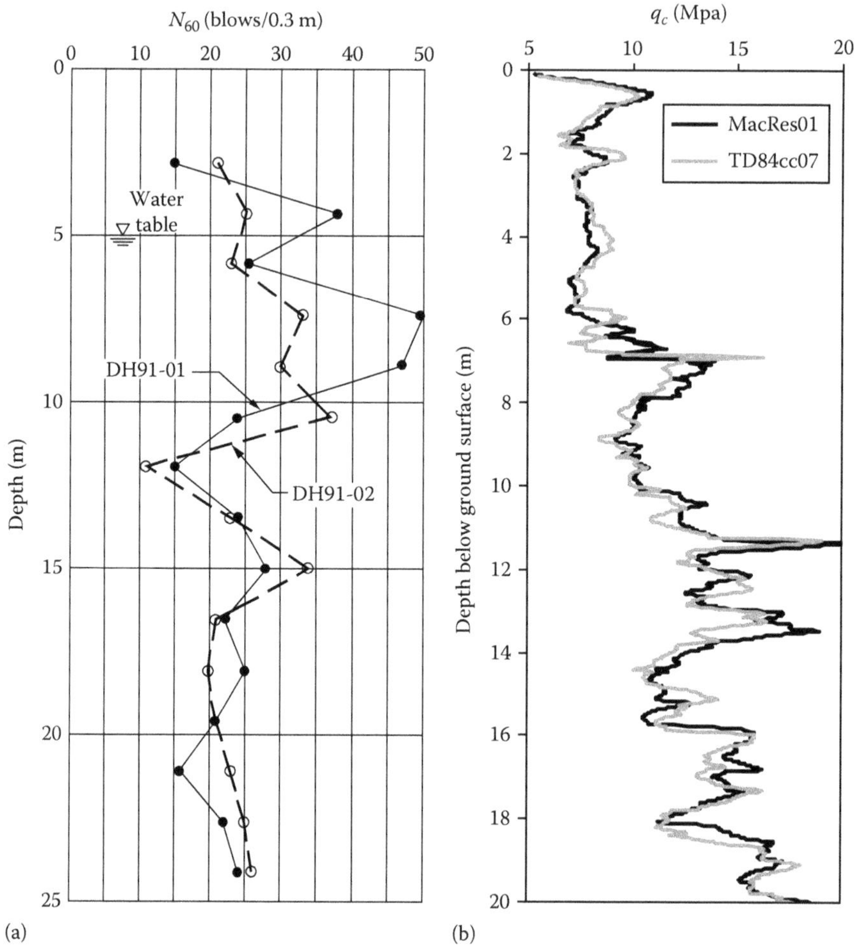

Figure 4.1 Comparison of SPT and CPT repeatability. (a) Two adjacent SPT borings (Richmond, British Columbia, Canada) and (b) two adjacent CPT soundings (Tarsiut, Beaufort Sea).

The particular advantage of the CPT is the interesting combination of a continuous data record with excellent repeatability and accuracy, all at relatively low cost. The same numbers are produced regardless of which contractor or driller carries out the work.

The repeatability of the CPT is illustrated in Figure 4.1b, which shows two soundings in sandfill 1.5 m apart at ground surface (but not the same site as Figure 4.1a). There is excellent repeatability, with only occasional slight vertical offset in the location of denser layers. The CPT shows combined accuracy, non-linearity and hysteresis that is close to the transducer limits of 0.5% full scale (or 0.25 MPa). Most modern CPTs will equal or better this performance. A standard for the CPT, ASTM D5778 (and this is an excellent standard that should be referred to in specifying CPT work), suggests that the standard deviation in measured q_t is approximately 2% of full-scale output.

In round numbers, the SPT, even with energy measurements and subsequent correction, is four to five times less repeatable than the CPT. The SPT also suffers from significant discretization error. Once N_{60} is greater than about 20, liquefaction problems have largely been avoided, so that the entire range from catastrophic through to adequate behaviour is spanned in something like 20 units of measurement. By contrast, even the crudest of modern CPT equipment will cover the same range in about 3000 discretization intervals to span 0–50 MPa (4096 points from 12-bit resolution with linear scaling) and more accurate CPT measurements are easily achieved by changing to a more sensitive transducer (e.g. a 0–20 MPa unit) and the possible enhancement of 18-bit discretization.

An argument sometimes put forward in favour of the SPT is that one usually obtains a geological sample of the stratum just tested, which of course is not obtained with the CPT. This argument is specious in two ways. First, the origin of the CPT is a stratigraphic logging tool, and there are reliable charts to estimate soil type (an example is given in Figure 4.2). Plus, the CPT sees soil layers in the order of 50 mm thickness, such detail being completely missed by the SPT N_{60} value. Second, interest lies in the mechanical behaviour of the soil, and the three data channels of the CPT measure just that. Exaggerating for the sake of making the relevant point, it really does not matter if a stratum is called silty sand or blue cheese with holes – what matters is its mechanical and hydraulic behaviour. The CPT, by having three data channels to the SPT's one, is obviously superior at providing useful data on these soil behaviours. And then there is the vastly better resolution, data density and accuracy of the CPT.

Before leaving the SPT, note that using the CPT still allows the use of any relevant SPT databases. Because both SPT and CPT are penetration tests, there is a relationship between

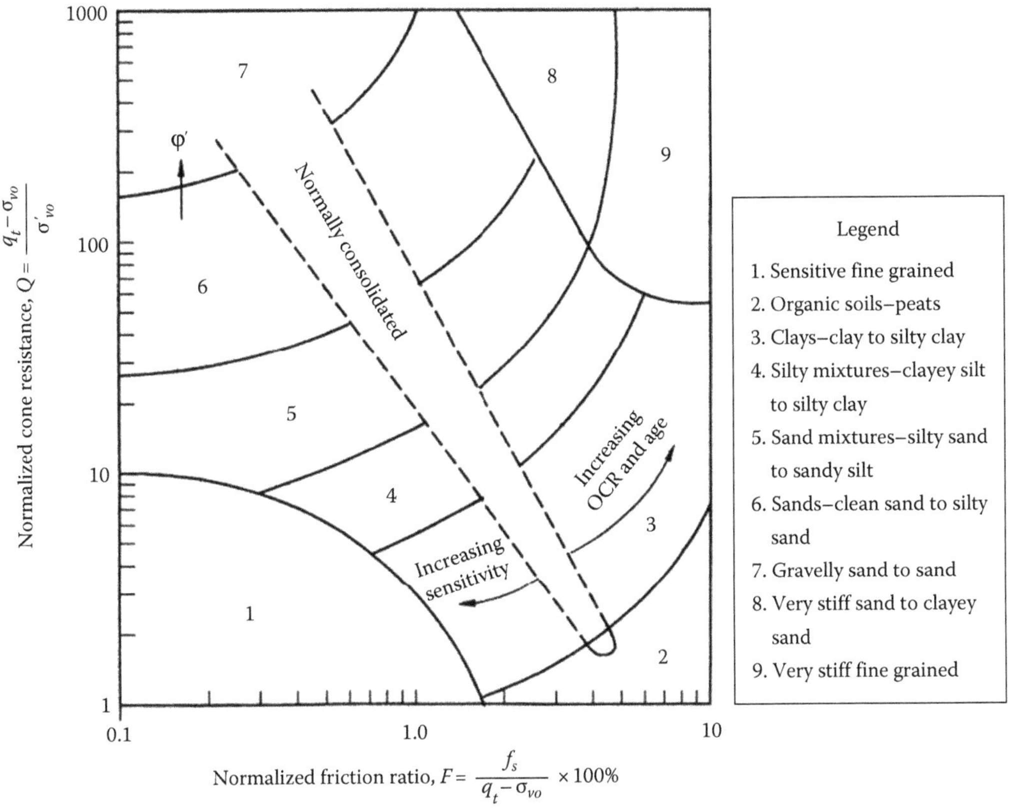

Figure 4.2 Illustration of soil-type classification chart using CPT data. (After Robertson, P.K., *Can. Geotech. J.*, 27(1), 151, 1990. With permission from the NRC, Ottawa, Ontario, Canada.)

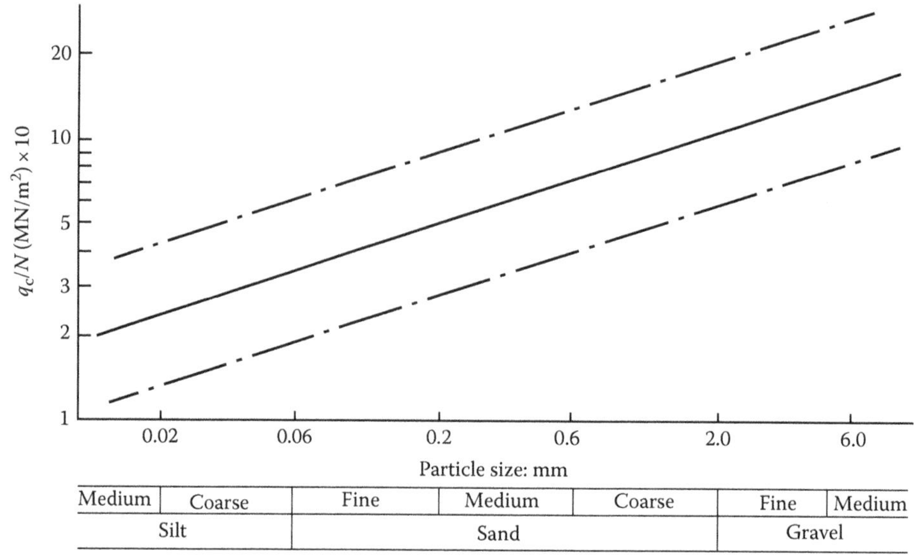

Figure 4.3 Relation between q_c/N and soil type. (From Burland, J.B. and Burbidge, M.C., *Proc. Inst. Civil Eng.*, 78(1), 1325, 1985. With permission from the Institution of Civil Engineers, London, U.K.)

the resistance measured in one and that seen in the other. Several people have looked into this, and the general form of the relationship is

$$q_t = \alpha N_{60} \tag{4.1}$$

The coefficient α (MPa/blow) depends on soil type. Initially, estimates of α were sought from the perspective of having SPT data in hand – that is one knew the geologic description of the strata, had a blow count and from this one wanted to estimate an equivalent CPT resistance. Burland and Burbridge (1985) give a good summary of the knowledge of α from geology, with detailed information being found in Robertson et al. (1983). Figure 4.3 summarizes the relationships developed. Here, the opposite situation is relevant as emphasis is on the CPT, and the requirement is to be able to look back and compare measurements to those sites where only SPT data are available. Because the CPT does not provide a sample to use in assessing grain size, reliance is placed on the friction ratio or soil behaviour type from the CPT. Jefferies and Davies (1993) as well as Robertson (2012) dealt with this and the methodology is given later after the general framework for evaluation of the CPT has been developed and explained. For now, accept that relying on the CPT does not mean abandoning those case histories for which only SPT data are available and that SPT-based experience can be used when only CPT soundings have been carried out.

There is also the important point, in choosing between whether to use the CPT or the SPT, that many case histories actually have CPT data, sometimes as well as the SPT and sometimes instead of the SPT. There is an erroneous impression given in the literature that the SPT should be the test of first resort because that is how liquefaction analysis started and it provides the precedent. Nothing could be further from the truth, as can be seen from scanning the case histories presented in Appendix F.

4.3 INVERSE PROBLEM: A SIMPLE FRAMEWORK

Although initial interest in the CPT derived from the geologic perspective of stratigraphic identification using a continuous record, it is the repeatability and accuracy of the CPT that is of most interest here. The determination of penetration resistance to an accuracy of typically better than 2% is certainly precise enough for soil testing and provides the basis for estimating the state parameter in-situ. However, this will not be achieved by simply introducing correlations. Rather, Wroth's strictures will be followed and a proper theoretical framework provided.

Determination of the in-situ state involves solving an inverse problem. The relationship between CPT resistance and the properties of the soil stratum is of the form

$$Q_p = f(\psi, G, \bar{\sigma}_v, M, \ldots) \tag{4.2}$$

whereas inversion of (4.2) is required to get the state from the CPT response:

$$\psi = f^{-1}(Q_p, G, \bar{\sigma}_v, M, \ldots) \tag{4.3}$$

Generally, there is an information loss with inversion. Q_p will be determined more accurately from (4.2), for a given range of uncertainties in material properties, than will be ψ using (4.3) and the same uncertainties. This is a fact of life and is the flip side of the ease of penetration testing in the field. Accurate parameter determination from the data requires a lot of care and attention to detail. Also, note an algebraic limitation. There is one item of information (q_t) and one equation. There is no possibility of learning five independent properties from the CPT penetration resistance alone, despite all the 'correlations' that have been published. Those properties appearing in the right-hand side of (4.3) must be known. This is a fundamental requirement based in algebra, not something that can be avoided by relying on correlations.

The approach taken here is to follow (4.3) rigorously. The starting point is, of course, to define the nature of the function f^{-1}. This is a little troublesome, as there is no closed-form solution for the resistance of the ground to a CPT. Obtaining a proper form for (4.2) and (4.3) needs large displacement analysis as the familiar small strain approximations (i.e. what you were taught in undergraduate engineering courses) do not give a stable limiting load as seen by the penetrometer. Further, the zone around the penetrometer is one of intense shearing with substantial changes in void ratio. A proper analysis needs a soil model whose behaviour reflects the effect of changing void ratio and stress, but such soil models are not tractable in closed-form solutions. What is more, these good soil models are difficult to implement in finite element codes with the soil flowing around a penetrometer.

As a first-order guide, the simplest acceptable approximation to the form of (4.2) is to combine spherical cavity expansion theory with a non-associated Mohr–Coulomb (NAMC) soil model. This combination does have a semi-closed-form solution and gives an analogue for the CPT in drained penetration of cohesionless soils. Spherical cavity expansion looks at the problem of expanding a spherical cavity of some initial radius (which may be zero) in an infinite soil medium, neglecting gravitational effects. An advantage of the cavity expansion analysis is that a difficult 3D problem is reduced to something viewed in terms of one-space variable, radius. The pressure required to expand a spherical cavity has a limiting value, given sufficient cavity expansion, and this limiting value is treated as an approximation of the pressure on the tip of the CPT as it displaces the soil in penetrating the ground.

The symmetry of cavity expansion has resulted in cavity expansion analysis becoming one of the classics of theoretical soil mechanics, and there is quite a history of progressive development of the analysis. Restricting our attention to cohesionless soils, the following are the key developments in an understanding of cavity expansion.

Chadwick (1959) gave a first solution for spherical cavity expansion in a frictional soil by using a Mohr–Coulomb soil model with a dilation angle equal to the friction angle. A natural strain approach accounted for large displacements. The solution was not closed form, requiring numerical evaluation of integrals. Vesic (1972) gave a delightfully simple solution for a Mohr–Coulomb soil using a large displacement formulation. However, simplicity was gained by writing the solution in terms of total volumetric change in the plastic zone – which, in general, is not known with sands, and this limited the applicability of the solution. A crucial aspect of soil behaviour was missing from what is otherwise a very elegant analysis. Baligh (1976) further developed Vesic's approach to include the effect of 'soil compressibility' (his terminology – it is actually suppression of dilation by mean stress) on the cavity pressure, illustrating how different penetration resistance profiles with depth can arise from this effect. Hughes et al. (1977) worked with only small strain expansion of a cylindrical cavity but were the first to introduce stress–dilatancy theory as the basic constitutive model. This removed dependence on prior uncertainties about plastic volume changes, as they were now calculated as part of the solution. However, the small strain basis prevented calculation of limit pressures. Carter et al. (1986) applied the NAMC model for plastic strains with a large strain formulation to both cylindrical and spherical cavity expansions. This was the first real result that is readily useable and expressed in terms of a few soil properties. Their solution was closed form, but implicit for the cavity expansion pressure. Subsequently, Yu and Houlsby (1991) adopted the same NAMC material idealization in an entirely different mathematical treatment and gave a series solution.

Thus, from the late 1980s, realistic guidance has been available in semi-closed form to indicate a basis for the proper interpretation of the CPT. This guidance is based on constant soil properties throughout the cavity expansion, which is a further simplification imposed on the already idealized case of radial symmetry. Since then, the increasing power of computers has become apparent, and more sophisticated soil models have been used that specifically include void ratio change as the cavity expands, with corresponding changes in the soil behaviour. These sophisticated models are not analytically tractable, and depend on numerical methods for solution even with radial symmetry in cavity expansion. Numerical solutions are examined later, but it is helpful first to use the simpler closed-form solutions as a guide. Of these solutions, that by Carter et al. is the easiest to use and is adequately thorough.

Stress–strain curves for the NAMC model used in the Carter et al. model are presented in Figure 4.4 and can be compared with examples of sand behaviour given in Chapter 2. In particular, notice that NAMC captures the basic dilatancy of sands, although in the simple sense dilatancy is constant once yielding starts. The NAMC model is elastic–plastic with four properties: two elastic (shear modulus G and Poisson's ratio ν) and two plastic (friction angle ϕ and dilation angle δ). There is no plastic hardening. The model is termed non-associated because the friction and dilation angles are different. The Mohr–Coulomb part is a conventional and familiar limiting strength criterion, which is taken as a yield surface in the NAMC's model. The Carter et al. solution for the limit spherical cavity expansion stress ratio Q_p in a NAMC soil is

$$\frac{2G}{p_0} = \frac{n-1}{n+2}\left[T\left(\frac{Q_p}{R}\right)^{\gamma} - Z\frac{Q_p}{R}\right] \tag{4.4}$$

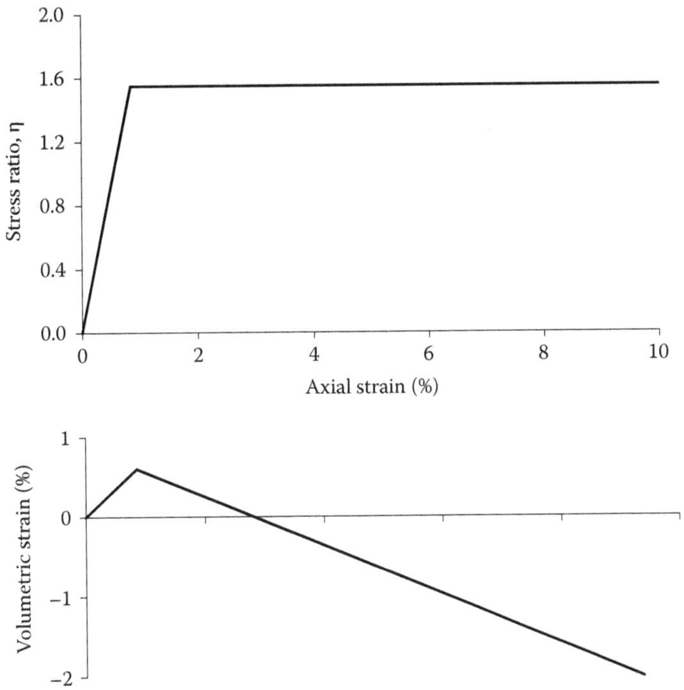

Figure 4.4 Example stress–strain behaviour of NAMC material in triaxial compression (properties for medium dense sand).

where the terms n, R, T and Z are functions of the four properties of the NAMC soil noted earlier. Notice that (4.4) is an inverse form for Q_p; one cannot simply rearrange the equation to recover Q_p given input of the four material properties. However, (4.4) is easily solved with the bisection algorithm in a spreadsheet; the downloadable spreadsheet *Carter.xls* returns Q_p for specified material properties using the bisection algorithm. Also notice that modulus in (4.4) has been normalized by far-field stress to give the dimensionless rigidity $I_r = G/p_o$; this parameter group arises routinely in mechanics-based understanding of soil behaviour, including penetration tests (note that p_o is effective stress in this context).

In the NAMC model, dilation and friction angles are treated as properties that can vary independently of each other, whereas in real soils friction angle links to dilatancy through stress–dilatancy behaviour. There are alternative theories describing different mathematical forms for the relationship between the friction angle ϕ and the dilation angle δ, but these are operationally quite similar. One common approach is to approximate Rowe's stress–dilatancy theory (Rowe, 1962) as suggested by Bolton (1986):

$$\phi = \phi_c + 0.8\delta \tag{4.5}$$

where ϕ_c is the critical state friction angle.

Figure 4.5 shows the relationship between Q_p and peak friction angle from (4.4) using Bolton's approximation for stress–dilatancy (4.5). The figure shows the Carter et al. solution as well as the results by Yu and Houlsby (1991, 1992). The two solutions, while not identical, are close. Also shown, as points, are four simulations using the large strain finite element method and the same NAMC model as Carter et al. The finite element results lie

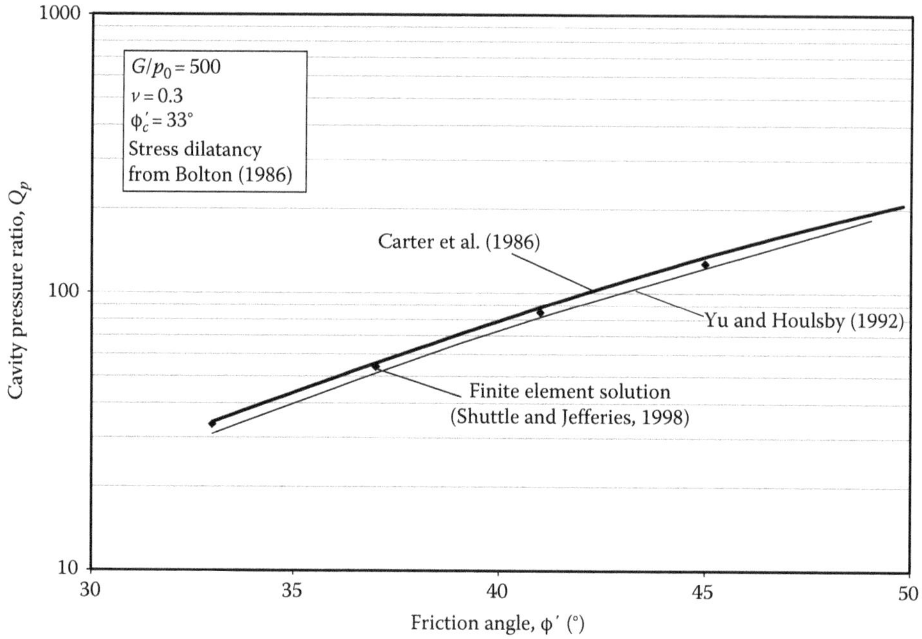

Figure 4.5 Spherical cavity limit pressure ratio versus friction angle for NAMC material with Bolton's approximation of stress–dilatancy.

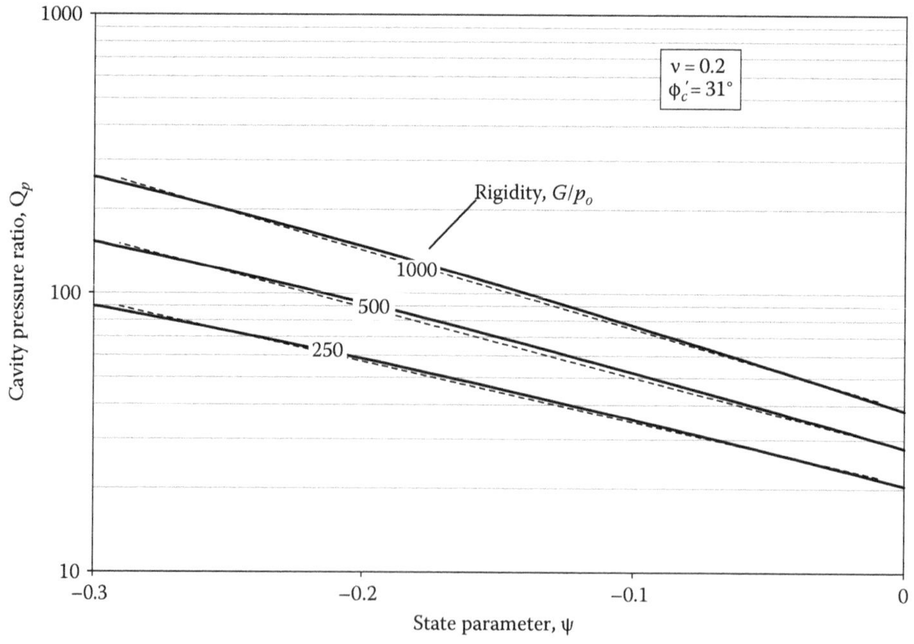

Figure 4.6 Spherical cavity limit pressure ratio versus state parameter (broken lines indicate linear approximation of Equation 4.7).

between the Carter et al. solution and the one by Yu and Houlsby. This finite element program is used later for developing detailed CPT simulations with the *NorSand* model, and the results for NAMC are shown here as a program verification.

The peak friction angle and the state parameter ψ are intimately related. Referring to Figure 2.7 the peak friction angle is given by

$$\phi = \phi_c\left(1 - \frac{5}{3}\psi\right) \quad \text{with}\, \phi_c = 32° \tag{4.6}$$

Based on the 20 sands tested, Equation 4.6 is broadly applicable regardless of grain size and shape, fines content or mineralogy. Introducing Equation 4.6 into 4.5 to determine δ from the chosen ψ, ϕ_c, and then substituting the results in (4.4), gives a closed-form relationship between Q_p and ψ. This relationship is plotted in Figure 4.6 for three values of rigidity.

Figure 4.6 is based on a solution for the penetration resistance given three material properties: I_r (which might also be regarded as a state variable since it changes with void ratio), ν (essentially a constant for any sand), and ϕ_c (also taken to be a constant for any sand); plus the state parameter ψ. However, the requirement is for the opposite problem of knowing Q_p from a CPT sounding and wishing to determine ψ. This can still be done by using the bisection algorithm to solve for ψ, but there is a simpler approximation. As shown by the dotted lines in Figure 4.6, the closed-form solution can be approximated by the very simple expression

$$Q_p = k\exp(-m\psi) \tag{4.7}$$

which is readily inverted to give the relationship sought:

$$\psi = \frac{-\ln(Q_p/k)}{m} \tag{4.8}$$

The interpretation parameters k and m are functions of the soil rigidity. This is then a basic expectation, anchored in theory, as to the form of the inversion sought for obtaining the state parameter from CPT data. Of course, there is a question as to the extent to which the simplifications in cavity expansion theory adequately represent the CPT. The relationship between spherical cavity expansion and penetrometer resistance in sand was experimentally tested by Ladanyi and Roy (1987), and their key result is presented in Figure 4.7. Penetrometer resistance and cavity pressure are similar functions of friction angle, but the two are not equal. Cavity expansion is only an analogue for the CPT, and calibration is required to obtain the appropriate parameter values.

As a final comment on closed-form solutions, Ladanyi and Foriero (1998) suggest that at least some of the large strain solutions discussed earlier err in combining a large strain measure (friction angle) with a ratio based on small strains (rigidity). So far, this proposition has not been further discussed in the literature, but the Carter et al. solution, being formed from integration of strain rates, should be acceptable.

4.4 CALIBRATION CHAMBERS

4.4.1 Description

Because theoretical and numerical solutions for the CPT are so difficult, and the spherical cavity solutions are approximations, most work at some point requires reference to

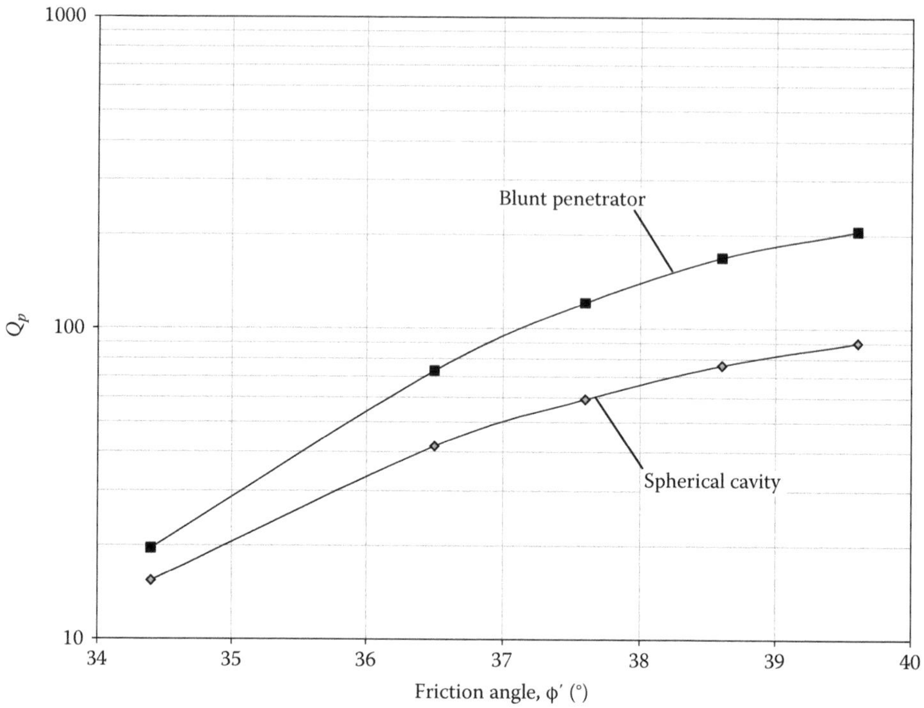

Figure 4.7 Comparison of experimental spherical cavity limit pressure with penetration resistance of blunt indenter. (After Ladanyi, B. and Roy, M., Point resistance of piles in sand. *Proceedings Ninth Southeast Asian Geotechnical Conference*, Bangkok, Thailand, 1987. With permission from Professor B. Ladanyi.)

calibration chamber (CC) studies. The CC is essentially a large triaxial cell, typically about 1 m in diameter, in which sand prepared to a known density can be loaded to a given stress level. A CPT is then carried out in the sand. Figure 4.8 shows a picture and a general arrangement of a CC.

CC studies involve carefully placing sand in the test chamber so that it is, to the maximum practical extent, of constant and a known density. Dry pluviation is commonly used. Then, a desired stress regime is applied in the same way as consolidating a triaxial sample. Vertical and radial stresses are independently controlled so that the effect of geostatic stress ratio can be evaluated. Setting up a sample of sand in the CC is not a trivial undertaking; over 2 tonnes of sand is involved. The CPT is pushed into the sand in the CC, just as in the field, with CPT data recorded in the usual way. Figure 4.9 shows examples of measured CPT parameters and the corresponding density measured throughout the various layers of a CC (after Been et al., 1987b). CC studies comprise many such tests over a range of densities, applying a range of confining stress levels. From this testing, it is possible to develop a mapping between the CPT penetration resistance q_c, initial confining stress, geostatic stress ratio and soil state (as measured by ψ, void ratio or relative density).

CCs are a finite size and, although many times the diameter of the CPT itself, the reality is that the far field from the CPT affects the penetration resistance. The CPT does not simply test soil a few tens of millimetres adjacent to the penetrometer. This is unsurprising, as the cavity expansion theory shows the elastic far field is important.

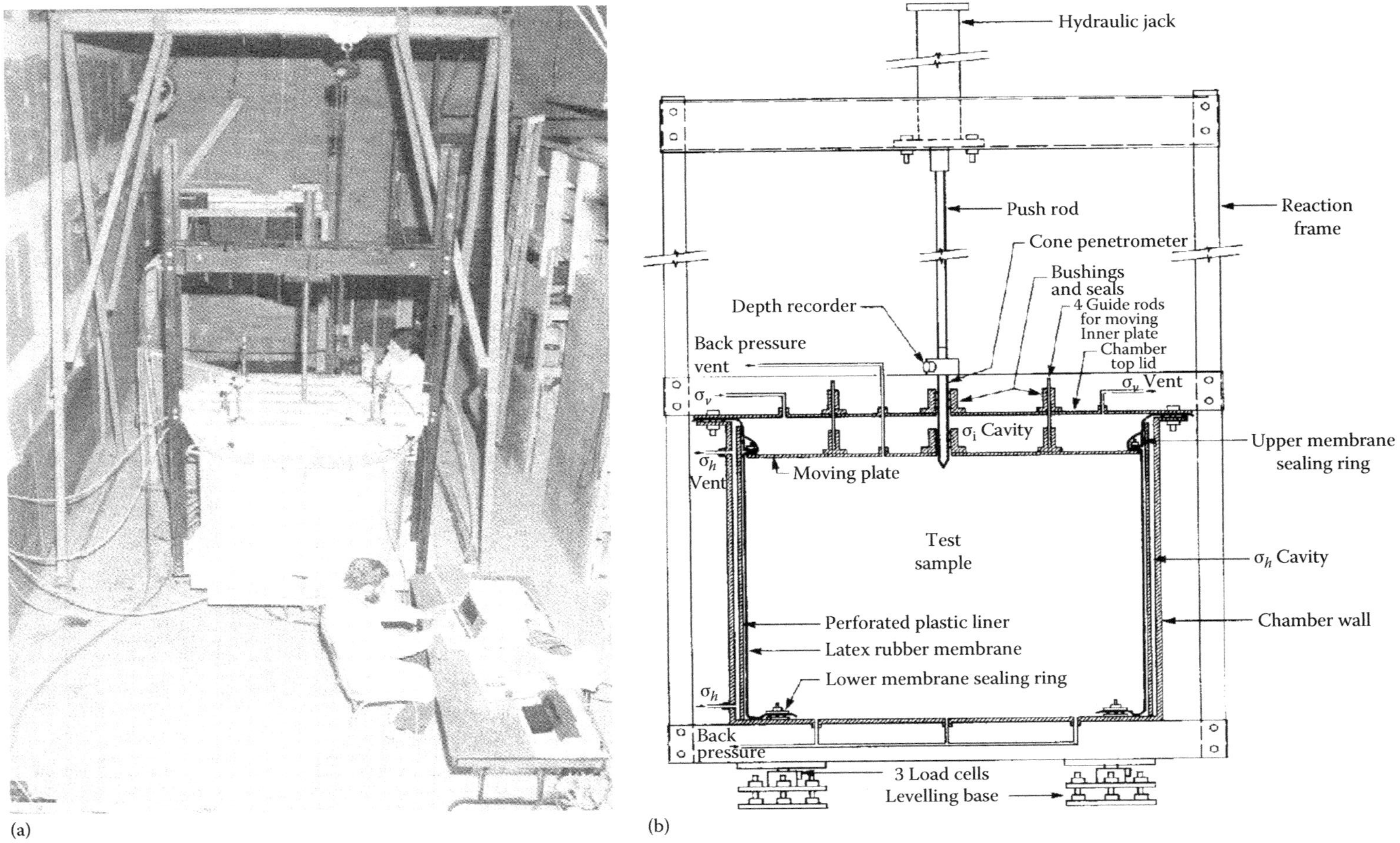

Figure 4.8 Example of CPT calibration chamber: (a) view of chamber and (b) general arrangement. (From Been, K. et al., *Can. Geotech. J.*, 24(4), 601, 1987b. With permission from the NRC, Ottawa, Ontario, Canada.)

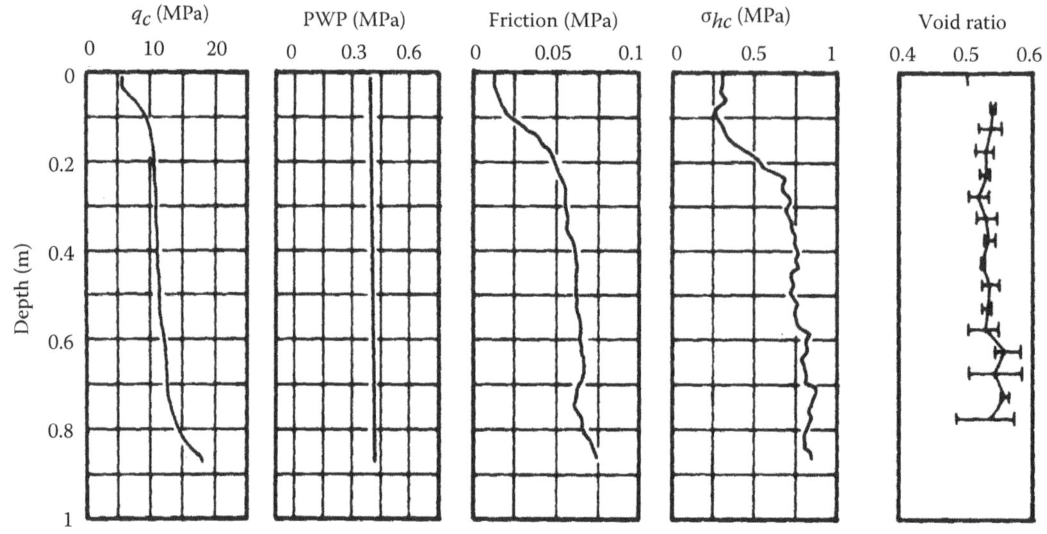

Figure 4.9 Example of CPT chamber test data. (Erksak sand, after Been, K. et al., *Can. Geotech. J.*, 24(4), 601, 1987b. With permission from the NRC, Ottawa, Ontario, Canada.)

Practically, this means chamber boundaries do influence the CPT, especially in dense, dilatant sands (Parkin et al., 1980; Parkin and Lunne, 1982; Ghionna, 1984). Chamber size corrections are, therefore, applied to the measured q_c data, depending on the diameter ratio between the cone and chamber, the density of the sand and the chamber boundary condition (constant stress or zero volume change). Correction factors are discussed in Section 4.4.3.

As will be appreciated, there is quite a bit of experimental scatter in chamber data because of difficulties in controlling (and measuring) sand density in the chamber as well as the influence of top and bottom boundaries. Picking the q_c values themselves off the traces also involves judgement, as it is rare to get a perfectly constant q_c value with penetration into the chamber. Nevertheless, reasonably constrained mappings can be developed given due diligence. These data become, in effect, the gold standard as it represents full-scale direct calibration of the CPT in sands.

4.4.2 Test programs and available data

CC studies have been carried out for some 35 years and in several laboratories, both academic and commercial. To date, at least 13 sands have been tested for which there is also a critical state line. Several groups have tested the same sand in different chambers, allowing comparison of equipment and repeatability. Table 4.2 summarizes CC studies of the CPT that have been published and for which CSLs are available (Table 2.1). Sands have been tested in both the normally consolidated (NC) and over-consolidated (OC) states to arrive at an indication of the effect of stress history on penetration resistance. Figure 4.10 shows the grain size distribution of the tested sands. Critical state parameters for CC sands were determined through the standard method of multiple undrained triaxial compression tests on each, while Figure 2.34 shows the CSL estimated for the sands using the semi-log idealization.

Table 4.2 Summary of CPT calibration chamber studies for which CSL is also known

References	*Sand tested*	*Tests carried out*	*Remarks*
Tringale (1983)	Monterey # 0	9 NC	
Huntsman (1985)	Monterey # 0	22 NC	Horiz. stress CPT
Baldi et al. (1982)	Ticino 1, 2, 4	83 NC, 63 OC	Each of the Ticino sands appears to be slightly different
Baldi et al. (1986)	Ticino 1, 2, 4	22 NC, 34 OC	
Golder Associates	Ticino 9	10 NC	
Parkin et al. (1980)	Hokksund	40 NC, 21 OC	
Baldi et al. (1986)	Hokksund	9 NC, 40 OC	
Lunne (1986)	Hokksund	30 NC, 20 OC	
Harman (1976)	Ottawa	30 NC	
Harman (1976)	Hilton mines	20 NC	
Lhuer (1976)	Reid Bedford	17 NC	
Been et al. (1987b)	Erksak	14 NC	Horiz. stress CPT
Brandon et al. (1990)	Yatesville silty sand	4 NC, 1 OC	42% fines content
Golder Associates	Syncrude Tailings	7 NC, 1 OC	
Fioravante et al. (1991)	Toyoura	23 NC, 5 OC	
Shen and Lee (1995)	West Kowloon	18 NC	Two tests on each sample
Lee (2001)	Chek Lap Kok	10 NC	Two tests on each sample
Pournaghiazar et al. (2012)	Sydney Sand	8 NC	

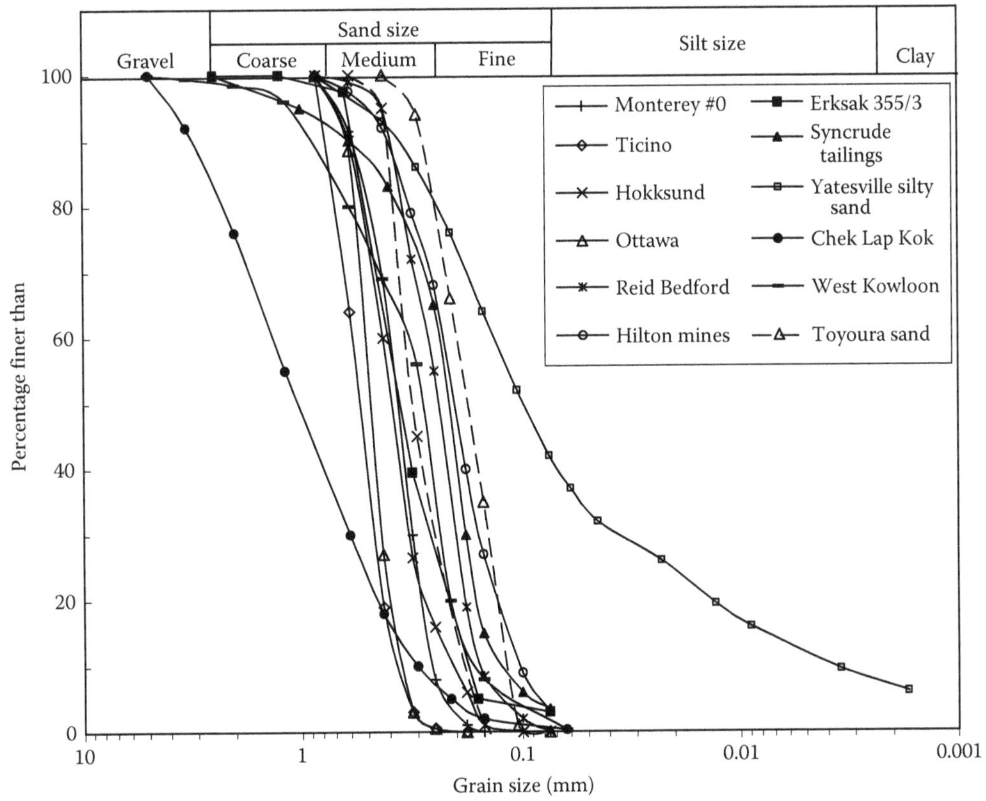

Figure 4.10 Grain size distribution curves for sands tested in calibration chambers.

Test data from these CC studies are given as a downloadable file *CPT_CC_data.xls* and in Appendix E as tables. There is one file for each of the sands tested. Each file gives source and reference information at the top of the main worksheet and then tabulates the individual test numbers with reported density, void ratio, initial stress conditions and q_c value. Added to these measured parameters are the computed chamber boundary correction factor, the Q_p determined using this boundary correction and ψ (calculated from the sample density and stress level). The information is provided because some of the references are obscure, while others are effectively only available from the records of Golder Associates.

4.4.3 Calibration chamber size effects

Correction factors for chamber size were initially established experimentally using cones and chambers with different combinations of diameters. Cones of different diameter (smaller than the standard 36.7 mm diameter) were used and normalized tip resistance plotted against the diameter ratio (CC diameter/cone diameter). The resistance at a large diameter ratio, typically 48–60, was taken as representative of the 'field' tip resistance. Loose sands showed minimal chamber size effects, while very dense sands showed a considerable effect. Been et al. (1986) provide useful summary of these empirical corrections and converted the data into state parameter terms.

The methodology to determine ψ from the CPT is largely based on these empirical corrections; however, considerable research has been carried out on chamber boundary effects since 1986. Analytical methods are of particular interest since the experimental data sets are small and show considerable scatter. Schnaid and Houlsby (1991) examined the potential size effects considering limit pressures in cylindrical cavity expansion and showed that cavity expansion theory was consistent with the experimental data. Salgado et al. (1997, 1998) describe a similar cavity expansion model including stress–dilatancy and calculate how the size effects vary as a function of vertical stress and relative density for Ticino sand. Hsu and Huang (1999) describe an innovative 'field simulator' in which the chamber lateral boundary stress is varied based on numerical simulations to represent field conditions. The focus of these studies was on the horizontal stress conditions at the chamber boundary, but Wesley (2002) shows (by rather simple equilibrium considerations) that the vertical stress arising from the downward force of the penetrometer may provide the best explanation for changes in cone resistance with chamber size. More recently, Pournaghiazar et al. (2012) elegantly pull together most of the aforementioned ideas in a single approach. Their method first corrects the CC vertical stress conditions as suggested by Wesley, and then uses drained spherical cavity expansion theory to compute the ratio between the chamber and field penetration resistance. The corrections using spherical cavity expansion corrected for vertical stress effects show a good agreement with physical tests, but are generally slightly less than empirical methods and better than those computed using cylindrical cavity expansion theory.

In summary, while CC data are indeed the gold standard for CPT inversion, size effects and boundary conditions are an important consideration when using the data. The corrections depend on the density, vertical and horizontal stress conditions, intrinsic properties of the sand, shear modulus, CC and cone diameter, and how boundary conditions are applied and controlled in the chamber. Correction factors applied to cone resistances used in this book are included with the data in Appendix E. These are state parameter–based correction factors, which we have used if the published sources do not provide corrections, and are consistent with current knowledge based on cavity expansion modelling although they are difficult to compare directly. The corrections could be increased or decreased by a smidgen, but there remains an unavoidable, although small, residual uncertainty.

4.5 STRESS NORMALIZATION

4.5.1 Effect of vertical and horizontal stresses

A striking feature of the CPT in sands is that the penetration resistance is a strong function of the stress level. This is illustrated in Figure 4.11, showing trend lines for the relationship between q_c and σ'_v from CC data on three sands (Hilton mines, Ticino, and Monterey) at two relative densities each. The Ticino sand data in Figure 4.11 are from the widely known and referenced paper by Baldi et al. (1982).

Because increasing stress level suppresses dilatancy, there is something less than a linear increase in q_c with σ'_v for a given relative density. As the state parameter gives a maximum dilatancy that is independent of mean stress level, one would expect a simpler trend of q_c with stress level for constant state ψ. This turns out to be true, as can be seen from Figure 4.12, which shows the data for Monterey sand on a q_c versus p' plot with individual values of ψ indicated. Simple radial trend lines fit the data in Figure 4.12 reasonably well,

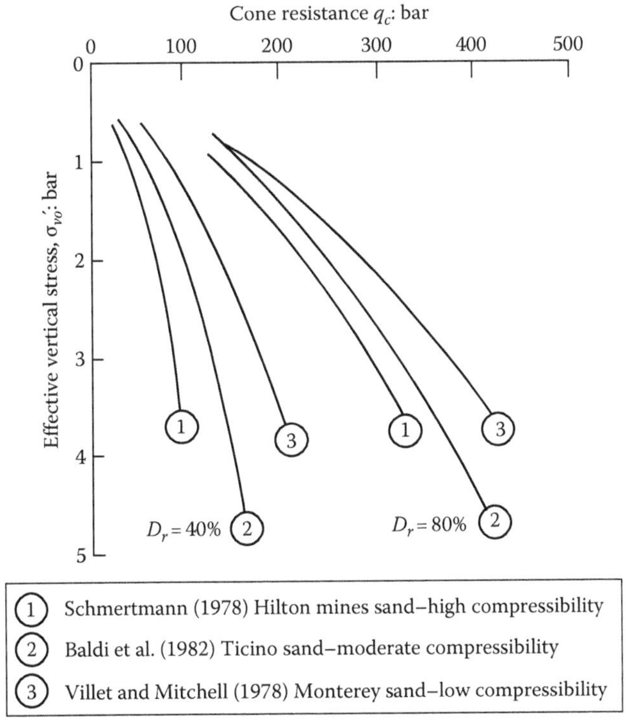

Figure 4.11 CPT resistance versus relative density for three sands. (After Robertson, P.K. and Campanella, R.G., *Can. Geotech. J.*, 20(4), 718, 1983. With permission from the NRC, Ottawa, Ontario, Canada.)

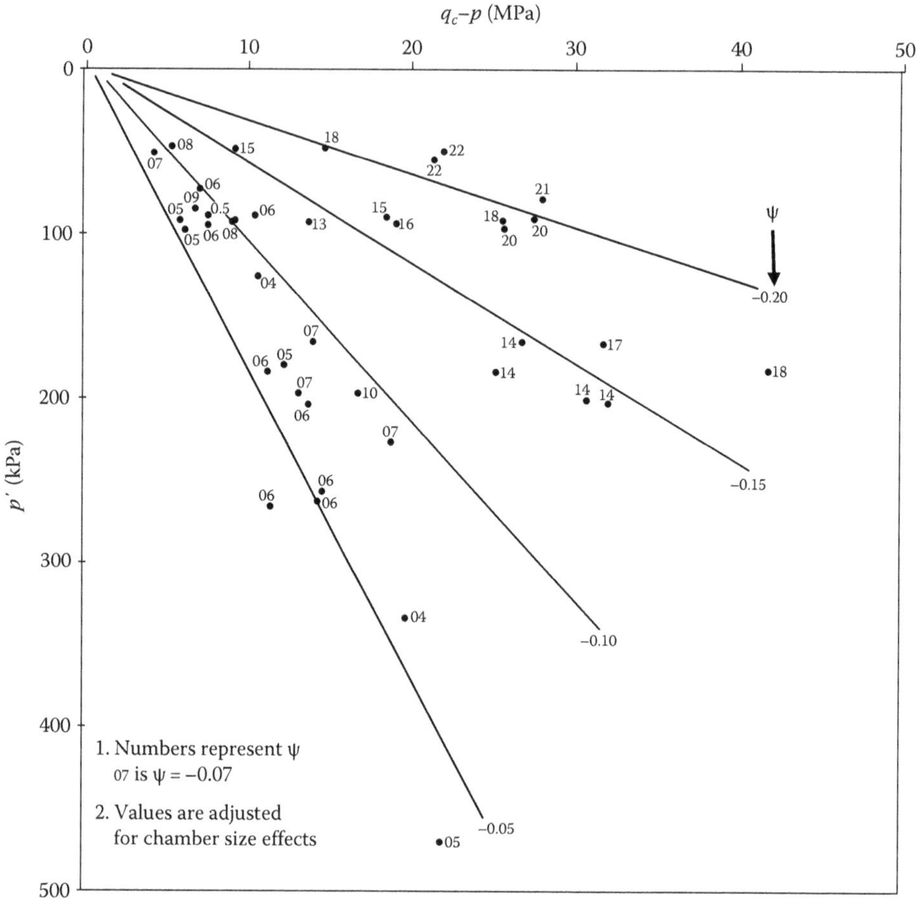

Figure 4.12 CPT resistance calibration for Monterey No. 0 sand. (Test data from Villet, W.C.B., Acoustic emissions during the static penetration of soils, PhD thesis, University of California at Berkeley, Berkeley, CA, 1981; Graph from Been, K. et al., *Géotechnique*, 36(2), 239, 1986. With permission from the Institution of Civil Engineers, London, U.K.)

indicating that CPT resistance is, broadly speaking, directly proportional to stress level at constant state parameter.

Given that stress level so dominates q_c, the first step in evaluating CPT data is to remove the effect of stress level. Before that, however, it is necessary to deal with the differing effects of vertical and horizontal stresses. Early experimental studies (Clayton et al., 1985) determined the relative influence of horizontal and vertical effective stresses by independently varying each in a series of CC-like studies using a dynamic penetrometer. Figure 4.13 shows their elegant results, from which it can be seen that the horizontal stress has twice the influence of the vertical stress. This immediately implies that the correct stress in normalizing penetration data is the mean effective stress, $p' = (\sigma'_v + 2\sigma'_h)/3$, as the two stresses are in exactly the required proportion.

The requirement for mean stress rather than vertical stress is certainly inconvenient. Generally, it becomes easiest to use the identity that $p' = \sigma'_v(1 + 2K_o)/3$, as this separates the

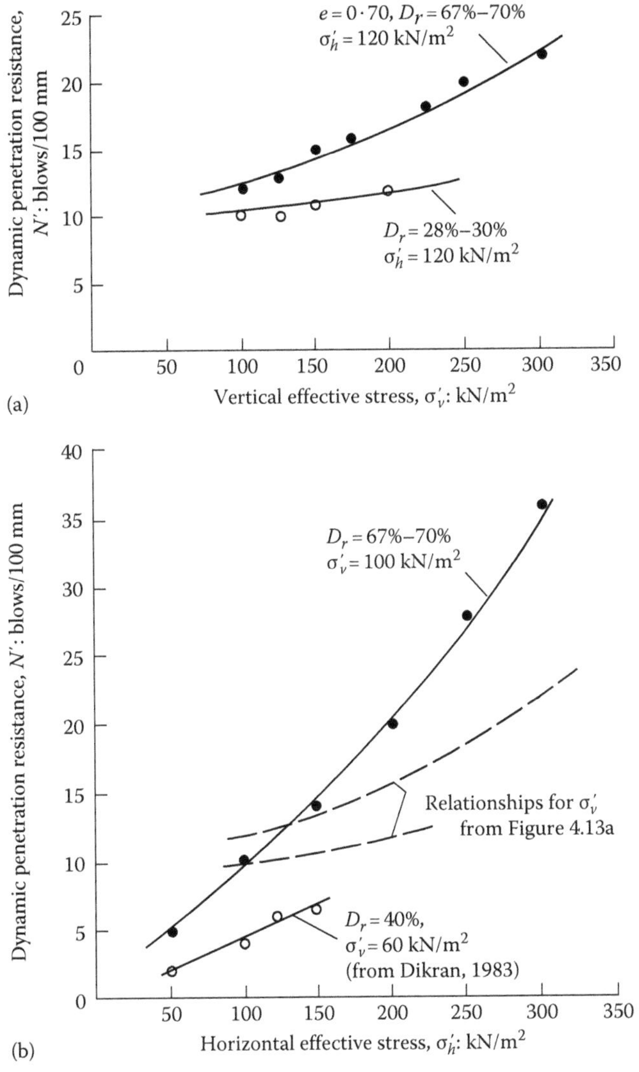

Figure 4.13 Effect of stress on penetration resistance in normally consolidated sand: (a) vertical stress and (b) horizontal stress. (From Clayton, C.R.I. et al., *Géotechnique*, 35(1), 19, 1985. With permission from the Institution of Civil Engineers, London, U.K.)

effect of stress level from the uncertainty in the geostatic stress ratio. Geostatic stress ratio and its measurement will be discussed after developing the framework for CPT evaluation based on mean stress.

4.5.2 Reference condition approach

The approach to normalizing the effect of stress on penetration test data developed for the SPT is the reference condition approach, in which penetration data at one stress level are mapped to an equivalent penetration resistance at a reference stress level. Commonly, the reference is taken as an effective stress σ_{ref} = 100 kPa (≈ 1 tsf or 1 atmosphere at sea level). Penetration resistances mapped back to this reference level are usually subscripted by a '1'; thus, $N_{(60)}$ becomes $N_{1(60)}$ for the SPT and q_c similarly becomes q_{c1} for the CPT. The mapping for the SPT is usually quoted in the form

$$N_1 = C_N N \tag{4.9}$$

with C_N being the mapping function. The key concept is obviously the meaning of equivalent. Equivalent in its original use of the reference condition approach meant the same relative density (e.g. see Figure 4.11; C_N is the function for a constant D_r line). This basis for original use has become lost, and many papers over the last two decades commonly used $N_{1(60)}$ without regard to the implicit constant void ratio condition.

The first work recognizing the effect of stress level on the determination of relative density was a study by Gibbs and Holtz (1957). Subsequently, Marcuson and Bieganousky (1977) reported on a reasonably comprehensive set of tests using the SPT in a relatively crude CC-like arrangement (there was no provision for control or measurement of horizontal stress), and their results are illustrated in Figure 4.14. Notice that the C_N function is different for a relative density of 40% compared to 60% or 80% for the same sand.

Various approximations for C_N were proposed from curve fitting the test data. However, the substantive next step was a comparative study by Liao and Whitman (1986) who

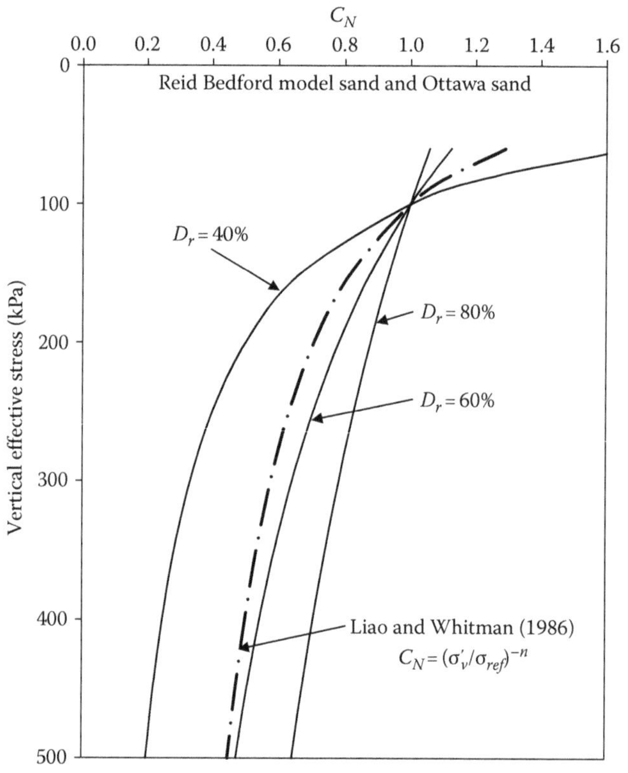

Figure 4.14 Experimentally determined C_N functions for Reid Bedford and Ottawa sand by Marcuson and Bieganowski (1977) and recommended C_N function by Liao and Whitman (1986).

suggested that there was not a great deal of difference between the various functions proposed for C_N and that a reasonable average relationship was the simple (and now widely used) equation

$$C_N = \left(\frac{\sigma'_v}{\sigma_{ref}}\right)^{-n} \quad \text{with } n = 0.5 \text{ for SPT} \tag{4.10}$$

This same approach has been widely applied to the CPT for addressing liquefaction problems. In the case of the CPT, the parameters are q_c, q_{c1} and C_Q, where $C_Q = C_N$ with a range for the exponent in $0.5 < n < 1$ that depends on the soil type.

There are many variations of the reference stress approach, for example Idriss and Boulanger (2004), Moss et al. (2006), Maki (2012) and Robertson (2012). These appear to provide small improvements in particular applications, for particular soils or for different state measures. A recent trend is to vary the stress-level exponent with the soil type, from $n = 0.5$ in sands to $n = 1$ in clay-like soils, but this then blurs the concept that the resistance normalized to the reference condition is an indication of relative density.

If done appropriately, the reference condition approach can indeed be considered dimensionless, since at a fundamental level it is indicative of a dimensionless parameter: void ratio or relative density. The difficulty with the reference condition approach is that relative density is not a good indicator of soil behaviour. Different soils do not behave the same way at the same relative density. What is more (and worse), because soil behaviour depends on the confining stress, the concept of mapping back to a reference stress loses a clear understanding of what is going on at the actual stress level in the target soil horizon. Behaviours are assessed at a stress which is not the in-situ stress, with additional adjustments then needed to map the strengths (e.g. cyclic behaviour in seismic analysis) estimated from the penetration test data back to the in-situ stress condition.

A further difficulty with the reference condition approach, as applied to CPT data, is the use of a variable exponent depending on soil type. While there is no argument that the two limits for the exponent, $n \approx 0.5$ for sands and $n \approx 1.0$ for clays, are reasonable, it is a huge leap to assume that silts would be intermediate and that there might be a gradual transition from one to the other. The simple cavity expansion model in NAMC soil indicates that the factor affecting the exponent is elastic shear modulus, not the soil type itself.

Overall, the widely used reference stress approach is a confusing framework requiring transposition of measured values to a situation which is not the in-situ condition, followed by correction of the computed strength back to the in-situ stress level using poorly understood factors. The confusion of the reference stress approach is further compounded because the known effect of horizontal stress is ignored, as are the effects of soil properties (e.g. compressibility). The reference stress approach also does not lend itself to a theoretically based framework, such as that developed from the Carter et al. cavity expansion analysis presented earlier, so improving on the poorly understood adjustment factors is difficult. Is the reference stress approach necessary? No. More rigorous results, which are readily understood within standard mechanics, are easily achieved using the state parameter with a simple linear stress normalization.

4.5.3 Linear stress normalization

The alternative to reference stress approach is to express the penetration resistance as a dimensionless number. This was introduced earlier and is simply the penetration resistance normalized to mean effective stress, Q_p or $(q_c - p_o)/p'_o \approx q_c/p'_o$. Radial lines from the origin on a plot of $(q_c - p)$ versus p' represent a constant Q_p, and Figure 4.12 shows a reasonable correspondence between radial lines and lines of constant ψ. This idea forms the basis for obtaining ψ from CPT data, which is explored in the following section.

4.6 DETERMINING Ψ FROM CPT

4.6.1 Original method

The methodology for using the CPT to determine ψ was put forward in two papers in the mid-1980s (Been et al., 1986, 1987c). Both papers were based on the then existing database of CC tests and the work became one of obtaining samples of the various chamber test sands to determine their respective CSLs (using the methods described in Chapter 2). With that done, it was then straightforward to reprocess the CC data to develop a method for the determination of ψ.

In the first paper, which used mainly the Monterey sand data, work concentrated on standardizing chamber size corrections to develop the form of plot as shown in Figure 4.12. Since radial lines approximated lines of constant state, stress level appeared to be removed when the data were viewed in terms of ψ. Plotting Q_p against ψ gave the desired result of a straightforward relationship between Q_p and ψ (Figure 4.15), with no further need to consider confining stress effects. The trend line in Figure 4.15 nicely captures all the data with rather little scatter. The trend line has a very simple exponential form:

$$Q_p = k\exp(-m\psi) \qquad (4.7\text{bis})$$

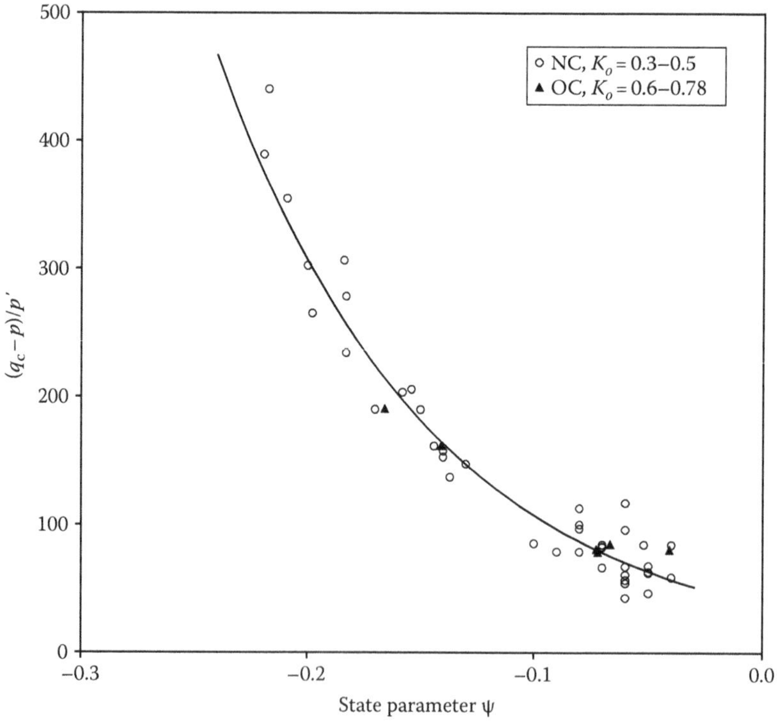

Figure 4.15 Dimensionless CPT resistance versus state parameter for Monterey sand (data from Figure 4.12).

As an historical note, the evaluation of CPT data for ψ was concurrent with the cavity expansion analysis pursued by Carter et al. (1986). Thus, while (4.7) can be developed either way, the development of (4.7) from the data did not follow Wroth's strictures faithfully (which also had not been published at the time), and was based solely on dimensional considerations.

The obvious question that followed from Figure 4.15 was whether the behaviour of other sands is the same as Monterey sand. The second paper took the established methodology and worked through the remaining chamber sands, giving the results shown in Figure 4.16, where

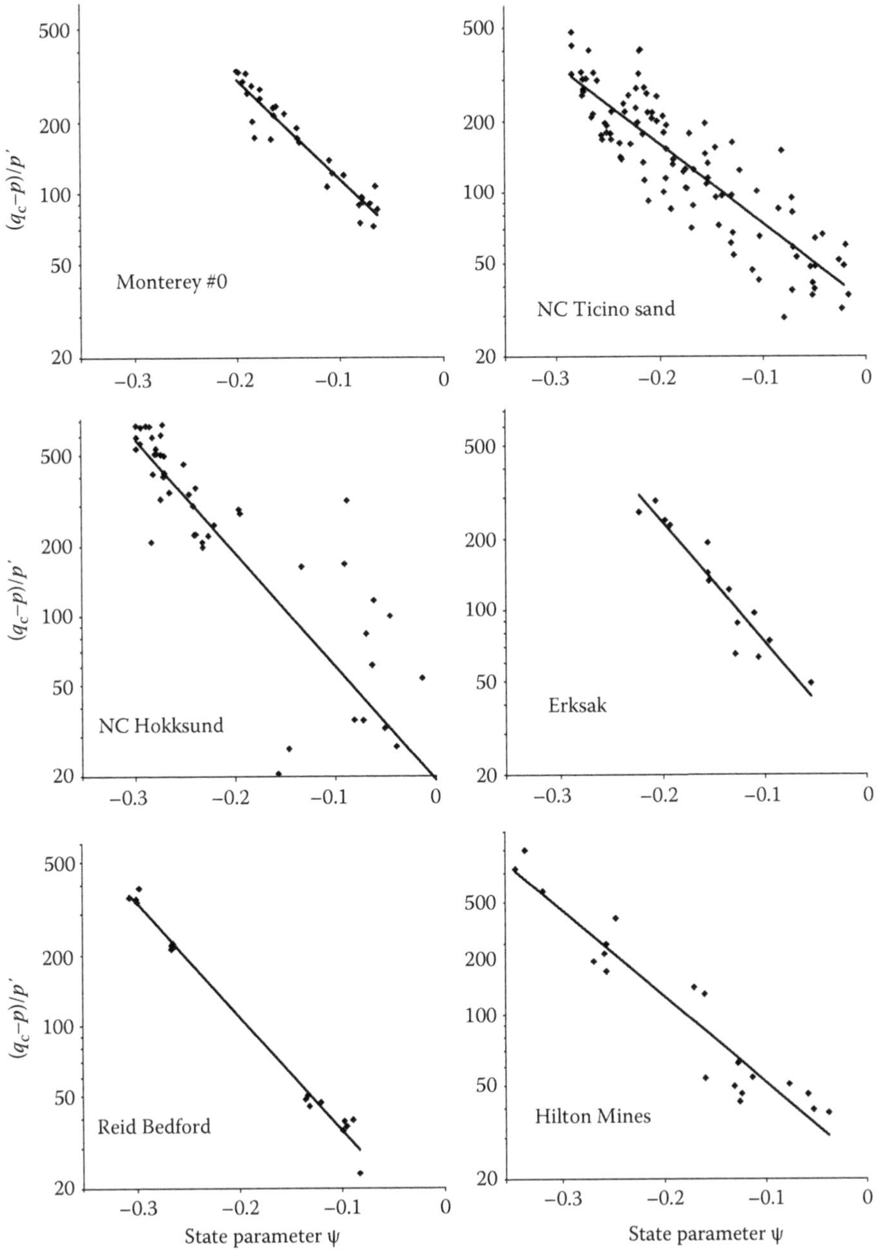

Figure 4.16 Normalized Q_p − ψ relationships from calibration chamber studies (NC = normally consolidated; OC = over-consolidated). *(Continued)*

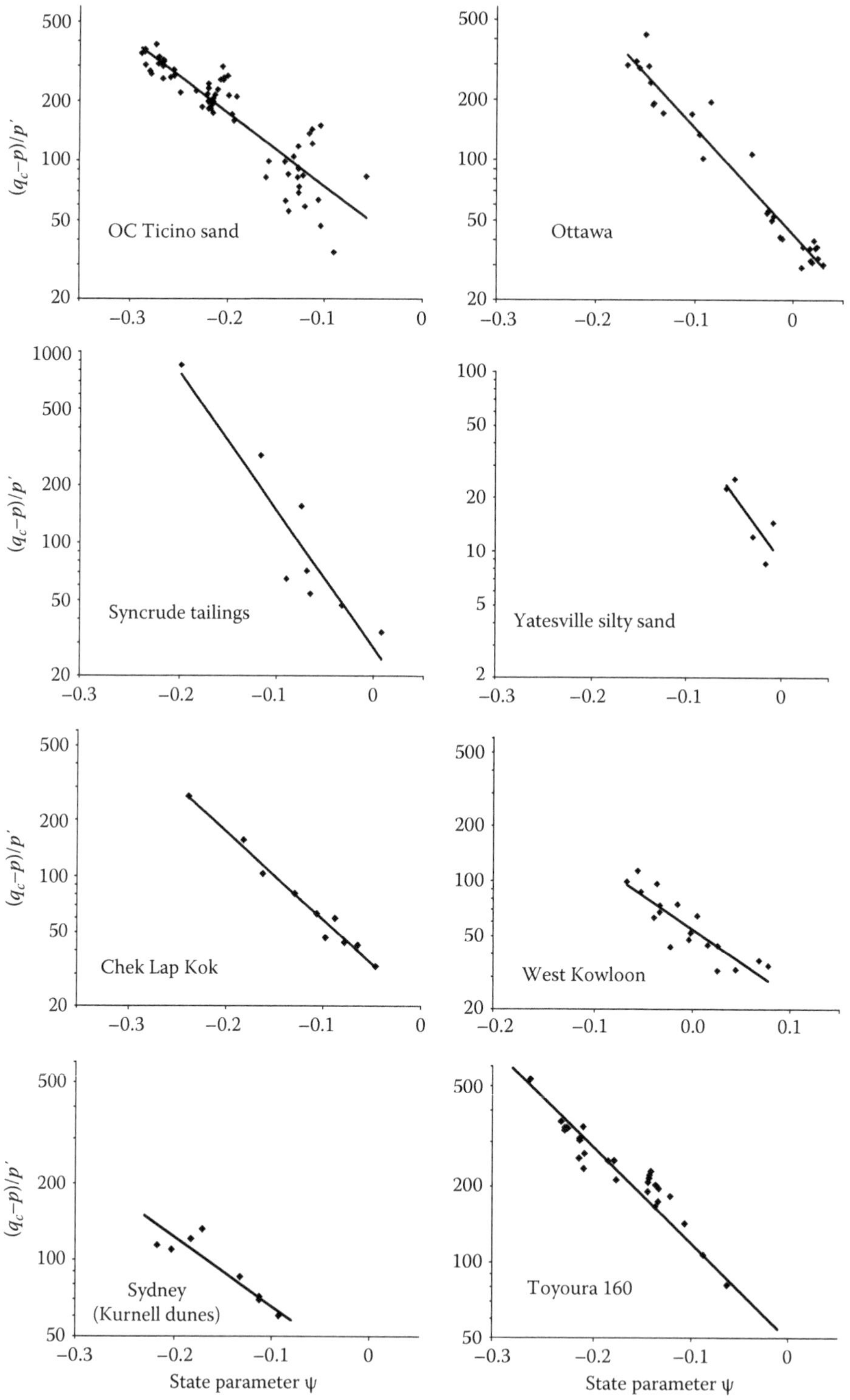

Figure 4.16 (Continued) Normalized Q_p − ψ relationships from calibration chamber studies (NC = normally consolidated; OC = over-consolidated).

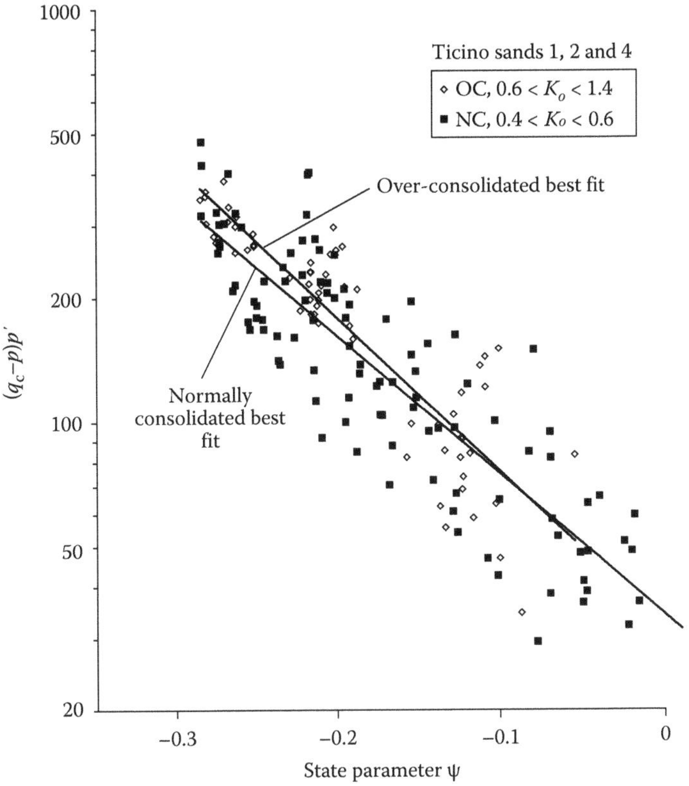

Figure 4.17 Normalized CPT resistance of normally consolidated and over-consolidated Ticino sand.

individual data can be seen. Since the paper was published, additional chamber testing has produced data on seven more sands: Erksak (Been et al., 1987b), Syncrude Tailings (Golder Associates files), Yatesville silty sand (Brandon et al., 1990), Chek Lap Kok (Lee, 2001), West Kowloon (Shen and Lee, 1995), Toyoura (Fioravante et al., 1991) and Sydney sand (Pournaghiazar et al., 2011). The data from these seven sands are also shown in Figure 4.16.

Most of the CC data is for NC sands. Several OC tests were carried out on Ticino and Hokksund sands, in addition to a few tests on Syncrude tailings, Toyoura sand and Yatesville silty sand (Table 4.2). Interestingly, Ticino sand shows very little difference between NC and OC results (Figure 4.17) because the effect of different values of K_o appears to be captured in the mean effective stress normalization. For Hokksund sand, there does appear to be a difference between NC and OC tests, but for loose sands only. This probably reflects the effect of elastic rigidity (which is higher for OC samples, and which is not captured in the stress normalization). The Hokksund data set is, however, not as good as that of Ticino. The tests were carried out at several different laboratories by several different people, and there is a lot of scatter in the data for loose samples. Figure 4.16 is necessarily compressed so as to show the information of the entire CPT calibration database in one glance. But, we do not expect these figures to be used for anything other than a brief visual comparison. The data, and the individual plots shown on Figure 4.16, are provided in the downloadable file *CPT_CC_data.xls* for further evaluation (if desired).

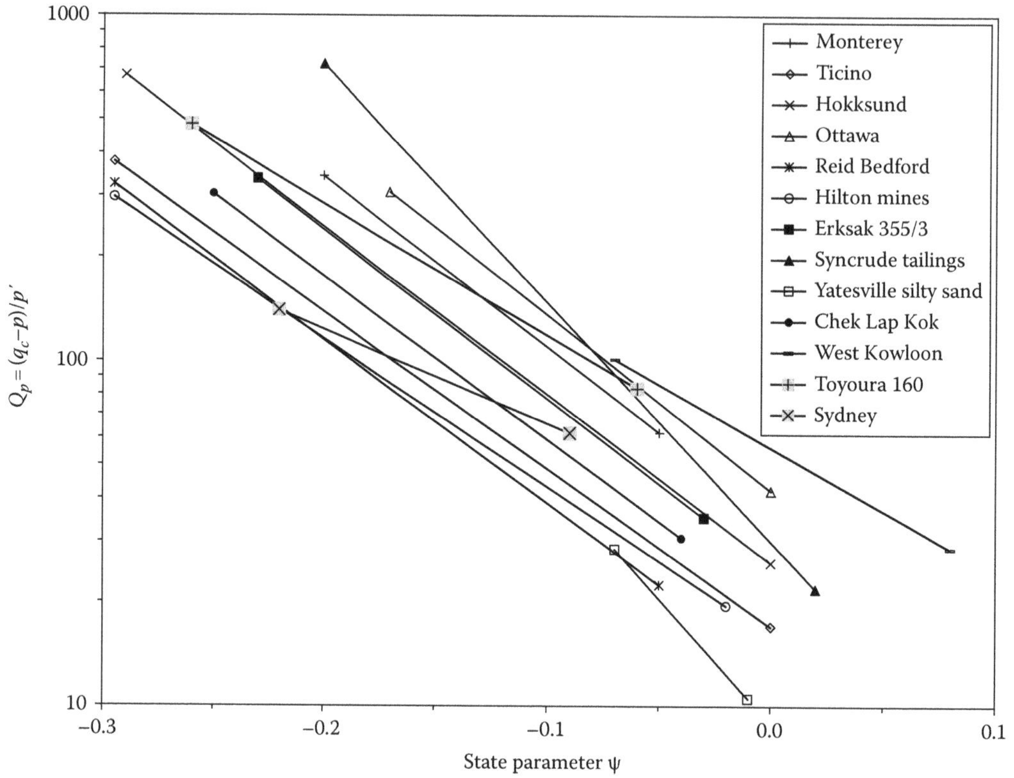

Figure 4.18 Comparison of $Q_p - \psi$ trends for different sands. *Note*: West Kowloon sand critical state line testing was not on exactly same sample as used for CC tests.

Figure 4.18 compares the best-fit trend lines for the various sands. Although the simple relationship between Q_p and sand state ψ fits each data set of Figure 4.16, it is clear from this comparison that the terms *k* and *m* in Equation 4.7 are sand-specific constants and as such are functions of other intrinsic properties of the sands.

Cavity expansion theory using NAMC indicates that the soil rigidity should affect the CPT behaviour parameters *k* and *m*. However, the NAMC model is perfectly plastic and has no concept of hardening, for which there would be a corresponding plastic modulus. Following from the principle of the Cam Clay model (Chapter 3) which states that plastic hardening is proportional to the inverse of the slope of the CSL as given by λ, it was suggested (Been et al., 1987c) that the sand-specific constants *k* and *m* were simple functions of plastic hardening, as plasticity dominates elasticity in large strain shear of sands. Figure 4.19 shows the result of plotting *k*, *m* against λ_{10}. Rather simple relationships are evident with

$$k = 8 + \frac{0.55}{\lambda_{10} - 0.01} \tag{4.11a}$$

$$m = 8.1 - 2.3 \log \lambda_{10} \tag{4.11b}$$

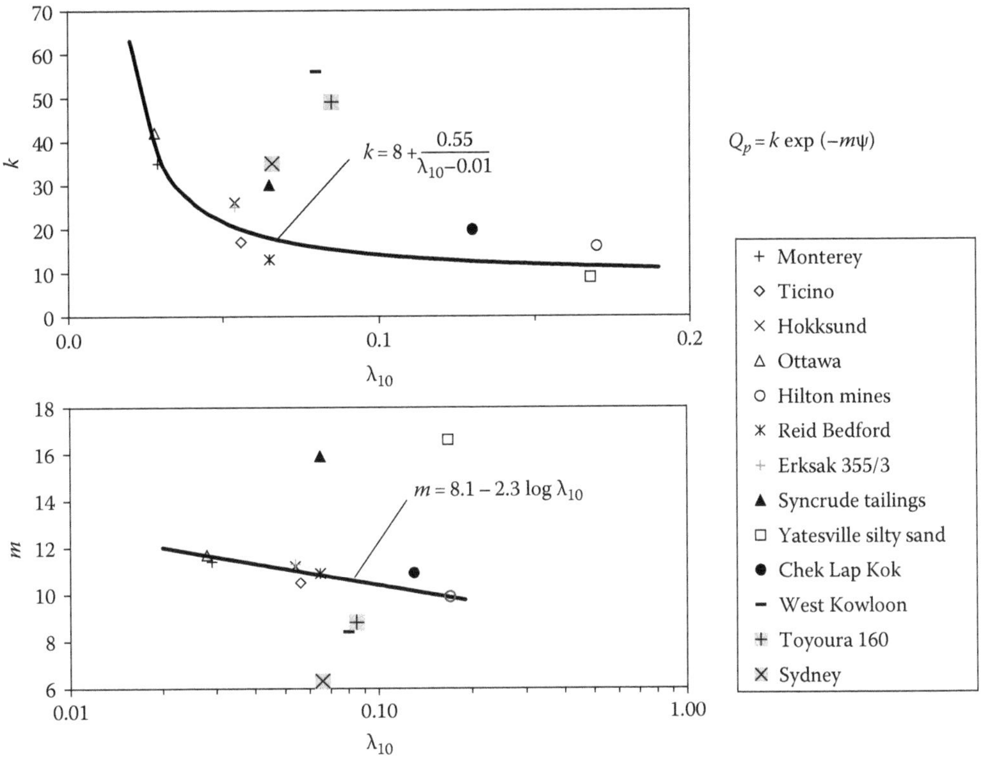

Figure 4.19 CPT inversion parameters versus slope of CSL, λ_{10}.

Equations 4.11a and b are readily combined with Equation 4.7 to give ψ from the measured data:

$$\psi = \frac{-\ln(Q_p/k)}{m} \tag{4.8bis}$$

This gives a method that recovers the initial state of any test in the CC database with an error of less than $\Delta\psi \pm 0.05$ at 90% confidence.

4.6.2 Stress-level bias

Although Equation 4.7 is both delightfully simple and consistent with a proper dimensionless approach, it has been criticized on the grounds that careful examination of the reference chamber test data indicates substantial bias with stress level. At least k, and possibly m, was a function of p' (Sladen, 1989a). If the data on Ticino sand are broken into six groups of approximately equal stress, then trend lines can be drawn through each group of data and these trends differ from each other. Figure 4.20 compares those individual trend lines with each other and with the original trend line drawn through the entire data set. As can be seen from Figure 4.20, for a particular value of Q_p, say $Q_p = 100$, the inferred state at a mean stress of 400 kPa would be approximately $\psi = -0.19$, whereas at a confining stress an order of magnitude less, 40 kPa, the same Q_p would give $\psi = -0.10$;

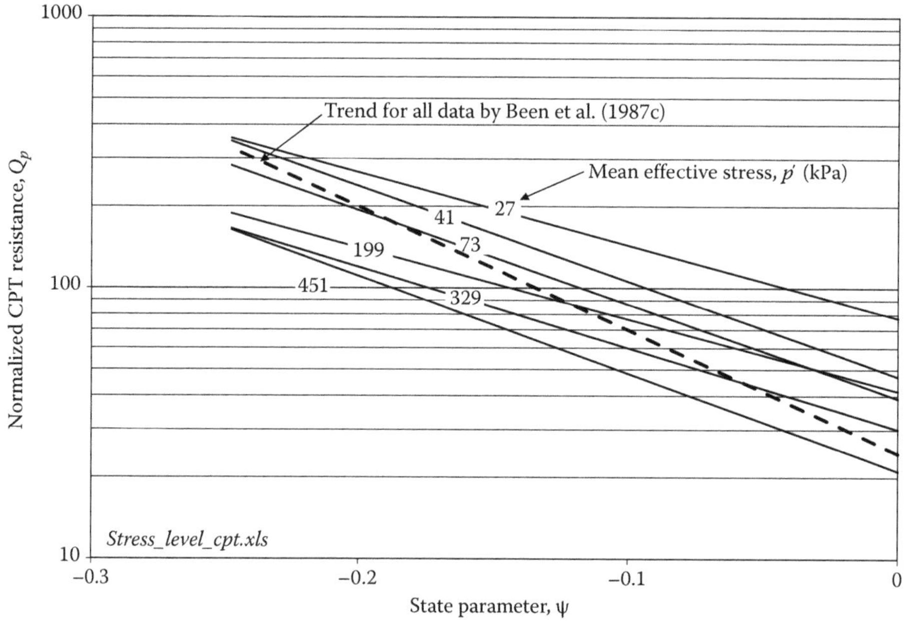

Figure 4.20 Summary of stress-level bias in $Q_p - \psi$ relationship for Ticino sand as suggested. (Reproduced from Sladen, J.A., *Géotechnique*, 39(2), 323, 1989a; Sladen, J.A., *Can. Geotech. J.*, 26(1), 173, 1989b. With permission from the NRC, Ottawa, Ontario, Canada.)

these values can be contrasted with $Q_p = 100$ giving $\psi = -0.14$ if the overall trend is used. Sladen suggested that 'an error in the assessment of in-situ void ratio of more than 0.2, that is more than 50% in terms of relative density, could not confidently be ruled out'. The actual error is $\Delta\psi = (-0.19 - 0.14) = -0.05$ with the 'more than 0.2' misrepresenting the data and inferred trends, but that misrepresentation does not obscure the fact the bias exists and causes an uncertainty in interpreted state parameter of $|\Delta\psi| < 0.05$. Something had been missed.

The existence of a stress-level bias is curious given the dimensionless formulation. At the time of the original work, there were no theoretical methods available to go further, but these became available some 10 years later and showed how to proceed. The key step was the development of the *NorSand* model, which allowed numerical simulations to explore the effect of the various material properties on the CPT resistance.

4.6.3 Simulations with *NorSand*

Because mathematical or numerical modelling of the drained penetration of the CPT is extremely difficult, Equation 4.7 was proposed on dimensional grounds and correlated to chamber test data. As shown earlier, this form of relationship is also suggested by the Carter et al. (1986) cavity expansion solution from which theoretical guidance can be obtained. A large-scale numerical examination of Equation 4.7 was undertaken by Shuttle and Jefferies (1998) using cavity expansion analysis and the *NorSand* model. This then developed into a universal framework for evaluating ψ from the CPT.

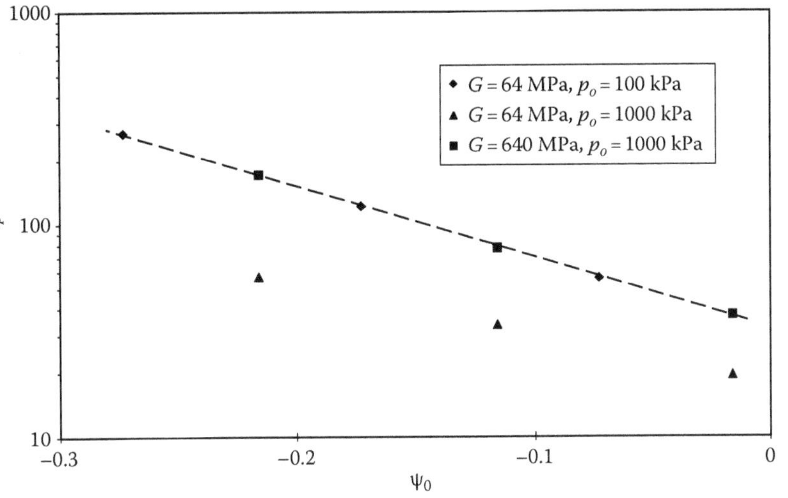

Figure 4.21 Numerical calculation of $Q_p - \psi_o$ relationship for Ticino sand, showing linearity and effect of elastic modulus as cause of stress-level bias. (From Shuttle, D.A. and Jefferies, M.G.: Dimensionless and unbiased CPT interpretation in sand. *Int. J. Numer. Anal. Methods Geomech.* 1998. 22. 351–391. Copyright Wiley-VCH Verlag GmbH & Co. KGaA. Reproduced with permission.)

An example of the results obtained by Shuttle and Jefferies is shown in Figure 4.21 for a single sand (computed using properties of Ticino sand). If Figure 4.21 is examined closely, a very small curvature can be detected in the trend line of the results, but this is much less than second order and for all practical purposes Equation 4.7 is a very good representation of CPT resistance at constant modulus. However, if the mean stress were changed and all other properties kept constant, different Q_p values were computed. Changing G, such that the ratio G/p_o was returned to its original value, put the computed results exactly back on the prior trend line. The explanation here is that the earlier work on determining the state parameter from the CPT missed a dimensionless group, the elastic soil rigidity $I_r = G/p_o$.

A similar conclusion can also be developed from the closed-form solutions in a NAMC soil, as discussed in Section 4.3. Rigidity varies substantially with depth, but was not captured in the $Q_p - \psi$ normalization because I_r scales roughly with the square root of mean stress (in sands) while Q_p was linearly stress normalized. Figure 4.22 shows the consequent variation of k and m for Ticino sand as a function of rigidity.

The elasticity of Ticino sand has been tested at some length, and Figure 4.23 shows the relationship between elastic shear modulus, void ratio and mean stress from Bellotti et al. (1996). The computed penetration resistance of the CPT in Ticino sand, using these data for soil elasticity and the previously computed trends for k and m, is compared with test results from the CC shown in Figure 4.24. The results are separated into the same six intervals of confining stress as used by Sladen (1989a) corresponding to the trend lines in Figure 4.20. The computed trend lines are an excellent fit to the data, with the possible exception of a second-order bias in the highest stress-level data (at 450 kPa). This bias is conservative (for inversion to state parameter) and is thought attributable to grain crushing, which is presently excluded from the *NorSand* model, and is unlikely to have much practical consequence.

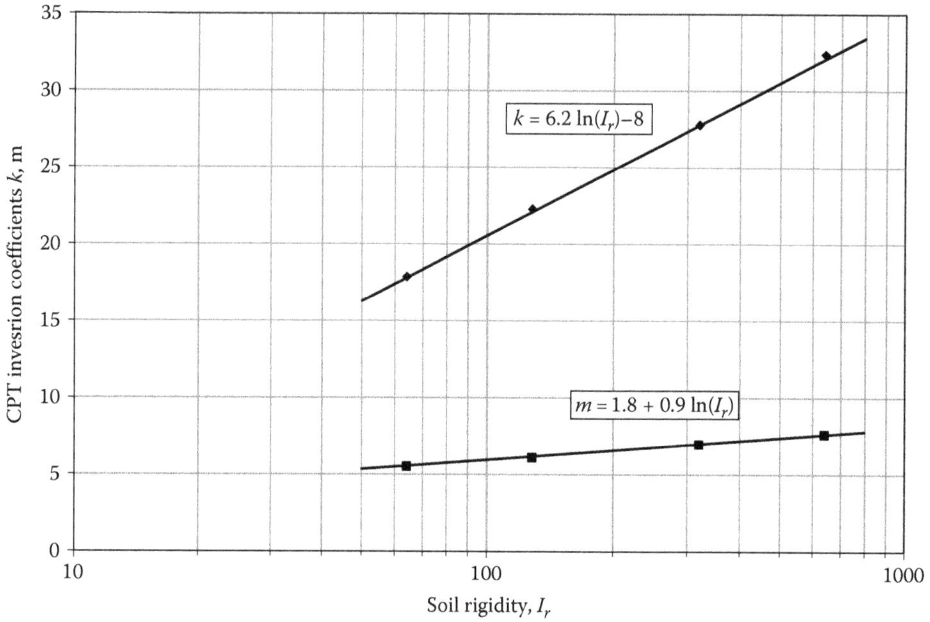

Figure 4.22 Computed effect of I_r on k, m coefficients for Ticino sand. (From Shuttle, D.A. and Jefferies, M.G.: Dimensionless and unbiased CPT interpretation in sand. *Int. J. Numer. Anal. Methods Geomech.* 1998. 22. 351–391. Copyright Wiley-VCH Verlag GmbH & Co. KGaA. Reproduced with permission.)

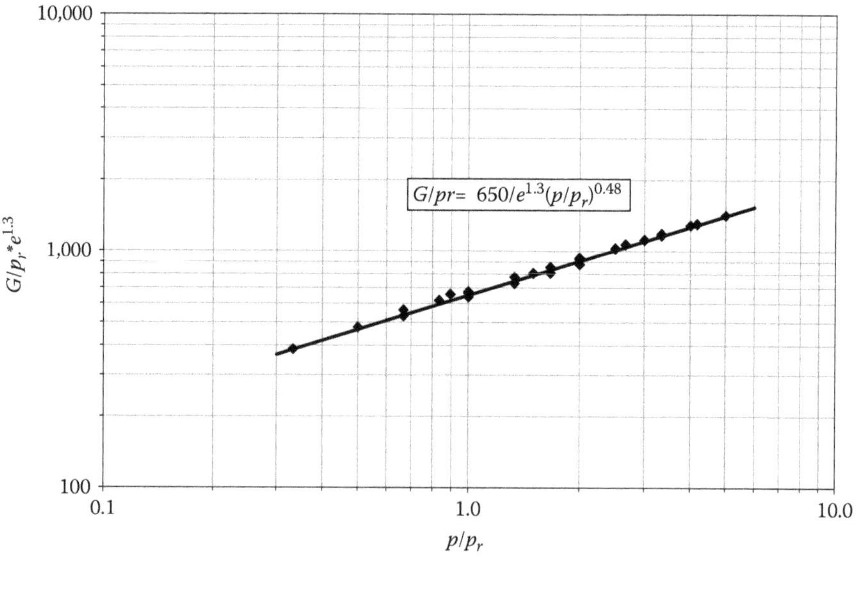

Figure 4.23 Shear modulus of Ticino sand versus confining stress: p_r is a reference stress level, here taken as 100 kPa. (From Shuttle, D.A. and Jefferies, M.G.: Dimensionless and unbiased CPT interpretation in sand. *Int. J. Numer. Anal. Methods Geomech.* 1998. 22. 351–391. Copyright Wiley-VCH Verlag GmbH & Co. KGaA. Reproduced with permission.)

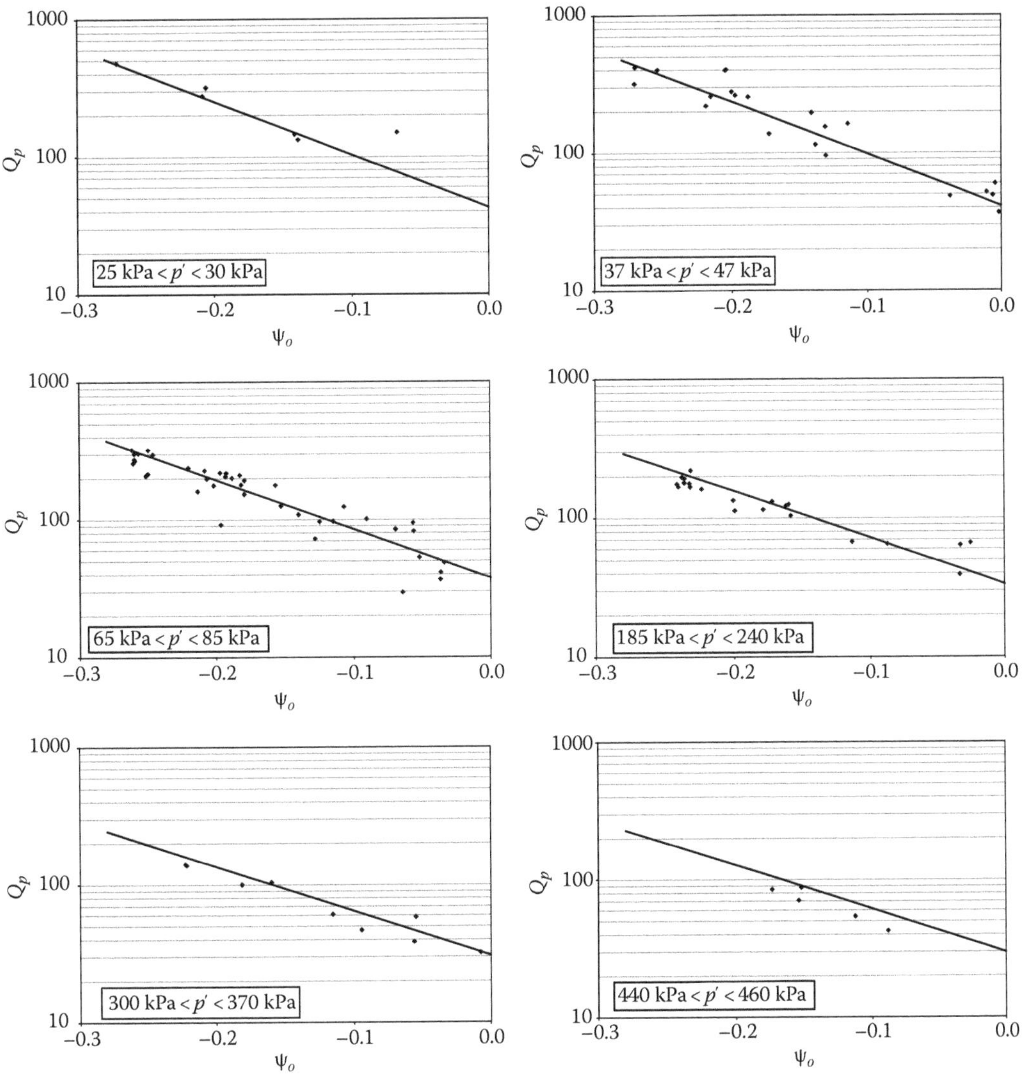

Figure 4.24 Computed Q_p versus ψ_o relationship for Ticino sand, shown as trend lines, compared to individual calibration chamber tests.

4.6.4 Complete framework

In order to evaluate the effect of soil properties in addition to rigidity on the penetration resistance, the simulations of Shuttle and Jefferies were extended to examine separately the key parameters of *NorSand*. These are as follows: M is the critical friction ratio, H is the plastic hardening modulus, N is a scaling parameter for the stress–dilatancy relationship and λ is the slope of the critical state locus in void ratio $\ln(p')$ space. All the properties are dimensionless. Figure 4.25 shows the results of simulations in which these properties are systematically varied one at a time around a central base case. In addition to the effect of rigidity, both intercept k and slope m are strong functions of the plastic hardening modulus H, as well as the critical friction ratio M. There was a much weaker influence of N and the soil compressibility λ. Poisson's ratio had essentially no effect.

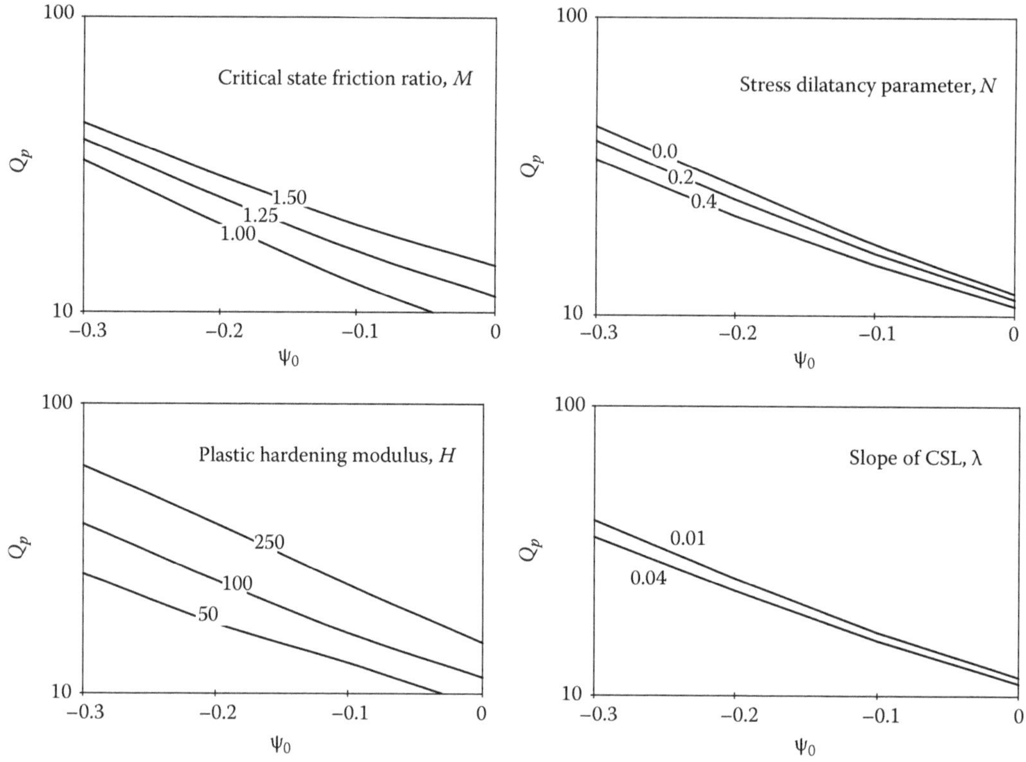

Figure 4.25 Effect of soil properties on spherical cavity expansion pressure ratio. (From Shuttle, D.A. and Jefferies, M.G.: Dimensionless and unbiased CPT interpretation in sand. *Int. J. Numer. Anal. Methods Geomech.* 1998. 22. 351–391. Copyright Wiley-VCH Verlag GmbH & Co. KGaA. Reproduced with permission.)

To determine the most accurate interpretation of the CPT in any soil, the methodology outlined by Shuttle and Jefferies requires detailed numerical simulations to ascertain the values of k and m as a function of rigidity and soil properties. This is fairly time consuming. However, they offered an approximate general inversion obtained by fitting trend lines to the offset of the computed results presented in Figure 4.25. The approximate inversion loses some accuracy, but is closed form and readily computable. The approximate inverse form is

$$k = \left(f_1\left(\frac{G}{p_o}\right) f_2(M) f_3(N) f_4(H) f_5(\lambda) f_6(\nu) \right)^{1.45} \tag{4.12a}$$

$$m = 1.45 f_7\left(\frac{G}{p_o}\right) f_8(M) f_9(N) f_{10}(H) f_{11}(\lambda) f_{12}(\nu) \tag{4.12b}$$

where the fitted functions f_1–f_{12} are given in Table 4.3. The performance of the proposed general inversion was verified by taking 10 sets of randomly generated soil properties/states and computing the Q_p value using the full numerical procedure. This computed Q_p was then input to the general inversion to recover the estimated value of ψ. Figure 4.26 compares true

Table 4.3 Approximate expressions for general inverse form $\psi = f(Q_p)$

Function	*Approximation*
$f_1(G/p_0)$	$3.79 + 1.12 \ln(G/p')$
$f_2(M)$	$1 + 1.06\ (M - 1.25)$
$f_3(N)$	$1 - 0.30\ (N - 0.2)$
$f_4(H)$	$(H/100)^{0.326}$
$f_5(\lambda)$	$1 - 1.55\ (\lambda - 0.01)$
$f_6(\nu)$	Unity
$f_7(G/p_0)$	$1.04 + 0.46 \ln(G/p')$
$f_8(M)$	$1 - 0.40\ (M - 1.25)$
$f_9(N)$	$1 - 0.30\ (N - 0.2)$
$f_{10}(H)$	$(H/100)^{0.15}$
$f_{11}(\lambda)$	$1 - 2.21\ (\lambda - 0.01)$
$f_{12}(\nu)$	Unity

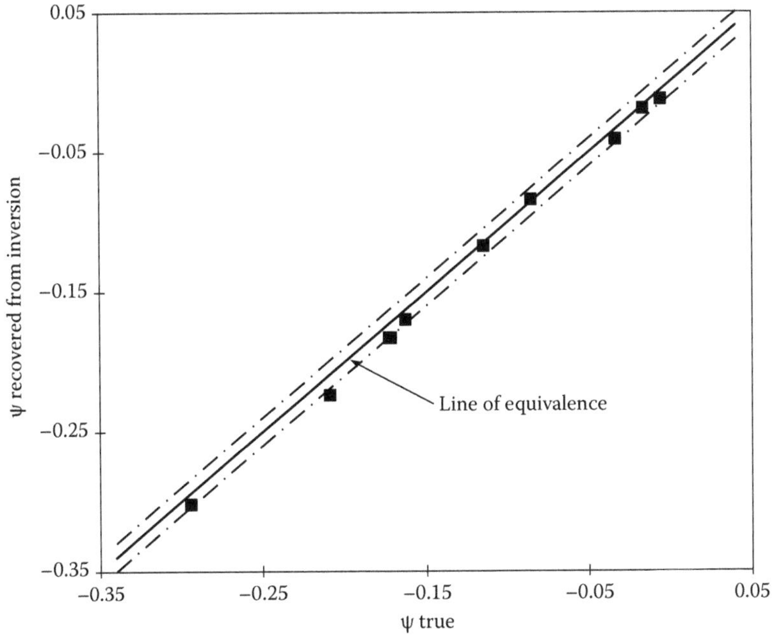

Figure 4.26 Performance of approximate general inversion on 10 sands with randomly chosen properties. (From Shuttle, D.A. and Jefferies, M.G.: Dimensionless and unbiased CPT interpretation in sand. *Int. J. Numer. Anal. Methods Geomech.* 1998. 22. 351–391. Copyright Wiley-VCH Verlag GmbH & Co. KGaA. Reproduced with permission.)

and estimated ψ values using this procedure. The proposed general inversion recovers ψ with an accuracy of ±0.02.

To apply the general inversion established by Equations 4.7 and 4.12a and b, several soil properties must be known. Most of these properties are general, not specific to *NorSand*. Three of the properties (M, N and λ) can be determined from triaxial compression tests on reconstituted samples, as described in Chapter 2. The shear modulus can be measured using the seismic piezocone as part of the CPT, by cross-hole shear wave velocity measurement or by SASW. The only *NorSand*-specific property is the plastic hardening modulus; other good soil models will include a comparable plastic hardening modulus, so it is not an issue of *NorSand* itself but how geotechnical engineering standardizes the representation of plastic hardening in shear (the analogous, accepted property in confined compression is the compressibility C_c). And that leads to a practical difficulty as plastic hardening appears affected by soil particle arrangement (fabric) and so plastic hardening should be measured in-situ (using data obtained with a self-bored pressuremeter (SBP), for example). For the moment, it is a situation of hoping that reconstituted laboratory samples replicate in-situ soil fabric and with H determined from such laboratory data by iterative forward modelling (IFM) as described in Chapter 3. Finally, of course, the horizontal stress in-situ must be known.

Equation 4.12 combined with Table 4.3 is a large step forward compared to standard practice, and the framework nicely validates against the CC data. Of course, the improvements come at the cost of requiring assessment of soil properties but that can hardly be viewed as a limitation since those properties do affect the CPT and it is better to be explicit (even if you end up estimating properties) than to neglect them. What is inescapable is that the engineering knowledge of the calibration of penetration tests is almost entirely restricted to clean quartz sands – not what is found in-situ most of the time.

4.7 MOVING FROM CALIBRATION CHAMBERS TO REAL SANDS

4.7.1 Effect of material variability and fines content

CC tests are carried out on uniform sand with constant intrinsic properties. Soils encountered in practical engineering, however, show considerable variability, for example layering with different grain sizes or silt contents in each layer. Silt content fluctuates between 1% and 7% even in uniform clean hydraulic fills placed under strict quality assurance procedures as indicated in Figure 1.5 for the Nerlerk berm. Fourie and Papageorgiou (2001) show that the variation for Merriespruit gold tailings was much wider (Figure 1.18). These variations matter.

In Section 2.6, data were presented showing how the slope of the CSL varied quite substantially with seemingly small changes in silt content. On a wider scale, moving from sands to clays, the critical friction ratio M decreases, the compressibility λ increases and the plastic hardening H decreases. In finer grained soils, these factors all decrease the CPT resistance for constant state parameter and over-consolidation ratio.

A key question arises. In inverting CPT data, how can the effects of soil type be differentiated from the effects of soil state? Crudely, a perfectly satisfactory (i.e. compact or dense) sandy silt can appear as a loose sand based on penetration resistance alone. This has led to various silt content 'corrections' being proposed for the SPT and the CPT.

Further, in considering finer grained cohesionless soils such as silts and mine tailings, the CPT is not fully drained and excess pore pressures are frequently measured during the test. In Chapters 2 and 3, it was shown that the state parameter and *NorSand* capture the stress–strain behaviour of silty sands well. The only real restriction on the CC tests and general methodology presented earlier is that the work so far has been based on drained penetration.

There are two approaches available to avoid the undesirable 'silt corrections' to the CPT, and these approaches also address the issue of partial drainage very well:

1. Measure the soil properties in the laboratory, as described earlier for drained penetration, and then calculate the CPT inversion coefficients using cavity expansion theory. When dealing with undrained penetration, an undrained implementation of cavity expansion in *NorSand* (Shuttle and Cunning, 2007) is required (a downloadable file).

2. Rely on the trend that relates the behavioural parameters of a soil to its soil type. Soil type can be determined from the CPT by using the friction and pore pressure data, as well as the tip resistance. Of course, relating parameters to soil type is only a first approximation, which is why the approach is sometimes called a screening-level assessment.

These approaches are discussed in more detail after first reviewing soil behaviour type and how this is obtained from the CPT.

4.7.2 Soil behaviour-type index from the CPT

The soil behaviour-type index (Jefferies and Davies, 1991) arose from the idea that the conventional CPT soil-type classification chart (Figure 4.2) could include the piezometric information directly by using $Q(1 - B_q)$, rather than Q, as the plot axis and thereby expanding the lower part of the chart and helping to distinguish between clays and silts. Actually, it is better to use $Q(1 - B_q) + 1$ which is the parameter grouping identified by Houlsby (1988), see Table 4.1. Although the '+1' term looks like a negligible difference, it in fact proves useful when dealing with loose silts. Distorting the soil classification chart by expanding the horizontal scale makes soil-type zone boundaries into approximate circles, as shown in Figure 4.27 (compare this to Figure 4.2). The radius of the circles can then become a material behaviour type index, I_c, where

$$I_c = \sqrt{(3 - \log(Q(1 - B_q) + 1))^2 + (1.5 + 1.3\log(F))^2} \tag{4.13a}$$

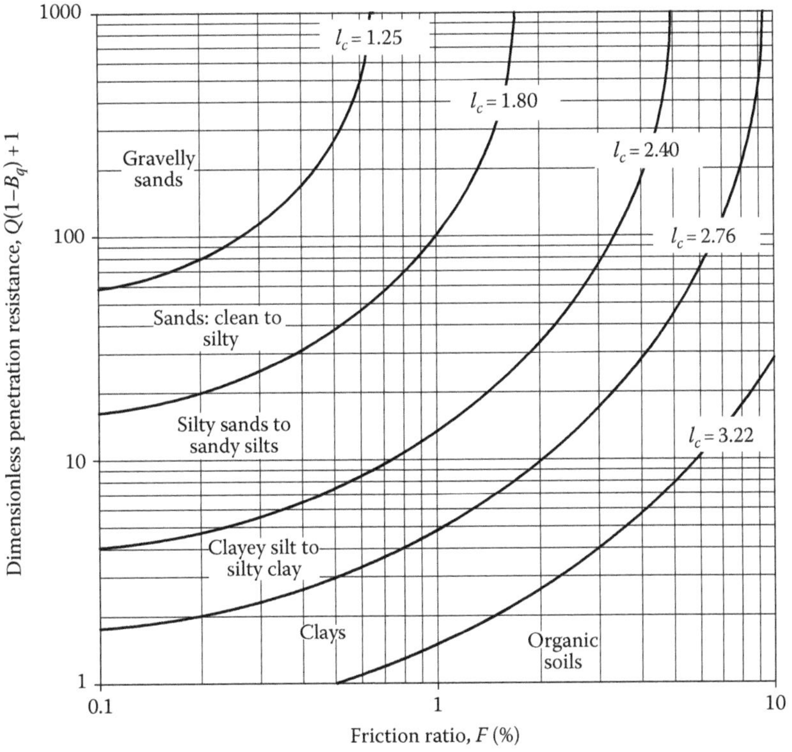

Figure 4.27 Soil-type classification chart showing constant I_c contours.

Table 4.4 Relationship of soil type to soil behaviour index I_c

Soil type	*Chart zone*	*Been and Jefferies Equation 4.13a*	*Robertson and Wride Equation 4.13b*
Gravelly sands	7	$I_c < 1.25$	$I_c < 1.31$
Sands: clean to silty	6	$1.25 < I_c < 1.80$	$1.31 < I_c < 2.05$
Silty sand to sandy silt	5	$1.80 < I_c < 2.40$	$2.05 < I_c < 2.60$
Clayey silt to silty clay	4	$2.40 < I_c < 2.76$	$2.60 < I_c < 2.95$
Clays	3	$2.76 < I_c < 3.22$	$2.95 < I_c < 3.60$
Organic soils	2	$I_c > 3.22$	$I_c > 3.60$

The relationship between soil behaviour type and I_c is given in Table 4.4. The advantage of I_c is that it is continuous and does not require artificial distinctions of soil-type zones. However, it is not strictly a soil classification as the CPT measures soil behaviour, not plasticity or grain size. For this reason, I_c is called a soil behaviour–type index.

Using I_c as a method of estimating soil behaviour is becoming quite common, but one needs to be aware of alternative definitions. Equation 4.13a was given by Been and Jefferies (1992) and obviously requires using a piezocone. Robertson and Wride (1998) suggested reverting to a $Q - F$ soil-type chart (they used that of Figure 4.2) and ignoring the piezometric data, leading to an alternative equation for I_c:

$$I_{c(R\&W)} = \sqrt{(3.47 - \log(Q))^2 + (1.22 + \log(F))^2} \tag{4.13b}$$

The inferred soil types using the Robertson and Wride equation for I_c are also given in Table 4.4. The Robertson and Wride definition of I_c ignores valuable information about the soil that is obtained from a piezocone and is best avoided. There is minimal (if any) additional cost in using a piezocone, and the extra data can be crucial when dealing with loose sandy silts to silts (as will be pervasively encountered with mine tailings, for example). A piezocone should always be used.

The soil behaviour type is now routinely calculated in commercial CPT software, and it has become part of standard CPT data processing (but do be aware of which I_c is being output so as to correctly associate the soil type as per Table 4.4). But we will now return to the problem at hand: determining ψ from the CPT for silty sands and silts for which penetration may also be undrained.

4.7.3 Theoretical approach using cavity expansion

A cavity expansion approach to determining CPT inversion coefficients is preferred for the most precise work. Samples must be obtained and tested in the laboratory, as described earlier, to understand the true values of H, M and λ. Then G_{max} and K_o are tested in-situ, discussed later in Sections 4.8 and 4.9. With these parameters known, the CPT inversion coefficients k, m can be calculated (Equation 4.12) and the state parameter determined from Equation 4.8. The method is not restricted to clean sands. In Chapters 2 and 3, it was shown that the state parameter and *NorSand* capture the stress–strain behaviour of silty sands well and the only real restriction on the general methodology is that the work so far has been based on drained penetration. So, the CPT data must be checked to confirm that penetration is indeed drained (say by limiting use of k, m to the portion of the sounding where $-0.02 < B_q < +0.02$).

For the case of excess pore water pressure during CPT sounding (easily seen in the measured data), the *NorSand* cavity expansion approach has been extended to undrained behaviour by Shuttle and Cunning (2007). An effective stress version of Equation 4.7 emerged:

$$Q_p(1 - B_q) + 1 = \bar{k}\exp(-\bar{m}\psi) \tag{4.14}$$

where the parameters $\bar{k}$ and $\bar{m}$ use the 'bar' notation to distinguish them from k and m for drained CPT penetration. Figure 4.28 shows the Shuttle and Cunning trend line for Rose Creek silt tailings, with a single line emerging for a range of effective stress, plastic modulus and rigidity index. As with drained penetration discussed in Sections 4.3 and 4.6, a 'spherical to CPT' correction is required and as a first approximation this can be taken as 2.

Another important result from the work of Shuttle and Cunning (2007) is that it provides a theoretical basis for the soil behaviour-type chart on Figure 4.27. There is of course the problem that Figure 4.27 is formulated in terms of Q and Equation 4.14 uses Q_p, so that variations in the value of K_o will cause scatter, but this scatter will generally be small and likely can be neglected in practical engineering. An inconvenience of the Shuttle & Cunning method is that, to date, it has not been reduced to trends in terms of soil properties – there is no equivalent of Equation 4.12/Table 4.3. We have therefore included a downloadable "widget" (executable) *CPT_state.exe* on the book website that computes the undrained CPT response in terms of initial state for the specified soil properties and which outputs a file for importing into Excel to give the coefficients in Equation 4.14. The source code for this widget is also downloadable should you wish to compile it as a DLL for working directly with Excel.

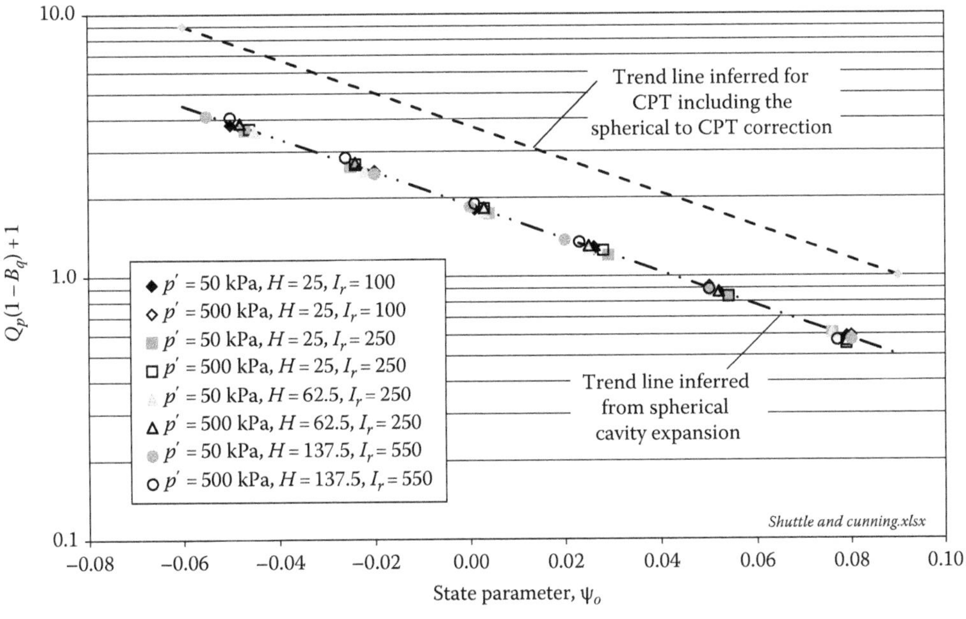

Figure 4.28 Spherical cavity expansion for undrained soil. (After Shuttle, D.A. and Cunning, J., *Can. Geotech. J.*, 44(1), 1, 2007.)

4.7.4 Screening-level assessment

The starting point for understanding a screening level assessment is the observation that CPT behaviour from sands to clays can be fitted by extending (4.7) to account for drained to undrained conditions. Been et al. (1988) first showed this semi-empirically from considerations of the similarity between ψ and OCR. The critical state model provides a basis for extending Q_p versus OCR relationships for clays (e.g. Sills et al., 1988; Wroth, 1988) into Q_p versus ψ relationships which are comparable to those for sands. For sands, ψ and OCR are independent quantities and such a transformation is not possible, although in clays it appears that the transformation is reasonable as the influence of OCR clearly dominates that of ψ.

An important difference between the Q_p – ψ relationships for sands and clays is drainage, or rather pore water pressure. CPT penetration in sands is drained, while in clays it is essentially undrained and in silts it is probably partially drained. This requires changing the normalized CPT resistance from Q_p to $Q_p(1 - B_q)$. Of course, for drained penetration with $B_q = 0$ the approach becomes identical to the framework for sands in Equations 4.7 and 4.11.

Before proceeding, we also note that in Equation 4.7 the parameter k is the normalized penetration resistance at the critical state ($\psi = 0$). Penetration at the critical state is expected to be a strong function of the critical state friction ratio M_{tc}, as well as the critical state volumetric hardening parameter λ. In addition, we prefer the addition of '+1' in the parameter grouping as suggested by Houlsby (1988) and which is underpinned by the cavity expansion theory of Shuttle and Cunning (2007). The relationship between Q_p and ψ can therefore be improved by incorporating M_{tc} into the interpretation, giving

$$Q_p(1 - B_q) + 1 = \bar{k}\exp(-\bar{m}\psi) \tag{4.14bis}$$

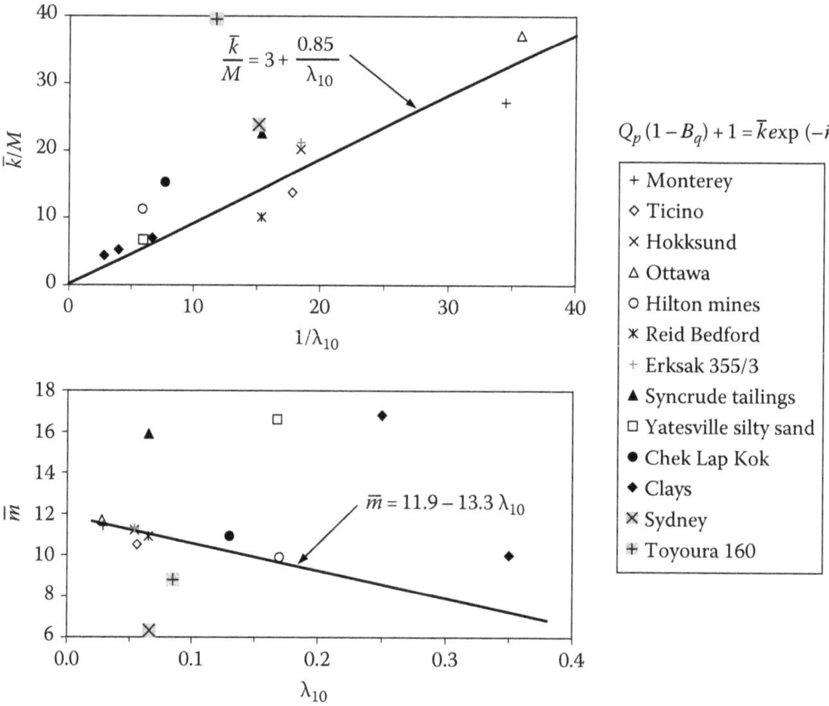

Figure 4.29 Trends in effective inversion parameters $\bar{k}$ and $\bar{m}$ with soil compressibility λ_{10}.

$$\frac{\bar{k}}{M} = 3 + \frac{0.85}{\lambda_{10}} \tag{4.15a}$$

$$\bar{m} = 11.9 - 13.3\lambda_{10} \tag{4.15b}$$

These equations were developed and improved successively by Been et al. (1988), Plewes et al. (1992) and Been and Jefferies (1992), with Figure 4.29 showing the current version in use with effective stress parameter grouping $Q_p(1 - B_q) + 1$.

The effective inversion coefficients $\bar{k}$ and $\bar{m}$ appear as functions of the soil type as indexed by compressibility λ_{10} illustrated in Figure 4.29. Soil compressibility λ is used as an index because there is presently insufficient experience or data to use H directly (which would be preferable given its large influence on Figure 4.25). Reliance is placed on H being related to λ, which is correct for Cam Clay and is certainly a trend with *NorSand* (choosing $H \sim 6/\lambda_{10}$ is a reasonable initial estimate, see Chapter 3). M_{tc} can be either treated independently as in (4.15) (preferable) or wrapped into a λ-trend, as high values of M_{tc} do not normally occur in clays, and low values do not normally occur in sands. This ties back to the original semi-empirical proposition in Been et al. (1987c) that k, m are related to λ.

The next step is to estimate λ from the CPT data, and here the approach of Plewes et al. (1992) differs from Been and Jefferies (1992). In the case of Plewes et al., a linear scaling between λ_{10} and F was suggested as a first approximation, and Figure 4.30 shows the data

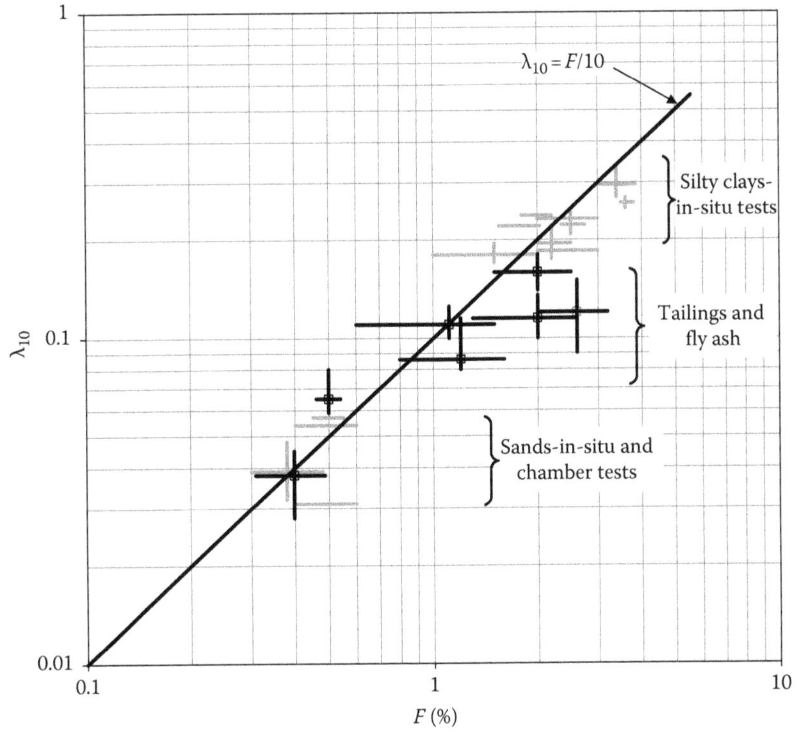

Figure 4.30 Relationship between λ_{10} and F suggested by Plewes et al. (1992) with additional data from Reid (2012) and authors' files.

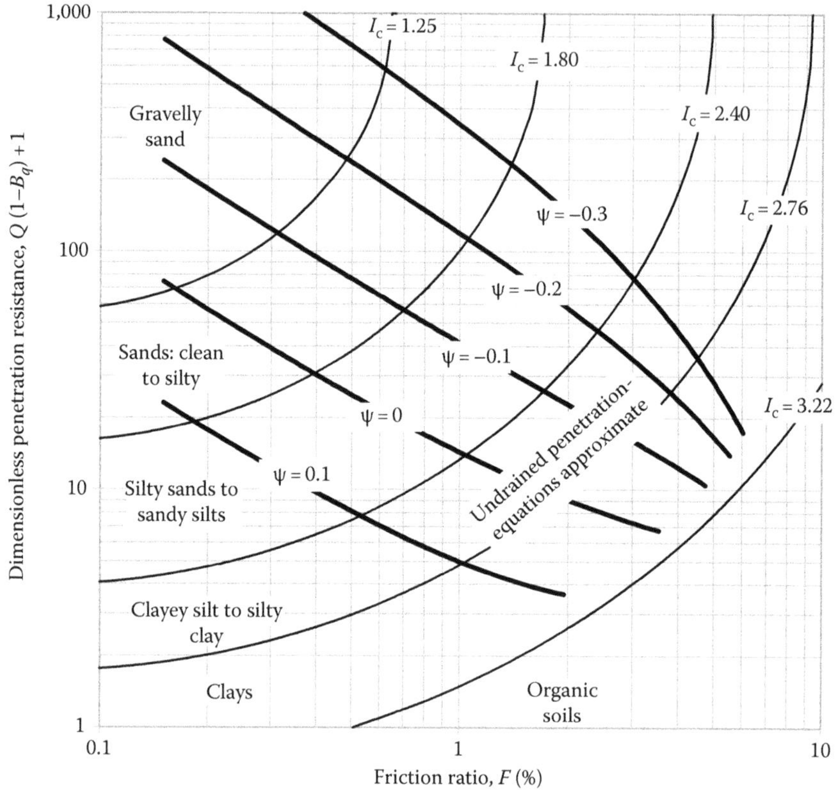

Figure 4.31 Screening-level liquefaction assessment chart for sands and silts showing I_c contours and approximate ψ contours calculated using Figure 4.30 and Equations 4.13 and 4.14.

to support this approximation. Figure 4.30 includes the data in Plewes et al. plus several additional sets of data from Reid (2012) where λ_{10} is known from laboratory testing and F has been measured by the CPT. There is no expectation that constant F indicates constant soil type on CPT soil-type classification charts (e.g. Figure 4.2), whereas λ is most certainly a soil property. However, Shuttle and Cunning (2008) note in passing that F appears to be a better indicator of soil properties than the soil behaviour–type index I_c, which was used to by Been and Jefferies (1992) to estimate λ_{10}. The Plewes et al. method has stood the test of time and has been improved with additional data (e.g. Reid, 2012), giving the final equation needed for the screening method:

$$\lambda_{10} = \frac{F}{10} \quad \text{(with } F \text{ in \%)} \tag{4.16}$$

The screening-level assessment can be considered as a form of soil behaviour classification that includes ψ as an indication of in-situ state of the material. Figure 4.31 illustrates the approach by showing the results of Equations 4.13a as well as 4.14 through 4.16 on a single chart. This form of relationship was first proposed by Plewes et al. (1992) and proves to be an extremely useful 'first look' at CPT data in practice.

4.7.5 Effect of interbedded strata

The CPT is affected by the soil above and below it as well as at the penetrometer itself. In particular, a denser/stiffer layer below the soil being tested will increase the penetration resistance, while a looser/softer layer will decrease it. These effects can be particularly marked if dealing with interbedded soils and when thin (say less than 0.5 m thick) strata have an impact on engineering solutions.

To date, there is no standard method to allow for these layering effects. Experimental studies and numerical simulations have been undertaken (e.g. Van den Berg, 1994) which suggest the effect is confined to within 150 mm of the boundary between the two strata. A slightly different problem is the 'correction' of penetration resistance in a thin soil layer embedded in stiffer materials. Vreugdenhil et al. (1994) used a simplified elastic solution to compute resistance changes around the interface between layers of different stiffness. They extended the analysis to show how the resistance within an included layer differs from the resistance in an infinite layer, but the National Centre for Earthquake Engineering Research (NCEER) workshop (Youd et al., 2001) recommended that more conservative 'corrections' be used for adjusting thin layer penetration resistances. This recommendation was based on analysis of field data by Castro and Robertson, although it is not clear how the 'true' value within the thin layer was estimated. Moss (2003) used field data from 23 sites for comparison with the Vreugdenhil solutions and suggested a set of corrections that are broadly consistent with the NCEER correction for a resistance ratio of 2 between the layers, but extended the recommendation to resistance ratios of 5 and 10.

Subsequent to the NCEER workshop, Berrill et al. (2003) carried out CC tests on Fontainbleau sand with different layerings to examine the applicability of Vreugenhil's elastic solutions. They note that although the CPT clearly involves plastic deformation, the elastic solution captures the effects of layering remarkably well and speculate that the stress state imposed on the plastic zone by the surrounding elastic region is very important. Their results were within 11% of the elastic solution for a thin dense included layer and within 22% for a single test on a thin loose included layer.

In practice, correcting the resistance is difficult and uncertain since there are two unknowns, layer thickness and true tip resistance of the soil. Vreugdenhil et al. (1994) illustrate a manual iterative approach for a single layer, in which the pore pressure from the CPT is used to determine the layer thickness and the layer stiffness, that is true tip resistance is determined by successive guesses. However, this procedure is not practical in most cases, and engineering practice is generally to neglect layering effects and develop estimates of liquefaction potential from data obtained away from the transition between strata. This approach will give satisfactory results provided layers are thicker than about 0.6 m and the stiffness ratio between the materials is less than five.

4.7.6 CPT inversion software

It should be appreciated by now that to retain reasonable precision in the recovered state estimate, inversion of CPT data to recover state is not as simple as it initially appears. Nevertheless, the entire methodology can be implemented in a spreadsheet and one is provided as a downloadable file (*CPT_plot.xls*) along with a typical data file to illustrate the results. Chapter 9 walks you through the process of using *CPT_plot* in detail. All routines are written in open code that is accessed through the VBA editor of Excel. The source code is 'plain English' with comments, so easy to read and follow.

Because the Plewes et al. (1992) screening-level evaluation based on F to determine λ_{10} (Equation 4.16) and $\bar{k}$ and $\bar{m}$ (Equation 4.15) has been found surprisingly accurate in estimating ψ, and because this method automatically senses changing soil type, the state profile from this method is always shown in *CPT_plot.xls*. The spreadsheet also allows estimation of ψ with three additional methods, with soil-specific calibrations, through the choice of the parameters k and m or $\bar{k}$ and $\bar{m}$. The methods for the inversion to state are as follows:

- Specify k, m directly in target horizon at a given stress level. This is the original method based on CC testing and Equations 4.8 and 4.11 for drained penetration (Been et al., 1987c).
- Specify soil properties M, N, H, ψ, G and compute k, m using the approach of Shuttle and Jefferies (1998) in Equation 4.12. In this approach, G varies as indicated in Equation 4.17, and any stress-level bias inherent in the CC data is considered. This method is preferred in the absence of chamber tests when the soil properties are outside the range of the chamber test soils.
- For silts, drained penetration no longer occurs. In this case, $\bar{k}$ and $\bar{m}$ need to be computed using the method of Shuttle and Cunning (2007) and then used in the inverted form of Equation 4.14 to obtain ψ from Q_p and B_q. Download the 'widget' to generate these CPT coefficients for your soil properties.

The general inversions include the effect of G on the CPT, and the basis for the method is that G can be expressed as (neglecting void ratio change effects):

$$G = G_{ref}\left(\frac{p'}{p_{ref}}\right)^{n} \tag{4.17}$$

where G_{ref} is a reference shear modulus at the reference stress level p_{ref}. The stress measure will commonly be either the vertical effective stress or the mean stress and the reference value will be 100 kPa by convention. Different soils will have different n; the valid range is $0 \le n \le 1$, with 0 being a soil of constant modulus, and 1 being a soil, typically clay, whose modulus increases in direct proportion to the stress level. For sands one typically finds $n \approx 0.5$ and this value can be used as a default. The worksheet provides a picture of the G versus stress profile input for checking that the chosen properties give the desired modulus profile, which leads to consideration of in-situ elastic properties.

4.8 ELASTICITY IN-SITU

The elastic shear modulus (strictly G, but often called G_{max} in practice) is needed to minimize uncertainty in deriving soil state from CPT data. How then should this shear modulus be measured or otherwise established? People have conducted laboratory studies using resonant column tests or bender elements to propagate a shear wave signal and in so doing developed rather well-controlled relationships. For example, Bellotti et al. (1996) reported an extensive test program on the elasticity of Ticino sand using bender elements, mainly investigating the cross-anisotropic elasticity. Their data can be averaged and fitted with the same form of functional dependence on void ratio and mean stress as they propose, and these are the data that were previously presented in Figure 4.23.

One might readily think from Figure 4.23 that in order to define the shear modulus as a function of density and stress level, all that is needed is a modest amount of laboratory testing. This is a delusion. Elastic properties are very dependent on the soil particle arrangement or fabric, and there is presently no relationship based in mechanics to ascertain the modification of shear modulus for the varying silt content and fabric that inevitably arise in-situ. The upshot is that precise work will require measurement of the shear modulus in-situ. The good news is that this task is not overly onerous. Two different methods that have been used to measure the shear modulus in-situ are unload–reload tests with a pressuremeter and geophysical tests to measure shear wave velocity.

Pressuremeters are quite straightforward because the slope of the expansion pressure versus hoop strain relationship is theoretically equal to $2G$ under elastic conditions. Elasticity is enforced by loading up the pressuremeter and then carrying out a gentle pressure reduction and subsequent reload. Although this is a very simple method, there are two subtleties. First, elastic strains are small and considerable care is needed with pressuremeter calibration and transducer quality. A reputable testing organization specializing in pressuremeter work must be involved. Second, pressuremeter testing gives only point estimates and if a series of tests are carried out in a borehole, there will usually be place-to-place variation in the shear modulus even in an apparently uniform fill.

The alternative method to determine shear modulus is to measure the travel velocity of a seismic shear wave. Modulus is then obtained using the standard relationship between shear modulus G, the shear wave velocity V_s and soil density ρ:

$$G = \rho V_s^2 \tag{4.18}$$

The shear wave velocity is determined by measuring the time taken to propagate a shear wave over a known distance. Because shear waves travel more slowly than compression waves, it is normal to use polarity reversal to clearly identify the shear wave arrival. Digital stacking oscilloscopes or seismographs are used to build up the signal and minimize the effects of noise on the data. There are three common arrangements for measuring seismic shear wave velocity directly: cross-hole testing, vertical seismic profiling in a borehole, and vertical seismic profiling with a seismic cone. Spectral analysis of surface waves, SASW (e.g. Nazarian et al., 1983; Stokoe et al., 1994), or MASW (Park et al., 1999) techniques provide non-intrusive methods to determine the shear wave velocity albeit only to 5–10 m depths.

Cross-hole testing, as the name suggests, involves drilling boreholes about 5 m apart in which plastic casing is grouted. A polarized seismic source, commonly a sliding hammer that can be struck in either an up or down direction, is lowered in one borehole and used to send a signal to a geophone in the other borehole. Because boreholes deviate from vertical, each borehole must be carefully surveyed throughout its full depth. The survey comprises taking multiple readings, not necessarily just horizontally, to construct a picture of the variation in shear modulus. Standard procedure D-4428, as published by ASTM, describes what is needed and how the testing should be carried out. This is a very useful standard and should be referred to even if testing with vertical methods (for which there is presently no corresponding ASTM standard). It is possible to reconstruct the place-to-place variation in shear modulus from tomographic inversion of cross-hole testing but that is a highly unusual level of effort and rarely encountered.

The need for multiple boreholes in cross-hole testing is inconvenient, as is the effort in precisely surveying them. Consequently, it is more common to use vertical seismic profiles (VSPs). VSPs comprise using a receiver placed down a cased borehole, much like in cross-hole testing, and then propagating a signal from the surface to the receiver. VSPs offer two

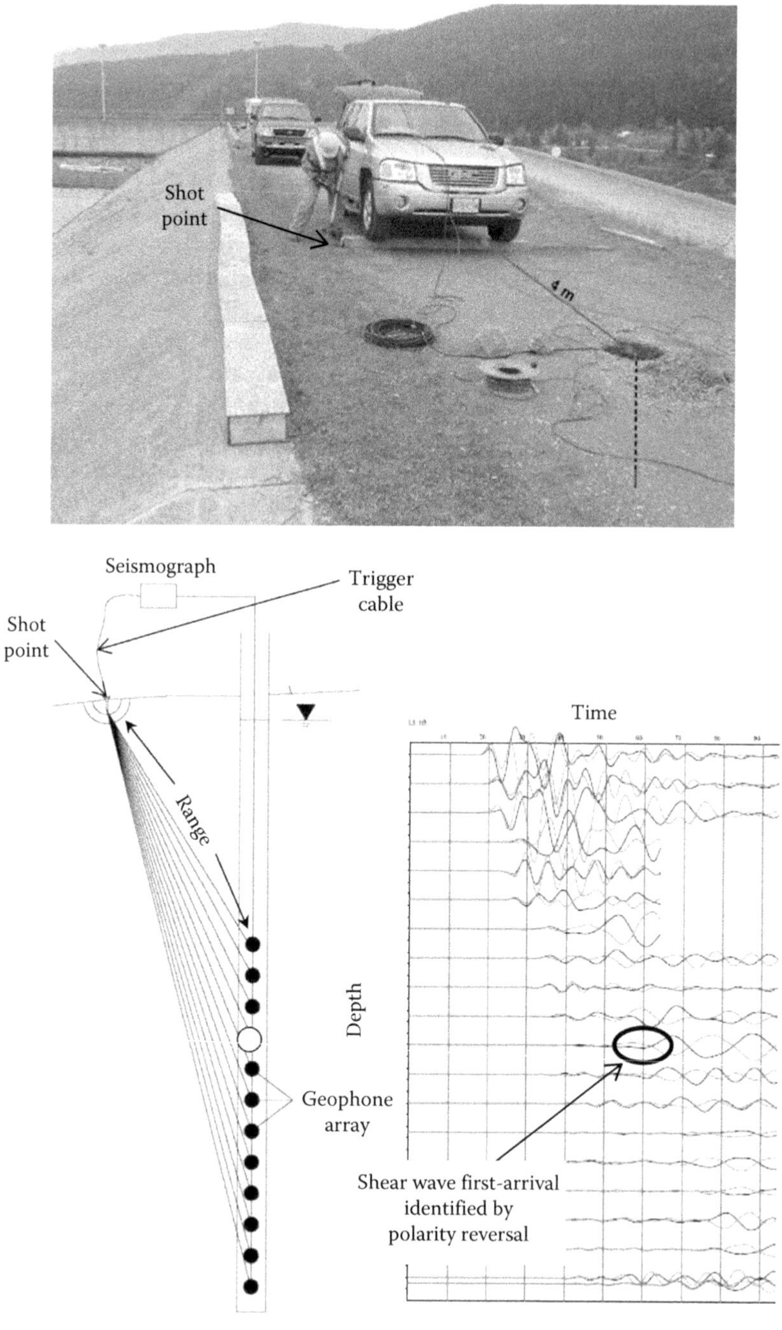

Figure 4.32 Vertical seismic profiling to determine in-situ shear wave velocity.

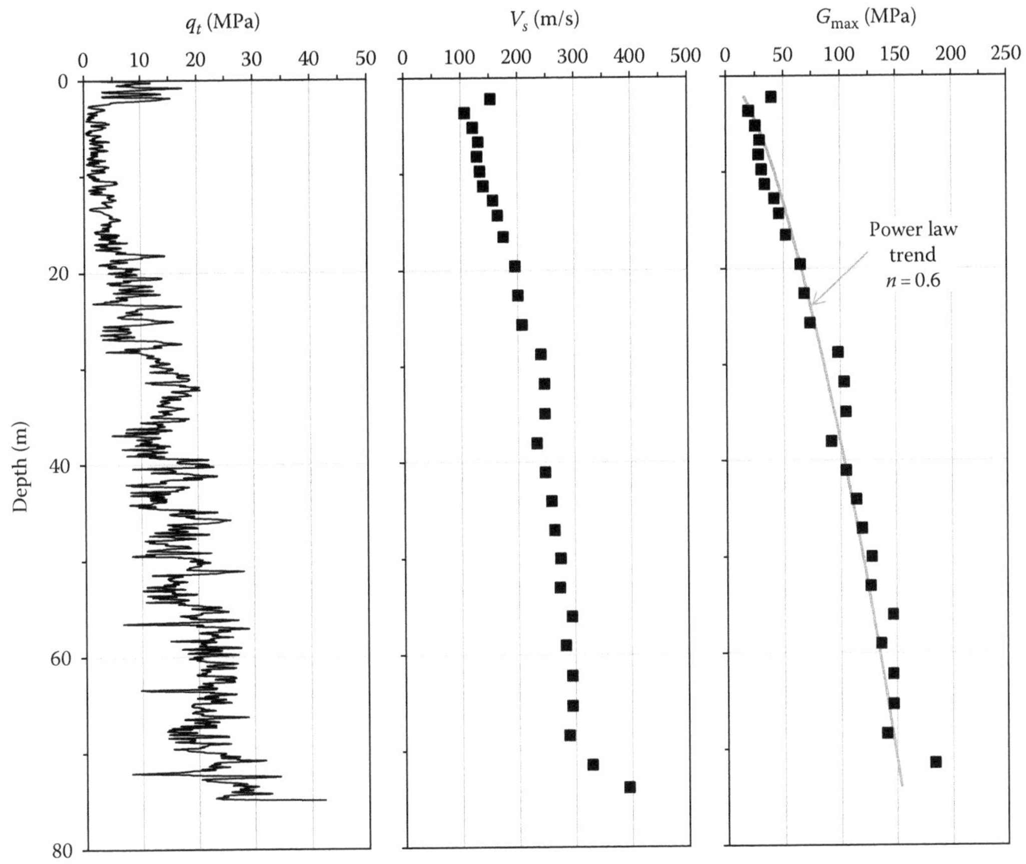

Figure 4.33 Example of seismic CPT in layered sand and silt tailings deposit; tip resistance, shear wave velocity and shear modulus.

advantages. First, because the signal travel path is predominantly vertical, it is common not to survey the borehole and accept the error so introduced in the estimated travel path. Second, the polarized source no longer has to be a downhole hammer, but now can be a plank of wood anchored to the ground (often using the wheels of a site vehicle parked on it), and simply hit with a sledge hammer as illustrated in Figure 4.32.

A convenient variation on the VSP approach is the seismic cone. The seismic cone is a standard CPT in which a geophone has been mounted as an extra transducer channel (Robertson et al., 1986). During each pause in the CPT sounding, usually at 1 m intervals when adding a rod, a shear wave is generated at the ground surface and the time required for the shear wave to reach the geophone in the cone is recorded. This is just like a standard VSP. Not only does this provide the needed data inexpensively, but there is the additional advantage that it is adjacent to the CPT, therefore relevant to the inference of state from that particular test. The method can also be improved by including two geophones, exactly 1 m apart within the CPT system. The difference in arrival time of the shear wave generated at the surface then gives the shear wave velocity between the two geophone depths.

Figure 4.33 shows the results of VSP testing with a seismic CPT in sand and silt tailings deposit. There is a strong layering evident in the tip resistance, which is typical of

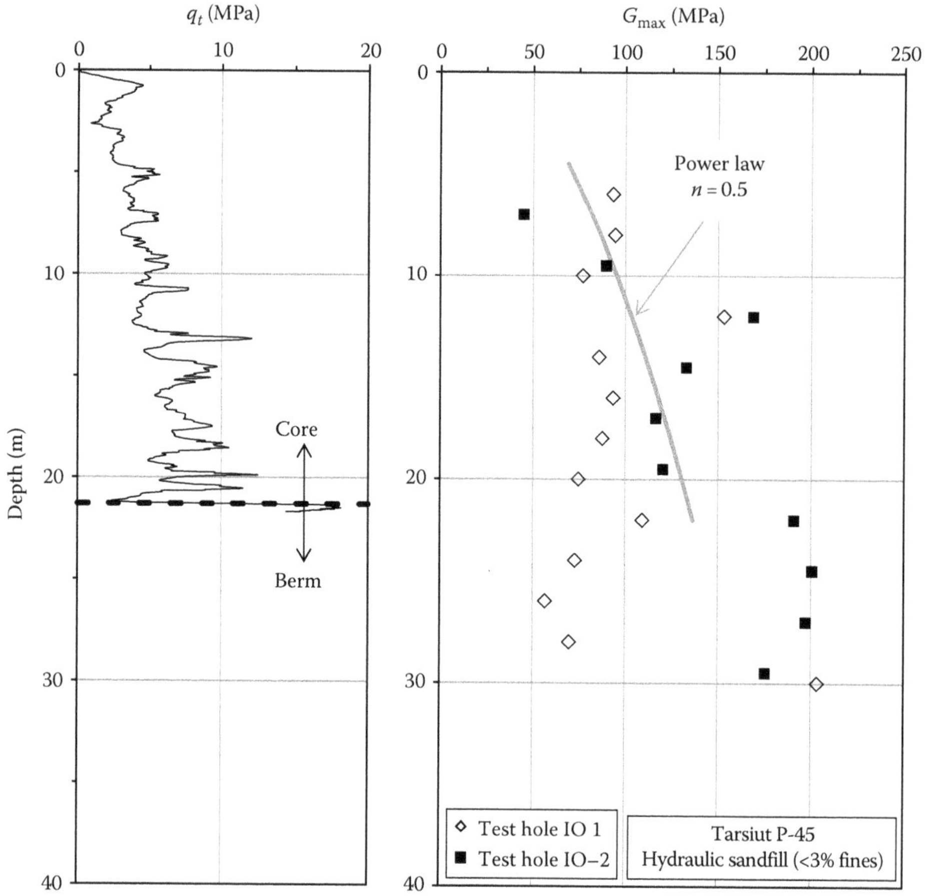

Figure 4.34 Shear modulus determined from VSP tests in hydraulically placed sandfill (Molikpaq core at Tarsiut P-45).

periodical hydraulic deposition of sandy and silty materials. The shear wave velocity profile averages properties over a metre and shows a very consistent trend of velocity increasing with depth. From this, the shear modulus G_{max} has been calculated and a power law trend from Equation 4.17 with $n = 0.6$ is shown for comparison. A CPT and two VSP testing boreholes in a uniform and hydraulically placed sandfill are shown in Figure 4.34. These data are from within the core of the Molikpaq at Tarsiut P-45. There is a substantial variation in modulus from place to place, about ±50% of the best-fit power law trend line for a uniform density material using the familiar power law (square root) trend. This variation is unsurprising when we look at how the properties vary (Chapter 5, Figures 5.5 and 5.6), and is the reason why shear modulus needs to be treated as an independent variable.

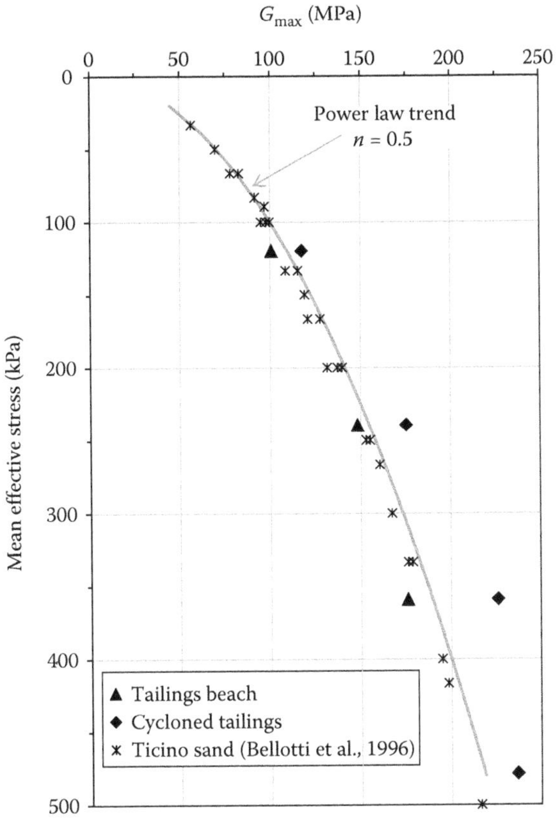

Figure 4.35 Bender element measurements of shear modulus in laboratory samples of Ticino sand and tailings, showing consistent power law trend with $n = 0.5$.

When introducing the stress normalization methods, it was noted that some workers were using an exponent of 0.7 to normalize CPT data in silt, apparently on the grounds that silt was a soil half-way between sand and clay. Figure 4.35 shows G_{max} data from laboratory bender element tests on a clean sand (Ticino sand, the same data from Bellotti et al., 1996, in Figure 4.23) a cycloned tailings (silty sand) and a tailings beach (sandy silt). For all intents and purposes, these materials show exactly the same trends in the laboratory and there is no basis for normalizing CPT results in silts using an exponent $n = 0.7$. Figure 4.36 shows in-situ G_{max} data on four silts; one a natural deltaic deposit (Fraser River Delta), one hydraulic fill (placed in Coquitlam dam core in 1910) and two tailings deposits. Here, there is a range from $n = 0.5$ to 0.8, but the range is attributed to different depositional environments and ageing (the bottom of the tailings deposits may be decades old, while the near surface material is a few years old). It is important to actually test the soil as opposed to relying on opinion of what the trend or exponent should be; if you test, then you know the facts.

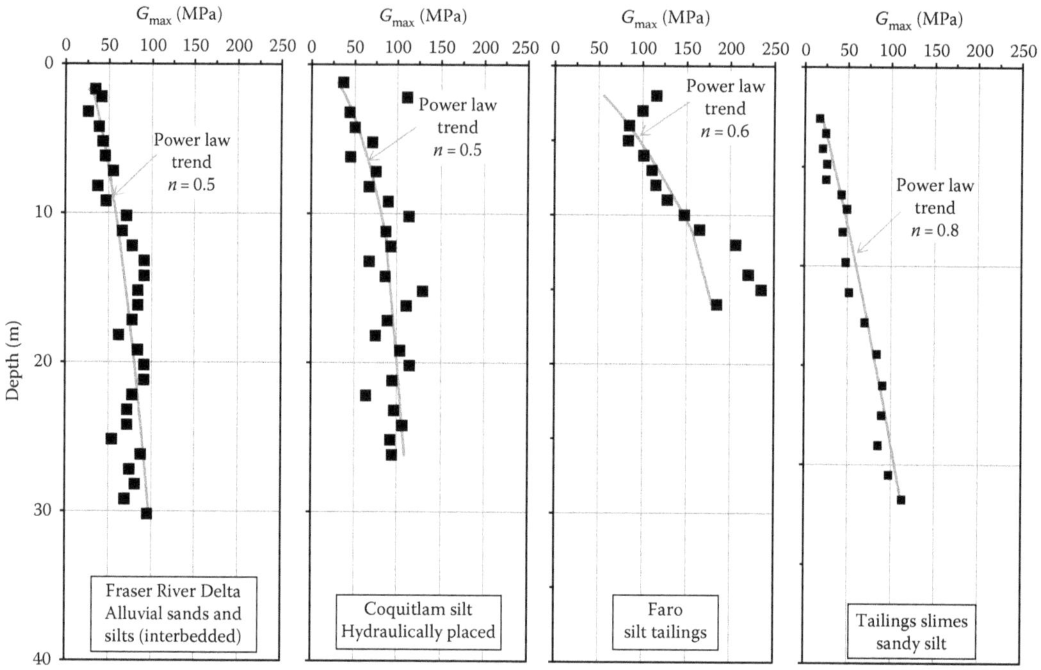

Figure 4.36 Comparison of shear modulus profiles for several sites, including sands and silts in different depositional environments.

4.9 HORIZONTAL GEOSTATIC STRESS

4.9.1 Geostatic stress ratio, K_o

Estimating the vertical effective stress is straightforward and generally can be done within a few percent of truth given a measured or otherwise known ground water pressure. But the difficulty is that penetration tests, whether CPT or SPT, respond to the mean stress and this requires knowing the horizontal stress to maximize information from penetration tests. A common way forward is to work through the geostatic stress ratio, $K_o = \sigma'_h/\sigma'_v$. This stress ratio is determined and then is usually taken as a constant in the stratum, allowing the mean effective stress to be related to the vertical effective stress.

The formula usually encountered for K_o in normally consolidated soils relates the geostatic stress ratio to the current friction angle through the Jaky equation:

$$K_o = 1 - \sin(\phi') \tag{4.19}$$

Besides the fact that Jaky developed this relationship for the stress on the axis of an accreting conical sand mass, there are fundamental problems with Equation 4.19. First, normally consolidated sands that are gently vibrated will densify and increase their friction angle, from which (4.19) would indicate that the horizontal stress decreases, since dense soils have a greater friction angle than loose soils. In reality, the horizontal stress increases with vibration instead of decreasing, so that Equation 4.19 has the wrong sense. Second, one would expect the geostatic stress ratio to be a function of soil stiffness and other small strain constitutive parameters, not friction angle, at least for the level ground case. The best that can be said of Equation 4.19 is that it indicates $K_o \approx 0.5$ under loose, normally consolidated conditions.

A better understanding of K_o can be developed from thinking about the confined (no lateral displacement) compression of an elastic solid under level ground conditions with the vertical effective stress as the maximum principal stress. Let elasticity follow the Hook's law ideal with properties E and ν. Increment the vertical stress by $\Delta\bar{\sigma}_1$, and from the definition of Poisson's ratio:

$$\dot{\varepsilon}_2 = \frac{\Delta\bar{\sigma}_2}{E} - \frac{\nu\Delta\bar{\sigma}_1}{E} - \frac{\nu\Delta\bar{\sigma}_3}{E} \tag{4.20}$$

Introducing the condition of no lateral displacement, $\dot{\varepsilon}_2 = 0$:

$$\Delta\bar{\sigma}_2 = \nu\Delta\bar{\sigma}_1 + \nu\Delta\bar{\sigma}_3 \tag{4.21}$$

From symmetry $\Delta\bar{\sigma}_2 = \Delta\bar{\sigma}_3$, which then gives the result:

$$K' = \frac{\Delta\bar{\sigma}_3}{\Delta\bar{\sigma}_1} = \frac{\nu}{1-\nu} \tag{4.22}$$

K' is often taken to be K_o in the literature; for example, Bishop and Henkel (1962) suggest that the incremental approach is the most accurate method for determining K_o. Those authors further remark that K' is experimentally constant over a wide range of stresses with cohesionless materials – hardly surprising given that we can see from (4.22) that K' is an alternative identity for Poisson's ratio and that elasticity is important in the confined compression of cohesionless materials. Reverting to the true definition of the geostatic stress ratio, for the changed stress state:

$$K_o = \frac{\bar{\sigma}_3 + \Delta\bar{\sigma}_3}{\bar{\sigma}_1 + \Delta\bar{\sigma}_1} \tag{4.23}$$

from which it can be seen that K_o is only equivalent to K' if starting from zero initial stress. Interestingly, Poisson's ratio for cohesionless soils commonly lies in the range $0.2<\nu<0.3$, giving corresponding values $0.25<K'<0.45$. This is close to the values obtained with the Jaky equation, which might be a reasonable approximation in laboratory experiments with reconstituted samples. Presumably this is what has led to the views of the equivalence of K' and K_o, as these views have been put forward by people with a largely laboratory experimental background. For real soils in-situ, relying on K' as an estimate for K_o entirely ignores geologic processes that may markedly change things, including many thousands of cycles of low-level cyclic stress, creep, and ageing.

There are data from two studies showing the effect of cyclic loading on K_o. Both studies tested clean sand in the void ratio range $0.5<e_o<0.6$ (which is not loose). In the first study, Youd and Craven (1975) used calibrated membranes in the cyclic simple shear test to measure how the horizontal stress evolved during several levels of near-constant cyclic straining. In the second study, Zhu and Clark (1994) used an oedometer, instrumented to measure lateral stress, that was mounted on a shaking table much as might be done for a minimum void ratio test. Figure 4.37 shows the data from both these studies illustrating how K_o evolves with cyclic loading. The two studies do not overlap. Youd and Craven limit themselves to 100 cycles whereas Zhu and Clark have 60 Hz loading and only show data after a few seconds of loading. Also, Youd and Craven nicely document the effect of cyclic strain amplitude whereas the tests by Zhu and Clark were a bit 'brute force' although taken to

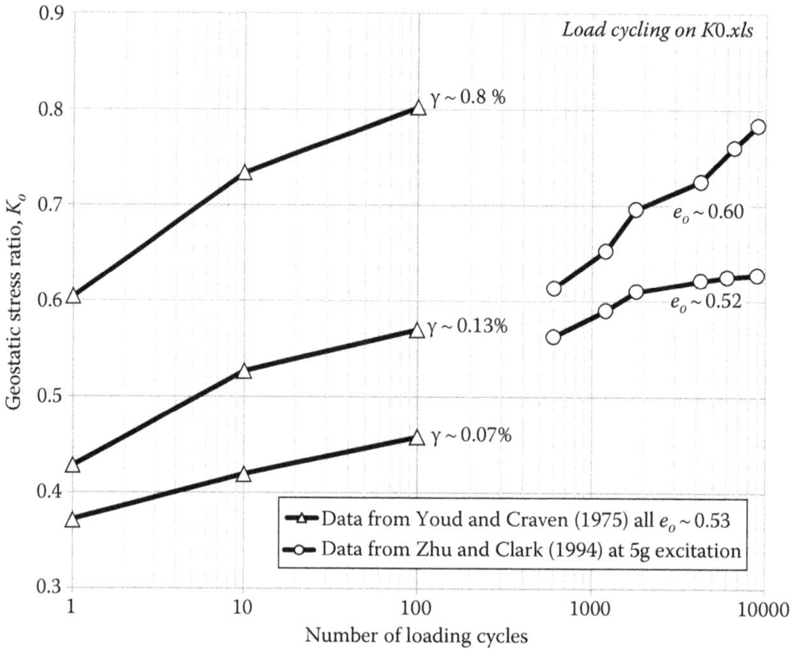

Figure 4.37 Effect of number of loading cycles on geostatic stress ratio (under laboratory conditions).

vastly more loading cycles. Despite these differences in test methodology, the data show that rather minor cyclic loading can easily double K_o from its as-placed value (i.e. the Jaky value).

One approach might be to somehow estimate the geological history of a stratum and what the effect of that history might be on the geostatic stress ratio. But that hardly seems sensible since we have no method to quantify geological history and our understanding of the mechanics as to why geostatic stress ratio increases with cyclic loading is limited. Practically, it is not possible to avoid treating K_o as an independent variable, and a variable that must be measured in-situ. How to measure K_o is the next question.

4.9.2 Measurement with SBP

Measuring K_o tends to rely on the SBP, although there are some alternatives. With the SBP there are, in principle, two approaches to determining K_o: initial lift off and formal solution of the boundary value problem. Some example data are examined before discussing these methods.

Figure 4.38 shows a composite plot of seven good SBP tests in a loose hydraulically placed sandfill where the measured cavity pressures have been normalized by the vertical effective stress at the test depth. (The case history is the Molikpaq offshore platform at the Tarsuit P-45 site, Jefferies et al., 1985, for which CPT and VSP profiles are shown in Figure 4.34.) Table 4.5 provides a summary of these tests as well as the interpreted results from the adjacent CPT profile for reference. The data plotted in Figure 4.38 for each test are the average of the three arms in the SBP, as this is required to compensate for possible pressuremeter movement in the ground as it is expanded. Only 7 out of 16 tests are shown; the other tests being abandoned as obviously grossly disturbed (e.g. for a large mismatch between the individual arm displacements). It seems to be common that only about half the SBP tests in sand remain relatively undisturbed as a result of the self-boring, possibly because the driller

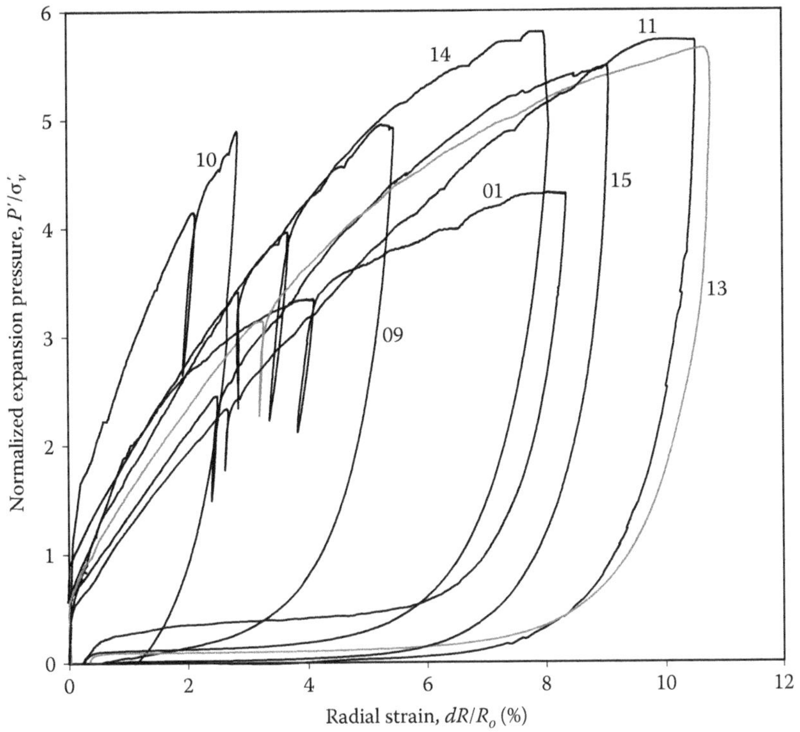

Figure 4.38 Results of SBP tests in hydraulically placed Erksak sand.

Table 4.5 Summary of near-undisturbed SBP tests in Tarsiut P-45 hydraulically placed sandfill and adjacent CPT data

			Lift-off method		*IFM (Ghafghazi and Shuttle, 2008)*		
Test	*Depth (m)*[a]	$\bar{\sigma}_v$ (kPa)[b]	$\bar{\sigma}_h$ (kPa)	K_o	K_o	q_c *(MPa)*[c]	Q_p
Spigotted core							
1	2.7	53	38	0.7	0.96±0.04	2.6±0.3	55
2	3.7	70	<120	–		≈4	57
8	7.7	109	Uncertain	–		≈6	55
9	10.7	139	75	0.5	0.65±0.11	9.2±0.7	87
10	11.5	146	181	1.2	0.9±0.10	14.7±5	107
11	16.2	193	130	0.7	0.8±0.20	12.3±5	79
13	18.2	212	105	0.5	0.8±0.20	8.0±2	45
14	20.0	230	145	0.6	0.63±0.10	12.7±2	74
15	21.0	240	217	0.9	0.88±0.13	12.4±3	57
Bottom dumped berm							
16	25.3	282	165	0.6		≈20	71
18	29.7	325	102	0.3		≈20	61

[a] Depths are quoted to the strain arm measurement axis of the SBP.
[b] Water table estimated at 3.5 m from surface from adjacent CPTu.
[c] Estimated characteristic value from adjacent CPT.

receives no feedback during self-boring of the pressuremeter as to whether the pushing rate is properly balancing the soil removal rate. However, as can be seen, the good tests show a consistent pattern of behaviour. The data files for these tests are included on the website with test details.

The lift-off method, as the name suggests, involves identifying when the expansion of the pressuremeter begins. The proposition behind this method is that the membrane remains pushed against the pressuremeter body until the net pressure balances the horizontal stress in the ground. Net pressure means that the effect of the SBP membrane must be subtracted from the total pressure. This offset (typically 18–30 kPa) needs to be measured as part of the pressuremeter calibration.

The boundary value method to determine horizontal stress involves fitting a model to the entire pressure versus displacement record. This method works because the pressure at any stage of the test is a function of the far-field horizontal stress, which is the geostatic stress sought. Ghafghazi and Shuttle (2008) used *NorSand* within a finite element model to analyze all of the SBP tests in Figure 4.38. They illustrate the use of an IFM approach, in which unknown parameters are estimated (in this case ψ and σ'_h), the pressuremeter expansion behaviour for these parameters is computed, and then the computed behaviour compared with what was measured. The estimate of properties is then revised and the process repeated until a good image match is obtained. Figure 4.39 shows the results of IFM for Test 01 from Ghafghazi and Shuttle (2008). Note that the model focuses on the cavity strains greater than about 1% as this reduces the effect of disturbance due to boring on the results. Although IFM results in a non-unique solution, as more than one combination of parameters may fit the data, this is not a limitation. How easily a fit is obtained, and the range of parameters that give a good fit, is an indication of robustness of the solution. The results of IFM for all seven tests which were modelled by Ghafghazi and Shuttle are included in Table 4.5 for comparison with the less accurate lift-off method. The data are consistently $0.6 < K_o < 1.0$ for IFM, with lift-off pressures giving a slightly wider range of K_o.

Figure 4.40 (after Graham and Jefferies, 1986) shows the estimated in-situ horizontal stress plotted against vertical effective stress for six case histories (including the data from

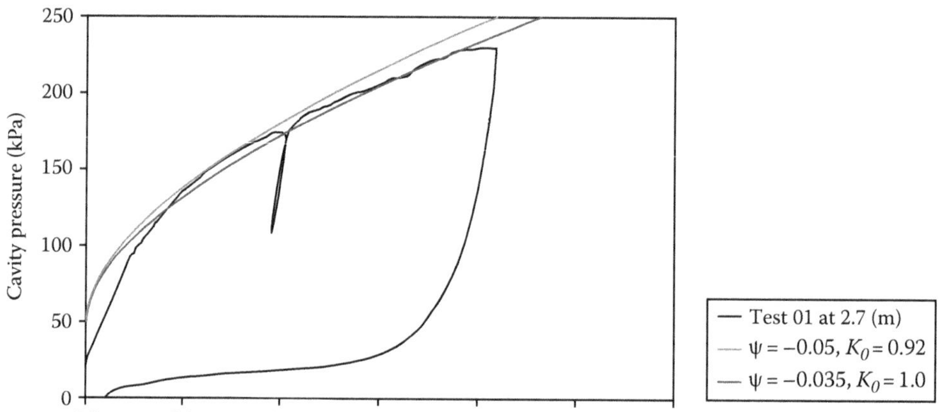

Figure 4.39 Iterative forward modelling (IFM) for interpretation of self-bored pressuremeter test in hydraulic sandfill. (From Ghafghazi, M. and Shuttle, D.A., *Can. Geotech. J.*, 45(6), 824, 2008. With permission from the NRC, Ottawa, Ontario, Canada.)

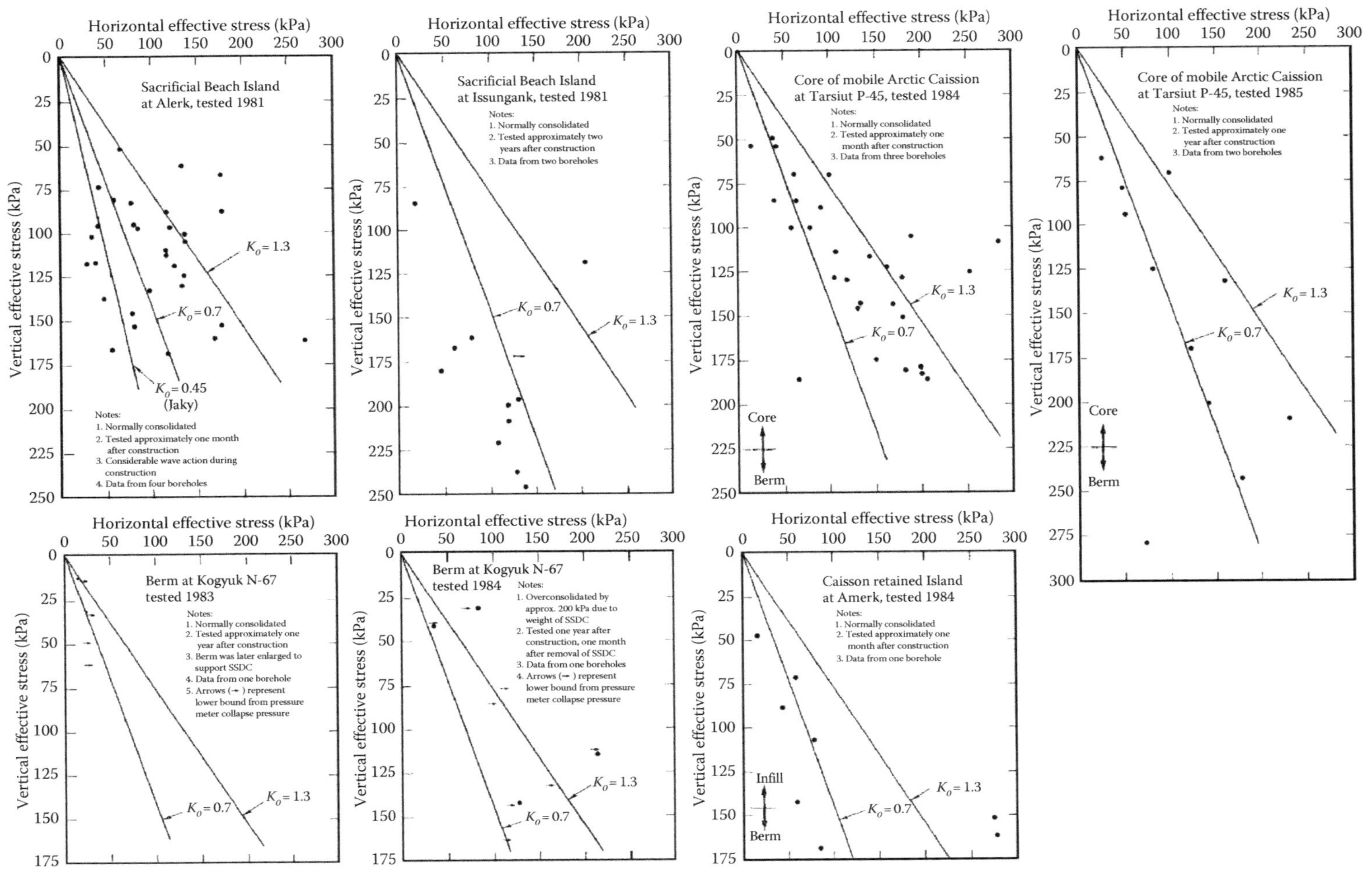

Figure 4.40 Horizontal geostatic stress in hydraulic fills. (From Graham, J.P. and Jefferies, M.G., Some examples of in-situ lateral stress determinations in hydraulic fills using the self-boring pressuremeter, *Proceedings of 39th Canadian Geotechnical Conference*, Ottawa, Ontario, Canada, 1986.)

Tarsiut P-45) of hydraulically placed sands in the Beaufort Sea. As can be seen, there is a common pattern to the lift-off pressures over these different case histories, which provides comfort in the repeatability of the procedures, as different testing contractors, field engineers and drillers were involved. Also, the Molikpaq core at Tarsiut P-45 was tested twice, once in 1984 and once in 1985, using different companies and different field engineers and yet very similar results were obtained. The average geostatic stress ratio across these six case histories was about $K_o \approx 0.7$. The Jaky equation gives a lower bound to the data, as expected, and something like $K_o \approx 1.3$ might be viewed as a sensible upper limit if a few values are excluded as possible outliers.

4.9.3 Measurement with horizontal stress CPT

The requirement to know the horizontal stress is certainly inconvenient for easy use of CPT data. However, it becomes a non-issue if the CPT itself measures the horizontal stress. There is also the interesting question of whether K_o is even constant in the light of the SBP data, and continuous measurement of σ_h is desirable.

Huntsman (1985) suggested installing strain gauges on the friction sleeve of the penetrometer as an additional data channel to measure hoop strain during CPT sounding. The horizontal stress sensing cone test was tried on an experimental basis in both the CC (Huntsman, 1985; Been et al., 1987b) and the field (Huntsman et al., 1986; Jefferies et al., 1987). Geotech AB in Sweden (www.geotech.se) (Ingenjörsfirman Geotech AB, 2014) manufactured the horizontal stress sensing cones that were used. The sensing element axis was 35 mm behind the shoulder of the penetrometer.

Because CPT penetration disturbs the ground, the horizontal stress measured on the sleeve is different from that in the ground. Inversion of the measured data is, therefore, essential to obtain the actual horizontal stress. Like other CPT work, the starting point is CC tests. Ground truth is known for these because the horizontal pressure is applied to the chamber. The measured effective radial stress on the CPT sleeve σ'_{hc} can be expressed as a ratio, called the amplification factor A, of the horizontal geostatic stress. Figure 4.41 shows the results of the measured data on A from CC tests on Monterey and Erksak sands, in each case comparing A with the state parameter as A is obviously related to soil dilation. There is a simple relationship that captures the data:

$$A = a\exp(-b\psi) \tag{4.24}$$

where a, b are coefficients analogous to k, m. There is now a small problem in that the inversion to recover ψ from CPT data depends on mean stress but to recover mean stress through (4.24) we need to know ψ. For the case that $m = b$, which approximates the test data, combining (4.24) with (4.7) and the relationship of K_o to the mean effective stress to eliminate ψ gives an equation (referred to as the 'linear algorithm'):

$$K_o = \frac{1}{(2 - ((3a/k)(q_c/\sigma'_{hc})))} \tag{4.25}$$

where measured parameters (q_c and σ'_{hc}) appear on the right-hand side. Jefferies et al. (1987) tested recovery σ'_h with (4.25) for the chamber test data (for which it was known). There was substantial scatter. However, applying (4.25) to the field data gave very much more stable results, shown in Figure 4.42. This figure compares the results using Equation 4.25 with

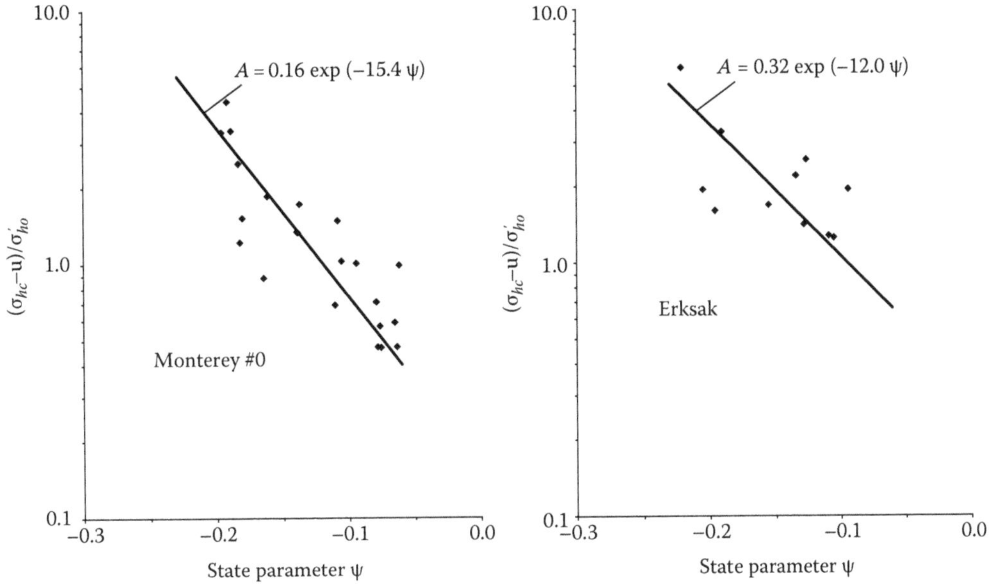

Figure 4.41 CPT horizontal stress amplification factor versus state. (From Jefferies, M.G. et al., *Géotechnique*, 37(4), 483, 1987. With permission from the Institution of Civil Engineers, London, U.K.)

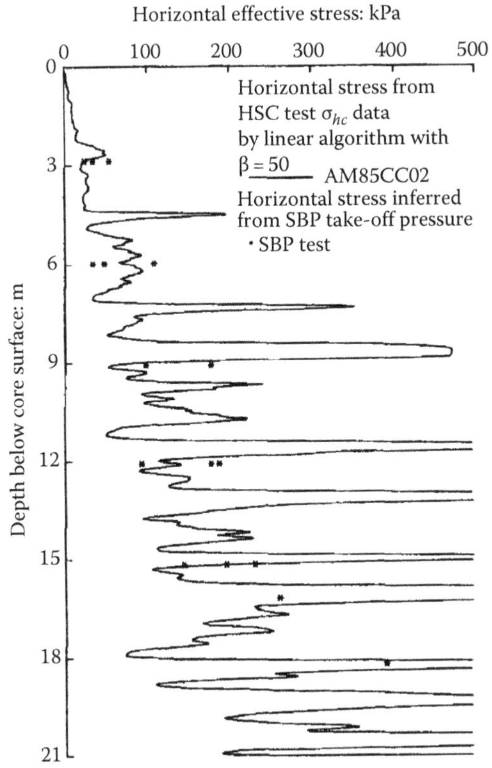

Figure 4.42 Comparison of geostatic stress from SBP and CPT in hydraulically placed sandfill. (From Jefferies, M.G. et al., *Géotechnique*, 37(4), 483, 1987.)

the estimated σ'_h from adjacent SBP tests (it is the same SBP data as appears in Figures 4.38 and 4.40). Rather good correspondence is evident, and the fluctuations in σ'_h determined through (4.25) looks intriguingly like the variation inferred from the pressuremeter data. One point not discussed in Jefferies et al. (1987) is that the chamber test data used to verify the recovery of σ'_h using (4.25) were based on data picks as representative of each chamber test. The apparently good performance of the field data was when the algorithm was applied to a data file continuously, on measured data at approximately 10 mm intervals.

At present, the horizontal stress CPT is very much an experimental device and not likely to be used in practice. It has been shown here for two reasons. First, the remarkable similarity of the K_o values to those from SBP data leads credence to the idea that the SBP data are reasonable and hopefully dispel any anxieties about rejection of the Jaky equation. Second, the horizontal stress CPT is an interesting device with substantial potential for improving precision in geotechnical engineering. Perhaps reading about it here might trigger further developments in transducer technology and the development of better inversion algorithms. Of course, the reality is that one really should conduct pressuremeter testing as well as using the CPT. The two are complimentary tools.

4.9.4 Importance of measuring K_o

Uncertainty in the magnitude of horizontal stress leads to uncertainty in the sand state inferred from the CPT. The importance of precision in the knowledge of σ'_h to the estimate state parameter is illustrated in Figure 4.43, which shows the uncertainty in state from Equations 4.8 and 4.11a and b as a function of uncertainty in horizontal stress. Defining an uncertainty factor , ξ, in horizontal stress as:

$$\frac{\sigma'_h}{1+\xi} < \sigma'_{h\xi} < \sigma'_h(1+\xi) \tag{4.26}$$

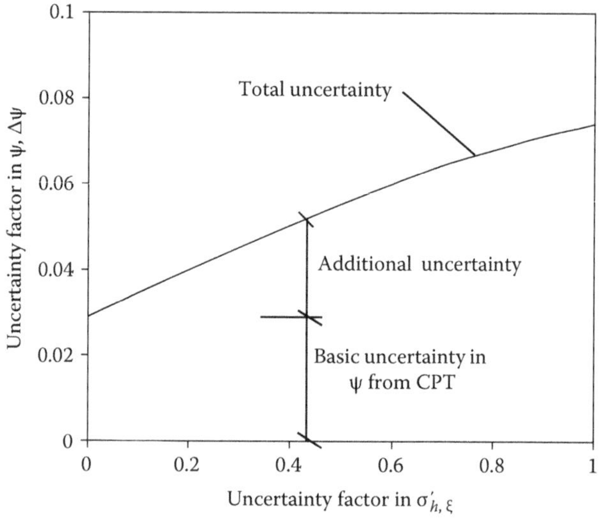

Figure 4.43 Effect of uncertainty in horizontal stress on uncertainty in estimated in-situ state parameter from CPT data. (From Jefferies, M.G. et al., *Géotechnique*, 37(4), 483, 1987.)

where

σ'_h is the true horizontal effective geostatic stress
$\sigma'_{h\xi}$ the estimate of it.

It can then be seen from Figure 4.43 that an uncertainty ξ of as much as 0.2 will not degrade significantly the estimated ψ.

However, once the uncertainty in σ'_h rises to $\xi = 0.6$, the uncertainty in ψ rises to ±0.05.

These numerical values for uncertainty can be put in perspective by considering the context of the liquefaction assessments. Soils show adequate behaviour to cyclic loading if $\psi < -0.2$, limited strain potential if $\psi < -0.1$, and may be prone to flowslides if $\psi > -0.05$. Confident engineering requires unarguable distinction between these possibilities and this requires determining ψ with a precision better than ±0.03. One, therefore, needs to know σ'_h to within $\xi < 0.2$ or 20%.

These test data presented show that K_o in-situ is unrelated to the Jaky expression. For hydraulically placed sands, the difference between K_o estimated from $(1 - \sin\phi)$ and reality is the difference between $K_o \approx 0.45$ and $K_o \approx 0.7$–0.9. This difference corresponds to $\xi \approx 1$, which is well outside the allowable tolerance. Further, the use of the Jaky equation is not conservative in determining ψ with the CPT.

Much of this chapter has been devoted to accounting for the effect of material properties and elastic modulus on the inferred ψ from the CPT. Although these aspects are important, the reality is that K_o alone is equally important and one is far more likely to estimate K_o poorly than any of the other parameters involved. So, if in doubt, measure K_o with an SBP or use some other reliable method to estimate K_o. Research on this topic is ongoing, with promise being shown (e.g.) in use of shear wave and compression wave anisotropy as an indication of in-situ stress anisotropy.

4.10 ALTERNATIVE IN-SITU TESTS TO THE CPT

4.10.1 Self-bored pressuremeter

The SBP is a remarkable test in that it is the only in-situ geotechnical test for which theory can be applied to the test data directly. Correlations are not needed. Additionally, perfect self-boring (no lateral displacement of the soil) is not needed for the data to be usable. So why is the CPT the mainstay of testing? Because SBP testing is expensive, frustrating in part and requires dedication to obtain good results. Nevertheless, the SBP is presently the most practical possibility for determining K_o and has the useful advantage of allowing estimation of dilation and state directly and entirely independently of penetration tests. The SBP will never replace the CPT, however, as the CPT is needed to understand the variability in ground conditions across a site. It is simply impractical to do sufficient SBP tests.

Determining dilation from SBP data is based on treating the SBP as the expansion of a cylindrical cavity, for which there are theoretical solutions for NAMC soil. The data from the SBP comprise effective cavity expansion pressure P' (after correcting the measured data for membrane tension and hydrostatic pressure) versus cavity displacement (which corresponds to hoop strain, ε_θ). Hughes et al. (1977) derived a relation between these two parameters by assuming the validity of Rowe's stress–dilatancy theory and small strain approximations that give the dilation angle δ directly:

$$\sin\delta = \frac{1 - R_c(1 - 2S)}{1 + R_c} \qquad (4.27a)$$

where S is the slope of the pressuremeter data when plotted logarithmically:

$$S = \frac{\Delta \log P'}{\Delta \log(\varepsilon_r + c)} \tag{4.27b}$$

and R_c is the critical stress ratio:

$$R_c = \frac{1 + \sin\phi_c}{1 - \sin\phi_c} \tag{4.27c}$$

The parameter c is a strain offset allowing for initial contraction before the onset of dilatancy.

Equation 4.27 has been widely used to estimate δ from pressuremeter data. But there are four limitations with this analysis that much degrade the validity of the estimated δ. First, c is somewhat subjective and, although arguable as a reasonable parameter to improve the NAMC model, there is precious little guidance on relating it to δ. Second, the analysis neglects elastic strains in the plastic region. Third, the analysis is small strain. Finally, the analysis assumes that a pressuremeter test is a cylindrical cavity expansion and neglects finite pressuremeter length effects.

Regarding the adequacy of the cylindrical cavity approximation, Ajalloeian and Yu (1998) reported on comprehensive chamber studies using pressuremeters with different geometries characterized by the ratio of pressuremeter length L to diameter D. Most commercially available pressuremeters have $L/D \approx 6$. Yu (1996) carried out numerical simulations and, based on these, suggests that the state parameter can be estimated directly from S^6 (where the superscript 6 denotes the standard commercial SBP geometry) by:

$$\psi = 0.59 - 2.2S^6 + 0.107S^6 \ln(I_r) \tag{4.28}$$

Although (4.28) is a useful first approximation, it is inferior to proper analysis of the SBP data using IFM as described in Section 4.9.2 and Figure 4.39. Concerns over the neglect of elasticity in the plastic zone and the reliance on small strain theory are automatically addressed within IFM using a full finite element simulation of the SBP test such as described in detail in Shuttle (2006) and Ghafghazi and Shuttle (2008). However, you must use a good soil model (e.g. *NorSand*).

4.10.2 Flat plate dilatometer

An additional possibility for determining the in-situ stress and state is to use the dilatometer, which as a test lies somewhere between the horizontal stress CPT and the SBP. The dilatometer imposes a fixed displacement on the soil, like the CPT, but then uses a further small increment of membrane displacement to measure a pressure response. It is a device that is less sophisticated than the SBP but which disturbs the soil in a manner that is repeatable. The dilatometer test gives two pressures: P_o, corresponding to initial lift-off of the sensing membrane, and P_1, corresponding to the pressure at full membrane displacement (1.1 mm).

Konrad (1988) carried out CC tests of the flat plate dilatometer (DMT) in Ottawa sand. The pressure ratio $(P_1 - P_o)/p'$ was found to be dependent on the state parameter. However, Konrad's chamber test involved placing the sand around the dilatometer and this does not replicate the conditions of the DMT in field use. Konrad also suggested that the results might be generalized from the particular case of Ottawa sands to other sands by normalizing the state parameter using e_{min} and e_{max} but, as shown in Chapter 2, this normalization is neither theoretically desirable nor does it unify triaxial test data. In short, Konrad's work indicates some potential for the DMT test in liquefaction assessments but much more needs to be done before there is a basis for relying on DMT data in engineering practice for this purpose.

4.10.3 Using the SPT database

The relative merits of SPT and CPT were discussed early in this chapter and it was noted then that a mapping of SPT from CPT allows the extensive existing experience of the SPT to be used while basing current testing on the much preferable CPT. This mapping was developed by Jefferies and Davies (1993). The methodology to recover N_{60} from the CPT uses soil classification derived from the CPT data itself. The CPT penetration resistance and SPT blow count are linearly proportional, Equation 4.1, and characterized by a coefficient α. The mapping treats α as a linear function of the soil behaviour–type index I_c given by (4.13a):

$$\alpha = 0.85\left(1 - \frac{I_c}{4.75}\right) \text{ MPa/blow} \tag{4.29}$$

The uncertainty in N_{60} recovered using Equations 4.1, 4.13a and 4.29 is less than the basic uncertainty in the SPT itself. This somewhat curious result arises because the intrinsic poor repeatability of the SPT is averaged out in the CPT–SPT correlation. If a reliable N value is wanted, it is best to avoid the SPT itself, carry out a CPT sounding and then use the mapping between the two test types to compute N from q_c. Robertson (2012) suggests that Equation 4.29 has been shown to be reasonable for a wide range of soils, but somewhat underpredicts N_{60} in clays.

4.11 COMMENTARY ON STATE DETERMINATION USING THE CPT

This chapter has presented and discussed the determination of the state parameter in-situ, which almost begs the question: why? This question is particularly pertinent since some approaches to liquefaction are directly based on using the penetration resistance without the intermediate step of computing the state parameter (or relative density). There are three very good reasons why the intermediate step of determining the in-situ state parameter is important:

1. The way soil properties affect the CPT is different from the way the same properties affect the liquefaction resistance, that is penetration data will not scale simply between case histories.
2. The framework for liquefaction, even in its most empirical form, involves at least an elemental concept of the critical state and working in terms of the state parameter is the simplest and best method to remove most of the effects of soil gradation on liquefaction strengths.
3. There is more to geotechnical engineering than liquefaction, and the state parameter underlies every good model for sand behaviour (which will be needed for any general evaluation of soil–structure interaction).

When critical state concepts were first put forward by Casagrande (1936) in the context of liquefaction resistance at Franklin Falls Dam, and then subsequently used in assessing the failure of Fort Peck Dam, interest focused on the in-situ void ratio. This approach has continued on and off for many decades and arguably culminated in the efforts to obtain undisturbed samples at Duncan Dam (in the 1990s). What has emerged from this work is that soil cannot be transferred from in-situ to an element test (e.g. triaxial or cyclic simple shear) in an undisturbed condition. The idea that if we are just clever enough we can avoid understanding soil and test under prototype stress paths is a chimera. Further, detailed measurement of void ratio misses how soil gradation affects the CSL. Knowing the void ratio is not enough, as you also need to know the corresponding CSL which defines the end state after shearing. All of which leads back to the need to know the state parameter, which is currently best measured using the CPT.

Once the CPT is accepted as the way forward, the next step is how to carry out CPTs. We have emphasized the accuracy and repeatability of the CPT and its independence from test operators. That is not quite the whole story as you need to choose the equipment to be used and ensure it is appropriate for the task. Section 9.5 provides the practical guidance on standards, selection of the right equipment and procedures to ensure you get the most out of the test.

A particular problem in the CPT industry is the persistence of the normalization of CPT data to a reference stress level, q_{c1}. This chapter has set out the case why q_{c1} is a misleading approach, but it is an approach that continues to have its advocates and which is found in most software packages. If in doubt, use the downloadable spreadsheet to verify the results obtained with the software you are proposing to use. As generally implemented, q_{c1} is an inaccurate index of void ratio and completely misleads on the state parameter.

Accurate evaluation of soil state from the CPT requires independent measurement of G_{max} and K_o. Of these, measuring G_{max} is arguably now seen as standard within the industry with many companies expecting every fifth or so CPT will be done using a seismic cone to measure the G_{max} profile; what we are asking of you is neither unusual nor difficult. The geostatic stress ratio is a very different matter, and there is no doubt that geotechnical engineering presently has a lot of trouble with K_o. But the facts are clear: K_o surfaces as an important variable in both CC studies and theoretical models of soil behaviour. Larger projects likely warrant detailed testing to determine K_o, but smaller projects may not be able to adsorb the costs of such testing. For smaller projects, we suggest 'engineering judgement' based in part on the data we have presented in this chapter, but do make sure you set K_o greater than (1–sin ϕ) which you would get from Jaky's equation!

Chapter 5

Soil variability and characteristic states

5.1 INTRODUCTION

So far, any particular soil has largely been assumed to be uniform and homogenous. This worked well until the discussion moved from calibration chambers to real soils and examined the effect of silt content. Implicit in the discussion was the fact that the CPT tip resistance, friction and pore pressure would vary as a function of soil type, as well as the in-situ state of the soil. Careful interpretation of the piezocone data will determine the state parameter, more or less continuously, throughout the soil at the sounding locations. But, experience shows that the soil state is rather variable laterally as well as vertically, even in a 'uniform' sand fill, that is, one that has been constructed or deposited geologically in a similar way and from the same source material. The question that now needs to be asked is: What state best characterizes the overall response of the soil to loading that may lead to liquefaction? This question may be divided into two issues:

1. What is the real distribution of state in-situ?
2. What value of the given distribution characterizes behaviour of the soil mass?

In the terminology of limit states design in civil engineering, as, for example, found in Eurocode 7, the question is more simply phrased as, 'What is the characteristic value of sand state?' Unfortunately there is no simple answer to this question, and the same issue is a major unresolved topic of discussion amongst developers of design codes.

This chapter describes the results of some important studies into the effects of soil variability on the response of sands to cyclic loading. It will also examine distributions of state in-situ, as well as analyses of how these distributions can influence performance in static loading cases. Finally, the chapter will round off with a brief discussion of characteristic values for liquefaction and limit states design codes.

5.2 EFFECT OF LOOSE POCKETS ON PERFORMANCE

During the 1960s and 1970s, the Dutch undertook major engineering works for sea defences in the Delta region where the Rhine, Meuse and Scheldt rivers discharge into the North Sea. They constructed many storm surge caissons founded on a sandy seabed in water depths of 15–35m, which had to resist both static loads from differential water levels on either side of the caissons and cyclic forces from waves hitting the caissons. In order to study liquefaction and the effectiveness of densification methods, a one-third scale field model caisson was built, and predictions of its performance were made, testing out methods of both analysis and design current at the time. One such prediction was made on the basis of centrifuge tests carried out at the University of Manchester and published by Rowe and Craig (1976). This study is interesting in that it highlights the effect that loose pockets of sand may have on dynamic performance. It is one of the few studies where this aspect has been actually tested, and despite the limitations of the centrifuge, there is much to be learnt from this study.

The centrifuge model caisson was 0.84 m ×0.455 m in plan and 0.2 m high. An acceleration of 110*g* in the centrifuge modelled a full size 50 m wide caisson; 33*g* modelled the test caisson (27.7 m long × 15 m wide × 6.5 m high). Table 5.1 summarizes the properties of the model, test and prototype caissons. The model caisson was placed on a sand bed prepared to a desired density index (relative density), either uniform or with a specific distribution of loose zones. The model was subjected to cyclic horizontal loading, with load levels increasing in steps to simulate wave loading on real caissons where the storm builds gradually over time. Table 5.2 summarizes the loadings used for the model. Because the loadings were periodic, and because of the issue of whether or not undrained behaviour would arise, care was taken with reproducing the same order of time factor in the model as in the prototype. The time factor calculations are summarized in Table 5.3. Although the test caisson time

Table 5.1 Dimensions and properties of model and prototype caissons

			Model	
	Test caisson	*Prototype*	*Test scale* $N_g = 33$	*Prototype scale* $N_g = 110$
Width (m)	15	50	0.45	0.455
Length (m)	27.7	100	0.84	0.84
Height of load action (m)	6.5	27	0.2	0.25
Submerged weight	13.75 MN	9.65 MN/m	0.38 kN	0.38 kN
EI/m (MN m²/m)	7.2×10^4	2.0×10^7	1.75	1.75
q (kN/m²)	33	193	33	193

Source: Data from Rowe, P.W. and Craig, W.H., *Design and Construction of Offshore Structures*, Institution of Civil Engineers, London, U.K., pp. 49–55, 1976.

Table 5.2 Cyclic loading stages in caisson models

	Field		Model			
Load parcel	*Static force H (kN)*	*Cyclic force ± ΔH (kN)*	*Static H (kN)*	*Cyclic ± ΔH (kN)*	*Maximum stress (H+ΔH)/A (kN/m²)*	H_{max}/W
P_0	200	±250	–	–		
P_1	400	±500	0.3	±0.4		
P_2	800	±1000	0.7	±0.9		
P_3	1200	±1500	1.1	±1.4		
P_4	1600	±2000	1.5	±1.8		
P_5	2000	±2500	1.8	±2.3	10.8	0.33
P_6	2400	±3000	2.2	±2.7		
P_7	2800	±3500	2.6	±3.2		
P_8	3200	±4000	2.9	±3.7		0.53

Source: Data from Rowe, P.W. and Craig, W.H., *Design and Construction of Offshore Structures*, Institution of Civil Engineers, London, U.K., pp. 49–55, 1976.

Table 5.3 Model time per cycle and time factors for centrifuge models

			As used on model	
	Field test caisson	*Prototype caisson*	*Mean*	*Lowest*
Number of cycles/parcel	300	300	300	
Number of parcels	5	5	4–8	
Total field time (2t)	75 min	250 min		
Time factor $T = c_v t/B^2$	1.5	0.45	0.21	0.04
N_t	23	2.1	3.3–1	0.66–0.2
Model time/cycle	69 s	21 s	10	2

Source: Data from Rowe, P.W. and Craig, W.H., *Design and Construction of Offshore Structures*, Institution of Civil Engineers, London, U.K., pp. 49–55, 1976.

factor exceeded that for the prototype, both were sufficiently high that most cyclic excess pore pressures would have been dissipated relatively quickly.

Both the uniform 50% and 70% density sand beds carried the full set of loading, with essentially drained displacement. The pore pressure generated during cyclic loading dissipated as quickly as it was generated, and the foundation behaviour was one of shakedown. Vertical and horizontal displacements arose with load cycling, eventually amounting to about 0.2% of the test caisson width prior to the onset of failure. Failure was always by sliding. Figure 5.1 shows the measured data. Interestingly, the field tests on the test caisson

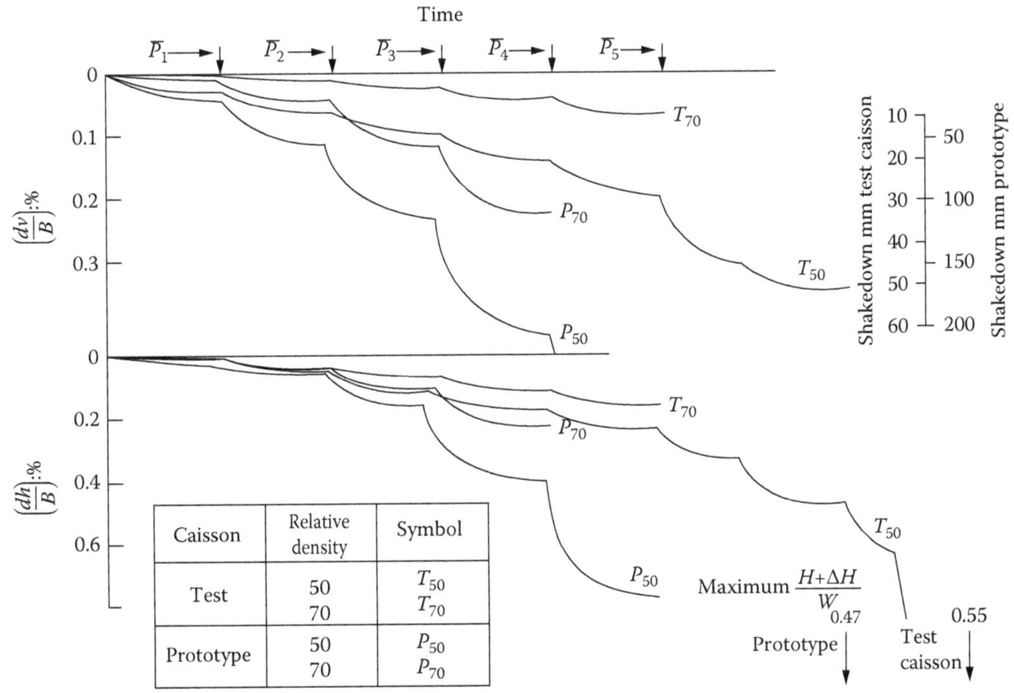

Caisson	Relative density	Symbol
Test	50 70	T_{50} T_{70}
Prototype	50 70	P_{50} P_{70}

Figure 5.1 Measured response of caissons subject to increasing stages of cyclic loading in centrifuge test. (From Rowe, P.W. and Craig, W.H., *Design and Construction of Offshore Structures*, Institution of Civil Engineers, London, U.K., pp. 49–55, 1976. With permission from the Institution of Civil Engineers.)

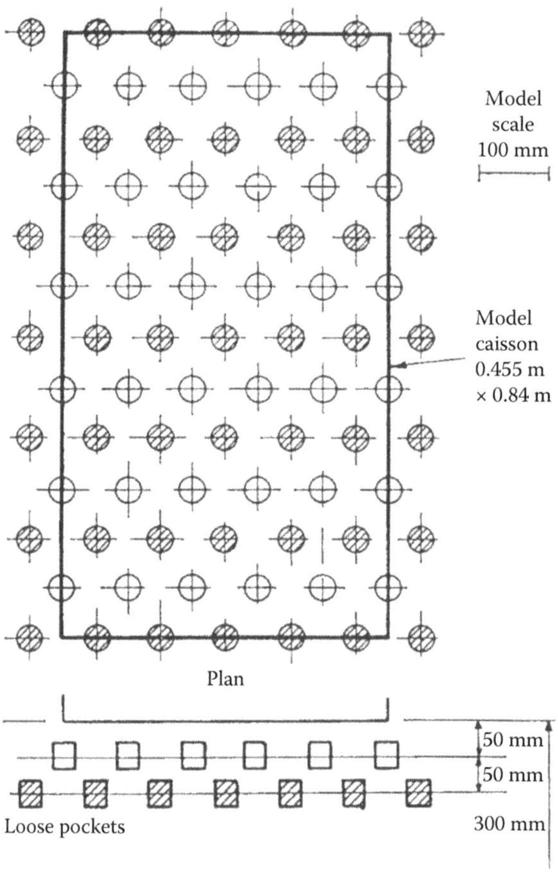

Figure 5.2 Layout of loose pockets below caissons. (From Rowe, P.W. and Craig, W.H., *Design and Construction of Offshore Structures*, Institution of Civil Engineers, London, U.K., pp. 49–55, 1976. With permission from the Institution of Civil Engineers.)

had shown rather different behaviour from that obtained in the uniform density foundation model. The field tests had shown early larger displacements and excess pore water pressures, although these had stabilized later in the loading cycle. Rowe and Craig therefore simulated non-uniform foundation conditions to ascertain if this was the cause of the difference between model and test caisson behaviours.

Figure 5.2 shows the distribution of loose pockets of sand below the model caisson. Two intensities of loose pockets were explored:

- Loose pockets of sand in the upper zone, representing one-third of the caisson width, and equal to 4% of volume in this zone.
- Loose pockets equal to 10% of volume in the same zone.

With 4% by volume loose pockets in an otherwise 50% relative density foundation, pore pressures developed in the loose pockets and spread to the denser zones, with significantly greater displacements after the third packet of 300 loading cycles. Pore pressure dissipation occurred with horizontal and vertical displacements of about 0.4% of test caisson width after five packets of loading. Figure 5.3 shows the displacement data and the piezometric

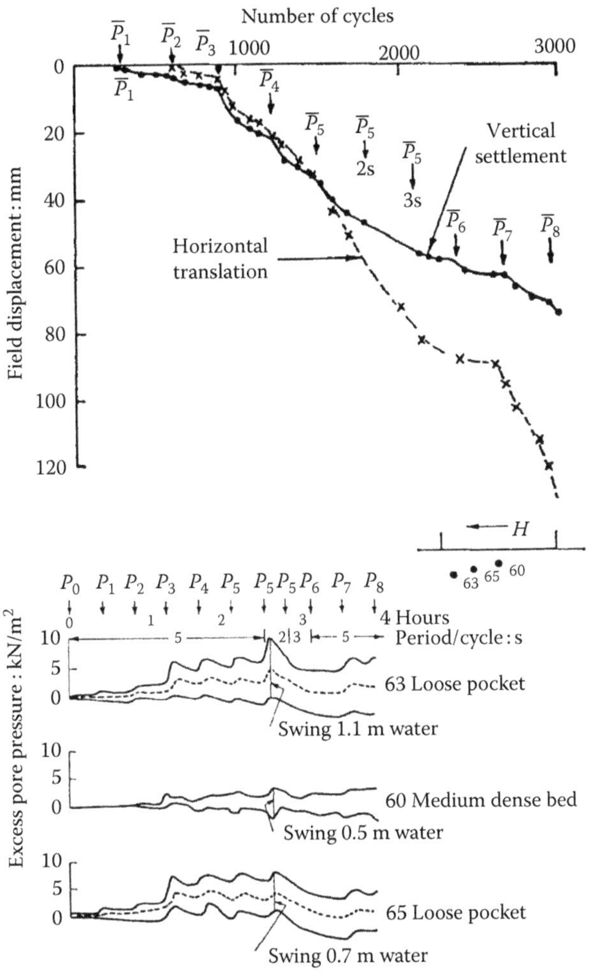

Figure 5.3 Scaled displacements and pore pressures observed in model with 4% loose zones in fill. (From Rowe, P.W. and Craig, W.H., *Design and Construction of Offshore Structures*, Institution of Civil Engineers, London, U.K., pp. 49–55, 1976. With permission from the Institution of Civil Engineers.)

data. Notice how the pore pressure generated in the loose pocket is also seen in the denser sand, not unexpected given the time factor of the loading. Given this type of behaviour, the loosest soil will begin to control what occurs.

In contrast to the benign behaviour with uniform and 4% loose zones, inclusion of 10% by volume loose pockets produced liquefaction failure during the second stage of loading. A few cycles into the third loading stage, the model reached its limits of displacement and the test was stopped and reset. High pore pressures, equivalent to the weight of the caisson, were observed in the sand. Figure 5.4 shows the displacement data and the piezometric data for these conditions.

This series of tests by Rowe and Craig illustrates that loose pockets have a disproportionate effect on the behaviour of sand foundations under cyclic loads. The effect is disproportionate to the volume of loose material as a result of re-distribution of pore pressures. High pore pressures generated in loose pockets dissipate into the surrounding denser material and, in effect, decrease the strength and stiffness of the denser material.

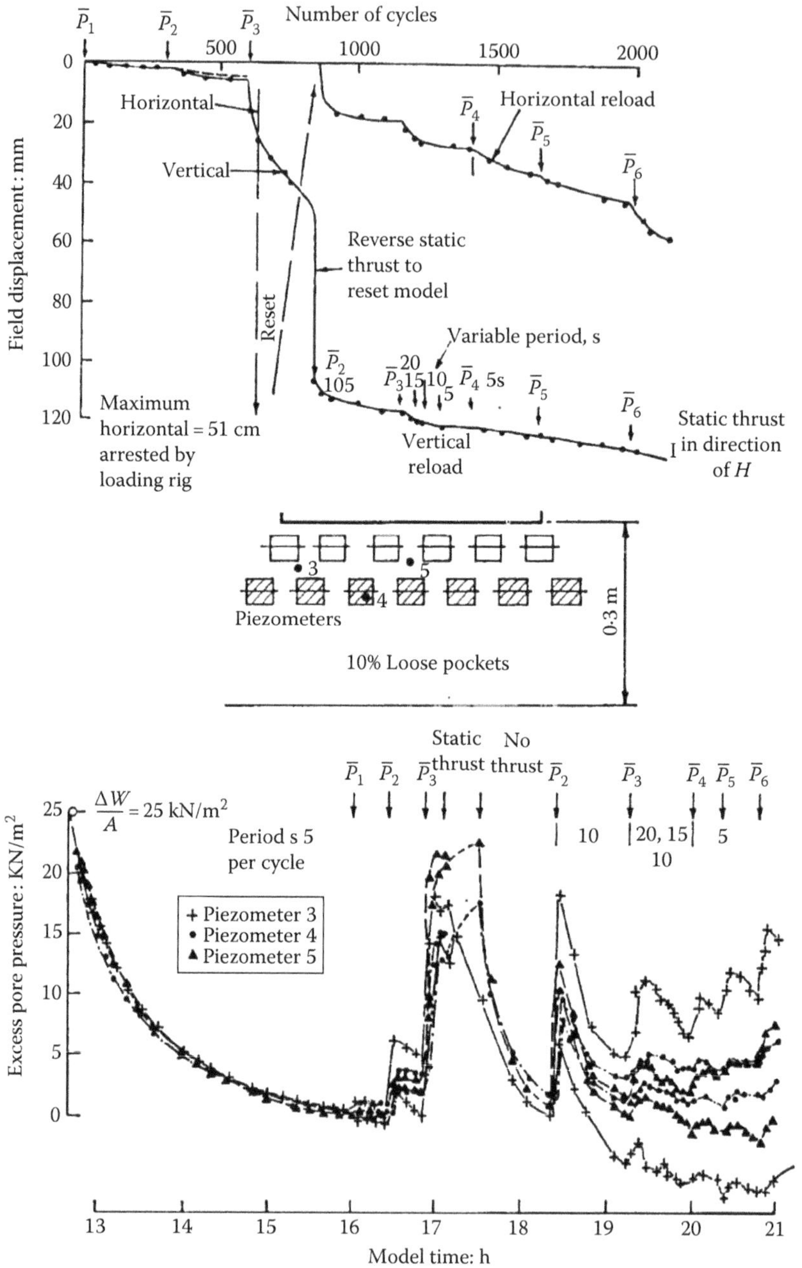

Figure 5.4 Scaled displacement and piezometric data for centrifuge model with 10% loose zones in fill. (From Rowe, P.W. and Craig, W.H., *Design and Construction of Offshore Structures*, Institution of Civil Engineers, London, U.K., pp. 49–55, 1976. With permission from the Institution of Civil Engineers.)

5.3 EFFECT OF VARIABILITY OF IN-SITU STATE ON CYCLIC PERFORMANCE

The analysis now shifts to some real in-situ data to examine its variability. First, the Tarsiut P-45 case history is examined. This data set was introduced in the previous chapter in the context of repeatability of the CPT and measurement of K_o with the self-bored pressuremeter test. Having determined the characteristics of the Tarsiut P-45 sand, and using numerical simulations of liquefaction, it is seen how sand with similar statistical characteristics performs under earthquake loading conditions. This work by Popescu (1995) was at the University of Princeton under the guidance of Prof. Prevost, and is an early example of what is now known as stochastic modelling.

5.3.1 Distribution of CPT resistance in Tarsiut P-45 fill

The Tarsiut P-45 fill was dredged Erksak sand. It was hydraulically placed within a caisson structure (0–21 m depth) and used to construct a sand berm (21–30 m depth), as illustrated in Figure 5.5. In Chapter 1 the Molikpaq structure and its performance at Amauligak I-65 when it experienced a large ice load event was discussed. The deployment of the Molikpaq

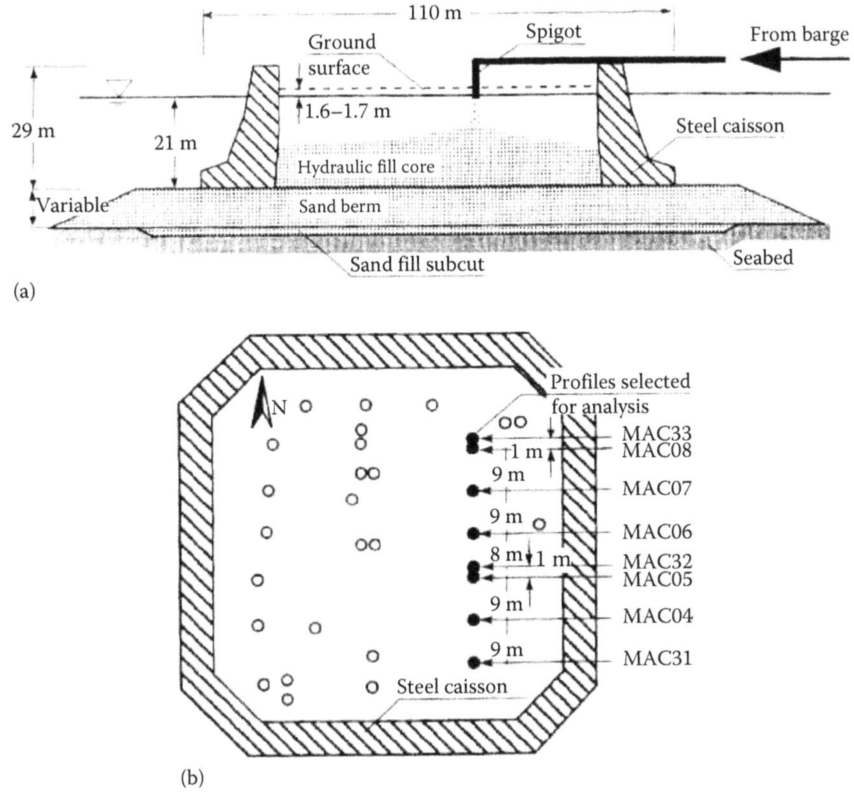

Figure 5.5 Schematic cross section of the Molikpaq at Tarsiut P-45 showing locations of CPTs to determine fill properties. (a) Cross section and (b) plan. (Adapted from Jefferies, M.G. et al., Molikpaq deployment at Tarsiut P-45, *Proceedings of ASCE Specialty Conference on Civil Engineering in the Arctic Offshore*, San Francisco, CA, pp. 1–27, 1985 by Popescu et al., 1997, reproduced with permission from the Institution of Civil Engineers, London, U.K. and Dr. Popescu.)

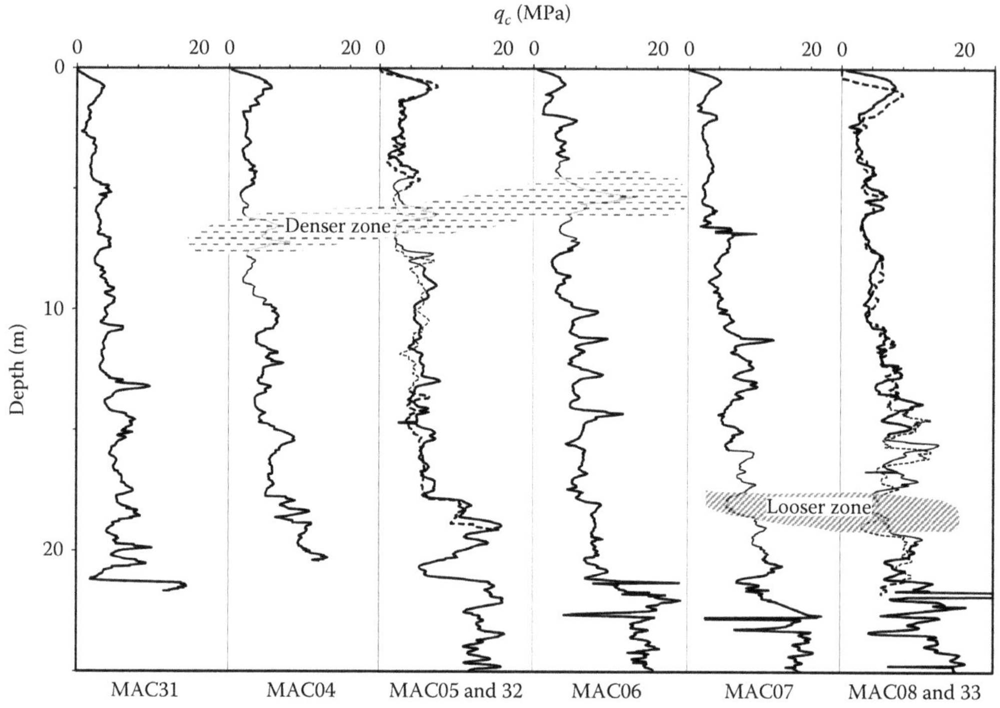

Figure 5.6 Examples of CPTs in Tarsiut P-45 fill. These CPTs are spaced about 9 m apart (see Figure 5.5). MAC 05 and 32 and MAC 08 and 33 are spaced 1 m apart to demonstrate repeatability of measurements. (Adapted from Jefferies, M.G. et al., Molikpaq deployment at Tarsiut P-45, *Proceedings of ASCE Specialty Conference on Civil Engineering in the Arctic Offshore*, San Francisco, CA, pp. 1–27, 1985; Popescu, R. et al., *Géotechnique*, 47, 1019, 1997.)

illustrated in Figure 1.20 was similar at Tarsiut P-45, although the conditions of the fill and the foundations were different.

The sandfill was comprehensively tested on two occasions: October 1984 and April 1985. In order to confirm the sand's adequacy before the structure was used for oil exploration, 32 CPTs (soundings TD84cc01–TD84cc33; # 29 is missing) were put down in October 1984 shortly after the fill was placed. A further five soundings (MacRes01–05) were put down in April 1985 to investigate the effect of ageing on the sand strength (there was none). The data are included on the website, including figures showing the location of the data.

The Tarsiut P-45 data gives us a good statistical data set in a geologically uniform material. Figure 5.6 shows examples of the CPT profiles. This data set represents a rather intense investigation, as the plan area of the site was just 72 m × 72 m. Typical spacing between CPT soundings was just 9 m, and several were done directly adjacent to each other to evaluate repeatability. There is excellent repeatability between CPTs conducted within 1 m of each other.

The CPT data showed a steady increase in tip resistance q_c with depth. Figure 5.7 shows the data from Figure 5.6 superimposed on the average trend in the data given by $q_c = 2.35 + 0.37z$ in the caisson core and $q_c = 6.16 + 0.44z$ in the sand berm where z is the depth below surface. A first statistical analysis of the CPT data was undertaken by dividing the fill into 1 m thick horizontal layers and, assuming that there was no effect of depth

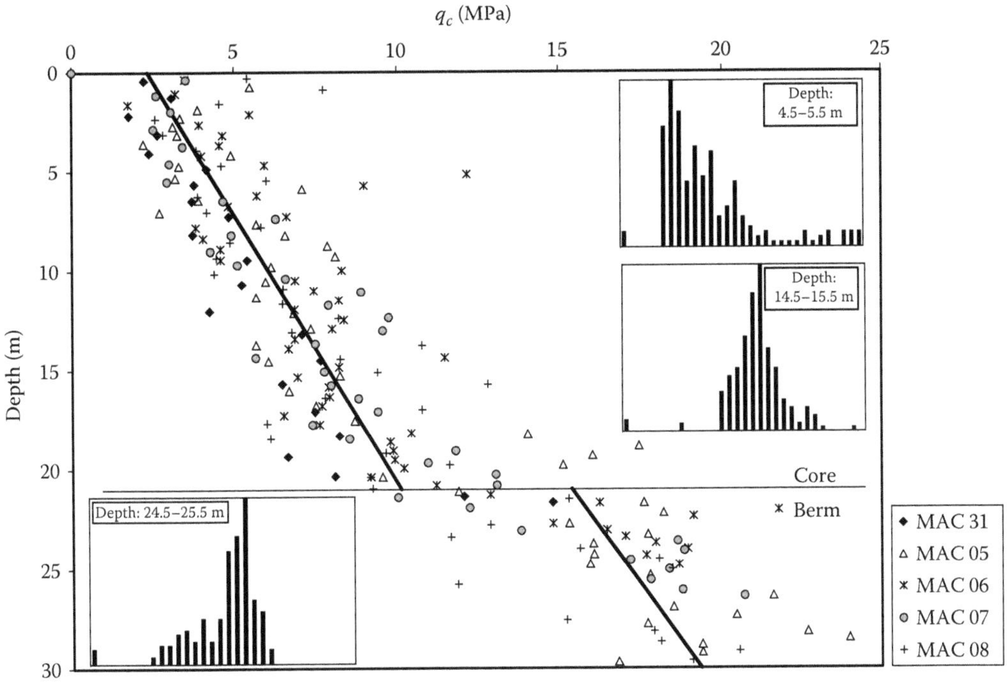

Figure 5.7 Selected Tarsiut P-45 CPTs plotted against depth with average trends in core and berm fill shown. Inset histograms are distributions of q_c values in 1 m depth intervals at depths of 5, 15 and 25 m.

within each 1 m layer, calculating the distribution of q_c values. A simple box-sort was used, in which the data range was taken as 0–50 MPa in 1 MPa increments, and each measured q_c value (typically there were 50 per m) was allocated to the appropriate box. This procedure gives the distribution of q_c at any particular depth (Figure 5.7 shows examples as insets), and can be plotted as a function of depth, as shown in Figure 5.8 in terms of the 50 and 80 percentiles (labelled as 'median' and 20% in Figure 5.8), together with the maximum and minimum values measured. There is quite a noticeable difference between the 50 and 80 percentiles. The distribution of q_c at any stress level appeared to approximate log normal, which would be a reasonable first expectation, since long normal distributions naturally arise where the variance is proportional to the mean. Such variance is not a bad approximation for soils, as both their stiffness and strength are proportional to the effective stress. Figure 5.8 also shows a similar statistical distribution of ψ interpreted from the q_c values at the same depths.

Popescu employed a more sophisticated approach to determining the distribution of tip resistance values around the mean trends, examining the correlation structure of the probability distribution functions he obtained. He made no *a priori* assumptions about the distributions, relying instead on curve fitting techniques to determine distributions. As it turns out, a skewed beta distribution fits the distribution of the tip resistance data around the depth normalized mean rather well. The correlation distances derived from this analysis of the data are:

- Core fill: 0.95 m vertical and 12.1 m horizontal
- Sand berm: 0.5 m vertical and 5.4 m horizontal

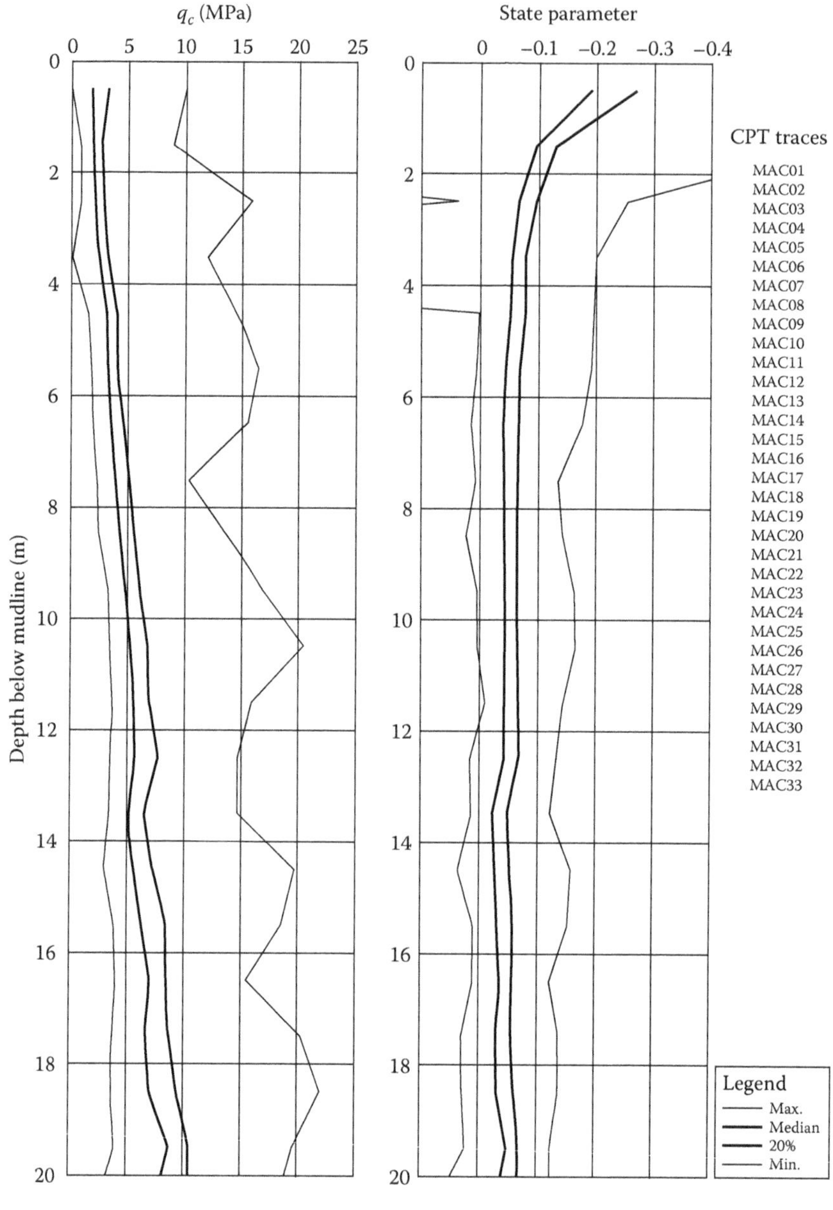

Figure 5.8 Statistical profile of penetration resistance q_c and state parameter at Tarsiut P-45.

Unfortunately, it also turns out that the horizontal correlation distance is of the same order of magnitude as the spacing between the CPTs (9 m), so there is some uncertainty regarding these values. Nevertheless, this analysis gives us some insight into the scale of variation of a 'uniform' sand deposit. As expected in a sedimentary deposit, there is a large difference in the scale of horizontal and vertical variability, in this case about an order of magnitude. The scale of horizontal variation is also such that one would expect significant variations under a typically sized man-made structure.

An interesting aspect of Popescu's work is that it did not stop with working out the statistics of the fill, but rather went on to develop stochastic simulations which illustrate the inferred

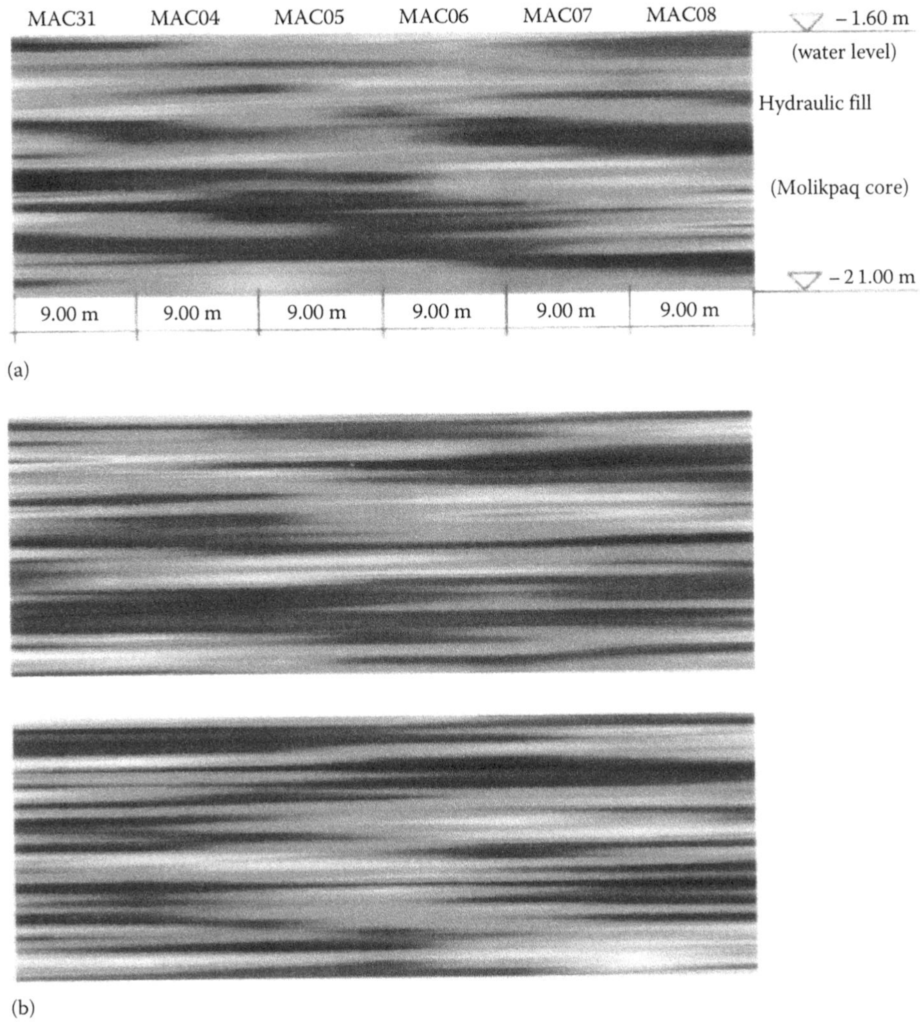

Figure 5.9 Stochastic reconstruction of Tarsiut P-45 fill by Popescu. (a) Normalized fluctuations of in-situ measured resistance (data from the hydraulic fill layer at Tarsiut P-45, represented on a 9.00 m × 0.25 m mesh) and (b) simulated sample fields of standardized cone resistance (the mid-point method has been used for data transfer). (From Popescu, R., Stochastic variability of soil properties: Data analysis, digital simulation, effects on system behaviour, PhD thesis, Princeton University, Princeton, NJ, 1995. Reproduced with permission from Dr. Popescu.)

relationship of the stratification in the fill. Figure 5.9 shows such simulations; the shading indicating deviation of q_c about the mean trend line value, looser zones being shown lighter and denser zones darker. The much layered structure with looser and denser zones is readily seen.

With this understanding of intrinsic variability in a 'uniform' soil, a question naturally arises as to just how much site investigation one needs in order to obtain a reasonable estimate of the soil's distribution. This question was addressed by Jefferies et al. (1988b), and the answer depends upon whether one assumes the form of the distribution, needing just its parameters (e.g. mean and variance), or whether one is 'starting from scratch', wanting to question the nature of the distribution. Obviously, much less effort is required in the former case, but about 15 CPT soundings will still be needed in a soil deposit for which there is no

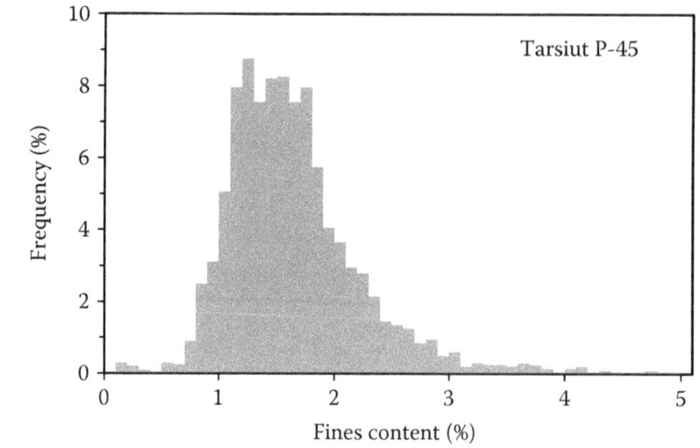

Figure 5.10 Distribution of fines content measured in Tarsiut P-45 fill. (From Jefferies, M.G. et al., Characterization of sandfills with the cone penetration test, *Proceedings of Conference on Penetration Testing in the U.K.*, Birmingham, England, pp. 73–76, 1988. With permission from the Institution of Civil Engineers, London, U.K.)

underlying geologic bias (e.g. buried channels). This means that two boreholes with SPTs simply cannot be relied upon to characterize a site. Even with the critical state locus (CSL), which provides something like 1–2 orders of magnitude more data than an SPT, 12–15 soundings are required on a well-controlled site. Geologic complexity can much increase this requirement. Fortunately, the CPT is inexpensive.

The variability in the silt content of this clean sand fill was also explored, as this influences the CSL, which in turn influences the inversion of the CPT data to state estimates. Samples were obtained routinely on the dredge and tested prior to discharge of the fill into the caisson. Once the fill was in place, boreholes were drilled and sampled. These samples were then also tested. Figure 5.10 shows the results – a log normal distribution of fines is indicated, with an average of about 1.5% silt, but with 90 percentile being as high as 2.5%–3.0% silt. Recall that, because of the hydraulic dredging used, this is in a thoroughly washed soil. The inferred variation of the CSL parameters is then quite large; see Chapter 2, Section 2.6.

5.3.2 Liquefaction analysis under earthquake loading

Popescu used the finite element computer program Dynaflow to compute the performance of a structure founded on a sand fill with the same statistical characteristics as the Tarsiut fill, under dynamic loading conditions equal to the Niigata earthquake in 1964. (Popescu did not use the state parameter approach, but rather related the tip resistance directly to the cyclic liquefaction resistance, and then used this relationship to calibrate constitutive parameters in a multi-yield surface constitutive model. This does not, however, reduce the relevance of his findings to this book and the current line of investigation – how cyclic liquefaction is simulated is described in Chapter 7.) He looked first at a so-called deterministic fill which used uniform mean values of tip resistance. He then compared the results of various stochastic realizations with distributions of tip resistance corresponding to the Tarsiut fill statistics. In addition, he looked at the effect of varying the correlation distance.

The important point from Popescu's work is the conclusion arising from the comparisons he made. The liquefaction index is defined as the ratio of the excess pore pressure to the vertical effective stress before loading, $\Delta u/\sigma'_{vo}$. In the uniform 'deterministic' deposit, the maximum excess pore pressure generated in the analysis was $\Delta u/\sigma'_{vo} = 0.6$, indicating that

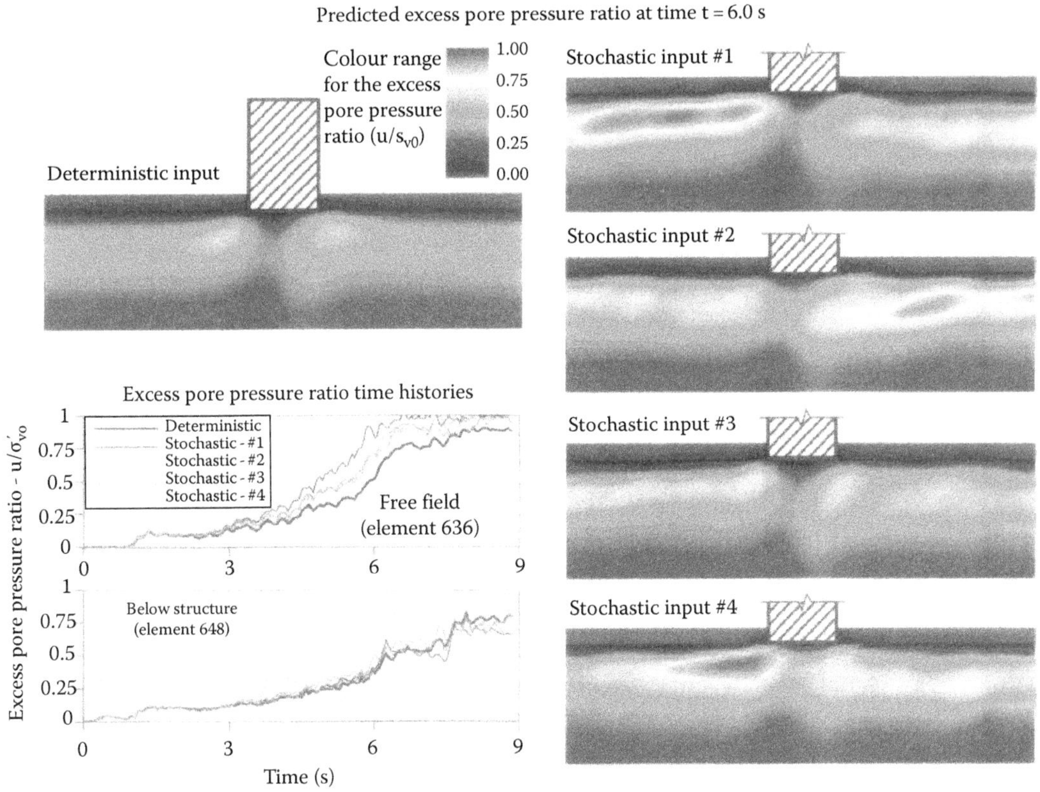

Figure 5.11 Liquefaction of variable fill computed by Popescu. (From Popescu, R., Stochastic variability of soil properties: Data analysis, digital simulation, effects on system behaviour, PhD thesis, Princeton University, Princeton, NJ, 1995. Reproduced with permission from Dr. Popescu.)

liquefaction did not occur. For the stochastically variable deposits, however, the maximum pore pressure was as high as $\Delta u/\sigma'_{vo} = 0.8 - 1.0$. Figure 5.11 shows some results of the simulations in terms of this pore pressure ratio. Consistent with the results of Rowe and Craig's centrifuge tests, there were higher pore pressures in looser zones, which then spread to denser materials. In particular, Popescu notes that the tails of the probability distribution (at the loose end) are especially important, and that pore pressures will be under-predicted if normal distributions are assumed.

Popescu et al. (1997) went one step further than Popescu's thesis work and carried out deterministic simulations with uniform properties, but with the uniform properties made systematically looser from one simulation to the next. In this case, their analysis was only for level ground under seismic loading. Figure 5.12 shows their results, presenting the pore pressures in six stochastic simulations, and six simulations with uniform properties. Comparing the results of the variable material with the uniform material, one can see that the 80 percentile uniform material approximates the extent of liquefaction in the stochastic simulations, whilst that of the 70 and 90 percentile simulations under- and overestimate it, respectively. Their logical conclusion is that the 80% value might be taken as characteristic for the conditions they were modelling.

Another conclusion of note from Popescu's thesis is that there is little influence on the pore pressure magnitude if horizontal correlation distances greater than those for Tarsiut-P45

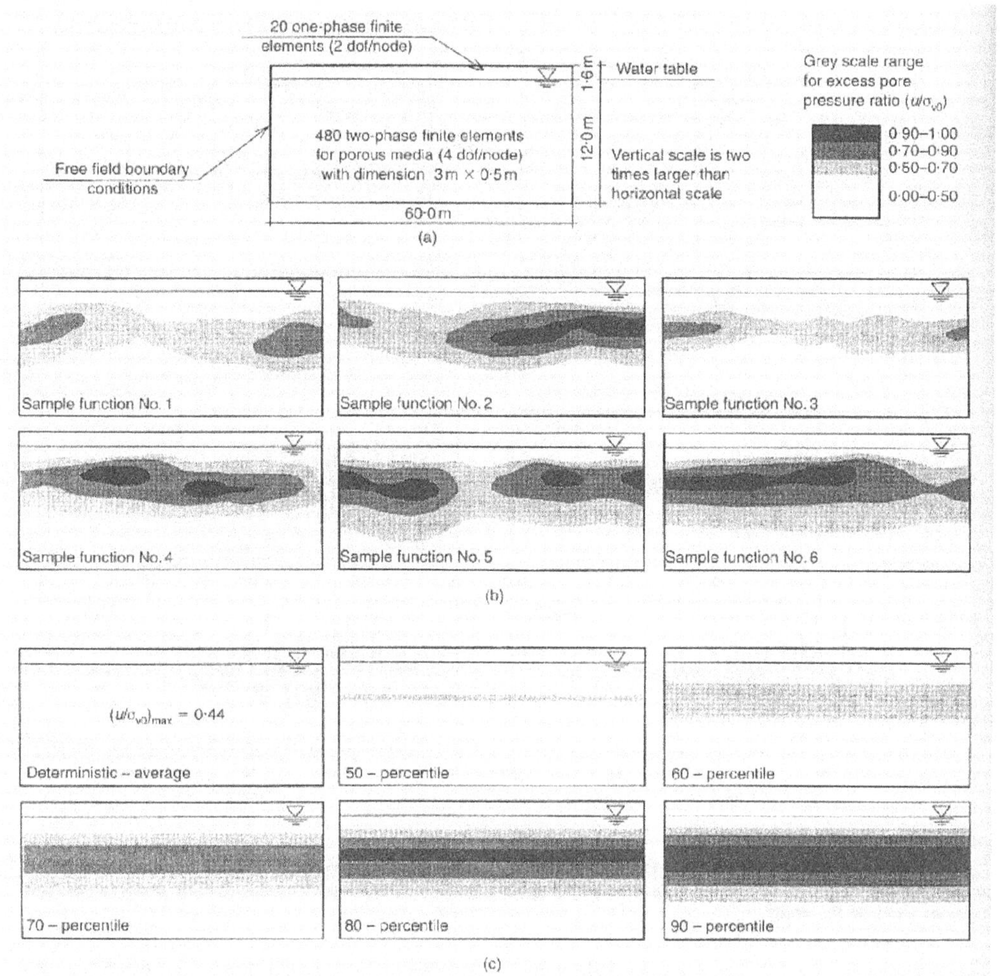

Figure 5.12 Comparison of uniform and variable fill results in Popescu et al. (a) Modelled geometry, (b) pore pressure ratios in six stochastic realizations and (c) pore pressure ratio in uniform layered fill. (Reproduced from Popescu, R. et al., *Géotechnique*, 47, 1019, 1997. With permission from the Institution of Civil Engineers, London, U.K. and Dr. Popescu.)

are input. In effect, in this example of a uniform sand, the scale of variability is such that loose zones are sufficiently large or close enough together to have an influence on the overall behaviour of the fill, even though on average the material would not be considered susceptible to liquefaction. The loose zones in sands cannot be ignored for liquefaction problems.

5.4 NERLERK CASE HISTORY

Spatial variability of density in real soils is not only an issue for cyclic loads. Discussion now returns to the Nerlerk berm failures, described in detail in Chapter 1, Section 1.3. Recall that the Nerlerk berm suffered several massive slope failures when the berm height was about 27 m above seabed, still 9 m short of its target height.

The Nerlerk engineers, as reported by Sladen et al. (1985a), worked from the morphology of the slides as defined by the bathymetry, and back analysed the failures assuming they occurred in the Nerlerk sand. They concluded that the slides were caused by liquefaction of the fill triggered by static loading. The problem with this explanation of the failures was that the sand fill had to be extremely loose ($\psi \sim +0.1$), but the CPTs did not indicate this to be the case.

Much discussion ensued about interpretation of the CPT and the state parameter approach, in particular whether the CPT over predicts state parameter for low stress levels and loose states (Been et al., 1989; Sladen, 1989a,b). The CPT–state parameter methodology presented in Chapter 4 has developed as a result of much of that discussion. It turns out that there was indeed a missing effect in the interpretation, the shear modulus, that manifests itself as a stress level effect. Figure 5.13 shows the current interpretation of the Nerlerk CPT data using the methodology of Chapter 4. The revised interpretation including the stress level effect in itself is nowhere near enough to explain the apparent discrepancy between the states interpreted from the CPTs and the looseness of ψ apparently needed for enaction of the Sladen et al. (1985a) liquefaction mechanism for the Nerlerk berm.

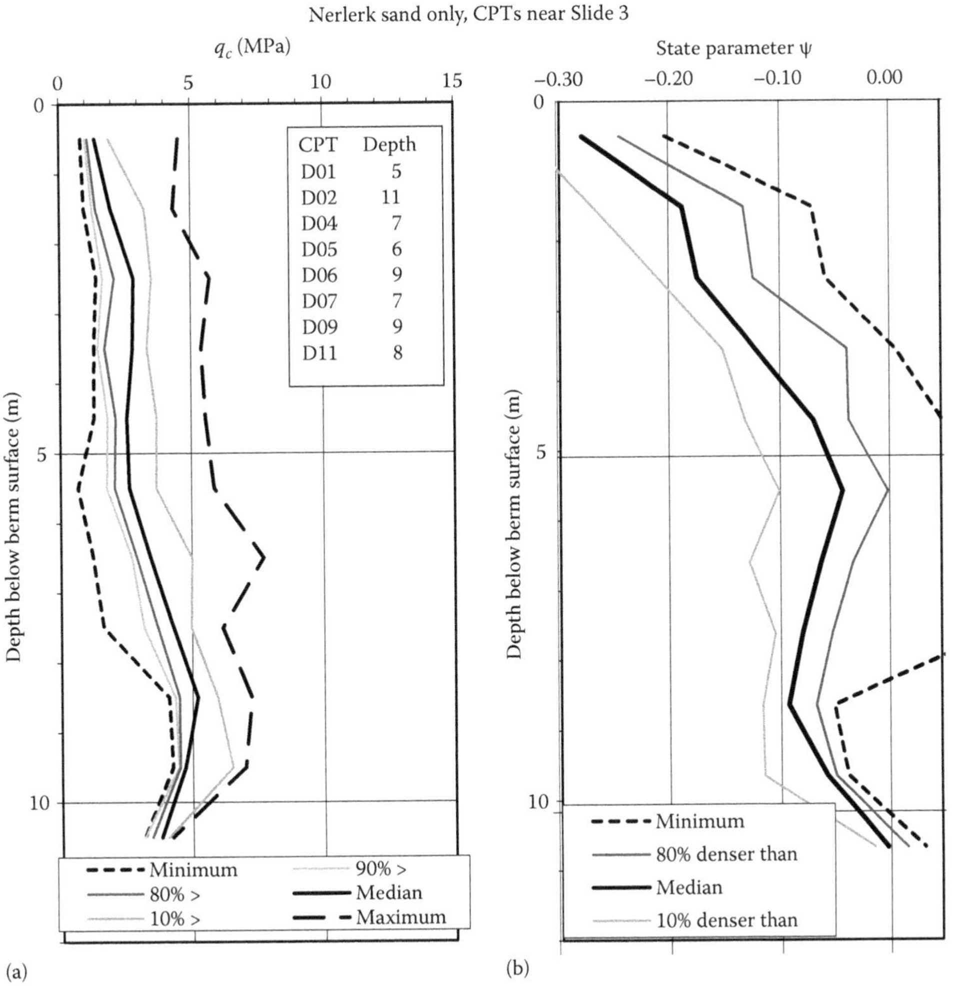

Figure 5.13 Summary of CPT statistics in the area of Slide 3 of Nerlerk berm. State parameter interpretation using variable shear modulus. (a) Penetration resistance and (b) state parameter.

Many other workers also examined this issue of the Nerlerk berm failures. Rogers et al. (1990) reported new data obtained during a Nerlerk site investigation in 1988 and considered the possibility of loose zones of fill at the interface between the 1982 and 1983 fill. Konrad (1991) picked up on the issues and proposed that the minimum undrained shear strength may be rather lower than the steady or critical state strength. Lade (1993) reassessed the Nerlerk berm stability by using the concept of an 'instability line' which occurs, in theory, for non-associated plasticity, and, in practice, is observed in laboratory tests on loose and dense sands. This approach requires a trigger mechanism, which Lade suggested could either have been rapid loading or movements in the underlying clay layer. Finally, Hicks and Boughrarou (1998) performed finite element analyses of the Nerlerk berm using a realistic, double hardening model for sands (Monot), which allowed them to look at the interaction between the different parts of the berm and the underlying clay. This work brought together much of the previous input by supporting the view that the underlying clay layer contributed to the Nerlerk slides (Been et al., 1987a; Rogers et al., 1990), while also agreeing with Sladen et al. (1987) and Lade (1993) that a translational slide through the clay alone was unlikely to explain the flow slides. Hicks and Boughrarou concluded that static liquefaction had occurred in the upper half of the Nerlerk fill, triggered by a combination of rapid sand deposition and limited movements in the weak underlying clay.

It is clear that the performance of the sand fill in the Nerlerk berm was unsatisfactory, but none of the studies mentioned earlier were totally conclusive. The main unresolved issue is whether significant pore pressures or liquefaction may be possible in the sand given the measured CPTs and the estimated state parameter. Figure 5.13 shows the distribution of CPT values in Nerlerk sand in the area of Slide 3, processed in 1 m depth intervals as described earlier for Tarsiut P-45. Between a depth of 3.5 and 10 m, the median $\psi \approx -0.08$. Why are the CPT values so high if the sand was collapsible? Assuming that the CPT data are not wrong, and that a dilatant fill would not liquefy, there was patently some other factor at work.

Onisiphorou (2000) undertook a static, random field analysis of a small part of the Nerlerk berm later published by Hicks and Onisiphorou (2005). This study was similar in a way to that of Popescu described earlier, except that Onisiphorou used ψ directly as the random field variable. She mapped the state parameter onto finite element mesh integration points, and then assigned material properties corresponding to these values of ψ to the integration points. Figure 5.14 shows an example of the result of this process. The spatial variability was generated by assuming a normal distribution of ψ scales of fluctuation that were typically 1 m vertically and 8 m horizontally.

Figure 5.15 shows the results of Onisiphorou's deterministic analysis of the Nerlerk berm based on uniform ψ. At values of $\psi > -0.02$ (i.e. looser than -0.02), a failure mechanism develops, while for $\psi < -0.08$ the berm is stable. At intermediate states there are significant strains but the slope can continue to be loaded. Now look at Figure 5.16, which shows the

Figure 5.14 Distribution of random ψ field mapped onto Nerlerk berm geometry, computed by Onisiphorou 2000. (Courtesy Dr. M. Hicks, University of Manchester, Manchester, England.)

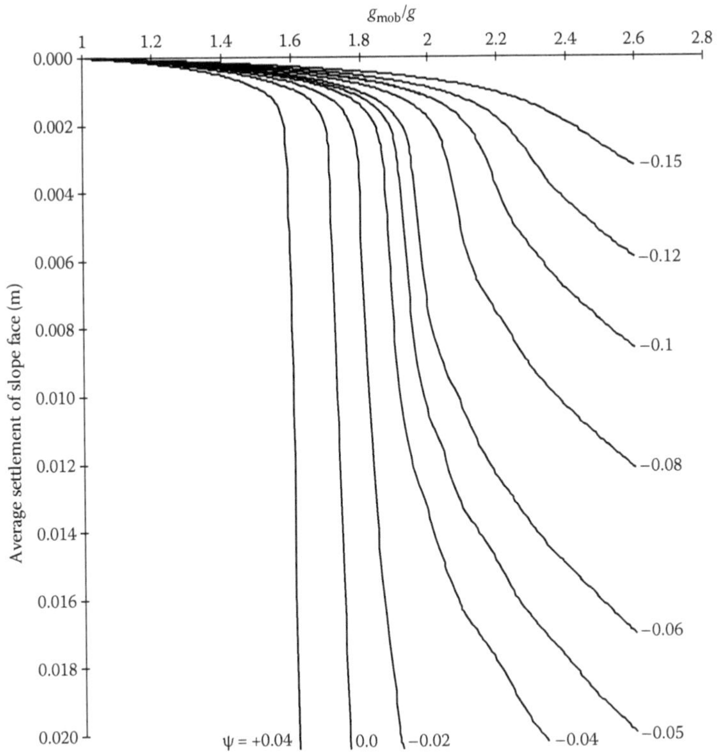

Figure 5.15 Results of Nerlerk berm analysis with uniform fill states. (After Onisiphorou, C., Stochastic analysis of saturated soils using finite elements, PhD thesis, University of Manchester, Manchester, England, 2000.)

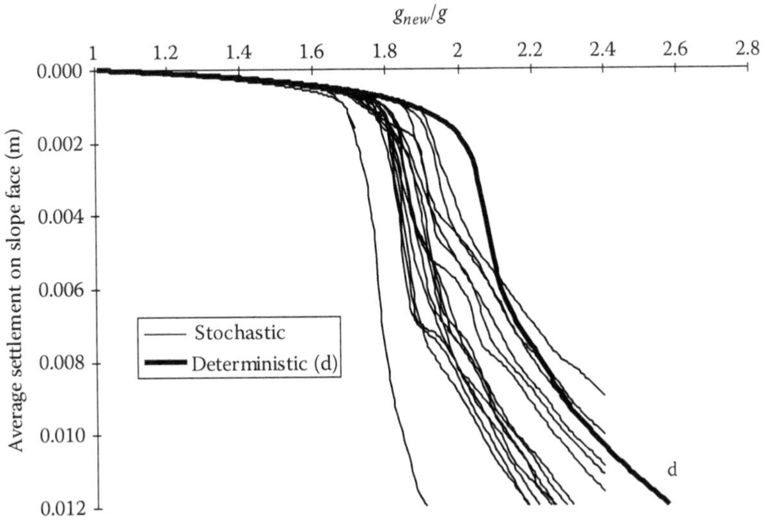

Figure 5.16 Results of analysis of Nerlerk berm with variable field ψ: μ = −0.08, σ = 0.05. (After Onisiphorou, C., Stochastic analysis of saturated soils using finite elements, PhD thesis, University of Manchester, Manchester, England, 2000.)

variable field analysis for a mean $\psi_\mu = -0.08$ and standard deviation $\psi_\sigma = 0.05$, compared to the uniform analysis results. The variable field berm shows a range of responses. The strongest response in the random field corresponds to the uniform field for $\psi = -0.08$, while the weakest response corresponds to the uniform field response for about $\psi = -0.02$ (seen by comparison with Figure 5.15). Based on a large number of sensitivity analyses, Onisiphorou concluded that Nerlerk type densities could liquefy as a result of variability. There was no conflict between the CPT evaluation and material behaviour – the missing parameter was variability, again stressing on the importance of understanding the layering and how the state is distributed within the berm.

So what is the distribution of state parameter in the Nerlerk berm, based on our current best interpretation of the CPT data? Figure 5.13a shows the distribution of q_c with depth for eight CPTs on the western side of the berm (near where Slide 3 occurred). Using the procedures in Chapter 4 to map these values of q_c to ψ gives the distribution in Figure 5.13b. Considering the depth interval from 3 to 7 m, we can see here that the median value of $\psi \approx -0.08$. Using this value in any analysis would indicate stable non-liquefiable behaviour. The 80 percentile value in the same depth interval is about $\psi \approx -0.03$, and there is an interval of 1 m in which the soil is looser than $\psi > -0.02$ which Onisiphorou reports as being the limit of clearly liquefiable behaviour.

5.5 ASSESSING THE CHARACTERISTIC STATE OF SANDS

For engineering, the first requirement of the characteristic strength of sand is that it must be a well-defined, measurable property of the sand. Usually it would be ϕ for sand, or undrained strength for clay. State parameter, ψ, is workable because there is a direct mapping to ϕ and it can be measured. Although a normalized cone penetration resistance could also be used directly, such an approach is less desirable as the influence of basic soil properties becomes less than clear which then limits how experience in one soil can be translated to the engineering of another. Direct use of CPT data is also unnecessary with a full framework for the mapping from the CPT to ψ described in Chapter 4. Thus, engineering of sand and silts should and can focus on the characteristic state, denoted as ψ_k, using the usual notation that the subscript k denotes *characteristic*. A characteristic value is simply the value to be used in a deterministic calculation to obtain the same outcome (performance prediction) as would be obtained with a full stochastic simulation and at a high level of confidence.

A second requirement of the characteristic strength is that it should recognize the shape of the underlying statistical distribution of the parameter. Sand strengths appear to be more log-normally than normally distributed, so that the arithmetic mean and standard deviation have little utility. It is more appropriate to define a percentile value, for example, the 80 percentile value, which means that 80% of the measurements will be 'stronger' than this value.

Finally, the characteristic strength must be statistically reasonable given the test procedure. It is not helpful to define the characteristic strength as the 95 percentile value of a measurement that can only practically be made 10 times. In addition, a precise test is required because the true strength distribution in the sand should not be confused with the distribution of the testing error. The SPT N value, for example, is a parameter which is unsuitable because the potential error in an individual value is much too large compared to the value itself. The CPT, by contrast, is a good test in this regard, as it displays a repeatability in tip resistance q_c commonly better than ±2%.

5.5.1 Characteristic state for liquefaction

Since ψ is measured with the CPT, an important question to be answered is what range of percentile values can be determined with a given level of confidence. The relationships between q_c and ψ (or ϕ) are dependent on sand type and have an accuracy of about $\Delta\psi = \pm 0.04$. This margin of potential error is perhaps typical of geotechnical strength measurement techniques. It is therefore simply not meaningful to work with the 99% exceedance value of ψ.

Existing CPT technology also does not in general allow a 90 percentile value of tip resistance or ψ to be determined reliably. While 50 percentile and 80 percentile profiles plotted against depth result in a relatively smooth profile, the 90 percentile value is generally erratic. This is partly the result of erratic readings (e.g. after rod changes), and partly the result of heterogeneity (e.g. locally higher silt content) within a natural sand mass (Jefferies et al., 1988b). One needs to look very carefully at the other CPT data channels (pore pressure and friction) once attention moves to these high confidence levels, as it is very easy to confuse a material type change with a soil state change.

The centrifuge testing of the Oosterscheldt caissons by Rowe and Craig indicated that as little as 4% by volume loose material in the critical part of the foundation could affect performance under cyclic loads. Certainly, 10% loose material had a major impact. The implication of that work is that the characteristic state for liquefaction analysis might be in the 90–95 percentile range. However, the Rowe and Craig result is likely to be a little on the conservative side, as the loose material was deliberately located in the part of the foundation that was most highly stressed.

The analyses of a hypothetical foundation subject to earthquake loading by Popescu indicates that a characteristic state lies in the 80 percentile range, but that factors like the scale of fluctuation and therefore the size and frequency of loose pockets are important.

The Nerlerk case history gives a static reference point. The depositional processes there resulted in loose layers or pockets that were sufficiently well connected ultimately to result in liquefaction when movement occurred in the underlying clay. Taking $\psi = -0.02$ as the boundary between acceptable and unacceptable performance, based on Onisiphorou's static deterministic analysis, the designers would have identified an inadequate design if they had adopted the 80 percentile from the data as ψ_k.

Practical experience is therefore that a reasonable characteristic value of state parameter lies between 80 percentile and 90 percentile values for liquefaction analyses. One should probably err closer to the 90 percentile value for cyclic loading cases, but can relax a little towards 80 percentile value for static design problems. In reality, the sensible designer will also look at the construction method and the scale of variation. For strongly layered systems such as Nerlerk, one should anticipate a greater likelihood of loose zones connecting together and causing a problem.

5.5.2 Characteristic strengths for foundation design

It is interesting now to see what characteristic strength is suggested by limit states design codes. The limit state design process starts with a given reliability, or probability of failure, of a structure. In order to achieve the target reliability, partial safety factors are used. Characteristic loads are defined to which load factors are applied and load effects calculated. Characteristic material strengths are also multiplied by partial factors, with the resistance to loads calculated and compared to the factored loads. Much has been published regarding appropriate values for the partial factors, but selection of the characteristic strength of soils is a frequently neglected aspect of geotechnical limit state design. The state parameter framework can help, firstly because there is a reliable method of making many measurements of state (the CPT), and secondly because there is a consistent mechanics approach from state parameter to engineering behaviours.

One code in which the process was well defined and the target probabilities explicitly stated was the Canadian 'Code for the Design, Construction and Installation of Fixed Offshore Production Structures' (Canadian Standards Association, 1992), which has now been withdrawn in favour of the ISO 19900 series of codes for offshore structure design. (The ISO codes broadly follow the Canadian Standards Association [CSA] model but are not explicit in stating target reliability levels.) Part 1 CAN/CSA-S471 indicated a target annual reliability level of 10^{-5} for Safety Class 1 structures. Safety Class 1 was defined as being when the consequences of failure are great risk to life or high potential for environmental damage or pollution. In addition, the annual probability of loads exceeding the factored loads was between 10^{-3} and 10^{-4}. This therefore meant that to achieve the target reliability level, the factored resistance should be approximately the 99% exceedance value.

It is clear that there are any number of combinations of characteristic strength and resistance factor that would result in a factored resistance having a 99% exceedance probability. Table 5.4 summarizes the results of some Monte Carlo simulations (Been and Jefferies, 1993) of an offshore structure, looking at the relationship between characteristic strength and partial factor of material strength to give a 99% probability of non-exceedance. An 80 percentile characteristic strength value combined with a resistance factor of 1.26 will result in the desired 99% probability. This corresponds almost exactly to ISO 19906 for Arctic Offshore Structures (ISO, 2010) which recommends that the material or resistance factor should not be less than 1.25 (see Section 9.4.3). Meyerhof (1984) also recommended $f_\phi = 0.8$ (corresponding here to a factor of 1.25) which is used in some design codes (e.g. the Canadian Foundation Engineering Manual, the Canadian Highway Bridge Design Code and the National Building Code of Canada).

A characteristic value is generally defined in ISO 19906 as a 'value assigned to a basic variable associated with a prescribed probability of being exceeded by unfavourable values during some reference period'. Eurocode 7 (or the adopted British version, BS EN 1997-1) defines characteristic value as 'the characteristic value of a soil or rock parameter shall be selected as a cautious estimate of the value affecting the occurrence of the limit state' (Clause 2.4.5.2(2)P) and goes on to point out that each word and phrase in this definition is important. Engineering judgment is required in the 'selection' of a value, conservatism is required in a 'cautious estimate' and the selected value must relate to a specific limit state and mode of possible failure. ISO 19906 (Section 9.3) is less specific than Eurocode 7 in some ways, but does provide guidance on factors that need to be considered in selection of the characteristic value, such as relevant soil layers, anisotropy, stress history, dilation, progressive failure, cyclic loading, stress path, thermal effects, etc. Revisions to the ISO standards (in progress)

Table 5.4 Resistance factors for characteristic strength percentiles for an offshore structure example

Percentile value of strength	*Calculated resistance (MN)*	*Resistance factor (to obtain 99% value of resistance – 631 MN)*
95	741	1.17
90	764	1.21
80	793	1.26
70	814	1.29
60	832	1.32
50	849	1.34

Source: Been, K. and Jefferies, M.G., Determination of sand strength for limit state design, *Proceedings of International Symposium on Limit State Design in Geotechnical Engineering*, Copenhagen, Denmark, Vol. 1, pp. 101–110, May 1993.

will encourage the use of statistical methods to determine the characteristic soil strength, while Eurocode 7 states that if statistical methods are to be used, the characteristic value should be such 'that the calculated probability of a worse value governing the occurrence of a limit state is not greater than 5%'. In addition, of course, there is a range of partial factors to be applied to the characteristic values to obtain design values of strength, and it is difficult to relate Eurocode 7 to the analysis mentioned earlier. It appears, however, that the statistical work described in this chapter, in particular the work carried out by Popescu, Prevost and their co-workers at Princeton, supports Eurocode 7, although the basis for the clause is not defended even in the commentary on the code by Simpson and Driscoll (1998).

5.6 SUMMARY

In summary, there is still much work to be done to understand the effect of soil variability on the performance of both soils and foundations under cyclic loads. This fact is reflected in modern design codes, mainly with respect to selection of characteristic values although stipulation of material or resistance factors appears (incorrectly) to be more certain. This chapter should have given you some appreciation of the issues involved and some background on which to base your engineering judgment which you will undoubtedly need in making these selections in practice.

Chapter 6

Static liquefaction and post-liquefaction strength

6.1 INTRODUCTION

This chapter considers undrained failure under monotonic conditions – a process often called static liquefaction when dealing with loose sands. Why start with static liquefaction? Because static liquefaction largely controls stability, even when the loading is cyclic (e.g. during earthquakes). If there is sufficient residual strength, then cyclic loading is going to manifest itself only as fatigue-like strains, which are unlikely to endanger anyone. Static liquefaction failures, on the other hand, have killed several hundred people on more than one occasion (Chapter 1). A second reason to start with static liquefaction is that it is relatively straightforward to understand, and there is no point dealing with more complex loadings until entirely comfortable with how excess pore water pressure is caused by plastic strain (and not the collapse of a metastable soil structure). No new models or properties for the soil are required as static liquefaction is an aspect of soil behaviour that fits simply within the state parameter framework.

Two situations of practical relevance arise with static liquefaction: undrained failure in monotonic shear and post-earthquake liquefaction. The two situations are similar in terms of how the soil behaviour evolves and the residual strength during the liquefaction event. The difference between the two is in the method of triggering. In the case of monotonic shear, what matters is the stress ratio η. If this stress ratio increases, either through an increase in deviator stress (e.g. slope steepening by erosion at the toe) or through a decrease in the mean effective stress through seepage pressures (as happened at Aberfan), then a static liquefaction can be triggered if the soil is loose enough. In the post-earthquake case, there will be cyclically induced excess pore pressures from the earthquake. These pore pressures may be sufficient to cause outright soil failure under the imposed loadings. Even if the pressures do not cause outright failure initially, their redistribution during dissipation can trigger further movement (as happened at Lower San Fernando Dam).

In these differing situations, the conventional view is that an undrained strength (or its equivalent) is appropriate and can be used in stability analysis. The fully softened (or large displacement) strength during monotonic liquefaction is usually denoted as s_r and is sometimes referred to as the residual strength. The peak strength is denoted as s_u. The difference between the two strengths indicates the brittleness of the soil. If s_r is less than the drained strength, then there is the potential for a flowslide. Not surprisingly, this leads to a substantial interest in s_r and how it can be determined for a soil in-situ.

Undrained behaviour is caused by an imposed boundary condition in the laboratory and by the drainage time in a field situation. The soil behaviour, however, continues to be the result of effective stresses, and previously established properties from drained tests apply.

This use of drained properties is crucial because, in a field situation, there will be some drainage in the short term and complete drainage in the long term. An 'understanding' based on undrained calibrations or undrained models alone is potentially misleading and certainly something that cannot be used in general.

The undrained monotonic behaviour of sands has received much attention in the liquefaction literature, particularly in relation to loose sands. Dense, dilatant sands are of little interest in undrained shear because the same sand can sustain much higher shear stresses undrained than drained: if the project at hand is stable in the long term, then it is stable in the short term. This is not the case with loose sands in which positive pore water pressures are developed during undrained shear, leading to possible liquefaction with a runaway slide. Under some circumstances, drained monotonic loading can transition into undrained liquefaction; the Aberfan flowslide is an example. However, far more soils than a limited range of sands can exhibit static liquefaction. To date, the range of experience includes rather coarse uniform sands through to nearly pure silt-sized soils and various combinations in between these gradational limits.

From a practical point of view, the undrained behaviour of loose sands is frequently only of academic interest. Soil that is sufficiently loose to fail in undrained monotonic shear poses such a risk of catastrophic failure that engineers will always specify ground treatment of some form to improve its density. It is simply not worth the risk to do otherwise. The true engineering problem is to identify when a soil is sufficiently dense that treatment is not required. A slightly different perspective may apply to mining projects, as the available space may be such that interest is in how far a potential flowslide will move. This clearly requires an understanding of liquefaction, and the phrase 'critical density' is not misplaced in terms of making a decision.

The question of 'how dense is sufficiently dense?' has been the subject of much discussion in the geotechnical engineering literature over the last 25 years (even though most engineers would probably agree on whether treatment is required for any particular project). Many 'concepts' for liquefaction have been put forward, some of which are not supported by mechanics and only confuse. The beauty of a plasticity model like *NorSand* is that it computes all behaviours of loose sands, starting from calibrations on dense samples under drained conditions: there is nothing magical about liquefaction. This ability to compute, using a model properly anchored in established mechanics, cuts through the confusion that has surrounded static liquefaction. Most importantly, the model is implemented within open-code software that is downloadable and runs within a spreadsheet environment. There is nothing hidden, and it is easy to develop both insight and an appreciation for the phenomenon.

The behaviour of loose sands in undrained triaxial compression and extension tests is the starting point and leads to an exploration of the uniqueness of the critical state line. Other stress paths and effects such as strain localization are then considered. Case histories are used to confirm model suitability and to address the issue of post-liquefaction strengths. Having gained this understanding from laboratory tests, *NorSand* and case histories, the answer to the question 'how dense?' is no longer mysterious and becomes accessible to all geotechnical engineers.

6.2 DATA FROM LABORATORY EXPERIMENTS

This section describes and discusses experimental data on static liquefaction. The data can be downloaded from the website for further manipulation, data being provided on some 20–30 soils in total (the number depends on how you count the same sand with different fines contents or grain size distributions). Most of these files are from tests carried out by Golder Associates using GDS software-controlled triaxial equipment, although there are a few others included. The downloadable files may be especially interesting when looking at particular details of soil behaviour, as plots can be expanded to show a point of interest. The plots in this chapter are annotated with the file name that produced them, allowing easy investigation of features that are highlighted, as well as plotting out at a larger scale. In particular, the downloadable files allow the behaviour of these different soils to be viewed in some detail to confirm that the behaviour presented in this chapter is characteristic of loose soils, regardless of mineralogy or gradation, fines content, etc.

6.2.1 Static liquefaction in triaxial compression tests

Figure 6.1 shows results from a typical series of undrained tests, in this case on reconstituted samples of Erksak sand prepared by the moist tamping method. The initial values of the state parameter for each test are shown in the figure, and the tested range is $-0.07 < \psi_o < +0.07$. All tests were load controlled, and the figure shows the stress paths followed by the samples together with their corresponding stress–strain response. The stress paths for samples with $\psi_o > 0$ (i.e. an initial state looser than the CSL) are similar. The corresponding stress–strain behaviour is brittle and initiates at the maximum deviator stress.

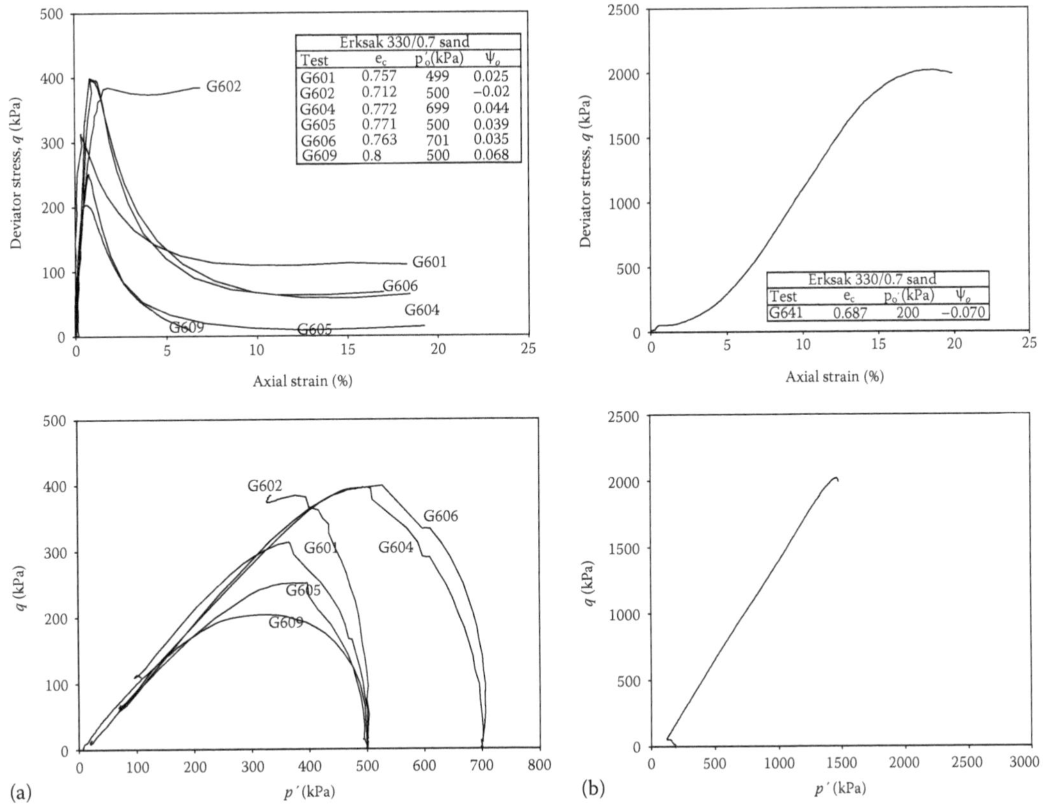

Figure 6.1 Undrained triaxial compression of Erksak 330/0.7 sand. (a) Contractive (ψ on −0.02) and (b) dilatant (ψ ~ −0.07).

The strength falls rapidly with strain and can reach a small fraction of the peak value for very loose samples (e.g. test G609 with $\psi_o = +0.068$). Usually, it takes less than 15% axial strain to establish critical conditions for very loose soils, but sometimes as little as 5% is sufficient as can be seen in Figure 6.1a. These strains are well within the capabilities of triaxial compression tests.

For initial states denser than critical ($\psi_o < 0$), dilation at large strains becomes dominant and creates negative excess pore pressure. Because the limit of dilation is not usually reached in the triaxial test (which is typically limited to 20% axial strain compared to the 50% or so needed to attain the critical state with dilatant soils), the maximum shear stress is ambiguous: the undrained strength measured depends on the strain to which the soil is taken in the test, and whether there is sufficient back pressure to allow the negative excess pore pressures to develop without being limited by cavitation.

Test G602 in Figure 6.1a had an initial state of $\psi_o = -0.02$ and hints at the onset of dilation to an apparently real maximum strength. Going to a slightly denser sample, Figure 6.1b shows Erksak sand test G641 at $\psi_o = -0.07$ where there is a strength plateau at 1%–2% axial strain before the onset of strong dilation, which then continues all the way to about 18% axial strain. The last part of the test suggests that dilation has ceased and the critical state has been reached. This may be real, as the excess pore pressure was −700 kPa and the sample back pressure was 1300 kPa. Moving to yet denser states makes it practically difficult to obtain sufficient back pressure in testing, and even in the case of G641 at $\psi_o = -0.07$, it is questionable whether any practical field case (except the deep offshore) is going to allow the generation of such high negative excess pore pressure.

Similar stress paths to those of Figure 6.1 are widely reported for loose sands in undrained laboratory tests within the literature, and there is nothing special about Erksak sand. To reinforce this point, Figure 6.2 shows the same behaviour with Ticino sand (a standard soil much used in laboratory research). Of course, it is not that surprising that two hard quartz sands of similar near-uniform gradation show similar behaviours, but what about well-graded soils and silts, rather than sands?

Liquefaction is by no means restricted to clean sands. Figure 6.3 shows the particle size distribution curve for both Erksak and Ticino sands as well as a well-graded silty sand (Bennett Dam Core) and a sandy silt (Guindon Dam Foundation). The fines fractions of these two additional soils are 31% and 65% respectively. Both of these soils were tested to determine their CSL, and examples of measured undrained behaviour of each soil are shown in Figure 6.4. Looking at the test data, there is little difference in behaviour between the silty soils and the clean sands, and someone shown Figure 6.4a would be unsurprised (and possibly expect) to be told that they were data from testing clean sand. These tests nevertheless had 31% fines. The sandy silt shown in Figure 6.4b exhibits less brittleness than readily achieved with clean sand. However, this should not be taken as proving high fines reduce liquefaction potential – the issue is rather laboratory sample reconstitution of high-fines soils, as much looser soil can be found in the field than can be prepared in the laboratory. If a soil has substantial positive state parameter, high fines content does not prevent liquefaction. The soil's state matters, not its fines fraction.

6.2.2 Triaxial extension

Soil behaviour depends on the stress path imposed, and so, a question naturally arises as to what extent triaxial compression test results, and the understanding developed from them, reflect other conditions. The opposite extreme in terms of stress combinations is triaxial extension, and it is helpful to look at undrained extension data to answer the question as it relates to stress path effects.

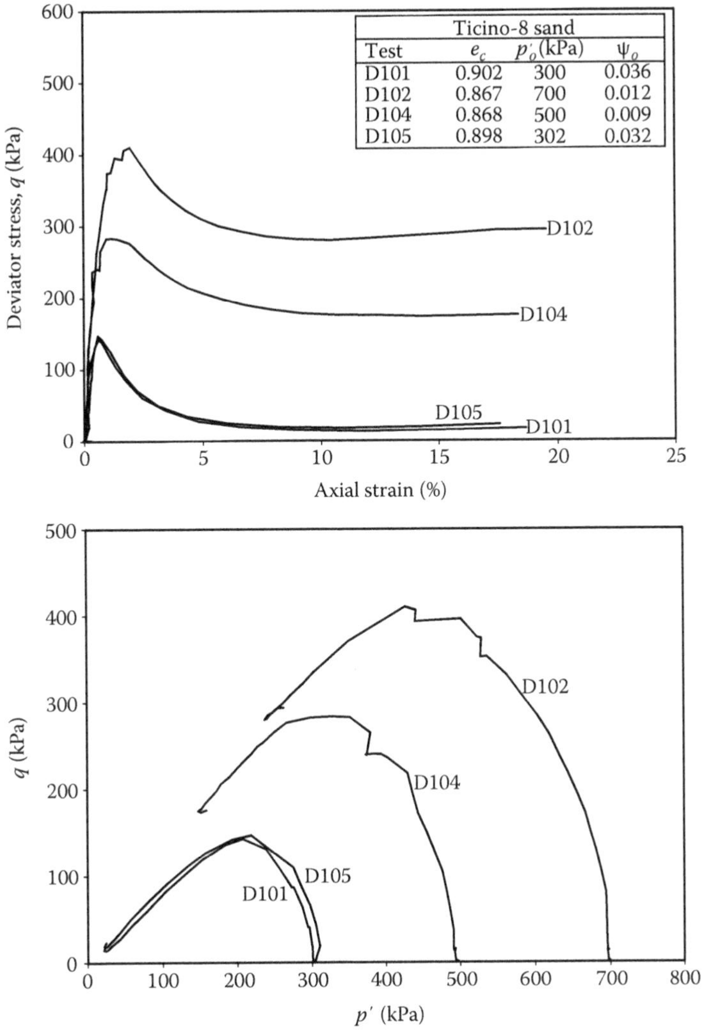

Figure 6.2 Loose Ticino sand in undrained triaxial compression.

Data for loose Erksak sand in triaxial extension are plotted in Figure 6.5, the tests having initial state parameters in the range $-0.03 < \psi_o < +0.05$. The figure shows the stress paths followed by the samples together with their corresponding stress–strain response. Some samples were anisotropically consolidated before being sheared undrained.

The stress paths for samples with $\psi_o > 0$ all show brittle post-peak strength reductions, just like compression tests, with samples further from the critical state producing the more brittle behaviour. The one sample slightly denser than the CSL shows an initial contraction followed by dilation to give an 'S'-shaped stress path, a behaviour entirely characteristic of soils marginally denser than critical in compression.

Triaxial extension behaviour is compared with compression, after normalizing out the effect of initial effective stress in Figure 6.6. Rather similar behaviour is evident, but the peak strength is markedly less in extension. It is also apparent that, although the effective stress paths have similar shape, the difference in peak strength flattens extension stress paths compared to those in compression. This flattening is hardly a surprise, however, as

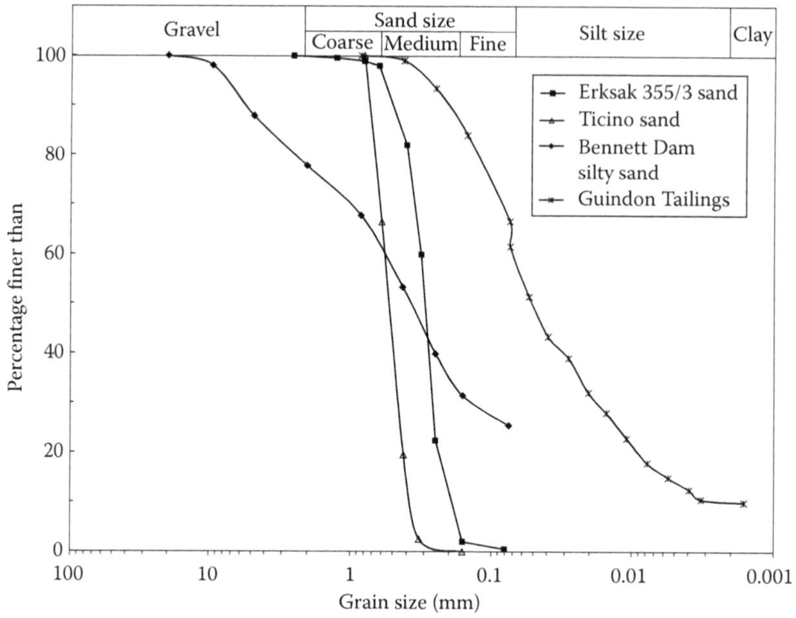

Figure 6.3 Particle size distribution curve for four liquefying soils.

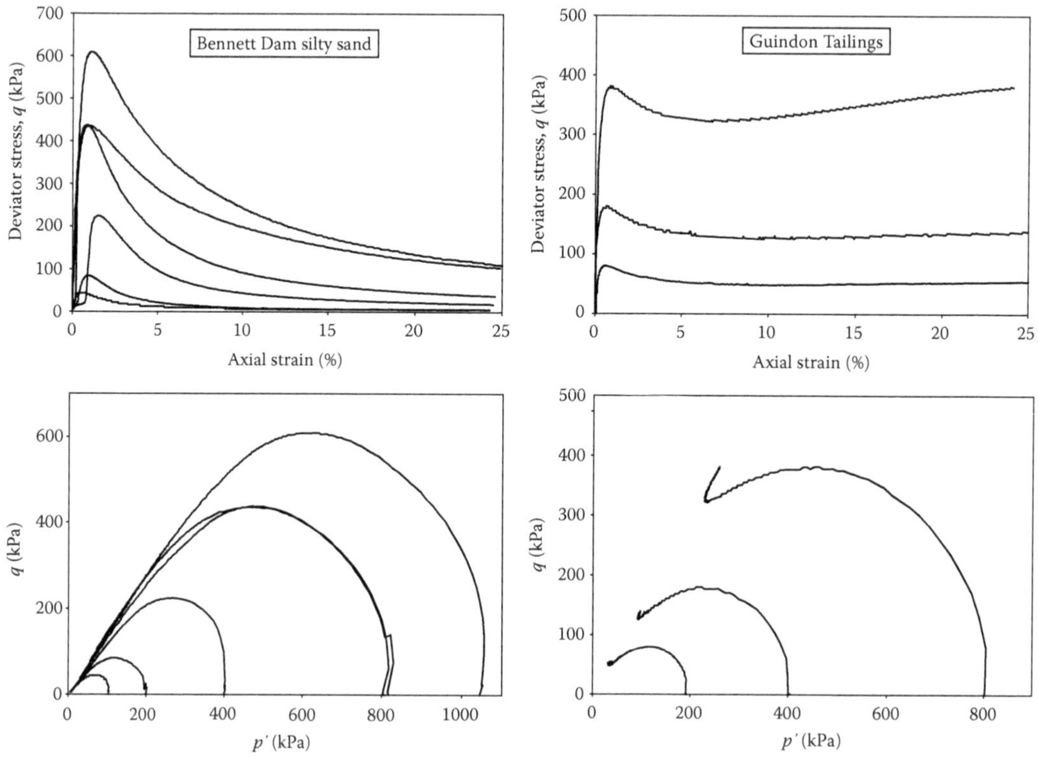

Figure 6.4 Loose silty sand (Bennett Dam) and sandy silt (Guindon Tailings) in undrained triaxial compression.

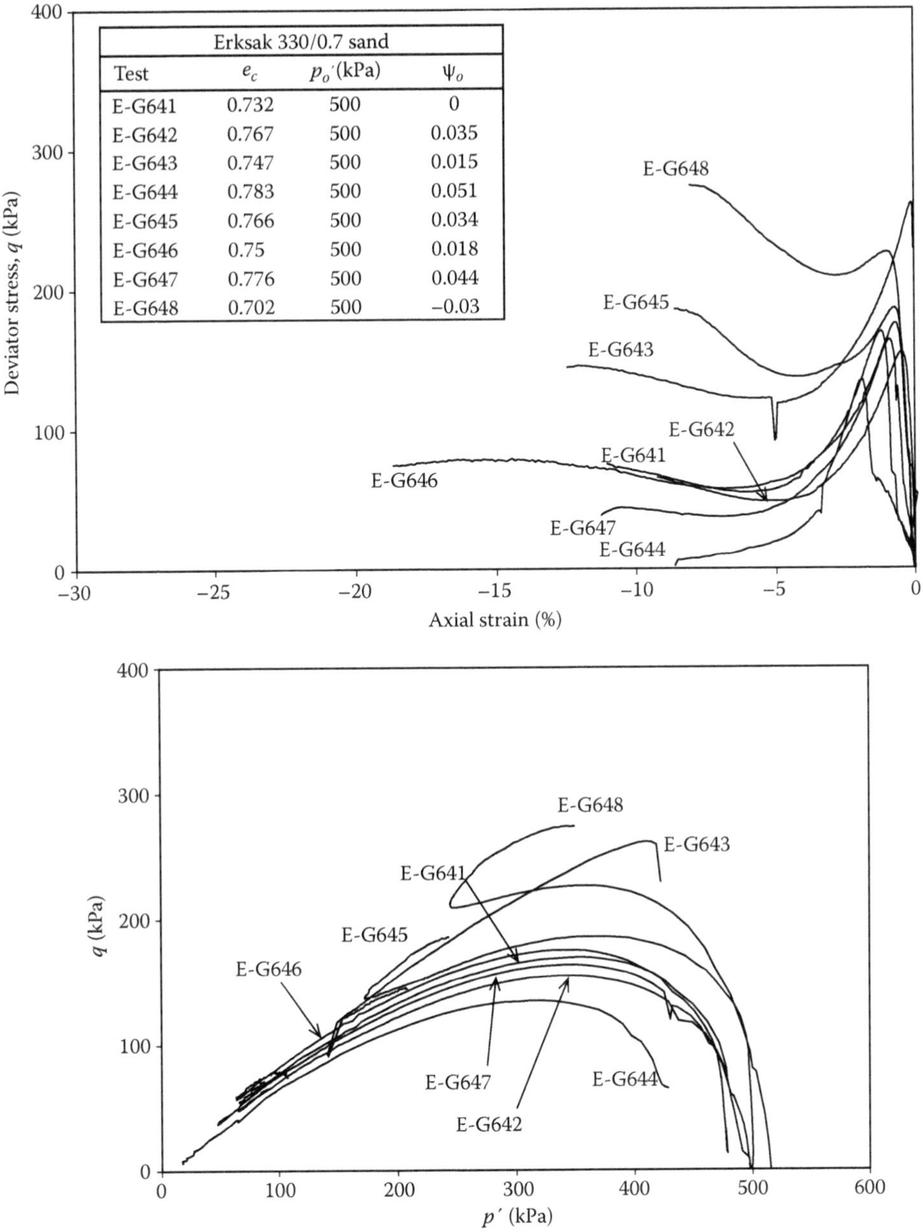

Erksak 330/0.7 sand			
Test	e_c	p_o'(kPa)	ψ_o
E-G641	0.732	500	0
E-G642	0.767	500	0.035
E-G643	0.747	500	0.015
E-G644	0.783	500	0.051
E-G645	0.766	500	0.034
E-G646	0.75	500	0.018
E-G647	0.776	500	0.044
E-G648	0.702	500	−0.03

Figure 6.5 Triaxial extension test data for Erksak 300/0.7 sand.

the critical state friction ratio M is markedly less in triaxial extension compared to triaxial compression (see Chapters 2, 3 and Appendix C).

Various workers have asserted that stress path affects the final critical state, in particular that the critical state in triaxial extension is much different from that in triaxial compression. While there is no issue with stress paths affecting M, the implication of a non-unique relationship between void ratio and mean effective stress (discussed in Section 2.5) strikes to the heart of a critical state approach. Is this implication actually correct? The critical state for

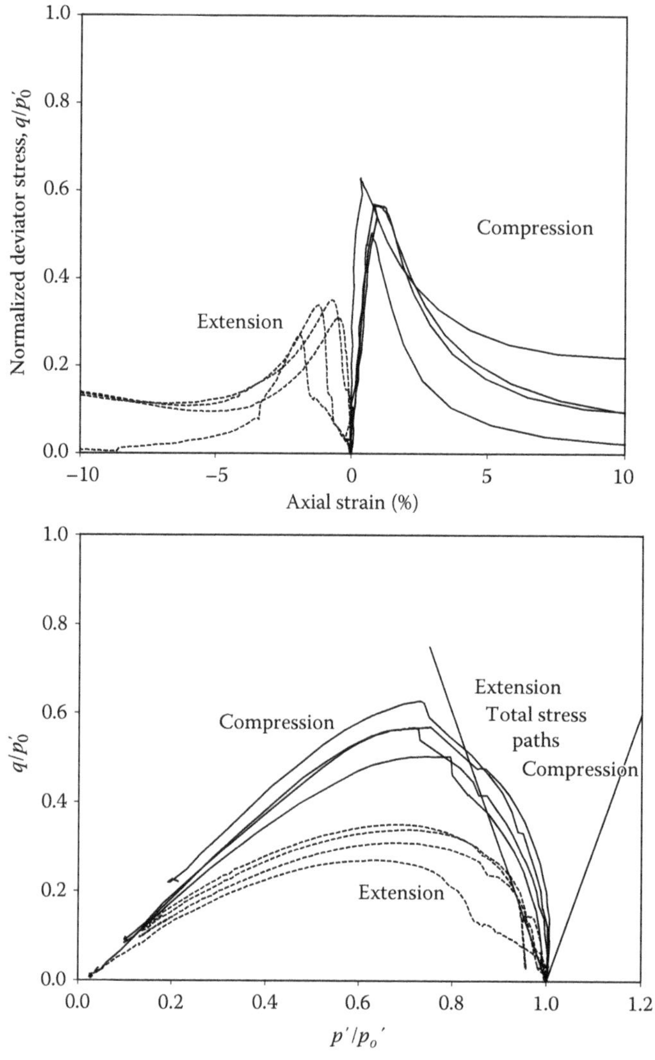

Figure 6.6 Comparison of extension and compression tests on Erksak sand (normalized).

the various stress paths on Erksak sand summarized in Figures 6.1 and 6.5 is plotted in void ratio space in Figure 6.7 at an expanded scale relative to Figure 2.26 where the same data are shown. As can be readily seen, within an experimental precision of about $\Delta e = \pm 0.005$, there is no effect of stress path on the critical state. Given that the extension test data show identical scatter to the compression data and that the trends of both sets are the same, it is arguable that stress path has no effect whatsoever on the critical state locus in $e - p'$ space.

6.2.3 Simple shear

Triaxial tests are the most commonly used in the laboratory, but the simple shear test is attractive as being a possibly better analogue of conditions in the ground when loaded in plane strain (a common situation). However, despite simple shear being a good analogue for an infinite slope, very few published liquefaction studies use the simple shear test, particularly

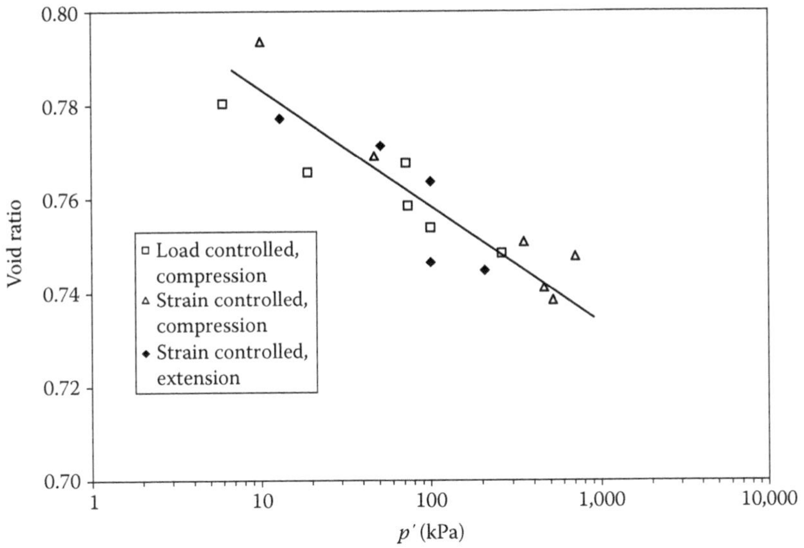

Figure 6.7 Effect of stress path on the critical state locus (at expanded scale).

tests on sands with initial states looser than critical. (The same does not apply to cyclic simple shear testing to examine seismic liquefaction on material with initial states denser than critical.) The VELACS project (Arulanandan and Scott, 1993) included several tests on Bonnie silt that allow comparison between simple shear and triaxial shearing of a loose material. Figure 6.8 shows five tests on Bonnie silt within a narrow void ratio range and an initial stress of 80 kPa. Two tests are simple shear with an 80 kPa vertical stress on the samples, two tests are triaxial compression and one is triaxial extension. In the triaxial tests, 80 kPa is the initial isotropic confining stress. Stress paths for all tests show an initial contraction, but in this case, the samples were sufficiently dense that dilation kicks in and determines the behaviour at higher strains. Simple shear and triaxial shear behaviours are similar, although the excess pore pressure is much greater initially in simple shear and with less intense subsequent dilation.

6.2.4 Plane strain compression

A substantial limitation of the simple shear test, with present equipment, is that the horizontal stress is not measured – which makes simple shear data problematic as a basis for assessing soil behaviour. Equally, plane strain is important, and details of plane strain apparatus at Imperial College and Nanyang Technological University are described in Section 2.7.2. (See also Figure 2.44 for a photograph of the NTU equipment, which allows intermediate principal stress, σ_2, to be measured.)

Wanatowski and Chu (2007) reported on the plane strain liquefaction behaviour of loose Changi sand using this equipment. This soil is a subangular marine-dredged silica sand used for the Changi land reclamation project in Singapore containing approximately 12% of shells (which affect its compressibility). Following various triaxial tests to determine the properties of Changi sand, three undrained tests were carried out in plane strain. These plane strain tests loaded the sand drained to a K_o condition of $0.41 < K_o < 0.48$ before increasing the vertical load undrained while holding σ_3 constant; these undrained data are shown in Figure 6.9. As can be seen in the figure, these three tests were close to their instability limit from the

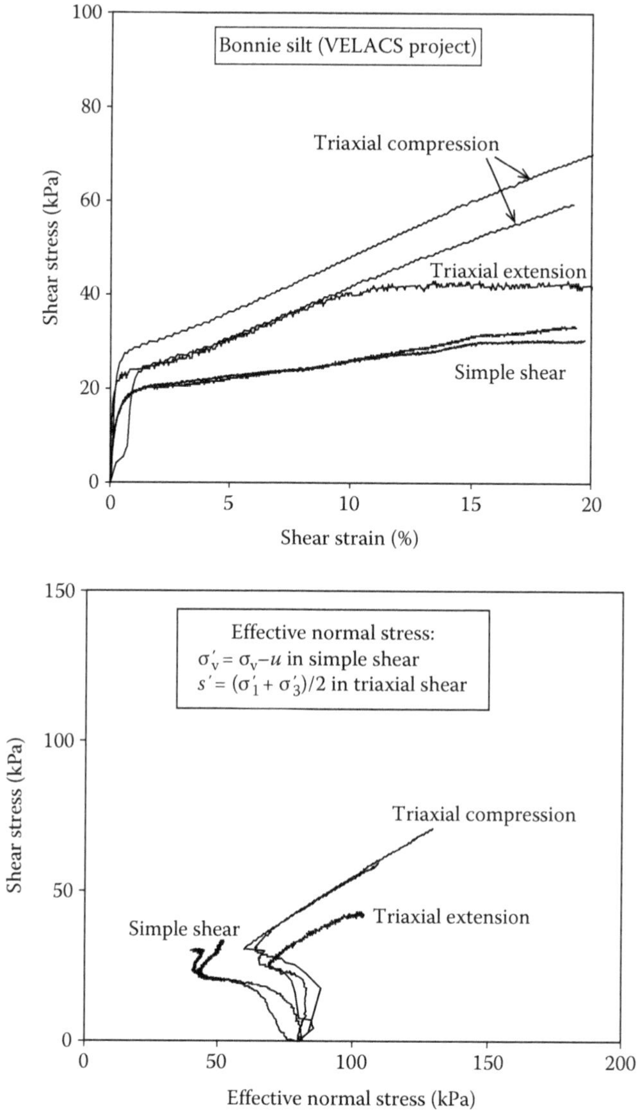

Figure 6.8 Comparison of Bonnie silt in simple shear, triaxial compression and triaxial extension (all tests at initial confining stress of 80 kPa, $0.683 < e_o < 0.753$).

stress state established by the drained loading. Once peak strength was reached, there was a strength loss that appears comparable to that seen in triaxial compression for similarly loose sand. The end stresses lie on a critical friction ratio $M_{ps} \sim 1.16$ (smaller than M_{tc} as expected because of the effect of the intermediate principal stress, see Chapter 3).

6.3 TRENDS IN LABORATORY DATA FOR s_u AND s_r

It is clear from the data presented so far that liquefaction may arise if the soil is looser than its critical state and that it does not matter a whole lot whether the soil is clean sand or has a high proportion of silt-sized particles. A useful first step to understanding liquefaction

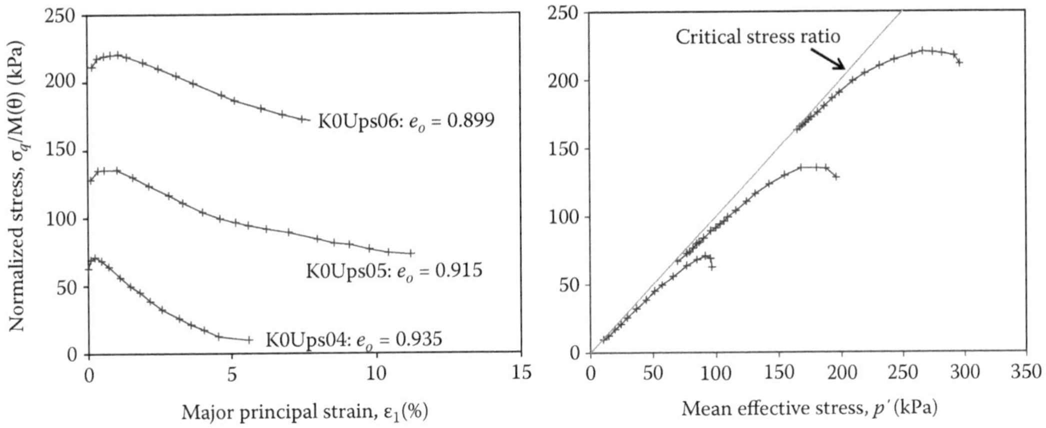

Figure 6.9 Response of loose Changi sand in undrained plane strain compression. (Data courtesy of Dr. Wanatowski.)

is to reduce these data on many liquefaction tests to fundamental trends. In what follows, attention concentrates on triaxial compression behaviour, so providing a convenient overview of trends and allowing ready comparison with soft clays. The effect of loading path is discussed later, in particular how the basic triaxial compression trends may be affected by a change to plane strain (which is the usual situation of practical interest).

Because the undrained response to the loading of loose soils is uniformly similar, the stress paths can be characterized by their peak undrained strength and the pore pressure at which this peak is reached. The end point of the stress path is the critical state, which also provides a convenient normalizing stress. The relationship between the end point and the peak is a measure of the soil's brittleness. Brittleness is extremely important as it is this aspect of soil behaviour that allows acceleration of slope movement into a potentially dangerous flowslide.

The undrained strength can be normalized in terms of the initial confining stress p'_o to give a s_u/p' ratio in the form

$$\frac{s_u}{p'_o} = \frac{(q_{\max}/2)}{\sigma'_{3c}} = \frac{(\sigma_1 - \sigma_3)_{\max}}{2\sigma'_{3c}} \tag{6.1}$$

and a pore pressure parameter A_f of the form

$$A_f = \frac{\Delta u}{\Delta \sigma'_1} \tag{6.2}$$

This puts the behaviour of loose sands and parameter usage in the same context as the undrained shearing of clays, with A_f being as defined by Skempton (1954).

Figure 6.10 shows the s_u/p'_o ratio for a number of different sands as a function of ψ. This figure is analogous to the familiar relationship of the undrained strength ratio in clays to the over-consolidation ratio (e.g. Wroth, 1984). There is some scatter, and the data form a band rather than a single trend in Figure 6.10, but nevertheless there is a strong relation between the undrained strength ratio and the state of the soil. Apart from Kogyuk sand, which is notable for anomalously high strengths, the data support a broad trend of strength ratio $s_u/p'_o > 0.15$ even when the sample is rather loose ($\psi = +0.05$), rising to $s_u/p'_o \approx 0.30$

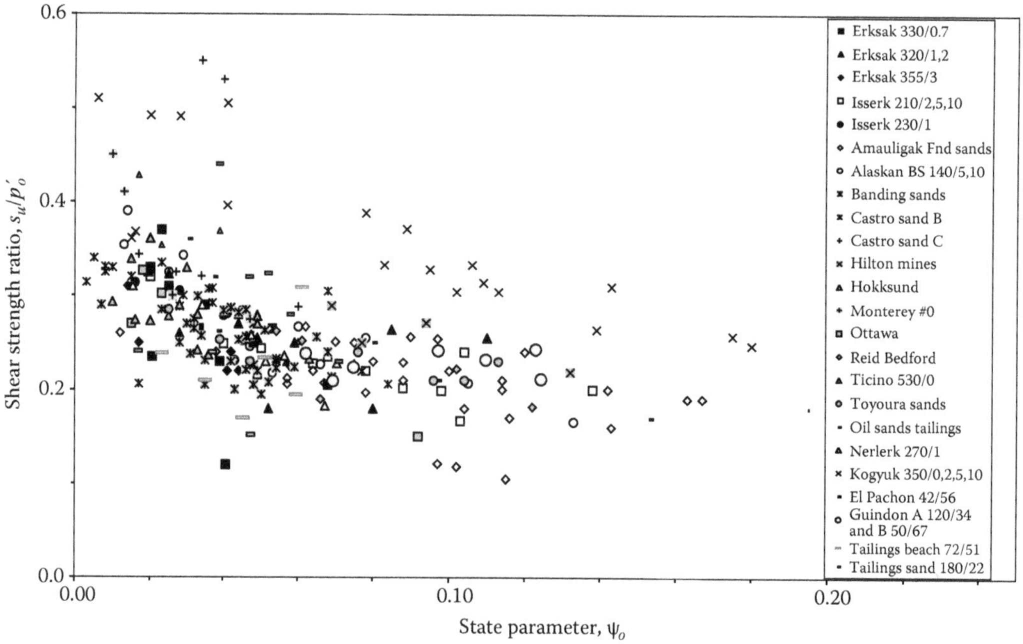

Figure 6.10 Normalized undrained strength of loose liquefiable sands, silty sands and silts.

at the critical state ($\psi = 0$). Strengths then rapidly increase with further increasing density, although this is not evident as the plot ends at $\psi = 0$.

There are two points of importance about these numerical values of the strength ratio s_u/p'_o. First, these ranges for the strength ratio s_u/p'_o have been known for some 40 years. Table 6.1 is taken from Bishop (1971), and similar s_u/p'_o values are quoted, although with only a crude indication of sample density. (Note that Huachipato Sand in Table 6.1 is Castro sand C in Figure 6.10 and Banding sand is Castro sand B.) Second, these undrained strengths are comparable to normally consolidated clays. Normalized clay strengths are much discussed in the literature, and Wroth (1984) presents a convenient summary of clay behaviour (this Rankine lecture ought to be mandatory reading for all geotechnical engineers). Wroth developed the following relationship:

$$\frac{s_u}{p'_o} = \frac{M}{2}\left(\frac{R}{r}\right)^{\Lambda} \approx \frac{M}{2}\left(\frac{1}{2}\right)^{0.8} \tag{6.3}$$

where

R is the over-consolidation ratio, with R=1 set for the normal compression being considered

r is the spacing ratio

$\lambda = (1 - \kappa/\lambda)$

Equation 6.3 follows from critical state theory, with insights from the Modified Cam Clay model in particular. The approximation assumes a spacing ratio of two between parallel NCL and CSL. The estimate that the exponent $\lambda = 0.8$ is based on calibrations to clays and their behaviour in isotropic compression, and this average calibration is discussed at some

Table 6.1 Observed values of the parameters $(s_u/p')_n$, ϕ' and I_B for consolidated undrained tests on cohesionless soils

Material	*State*	*Type of test*	*P*	$(s_u/p')_n$	ϕ'_f	ϕ'_r	I_B (%)	*References*
Banding sand (Ottawa sand)	Very loose	comp.	4 kg/cm²	0.206	17.0	30	94	Castro (1969)
Banding sand (Ottawa sand)	Medium loose	comp.	4 kg/cm²	0.288	21.8	30	65	Castro (1969)
Banding sand (Ottawa sand)	Medium loose	comp.	4 kg/cm²	0.333	24.3	30	10	Castro (1969)
Huachipato sand (Chile)	Very loose	comp.	4 kg/cm²	0.288	26.8	35	91	Castro (1969)
Huachipato sand (Chile)	Loose	comp.	4 kg/cm²	0.333	29.4	35	65	Castro (1969)
Ham River Sand	Loose	comp.	990 psi	0.249	21.3	33	94	Bishop et al. (1965)
Ham River Sand	Dense	comp.	990 psi	0.596	33.8	33	–	Bishop (1966a)
Ham River Sand	Loose	comp.	1840 psi	0.251	20.1	33	–	Bishop et al. (1965)
Ham River Sand	Loose	ext.	99.4 psi	0.119	12.0	33	~75	Reades (1971)
Ham River Sand	Loose	ext.	90.0 psi	0.052	11.5	33	~75	Reades (1971)
Brasted Sand	Loose	comp.	10.6 psi	0.825	29.2	33	–	Eldin (1951)
Brasted Sand	Loose	comp.	30.4 psi	0.457	31.4	33	–	Eldin (1951)
Mississippi River Sand	Very loose	comp.	29.4 psi	0.230	17.2	31	–	Hvorslev (1950)
Mississippi River Sand	Medium loose	comp.	29.4 psi	0.415	28.9	31	–	Hvorslev (1950)

Source: Data from Bishop, A.W., Shear strength parameters for undisturbed and remoulded soil specimens, *Stress–Strain Behaviour of Soils: Proceedings of the Roscoe Memorial Symposium*, Cambridge, U.K., R.H.G. Parry (ed.), Foulis, pp. 3–58, 1971.

length in Wroth's Rankine lecture. Equation 6.3 is plotted in Figure 6.11 and illustrates that it is reasonable to expect a normally consolidated undrained strength ratio of about $(s_u/p'_o)_{nc} = 0.35$ for a critical friction ratio $M = 1.25$ (where the subscript nc has been used, following Wroth, to indicate specifically the normally consolidated strength).

Although the state parameter and over-consolidation are different entities, and there is no unique normal consolidation locus, what Wroth had in mind with normal consolidation was a state a little above the CSL. Wroth's use of a spacing ratio $r = 2$ in Equation 6.3 is actually the same thing in terms of void ratio as $\psi_o = +0.69\lambda_e$, corresponding to $\psi_o \approx +0.015$ for most clean sands. Looking at Figure 6.10, it is apparent that Wroth's normally consolidated strength ratio is in the bandwidth for the trends from the many triaxial measurements on loose sands at this state. There is no substantive difference between the undrained peak strength of loose sands and 'normally consolidated' clays.

Turning to the excess pore pressure at peak strength, Figure 6.12 plots the pore pressure parameter A_f. While the undrained strength data lie in a reasonably narrow band, A_f scatters widely around a backbone trend. This is probably caused by a combination of the difficulty in measuring pore pressure accurately as the sample fails and the pore pressure at failure being influenced by small differences in the fabric or structure of the sand. These values of A_f for sands are again familiar in the context of clays. For example, Bishop and Henkel (1962) plot detailed results for A_f versus over-consolidation ratio for Weald and London clay, both clays showing $A_f \approx 0.9$ when normally consolidated, which are mid-band for the trend in Figure 6.12 at $\psi_o \approx +0.015$ (the equivalent of normal consolidation as

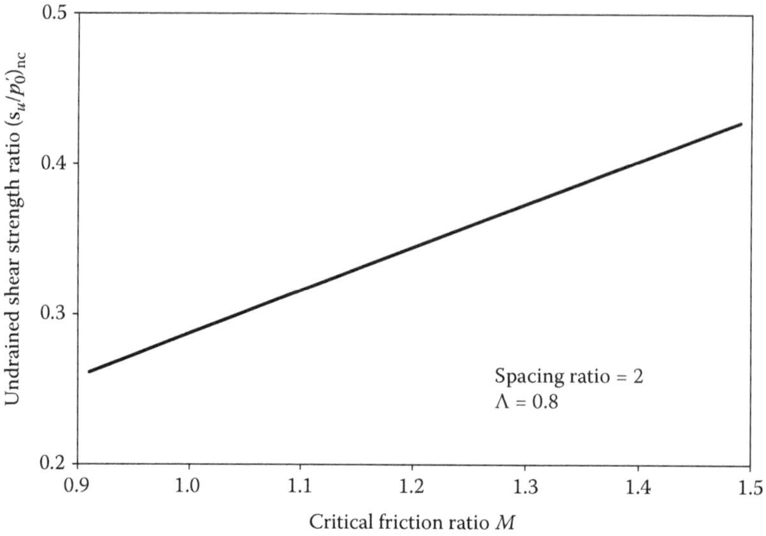

Figure 6.11 Undrained strength ratio of normally consolidated clay. (Equation in Wroth, C.P., *Géotechnique*, 34(4), 449, 1984.)

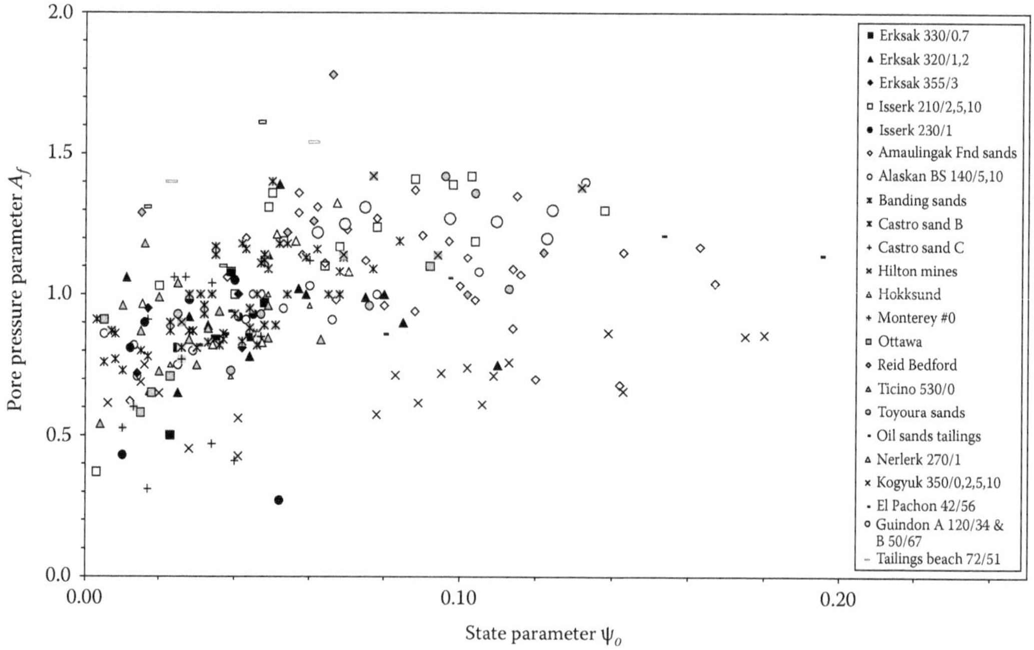

Figure 6.12 Pore pressure ratio A_f of loose sands at peak strength.

discussed earlier). The similarity of A_f to common normally consolidated clay experience is unsurprising given the similarity in strength ratios between sands and clays.

Wroth's framework is restricted to states close to the CSL, whereas there is no difficulty in getting sands and silts to states markedly looser or denser. However, Wroth's framework can be pushed a little further using ψ_o/λ as an alternative normalizing group. Although ψ_o/λ has no application to drained stress paths, ψ_o/λ is useful in the context of residual strength during undrained liquefaction, s_r, since

$$\frac{s_r}{p'_o} = \frac{M}{2}\exp\left(\frac{\psi_o}{\lambda}\right) \tag{6.4}$$

where (6.4) follows from the conventional critical state idealization (i.e. $\ln(p'/p'_c) = \psi_o/\lambda$), the definition of the state parameter (Figure 1.1) and an undrained stress path.

Figure 6.13 shows the same peak strength ratio information s_u/p'_o as Figure 6.10, but changes the state measure to the parameter group ψ_o/λ. There is no apparent reduction in the scatter of the data because the peak undrained shear strength is not significantly affected by the λ of the soil. However, this new figure illustrates the relationship of peak to residual strength. Equation 6.4 is plotted in Figure 6.13 (using $M = 1.25$, which is typical for the sands shown) to illustrate how brittleness develops with more positive states. There are several interesting aspects to this figure.

The conventional understanding of 'normal consolidation' gives $\psi_o \approx 0.7\lambda$. At this state, it is readily apparent from Figure 6.13 that the peak and residual undrained strengths are approximately equal. This is exactly what the Cam Clay model represents, and it is what is widely understood to be the behaviour of insensitive soft clays. As the state parameter becomes more positive, then the potential for a marked strength reduction post-peak (i.e. from s_u to s_r) becomes ever greater. The peak strengths measured in loose liquefying soils are not necessarily

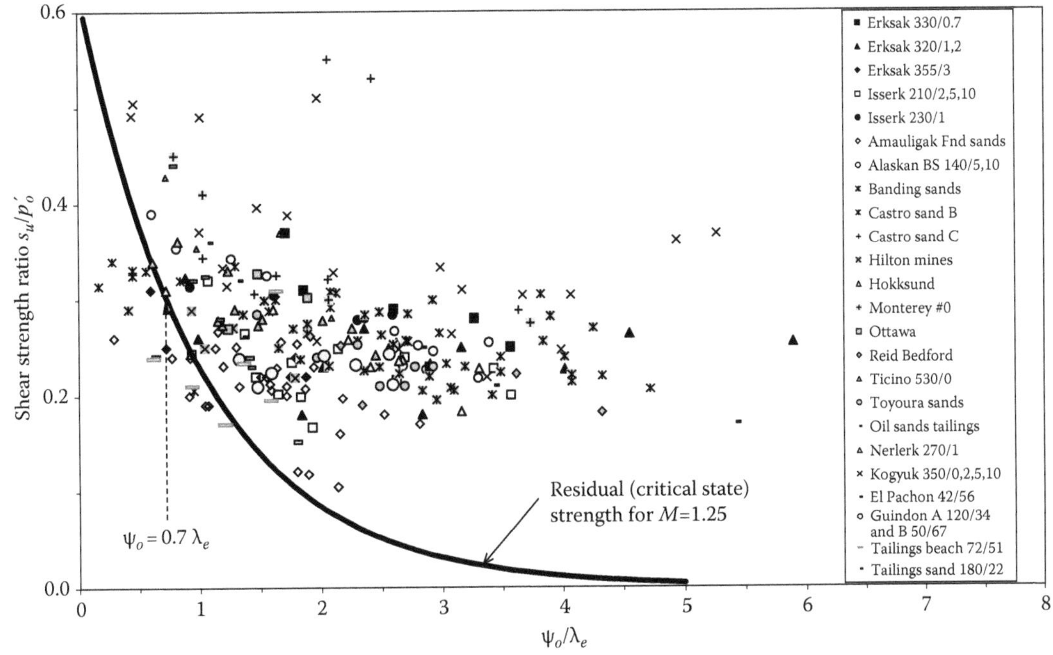

Figure 6.13 Comparison of normalized peak and residual undrained strengths.

that small, but it is the subsequent brittle strength loss that makes static liquefaction so very dangerous. This stress drop can accelerate an incipient failure into a rapidly moving flowslide.

Bishop used a brittleness index I_B to describe the strength drop as a dimensionless ratio:

$$I_B = \frac{s_u - s_r}{s_u} \tag{6.5}$$

The brittleness index has been calculated for the many strength tests summarized earlier and is plotted against ψ_o/λ in Figure 6.14. This type of plot was used by Hird and Hassona (1990), and some very simple relationships arising from critical state soil mechanics bring clarity to the empirical brittleness index plot. Brittleness arises only if $\psi_o/\lambda_e > 0.7$, which is a restatement of the limit of conventionally understood normally consolidated behaviour. Brittleness rapidly develops with increasing ψ_o/λ.

So far, a substantial body of data has been brought into dimensionless ratio form. This dimensionless form makes it apparent that the peak strength of even very loose sands is not unusually low. What is unusual about liquefying soils is their low critical state (residual) strength compared to their peak strength, best illustrated by brittleness index. There are well-defined trends in terms of ψ and ψ_o for the laboratory triaxial compression

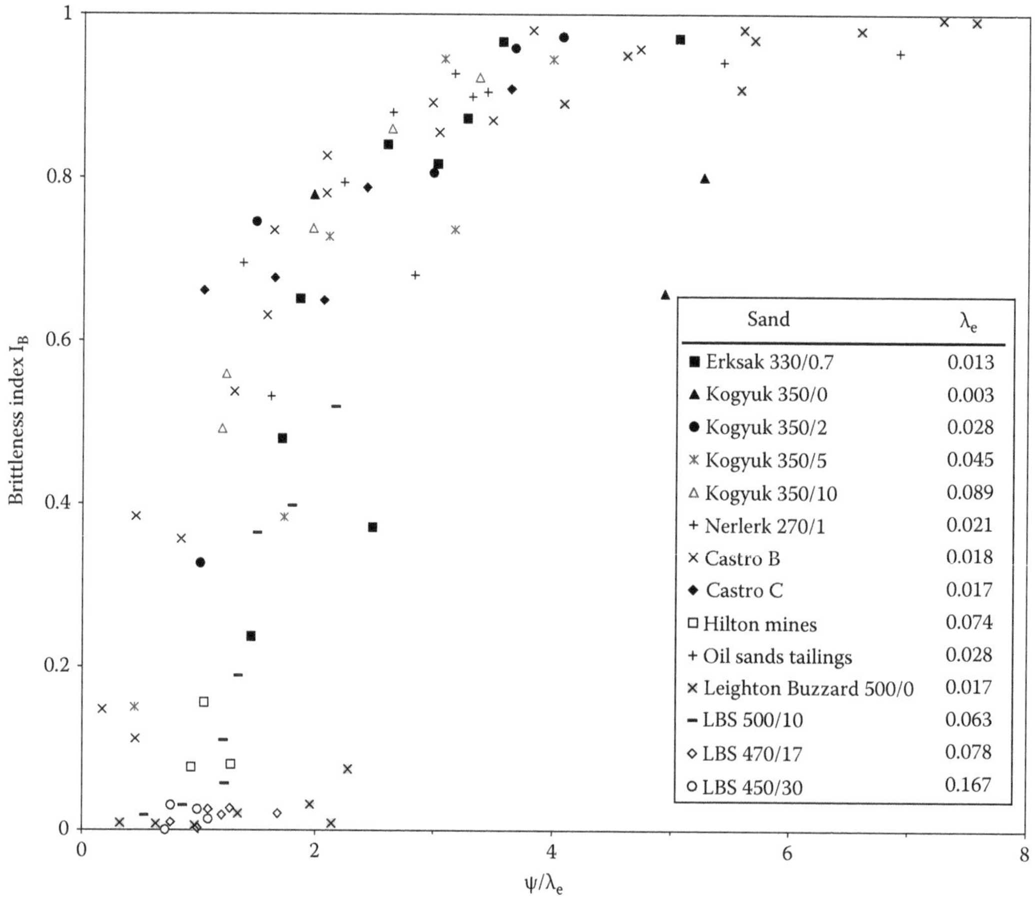

Figure 6.14 Brittleness index of sands with a range of λ. (Filled symbols represent low λ and open symbols higher λ values.)

tests considered, but this then leads to the interesting question: what are the underlying physical processes that lead to static liquefaction?

6.4 NATURE OF STATIC LIQUEFACTION

The observation that static liquefaction arose with a substantial brittle strength reduction from peak to residual led many workers to view liquefaction as caused by the collapse of a metastable arrangement of soil particles. It has become common to represent the transition from stable work hardening behaviour to sudden strength loss in undrained tests on loose sands with a collapse surface, 'collapse' being associated with a sudden rearrangement of the supposed metastable arrangement of the soil particles in a liquefiable soil. Sladen et al. (1985b) noted that if several liquefiable samples with the same initial void ratio are tested from differing initial confining stress conditions, the locus of peak strengths forms a line in effective stress space, which they termed the collapse surface. Others have introduced and used a variation on the collapse surface called the instability line (e.g. Lade and Pradel, 1990; Ishihara, 1993; Chu and Leong, 2002; Lade and Yamamuro, 2010). Figure 6.15 compares

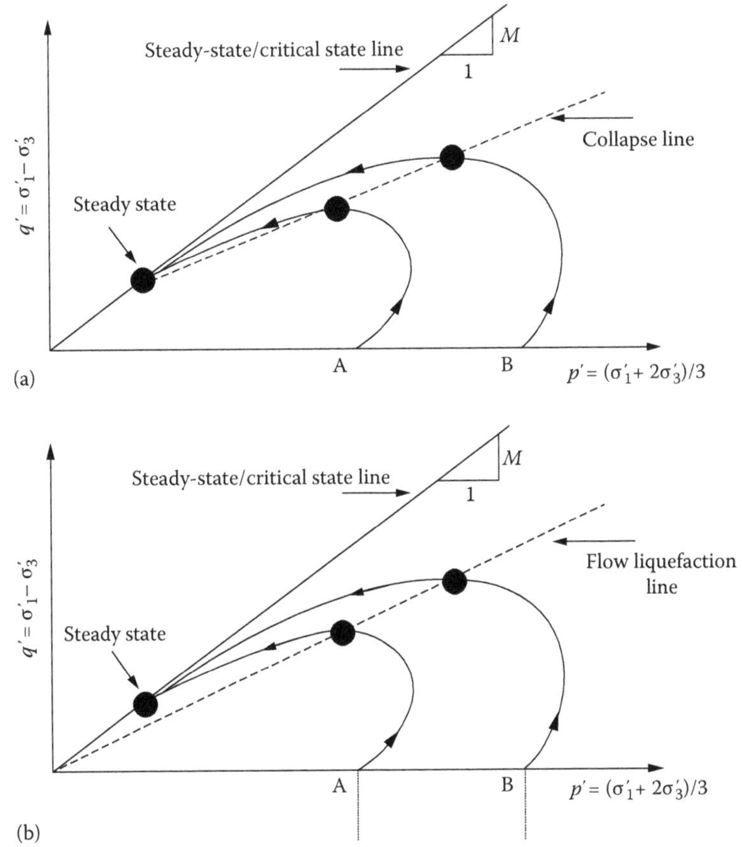

Figure 6.15 Comparison of collapse surface and instability or flow liquefaction line representations for the onset of liquefaction. (a) Collapse surface or line and (b) instability or flow liquefaction line. (After Yang, J., *Géotechnique*, 52(10), 757, 2002. With permission from the Institution of Civil Engineers, Londom, U.K. and Dr. Yang.)

the collapse surface and instability line frameworks. Both frameworks invoke the idea that there is a 'soil structure collapse' at the peak strength, and the only difference is in whether there is a cohesion-like intercept to the instability line.

While there is nothing wrong with normalizing peak undrained strength data in the form of a stress ratio, such as s_u/p'_o, difficulties arise once either the instability line or collapse surface is given physical significance in terms of an effective stress basis for soil behaviour (i.e. the slope of the collapse or instability line in Figure 6.15 becoming a property for a soil model). Examples of such use include Alarcon et al. (1988), Skopek et al. (1994), Yamamuro and Lade (1997), Lade (1999) and Chu et al. (2003). An effective stress instability criterion raises the question: how can the mobilized stress ratio at the onset of flow liquefaction (η_L at collapse) be much less than the ratio M that the sample is subsequently able to sustain at the critical state even though there is no densification of the sample (because of the imposed undrained boundary conditions)? Could it be that the concept of soil structure collapse is erroneous and that other mechanisms are involved?

The observation that the mobilized stress ratio at the onset of flow failure is much smaller than the critical friction predates Sladen et al. (1985b), being first suggested by Bishop (1971, 1973) from his investigations into soil behaviour during the Aberfan slide. However, Bishop also realized that the mobilized friction angle at liquefaction had no intrinsic significance in itself. It was a soil behaviour, not a soil property.

To progress the understanding of liquefaction, it is helpful to look at a liquefaction test in detail. Test G609 on Erksak sand is the loosest liquefaction test on a clean sand in our records (with an initial void ratio $e_o = 0.800$ and a corresponding state parameter of $\psi = +0.068$) and was carried out using displacement control with computerized data acquisition. Figure 6.16 shows the standard view of the test data, illustrating both an 80% drop in strength post-peak and an effective stress ratio at 'collapse' of $\eta_L = 0.62$. Also shown is the mobilization of excess pore water pressure with strain, with the peak deviator stress point highlighted. Excess pressure is smoothly generated throughout the stress path, and in particular, there is no inflection point at peak strength that might be associated with a collapse of a metastable arrangement of soil particles.

Going further, Figure 6.16 also shows the measured data transformed into the mobilization of η with strain. Like excess pore pressure, shear stress ratio is smoothly mobilized with strain, and the soil is indifferent to the 'collapse surface'. This particular test shows that there is no sudden change in soil behaviour or collapse during shear, but rather the loose state of the sample sets a limit on the shear hardening (discussed in Chapter 3), while the ongoing plastic volumetric strains continue to cause further excess pore water pressure. This is the opposite of the postulated physics of a collapse or instability surface.

Test G609 was not selected because it is in any way special other than being the loosest in our database. The physics of liquefaction illustrated by this test is exactly the same as for every other liquefying sample we have tested. The data files can be downloaded, and this aspect verified, as it really is crucial to understand the misleading nature of collapse surface idea. Soil liquefaction is not the result of metastable arrangements of soil particles.

The erroneous idealization of liquefaction as a metastable collapse event with a low limiting stress ratio (or friction angle) is further confirmed by drained tests on very loose sand samples, such as sample D684. This particular sample was at $e = 0.820$, which is significantly looser than $e_{max} = 0.76$ of the sand and also looser than the very loose liquefiable sample G609 just discussed. The initial confining stress of the drained sample is less, but the state parameter remains greater (more contractive) than the undrained sample. Figure 6.17 plots the loose drained test behaviour and shows the intersection of the 'collapse surface' friction ratio $\eta_L = 0.62$ from G609 (which will be an upper bound since D684 is both looser than

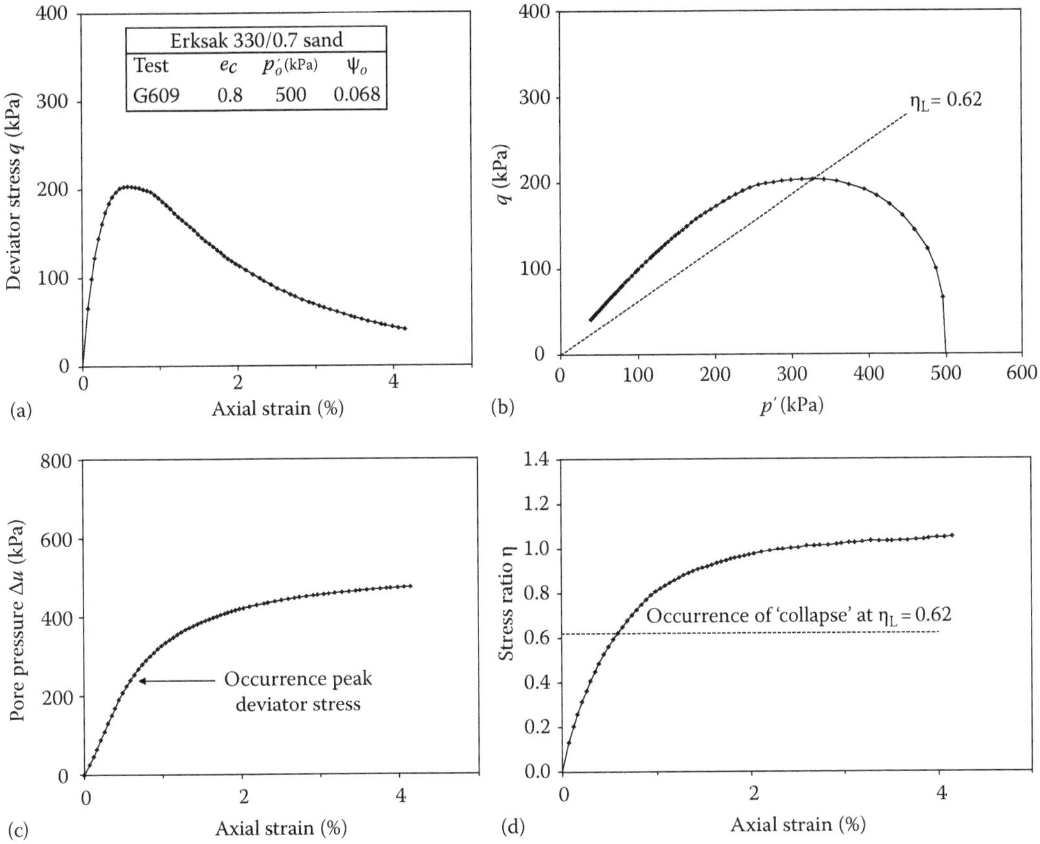

Figure 6.16 Erksak sand test G609 illustrating the nature of static liquefaction and collapse surface at η_L. (a) Stress strain, (b) stress path, (c) excess pore pressure and (d) stress ratio.

G609 and is also more positive ψ_o) with the drained stress path. As can be seen, not only does the sand not collapse at the shear stress ratio corresponding to the 'collapse surface', but also there is no change in the sand behaviour at this shear stress ratio. As with mobilized η in the undrained test, the drained soil behaviour is utterly indifferent to the fiction of a 'collapse surface'.

Of course, there is now an interesting question of physics. If the undrained strengths of liquefying sands are not unusual (as shown earlier), and it is now clear that liquefaction does not involve collapse of a metastable particle arrangement, then just what is the mechanism? A return to the discussion of *NorSand* will provide the insight.

6.5 UNDRAINED *NorSand*

6.5.1 Representing the undrained condition

The early literature on modelling undrained soil behaviour concentrated on a total stress approach, arriving at the undrained conditions through using a Poisson's ratio to impose constant volume. Such an approach misrepresents the nature of undrained loading of soil.

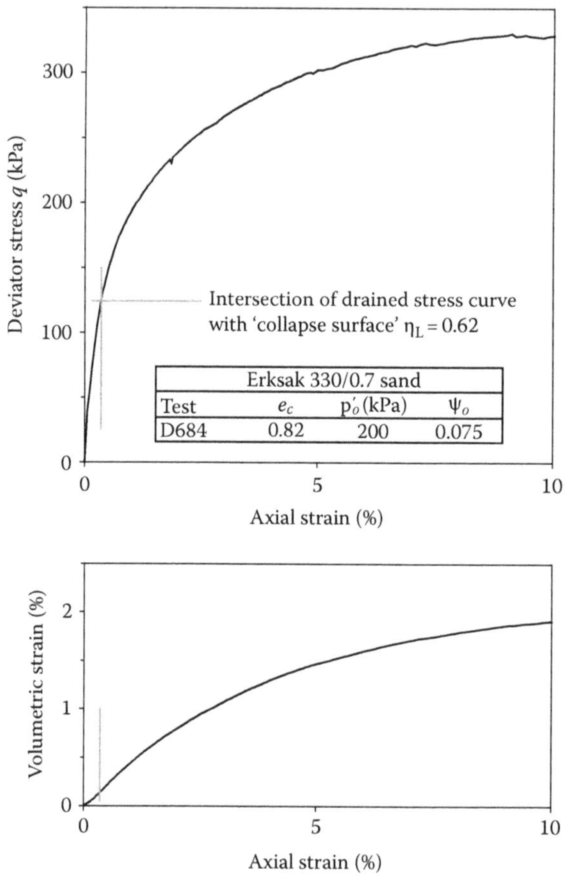

Figure 6.17 Very loose Erksak drained test D684 showing intersection of stress path with 'collapse surface'.

The basic condition for undrained loading, neglecting the minimal elastic compressibility of soil particles and the pore water, is that

$$\dot{\varepsilon}_v = 0 \Leftrightarrow \dot{\varepsilon}_v^p = -\dot{\varepsilon}_v^e \tag{6.6}$$

where the implication for the individual volumetric strain components has followed simply from invoking the fundamental strain partition of (3.1). Zero volumetric strain rate overall does not mean zero volumetric plastic strain and zero elastic strain – the two strain components can be non-zero provided that they balance each other. Undrained loading is a boundary condition effect, not a fundamental aspect of soil behaviour. Plasticity models capture undrained conditions by calculating plastic strains exactly as the drained case and then invoking (6.6) to obtain the required no-volume-change condition. The change in mean effective stress immediately follows by using the elastic bulk modulus K:

$$\dot{p} = -\dot{\varepsilon}_v^p K \tag{6.7}$$

The effective stress change in undrained loading responds only to the shear component of load. An external load increment that increases the total mean stress produces an equal response in the pore water pressure for fully saturated soils (i.e. with $B = 1$). Partial saturation effects or finite pore water compressibility can be added in easily enough using $B<1$ and computed from compatibility of volumetric compressive strain, but doing so provides no further insight and is not discussed further.

6.5.2 Simulation of undrained behaviour

Chapter 3 introduced *NorSand* and then considered drained loading. The drained derivation is now extended to undrained conditions before being applied to static liquefaction.

The steps in computing an undrained response with *NorSand* follow much the same procedure as for a drained response. A plastic shear strain increment is imposed, and the corresponding plastic volumetric strain computed immediately from the stress–dilatancy relationship. Mean effective stress change follows from (6.7). The change in shear stress then follows through the consistency condition:

$$\dot{\eta} = \eta \frac{\dot{M}_i}{M_i} + M_i \left(\frac{\dot{p}_i}{p_i} - \frac{\dot{p}}{p} \right) \quad \text{(C.51bis)}$$

There are two terms in the brackets on the right-hand side of (C.51). The first of these is given by the hardening law (3.37), which depends only on the current stress state, the state parameter and the plastic shear strain increment. The second term on the right-hand side is what has been computed from (6.7). It then remains to add the elastic shear strain increment using the computed stress change. In stress paths with rapidly decreasing p', the hardening law needs an additional term to prevent the stress path crossing the internal cap to the yield surface. This addition to the hardening law is detailed in Appendix C and applies to all such stress paths whether drained, undrained or partially drained (although with negligible effect in normal drained loading). The routine *NorTxlU* of the downloadable *NorSandM.xls* spreadsheet implements the undrained calculation for the triaxial test using commented code written in 'plain English'; do look at routine as there really is no magic to the mechanics of liquefaction.

Because undrained behaviour is a boundary condition effect, it is usual to carry out the base calibration of models using drained behaviour. Care is needed, however, because undrained behaviour depends very strongly on the elastic bulk modulus, which is both sensitive to sample grain contact arrangement and somewhat variable on a test-to-test basis. Ideally, bender elements are used to measure the shear modulus of the sample after it has been set up in the test cell, with the bulk modulus then inferred assuming a constant Poisson's ratio using the standard relation:

$$K = \frac{2(1+\nu)}{3(1-2\nu)} G \quad (6.8)$$

Bender elements are common in research laboratories these days, and also found in many commercial testing facilities. But bender elements were not common practice in the past, and the wealth of available data in the literature usually does not have direct measurement of elastic properties. The absence of directly measured moduli makes fitting *NorSand* to undrained behaviour less than a prediction and more like a calibration. Nevertheless, the starting point remains drained calibration.

The extensive series of tests on Erksak 330/0.7 sand was outlined in Chapter 2 in the context of CSL determination, and the drained data were then used to illustrate how well *NorSand* captured sand behaviour in Chapter 3. Comparison of *NorSand* with undrained

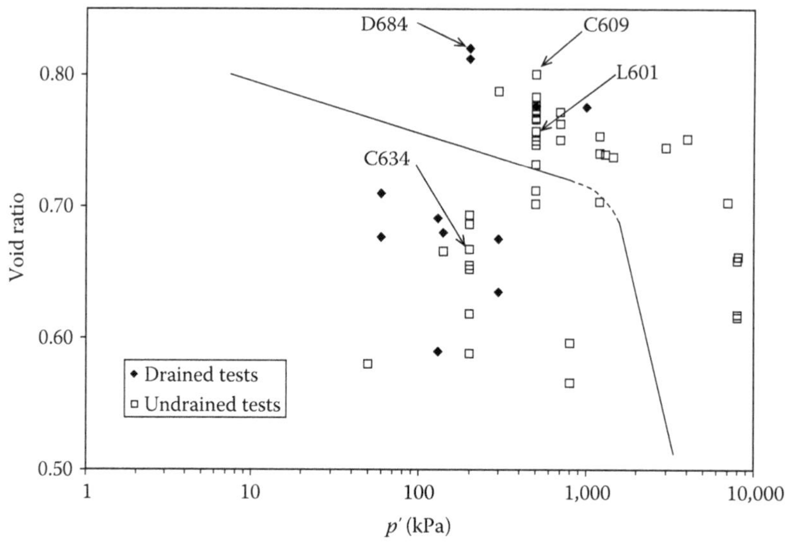

Figure 6.18 Initial state diagram for the series of triaxial tests on Erksak 330/0.7 sand.

experimental data for Erksak 330/0.7 sand will now take properties from this drained calibration as constant (i.e. $M_{tc}=1.26$, $\Gamma=0.817$, $\lambda_{10}=0.031$, $\chi_{tc}=3.8$). The hardening modulus H varies with ψ_o (see Figure 3.17) and averages $H=103\text{–}980\psi_o$. The corresponding undrained data from the series of tests on Erksak sand were presented at the beginning of this chapter. The state diagram illustrating the range of void ratios and stresses tested is shown in Figure 6.18, with tests of particular interest highlighted.

Elasticity of Erksak sand in laboratory-reconstituted samples was measured as a bulk modulus using unload/reload stages (Section 3.7.1) giving the shear rigidity in Equation 3.41. This equation can be simplified, assuming Poisson's ratio of 0.2 and reasonable values of the other constants to $I_r=980\ (p'/p_{ref})^{0.5}$, which has been used for comparison (the reference stress level has been taken as 100 kPa as is conventional). Figure 3.16 presented the fit *NorSand* achieves to drained triaxial compression data, with good-to-excellent fits being evident. Turning to undrained behaviour, Figure 6.19 compares *NorSand* with three undrained liquefaction tests representing the spectrum from highly contractive to lightly dilatant. In fitting the tests, H and I_r have been varied around the mean trends to obtain the best fit to the stress strain data, with the parameters used being shown for each simulation. Everything else is as per the dense calibration. These simulations can readily be redone using the downloadable *NorSandM.xls*. It is emphasized that the good fits are obtained using parameters from dense drained triaxial tests and without a collapse surface. Detailed comments now follow.

Test L601 was a lightly contractive sample with an initial state $\psi_o=+0.025$. A rather good fit to the stress path is computed with *NorSand*, although using a slightly stiffer plastic modulus than trend. Importantly, the peak strength and the onset of liquefaction are nicely predicted. The subsequent strength drop with increasing strain is less dramatic in *NorSand* than the experiment, but this may be a rate or inertial effect as the experiment was load controlled. The residual strength is well predicted, which confirms the CSL and initial void ratio determinations for this sample. A little over-consolidation ($R=1.15$) was introduced to replicate soil structure effects that affect the initial shape of the stress path.

Perhaps the most dramatic aspect of the undrained response of sand is the static liquefaction of very loose samples. The loosest sample tested in undrained compression was sample

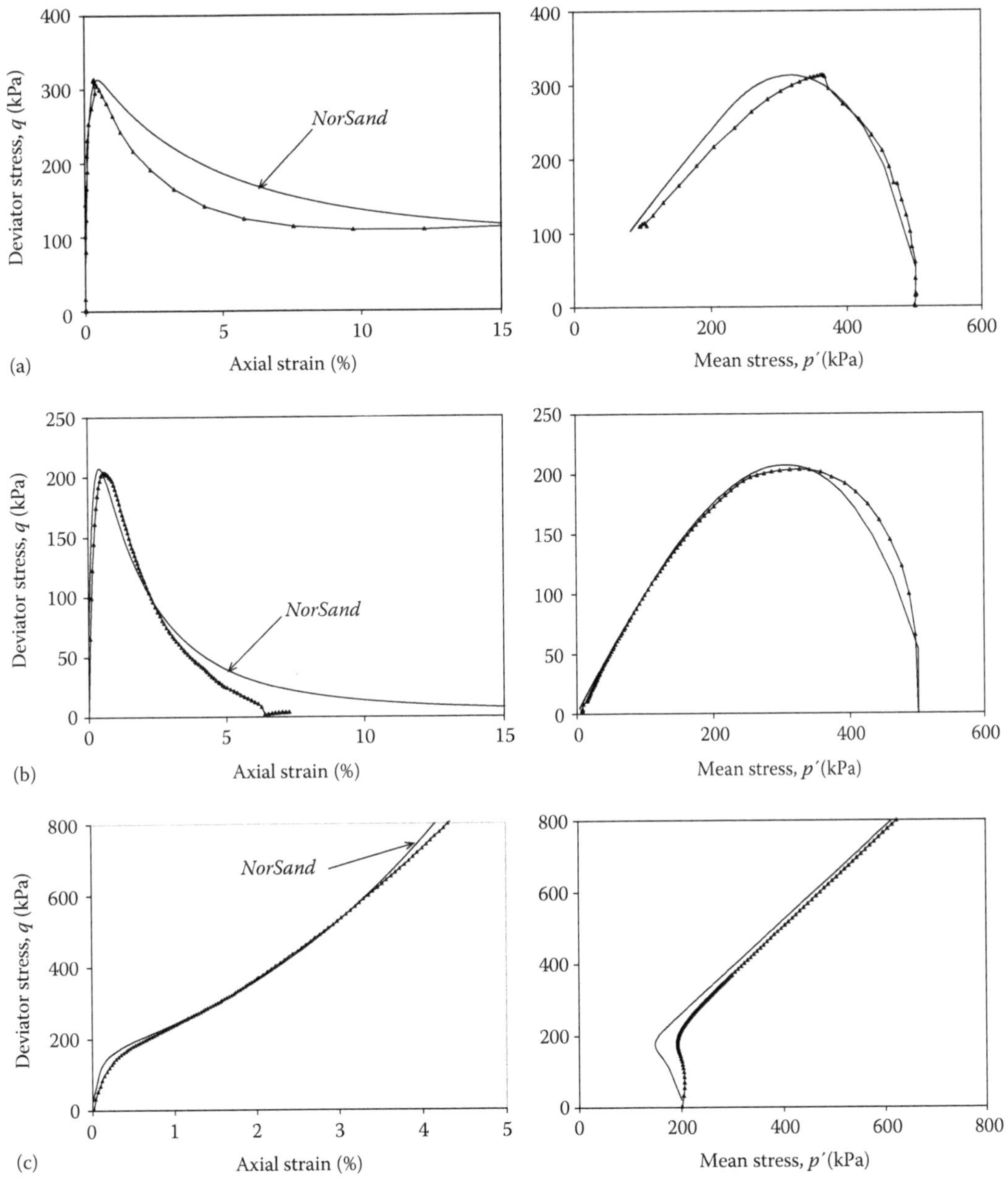

Figure 6.19 Triaxial compression static liquefaction – *NorSand* compared to Erksak sand data. (a) Test L601 $\Psi_o = +0.025$, modelled with $H = 250$, $I_r = 170$, $R = 1.15$, (b) test L609 $\psi_o = +0.07$, modelled with $H = 140$, $I_r = 140$, $R = 1.15$ and (c) test C634 $\psi_o = -0.08$, modelled with $H = 300$, $I_r = 250$, $R = 1.20$.

C609 with $\psi_o = +0.068$ (the sample discussed in Section 6.4). Using unchanged material properties, a good fit to the measured behaviour is computed including the characteristic extreme post-peak strength drop of very loose samples as illustrated in Figure 6.19. A less dramatic, but more interesting, behaviour is the undrained response of lightly dilatant sand. Figure 6.19c shows the response of sample C634 with $\psi_o = -0.08$. Again, without changing any material properties, a reasonable match to this behaviour is computed, including the reversal of curvature of the stress path. Overall, *NorSand* calibrates well to the experimental data on Erksak sand for drained loading (Chapter 3), and these dense drained calibrations then closely replicate undrained loading and in particular static liquefaction of very loose samples.

6.5.3 How *NorSand* models liquefaction

Having demonstrated that *NorSand* predicts the measured response of sand in undrained triaxial shear, including what are conventionally termed liquefaction tests, it is now possible to examine the nature of static liquefaction by looking into the details of *NorSand* when a loose sample is loaded undrained.

The first point is that contractive strain is not the difference between normally consolidated behaviour of clays and liquefaction of sands. Contractive strains occur in both cases. What happens in liquefaction is that there is proportionately less shear hardening. Normally consolidated clay behaviour is one in which the contractive volumetric strain produces just enough hardening to balance the decrease in mean stress with an increase in shear stress ratio, and overall, there is a monotonic increase in deviator stress to the critical state. In liquefying samples, however, the volumetric contraction has the effect of reducing the mean effective stress by more than the shear strength gain induced by the hardening of the yield surface. These processes are apparent in the test data, but can now be looked in more detail with *NorSand*.

Yield surface size in *NorSand* is characterized by two parameters, M_i and p_i. M_i sets the relative size of the yield surface in the deviatoric direction, while the image mean effective stress p_i determines its absolute size. Recall that *NorSand* controls hardening (i.e. sets the maximum yield surface size) using a limit on p_i expressed by

$$\left(\frac{p_i}{p}\right)_{max} = \exp\left(\frac{-\chi_{tc}\psi_i}{M_{tc}}\right) \qquad (3.35\text{ bis})$$

For loose soils, ψ is positive, and the term on the right-hand side is less than unity. This hardening limit on p_i has the effect of allowing excess pore pressure while limiting the hardening rate substantially. Although stress ratio η is still increasing, as the peak strength is approached, the rate of hardening slows, and the deviator stress drops because the rate of excess pore pressure increase exceeds the rate of η increase. Fundamentally, the yield surface has to soften (contract) to the critical state. This can be visualized by looking at a yield surface for a liquefiable soil in Figure 3.13. Note in particular that Equation 3.35 is not remotely like the concept of a collapse surface and, as a hardening limit, suffers from none of the theoretical deficiencies associated with collapse or instability surfaces. It is also the same hardening limit that controls the dilatancy of dense soil and is not something introduced just to simulate liquefaction.

6.5.4 Effect of soil properties and state on liquefaction

It is helpful to look at systematic parameter changes to understand how the *NorSand* model responds to different parameters in undrained loading. Simulations are presented using the Erksak 330/0.7 sand CSL parameters for both loose (ψ = +0.05) and lightly dilatant (ψ = −0.05) states, as interest is mainly on soil behaviour around the critical state.

Figure 6.20 shows the effect of elastic rigidity, I_r, on the computed undrained behaviour, all other properties being held constant. For the loose sample (ψ = +0.05), increased elastic stiffness significantly decreases the strength at the onset of liquefaction. There is about a 30% reduction in strength for the case of a doubling of elastic stiffness. There is

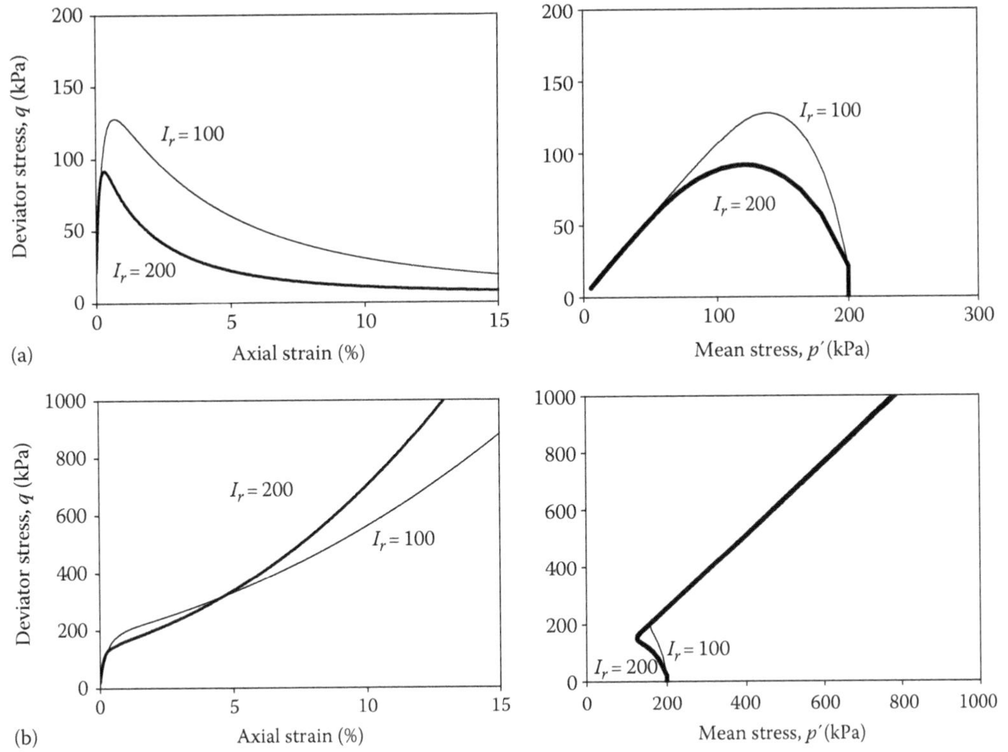

Figure 6.20 *NorSand* simulations showing the effect of elastic modulus on undrained behaviour. (a) Loose $\psi_o = +0.05$ and (b) compact $\psi_o = -0.05$.

also an effect on the stress–strain curve and the brittleness with strain. The effect of I_r on the denser sample is more subtle in that increasing modulus decreases the shear stress of the transition to dilatant behaviour, but then the stiffer soil gains dilatant strength more quickly.

Figure 6.21 shows the corresponding effect of plastic modulus H on the computed undrained behaviour, again with all other properties held constant. In the case of the loose soil, the stiffer plastic modulus (large value of H) acts in the opposite sense to elastic rigidity and produces the greater strength. Unlike elasticity, plastic modulus has quite an effect on the post-peak behaviour, with the stiff modulus producing a markedly more brittle post-peak regime. In the case of the denser soil, a stiff plastic modulus increases both the shear stress at the transition from contractive to dilative behaviour and the effectiveness of dilation after this transition.

Figures 6.20 and 6.21 show that peak undrained strength (i.e. at onset of the liquefaction) is influenced by the plastic hardening modulus and the elastic rigidity. Peak strength is not simply a frictional property, although of course it scales with the critical stress ratio, M_{tc}, for any fixed combinations of other soil properties.

Figure 6.22 illustrates how the computed peak undrained strengths match the spectrum of test data, with trend lines for various H/I_r ratios. These simulations are for a common $M_{tc} = 1.25$, $\chi_{tc} = 3.8$ and $\nu = 0.15$, which are reasonable values for most test data on the plot. *NorSand* replicates the trend in a large body of data. The variations within this trend are easily understood as being caused by variations in the ratio of elastic-to-plastic modulus with some contribution from the critical friction ratio.

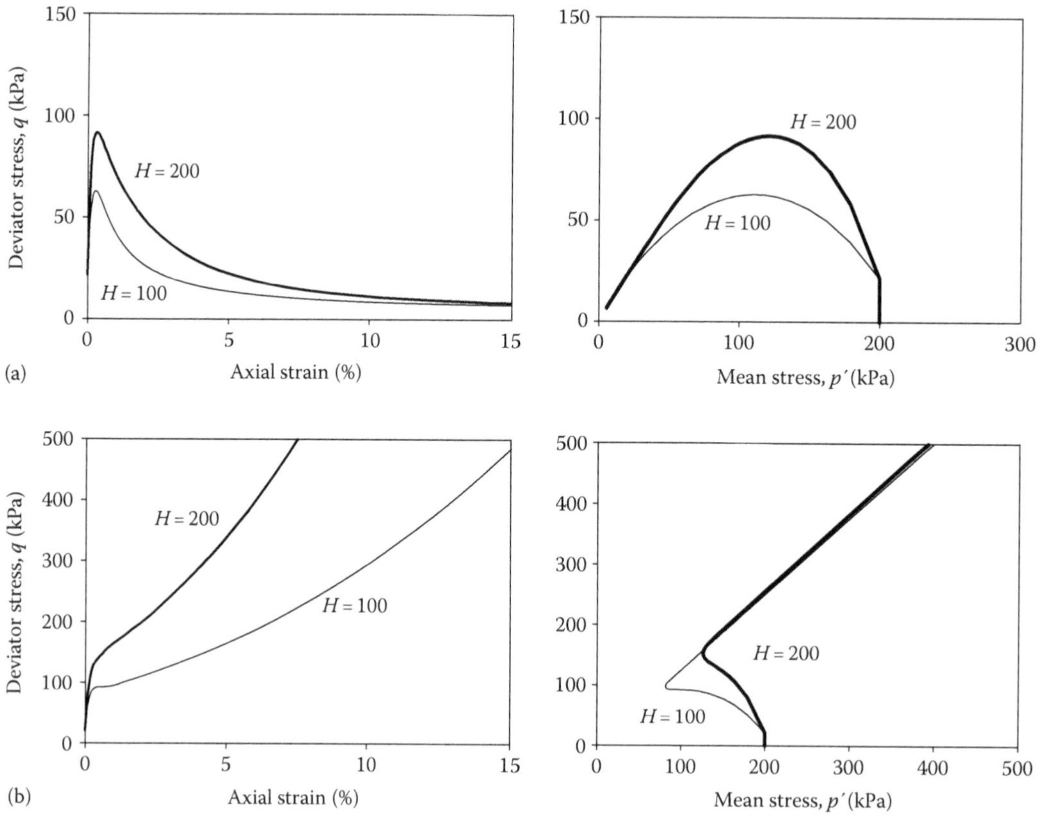

Figure 6.21 *NorSand* simulations showing the effect of plastic modulus on undrained behaviour. (a) Loose $\psi_o = +0.05$ and (b) compact $\psi_o = -0.05$.

6.6 UNDERSTANDING FROM *NorSand*

Having established that *NorSand* fits experimental test results, attention now turns to what might be regarded as more controversial issues surrounding liquefaction and to the clarification that can be gained from *NorSand*. It may be helpful to open the *NorSandM.xls* spreadsheet and run the simulations while reading this section. While this section provides a guide to soil behaviour, it is easy to run far more simulations than can be shown here, and much more insight will be gained by looking at how changing properties and state changes soil response.

6.6.1 Uniqueness of critical state

In Chapter 2, the critical state and its measurement were introduced, together with how confusion in identifying the critical state from test data has led people to question whether the critical state is unique. Uniqueness of the critical state is now considered within the general understanding of undrained behaviour. An issue with uniqueness of the CSL concerned the F (for fast) and S (for slow) steady-state lines (Casagrande, 1975; Alarcon et al., 1988; Hird and Hassona, 1990) and just what each line implies. Triaxial tests are limited to an axial strain of about 20%, so the state of samples at 20% strain represents the closest the samples get to the critical state in a test. If these states are plotted on an *e*–log *p*′ diagram,

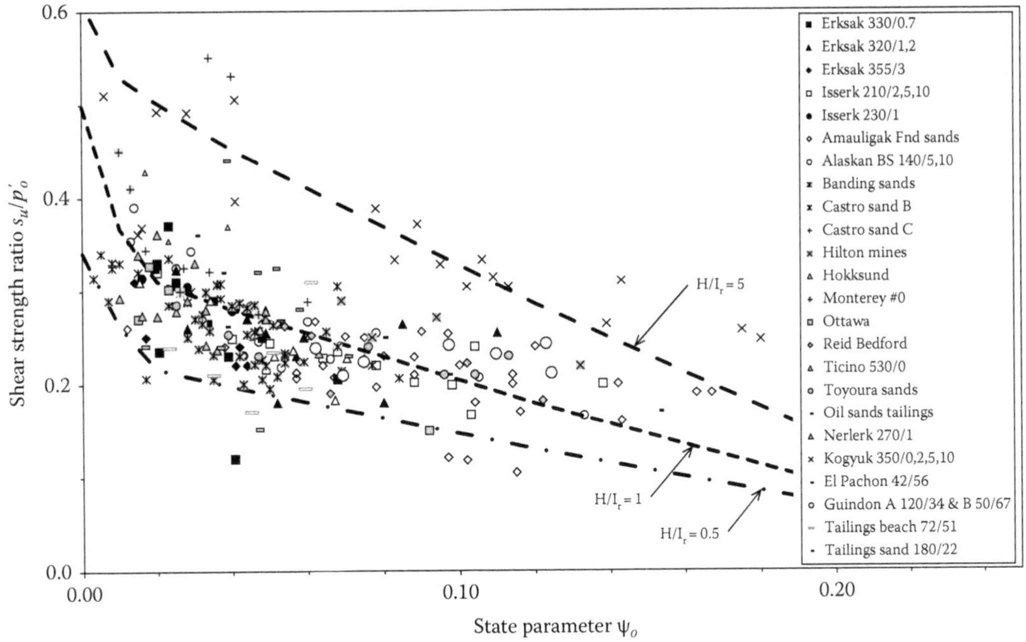

Figure 6.22 Peak undrained triaxial compression strength for liquefying sands with trends from *NorSand*.

the S and F lines from *NorSand* that are equivalent to the test data presented in the literature can be defined. Figure 6.23 is just such a plot for Erksak sand, showing very clearly how drained and undrained tests suggest different critical state lines even though they have the same intrinsic CSL. The only difference is the initial state and stress path followed. At 20% strain, the critical state has not been reached by most of the samples.

The *NorSand* behaviour shown in Figure 6.23 is consistent with the experimental data reported in the literature and illustrates that part of the difficulty in interpreting laboratory tests is the practical maximum strain limitations of the triaxial apparatus. Theoretically, shear strain to the critical state for denser soils may exceed 50%. The potential for confusion is large when measurements use test equipment where strains greater than 15% are problematic. This is why CSL determination is best done with contractive samples, as only these samples get to critical conditions within the limits of the test apparatus.

Sample preparation effects are sometimes invoked as to why the critical state cannot be unique. This view is readily shown false using *NorSand*. There are two ways soil fabric (particle orientation) effects are captured within *NorSand*. First, the dimensionless plastic hardening modulus (H) can be expected to be a function of particle arrangement – certainly whether a reconstituted sample was prepared by water pluviation or moist tamping affects H. Second, elastic rigidity is as much a function of particle contact arrangement as it is of stress level. The effect of H and I_r on soil behaviour was examined earlier (Figures 6.20 and 6.21), and a wide range of soil behaviour was simulated with an explicitly unique CSL. But it is now time to bury the supposed non-uniqueness of the CSL.

To illustrate the reality of parameter variation as the explanation of sample preparation effects, Figure 6.24 shows simulations of the measured behaviour of the moist-tamped and water-pluviated sands reported in Section 2.5 and Figure 2.31, and an excellent fit of *NorSand* to the test data is apparent for either sample preparation technique. This excellent fit was

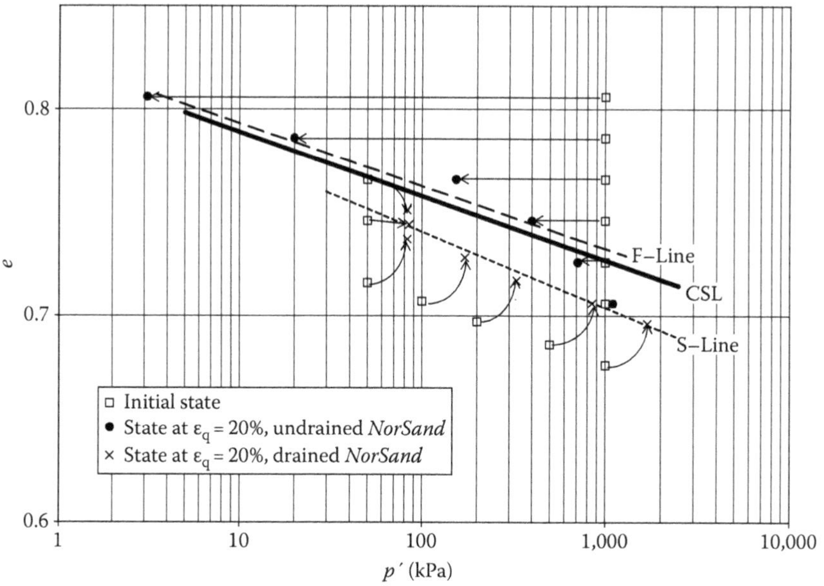

Figure 6.23 S and F lines from *NorSand* simulations of triaxial compression tests. (From Jefferies, M.G. and Been, K., Undrained Response of Norsand, In *Proceedings of the 45th Canadian Geotechnical Conference*, Toronto, October 26–28, 1992).

achieved by modest changes in the elastic and plastic model parameters (I_r, H and χ) and with remaining parameters (Γ, λ, ν) constant. All changes in behaviour can be attributed to soil structure, in as much as both elastic and plastic parameters move together. Softer elasticity goes with softer plasticity, and the fit of the model to the test data is excellent in both instances.

The importance of capturing fabric effects with *NorSand* is not so much that it can be done, but that it disproves the assertion that different stress paths from different sample preparation methods (but with the same state) require a non-unique CSL. These simulations had an explicitly unique CSL, and yet they capture the details of both stress path and stress–strain changes caused by differing initial soil structures.

It was noted earlier that simulation of soil fabric with an isotropic model would be illustrated. This is what has just been done, as it is the different sample preparation techniques that produce the different fabrics.

Of course, it is easy enough to show how rather small variations in H and I_r can be used to explain laboratory test data, but this then raises the question of how to apply either laboratory data or simulations to in-situ soils. Presently, there is no way of measuring soil particle arrangement that can be used in practice. This is why, in Chapter 4, the indirect approach of ignoring fabric and instead determining I_r with shear wave velocity measurements and H with the pressuremeter test was advocated (although a long way from a routine procedure). Good determination of plastic hardening modulus, which seemingly must be in-situ as it is clear that different laboratory procedures produce different soil fabrics each of which has its own hardening modulus, is an area requiring further work regardless of constitutive model choice. Shuttle (2008) suggests that a way forward may be to work with the observation that H and I_r appear correlated, which is physically plausible, and thus measured in-situ G_{max} could be used to develop the best estimate for H. Of course, calibration of H to reconstituted samples in the laboratory is more straightforward and likely to remain so for some time. Although advocating measurement of H in-situ is counsel of perfection.

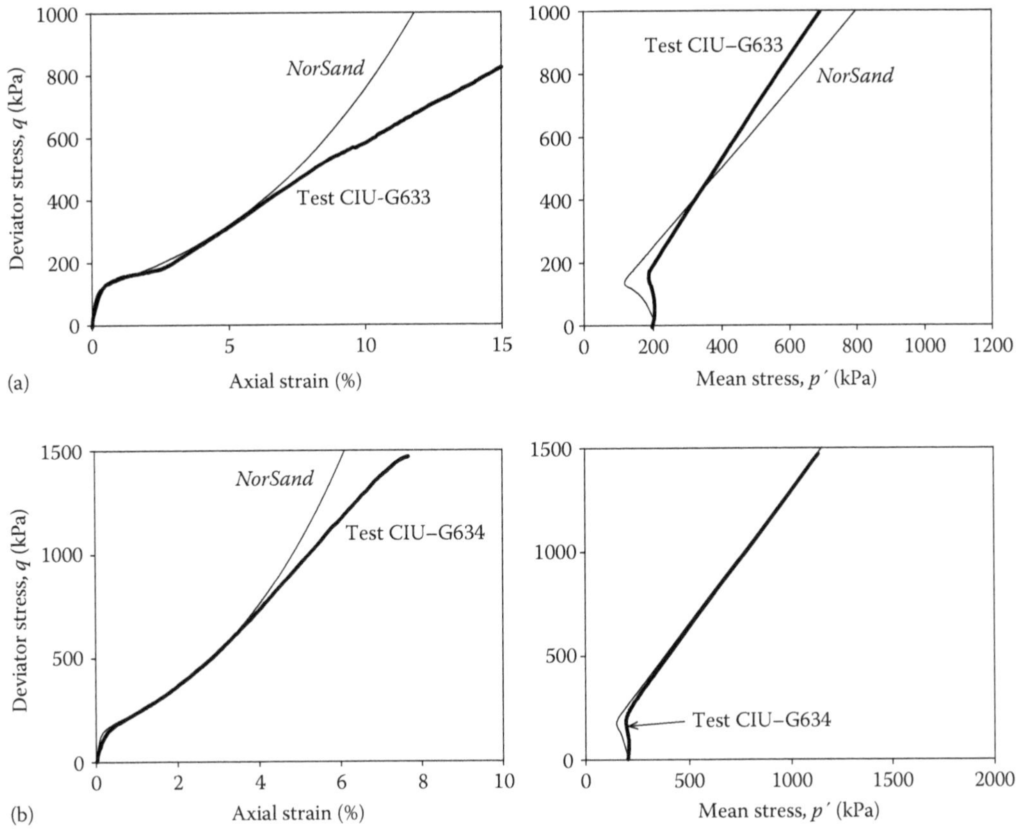

Figure 6.24 Simulations showing modelling of sample preparation effects (for both samples, $\Gamma = 0.816$, $\lambda_{10} = 0.031$, $\nu = 0.2$). (a) Test CIU-G633, moist tamped ($\psi_o = -0.089$, $H = 150$, $I_r = 200$, $\chi_{tc} = 2.5$) and (b) test CIU-G634, pluviated in water ($\psi_o = -0.077$, $H = 300$, $I_r = 250$, $\chi_{tc} = 3.8$).

6.6.2 Instability locus

The instability locus (IL; also the collapse surface or flow liquefaction line) is an accepted view of the nature of liquefaction triggering based on trends seen in undrained tests. However, in Section 6.4, we looked at direct experimental evidence for a collapse surface or IL that would trigger liquefaction in sand samples and found none. So, if there is no collapse or breakdown of a metastable particle arrangement, what can *NorSand* tell us about the instability mechanism that then leads to brittle collapse?

Brittle collapse can arise with any strain softening stress–strain behaviour. Whether collapse actually occurs depends on load redistribution and drainage of excess pore water pressures.

Consider the undrained strength that develops from differing (drained) K_o conditions, a set of scenarios that illustrate how the mobilized shear stress at failure might change depending on how much of the loading path is drained and how much is undrained. The following simulations were carried out using the Erksak 330/0.7 calibration of *NorSand* and for an initial state parameter of $\psi_o = +0.05$. This is actually rather loose, as will be seen from the computed stress–strain curves. A common mean stress level of 200 kPa was chosen for the simulations, which started from a range of K_o conditions equivalent to the end of a drained loading path before transition into undrained loading. The results are shown in Figure 6.25.

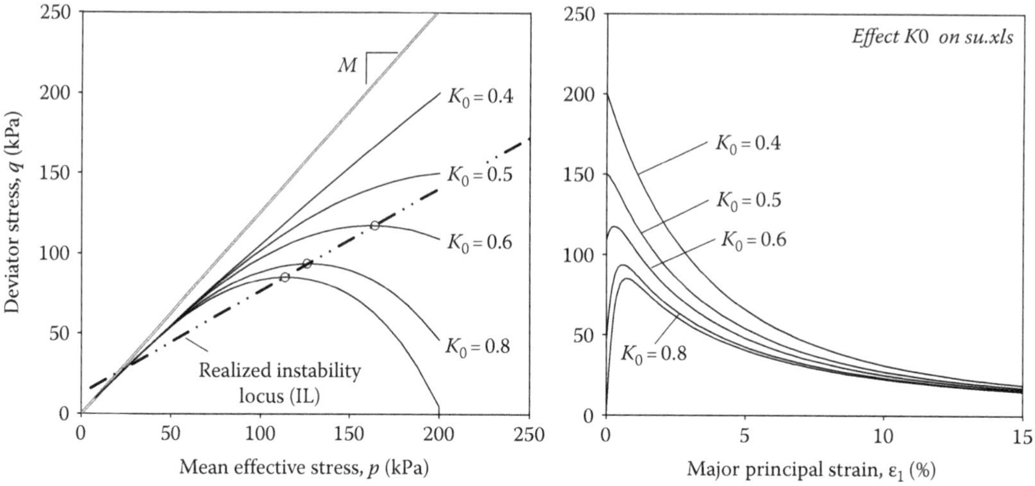

Figure 6.25 Effect of K_o on undrained triaxial compression strength (Erksak sand properties).

Several interesting behaviours are apparent in Figure 6.25. First, there is an increase in peak undrained strength as K_o decreases (increased initial shear stress), because a smaller part of the strain path is in the weaker undrained mode. Second, this process goes only so far. In the case of the simulations shown in Figure 6.25, once K_o reaches about 0.55, the strength immediately and rapidly falls after the switch to the undrained loading. This transition to an immediate strength drop arises at far below the drained strength.

Figure 6.25 shows that a stress perturbation above the IL could result in immediate undrained behaviour and rapid strength loss. Why did this not arise in the loose drained laboratory tests we discussed earlier? Because those laboratory tests were carried out using displacement control to demonstrate that there was no metastable collapse in the arrangement of sand particles. What we are exploring here using *NorSand* is the effect of drainage conditions when simulating load control. And what is apparent is that despite *NorSand* having no concept of an IL in its formulation, the locus through peak strengths (circled on the respective stress paths) is a straight line. Instability is caused by drainage conditions, not metastable collapse of the soil fabric. In addition, the IL is not a soil property but changes both with the state parameter of the soil and with the soil's elastic and plastic properties.

Further examination of *NorSand* simulations and data for a silty sand material is instructive to reinforce this point. The material considered is Guindon silty sand with 33% fines for which the CSL is shown in Figure 2.38. Figure 6.26 shows measured data for an undrained test on this silty sand with an initial state parameter $\psi_o = +0.065$ and how small variations of the elastic and plastic moduli (i.e. I_r and H) affect the undrained stress–strain curves and stress paths within *NorSand*. Both have a small but significant effect on pore pressure response and post-peak liquefaction behaviour. Increasing I_r results in a higher pore pressure, all else being equal (Figure 6.26a), since a larger elastic volumetric stain is required to compensate for shear-induced contraction in constant volume (undrained) shearing. Increasing the plastic modulus has a similar effect on pore pressure (Figure 6.26b). The key point is that the stress paths in Figure 6.26 show that the undrained shear strength, and thus also the IL, changes location in q–p' space as these moduli change.

We can take this a step further and use *NorSand* to predict the stress ratio η_L at instability, or the IL, as a function of changes in the elastic and plastic moduli. Figure 6.27 shows the results when this is done, for a particular set of soil parameters and one initial state

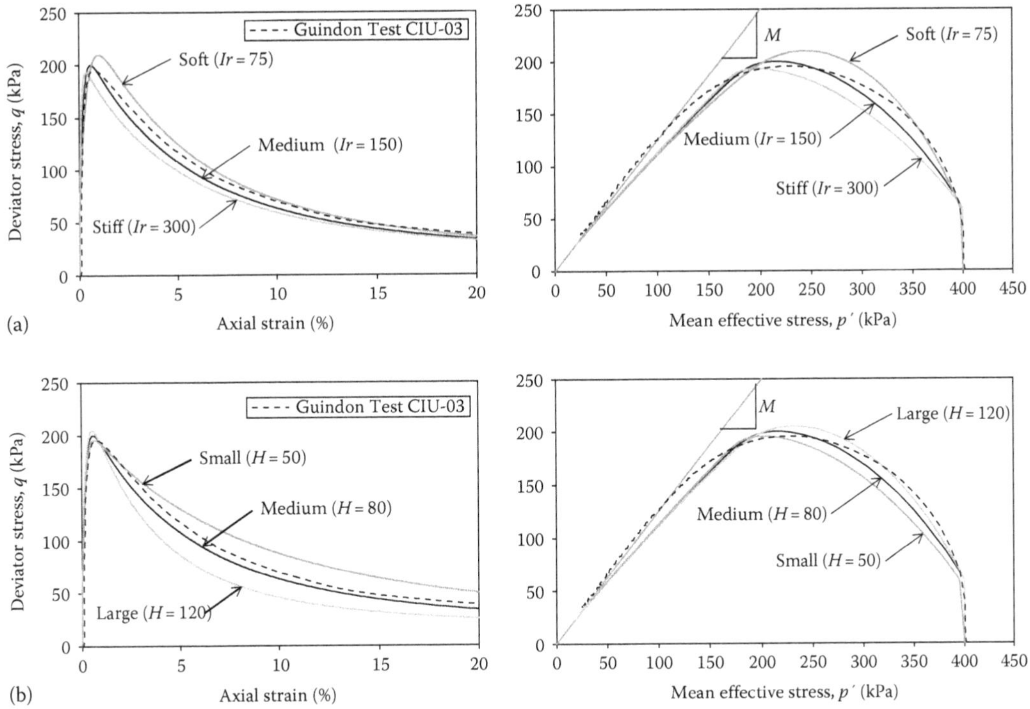

Figure 6.26 Influence of soil moduli (H, I_r) on liquefaction of Guindon silty sand and comparison with test CIU-103 ($\lambda_{10}=0.046$, $M_{tc}=1.25$, $N=0.5$, $\chi=6$, OCR = 1.15, $K_o=1$, $\nu=0.2$ and $\psi=+0.065$). (a) Effect of elastic modulus ($H=80$) and (b) effect of plastic hardening modulus ($I_r=150$).

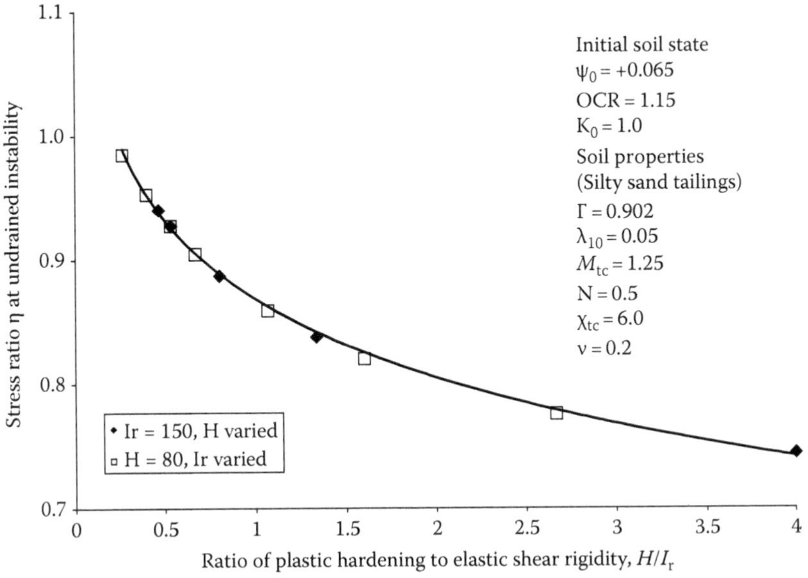

Figure 6.27 Variation of stress ratio at undrained instability or collapse surface as a function of elastic and plastic soil moduli (Guindon silty sand parameters).

($\psi_o = 0.065$). The horizontal axis is the same ratio H/I_r as shown in Figure 6.22, but Figure 6.27 shows clearly how the instability stress ratio can vary easily from $\eta_L = 0.75$ to $\eta_L = 1.0$ for the same value of critical friction ratio M_{tc}. The peak undrained shear strength and pore pressure generation, or triggering of liquefaction, cannot be represented by an effective stress friction ratio or instability line. The effective stress ratio is strongly dependent on the elastic stiffness and plastic hardening (or softening) of the sand during shear, and so, the IL or peak undrained shear strength is not simply a frictional soil property.

If we refer back to our experience with the undrained behaviour of clay soils, in particular, the observation that the undrained shear strength depends significantly on how it is measured, this observation that the undrained shear strength of sands is not a frictional property should come as no surprise. Just as with clays, the correct approach is to work with an undrained shear strength ratio s_u/σ'_v or s_u/p', and we can use the appropriate laboratory testing, supplemented by *NorSand* if necessary, to tell us what ratio is appropriate depending on the initial state parameter, the soil properties, the loading path and the current geostatic stress state.

The downloadable spreadsheet *NorSandM.xls* outputs the strength ratio s_u/p' and IL slope η_L automatically as a result from any simulation. You can run this spreadsheet as there is no substitute for seeing soil response change as the initial conditions and soil properties change (this is one of the beauties of constitutive modelling).

6.6.3 Effect of silt (fines) content on liquefaction

Fines content has long been an issue in liquefaction. Much of the field experience of liquefaction has been viewed in the context of penetration test data as the index of soil state. All else being equal, however, an increasing fraction of finer soil particles lowers the penetration resistance because increasing fines gives a more compressible soil. Correspondingly, how to evaluate the effect of fines on liquefaction susceptibility is a genuine concern.

Laboratory experiments in which the silt content has been systematically varied include Kuerbis et al. (1988), Pitman et al. (1994) and Lade and Yamamuro (1998). The first two papers indicated that fines tended to make a soil more resistant to liquefaction by filling the void space, in effect reducing the granular void ratio and making the soil appear denser than it was. Yamamuro and Lade obtained the exact opposite effect (and indeed they called it 'reverse behaviour') with increasing fines content producing an increasing liquefaction potential. The conflict between these results is caused by working in terms of relative density, but a clear understanding emerges using ψ.

The relationship between peak dilatancy and state parameter presented in Chapter 2 was based on testing many sands with silt contents as great as 35%. There is a sensibly unique relationship independent of soil gradation (Figure 2.6). Although the overall grain size distribution, which is more than just fines or silt content, affects the CSL (Figures 2.36 and 2.37), this does not affect the state parameter framework. Because dilatancy and ψ are simply related, dealing with the effect of fines content (or, more generally, soil gradation) on soil behaviour is straightforward – just use the correct CSL.

The adequacy of the state parameter in capturing silt content effects is illustrated by two examples. Silty sands often arise when dealing with tailings, and interesting data were obtained for the Rose Creek impoundment (Shuttle and Cunning, 2007). Various samples were obtained from the site, and two representative gradations were chosen, referred to as A and B, comprising 67% and 30% fines, respectively. Both gradations were tested to define the CSL using the usual test protocols on reconstituted samples. Figure 6.28 shows an example of the static liquefaction of each gradation. Also shown in the figure is the behaviour computed using *NorSand* (and the parameters used).

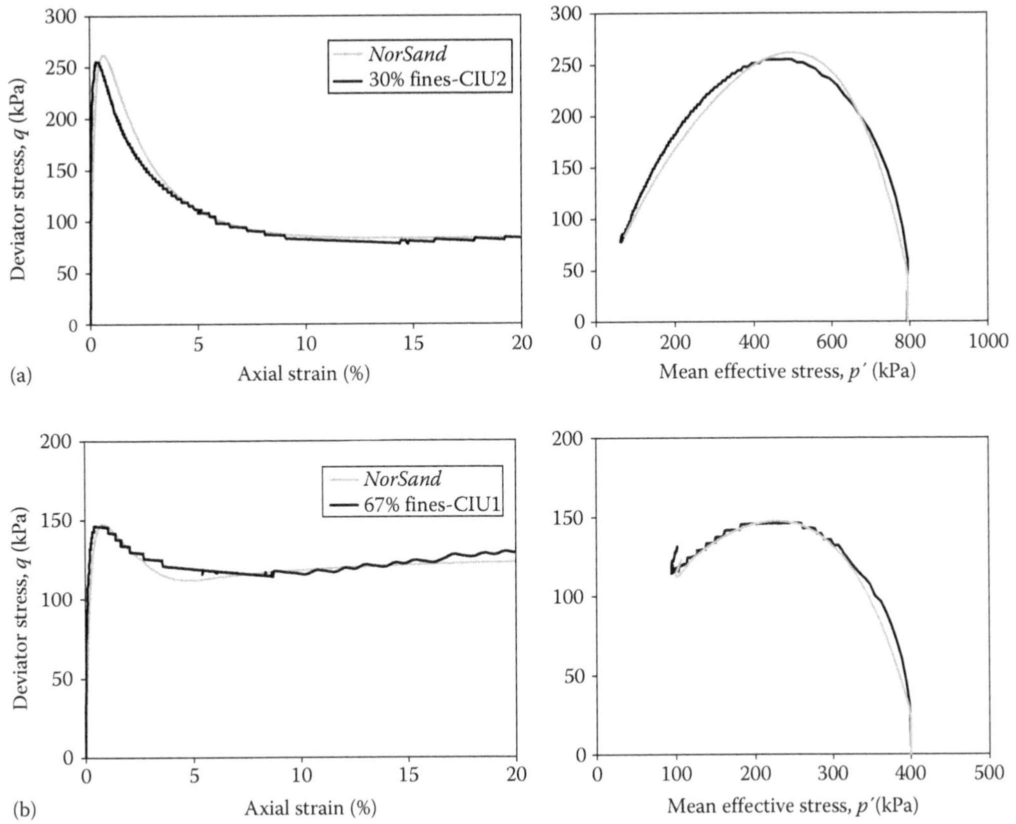

Figure 6.28 Simulation of liquefaction of loose sandy silts from Rose Creek Impoundment. (a) 30% fines (ψ_o = 0.085, R = 1.06, H = 60, I_r = 107, χ_{tc} = 5, M_{tc} = 1.15, N = 0.5, λ_{10} = 0.082) and (b) 67% fines (ψ_o = 0.09, R = 1.06, H = 40, I_r = 100, χ_{tc} = 5, M_{tc} = 1.15, N = 0.5, λ_{10} = 0.159).

Taking the 30% fines sample first, this test was quite loose with $\psi_o \approx +0.07$ based on the sample void ratio (after the saturation and consolidation stages) and the CSL determined from the group of tests on this gradation. The slope of the CSL is intermediate with λ_{10} = 0.082. Liquefaction was extreme, with no hint that fines in anyway suppressed liquefaction. *NorSand* was fit to the data readily.

Turning to the 67% fines gradation, this sample was tested at $\psi_o \approx +0.09$. The sample did liquefy, but had less brittleness than the 30% fines sample, which was the effect of increasing compressibility (greater λ) with the 67% fines sample being twice as compressible as the 30% fines sample. Despite this much greater compressibility, *NorSand* readily fit this test using the measured ψ_o, with typical soil properties (but the higher λ_{10} = 0.159).

At this point, a view might form that the comparison of the two tests shown in Figure 6.28 justifies the idea that increasing fines reduces brittleness. This would be wrong. Higher λ does mean the brittleness is reduced at the same ψ, but the problem in the laboratory is that samples could not be prepared for triaxial testing as loose as found in-situ. This situation is much like sand samples that cannot be set up by underwater pluviation in the laboratory anywhere near as loose as occurs during hydraulic deposition in the field. The 67% fines gradation was encountered in-situ at states estimated to be $\psi_o > +0.15$, the CPT data in some areas indicating local undrained liquefaction around the CPT tip to depths as much as

20 m with minimal soil strength. Such loose states are extremely brittle, even with the rather high compressibility of this particular silt.

That ψ characterizes such silty soils very well, with *NorSand* closely fitting the measured test data, is not a minor academic point. The tacit assumption in adjusting penetration test data for fines content in liquefaction analysis is that the high fines content soil is stronger than would be expected based on clean sand experience. But high-fines soils are accurately modelled in a critical state framework with ψ having exactly the same role and with the same trends as for clean sands. What is needed is to properly understand how penetration resistance varies with soil properties (Chapter 4), not to pretend that the soil is a different material from what it actually is and apply 'fines content corrections' to penetration test data.

6.6.4 Liquefaction in triaxial extension

Liquefaction can be much affected by loading direction, in particular triaxial extension as opposed to compression. The comparison in Figure 6.6 showed the differences in behaviour for Erksak sand in extension and compression, and Figure 6.8 compares compression, extension and simple shear data for Bonnie silt. More fundamentally, Vaid et al. (1990) asserted that the critical state is different in extension and compression, and is therefore not uniquely related to void ratio. The question then becomes whether this change in behaviour is predicted and understood.

Figure 6.29 compares *NorSand* with the experimental data of an extension test (CIUE-G642) on Erksak sand. The *NorSand* simulation used the calibration developed in drained triaxial compression. No changes were made to the model parameters, and a reasonable fit of model to data is evident despite the change from compression to extension and drained to undrained boundary conditions. This has been achieved with the explicit assumption of a unique CSL in the model.

There are two effects captured with *NorSand*, which give the good prediction of extension behaviour from compression data but that eluded the experimental workers. First, M varies with stress conditions as discussed in Chapter 2 and Appendix C. Second, dilatancy is different in compression compared to extension because there is a change in the symmetry. In triaxial compression, $\varepsilon_2 = \varepsilon_3$, whereas in extension $\varepsilon_2 = \varepsilon_1$. This affects the strain invariants substantially and causes a marked change in stiffness. Experimental work, while correctly drawing attention to the different curves in extension and compression, was extrapolated to

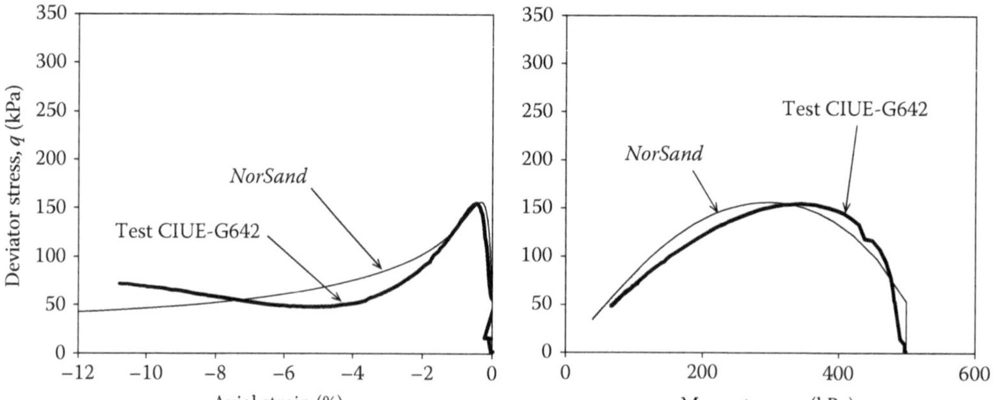

Figure 6.29 Simulation of triaxial extension, Erksak 330 test CIUE-G642.

draw erroneous conclusions about the critical state. Despite the claim of some experimentalists, soil behaviour cannot be understood from tests alone, and measurements must be put in the context of a proper mechanics framework (and which today means work hardening plasticity with a CSL and ψ).

6.6.5 Liquefaction with constant deviator and reducing mean stress

Soils are brought to failure as the ratio of deviatoric stress to mean effective stress ($\eta = q/p'$) increases. Conventional laboratory testing increases η by increasing q, and most current understanding of soil behaviour has followed from such testing (essentially all of 'taught' soil mechanics). But η can be increased by decreasing p'. Practically, this situation arises with water infiltration into slopes with an increasing pore pressure causing the mean effective stress to decrease while the shear stress remains near constant. This was the trigger mechanism for the flowside at Aberfan (described in Chapter 1, Figure 1.15). A feature of the Aberfan failure was that it occurred rapidly and without warning. The slope was inspected just one day before the failure with no evidence of impending instability (e.g. tension cracks). Subsequent investigations into the cause of the Aberfan failure (Bishop and Penman, 1969) recognized that the failure mechanism was related to the critical void ratio, but that was broadly as far the investigators were able to go without the required constitutive theory (which did not exist at the time). Exactly, this mechanism has also been reported in physical model testing by Take and Beddoe (2014), in which they gradually increased the pore pressure at the toe of a loose sand slope to trigger a liquefaction flowslide.

Although Bishop and Henkel (1957) discuss soil failure caused by increasing pore water pressure, further interest in this loading path appears scant until workers at the University of Alberta (Sasitharan et al., 1993) applied this path to investigate aspects of the collapse surface (Section 6.6.2), then viewed by some as the causative mechanism for liquefaction. The test on loose Ottawa sand by Sasitharan et al. (1993) exactly echoes the Aberfan disaster: (1) 'the axial strain ... at initiation of collapse was only 0.4%' and (2) 'the sample reached such a large momentum during collapse that when the loading head hit the restricting nuts the entire laboratory felt the vibration'. Other workers (e.g. Matiotti et al., 1995; Di Prisco and Imposimato, 1997; Gajo et al., 2000) have since confirmed and expanded on the results obtained by Sasitharan et al. (1993). An interesting set of data that was reported by Junaideen et al. (2010) used a servo-controlled system to carry out constant shear drained (CSD) loading of loose and dense decomposed granite from two sites in Hong Kong, their results being shown in Figure 6.30; the onset of yielding (occurrence of plastic strain) and the onset of collapse are shown on the various paths plotted for ease of understanding. All these workers considered triaxial conditions, and Wanatowski and Chu (2012) then extended knowledge of CSD behaviour from triaxial to plane strain.

While the body of test data and key observations is consistent across the various studies, the measured behaviour was not considered from the perspective of computational soil mechanics other than a preliminary effort we reported in the first edition of this book. The CSD path is theoretically interesting as it offers insights into yielding, and a variant on the CSD path was used in our laboratory to measure sand behaviour. The soil tested was Fraser River sand, a soil whose properties were discussed in Chapter 2. The CSD test was controlled by specifying the target pressure and a time to reach this target pressure, with the software-controlled pumps adjusting to maintain the required stress path; in essence, the same approach as adopted by Junaideen et al. (2010), Daouadji et al. (2010) and Chu et al. (2012). However, we modified the usual CSD path to include a slightly reducing deviator stress as that offers clearer identification of the elastic limit. The test

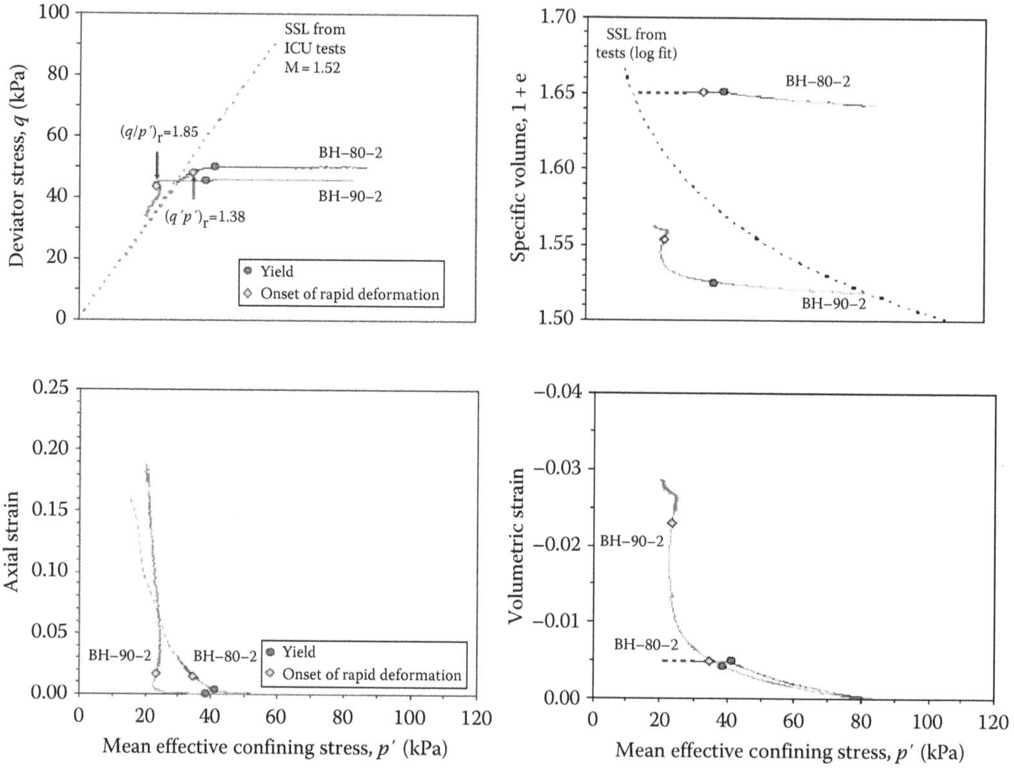

Figure 6.30 Comparison of CSD behaviour of loose and dense samples with the onset of plastic yielding and transition into failure identified. (After Junaideen, S.M. et al., *Can. Geotech. J.*, 47(4), 648, 2010. With permission from the NRC, Ottawa, Ontario, Canada.)

was on a dense sample ($\psi_o \sim -0.13$), and the data are shown in Figure 6.31 with three key points identified.

Initially, the sample was loaded to a stress ratio $\eta = 0.5$ under constant mean stress, denoted as Point A in Figure 6.31, and held overnight (which allowed some creep strain to develop). CSD loading was carried out by reducing the confining stress at the same time as a much smaller proportional reduction in the deviator stress, the stress path being plotted in the figure. The elastic limit is shown as Point B and was identified by the change in the q–ε_1 behaviour, although a matching change in the pattern of response is seen in the volumetric strain behaviour; the mobilized stress ratio at the onset of the transition from elastic unloading to plastic yielding was $\eta = 1.08$, well below the critical state stress ratio M. Continuing reduction of mean stress from Point B initiated yield in 'unloading', with reversal of axial strain from as p decreased to axial strain now increasing as p decreased; there was strong dilation associated with this yielding. There was no obvious change in the pattern of the soil response when the stress path crossed the critical state with smooth soil response until the limiting stress ratio was reached, Point C. When the limiting stress ratio $\eta = 1.84$ was reached, the stress path could no longer be held, with q now reducing at what appears to be constant p as the sample dilated towards its critical state ('appears' because the servo in the test equipment had great difficulty in controlling the test; it would also be plausible to suggest that the true behaviour was failure at the current η_{max} value, something that reduced to M_{tc} as the sample dilated). The axial strain rapidly increased past Point C despite the reducing load. Comparing Figures 6.30 and 6.31, it is clear that Fraser River sand in the CSD

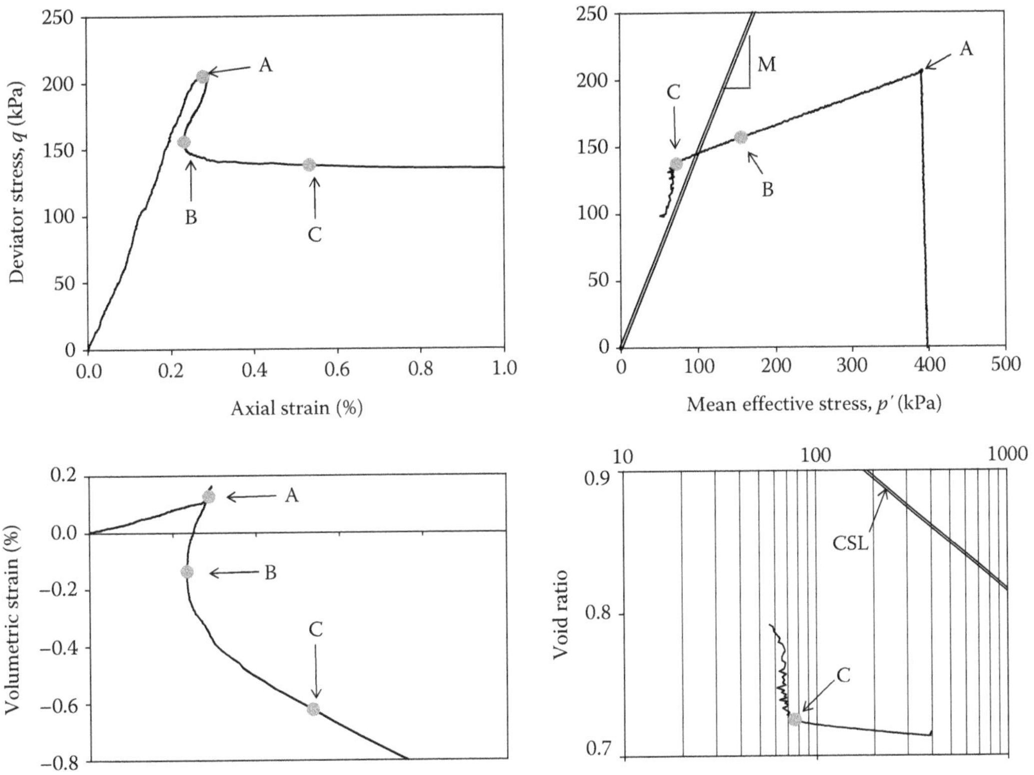

Figure 6.31 Dense Fraser River sand behaviour during CSD test.

path shows the same behaviour, and details in that behaviour, as reported by Junaideen et al. (2010) in their tests on a decomposed granite. This provides independent corroboration of the nature of soil response during CSD loading. We now turn to understanding this behaviour using computational soil mechanics.

Self-consistency in the derivation of *NorSand* required an internal cap to the yield surface (see Chapter 3); this internal cap results in *NorSand* yielding in what might be viewed as 'unloading', an aspect of soil behaviour explored in Chapter 3 and presented in Appendix C where it is seen to provide an excellent prediction of the hysteresis loops measured in conventional drained triaxial loading that include unload–reload cycles. Exactly, this aspect of *NorSand* is relevant to understanding soil behaviour in a CSD path.

Two *NorSand* yield surfaces are superimposed over the measured CSD path of the Fraser River sand test in Figure 6.32 (computed using the properties of Fraser River sand, see Section 7.3). The larger of the two yield surfaces is that established by the initial loading to Point A; the size of this yield surface sets the elastic zone for the initial part of the CSD path. The observed elastic limit in the test data, Point B, lies very close to the *NorSand* inner cap (which is an elastic limit). The subsequent yield under decreasing mean stress drags the yield surface with it (because of the consistency condition) with yield surface softening (shrinking) as that is physically required by cap yielding (see Appendix C). Softening is stable while the stress state moves up the cap, but when the shear stress ratio η reaches its limiting value at the intersection of the cap with the outer bullet, then the soil is in a situation of decreasing strength as it shears to the critical state. The coincidence of the observed 'runaway', Point C, in the test with the softened *NorSand* yield surface is apparent in Figure 6.32.

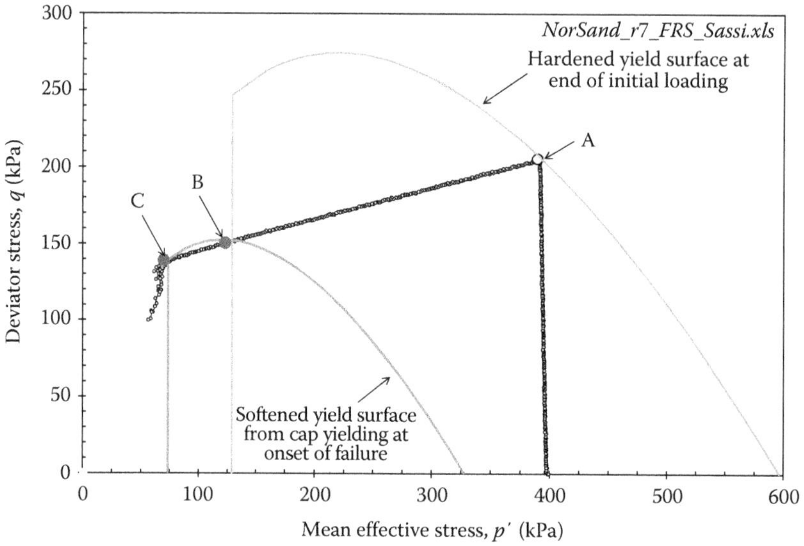

Figure 6.32 Comparison of *NorSand* yield surfaces with the onset of differing behaviours in unloading (Fraser River sand).

So far, so easy and accurate, using soil properties determined by calibration to standard drained triaxial compression tests. But there are two theoretical aspects of interest and importance: stress dilatancy during cap yielding and the softening of the plastic modulus.

In the case of stress dilatancy, the overall governing equation during loading (i.e. for η increasing) is $D^p = (M - \eta)/(1 - N)$. *NorSand*, along with Cam Clay, is derived from theoretical axioms and with an idealized mechanism for dissipating plastic work as heat: M is a soil property capturing the work done in distorting the arrangement of the soil particles, while N is a soil property that captures inelastic energy storage as the void ratio changes (Jefferies, 1997). Now consider the transition at Point B. It has $D^p < 0$ (i.e. dilation), but $\eta < M$, which implies that the operating work dissipation on the cap is a lot less than on the outer bullet. To date, the general nature of stress dilatancy during yield in unloading is open to a range of idealizations; *NorSand* uses one plausible idealization that involves imaging the stress q on the cap across to the outer 'bullet' yield surface.

In the case of the plastic modulus softening, one might anticipate that the change in size of the yield surface as a consequence of plastic strain might be controlled by a single hardening law. However, such a law has not been derived (say via self-consistency arguments), and presently, the plastic modulus for cap softening must be treated as another soil property – which means fitting the test, not predicting its stiffness during yield in unloading.

The upshot of the adopted work dissipation function and fitting the cap softening modulus is presented in Figure 6.33 where the measured stress–strain behaviour of the CSD test is compared to the *NorSand* simulation. A reasonable fit is evident, suggesting that the underlying extension of the basic Cambridge-type idealized work mechanism is plausible. However, the theoretical implications in terms of plastic work dissipation and recovery of stored inelastic energy warrant more consideration than achieved to date (see Appendix C). This may be an aspect of soil behaviour better understood by micromechanical modelling.

Despite the limited theoretical understanding of plastic work dissipation during a CSD path, the practical implications are very obvious and simple: be very, very careful with the

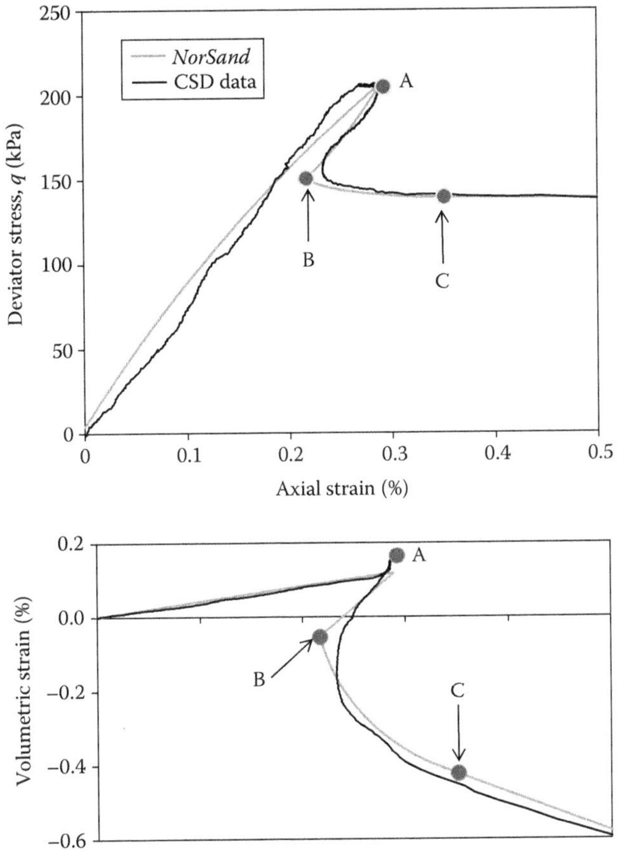

Figure 6.33 Measured and computed behaviour in CSD loading of Fraser River sand.

observational method for an in-situ situation with anything near a CSD path. There will be very little, if any, observable movements before catastrophic failure – exactly what happened at Aberfan. By all means, measure pore pressure, but do not allow the stress path to extend past 'Point B' on the CSD path (which is readily computed from drained triaxial compression data).

6.6.6 Pseudo–steady state

An aspect of liquefaction that has attracted considerable discussion is the pseudo–steady state, also called the phase change condition. This pseudo–steady state is the temporary condition when the rate of excess pore pressure change becomes zero (Chapter 2). Several workers have proposed this pseudo–steady state as a reference condition, including Ishihara (1993) and Konrad (1990b).

Figure 6.34 shows the stress–strain behaviour and stress path from compact (dense of critical) simulations of undrained triaxial compression. The two simulations differ only in their values of shear modulus or elastic rigidity, I_r. Both exhibit a pseudo–steady state between 1% and 2.5% axial strain, and the corresponding stress paths show a sharp vertex indicative of a transient zero dilation rate (between contraction and dilation) sometimes called the phase change condition.

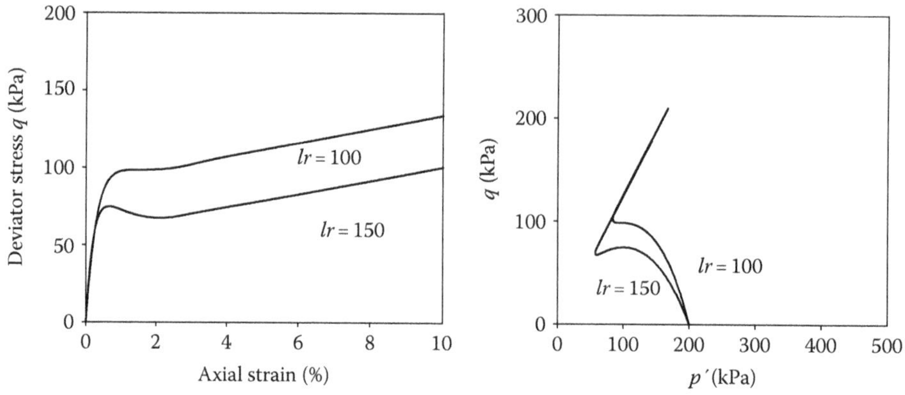

Figure 6.34 NorSand simulations of pseudo-steady-state condition.

Two points follow from Figure 6.34. First, both simulations had an explicit unique critical state $p'_c = 502$ kPa, but the pseudo–steady states are at $p' = 60$ kPa and $p' = 85$ kPa for the two simulations. The difference between the two simulations is caused by a change in I_r. Confusing the pseudo condition with the real critical state causes an error of a factor of 6–8 in estimating p'_c in this instance. Second, the pseudo–steady state is not a property or otherwise unique behaviour. A relatively minor change in the elastic modulus, which is unrelated to any plastic behaviour, caused a marked change in the pseudo–steady state. Claimed proof of non-uniqueness of the CSL based on estimating p'_c from the pseudo condition is untenable.

Because the pseudo condition depends on soil fabric, it is a conceptual error to use the pseudo condition as a reference state. This error is compounded by the pseudo condition being a near small-strain behaviour, not a large strain end state.

6.7 PLANE STRAIN VERSUS TRIAXIAL CONDITIONS

So far, the understanding of liquefaction presented has been based on triaxial conditions, which have a special symmetry. Triaxial tests also impose conditions that are largely unrelated to field circumstances. So, it is necessary to examine how this laboratory-based understanding should be carried into real field situations. Simple shear is a useful idealization for this purpose, but how well is simple shear behaviour predicted by triaxial compression calibrations? How do strengths in simple shear scale as s_u/p' ratios?

Simple shear is also used as the basis of a laboratory test, so why not just work with simple shear tests? The difficulty with the simple shear test is that only two of the three stresses needed to resolve the principal stress components and invariants are measured, and the test equipment provides only an approximation to simple shear. Although true simple shear is difficult experimentally, it remains easy to look at in theory as it is another simple element test for which direct integration methods can be used. Appendix C contains the derivations for *NorSand* in simple shear, and it is implemented for undrained conditions in *NorSandSS.xls* as the VBA subroutine *NorSSu*.

Figure 6.35 compares the undrained simple shear behaviour of an 'undisturbed' silt sample from the core of Coquitlam Dam with the computed *NorSand* response. An excellent fit of model to data is evident, but there are real problems with the simple shear apparatus as a method of obtaining soil properties and strength, and the fit shown in Figure 6.35 obscures these issues.

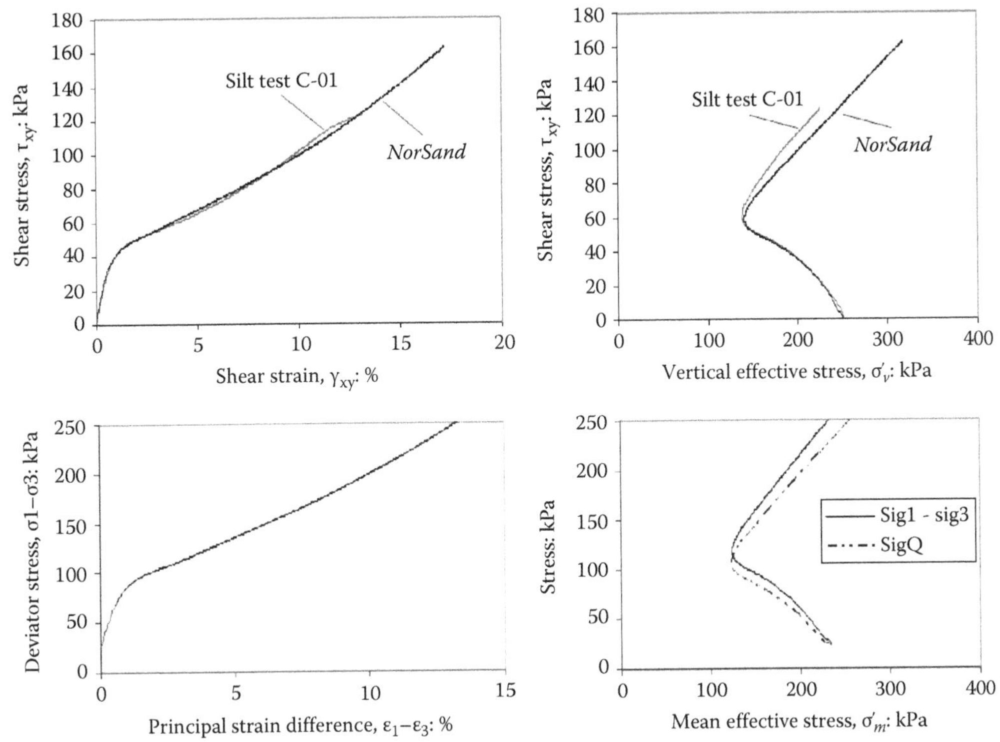

Figure 6.35 Comparison of *NorSand* with measured undrained silt behaviour in simple shear.

Simple shear results depend on the initial geostatic stress ratio, K_o. The value of K_o was varied to obtain the fit of *NorSand* to the test data in Figure 6.35, but this ratio is not measured in the simple shear equipment. So, K_o was treated as a 'free' parameter in fitting model to data, the achieved fit adopting $K_o=0.95$, which is consistent with the sample slumping during transfer from its in-situ state to sitting in the cell of the Geonor equipment. More generally, Figure 6.36 illustrates how the stress–stain behaviour of a loose sand differs between $K_o=0.5$ and $K_o=0.9$ in simple shear. The former value is perhaps at the low end of what is encountered in-situ, while the second value is towards the upper end unless the soil has experienced ageing, cyclic loading or disturbance during set-up for a test.

A consequence of the behaviour illustrated in Figure 6.36 is that simple shear tests are of limited use for calibrating any constitutive model. Because K_o is unknown in the test, K_o can be varied arbitrarily within rather wide limits to match model to data, leaving substantial uncertainty over what are the true soil properties as equally good fits can usually be achieved with markedly different property sets simply by using different values for the unknown geostatic stress ratio. However, if the soil properties are known from conventional drained triaxial tests, then we can use the constitutive model to assess what is likely in-situ under plane strain conditions.

Carrying out repeated simulations of simple shear behaviour allows the development of relations between s_u/p' and ψ. The results of such simulations using properties that are not unusual for natural sand deposits are presented in Figure 6.37. Properties are 'not unusual' in the sense of being what might be commonly found, the word 'typical' being avoided as it implies unwarranted confidence that other parameter combinations can be neglected. Specifically, this parameter set is directed at what might be found in natural deltaic sand in

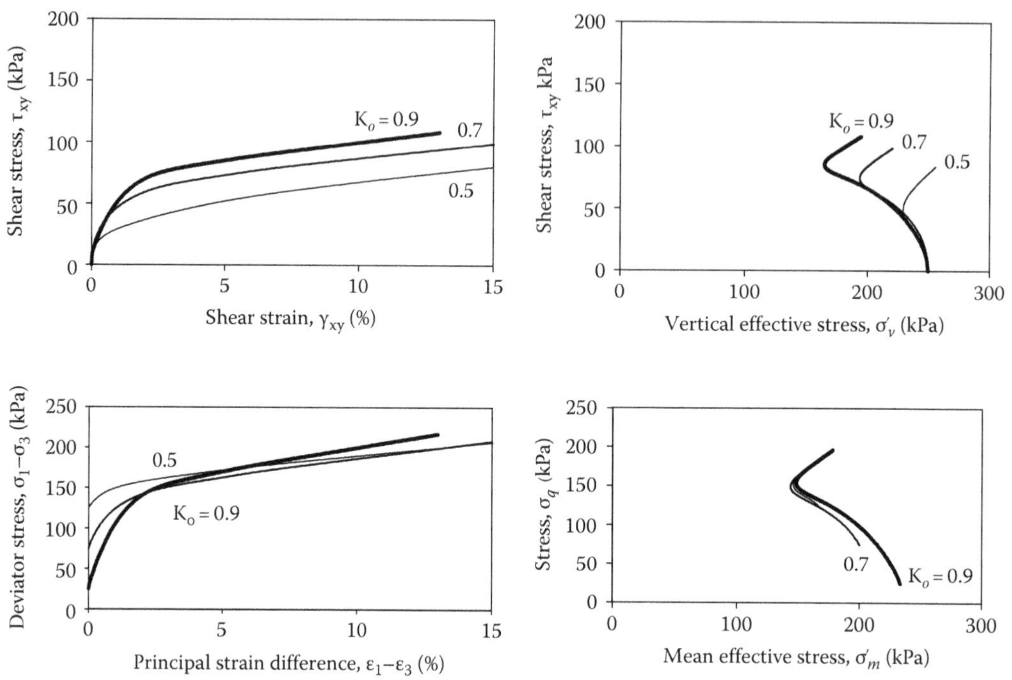

Figure 6.36 Example of effect of initial geostatic stress ratio K_o on undrained strength in simple shear.

contrast to the more familiar properties for laboratory quartz sands devoid of any silt particles. It is trivial to create similar figures for different combinations of soil properties using the *NorSandSS*.xls spreadsheet. Both peak and residual strengths are shown in Figure 6.37, and the trends are shown for two K_o values that span the usual range of loose soils. Trends have been drawn only for positive state parameter because 'S'-shaped stress paths develop once $\psi < 0$, and no peak strength is observed (the maximum is in fact the critical state in dilating soils, and this is attained only at high strains).

Simple shear tests measure τ_{xy} rather than τ_{max}, and rotation of principal stress in simple shear means that $(\tau_{xy})_{max} \neq \tau_{max}$. In addition, the deviator stress $\sigma_1-\sigma_3$ no longer corresponds to the stress invariant σ_q (because of σ_2) so just what the 'strength' is in a simple shear test requires caution. In general, the measured value of $(\tau_{xy})_{max}$ is reported as the strength by laboratories, but this is not directly comparable to the triaxial strength $s_u = (\sigma_1-\sigma_3)/2$. Since most undrained strengths are used in limit state calculations in which the effect of intermediate principal stress is neglected, the simple shear strength plotted in Figure 6.37 is also the measure $s_u = (\sigma_1-\sigma_3)/2$. This value is easy to obtain from numerical simulation of simple shear but cannot be obtained from laboratory tests.

The offset between triaxial compression strengths and those in simple shear is less than about 5% for the simulations carried out. This is perhaps less than that might have been expected from various comparisons found in the literature, although it should not be entirely surprising as the Lode angle in simple shear is far closer to that of triaxial compression than triaxial extension. This greater similarity has been obtained mainly because the comparison is in terms of the same undrained strength parameter (normalized by mean effective stress). A rather different picture emerges when strength is expressed in terms of the ratios with vertical effective stress and τ_{xy} or τ_{max} as illustrated in Figure 6.38, which are the same data as just shown in Figure 6.37.

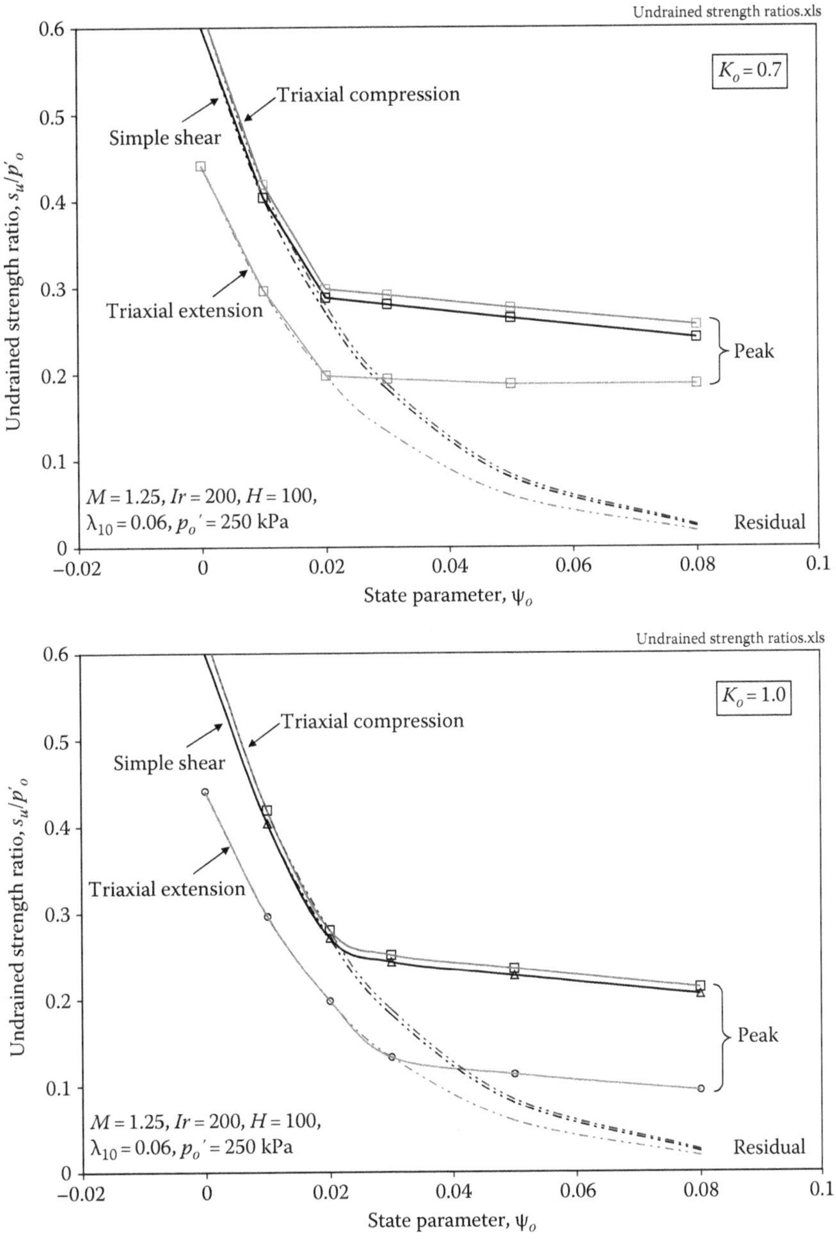

Figure 6.37 Effect of initial state on peak undrained strength in simple shear versus triaxial conditions. (Use the downloadable spreadsheet *NorSand.xls* for further simulations using different soil properties, as these graphs are property dependent and so not unique.)

The reduction of undrained strength in simple shear was addressed by Wroth (1984) in his Rankine Lecture. Two of Wroth's plots (his Figures 7 and 27) have been combined to produce Figure 6.39, and the friction angle axis has been inverted so that his assessment of trends in laboratory data can be compared directly to those from the state parameter approach. For states $\psi > +0.02$, the same trend of undrained strength ratio is seen in Figures 6.38 and 6.39. The strength ratio in isotropically consolidated triaxial

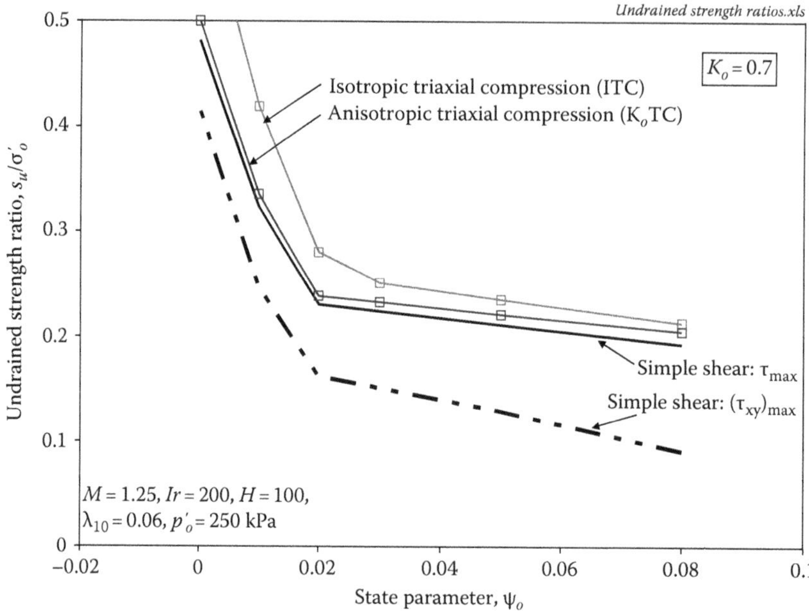

Figure 6.38 Computed peak undrained strength in simple shear versus triaxial conditions when normalized by the initial vertical effective stress.

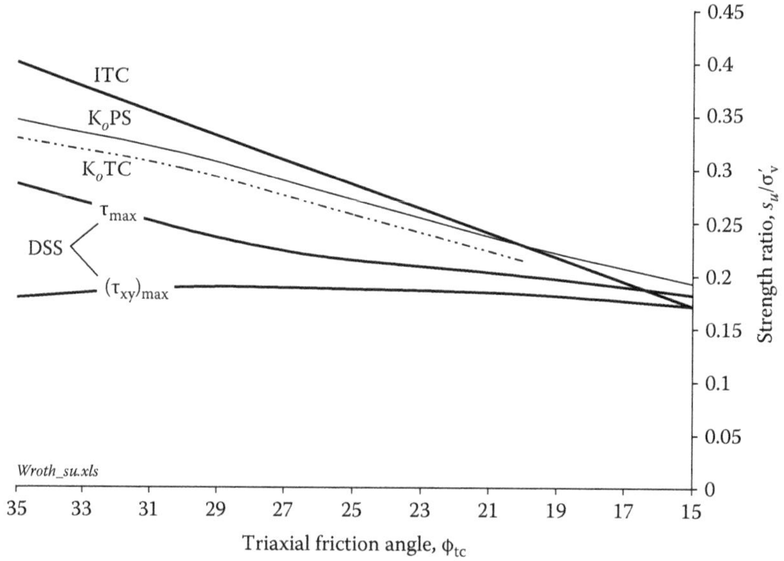

Figure 6.39 Peak undrained strength of normally consolidated clay in different strain conditions. (Based on Figures 7 and 27 of Wroth, C.P., *Géotechnique*, 34(4), 449, 1984.)

compression is greater than that for anisotropic consolidation, which is in turn greater than under simple shear conditions. The strength also reduces somewhat with increasing ψ and decreasing ϕ_{tc}. But whereas Wroth was inferring trends for normally consolidated clay, the relationships shown in Figure 6.38 were calculated directly using *NorSand* and soil properties that might be found in natural deltaic sands. The large increase in undrained strength ratio when $\psi < +0.02$ is not apparent in Wroth's trends as this would represent an over-consolidated material.

Figure 6.38 can be used only as a guide to soil behaviour because the normalized strength depends on the soil properties and K_o. The downloadable spreadsheet *NorSandSS.xls* implements undrained simple shear and allows rapid updating of Figure 6.38 for any choice of soil properties, initial geostatic stress condition and initial state.

A particularly important aspect of Figures 6.38 and 6.39 in the context of simple shear testing is the difference between $(\tau_{xy})_{max}$ and τ_{max} because for $K_o \neq 1$, the geostatic stress imposes an initial shear stress that reduces the possible increase in shear stress on the horizontal plane before the limiting strength is encountered. This geostatic condition is unknown in the simple shear test (as the horizontal stress is not measured) although it is $(\tau_{xy})_{max}$ that comprises the test result.

6.8 STEADY-STATE APPROACH TO LIQUEFACTION

6.8.1 Basic premise of steady-state school

Although a critical state–based view has been followed so far, North American interest in liquefaction and the critical state took the simpler approach of what we call the *steady-state school*. This approach was an important step in engineering practice and was dominant in the late 1970s to perhaps mid-1980s. Aspects of the approach remain today with the concept of post-liquefaction residual strengths. The ideas developed over many years, but the approach was clearly stated by Poulos (1981) with a more detailed exposition in Poulos et al. (1985). The ideas are as follows.

The key doctrine of the steady-state school is that the undrained steady-state strength is an assured minimum. This assured minimum cohesive-like strength finds its application in a simple total stress limit equilibrium calculation. If the structure or slope is stable under the total stress steady-state condition, then it is not going to liquefy statically, and even cyclic loading would produce only modest strains. The steady-state doctrine gives an apparently attractive and expeditious way of finding out whether or not liquefaction might be a cause for concern.

This steady-state approach is based directly on the two axioms of critical state theory: that the CSL is unique and that all state paths end up on the CSL. This becomes especially interesting for the residual strength in undrained shear. If the CSL and the current void ratio are known, then the mean effective stress at the limiting critical condition is known. Multiplying this by the critical friction ratio (or rather $M/2$) gives the undrained shear strength, usually denoted by s_r (the subscript r denotes residual), that is

$$s_r = p'_c \cdot \frac{M}{2} \tag{6.9}$$

The steady-state school, in various publications, invoked the idea of a 'flow structure' as a key feature of the soil steady-state strength which distinguished the steady-state strength from the related critical state strength for the soil's current void ratio given by (6.9). This flow structure of the steady state is never defined and cannot be measured, so it remains a conjecture. Looking at the testing procedures for the steady state, what is determined is one and the same thing as the critical state. Certainly, neither mathematics nor physics has been produced to support a unique flow structure. This leaves the steady-state school as an application of critical state principles using particular measurement methods, methods which were indeed unique to the steady-state school.

The devil is in the details, as always, and there are two challenges that must be faced in applying the steady-state concept: determining the CSL and determining the in-situ void ratio (the steady-state approach predated the state parameter and interest centred on void ratio rather than ψ). The documented procedure (Poulos et al., 1985) goes about estimating the steady state in a rather convoluted manner, as they suggest that the in-situ CSL is not the one measured on reconstituted samples. A five-step procedure is used:

1. Determine the in-situ void ratio
2. Determine the CSL using reconstituted specimens
3. Determine the steady state for undisturbed specimens
4. Correct the 'undisturbed' strengths to the in-situ void ratio
5. Calculate the factor of safety in a limit equilibrium analysis

The limit equilibrium analysis of the last step is a standard geotechnical calculation and will not be considered further. Attention is focused on how the undrained strength for this calculation is derived.

For the first step of void ratio measurement, three methods are suggested: fixed piston sampling, ground freezing and coring and test pit sampling. Poulos et al. (1985) assert that of the methods, fixed piston sampling, provides reasonable results provided that it is carefully done. Carefully done means the least possible clearance ratio which retains the sample in the sample tube that the sample must be pushed slowly, that the mud level in the hole must be above the ground water table and that the sample must be carefully removed from the hole. Void ratio is then calculated from measurements of sample length in the tube and the measured dry weight of soil recovered.

Ground freezing and coring is always a possible method of sampling sands, if it can be afforded. However, our experience concurs with Poulos et al. that careful fixed piston sampling can give repeatable (and hence presumably accurate) estimates of void ratio. It is nevertheless simpler to estimate void ratio from measured water contents, assuming full saturation, than calculating density from measurements of recovered sample size and weight. With the water content method, any water escaping from the sample tube is collected during extrusion and then added back into the sample during water content determination. A stationary piston sampler is not always necessary as surprisingly good (repeatable) results can be obtained, especially in silty sands to silts, with hydraulically pushed thin wall sampling tubes with a sharp cutting edge (in our case, typically 60 mm diameter stainless steel). The trick is to have an end cap ready, as the sample tends to be lost shortly after breaking the surface of the mud in the borehole. A careful driller pulls the sampler the last few feet out of the borehole without the sampler banging on the sides of the drill string with the soil technician in close proximity. The technician puts an end cap over the sample tube as soon as possible, usually while the driller holds onto the sampling string to stop it banging against the casing (and definitely the end cap must be there before the tube is removed from the sampling sub).

The second step, determining the CSL using reconstituted samples, follows the procedures set out in Chapter 2 and Appendix B. The soil used in these tests should be that found in the ground, and the reconstituted tests should all use the same soil. It is common to use several samples and thoroughly mix the soil to provide sufficient material for a full suite of undrained tests. However, there is a wrinkle in that the data are usually presented as a plot of void ratio versus effective minor principal stress, or undrained strength, rather than mean effective stress at the critical state (i.e. σ'_3 and $q_{ss}/2$ are plotted, rather than p'). The measured shear strength in triaxial compression is then also reduced by $\cos(\phi_s)$, where ϕ_s is the steady-state friction angle, to obtain the shear stress on the plane of failure. This $\cos(\phi_s)$ factor typically reduces the measured strength by about 30%, since ϕ_s is about 30°.

The third step is to take as undisturbed a sample as possible and to carry out consolidated undrained tests. What is being recognized in this step is that minor changes in soil gradation and mineralogy do affect the CSL, and the aim in testing an 'undisturbed' sample is to derive one point that is on the CSL of the actual soil. Inevitably, there will be sample disturbance, and this will show up as densification during consolidation. Because of the desire to work with contractive samples, the consolidation pressure used in testing may be markedly greater than in-situ. The actual test pressure is a matter of balance between the desire to induce contractive behaviour and the desire not to apply confining stress more than double that in-situ. At the end of the consolidated undrained test, a steady-state strength will have been measured. As this corresponds to a void ratio that is denser, and perhaps a lot denser than in-situ, the measured strength must be corrected to that corresponding to the in-situ void ratio.

Correcting the measured strengths, the fourth step, proceeds by assuming that the slope of the CSL of the 'undisturbed' sample is the same as that of the CSL for reconstituted samples. A line is drawn through the measured undisturbed strength parallel to the CSL, projecting this to intercept the measured in-situ void ratio. The estimated in-situ steady-state strength is then read off the graph. This overall procedure is illustrated in Figure 6.40, which is taken from Poulos et al. (1985).

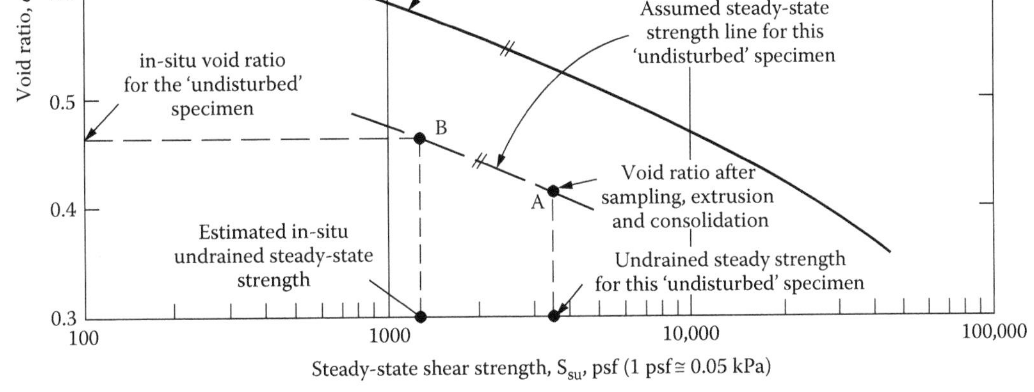

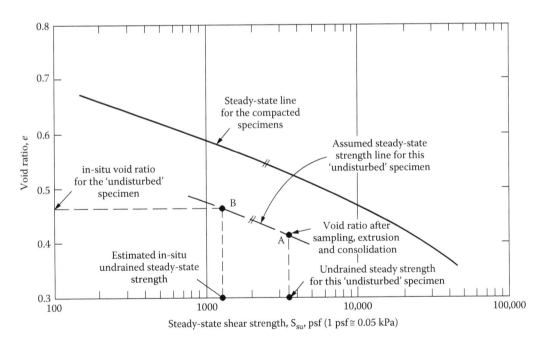

Figure 6.40 Steady-state school method to determine steady-state strength of soil at in-situ void ratio. (After Poulos, S.J. et al., *J. Geotech. Eng. Div., ASCE*, 111(GT6), 772. With permission from the ASCE, Reston, VA.)

6.8.2 Validation of the steady-state approach

The steady-state procedures were thoroughly evaluated in a comparative back-analysis of the Lower San Fernando liquefaction failure during the 1971 San Fernando earthquake. The Lower San Fernando Dam failure is one of the great case histories of liquefaction and the near-catastrophic consequences of the post-liquefaction movement (Figure 1.13) more than any other liquefaction event caused engineers and officials in California to begin thinking about liquefaction hazards.

Lower San Fernando Dam was substantially investigated immediately after the failure. Interest here, however, focuses on a further investigation during 1985 which was directed at examining the validity of steady-state concepts. The project involved a comparative back-analysis of the failure by three groups: Geotechnical Engineers Inc (GEI), which is the consulting company of Poulos and Castro; a group from the University of Berkeley led by Seed and a group at Rensselaer Polytechnic Institute led by Dobry. It was sponsored by the Waterways Experiment Station of the U.S. Army Corps of Engineers. The documentation of the 1985 investigation is public domain as a three-volume set of reports by Castro et al. (1989), published by the Department of The Army and available through the U.S. National Technical Information Service. The companion report by Seed et al. (1988) is an Earthquake Engineering Research Centre document. An easily accessible summary is given by Marcuson et al. (1990).

Back-analysis of the Lower San Fernando Dam failure for operating strengths at the onset of the sliding failure (Figure 6.41a) gave an average shear stress in the zone that liquefied of about 850 psf (pounds per square foot will be used here since that is what all the studies report). Estimating the residual strength requires use of the failed geometry and allowance for inertial effects as the failed mass was slowed to a stop. Figure 6.41b shows the failed geometry. Of the groups in the comparative study, the Seed group estimated the residual strength of the liquefied zone as 400 ± 100 psf. The GEI group estimated that the residual strength was about 500 psf.

The steady-state strength was also estimated using the steady-state procedures exactly as presented in the previous section. However, the testing was done in several different laboratories, and the in-situ steady-state strength was estimated by the three principal groups of investigators. Table 6.2 summarizes the estimated steady-state strengths. Marcuson et al. viewed the steady-state approach as validated based on the San Fernando Dam analysis. Seed et al. (1988) drew a somewhat different conclusion, as in their view, the residual strength based on the 35th percentile value (580 psf) was greater than the greatest back-calculated value during the failure (500 psf). Looking at average values in Table 6.2, the steady-state procedures can result in a substantial overestimation of the operating strength during the post-earthquake slide.

Some of the steady-state data are summarized in Figure 6.42 and reveal that some very large 'corrections' were used to change what was measured to what was estimated to have existed in-situ. Remember that these adjustments were not being applied as a 'Class A' test of the method. The relevance of the method in this case depended on how the adjustments were calculated or estimated.

A significant adjustment in the case of this back-analysis is necessary because the samples retrieved for testing were not from the failed soil prior to the earthquake. The samples were retrieved from the unfailed downstream part of the dam, with assumptions made about the construction procedures to extrapolate to the upstream part of the dam. Further, the void ratio of these retrieved samples was adjusted to allow for the effect of settlements induced in these non-failed soils by the 1971 earthquake. These pre-1985 adjustments amounted to about a 30% or so reduction in the estimated steady-state

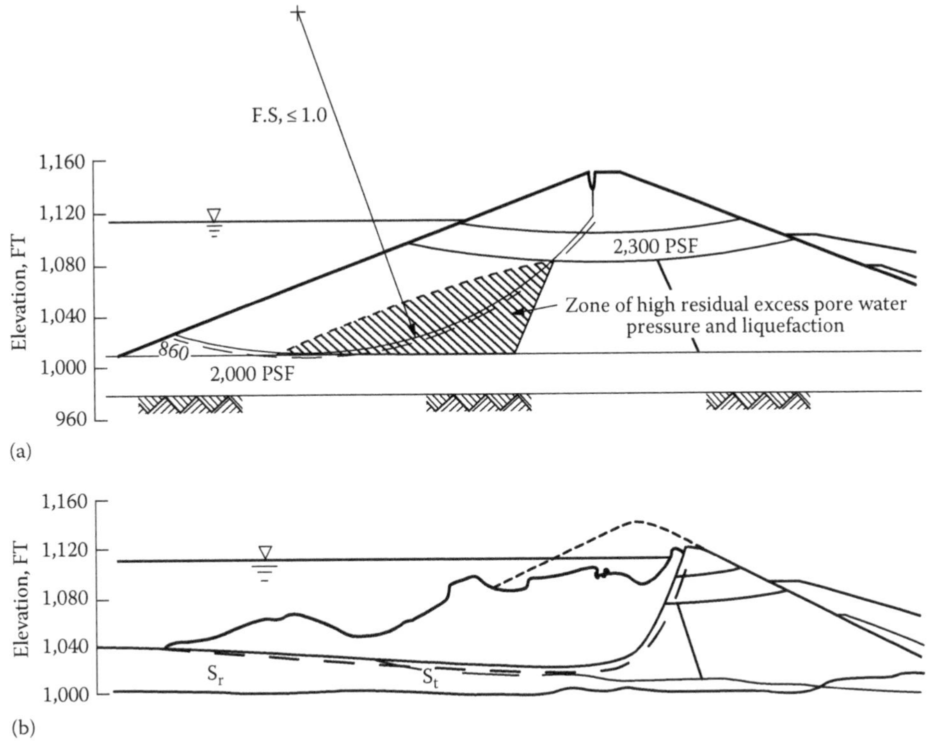

Figure 6.41 Cross sections of Lower San Fernando Dam. (a) Idealization of pre-failure geometry for slip circle analysis and (b) estimated section after the earthquake and after sliding of the upstream shell. (From Seed, H.B. et al., Re-evaluation of the slide in the Lower San Fernando Dam in the earthquake of February 9, 1971, *Report UCB/EERC-88/04*, Earthquake Engineering Research Centre, University of California at Berkeley, Berkeley, CA, 1988. With permission from the EERC.)

Table 6.2 Summary of steady-state strength determinations from laboratory tests for Lower San Fernando Dam

	Steady-state strength in upstream area at time of earthquake (psf)			
Group	*Average*	*Range*	*33 or 35 Percentile value*	*Remarks*
GEI Group	630	260–940	520 ± 100	GEI data
Seed Group	800		580	GEI and Stanford data
Dobry Group	700	150–2000	420	RPI data

Source: After Marcuson, W.F.H. et al., *Earthquake Spectra*, 6(3), 529, 1990.

strength, as can be seen in Figure 6.42. Given all of these corrections to obtain a reasonable agreement, it is difficult to concur that the steady-state procedures have been validated by Lower San Fernando Dam.

In Chapter 5, the discussion of the characteristic state and what might be the appropriate value for liquefaction analysis suggested that something like the loosest 15th percentile (85% denser) is appropriate. Figure 6.43 shows the range of in-situ void ratio measured at San Fernando. If the CSL of the blended and remoulded samples is compared to the in-situ void

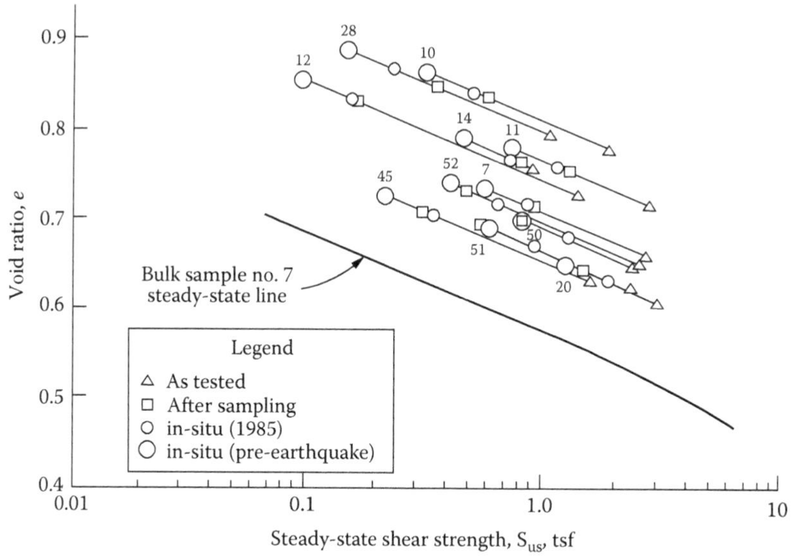

Figure 6.42 Adjustments of measured undrained steady-state strengths to in-situ conditions at Lower San Fernando Dam. (From Seed, H.B. et al., Re-evaluation of the slide in the Lower San Fernando Dam in the earthquake of February 9, 1971, *Report UCB/EERC-88/04*, Earthquake Engineering Research Centre, University of California at Berkeley, Berkeley, CA, 1988. With permission from the EERC.)

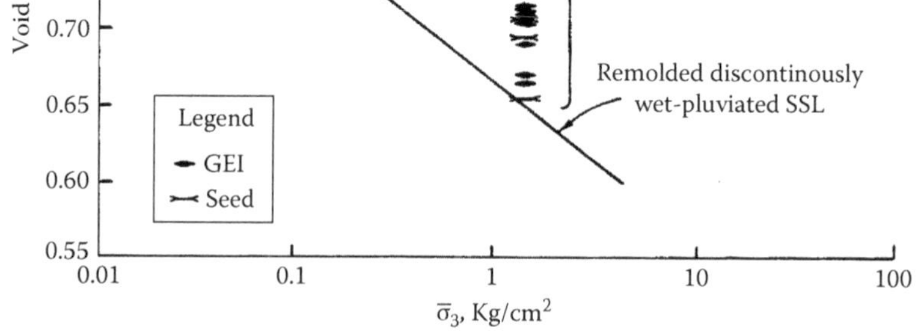

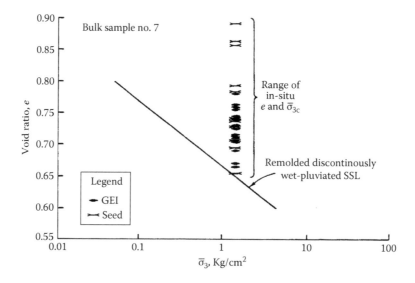

Figure 6.43 Comparison of in-situ void ratios and remoulded SSL for Lower Sand Fernando Dam determined by Vasquez-Herrera and Dobry. (Reproduced from Vasquez-Herrera, A. and Dobry, R., *The Behavior of Undrained Contractive Sand and Its Effect on Seismic Liquefaction Flow Failures of Earth Structures*, Rensselaer Polytechnic Institute, Troy, NY, 1988. With permission of Professor Dobry.)

ratio, and undrained conditions assumed, then the expected 15th percentile characteristic undrained residual strength is about 200 psf. To arrive at this value, take Figure 6.43 and a characteristic in-situ void ratio of about $e \approx 0.77$ to derive $\sigma'_3 \approx 200$ psf or 0.1 kg/cm^2 for the critical state. The stress ratio $\sigma'_1/\sigma'_3 \approx 3$ at the critical state. The residual undrained strength is then $s_r = (\sigma'_1 - \sigma'_3)/2 \approx 200$ psf, which is somewhat lower than the back-figured strengths, and illustrates how easy it is to obtain apparent validation in a back-analysis.

6.8.3 Deficiencies of the steady-state approach

On closer examination, the steady-state method has several shortcomings all of which bear on the method over-estimating apparent strength during post-liquefaction slides.

The first deficiency is blending several samples of possibly differing gradation to produce a composite sample for the reconstituted testing. As was shown in Chapter 2, the CSL is very sensitive to even small variations in silt content, so there is no assurance that a blended sample is relevant to in-situ conditions. The assumption of essentially constant CSL slope for a geological stratum is too simple, even for the minor variations in silt content that are found in processed hydraulic fills. Natural deposits are more varied, and the assumption of constant CSL slope will be generally wrong.

The second deficiency is reducing the measured strengths by about 30%, the idea being that it is the shear stress on the failure plane that matters. The undrained limit equilibrium analysis, however, proceeds in terms of the difference between the principal stresses and does not recognize that the strength being used is anything other than the undrained strength $(\sigma'_1 - \sigma'_3)/2$. The cosine factor invoked to give this 30% strength reduction is misguided and incorrect.

The third deficiency is that the critical friction ratio M, which affects the residual post-liquefaction strength, depends on the relative level of the intermediate principal stress. Most case histories involve near plane strain failure, and the simulations presented earlier suggest that allowing for differing M from the triaxial conditions of laboratory tests would reduce the field strengths by about 5% (Figure 6.38).

There is also the practicality of testing 'undisturbed' samples. In the case of sands, and even silty sands, piston samples cannot be extruded and mounted in a test cell – they fall apart when extruded, never mind trimming the ends and mounting in a test cell. One has to resort to ground freezing and coring. This is inconvenient, but not impossible if the tester (or rather the project owner) is sufficiently motivated. With frozen samples, a direct measure of the in-situ behaviour can be obtained, although strains during thawing will have to be monitored, and the in-situ stress state will have to be known so that the original stress state can be imposed on the sample while it is thawing and before it is tested; frozen samples are not assured to be representative.

Although seemingly based on a powerful theory, the most fundamental objection to the steady-state approach is that it is based on the assumption that there is a minimum assured strength. This is a theoretical error unless the soil is totally contractive. This error becomes important, practically, with lightly dilatant soils. Lightly dilatant soils are initially contractive before dilating to the critical state, for example, Figure 6.20b. Roughly, these are soils in the range $-0.1 < \psi < 0$. Looser soils with $\psi > \sim 0.03$ are completely contractive, and the steady-state minimum strength is entirely applicable to such loose soils, although not many such soils are found in nature. It is the lightly dilatant soils that tend to cause the most trouble for engineering predictions, design and remediation.

An implicit assumption in the steady-state approach is that undrained conditions prevail during liquefaction. Of course, liquefaction events tend to be quick, in the range of a few minutes to perhaps a few tens of minutes (the exception is offshore structures, when environmental loadings may cause cyclic stresses in the foundations of the structures extending over a few hours). Are a few minutes short enough to hold undrained loading in a large structure even in sandy soils? It turns out not to be.

Figure 1.21 showed the measured decay of excess pore water pressure in a large sand fill hydraulically constructed using Erksak sand with about 2%–3% silt. The data are from a piezometer at a depth of 15 m with double drainage of approximately 15 m either way. There was a 20% reduction in the excess pore pressure in 2 min after the load cycling stopped, and it took about 8 min for the pore pressures to dissipate completely. By way of comparison, Lower San Fernando did not fail during the earthquake, but started moving about 30 s later (Section 1.3.4). Undrained conditions cannot be assumed without considering the consolidation time factor of the soil and structure in question.

The time factor then raises a second question: what drainage path length should be used in determining the time factor? In the case of contractive soils, all the soil is trying to expel excess pore water, and the applicable length is the distance to the drainage boundary. In the case of dilatant soils, the distance is very much shorter. The reason it is shorter is strain localization. Strain localization refers to the tendency of soil failing in shear to form shear bands rather than to distort uniformly. Because the stress–dilatancy relationship remains the same both inside and outside the shear band, the increased shear strain rate in the shear band produces an increased volumetric strain, and the shear band develops a locally looser void ratio than the soil around it. For samples that get their undrained strength from dilating to critical, this means that the strengths seen in small laboratory samples will not be seen at field scale, as at field scale, there is a much larger volume of soil to feed water into the shear band.

Experimental data on strain localization and void ratio have been produced by several researchers, and a particularly fascinating study is that of Oda and Kazama (1998). Figure 6.44 is taken from this study and shows x-ray photographs of two sand samples after the formation of shear bands. The shear bands are identified by the lighter colour on

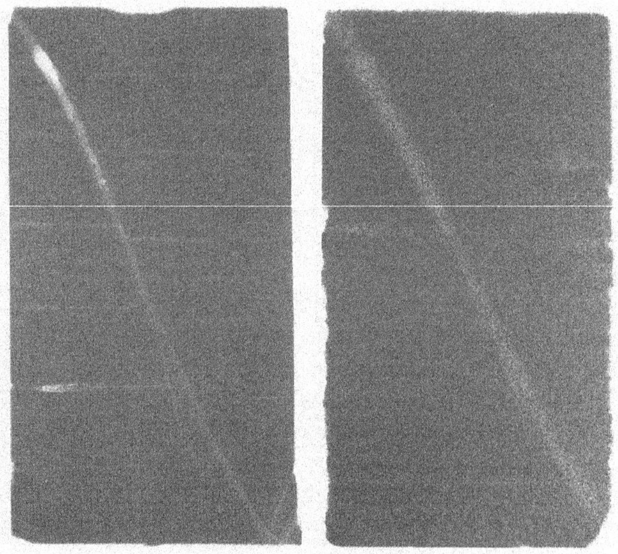

Figure 6.44 X-ray images of shear bands in triaxial samples. (From Oda, M. and Kazama, H., *Géotechnique*, 48(4), 465, 1998. With permission from the Institution of Civil Engineers, London, U.K. and Dr. Oda.)

the photograph, arising from the lesser absorption caused by the higher void ratios. Strain localization means that the distance required for water to migrate into the shear bands falls to the order of a metre (or the spacing between shear bands). From the statistical analysis of real (variable) data in Chapter 5, the scale of fluctuation of state might be in the order of a few metres, or more specifically ~1 m vertically and ~10 m laterally for a man-made fill.

Besides being a poor assumption, localization is a theoretical error for the steady-state approach. Critical state theory is based on Drucker's stability postulate (Chapter 3), which is in turn based on work hardening conditions. *NorSand* work hardens until the soil reaches the maximum plastic dilatancy, at which point the yield surface softens to the critical state. The steady-state approach is mathematically consistent for $\psi > 0$, as that regime is entirely work hardening with *NorSand*. However, for $\psi < 0$, *NorSand* becomes work softening after peak strength, and any assurance of getting to the critical state at the outset void ratio is lost; the doctrine of 'minimum assured strength' correspondingly also loses its theoretical guarantee. The steady-state approach is rescued once soils become sufficiently dense because the undrained strength at the onset of plastic dilation becomes greater than the drained strength, which arises when the volumetric strain is negative at the onset of dilation. This volumetric strain condition was shown for some 20 sands in Figure 2.9 with about $\psi \approx -0.06$ being the controlling state parameter on average, but with a substantial variation that is related to the soil's properties.

Localization is currently a subject of intense research, and it will become something to incorporate in practice as the theory evolves. What should be done in the meantime? Full-scale data will be examined before attempting to answer this question.

6.9 TRENDS FROM FULL-SCALE EXPERIENCE

6.9.1 Background to the empirical approach

The father of critical state theory, Arthur Casagrande, was well aware of its theoretical and experimental limitations half a century ago and wrote (Casagrande, 1950):

> Ever since I presented in a lecture before this Society (the Boston Society of Civil Engineers) in 1935 the concept of critical void ratio in an attempt to define a minimum density to which a sand should be compacted to be safe against liquefaction, much laboratory research has been conducted by a number of investigators on this question. Instead of coming closer to a solution of this problem, recent investigations have raised serious doubts whether any laboratory tests so far used or suggested will permit positive identification as to whether or not a given sand deposit is susceptible to liquefaction. I now believe that the original procedure by means of triaxial tests, which I developed in 1937, gives results which are on the unsafe side. The best we can hope for is that a long-range program of laboratory research combined with field investigations will eventually lead to a laboratory or field test which will identify reasonably well critical conditions. However, at the present time we are obliged to rely on empirical criteria, which are derived directly from a study of flow slides.

Of course, history is often neglected, and it was Seed some 35 years later who is largely attributed with instigating the case history approach to determine residual or steady-state strengths. Seed (1987) looked at seven liquefaction-related failures, estimated a residual strength for each and combined these strengths with estimated SPT resistances to produce an apparent correlation between the two. Marcuson et al. (1990) noted that there was no theoretical basis for the proposal and that it was based on rather few case histories.

Nevertheless, it had the attraction of being expedient, such that a site could be properly tested to account for variability in soil conditions.

Many workers have added to the database and methodology since 1987, including Seed and Harder (1990), Stark and Mesri (1992), Ishihara (1993), Jefferies (1998), Wride et al (1999), Olson (2001), Olson and Stark (2001), Olson and Stark (2002), Idriss and Boulanger (2007) and Robertson (2010). Of course, developing design relationships directly from field performance experience is attractive, as such an approach in principle captures scale effects, modelling uncertainty and so forth. This has become the predominant methodology, although there were substantive disagreements between different groups as to the appropriate framework in which to develop design strengths from the experience base of the case histories (e.g. Seed, 1987, has rather basic errors in mechanics). Today, there is closer to a consensus, anchored in mechanics, for which we claim some credit as a result of the first edition of this book.

There is also the difficulty that several of the case histories are short on information, some being based on hearsay evidence about the initial soil density, or as little as a single SPT (i.e. one SPT, not one borehole with several SPTs). Contrast this with the statistical assessment of penetration data in Chapter 5, which was needed to derive a proper estimate of characteristic properties. In addition, most case histories have minimal (or even no) information on the basic properties of the soils that failed. There is a correspondingly large degree of uncertainty.

Several of the case histories involve underwater slopes. The precision with such post-failure slopes are measured is often exaggerated, and there is much unwarranted confidence in this aspect. This exaggeration becomes important because these underwater slides lead to some of the lower estimated residual strengths. Some workers have apparently, and incorrectly, omitted the stabilizing effect of water pressure on the failing slope.

The case history information is summarized later, with detailed notes on the important histories being given in Appendix F. A striking feature of the case histories is the range in values reported, both for the estimated operating strengths and for the estimated characteristic penetration resistances. These ranges are indicated in the summary, while Appendix F contains detailed evaluations traced back to the source data. Some aspects of the soil strengths and state evaluated and reported in the literature might best be described as guesses.

Centrifuge studies are excluded from the assessment. Although the centrifuge would allow liquefaction mechanisms to be conveniently tested, there are three basic problems with current centrifuge testing. First, the soil state of the in-flight models is presently very uncertain. The as-prepared density on the workbench is measured, but not how this changes as the model is spun up to speed. Neither is the horizontal geostatic stress 'in flight' known. Second, there are questions about edge effects: whether the models are big enough and whether the time factor is properly scaled. Third, in one 'Class A' prediction of prototype experience, the centrifuge results were completely erroneous (Jefferies et al., 1988a). So, full-scale case history experience provides the basis for learning, despite the limitations of partial or incomplete data.

6.9.2 Strength (stability) assessments

Estimating strengths from failure case histories requires stability calculations. Such calculations, to date, dismiss model uncertainty and assume that a factor of safety of unity on soil strengths implies failure. Thus, despite the abundant evidence in the literature that slopes and foundations have failed at factors of safety other than unity (both greater and less), complete confidence is assumed in the calculation methods (of which a range is used). This is not a trivial point as the calculation methods for stability have their own assumptions and idealizations. There certainly is no a priori basis for assuming that calculation methods are perfect, although it is fair to note that provided the estimated strengths are used in the same calculation method for a prototype application, then issues of model uncertainty are at least partially overcome.

Calculations almost invariably adopt limit equilibrium methods. Four approaches are found, with the choice of which to use depending on the nature of the apparent failure mechanism. The approaches used are as follows:

- Infinite slope analogy
- Wedge analysis
- Stability charts (simplified total stress)
- Method of slices (especially with non-circular slip)

All methods can be used with pre- or post-failure geometry. Often, the pre-failure geometry is used to give s_u and the post-failure geometry is used to obtain s_r. Inertial effects may or may not be included, depending on a particular worker's preference. However, since a flowslide is necessarily decelerating to its final resting position, more than self-weight forces are involved. It is also useful to note that while several case histories are recorded as having developed over several hours with retrogressive sliding, some workers use the overall geometry rather than the locally failing one in the back-analysis.

All of these methods for estimating mobilized strengths are found in the various case histories. All are tacitly assumed to provide comparable strength estimates, and choosing which method to use is only a matter of what is most convenient for the situation. Examples of each method will be found in the case histories presented in Appendix F.

6.9.3 Summary of full-scale experience

Chapter 1 introduced some liquefaction failures which, at least to the profession, are the large and now well-known cases. However, many smaller failures have been recorded, and there are now some 30 cases of liquefaction-related failures which have been investigated or documented and are available for analysis. These case histories range from mine tailings slides to post-earthquake failure of structure foundations. Much of the case history record has been gathered with geology rather than mechanics as the starting point. Case histories are classified based on 'geological' observations into:

- Post-earthquake slide versus static failure
- Tailings, natural soils or fills
- Mass movement versus basal failure

These three categories break the case histories down broadly into whether or not there might have been residual excess pore pressure before the soil started moving, whether we might expect geologic bedding to be present and whether the failure was likely caused by some preferentially weaker layer. Getting a little ahead of the story, basal failure seems especially dangerous, as it induces a decreasing mean effective stress higher in the soil column, and it has been shown in the laboratory (triaxial extension tests, and in particular the experiments of Sasitharan et al., 1993, mentioned earlier) that this is an easy way to cause soil to liquefy catastrophically.

Table 6.3 summarizes the important case histories used for back-analyzing steady-state strengths and shows the type of structure involved, the height and slope, its classification as just described and the original (and any supplementary) reference reporting

Table 6.3 Some important case histories giving insight to full-scale post-liquefaction strength

Case history	Year	Soils	Initial slope		Classification	References
			Height	*Slope*		
Zeeland coast	1881 onwards	Holocene fine uniform sand	< 15 m	< 27°	Coastal static flowslides of river foreshores. Mass movement.	Koppejan et al. (1948)
Wachusett Dam	1907	Sandy silt to silty sand fill	25 m	26.5°	Static liquefaction slide during first impoundment of reservoir. Large mass movement.	Olsen et al. (2000)
Calaveras Dam	1918	Sandy silt fill	60 m	18.4°	Static liquefaction slide during construction. Likely mass movement on limited zone.	Hazen and Metcalf (1918); Hazen (1918, 1920)
Sheffield Dam	1924	Silty sand to sandy silt fill	7.6 m	21.8°	Static liquefaction after earthquake induced excess pore pressure. Slide on weak layer.	USACE (1949); Seed et al. (1969a,b)
Fort Peck Dam	1938	Sandy silt fill	61 m	14°	Mass movement static liquefaction slide triggered by yielding foundation (see Fig 1.2).	Middlebrooks (1940); USACE (1939); Casagrande (1965).
Kawagishi-Cho Apartments	1964	Holocene sands	NA	NA	Building foundation failure caused by earthquake-induced liquefaction (see Fig 1.8).	Ishihara and Koga (1981)
Aberfan Tip	1966	Coal tailings, about 10% silt sized	67 m	36°	Mass movement liquefaction slide triggered by increasing pore pressure (see Fig 1.15).	Bishop (1973
Hokkaido Dam	1968	Silty sand tailings	7–9 m	18.4°	Static liquefaction after earthquake induced excess pore pressure. Slide on weak zone.	Ishihara et al. (1990)
Lower San Fernando Dam	1978	Sandy silt tailings (fines > 50%)	43 m	21.8°	Liquefaction slide in weak zone after earthquake induced excess pore pressure (Fig 1.13).	Marcuson et al. (1990)

(*Continued*)

Table 6.3 (Continued) Some important case histories giving insight to full-scale post-liquefaction strength

Case history	*Year*	*Soils*	*Initial slope*		*Classification*	*References*
			Height	*Slope*		
Mochikoshi Dams (no. 1 and no. 2)	1978	Sandy silt tailings (≈ 50% fines)	14, 18 m	18.4°	Static liquefaction mass slides triggered by earthquake-induced excess pore pressures.	Ishihara et al. (1990)
Nerlerk Berm	1982/1983	Sand fill, trace fines	24 m	12.7°	Multiple static liquefaction slides triggered by yield in underlying clay foundation.	Sladen et al. (1985a); Been et al. (1987a)
La Marquesa Dam	1985	Silty sand foundation	7, 8m	27°, 34°	Slides of both dam shells on basal silty sand, triggered by an earthquake.	De Alba et al. (1988)
La Palma Dam	1985	Silty sand foundation	11 m	34°	Slide of upstream shell on liquefied foundation after an earthquake.	De Alba et al. (1988)
Lake Ackerman Embankment	1987	M to F sand fill, trace silt	6.5 m	26.6°	Machine vibration-induced mass flowslide of embankment (Fig 1.23).	Hryciw et al. (1990)
Sullivan Dam	1991	Sandy silt tailings (fines > 50%)	12 m	≈18°	Static liquefaction induced by dike raising, mass movement but not flowslide.	Klohn Crippen (1992); Davies (1999)
Jamuna River Banks	1994	Holocene fine sand, trace silt	≈20 m	≈14°	Mass flow liquefaction slide of river band induced by dredging to steepen slope.	Yoshimine et al. (1999)
Merriespriut	1994	Sandy silt to silt tailings	31 m	Not given	Overtopping by retained water lead to retrogressive liquefaction failure.	Wagener et al. (1998); Fourie et al. (2001)

the data. Table 6.3 does not contain all the cases cited by various workers investigating post-liquefaction strength, as some studied by Seed and co-workers have little real data. Appendix F presents detailed and extensive information for many of the case histories of Table 6.3, which have been used to develop the methodology of this section, while the Lower San Fernando case history is considered at length at the end of this chapter both as an example and because of its prominence as a case history to validate the approach.

The empirical approach to post-liquefaction strengths relates residual strength to a pre-liquefaction penetration resistance. Some of these case histories are old, and penetration resistance data are usually from SPT and done with old equipment for which the energy content is unknown. Sometimes, the resistance was measured by non-standard static penetration cones. Some case histories have no penetration data at all, and for these people have estimated (i.e. guessed) the penetration resistance based on the reported relative density or construction method. The more recent case histories either have both CPT and SPT, or sometimes CPT alone. Some of the older case histories were actually tested for penetration resistance between several years and several decades after the failure, with these tests being carried out on the material still in place and which is thought representative of the material that failed.

Penetration resistance values vary naturally within any stratum. Different workers developed different views on what was characteristic. In developing these views, formal statistical analysis of the data is rare, and in many cases, there are little data to process in any statistically meaningful way. There is a wide discrepancy on what is the characteristic penetration resistance for any case history.

There is also a question as to the type of material involved in the liquefaction. Few case histories have published grain size data, and even fewer have triaxial data from which properties can be estimated. It is usual to take the fines content, defined as the fraction by weight passing the #200 sieve, as an indicator of soil type.

Table 6.4 Comparison of post-liquefaction residual strength s_r (*psf*) from back-analysis of failure as reported by various investigators

	Investigator and their quoted mobilized residual strength range (some best estimate only) in psf						
Case history	*(a)*	*(b)*	*(c)*	*(d)*	*(e)*	*(f)*	*(g)*
Calaveras Dam	750	600–1100	700	600–700	—	600–700	600–750
Fort Peck Dam	600–700	700	700	250–450	—	250–450	250–700
Juvenile Hall	140	200	200	60–200	—	60–200	60–200
Lower Sand Fernando	750	500–1000	510	300–500	—	300–500	300–750
Mochikoshi Dam No. 2	250	75–200	250	100–400	230	100–400	100–400
	Corresponding characteristic normalized SPT blowcount $(N^1)_{60}$						
Calaveras Dam	*12*	2	2	*12*	—	*12*	2
Fort Peck Dam	*11*	*5.5*	*5.5*	*10*	—	*10*	5
Juvenile Hall	2	—	4	6	—	6	2
Lower Sand Fernando	11.5		8.5	11.5	—	11.5	6
Mochikoshi Dam No. 2	1	0	0	0	Weight rods	0	0

Notes: (1) Strengths are presented here in *psf* as this is the form found in most of the references; (2) Key to investigators: (a) Seed (1987), (b) Poulos (1988), (c) Davis et al. (1988), (d) Seed and Harder (1990), (e) Ishihara et al. (1990), (f) Stark and Mesri (1992) and (g) Wride et al. (1998); (3) Penetration resistances shown in italics are estimated from fill description, rather than being based on measured resistances.

Finally, there are the back-figured strengths. These depend on assumptions about inertial effects, location of failure planes and several other factors in the back-analyses, and again there are differences among different analysts. Most cases were analyzed using the conventional method of slices, but infinite slope analysis was used in some instances.

The combined effect of the various uncertainties in the back-analysis of the failure case histories is that there is no unanimous agreement on what strength was mobilized, in what material and at what penetration resistance. This lack of consensus is illustrated in Table 6.4 for five of the case histories. There is an uncertainty of easily ±30% in the calculated mobilized s_r (which is a little surprising, given that the geometry of the post-failure slope is often the best-known aspect of any case history). There is an even larger uncertainty in the characteristic penetration resistance.

6.9.4 Residual (post-liquefaction) strength

6.9.4.1 Background

A fundamental requirement is that equations must be 'dimensionally consistent', which in plain language means every group of terms in an equation must have the same physical units; for example, an equation will be consistent if all the terms that are added, subtracted or taken as equivalent have the dimensions of 'stress', which has the units of *Force/Length*^{2} (or FL^{-2} as you will find it annotated in mechanics texts). In the case of soils, strength increases with confining stress so that the obvious parameter group to capture the experience derived from analysis of the case history record is the ratio s_r/σ'_{vo} (which, as a ratio of stresses, is dimensionless). The requirement of dimensional consistency means that a stress-normalized penetration resistance will be an acceptable measure to compare in-situ conditions to this stress ratio. A single trend will not necessarily exist between these two dimensionless groups as there are almost certainly other factors involved, for example soil compressibility, but these other factors should be reduced to one or more groups of the same dimensions as the governing relationship and which will then produce a family of trend lines. Empirically derived relationships from experimental data have the simplest form (least number of terms) if all groups are dimensionless – the Buckingham 'Pi' theorem.

6.9.4.2 History

What was 'accepted', or at least 'widely used', practice until at least 2001 is derived from the proposal by Seed (1987). There was a strong preference for the SPT. Measured penetration resistance was adjusted to what might have been obtained at a vertical effective stress of 1 tsf (near enough 1 kg/cm^2 or 100 kPa) and as if the test had been done in clean sand rather than the actual soil that liquefied. The adjustment of energy-corrected penetration resistance N_{60} for stress level is the familiar c_n approach that was discussed in Section 4.5 and gives the adjusted resistance $(N_1)_{60}$. The adjustment of the measured data to what might have been obtained if clean sand were tested $(N_1)_{60,ecs}$ was based solely on fines content (the subscript *ecs* is there to denote '*equivalent clean sand*') and was strictly speculative with no explanation of any underlying and postulated normalized soil behaviour. Seed (1987) and other subsequent workers (e.g. Seed and Harder, 1990) who followed this framework developed strength trends of the following form:

$$s_r = f((N_1)_{60,ecs}) \tag{6.10}$$

A fundamental error in (6.10) is that it fails the requirement of 'dimensional consistency', and having failed that requirement, the assured consequence is that the derived strength trends are wrong for general engineering (and we mean 'wrong', there is no wiggle room). There are also real issues between what we will call 'engineering geology' and 'soil mechanics'. By engineering geology, we make reference to the fact that the soil is characterized by its fines content and similar geological descriptions. In contrast in soil mechanics, the soil is characterized by compressibility, permeability, friction angle or similar mechanical parameters.

Various groups opposed the form of (6.10) going back many years, but arguably, it was Olson and Stark (2002) who brought the ratio s_r/σ'_{vo} into widespread acceptance as the most appropriate basis for assessing trends from the case history record.

There has also been a shift away from the SPT. Partly, this came about from recognition that the arguments in favour of the SPT were misleading, partly from the recognition that the CPT is a better test in all respects and partly because there are now lots of testing contractors with CPT equipment. And, there is robust software to process CPT data – a convincing argument for consulting engineers.

Thus, in the 10 years since the first edition of this book was written, the situation has evolved to general acceptance that the case history record should be assessed in the context of s_r/σ'_{vo} versus stress-level normalized CPT data. However, there remain disagreements on the best way of dealing soil compressibility (often expressed as fines content) and the nature of the stress normalization.

6.9.4.3 Current best practice in the United States

The current best practice within the United States is that set out by Robertson (2010) and which represents a further evolution from Olson and Stark (2002). A dominant idea continues to be that the appropriate normalization of the CPT data is the reference stress method (i.e. measured q_t is transformed to q_{t1} as an index of soil state) with developments centring on how to account for the effect of soil type within the reference stress transformation. The general view is that the stress-normalized CPT resistance Q_{tn} is given by

$$Q_{tn} = \left[\frac{(q_t - \sigma_{vo})}{p_a}\right]\left(\frac{p_a}{\sigma'_{vo}}\right)^n \tag{6.11}$$

where

p_a is atmospheric pressure ($\approx$100 kPa by convention as the reference pressure)
n is an exponent that depends on soil type, notionally moving from $n \approx 0.5$ in clean quartz sands to $n \approx 1.0$ in clays

The normalization exponent n is related to the soil-type index (Section 4.7.2) with various workers contributing to the present position (after Robertson, 2010) that

$$n = 0.381(I_{cRW}) + 0.05\left(\frac{\sigma'_{vo}}{p_a}\right) - 0.15 \tag{6.12}$$

In terms of including the effect of soil type on the strength ratio s_r/σ'_{vo}, the original Berkeley School (see Chapter 7) ideas continue with test data being transformed to *equivalent clean sand* values via a factor K_c, thus

$$Q_{tn,cs} = K_c Q_{tn} \tag{6.13}$$

This transformation factor is now also computed from the soil-type index, not fines content as measured in a laboratory test, so that

$$\text{If} \quad I_{cRW} \le 1.64 \Rightarrow K_c = 1.0$$

$$\text{else} \tag{6.14}$$

$$K_c = -0.403(I_{cRW})^4 + 5.581(I_{cRW})^3 - 21.63(I_{cRW})^2 + 33.75 I_{cRW} - 17.88$$

Robertson (2010) reviewed the case histories listed in Table 6.3, amongst others, and considered the range of views about each reported by various investigators. Reliability factors from A to D were assigned to the reported information, with A being tolerably reliable to D being speculative (our words). A best estimate was quoted for strengths and penetration resistances, although the rationale for this best estimate judgement is unclear as some of the quoted values do not capture the issues documented in Appendix F. The best estimate for the category A and B data was plotted to infer operating post-liquefaction strengths as they might be directly derived from CPT data (Figure 6.45). Noting Roberson's assertion that $Q_{tn,cs}$ is equivalent to ψ, an exponential trend would be expected on theoretical grounds as per (6.4) with the best fit of this type of equation to a lower bound of the plotted data being

$$\frac{s_r}{\sigma'_{vo}} = 0.0055\exp(0.05Q_{tn,cs}) \tag{6.15}$$

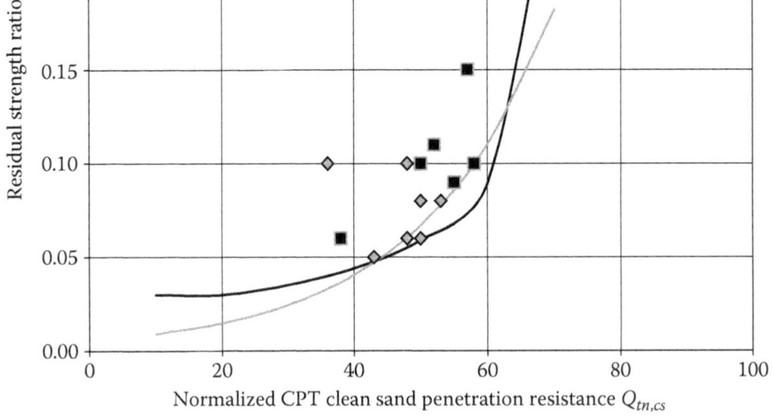

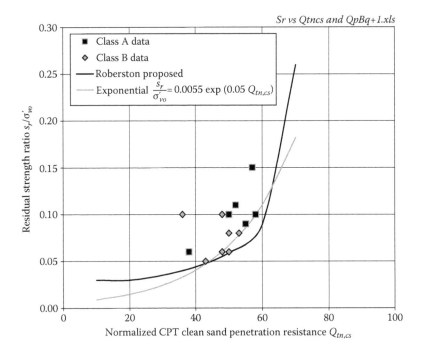

Figure 6.45 Common U.S. practice to estimate post-liquefaction undrained shear strength from CPT. (Adapted from Robertson, P.K., *J. Geotech. Geoenviron. Eng., ASCE*, 136(6), 842, 2010.)

This equation differs from Robertson's curve fit equation as shown in Figure 6.45, which also has the proviso that $0.03 < s_r/\sigma'_{vo} < \tan\phi'$, but Equation 6.15 is implemented as one option for the determination of post-liquefaction strength in the CPT processing spreadsheet that can be downloaded.

While Roberson (2010) represents present best practice in California, there remains a fundamental issue that the concept of 'equivalent clean sand' behaviour is poorly defined and not demonstrated to unify basic laboratory or triaxial data. It is questionable mechanics. This issue of appropriate applied mechanics is compounded by using a geological description of soil type rather than soil mechanics properties (e.g. compressibility, G_{max}), although the use of a soil behaviour type from the CPT rather than 'fines' itself might be argued as a partial step in the correct direction. Robertson (2010) argues that the trends so developed are consistent with critical state principles, but that then begs the question: why not assess operating strengths with a method anchored in applied mechanics directly? That, of course, leads to the state parameter method.

6.9.5 State parameter approach

If undrained conditions occur in the small scale (i.e. no bifurcation or strain localization), then for the usual semi-log idealization of the CSL, Equation 2.1, the mean effective stress in the critical state is related to the current mean effective stress through the definition of the state parameter, Equation 2.6, to give

$$p'_c = p'_o \exp\left(\frac{-\psi_o}{\lambda_e}\right) \tag{6.16}$$

where

the subscript 'o' denotes initial conditions as usual

λ is on a natural logarithm base

Equation 6.16 is independent of stress path or strain conditions (plane strain, extension, etc.) other than the assumption of undrained (constant volume) conditions. The post-liquefaction strength s_r is obtained by combining (6.9) and (6.16):

$$\frac{s_r}{p'_o} = \frac{M}{2}\exp\left(\frac{-\psi_o}{\lambda_e}\right) \tag{6.4bis}$$

or, in terms of vertical effective stress,

$$\frac{s_r}{\sigma'_{vo}} = \frac{1+2K_o}{3}\frac{M}{2}\exp\left(\frac{-\psi_o}{\lambda_e}\right) \tag{6.17}$$

The factor of 2 has arisen because shear strength is half the deviator stress. Some ambiguity will arise in practice as M depends on the intermediate principal stress, but as shown earlier with numerical simulations of simple shear, this ambiguity is unlikely to overestimate strength by more than about 5%. Now, recall that the in-situ state can be measured with the CPT and that the general equation is

$$\psi = -\ln\frac{(Q_p/k)}{m} \tag{4.8bis}$$

Substituting (4.8) in (6.17) and changing from mean to vertical effective stress gives

$$\frac{s_r}{\sigma'_{vo}} = \left(\frac{1+2K_o}{3}\right)^{1-1/m\lambda} \frac{M}{2}\left(\frac{Q}{k}\right)^{1/m\lambda} \tag{6.18}$$

The term including K_o is close to unity, so that Equation 6.18 simplifies to

$$\frac{s_r}{\sigma'_{vo}} = \frac{M}{2}\left(\frac{Q}{k}\right)^{1/m\lambda} \tag{6.19}$$

This equation shows that critical state theory requires that field experience be viewed in terms of the dimensionless framework of the ratio s_r/σ'_{vo} versus the dimensionless resistance CPT resistance Q. There is no uncertainty about dimensions in (6.19) or which variables should be in it. The equation follows from cavity expansion studies whether using the either simple NAMC or more sophisticated *NorSand* constitutive models. This is a particular advantage of the CPT over the SPT – because the CPT is quasi-static and with known geometry (the SPT is affected by soil moving up inside the sampler during the test), it is trivial to understand CPT data in terms of stress and to change those stresses into dimensionless ratios. It is then equally straightforward to understand the dimensionless penetration resistance in terms of applied mechanics using fundamental soil constitutive models.

Several workers (e.g. Maki et al., 2014) have criticized the framework of (6.19) on the grounds that it normalizes sand behaviour by what would be an unusual exponent $n=1.0$ in sand if the data were viewed in California-based approach of Equation 6.11. That criticism is wrong in several regards. First, if calibration chamber data are processed into state parameter form, then indeed it is very close to a perfect semi-log relationship (see Figure 4.16 and download the data from the website to check), and these semi-log trends are also found by formal cavity expansion analysis for constant dilation (Figure 4.5) or constant state (Figure 4.21). The problem lies with Equation 6.11 and the loose idea of 'equivalent' behaviour versus the precise definition of ψ. Second, there is a second-order bias with stress level that is not reflected in (6.19) (see Figure 4.24), but that bias is a consequence of neglecting the effect of in-situ shear modulus. G_{max} is readily included in a general CPT framework if the extra precision is desired (see (4.12), which modifies the parameter k, m in (6.19)) and the in-situ modulus is measured. The California-based approach itself neglects G_{max}. Third, soil type (fines content) is itself a poor predictor of the soil compressibility λ that affects both the CPT resistance and the post-liquefaction strength ratio (see Figures 2.36 and 2.37).

Accepting that leaving out G_{max} loses a little precision, the 12 case histories of liquefaction data in Appendix F are evaluated using this fundamental state parameter framework. The Lower San Fernando Dam is considered on its own in the next section of this chapter. In evaluating these case histories, we have considered the views of various workers on the mobilized strength in each case, the stress levels involved and the characteristic penetration resistances. Penetration resistance data are always given in terms of CPT values, which have been measured in most instances. Soil properties have been estimated using either measured data (available in several cases) or from the CPT using the methods presented in Chapter 4. The way this has been done is discussed for each of the case histories in Appendix F. Table 6.5 summarizes the numerical results of these back-analyses, generally giving a range of values indicative of the uncertainty in the estimates. Figure 6.46a shows a dimensionally correct form of the case history record for undrained shear strength as a function of normalized penetration resistance. Bands show the uncertainty in back-analyzed strengths and

Table 6.5 Summary of case history data for mobilized post-liquefaction strength

Case history	Q_k[a]	λ_{10}	s_r *(kPa)*	σ'_{vo} *(kPa)*	s_r/σ'_{vo}	ψ_k	ψ/λ
Zeeland coast	30–50	~0.06	—	—	~0.13	−0.09 to −0.02	−3.4 to −0.8
Wachusett Dam	10–30	0.06–0.10	10.4–19.1	—	0.07–0.13	−0.05 to +0.07	−1.4 to +2.0
Calaveras Dam	4–8	0.1–0.15	38–56	110–180	0.31–0.35	+0.11 to +0.14	+2.1 to +2.5
Sheffield Dam	6–12	0.1–0.15	3–5	~70	0.04–0.07	+0.04 to +0.15	+0.9 to +2.3
Fort Peck Dam	Not available	0.19	10–30	400–530	0.04–0.06	−0.05 to −0.01	−0.6 to −0.1
Hokkaido Dam	5–7	0.1–0.2	—	—	0.08–0.12	+0.07 to +0.12	+1.6 to +2.3
Lower San Fernando Dam	5–8	~0.1	15–25	200–220	0.07–0.12	0.0 to +0.07	0 to +0.16
Mochikoshi Dam No. 1	3–5	0.15–0.25	~15	~195	~0.08	+0.13 to +0.25	+2.0 to +2.3
Mochikoshi Dam No. 2	3–5	0.15–0.25	18–21	~130	0.14–0.16	+0.13 to +0.25	+2.0 to +2.3
Nerlerk Berm	44–52	0.04–0.05	—	—	0.09–0.15	−0.05 to −0.03	−2.9 to −1.4
La Marquesa Dam	15–25	15–25	4–13	50–85	0.08–0.10	−0.05 to +0.05	−1.0 to +1.0
La Palma Dam	9–15	9–15	10–12	~80	0.12–0.15	+0.01 to +0.08	+0.4 to +1.5
Sullivan Dam	10–14	0.1–0.2	~ 10	80–140	0.07–0.13	+0.05 to +0.1	0.9 to +1.8
Jamuna River Bank	14–16	0.1–0.2	—	150–300	0.12–0.20	−0.04 to +0.05	−0.7 to +0.9

[a] *k* subscript denotes characteristic value (80–90 percentile). *Q* is defined on σ_v and $K_o = 0.7–0.8$ is assumed.

inferred characteristic *Q*. If the full-scale case history data are considered in terms of *Q*/*k* rather than *Q*, then effects of critical friction angle and shear modulus are incorporated through *k* as well as stress level through normalization of *Q*. Figure 6.46b shows the same information as Figure 6.46a, but in terms of *Q*/*k*. A somewhat tighter distribution around the trend line is apparent in Figure 6.46b. Unfortunately, *k* is unmeasured for most case histories, so that crude estimates have to be used at present, although *m* has little effect at the critical state and varies rather gently with soil behavioural parameters (Figure 4.19). In a way, using *Q*/*k* could be viewed as the proper fines content 'correction' to the dimensionless penetration resistance *Q*. The value of *m* will depend on soil type, the initial conditions and the subsequent stress path during shear. So, there never will be a single trend line relating normalized penetration resistance to an undrained strength ratio. The trend line is affected by the properties of the soil.

This point is well illustrated by the Jamuna River Bank case history. The sand is reported by Yoshimine et al. (1999) to be micaceous. Based on laboratory tests on Leighton Buzzard sand with added mica (Hird and Hassona, 1986), relatively small quantities of mica have a potentially large effect on the critical state of sand. For this case history, a relatively high value of CSL slope, $0.1 < \lambda_{10} < 0.2$, was estimated, leading to a CPT inversion coefficient *k* of 10–12. As a result, the Jamuna data in terms of *Q*/*k* lie close to the trend line, while it is well above the trend in terms of *Q* alone.

The Calaveras Dam case history data lie nowhere near the trend from the other case histories. It appears that the undrained strength ratio is too high for the apparent penetration resistance. However, for this case record, penetration resistances were not measured at the time of the failure. As discussed in Appendix F, SPT *N* values were estimated based on descriptions of the construction methods, and these estimates ranged from 2 (Poulos, 1988) to 12 (Seed, 1987). This difference is partly the result of interpretations of whether the

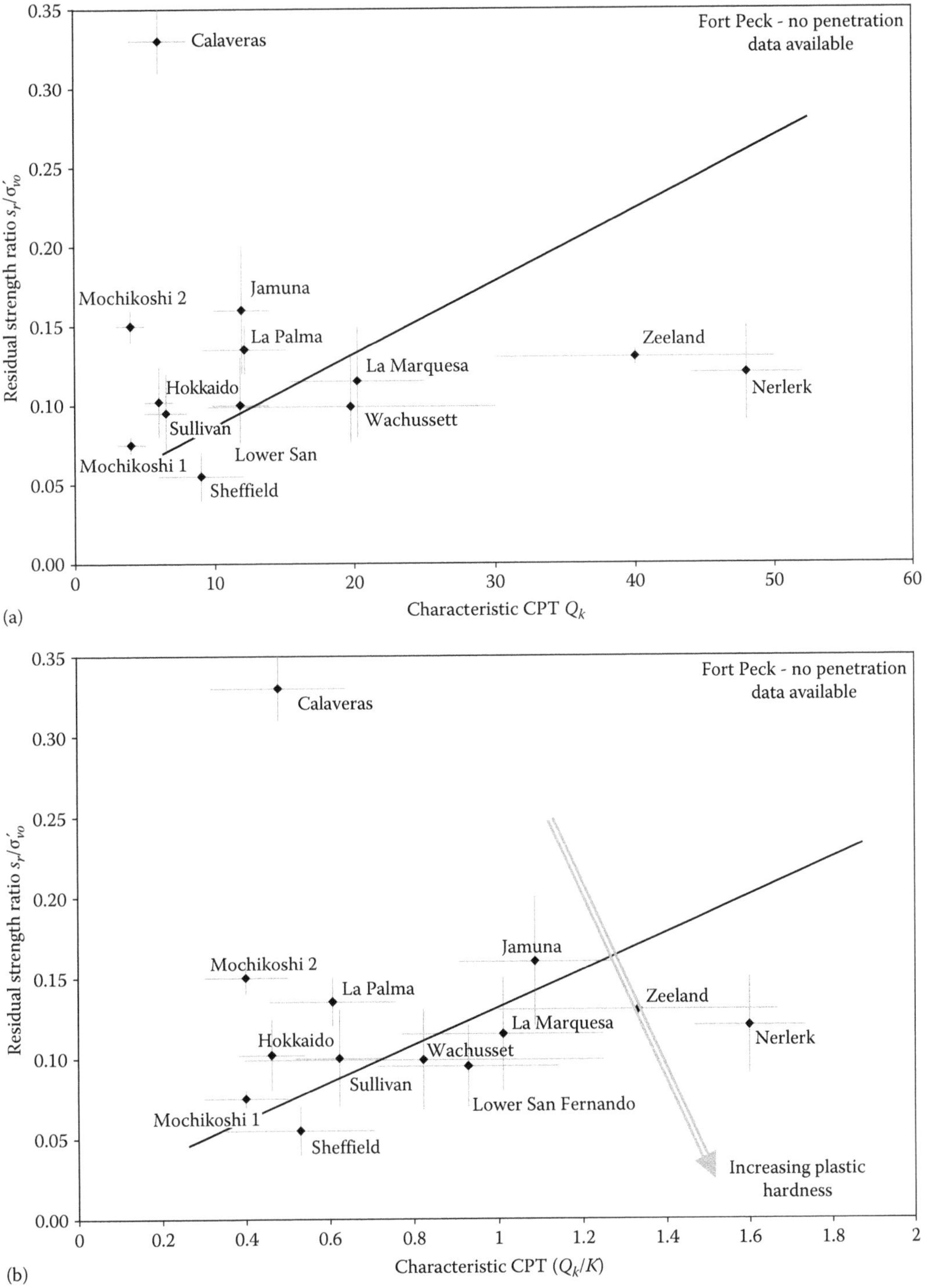

Figure 6.46 Residual undrained strength ratio versus penetration resistance from liquefaction case histories. (a) Dimensionless CPT resistance Q and (b) dimensionless and normalized CPT resistance ratio Q/k.

material controlling liquefaction is a silty sand or a loose silt. Our assessment based on the available descriptions is that the silts would control liquefaction, and Figure 6.46 reflects this assessment, which appears to be incorrect. Given the lack of definitive data, Calaveras Dam should probably be removed from the case record, as any further attempts to interpret which material controls liquefaction and its characteristic penetration resistance are bound to be biased by knowledge of what the answer should be.

Going directly to penetration resistance in Figure 6.46 compounds two factors: how state parameter is related to penetration resistance and how steady-state strength is related to state parameter. In particular, as seen in Equations 6.12 and 6.14, λ is important in going from ψ or Q to s_r. The preferable approach is to estimate ψ from the CPT directly and then to understand how s_r at field scale correlates with ψ. Such an approach separates the uncertainty in inverting penetration test data (i.e. the estimates of k, m for the soils in question) from the uncertainty of soil behaviour at field scale (which also needs an estimate for λ, amongst other mechanical parameters).

This preferred approach is shown in Figure 6.47, where the strength ratio is plotted against the state parameter, but since strength depends on soil properties, the data are separated into three ranges. These ranges represent low compressibility (= low λ clean sands), intermediate (silty sands) and high compressibility (= high λ, sandy silts to silts). The uncertainty in the assessed strengths (mainly arising from difference between the various investigators) and the uncertainty in assessed state (because soil properties have generally had to be estimated, with only Nerlerk having good data) are both shown as error bars in the plot. These uncertainties are discussed in detail in Appendix F for each of the case histories. Generally, this preferred approach of Figure 6.47 shows rather nice trends in the data, with the effect of compressibility obvious by comparing trends between the three ranges. Mochikoshi no. 2 is the only real outlier, being apparently far stronger than might reasonably have been expected from the measured CPT resistance. Triaxial compression test data are particularly missing for this case history, with the implication that the retained tailings at Mochikoshi no. 2 dam might have been markedly siltier than those retained by the nearby Mochikoshi no. 1 dam.

Two sets of trend lines are shown in Figure 6.47 in addition to a lower bound to the data. One trend line is from simple theory (the critical state strength) and the other from numerical simulations allowing for additional effects. The simple theory is the same as the steady-state school, which formally has no pore water movement at all (undrained throughout the soil mass) and which precludes localization effects. This line is plotted in Figure 6.47, denoted as 'steady-state strength', computed using a typical sand $M_{tc} = 1.25$, a typical loose in-situ geostatic stress ratio $K_o = 0.7$, and λ values that are mid-range for each of the data sets. This steady-state strength does not fit the field case data very well except at extremely loose states. What is going on? *NorSand* offers insight and another set of trend lines in Figure 6.47.

The mathematical derivation of *NorSand* is based on Drucker's (1959) stability postulate. The introduction of instability criteria to geomechanical constitutive models is a very new and active subject, so definitive conclusions are as yet difficult. What we are presenting here is an engineering assessment based on a simple idealization that seems to match large-scale experience. Drucker's postulate leads us to expect instability in *NorSand*, or any other proper constitutive model of soil, when (with total stresses)

$$\dot{\sigma}_q \dot{\varepsilon}_q + \dot{\sigma}_m \dot{\varepsilon}_m < 0 \tag{6.20}$$

When (6.20) applies, undrained conditions can no longer be maintained locally. For lightly dilatant soils, the post-peak strength drop corresponds to the occurrence of the condition described by (6.20). At such time, the steady-state doctrine that the soil will proceed to critical state conditions at the same void ratio as existed prior to loading is false. Keep in mind

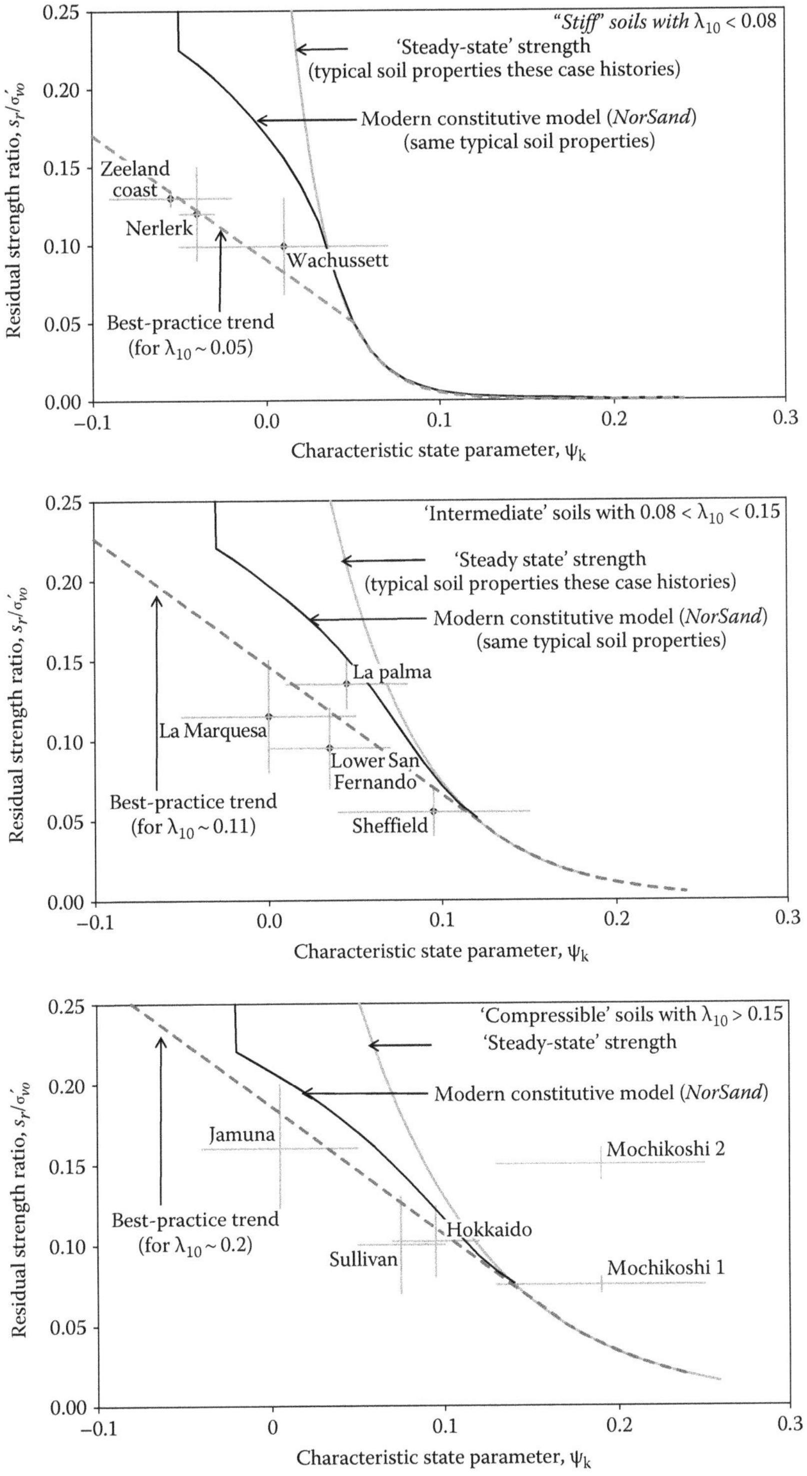

Figure 6.47 Relationship between initial in-situ state parameter and mobilized steady-state strength from case history data, for ranges of hardening (strong, intermediate, low).

that in laboratory tests, undrained conditions are imposed by shutting a valve, but that is not what happens in-situ. In-situ, small amounts of water may easily be drawn into a shear band from the surrounding large soil mass leading to a significant void ratio change in the shear band. The consequence is that lightly dilatant soils will not have the same 'steady-state' strength in large-scale situations.

The second trend plotted in Figure 6.47 is for the minimum post-peak stress predicted using the undrained simple shear model of *NorSand*, that is approximately the condition when (6.20) occurs. It is shown on each of the three compressibility bands of Figure 6.47. These trends are computed using a mid-point value of λ for each band and with $H/I_r = 0.5$. There is a remarkable coincidence of the minimum undrained strength predicted by *NorSand* and the case history data. Interestingly, *NorSand* does not show the gentle upward curve that might have been expected as a variation on (6.4), but instead shows only modestly increasing residual strengths with density until the state parameter becomes denser than about $\psi_o < -0.05$ and at which point the undrained strength becomes asymptotically large with further decreases in the state parameter.

By now, it will no doubt have been noticed that this minimum strength corresponding to the transition from work softening to work hardening is pretty much the same as using the phase transition strength advocated for design by Konrad (1990b), Vaid and Eliadorani (1998) and Yoshimine et al. (1999), amongst others. Some of their arguments are based on a misconception about the CSL, but the result remains the same. Of course, now that it is apparent that localization needs to be considered, a lot of simplicity is lost. Soil properties now matter, especially the in-situ plastic hardening and the in-situ elastic modulus. The ratio *H/Ir* is rather important, exactly the point Wroth made in his Rankine lecture when dealing with clays. There will never be a single trend line for post-liquefaction strength in terms of either normalized penetration resistance or state parameter. So, what is a practical engineer to do? The answer is to follow Wroth's strictures and use the theory to guide the assessment of experience, which comprises the final set of lines shown as 'best practice trend' in Figure 6.47. These best practice trend lines are a conservative fit to the case histories. They

```
Public Function sr_over_sigV(Mtc, k0, lambda, psi)
'returns the best fit of the strength ratio sr/sigv' to the case-history data base
'for post-liquefaction strength:
Dim lambda_e
Dim sr_sigV_origin
Dim k0_factor
Const Slope = 0.8

k0_factor = (1 + 2 * k0)/3

Select Case psi
  Case Is > lambda 'use critical state theory for very loose soils
    lambda_e = lambda/2.3
    sr_over_sigV = k0_factor * 0.5 * Mtc * Exp(-psi / lambda_e)

Case Else          'use case history trends
    sr_sigV_origin = k0_factor * 0.0501 * Mtc ' anchor the trends to the theoretical strength at psi = + lambda10
    sr_over_sigV = sr_sigV_origin + Slope * (lambda - psi)
End Select
End Function
```

Figure 6.48 VBA function for proposed minimum undrained shear strength after liquefaction as a function of ψ and λ_{10}.

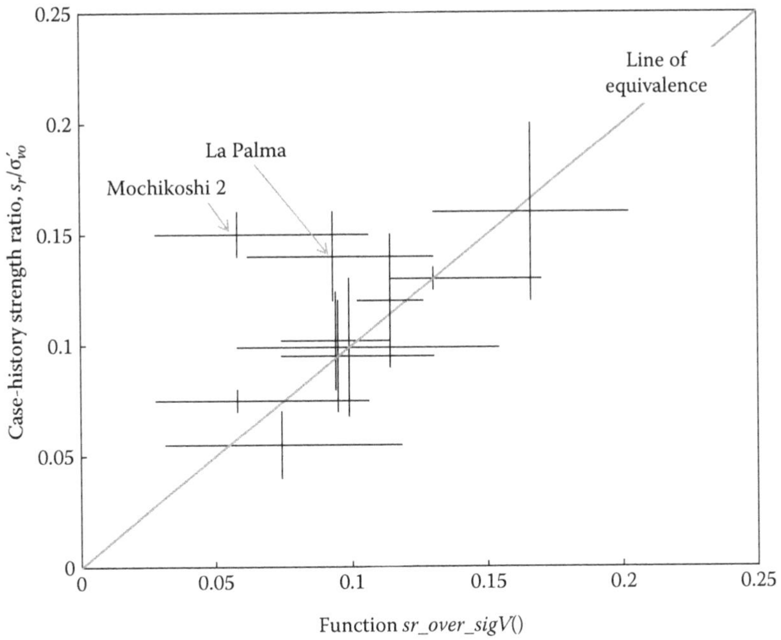

Figure 6.49 Demonstration of the adequacy of VBA function (Figure 6.48) to match case history post-liquefaction strengths.

are roughly parallel in Figure 6.47 with a slope of 0.8, and a computable form comprising the VBA function listed in Figure 6.48 is implanted as an alternative (i.e. selectable option) to (6.15) for deriving post-liquefaction strength in the spreadsheet *CPTplot.xls*.

The adequacy of the derived function was validated by directly comparing its output with the estimated operating strength for each of the case histories, the result being shown in Figure 6.49. The uncertainties in the operating strength derived from back-analysis are shown, and the uncertainty in the function output is a consequence of the uncertainty in the assessed soil state and compressibility. As can be seen, the function is essentially an unbiased predictor of operating strength with the central estimate of the uncertainty range lying very close to the line of equivalence; the two outliers are Mochikoshi no. 2 (already noted as seemingly too strong) and La Palma (which also shows are being a little stronger than predicted).

6.10 LOWER SAN FERNANDO DAM REVISITED

The comparative back-analysis of the Lower San Fernando Dam to validate the steady-state concepts was described in some detail in Section 6.8. The development of an empirically based residual strength framework occurred later (and was presumably spurred on by results of the Lower San Fernando Dam study). It is worth revisiting this important case history and examining it in the light of what is now known.

The 1985 investigation included 12 piezocone CPTs. Digital CPT data, however, are not part of the archive set, and we are indebted to Dr. Scott Olson (University of Illinois at Urbana-Champaign) for these records. Data files were recovered only for 9 of the 12 CPTs (C101–109 inclusive), but these largely cover the area of interest. Note that in presenting the information, the CPT data have been transformed to the familiar SI units, but the depth and elevation scales have been left in feet for ease of comparison with the published record and dam cross sections.

Lower San Fernando Dam was reconstructed in 1975 to act as a back-up for a new dam constructed upstream, but has not retained water since 1971. It is this reconstructed back-up dam that was investigated in 1985, and care was taken during the investigation to test materials that were as representative as possible of those that failed during 1971. The pre-1971 and reconstructed configuration that was tested in 1985 are compared in Figure 6.50. The 1985 investigation has relied on the fact that what remained downstream is to some extent representative of the failed soils. The case for similarity of materials tested in 1985 with those that failed in 1971 is based on two points: the material in both shells of the dam came from the same source (the reservoir floor) and construction photographs indicate that the dam was raised symmetrically with the same hydraulic filling method for the upstream and downstream shells. Two SPT borings 350 ft apart (S103 and S111) show similar results indicating similar zonation of the dam parallel to its centreline, which leads credence to systematic construction under engineering control.

Of course, failure in only an upstream direction suggests that the downstream soils were not quite representative of the upstream, but this difference in behaviour appears likely to have been caused by the differing water tables and saturations (high excess pore pressures do not occur in partially saturated soils during earthquake shaking). Engineers have been cognizant of the difference between upstream and downstream and also that both the earthquake and the subsequent dewatering and reconstruction of the dam may have densified the downstream soils. Some adjustments are needed to nearly all of the investigation measurements to account for the differences between 1971 and 1985 and between upstream and downstream, discussed later.

A plan of the 1985 investigation is shown in Figure 6.51, and as can be seen, the investigation concentrated on what was originally the downstream shell. Four upstream–downstream sections through the dam were tested. Three of these upstream–downstream sections had duplicate SPT–CPT pairs, with the fourth section being only CPTs. The key section, in terms of understanding the results of the investigation, is that through the centreline of the sliding mass, which is at Station 09 + 35, and comprises SPTs S103, S104, S105, the corresponding CPTs C103, C104, C105 and a test shaft. This cross section is shown as Figure 6.52 with the SPT resistance profiles superimposed. The fill-type boundaries shown in Figure 6.52 are those shown by Castro et al. (1989) in their report, and it appears that this is a consensus view of the internal zonation of the dam. The soil identified from the boring logs as Zone 5 was characterized by Castro et al. (1989) as the 'critical' soil unit (their term) for the dam from the liquefaction standpoint for the following reasons:

- Zone 5 is at approximately the same elevation as the upstream zone that experienced large strains during the post-earthquake slide, and most of the failure surface in the back-analysis lies within this zone.
- The static shear stress was greatest at the base of the hydraulic fill.
- SPT penetration resistance values are generally lower in Zone 5 than higher in the fill, and this difference becomes even more pronounced when the data are adjusted for overburden pressure.

Of course, the second point assumes that it is s_r that matters rather than s_r/σ'_{vo}, but the telling point really is the penetration resistance profile. The weakest soils were at the base of the fill according to the SPT. The laboratory testing for the evaluation of the steady-state approach concentrated in these Zone 5 soils.

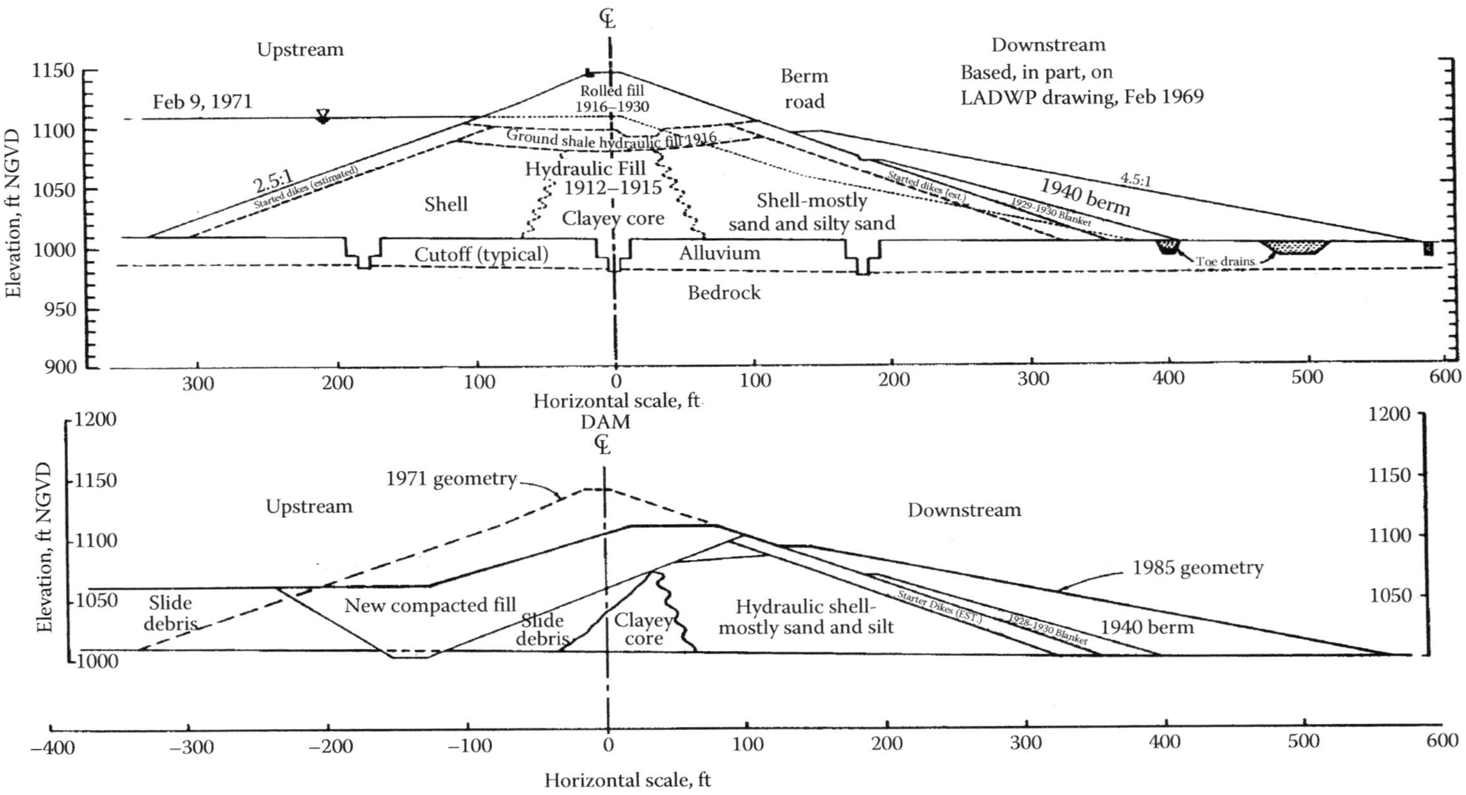

Figure 6.50 Lower San Fernando Dam showing as-constructed section (above) and section during the 1985 investigation (below). (After Castro, G. et al., *Re-Evaluation of the Lower San Fernando Dam, GEI Consultants*, Inc., Contract Report GL-89-2 Volume 1, US Army Corps of Engineers, Washington, DC. Reproduced with permission from Dr. Castro.)

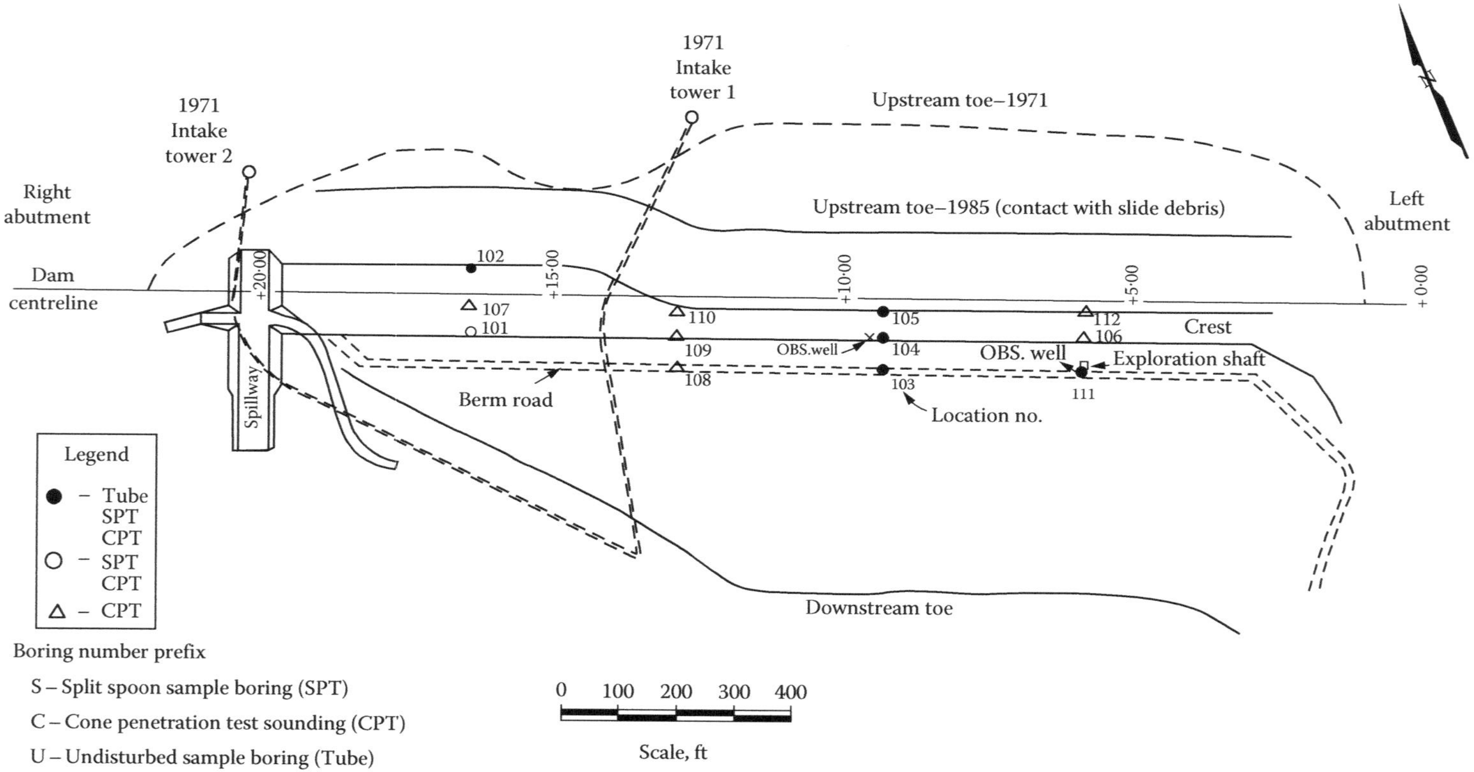

Figure 6.51 Plan of Lower San Fernando Dam in 1985 and showing location of investigation borings/soundings. (After Castro, G. et al., *Re-Evaluation of the Lower San Fernando Dam, GEI Consultants*, Inc., Contract Report GL-89-2 Volume 1, US Army Corps of Engineers, Washington, DC, 1989. Reproduced with permission from Dr. Castro.)

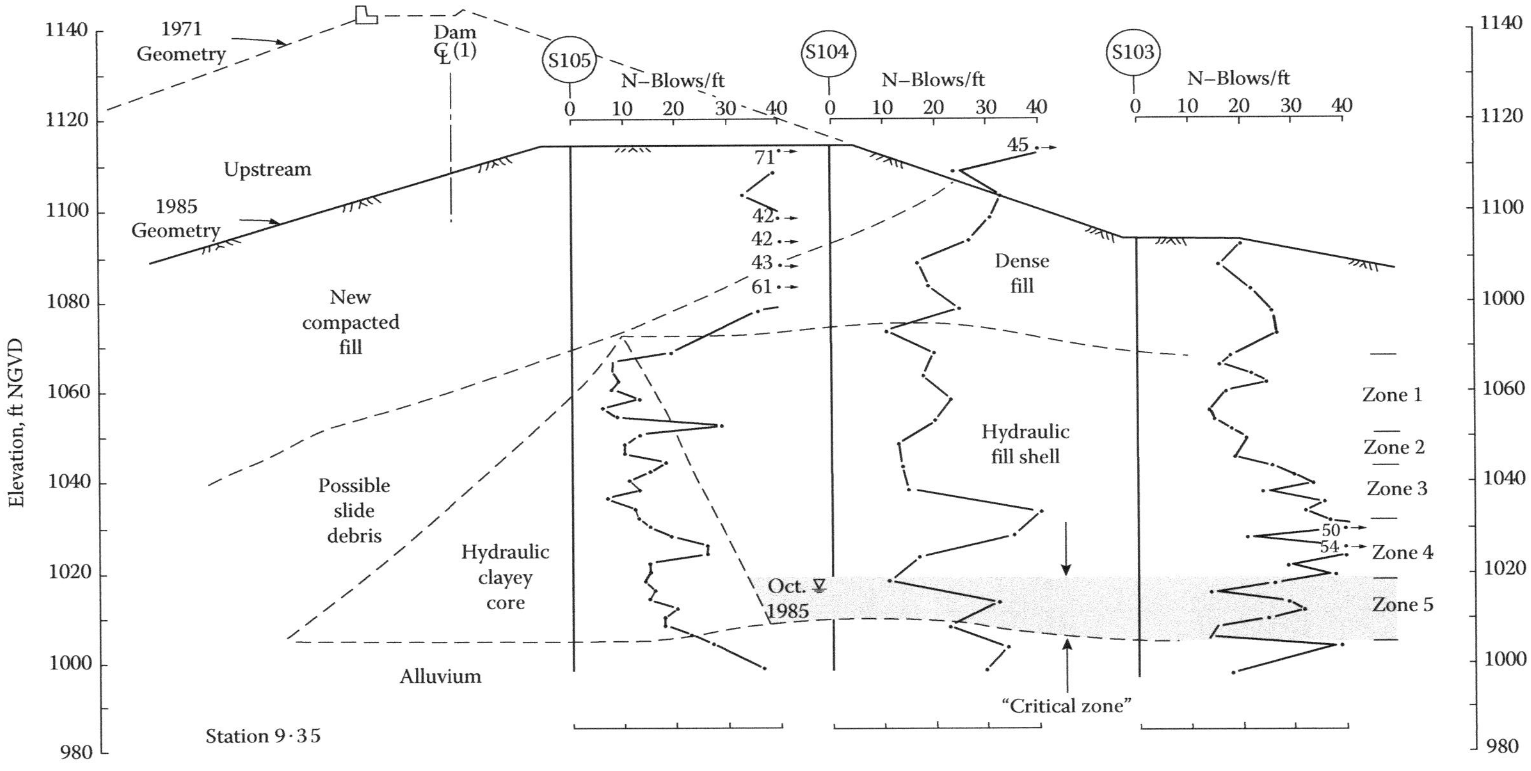

Figure 6.52 Cross section through Lower San Fernando Dam at St. 09+35 (approximately centreline of sliding mass) showing inferred zonation of dam from 1985 study. (Reproduced with permission from Dr. Castro.)

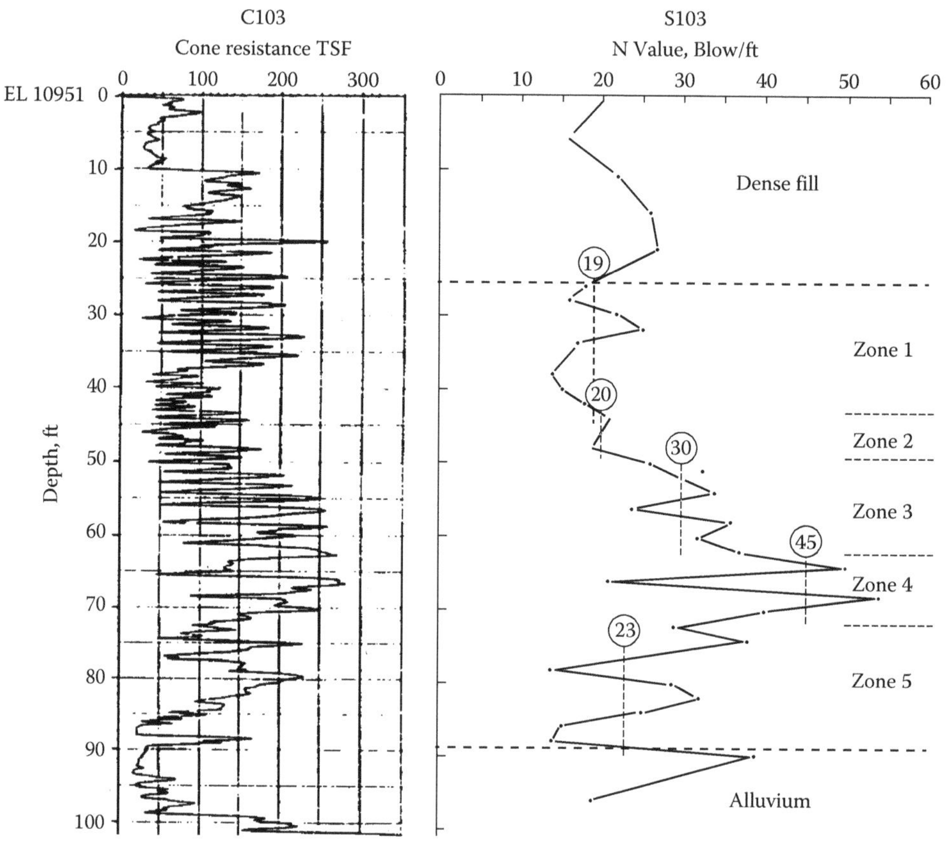

Figure 6.53 Comparison of CPT and SPT resistances at Lower San Fernando Dam. (From Castro, G. et al., *Re-Evaluation of the Lower San Fernando Dam, GEI Consultants*, Inc., Contract Report GL-89-2 Volume I, US Army Corps of Engineers, Washington, DC, 1989. Reproduced with permission from Dr. Castro.)

The CPT data do not show the same pattern of soil state as that estimated from the SPT, Figure 6.53 showing one of the comparisons presented by Castro et al. This comparison between S103 (the SPT) and the C103 (the CPT) is on the centreline of the slumped mass movement. Within the hydraulically placed shell soils, the SPT profile was subdivided into Zones 1 through 5, and a characteristic penetration resistance was assigned to each zone (shown as the circled number in Figure 6.53). The CPT tip resistance profile has a superficial resemblance to the SPT profile in that the average trend in q_c has a greatest value around 60 ft depth and the average clearly decreases below that. However, the CPT also shows a far more layered deposit. This layering was referred to as macro layers by Castro et al., who also reported that it could be seen to a lesser extent in the SPT split-spoon samples. This layering was found in a test shaft excavated through the dam fill, and it is also evident in the range of gradations measured in the Zone 5 soils which are shown in Figure 6.54. The fines content ranges from a low of about 20% to a high of about 70%.

Rather more detail emerges if all the CPT data channels are processed, Figure 6.55 showing the CPT C103 record including a plot of the soil-type index I_c. The water table elevation was reported to be low during the 1985 investigation (at about elevation 1012 ft or a depth of 83 ft in CPT 103), but rather large values of pore pressure were measured above the

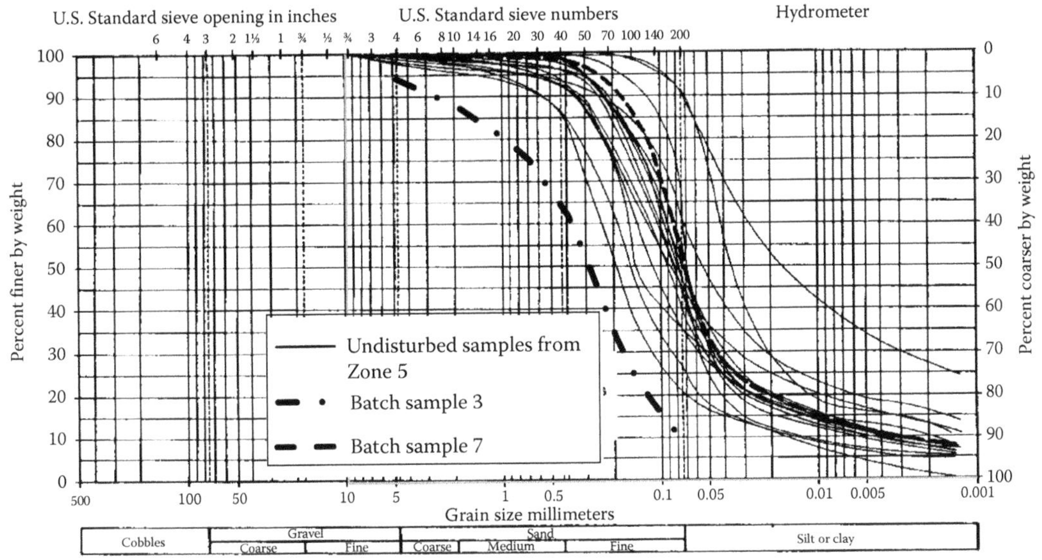

Figure 6.54 Particle size distribution of soils within Zone 5 of Lower San Fernando Dam ('Batch Samples' 3 and 7 were used for steady-state triaxial tests). (Modified from Castro, G. et al., *Re-Evaluation of the Lower San Fernando Dam, GEI Consultants*, Inc., Contract Report GL-89-2 Volume I, US Army Corps of Engineers, Washington, DC, 1989. Reproduced with permission from Dr. Castro.)

water table. This results in a notable difference between the I_c values following Robertson and Wride (1998) and Been and Jefferies (1992) (see Table 4.4), both of which are shown in Figure 6.55.

The processed CPT data (primarily the dimensionless tip resistance) suggest that the dense dam fill extends to a slightly greater depth, about 38 ft, than assessed from the SPT (i.e. the upper part of Zone 1 is actually a compacted fill). Below this dense fill, down to 85 ft, there is a variable and interlayered fill, which the I_c values suggest may range sand to silty sand (at the coarse end) to sandy silt or even clayey silt in the finer layers. The Robertson and Wride I_c indicates slightly coarser material than the Been and Jefferies I_c, but in fact both forms of I_c give a reasonably accurate picture of the gradation curves and more importantly how these gradational differences occur in-situ. Below 85 ft, there are silty clay or clay materials representing the underlying alluvial soils.

Going further, nothing like Zones 2 through 5 can be discerned in the I_c plot. There is a case for treating the hydraulic fill as simply one unit from about 38 to 85 ft, and the variability of this hydraulic fill can be evaluated over the shell. Figure 6.56 shows the q_c profiles for four CPTs. The data are aligned by elevation and windowed to lie entirely within the hydraulically filled shell. Taking the profiles (C103 and C104) on Station 09+35 first, which are near the centre of the sliding mass, thicker and systematic layering is evident on the more downstream CPT (C103) than on the upstream CPT (C104). This is entirely usual for a hydraulic fill dam constructed in the early twentieth century as is the case for Lower San Fernando Dam (in fact, much of the construction is like Calaveras and Fort Peck, two other classic dam failures). In such dams, fill was spigotted from the shells inwards so as to preferentially separate the coarser sands to the outside of the dam and the finer material as close to the dam centreline as possible. A similar but less pronounced pattern is seen on C108, which is downstream of the crest (west of C103), and C106, which is at the crest (east of C104).

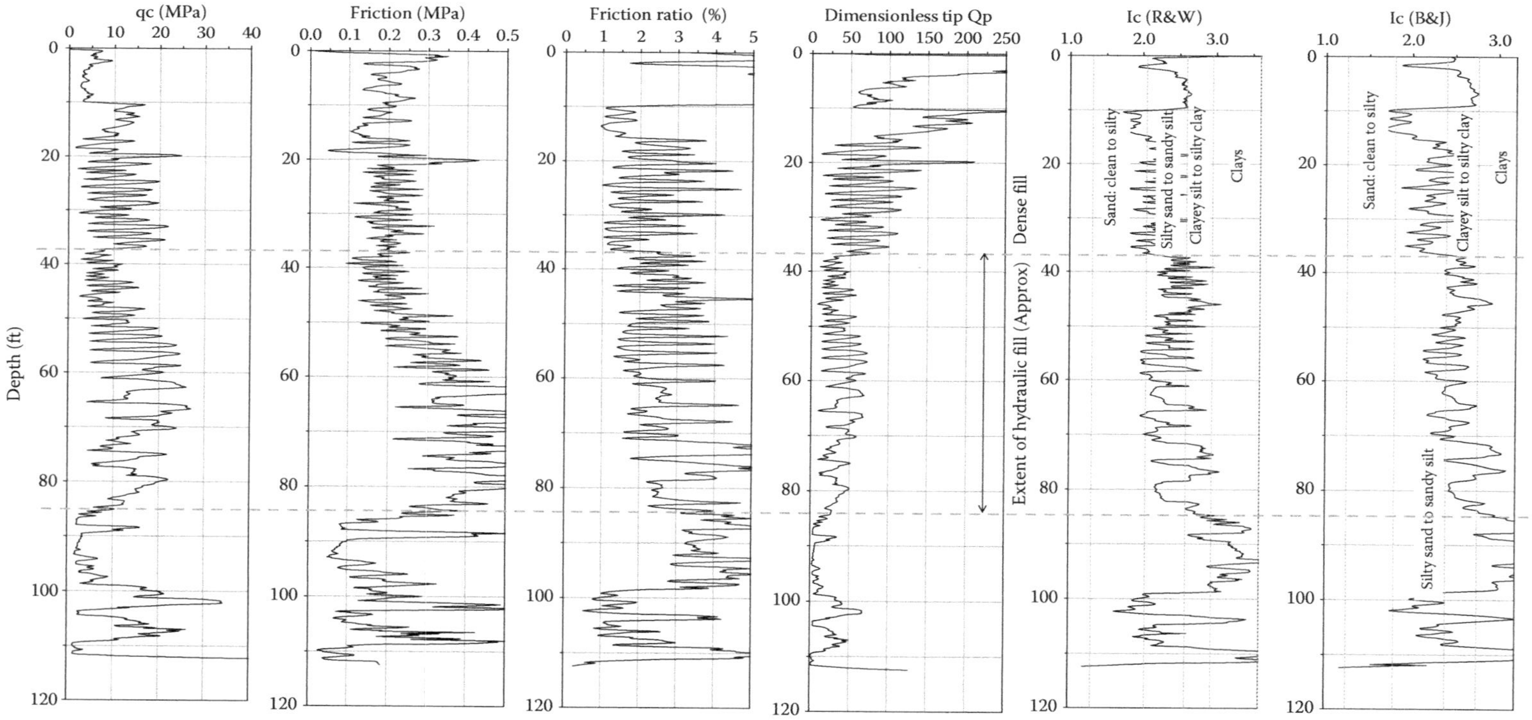

Figure 6.55 CPT C103 from Lower San Fernando Dam investigation in 1985.

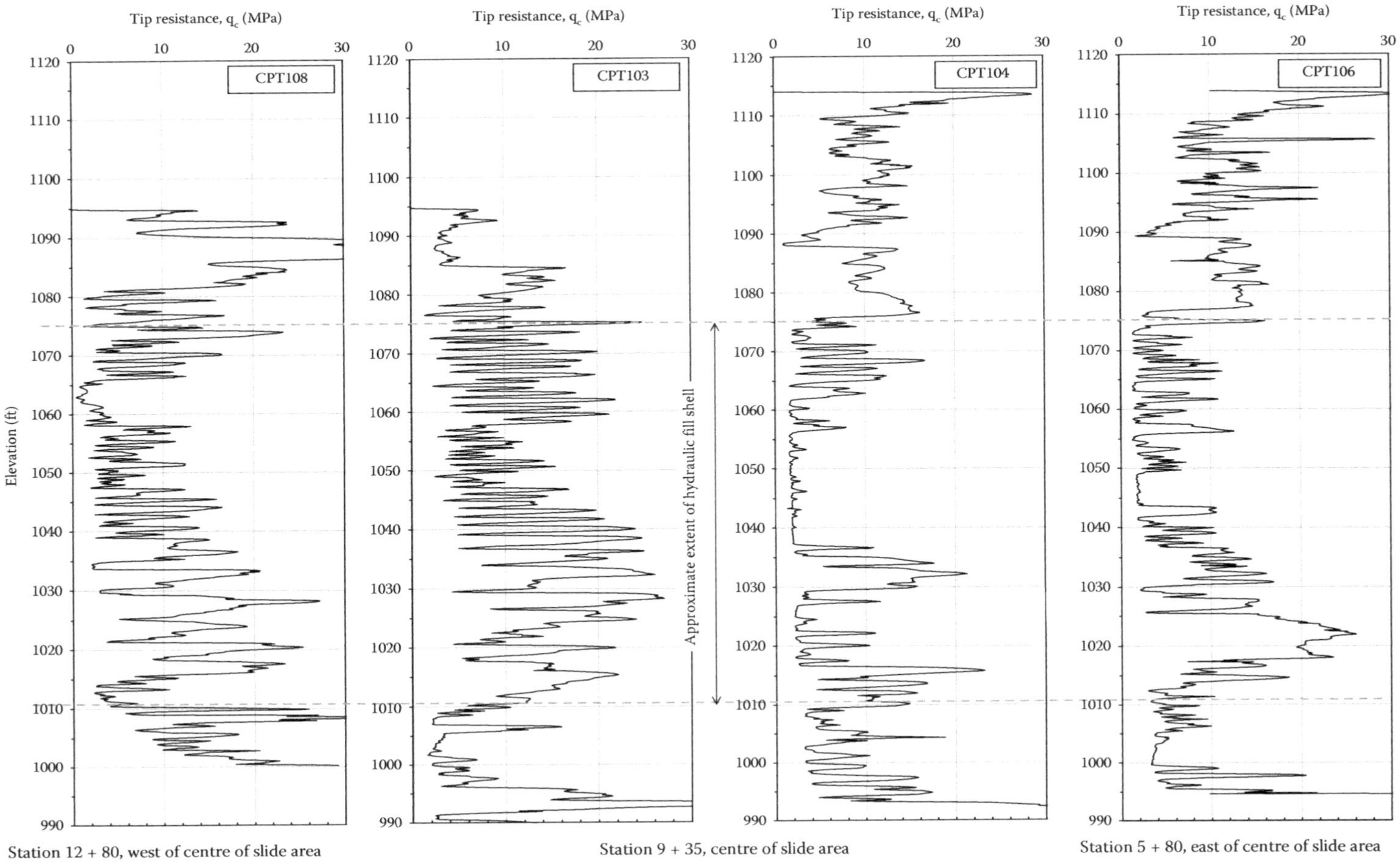

Figure 6.56 Comparison of CPT tip resistance profiles in hydraulic fill.

The question now becomes the following: what is the characteristic penetration resistance for the shell? The zonation adopted by participants in the 1985 study is not supported by the CPT data, although the CPT data are consistent with construction practice of the era. There is a difficult situation here as there is a layered system with two soils. One is primarily a sandy silt and the other is primarily a silty sand. Within each of these soil types, there is also a variation in state, sometimes at quite fine scale.

Two batch samples with gradations shown in Figure 6.54 were tested for steady-state line and incidentally resulted in parallel SSLs with $\lambda_{10} = 0.10$. The testing of sands at various silt contents that was presented in Chapter 2 indicates this is somewhat unlikely, but it illustrates that you really should measure the soil properties and not make assumptions based on grain size distribution. This similarity in λ_{10} across grain sizes that reasonably represent the potential range in-situ makes processing of the CPT in terms of state parameter rather straightforward. Indeed, in this instance, the screening-level method of Section 4.7 turns out to give a rather optimistic picture, because the friction ratio F (Figure 6.55) is in the range of 2%–4%, which would result in typical λ_{10} values in the range of 0.2–0.4, compared to the measured 0.1.

Using Equations 4.15 and 4.16 and $\lambda_{10} = 0.1$, the CPT evaluation coefficients k,m are estimated as $k = 14.1$ and $m = 10.4$. The q_c data (Figure 6.56) are then transformed into ψ profiles, Figure 6.57, windowed to show only the hydraulic fill between elevations 1010 and 1070 ft. The state parameter values in the hydraulically placed fill tend to split according to gradation and also according to distance from the dam crest. The peaks of ψ represent loose layers, and troughs of ψ represent denser layers. At the dam crest, C103 and C108, the sandier soils show $\psi \approx -0.10$ to -0.2, while the siltier soils show $\psi \approx 0.0$ (or between about -0.05 and $+0.03$), and there is a systematic layering with a thickness typically less than about 1 ft (count the

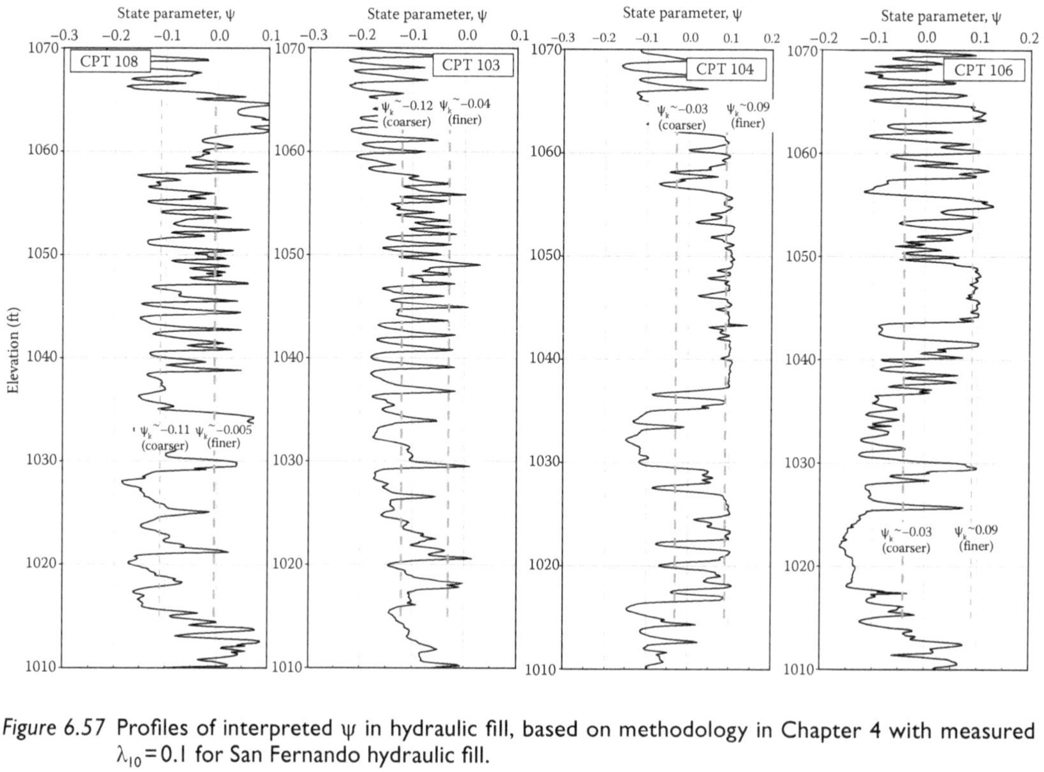

Figure 6.57 Profiles of interpreted ψ in hydraulic fill, based on methodology in Chapter 4 with measured $\lambda_{10} = 0.1$ for San Fernando hydraulic fill.

Table 6.6 Characteristic values of state parameter ψ_k and post-liquefaction residual strength ratio s_r/σ'_{vo} for Lower San Fernando CPTs

	Sandy, denser layers		Silty, looser layers	
CPT	ψ_k	s_r/σ'_{vo}	ψ_k	s_r/σ'_{vo}
C108, crest	−0.11	0.22	−0.005	0.13
C103, crest	−0.12	0.23	−0.03	0.15
C104, berm road	−0.03	0.15	+0.09	0.06
C106, berm road	−0.04	0.16	+0.09	0.06

peaks and troughs in a 10 ft interval anywhere between elevation 1020 and 1070 ft to arrive at this estimate). Caution is needed in the state parameter estimates, as no 'thin-layer' corrections have been made in this analysis. Away from the crest below the berm road (C104 and C106), the layering is markedly different with substantial thicknesses of loose and dense material evident from the CPT. Also, the silty layers are looser $\psi \approx +0.05$ to $+0.1$ than at the dam crest, while the denser sandy material has ψ between about −0.15 and −0.05.

Looking at the 90 percentile as characteristic, the broken lines indicate visually estimated ψ_k values for each CPT which are shown in Table 6.6. For the looser, silty layers, $-0.03 < \psi_k < +0.09$ and for the denser sandy layers $-0.12 < \psi_k < -0.03$.

The simplistic approach is next to determine the residual stress ratio s_r/σ'_{vo} using Equation 6.17. However, this equation makes no allowance for localization and fits field data for only very loose soils as shown in Figure 6.47. Ideally, *NorSand* simulations would be used to define the locus of the localization points (which correspond to the pseudo-steady-state strength). But such calculations require calibration of *NorSand* to triaxial tests, some of which must be dense drained tests (to determine χ). The required tests are not available for the Lower Sand Fernando fill. Therefore, the best practice trends to the case history data discussed earlier and the equation embodied in the VBA function in Figure 6.48 are used to calculate liquefied strength ratios shown in Table 6.6. Overall, $s_r/\sigma'_{vo} \approx 0.06–0.15$ might be taken as representative of the looser silty fill as tested in 1985 and $s_r/\sigma'_{vo} \approx 0.15–0.23$ of the denser layers of fill.

The next question is this: how is the strength ratio of 1985 related to what the soil actually possessed during the 1971 earthquake? The answer is, to practical precision, the same strength ratio. First, although the drainage during dewatering will have produced some consolidation of the fill, such void ratio changes are minor compared to the range of state possible. Increasing stress also tends to increase (make more positive) the state parameter because the consolidation-related void ratio changes are less than the change in the critical void ratio for the same stress increase. For this reason, the assessed strength from the 1985 data may underestimate the actual strength. Second, because the strength is expressed as a strength ratio of the initial effective stress, the very different effective stress conditions in the upstream shell are automatically accounted for. Third, densification of the downstream shell from cyclic strains (even though drained) can hardly be more than about 0.5%, which corresponds to a possible shift in the state parameter of about 0.01 and in the opposite sense to the change induced by the dewatering. Putting these factors together, the best judgement is that the strength ratios evaluated in 1985 might reasonably have been expected to be operative in the post-liquefaction situation.

How does this strength estimated from the CPT data compare to reality? The strength ratio to prevent large-scale slope movement after cyclic liquefaction would have needed to be $s_r/\sigma'_{vo} \approx 0.4$, which far exceeds the ratios shown in Table 6.6. In terms of when the slip stabilized, a reasonable best characterization of an overall post-slip slope angle for the upstream shell is 1V:5.8H. The various back-analyses of this post-failure slope presented in the literature give the range $0.06 < s_r/\sigma'_{vo} < 0.12$ (Table 6.5). This is remarkably close to the estimated residual strength ratio from applying the state parameter approach to the CPT data, a strength $s_r/\sigma'_{vo} \approx 0.06\text{–}0.15$ having been computed as characteristic for the loose silty layers in the downstream shell in 1985.

Too much should not be read into how closely the state parameter approach matches the operating residual strength ratio at Lower San Fernando Dam, as there are likely compensating errors. Using average trends on only a few CPTs, without a detailed assessment of the grain size, distribution ranges and CSL parameters would normally be considered just one step better than a screening-level assessment. Recall that there is significant layering, and how such layering affects CPT values alone is difficult, never mind how such layering would affect the liquefaction behaviour. There have also been no detailed *NorSand* simulations of the dam fill because the required triaxial data (dense, drained tests) are not available to calibrate the model. Finally, the time delay between the earthquake and the failure has not been explained.

6.11 HOW DENSE IS DENSE ENOUGH?

At the beginning of the chapter, a promise was made to answer the question, 'How dense must a sand be to avoid catastrophic failure?' as well as to address the residual strength after liquefaction. Engineers must assume that there is always the possibility of some strain in any structure, for example movements in underlying weaker clays, wave loading or erosion on shorelines or simply more fill placement at the top of a slope. If the soil is loose, liquefaction will be triggered. If the soil is sufficiently dense, there is very little strength drop as a result of increased pore pressures. In short, what is a threshold ψ at which the discussion needs to change to ground improvement (or comparable remediation) rather than consideration of acceptable displacements? There are three levels to answering this question: (1) aids to judgement directly from laboratory test data, (2) normalized CPT charts capturing case history trends and (3) in-situ and laboratory testing to support detailed numerical simulations using *NorSand* or a comparable model in project-specific studies.

6.11.1 Basis for judgement from laboratory data

In terms of developing engineering judgement, a few observations can be made from the laboratory data presented in this chapter:

- Strength drop is caused by pore pressures generated by volumetric compression. If the sand is dense enough that shear dilation dominates volumetric compression, it will not lose strength. This occurs at about $\psi < -0.08$ in isotropic triaxial compression and at about $\psi < -0.05$ in simple shear tests (where the starting geostatic stress state is unknown, but almost certainly $K_o < 1.0$).
- Undrained stress paths for sands tested in CIU triaxial compression that are denser than about $\psi < -0.08$ do not show a phase transformation or pseudo–steady state. They dilate continuously under monotonic loading.
- A lower bound trend to the undrained strength data in Figure 6.22 would extrapolate to negative states and intersect the drained shear strength ratio ($\tan \phi'$ or $M/2$) at around $\psi = -0.06$.

The laboratory tests showing these trends are nearly all from isotropic initial conditions so that the entire stress path is undrained. In-situ, it would be very unusual to encounter $K_o = 1$ with loose soils with the general expectation being in the range $0.5 < K_o < 0.8$. This range of in-situ K_o translates to perhaps only two-thirds of a potential loading path being undrained, and thus less excess pore water pressure at the onset of instability: looser states than the laboratory tests will generally be stable. A judgement criterion of $\psi < -0.05$ from laboratory testing should be conservative in practice. Let us now see how well this compares with CPTs and the case history record.

6.11.2 CPT charts and case history trends

The calibration chamber is the basis for evaluating CPT data in sands, but poses a challenge in how to quantify the effects of soil properties between data from one sand to another. Chapter 4 considered this issue and presented the results from detailed cavity expansion analyses by Shuttle and Jefferies (1998) showing the effects of sand compressibility and critical friction on the normalized penetration resistance. The situation is more difficult with silts as there are no calibration chamber studies in silt. However, cavity expansion is used as an analogy for the CPT regardless of whether penetration is drained or undrained. Noting this, Shuttle and Cunning (2007) used measured parameters and cavity expansion modelling with *NorSand* to show what we should expect from silts with the CPT. Their paper produced an interesting exchange with Robertson (2008) commenting on their work, and Shuttle and Cunning (2008) in reply suggested that there was indeed a common view of the threshold between large-scale movements or limiting softening that can be expressed in a normalized chart for plotting CPT data. Robertson (2010) further expanded on the case history data supporting such a view.

Before comparing these approaches, the key differences should be noted. Shuttle and Cunning use a normalized penetration resistance that includes the measured excess pore water pressure in the parameter grouping $Q_p\,(1 - B_q) + 1$, while Robertson uses the reference stress approach and an fines content correction in the parameter $Q_{tn,cs}$ (i.e. what would be measured in a clean sand at 1 tsf, 1 atmosphere or 100 kPa). The two approaches can best be compared then for clean sands (i.e. with F less than 1.5%) and when the vertical stress is about 100 kPa. In that case, the pore pressure is approximately zero so that $B_q = 0$, and there are essentially no 'corrections' in the Robertson approach.

Regardless of background, all workers agree, and all the data support, that the upper part of a Q–F plot (or related variant) represents soils that may lose stiffness during cyclic loading but which never reach the condition of the undrained strength being less than the drained. These soils have become labelled as '*dilatant*' as a catch-all name. Conversely, all agree, and the data support, that the lower part of the diagram represents soils where liquefaction (however caused) will result in the short-term undrained strength being less than the drained strength. These soils have become labelled as *contractive*. The question is the location of the boundary between these two very different classes of soil behaviour, and how that boundary varies by soil type.

Shuttle and Cunning (2008) took the *contractive/dilatant* boundary at large scale to correspond to $\psi < -0.05$, which is the state parameter criterion that emerges from laboratory tests (Section 6.11.1) and which appears to be the limiting situation for various flowslide case histories in sands and silts (Figure 6.58) when assessed in a state parameter context. In developing their trend line, Shuttle and Cunning in essence drew a smooth curve between calibrations in sand (Erksak, Ticino) and in a very weak silt (Rose Creek).

Robertson (2010) considered largely the same flowslide case histories that have been discussed in this chapter, albeit taking a different view on what comprised the best estimate of the representative values and concluded that $Q_{tn,cs} = 70$ would be a tolerably conservative representation of the *contractive/dilatant* boundary above which flowslides have never been observed ($Q_{tn,cs} = 50$ might be better fit to the data ranked as reliable).

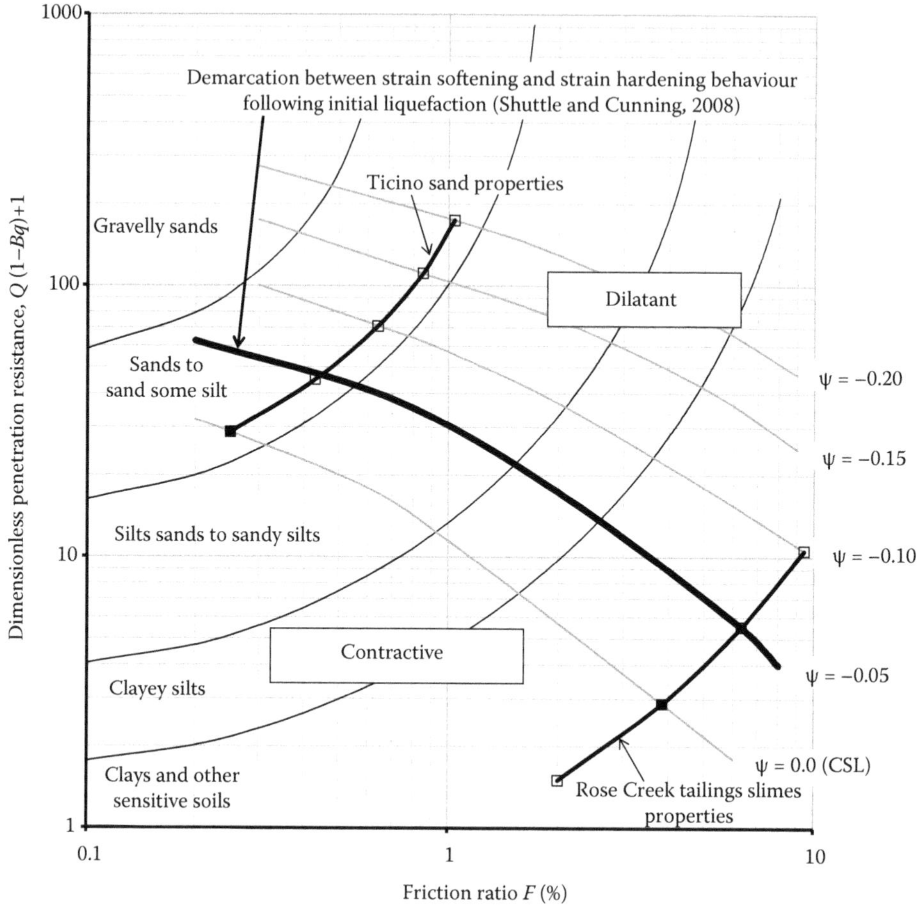

Figure 6.58 CPT screening tool: dimensionless penetration resistance versus friction ratio showing approximate 'flowslide demarcation' for sands and silts. (Adapted from Shuttle, D.A. and Cunning, J., *Can. Geotech. J.*, 45(1), 142, 2008.)

Figure 6.59 compares the Shuttle and Cunning and Robertson criteria for the location of the *dilatant/contractive* boundary. Despite the very different backgrounds and approaches of these authors, there is a considerable commonalty in the end result as it would affect a practical engineer, and they do coincide where expected for clean sands.

6.11.3 Project-specific studies

Figure 6.59 summarizes the current knowledge of what will be 'adequately dense' at what might be viewed as a 'screening level'. Case history information and laboratory trends have been included, and the CPT captures the in-situ state. But soil properties such as G_{max} and λ_{10} are included only to the extent that they are represented by the soil behaviour type index. The location of the *contractive/dilatant* boundary can be refined using project-specific tests and calculations that fully capture the details of these factors. This has been done for a few 'high-value' projects where the scale of the works is sufficient to warrant the engineering effort.

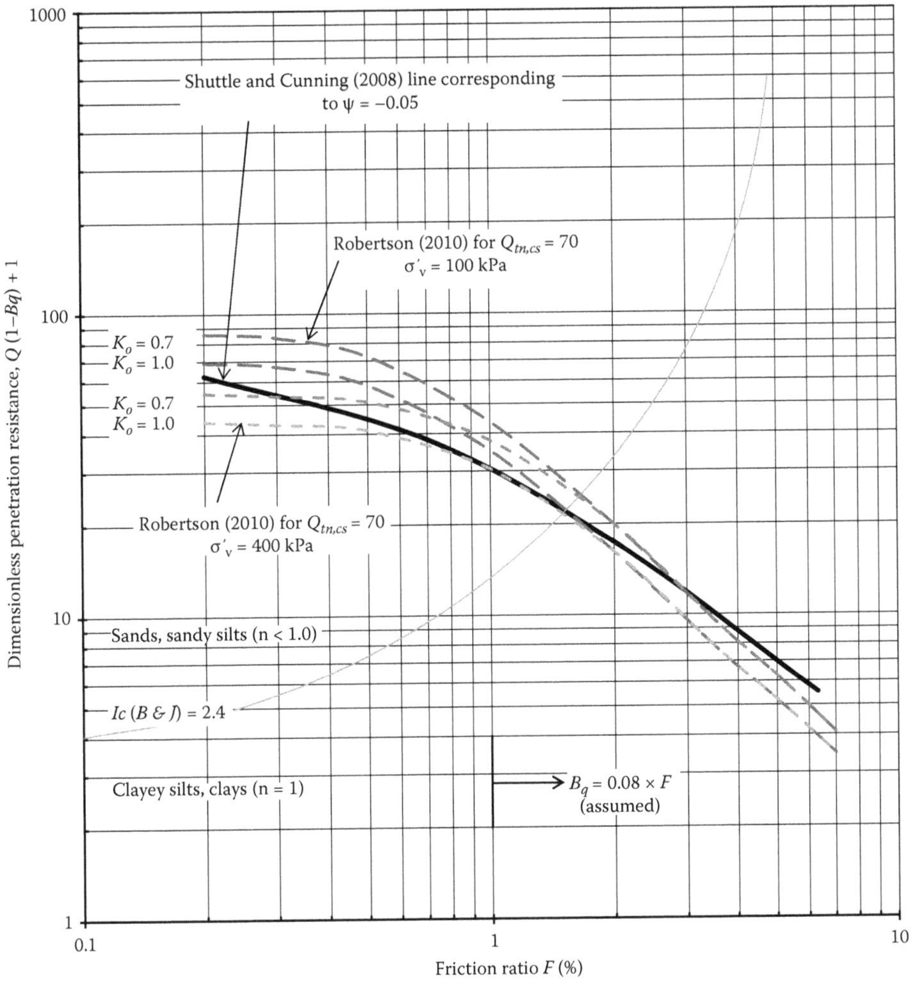

Figure 6.59 Comparison between Shuttle and Cunning (2008) and Robertson (2010) boundaries between satisfactory and unsatisfactory undrained performance of sands in terms of CPT penetration resistance.

It is possible to use *NorSand* (or similar state parameter–based model) in finite element analyses to assess the likelihood of a flowslide developing for a given project, and this has been done for at least three large dams to our knowledge. However, civil engineers are conservative people, and the difficulty becomes one of 'model uncertainty', or to what extent you trust the calculations and the data the calculations are based on.

In terms of model uncertainty, only the Nerlerk case history has a reasonably comprehensive set of CPT and supporting triaxial data, but the knowledge of slide geometry is quite poor. Lower Sand Fernando Dam has reasonable CPT data and reasonable knowledge of geometry, but the supporting laboratory testing is inadequate to assess the basic properties of the sands and silts involved. The remaining case histories are markedly inferior to both of these regarding data quality. So, even if we have perfection in the finite element work, how can we validate the analysis?

Broadly, until our profession as a whole does a better job of documenting and testing those case histories that do arise, we are not going to make progress beyond $\psi > -0.05$ as the criterion for flowslides. When the soil is looser, the discussion with the owner should be about 'ground improvement'. When the soil is denser, the discussion can be about 'tolerable movements'.

6.12 POST-LIQUEFACTION RESIDUAL STRENGTH

If the 'how dense' assessment leads to a conclusion that the site is *contractive*, then the next question becomes whether or not the residual (post-liquefaction) strength is sufficient. Dams in particular may be sufficiently movement tolerant to allow residual strengths to become the basis of design. This often arises in tailings engineering, which differs from the much of the case history record in two important aspects: (1) the materials are typically silts, sandy silts or silty sands; and (2) it is practically given that hydraulic deposition of tailings or thickened tailings will result in an in-situ state that is susceptible to liquefaction under seismic loads (and the mines producing these tailings always seem to be in earthquake-prone areas). Many tailings impoundments seem to result in stability being assessed on a post-earthquake basis using residual strength, but it is not just tailings dams. The 'high-consequence' Duncan Dam, part of the Columbia River Treaty works straddling the U.S./Canada border, was explicitly assessed as 'safe' using residual strengths, and this basis was accepted by the regulatory authorities. The issue for the practical engineer is what residual strength to use.

6.12.1 Residual strengths guided by case histories and penetration resistance

Earlier in this chapter, we considered strengths derived from the case history record as indexed by penetration tests. The two approaches that emerge to synthesize this case history data into trends for engineering use are (1) penetration resistance that has been adjusted using the concept of 'equivalent clean sand' and (2) the state parameter.

The 'equivalent clean sand' characterization of soil has long been advocated by workers familiar with the simplified Seed method for seismic liquefaction triggering. Roberson (2010) is the current best practice version of the method (Section 6.9.4). Although this 'equivalent clean sand' characterization is popular with the U.S. regulators, especially for the effect of earthquakes, any reasonable engineer ought to baulk at the application to sandy silts and silts. The adjustments for fines content are strictly speculative and are not substantiated by laboratory tests of soil strength and compressibility.

State parameter–based strengths (Section 6.9.5) are built on a line of development in soil mechanics that goes back more than 70 years but which appears to have been contaminated by the poor performance of the 'steady-state' school at Lower San Fernando Dam (Section 6.10). Hopefully, we have adequately documented and explained the various misconceptions involved with Lower San Fernando Dam, and the state parameter method fits that case history rather nicely. The particular advantages of the state parameter approach are that it is consistent applied mechanics, it captures the entire stress–strain behaviour, not just 'strength' and it characterizes soil using standard, widely used properties.

Clearly, we have a preference for the state parameter approach. However, both approaches have been implemented in the downloadable CPT processing spreadsheet *CPT_plot.xls* as user-selectable alternatives. The judgement on what to use is yours; the basis for such judgements is what we have set out in this chapter. Do you prefer *Engineering Geology*, or do you prefer *Soil Mechanics*?

6.12.2 Residual strengths by numerics

We have shown that *NorSand* captures all aspects of static liquefaction at laboratory scale, whether in triaxial compression, simple shear or triaxial extension. The various spreadsheets are downloadable, and this view can be validated against the test data. However, in considering the case history record, the evidence is that undrained conditions of the laboratory cannot be maintained at field scale. Thus, examination of the case history data resulted in the working hypothesis that the post-liquefaction residual strength corresponds to the undrained shear strength at which localization occurs, which in *NorSand* is predicted when Equation 6.20 occurs. The idea is expanded upon in Figure 6.60 in which the state path, stress–strain curve and p'–q stress path for a lightly dilatant material are shown. The state path shows that a lightly dilatant material (starting at a state close to the critical state line, indicated by the black square) will initially follow the undrained constant void ratio state path to the black triangle. The state path moves to the left as a result of increasing pore water pressures causing decreasing mean effective stress (p'). At the low point in the stress–strain curve (i.e. at the black triangle), localization occurs. A narrow shear zone is formed, and the material in the shear zone is able to dilate by flow of water from the surrounding material into the shear zone. Within the shear zone, the undrained or constant volume condition no longer holds true. The state path moves upward towards the critical state at the black circle in Figure 6.60.

The state path in Figure 6.60 implies that the material in question will reach the critical state at a lower mean effective stress (and higher void ratio) than if it were undrained. We now have a method to estimate the residual strength following strain localization or a theoretical basis on which to compute the post-liquefaction residual shear strength of sands and silts. The *NorSandSS.xls* spreadsheet has been programmed so that the localization strength is automatically captured and reported for the simulation of any parameter set; the hypothesis is that it is this strength that captures the full-scale experience. The localization strength depends not only on the state parameter, but the complete suite of soil properties and the geostatic stress ratio as well.

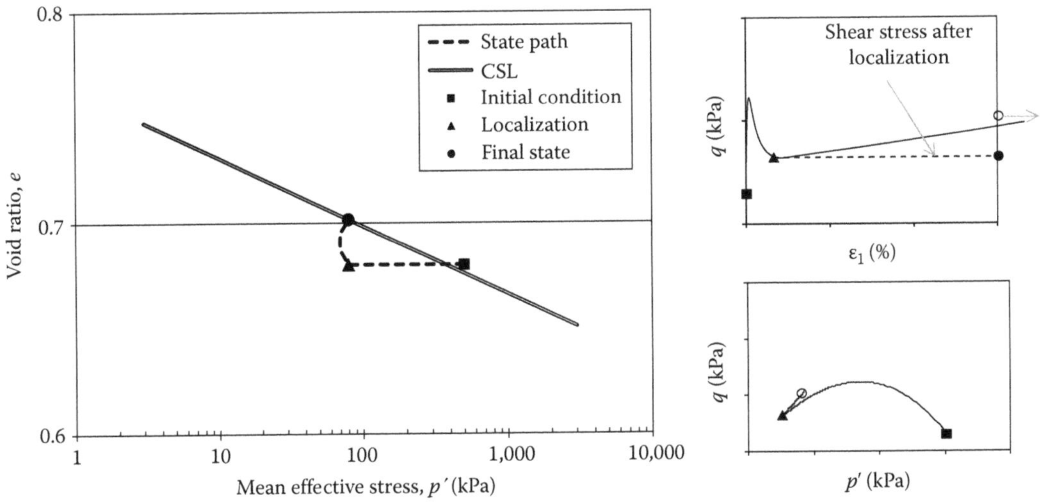

Figure 6.60 Possible state path for a lightly dilatant material in undrained loading. Localization occurs at the black triangle and the shear zone dilates to the critical state at the black circle.

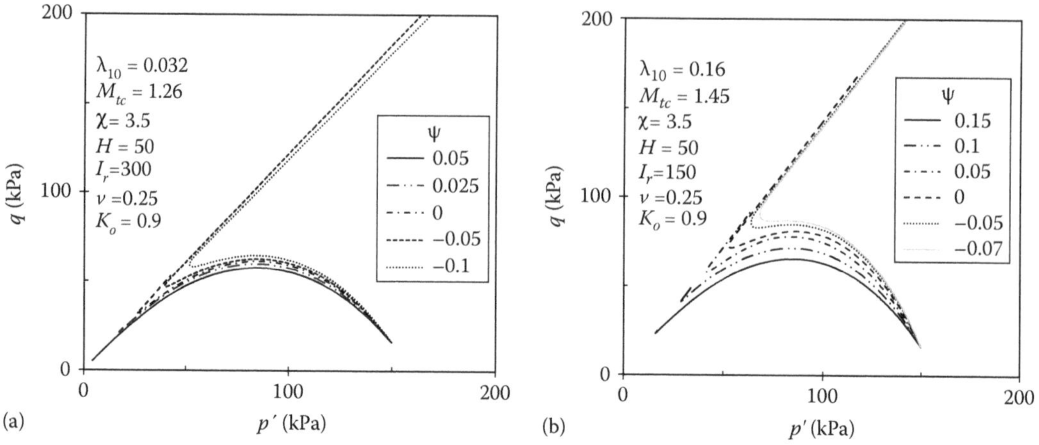

Figure 6.61 *NorSand* simulations of undrained stress paths for (a) clean sand and (b) sandy silt tailings.

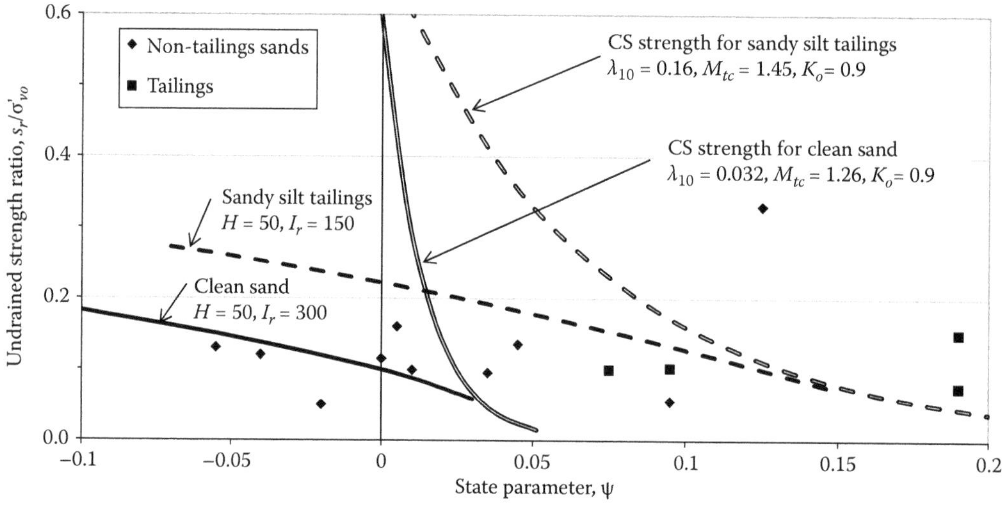

Figure 6.62 Minimum undrained shear strength for clean sand and for sandy silt tailings from *NorSand* simulations in Figure 6.61.

Some *NorSand* simulations for two sets of parameters, representing a clean sand and a sandy silt tailings, are shown in Figure 6.61 in terms of the p'–q stress paths. In reality, only three parameters differ between the sets of simulations: λ, M_{tx} and the elastic stiffness I_r (which of course changes the ratio of H/I_r). The hooks or low points on the stress paths are when localization occurs and the undrained shear strength at these points is used to compute a residual strength ratio, which is in turn plotted in Figure 6.62. Unsurprisingly, the residual strength at localization is different for clean and silty sands, and Figure 6.62 provides the needed insight into the empirically observed difference between post-liquefaction strength of clean sands and silts. Note that the presentation in Figure 6.62 is entirely consistent with the back-calculated residual strengths in Table 6.5 and Figure 6.47, and the approach uses engineering parameters that can be measured in the laboratory to estimate the residual strength for any particular material.

A word of caution about this approach is still needed. We started off calling it a 'working hypothesis' that the post-liquefaction residual strength observed from case histories is the result of localization so that behaviour is no longer undrained locally. While the argument is strong, it has certainly not been proven, and it is unlikely that we will be able to observe this mechanism in full-scale events, although detailed CPT soundings through a slumped mass of a future case history may be helpful. Practical geotechnical engineering could beneficially use further theoretical development to support what appear to be consistent trends at field scale.

6.13 LIQUEFACTION ASSESSMENT FOR SILTS

We have touched on the behaviour of silts a few times in this chapter, and silts are certainly important in practice – especially with the mining industry. Metal mining creates a lot of ground rock (tailings), and the most common method of tailings disposal used to be slurry deposition into an engineered containment (depending on site topography, variously valleys closed by dams or facilities with perimeter containment dykes). Tailings management is now changing. In the last decade or two, there has been a move away from slurry deposition towards creating non-segregating tailings, or 'paste', which can then be deposited on a slope with minimal containment other than a starter dyke. This means that stability of the tailings itself becomes an engineering issue.

Tailings are largely silts, and many of the big mining projects using paste tailings are situated in high seismicity parts of the world. Post-earthquake flowslides are an obvious concern, but static liquefaction also needs to be considered for some of these extraordinarily large schemes.

The approach put forward throughout this book is to look at measured soil properties, constitutive theory and evaluation of the case history record, with soil liquefaction just another facet of soil behaviour. Silts are just another soil, and tailings are usually ground-up rock without any clay minerals so can be considered as very fine sands. Earlier in this chapter, we showed how *NorSand* was perfectly able to reproduce the liquefaction behaviour of silt tailings from the Rose Creek and Guindon impoundments. We also showed that the liquefaction instability line and residual strength after liquefaction (or rather the strength at which shear localization occurs, which we suggested is the appropriate residual strength to consider) were soil behaviours that depend on three key properties in *NorSand*. These are the slope of the critical state line (λ), the shear modulus or elastic rigidity ($I_r = G/p'_o$) and plastic hardening parameter (H). The state parameter approach works very nicely for silts at laboratory test scale, but that leaves the issue of scaling from laboratory to full-scale construction.

The case history record (Appendix F) includes a few tailings slopes, the two 'upstream raised' Mochikoshi failures in particular. If anything, back-analyses of these two failures suggest stronger residual strengths than the trend in the rest of the case history record. This is not entirely surprising given the working hypothesis that it is local pore water migration that is the cause of deviation between undrained laboratory tests and full-scale experience. As the soil becomes more silty, the diffusion coefficient (i.e. c_v and/or c_h) controlling pore water movement becomes exponentially smaller, which means less pore water migration even at the local scale.

The upshot of this consideration comes back to the importance of accurate determination of the in-situ state parameter, which takes us back to CPT testing, the subject explored in Chapter 4. Of course, the difficulty is that so much of the CPT depends on the calibration chamber, and there are very little such data for silts (in fact seemingly none with the exception of limited small-scale work by Baxter et al. (2010), but that study is devoid of soil property measurements). We can use numerical simulation of the CPT (using *NorSand* or similar), but there is some uncertainty in such efforts because cavity expansion is an imperfect analogue for CPT resistance. Adding in measured excess pore pressure data through B_q helps constrain the uncertainty. What is also interesting is the idea that the shear stress on the CPT friction sleeve could be a measure of s_r – a plausible approach because soil immediately adjacent to the CPT has been sheared to the critical state (theoretically, at least).

Overall, assessing s_r in tailings requires working with all the aspects discussed and then developing a judgement by weighting the various factors – the familiar and pervasive situation of geotechnical engineering.

What about well-graded silts and other non-tailings silts? In Chapter 2, we looked at critical state lines and dilatancy. Increasing the silt content of a sand tends to increase the slope of the critical state line, but only in a very broad sense. In fact, we showed that for two silty sands (with ~34% silt content), it was whether the material was uniformly or well graded that mattered, not the amount of 'fines' in itself. Overall, there really is not much difference between the critical state line and stress–dilatancy behaviour of silts compared to sands. The numerical values of the properties change a bit, and that is it.

A reasonable approach for silts in general is to adopt the best practice trends of Figure 6.47 (and the associated algorithm, listed in Figure 6.48) and focus on assessing the in-situ state parameter from the CPT. For silts that are relatively incompressible, CPT penetration may be close to drained (easily determined as $B_q \sim 0$), and in those situations, the CPT – ψ relationship is well constrained as per Chapter 4. When measured $B_q > 0.1$, revert to treating the silt as similar to tailings as just discussed. When $B_q < 0$, you do not have a problem as the soil is self-evidently substantially dilatant.

6.14 SUMMARY

This chapter has made the following points:

- Looked at laboratory tests on undrained samples of loose, lightly dilatant and compact sands and shown how density affects undrained behaviour, as well as the differences between triaxial compression, extension and simple shear.
- Run *NorSand* simulations with undrained boundary conditions and neatly captured undrained liquefaction behaviour using calibrations from dense drained tests.
- Used the insight from *NorSand* to explain many of the confusing interpretations of test data and observations related to liquefaction reported in the literature.
- Described the steady-state approach to liquefaction assessment and pointed out its theoretical limitations.
- Looked through most of the full-scale case histories of experience with static liquefaction with a reasonable volume and quality of data.
- Shown that this case history experience can be computed rationally with *NorSand.*
- Included the liquefaction behaviour of silts in the discussion to show that silts can be treated just the same as sands without massive 'fines corrections'.
- Pointed out that there is much more to be done to understand the implications of strain localization (and Drucker's stability postulate) and how it impacts the behaviour of lightly contractive or lightly dilatant soils.

Finally, the similarity in some regards to the steady-state school is noted, but what is seen now is that while the critical state framework is indeed reliable and something that allows understanding of static liquefaction in practical situations, the simplicity of the steady-state school is an illusion. Rather it is essential to consider the range of in-situ conditions, the actual soil properties, the likely effects of drainage and the corresponding material behaviour. What has been provided allows this to be done (and you can download the spreadsheets rather than program it all from scratch).

Chapter 7

Cyclic stress–induced liquefaction (cyclic mobility and softening)

7.1 INTRODUCTION

7.1.1 Cyclic mobility

The last chapter explored static liquefaction in which large undrained strength reductions can be caused by an increase in pore water pressure. Although static liquefaction can be very dramatic, it is rather different from the failures during the Niigata and Alaskan earthquakes of 1964 that brought earthquake-induced liquefaction to the forefront of geotechnical engineering. The key feature of earthquakes is, somewhat obviously, the ground shakes. Shaking varies loads and stresses cyclically, and it is this cyclic action that can cause liquefaction.

In one respect, static liquefaction and cyclic liquefaction are caused by the same condition – there is a plastic volumetric strain that arises sufficiently quickly that the pore fluid cannot escape as fast as the plastic strain accumulates. This leads to increasing excess pore pressure, a reduction in mean effective stress and a corresponding reduction in shear stiffness and strength. The difference between static- and cyclic-induced liquefaction is the way in which plastic volumetric strains are generated. In the case of static liquefaction, a necessary condition is that the soil be loose so that plastic volumetric strain through the usual stress–dilatancy response is greater than the corresponding work hardening of the skeleton to support the increased stress. Any soil loose of the critical state (i.e. $\psi > 0$) may show static liquefaction to a greater or lesser extent. In the case of cyclic-induced liquefaction, the plastic volumetric strains arise through densification brought on by the cyclic stress changes that tend to pack the soil particles closer together. Cyclic-induced densification affects any soil, including dense sands and overconsolidated clays.

Where cyclic liquefaction and static liquefaction differ in that, for dense soils, cyclic liquefaction will be strain limited. Once things start moving, the shear-related dilation of dense soils rolls in and acts to offset the densification-induced excess pore pressure. Cyclic liquefaction then tends to be a softening of the ground in the case of all dense soils rather than the outright brittle (runaway) collapse that can arise with loose soils. This difference in the consequence of liquefaction created quite a furore in the literature, echoes of which still exist today.

There were really two issues that confused the understanding of liquefaction. First, many of the early workers in earthquake-induced liquefaction took a geologic rather than a mechanics perspective. These workers focused on the case history record. They tended to classify things rather than look to physics, and indeed some were dismissive of mechanics. To this group of workers, whether the displacements were large or small was a key observation that related to the phenomena. Second, there was a general lack of consideration of dilation. This resulted in a fixation on liquefaction being a zero effective stress condition, which in reality can only arise for a brief transient in a cyclically loaded dense sample (if even then).

The jargon that became accepted is that brittle collapse–type failures were called liquefaction, whereas those involving cyclic softening without runaway strains were called cyclic mobility. Note that despite the apparently limited strains, cyclic mobility is not benign; it has caused billions of dollars of damage as illustrated in Chapter 1.

In static liquefaction, it is the loosest soils that create the greatest excess pore pressures, and drainage will improve their strength. This means that the overall behaviour can usually be assessed by thinking in terms of undrained strengths without having to calculate the effects of pore water migration as the excess pore pressures dissipate. Cyclic mobility is rather different. In cyclic mobility, the zone of maximum excess pore pressure generation may not be the loosest soil but rather the soil that was in the most stressed location. As excess pore water migrates during dissipation, it may cause strength or stiffness reductions elsewhere and lead to delayed failure. (The Lower San Fernando Dam is a case in point. It did not start slipping until some 30 s after the earthquake shaking stopped.) Cyclic mobility ought to be viewed as something happening to the whole domain and not viewed as strength or stiffness of various soil elements. This is properly a boundary value problem requiring a fully coupled stress analysis. However, it is exceedingly rare to see such an approach outside a few leading universities. The state of engineering practice is much simpler, as will be seen in a moment, but this then should lead to considerable caution in extrapolating experience.

Another aspect of cyclic mobility is that it has been dominated in the literature by earthquake hazard mitigation concerns. To some extent, this is a consequence of the earthquake hazards reduction programme of the U.S. National Science Foundation that put much money into earthquake-related research. However, the fact remains that cyclic mobility, and the less dramatic but related cyclic softening, can arise in other situations. Machine-induced vibrations are almost self-evident with a good example being the collapse of a road by cyclic liquefaction induced by vibroseis trucks (Figure 1.23). Wave-induced cyclic mobility or softening arises in the context of piled and gravity-based offshore structures. Storms can cause rather large cyclic loads, and corresponding strains have been an issue for both near shore harbour works and offshore platforms. Ice is perhaps a surprise in this context, but ice moving past a structure (bridge pier or offshore platform) often forces substantial vibrations in the structure.

These non-earthquake loadings tend to involve many cycles of loading. Whereas an earthquake may last in the order of a minute or so, with a few to barely 15 *significant* load cycles (although there are many smaller cycles), these other load types usually have several hundred to thousands of cycles of rather similar amplitude. Soil shows fatigue-like behaviour in that much the same result is obtained with many low amplitude cycles as a few large cycles. Figure 7.1 shows some drained cyclic shear data illustrating the induced settlement of a sample of dense sand when loaded to 10,000 cycles. It is evident that there is a trend of soil behaviour with the number of cycles imposed, and it is helpful to think of the whole spectrum of high- and low-cycle response when understanding cyclic mobility/softening. This avoids artificial distinctions about the type of cyclic loading. One framework should suffice for all soils whether loose or dense sands and normally or overconsolidated clays.

It is possible to understand the way in which soils respond to load cycling in terms of fatigue using some simple mathematical idealizations of consistent physics, an approach that was used by van Eekelen (1977). The first-order expectation is that cyclic behaviour should be linearly related to the logarithm of the number of cycles N. The basis of this approximation is that it is a plausible function midway between the two mathematical extreme idealizations (proportional or infinite) where the fatigue approaches become inconsistent. From inspection of Figure 7.1 it is seen that this $\log(N)$ view is a pretty good first approximation to the cyclic densification of the soil.

A word of caution on the loading type is needed at this point. One of the features of cyclic loading in many civil engineering situations is that it is usually associated with repetitions

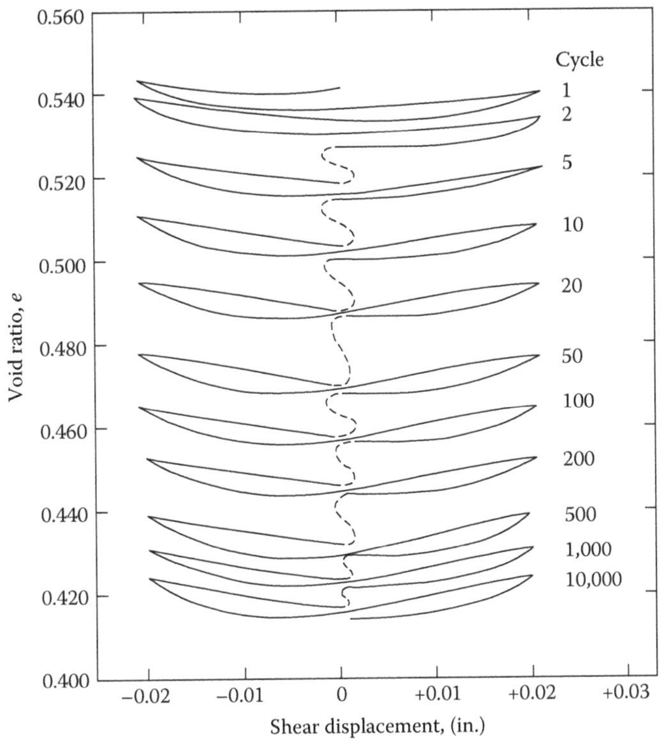

Figure 7.1 Void ratio reduction induced by cyclic shear. (After Youd, T.L., *ASCE J. Soil Mech. Found.*, 98(SM7), 709, 1972. With permission from the ASCE, Reston, VA.)

not that far away from the fundamental period of the structure of interest (typically in the order of 1 s). This means that there may be amplification or attenuation of the cyclic loads because of inertial effects. Calculation of stresses allowing for inertial effects is a subject in itself. What is presented here is the stress–strain behaviour.

This chapter explores a range of issues with cyclic loading of soils. Laboratory data will be examined so see how soils behave under cyclic loading. We start with cyclic triaxial test data as that was the historical basis for understanding soil behaviour under cyclic loading. It turns out, however, that cyclic simple shear tests are a better analogue for earthquake and many other cyclic loading problems. Developments over the last 20–30 years in laboratory testing equipment and computer logging and control have brought simple shear testing within reach of mainstream geotechnical engineering, and our primary laboratory database for this chapter is the cyclic simple shear testing on Fraser River sand (FRS). As with static liquefaction, full-scale field data are the gold standard for engineering and the Berkeley methodology based on case histories is described. Once this case history-based data has been presented, applied mechanics then links the laboratory and field data through the state parameter approach. This leads to a complete framework that is consistent with laboratory and full-scale experience.

7.1.2 Alternative forms of cyclic loading

In introducing cyclic mobility, it has been assumed so far that *cyclic* has a precise meaning. This is actually a huge step and there are three different cyclic loadings that must be distinguished: cyclic variation in the imposed deviator stress (or the stress ratio η), principal

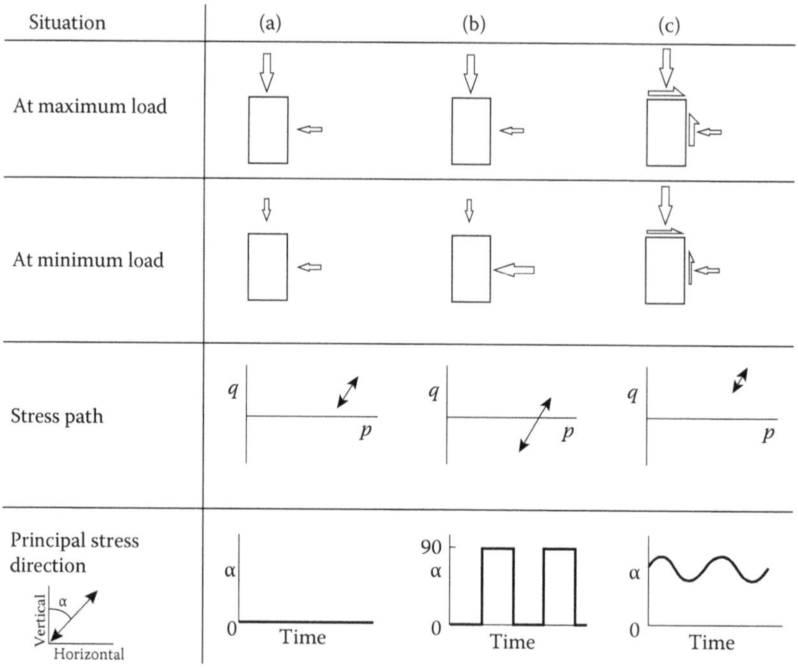

Figure 7.2 Schematic illustration of the different forms of cyclic loading. (a) One way (repeated) loading, (b) two way (cyclic) loading and (c) cyclic principal stress rotation.

stress flips (jumps) and principal stress rotation. All three types of cyclic loading arise both in-situ and in the laboratory. The difference and similarities are illustrated in Figure 7.2 which shows elements of soil in each condition, examples of the imposed stress paths and the variation in the direction of major principal stress with time. For simplicity, this figure is for plane strain in which the intermediate principal stress is out of plane and remains fixed in magnitude and direction. Similarly, the initial vertical stress is taken to be the major principal stress σ_1, and the vertical direction is taken to be the frame of reference defining the angle α that measures the orientation of σ_1.

The first case is illustrated as Figure 7.2a and is often referred to as repeated loading. In this situation, the stresses fluctuate and the stress path oscillates in q–p space without ever crossing the $q=0$ axis. The direction of σ_1 remains fixed (taken to be vertical throughout time here). Conceptually, this is a triaxial compression test with variable deviator stress (although variable mean stress or some combination of mean and deviator stress is also an admissible example).

The second case, Figure 7.2b, also keeps the orientation of the stresses fixed, but now the variation in magnitude is such that for part of the load cycle it is the vertical stress that is the major principal stress and for part of the load cycle it is the horizontal stress that is the major principal stress. This form of loading produces jumps of 90° in the direction of σ_1 as the loading varies. The time history of principal stress direction is like a square wave (although not necessarily symmetric) switching between 0° and 90°. Conceptually, this is the cyclic triaxial test with repeated excursions between triaxial compression and triaxial extension.

The third case, Figure 7.2c, has both vertical and horizontal stress as earlier, but also a shear stress. It is this shear stress that is the principal variation of the loading with time. Because of shear, σ_1 is no longer aligned with the reference vertical direction, and as the shear stress varies with time, so do the directions of σ_1 and σ_3, that is the principal stresses rotate. An example of this situation is the cyclic simple shear test, and it is also a close analogy to the vertical propagation of shear waves through the soil in an earthquake. In plane strain, principal stress rotation is restricted to one plane. In general, however, principal stress rotation may arise around all three axes.

Cyclic stresses are imposed on an initial condition. Current terminology tacitly takes a particular plane as *special* with the shear stress state in that initial condition referred to as the *static bias* characterized by the ratio τ_{st}/σ'_{vo} and where τ_{st} acts on the *special* plane. Similarly, the cyclic component is characterized by a representative amplitude τ_{cyc}/σ'_{vo} (what is meant by representative will be dealt with later) with τ_{cyc} acting on the same *special* plane as τ_{st}. These normalized stress measures have their origin in triaxial testing in which situation the *special* plane is the plane of *maximum shear stress* (i.e. 45 inclination in the sample), and the difference between the situations shown in Figure 7.2a and b is the ratio between these two terms. If $\tau_{cyc}/\sigma'_{vo} < \tau_{st}/\sigma'_{vo}$ then it is situation (a) and repetitive loading. If it is the opposite, then it is situation (b) and a case of cyclic loading. The relative proportions of these two parameters indicate the differing intensity of the cyclic loading, although the principal stress always flips through 90° in the cyclic loading case.

Static bias is rather meaningless in the case of more general loading conditions of Figure 7.2c, which has a horizontal *special* plane, as here the effect of static bias is merely to incline the initial direction of σ_1. The variation in this direction is entirely attributable to the variation in the applied shear, although there is usually a variation in the stress ratio η as well. It is also possible to contrive a laboratory test in which the deviator and mean stresses are kept constant while the principal stress direction rotates (and, yes, soil can be liquefied by such a path). However, despite its slightly misleading notion, current notation for the cyclic simple shear test treats the *horizontal* plane as the *special* situation.

It is difficult to underestimate the importance of the difference between periodic (cyclic) variation in the shear stress ratio η and the periodic variation in principal stress direction. Most of the literature on earthquake-induced liquefaction, at least in North America, has ignored principal stress rotation despite it being readily shown in the laboratory to be the principal (forgive the pun) mechanism causing cyclic mobility. Cyclic variation in stress magnitude η is not unimportant; however, it is just a lesser effect in terms of soil behaviour. Cyclic test laboratory data illustrate the relative importance of stress ratio and principal stress direction.

7.2 EXPERIMENTAL DATA

7.2.1 Laboratory cyclic test methods

A variety of laboratory tests have been used to investigate the cyclic loading of soils, the most important of which are triaxial, simple shear and torsional shear (or hollow cylinder) testing (Figure 7.3). All laboratory tests are directed at testing an element of soil, by which it is meant that the stress and deformation conditions are uniform so that the constitutive behaviour can be derived directly from the measurements made.

Cyclic triaxial tests have been most commonly used to characterize cyclic behaviour of sands since the early work of Seed and Lee (1966). However, Seed recognized early on that triaxial tests do not in general duplicate in-situ stress conditions very well and he focused more on simple shear. In the simple shear apparatus, the vertical normal stress and shear stress on the horizontal plane are controlled and a zero lateral strain condition is imposed, which approximates conditions in the ground during earthquake loading. As discussed in Section 2.7, there are problems with complementary shear stresses on the edges of simple

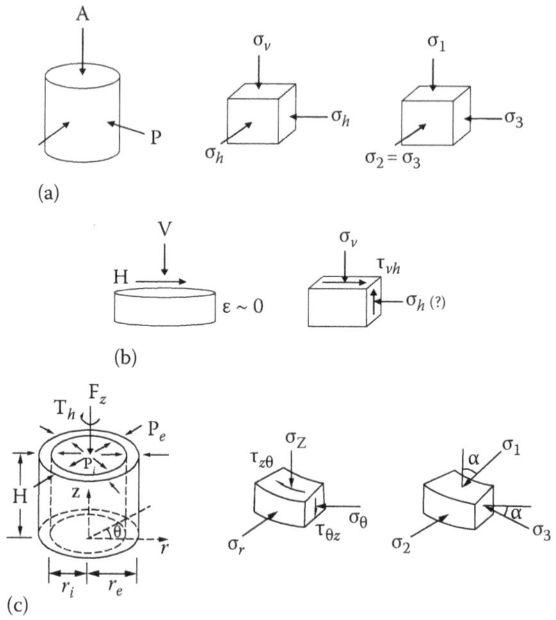

Figure 7.3 Stress conditions in (a) triaxial, (b) simple shear and (c) hollow cylinder tests.

shear samples as well as uniformity of stress conditions within the sample. In addition, the horizontal stress is not measured in the cyclic simple shear test, making calibration of constitutive models to the measured results problematic.

The shaking table (De Alba et al., 1976) is a variation of the simple shear device where the influence of the sample vertical boundaries is minimized by the very large plan area of the sample (2.2 m × 1.1 m for a 100 mm thick sample). Only the central portion of the sample where uniform conditions are expected to exist is instrumented and considered in the analysis of the test data. The cyclic loading is applied by a strain-controlled oscillation of the apparatus rather than by stress control. In this way, multidirectional shaking is applied that is more representative of real earthquake conditions.

Arthur and co-workers showed in two seminal papers (Arthur et al., 1979, 1980) the importance of changes in principal stress direction to soil behaviour. Their contribution was an important step forward as, until then, it was regarded as sufficient to work with stress invariants to study the constitutive behaviour of soil (similar to the approach with metals). Demonstrating the importance of principal stress rotation requires a different type of test from those discussed so far. Although principal stresses rotate in the simple shear test, this test is inadequate for investigating constitutive behaviour because the horizontal stress is not measured which leads to large ambiguities in what the test data actually mean. Most workers now use the hollow cylinder test to investigate the effects of changes in principal stress direction, Figure 7.4 showing an example of the hollow cylinder apparatus.

The hollow cylinder test is much like a triaxial test except that, as the test name suggests, the sample is hollow and torsion is applied to the top and bottom of the sample as well as more usual axial load. When the internal and external cell pressures are different, the hollow cylindrical sample is subject to hoop or tangential stresses. This combination of radial, axial and tangential stresses, combined with torsional shear, gives the hollow cylinder test a lot of versatility. For example, cyclic tests have been carried out by maintaining a constant deviator stress and continuously rotating principal stress directions. More commonly, the

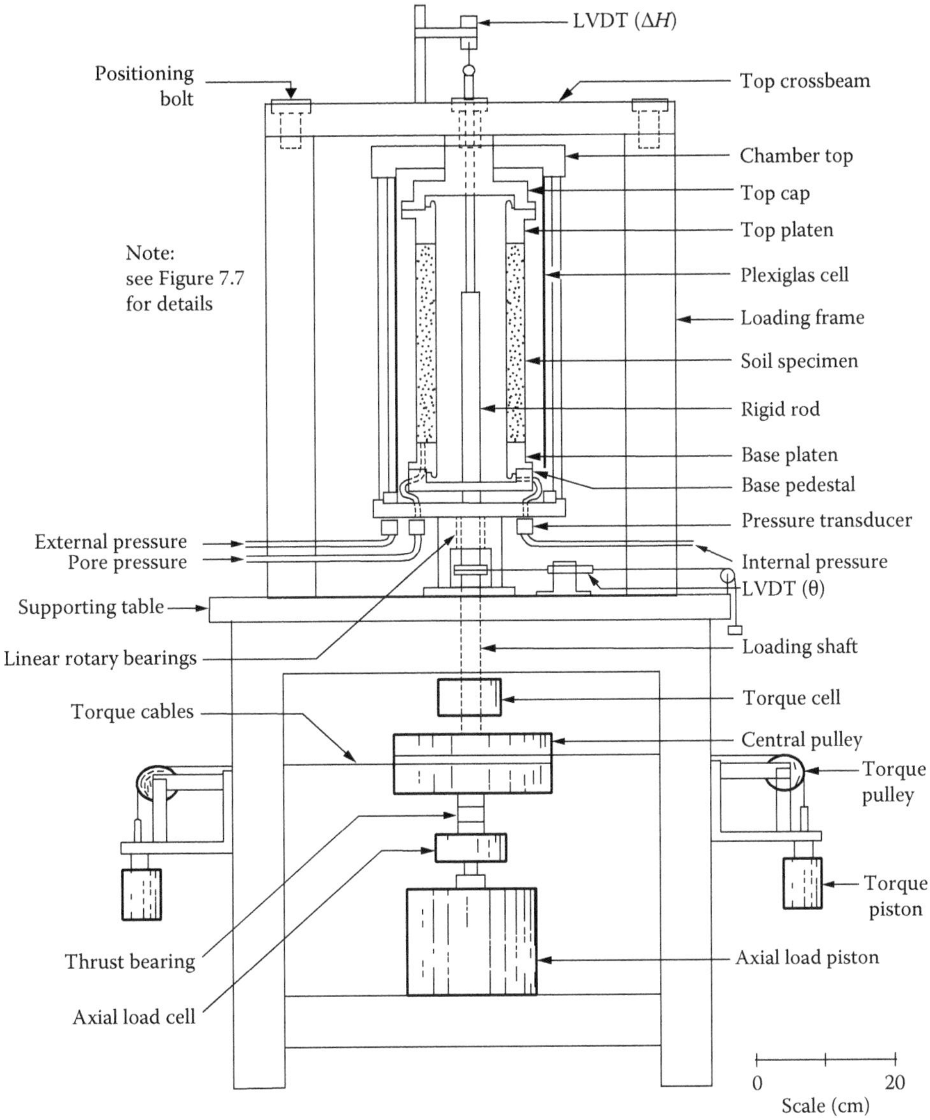

Figure 7.4 The hollow cylinder test apparatus at the UBC. (From Vaid, Y.P. et al., *Can. Geotech. J.*, 27(1), 1, 1990. With permission from the NRC, Ottawa, Ontario, Canada.)

internal and external cell pressures are kept equal, resulting in stress conditions close to simple shear (they are actually closer to plane stress rather than plane strain) but without the problem of complementary shear, although stress non-uniformities do occur on certain stress paths (Wijewickreme and Vaid, 1991). Importantly, in the hollow cylinder tests all principal stresses are known.

The hollow cylinder test has existed for more than 30 years. However, earlier equipment had rather small samples and the conditions in the sample departed significantly from the ideal of a uniformly stressed element. Substantive improvements in the hollow cylinder test were obtained at Imperial College (Hight et al., 1983) and at the University of

British Columbia (UBC) (Vaid et al., 1990b) by increasing the diameter of the sample to 250 mm (internal) and 300 mm (external).

Of the test types mentioned earlier, only the cyclic triaxial and cyclic simple shear tests are carried out in commercial soil testing laboratories. Building and carrying out tests with the more complex devices are usually topics for PhD theses, and the variations and refinements are almost limitless. Microcomputer technology has made many of the tests much easier to carry out and advanced designs of control systems are readily available from suppliers such as GDS (United Kingdom) and Trautwein (United States). There is also a difference between testing for insight and testing to calibrate models. While there is no present substitute for the modern hollow cylinder to evaluate constitutive behaviour, it is a different matter when that behaviour is given by a model and interest then centres on evaluation of plastic softening modulus (or equivalent) during principal stress rotation for a particular soil. Calibration of stress rotation-induced softening of *NorSand* has used cyclic simple shear, shown later in this chapter. However, we start with cyclic triaxial test data as these are the historical bases of understanding cyclic behaviour and liquefaction assessments.

7.2.2 Trends in cyclic triaxial test data on sands

Nevada sand at a relative density of 60% was used for the VELACS project (Arulanandan and Scott, 1993). Figure 7.5 shows a cyclic triaxial test on a medium dense sand sample ($\psi \sim -0.18$) which failed after about 11 load cycles. The small plots show the measurements of deviator stress, axial strain and pore pressure as a function of time during the test. After application of an initial deviator stress of 8 kPa, the stress was cycled ±26 kPa once a second in a sinusoidal form. Axial strains are initially small during the first nine cycles and then rapidly increase during the last two cycles when failure occurs. The pore pressure builds up gradually during each cycle until it is close to the confining stress of 40 kPa used in this test. Note, however, that the large cyclic strains occur when the pore pressure is about 30 kPa. Also, the pore pressure never actually reaches 40 kPa, except possibly at the peak of the last loading cycle.

The deviator stress–strain graph is also shown in Figure 7.5. For the first eight cycles the curves are very close together, but as the sample approaches failure, the strains increase and the hysteresis loops open up quickly. The q–p' graph reflects the gradual build-up of pore pressure as the effective stress (p') reduces until the stress path approaches the critical stress ratio at which time the sample starts failing and the shape of the stress path changes completely. The stress path becomes *hooked* towards the later stages of the test. Because of the high pore pressures, the sample reaches the critical stress ratio at low q values but as the load increases dilation moves the stress path up the failure line. When the stress reverses, dilation ceases and volumetric contraction drives the stress path back down towards the origin until the critical stress ratio is encountered in the opposite (extension) direction.

Two important points are evident from Figure 7.5. First, there is no sustained zero effective stress, or liquefaction, condition. A transient zero effective stress condition may occur as the deviator stress crosses the zero axis (but this does not occur in this test). Second, strains accumulate rapidly once the soil reaches this condition of transient dilation–contraction. Note also how the stress path on the extension side lies well above the critical stress ratio line and suggests a *cohesion* component in extension. This is a test artefact, possibly membrane extension, rather than a real property of the sand.

Although there is much detail in the stress–strain response under cyclic loading, the behaviour is commonly reported in terms of a fatigue model in which the number of cycles N to a particular failure criterion is taken as the basic result of the cyclic triaxial test. Commonly, data are presented in terms of the cyclic stress ratio (CSR) against the number of cycles of loading to cause 5% double amplitude strain, with most other factors kept equal. Sometimes the

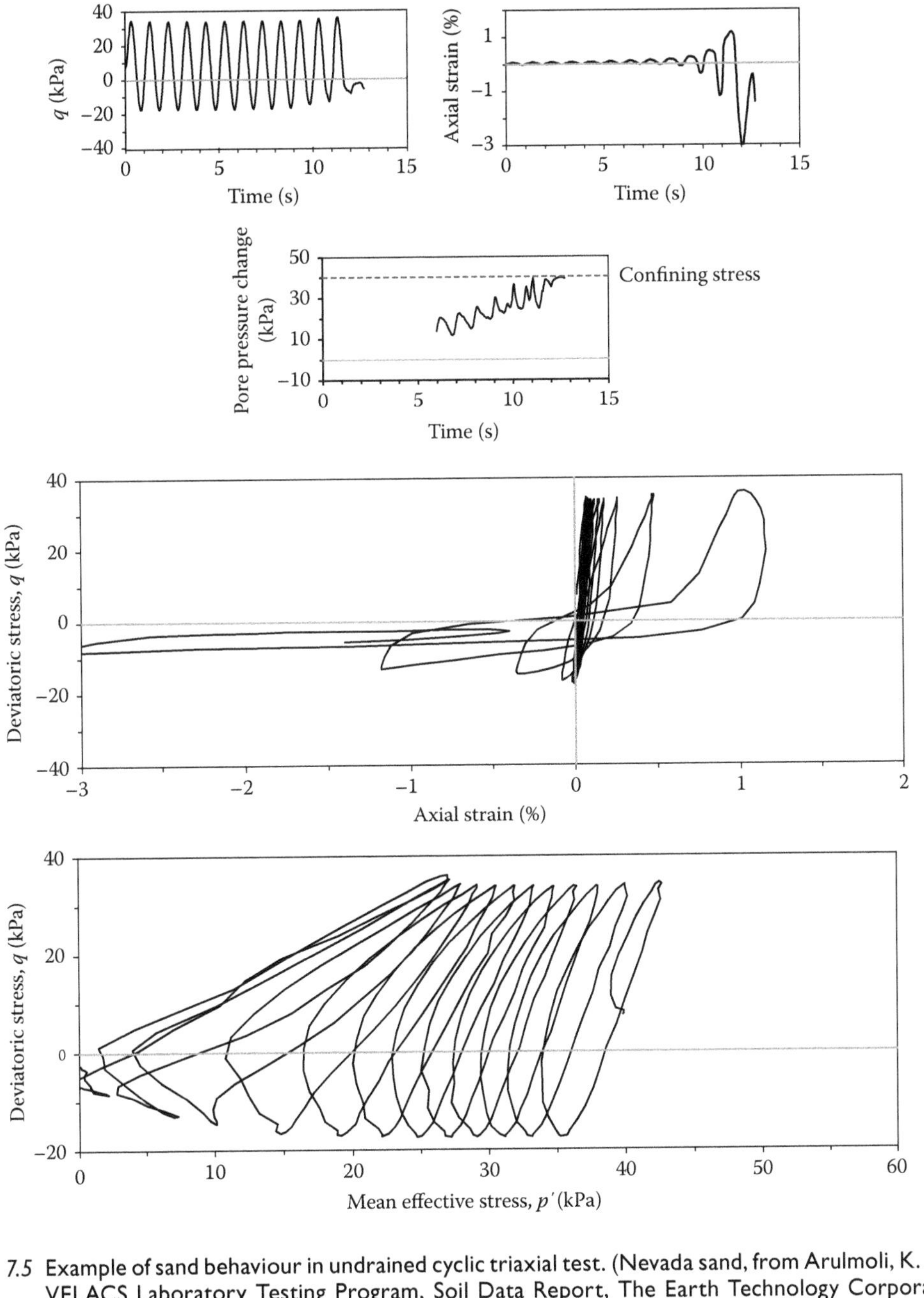

Figure 7.5 Example of sand behaviour in undrained cyclic triaxial test. (Nevada sand, from Arulmoli, K. et al., VELACS Laboratory Testing Program, Soil Data Report, The Earth Technology Corporation, Irvine, CA, Report to the National Science Foundation, Washington, DC, 1992.)

condition when the excess pore water pressure first equals the initial mean effective confining stress is taken as an alternative failure condition to the double amplitude strain. In the test shown in Figure 7.5, this pore pressure condition is reached after 13 cycles of loading, while the 5% double amplitude failure condition is not in fact reached as the strain accumulates in the extension direction with a cyclic component of about 2%. In loose sands, it is found that excess pore pressure and strain amplitude failure criteria are equivalent, but the criteria tend to diverge as density increases. Denser samples can be cycled through the transient zero effective stress condition with a relatively slow accumulation of strains.

Cyclic triaxial test data are reasonably repeatable. Figure 7.6 shows a set of results from cyclic triaxial testing of dense and loose samples of Toyoura sand. These data were generated as part of a cooperative test programme by five laboratories in Japan and are reported in detail by Toki et al. (1986). Samples were tested either dense ($e=0.669$–0.702, $D_r \sim 80\%$) or loose ($e=0.765$–0.823, $D_r \sim 50\%$), at an initial effective confining stress of 98.1 kPa (1 kgf/cm^2). Failure is defined as 5% double amplitude strain for the graphs in Figure 7.6. The cyclic stress is normalized to a CSR for triaxial conditions as $CSR = q/2\sigma'_{3c}$, that is the shear stress is divided by the initial confining stress. Loose Toyoura sand is able to sustain a

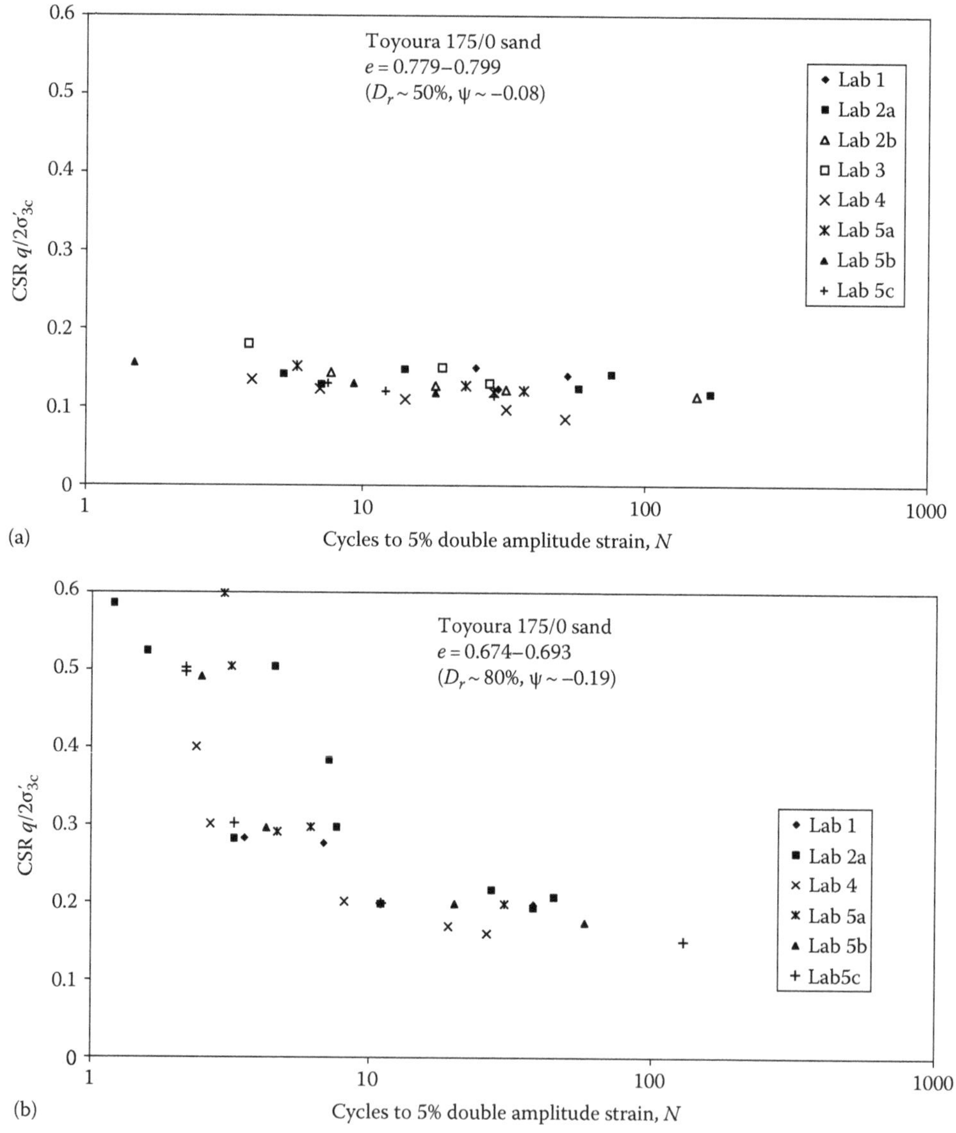

Figure 7.6 Cyclic strength of Toyoura sand in triaxial compression. (a) Loose (D_r = 50%) and (b) dense (D_r = 80%). (Data from Toki, S. et al., *Soils Found.*, 26(3), 117, 1986.)

maximum CSR of 0.2 for only one cycle of loading, but a stress ratio of less than 0.1 causes little problem even for hundreds of load cycles. Dense Toyoura sand similarly can sustain hundreds of cycles if the stress ratio is less than about 0.15, but very few cycles if the stress ratio is greater than about 0.3. To put these stress ratios in perspective, during the first loading cycle when pore pressures can be assumed to be zero, a CSR of 0.3 corresponds to a mobilized friction angle of only 13.3° while a CSR of 0.1 corresponds to 5.2°. The mobilized stress ratio increases during the tests as pore pressures are generated but the point is that relatively modest cyclic loads can be very damaging to sands.

Despite the large body of cyclic triaxial test data available, comparison of data between different soils and different testing laboratories is confusing as there are numerous testing factors that influence the test results. There is also the issue of how data should be normalized for void ratio (density) and soil type (gradation, D_{50}). During the 1970s these issues were explored in detail, and the American Society for Testing and Materials (ASTM), as it was then known even held a symposium with a Special Technical Publication on Dynamic Geotechnical Testing (STP 654) in 1978. Townsend (1978) summarized the factors affecting cyclic triaxial testing as follows:

- Specimen preparation (i.e. initial fabric)
- Confining stress
- Loading wave form
- Density
- Prestraining
- Consolidation ratio (i.e., $\sigma'_{vo}/\sigma'_{ho}$)

Figure 7.7 is a summary of cyclic triaxial test results on 13 sands plotted simply as CSR to reach failure (the failure criteria vary from 5% double amplitude strain to initial liquefaction or 100% pore pressure). No attempt has been made to normalize these data for relative density, although in general the relative density lies between 30% and 80%. All the data in Figure 7.7 are for isotropically consolidated samples and thus include 90° jumps in principal stress direction as conditions change from compression to extension during each half cycle of loading. Also shown are the results, normalized to $D_r = 50\%$, of Garga and McKay (1984) who presented a comprehensive collation of data for 20 different tailings and 13 non-tailings sands. It is interesting that the data normalized to a relative density of 50% show much the same range as the un-normalized data.

A striking feature of Figure 7.7 is the extremely wide range in data. A CSR as low as 0.1, or as high as 0.35, may be required to induce failure in 15 cycles in 2 different sands at the same relative density. This large variation is presumed to be due to a combination of the factors listed earlier, particularly fabric, stress level and relative density. For these test conditions with principal stress flips each cycle, it may be concluded that if the CSR is less than 0.09, failure is unlikely to occur in several hundred cycles. Conversely, if the CSR is greater than 0.2, failure is bound to occur within 100 cycles.

Despite the large range in behaviour, shown in Figure 7.7, there is actually a simple underlying behavioural trend. For any given sand, state parameter and sample preparation procedure, the CSR is normalized by the CSR that results in liquefaction failure after 15 cycles of loading, CRR_{15}, giving Figure 7.8. (Here we have introduced the strength of the soil as the cyclic resistance ratio (CRR) which is the CSR that causes failure in a given number of cycles.) A relatively narrow band now fits the data, consistent with the fatigue framework suggested by van Eekelen (1977) referred to at the start of this chapter, and the trend smoothly extends from a few cycles to nearly a thousand cycles. There is no difference in the pattern of soil behaviour between low or high cycles. Of course, specific material testing

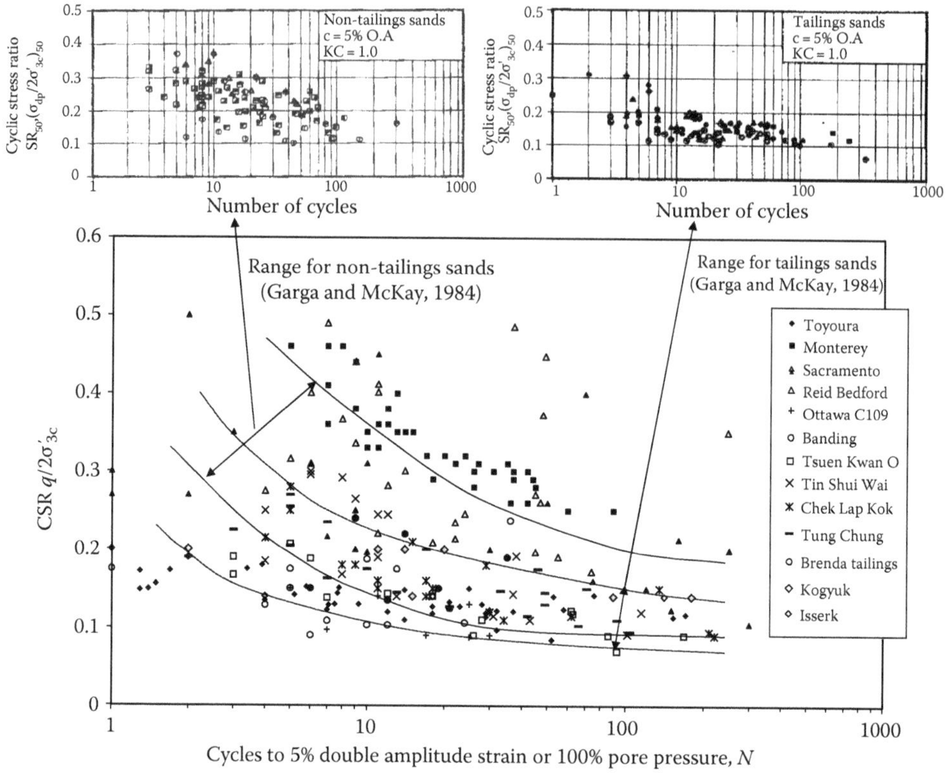

Figure 7.7 Cyclic triaxial test data on 13 sands for which CSL is known, with ranges obtained by Garga and McKay for sands and tailings sands shown. (From Garga, V.K. and McKay, L.D., *J. Geotech. Eng. Div., ASCE*, 110(8), 1091, 1984. With permission from the ASCE, Reston, VA.)

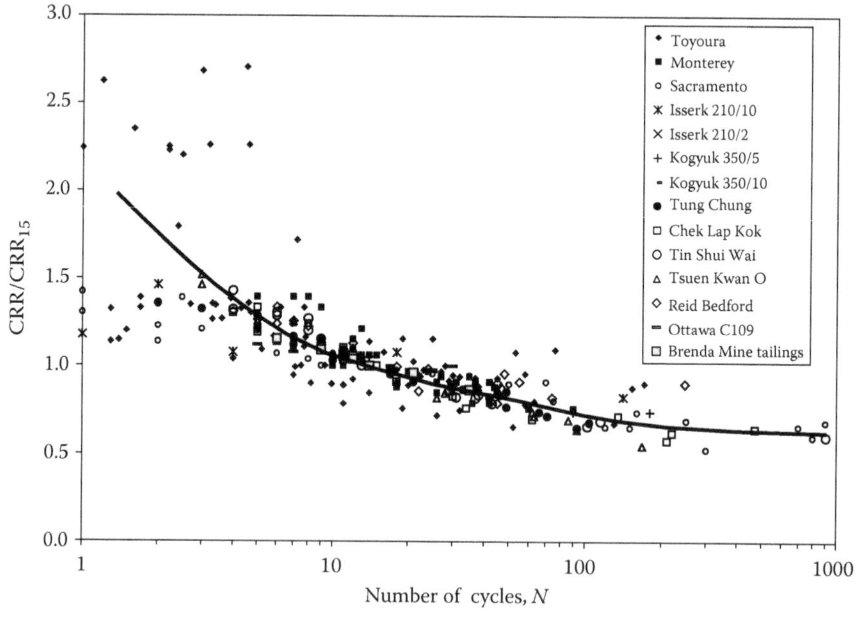

Figure 7.8 Cyclic triaxial test data from Figure 7.7 normalized to CRR for 15 cycles, CRR_{15}.

is still necessary for the definition of cyclic strength of any particular sand but the situation has now been largely reduced to understanding how state (density and stress level) and fabric control CRR_{15}.

In many cases of cyclic triaxial testing, anisotropic consolidation is used giving a static bias. Static bias, usually applied as a greater vertical than radial stress in the triaxial sample, reduces the excursion into triaxial extension conditions during each principal stress change cycle and generally gives higher resistance to cyclic triaxial loading than a corresponding isotropic initial condition sample. The CSR is commonly plotted against the number of cycles to 2.5% strain in compression for anisotropically consolidated cyclic tests, with different lines shown for different consolidation principal stress ratios, K_c (= σ'_1/σ'_3). Garga and McKay (1984) show that for several tailings materials, the different K_c lines can be normalized to a single line by division of the CSR by K_c. Figure 7.9 shows some data normalized in this manner, which in effect means the cyclic shear stress is divided by σ_1 rather than σ_3. All the data lie between a modified stress ratio of 0.1 and 0.2, but there is otherwise no apparent trend between stress ratio and number of cycles. Figure 7.9 is not very helpful in practice, but it does illustrate why researchers have struggled to reach any agreement on the effect of static bias on cyclic strength.

An explanation for the behaviour implicit in this plot has recently been shown by Baki et al. (2012). They did careful testing of matched pairs of static and cyclic testing, both the one-way and two-way variety in Figure 7.2a and b, and showed that triggering of cyclic instability occurs shortly after the cyclic effective stress path crosses the instability *zone* as determined

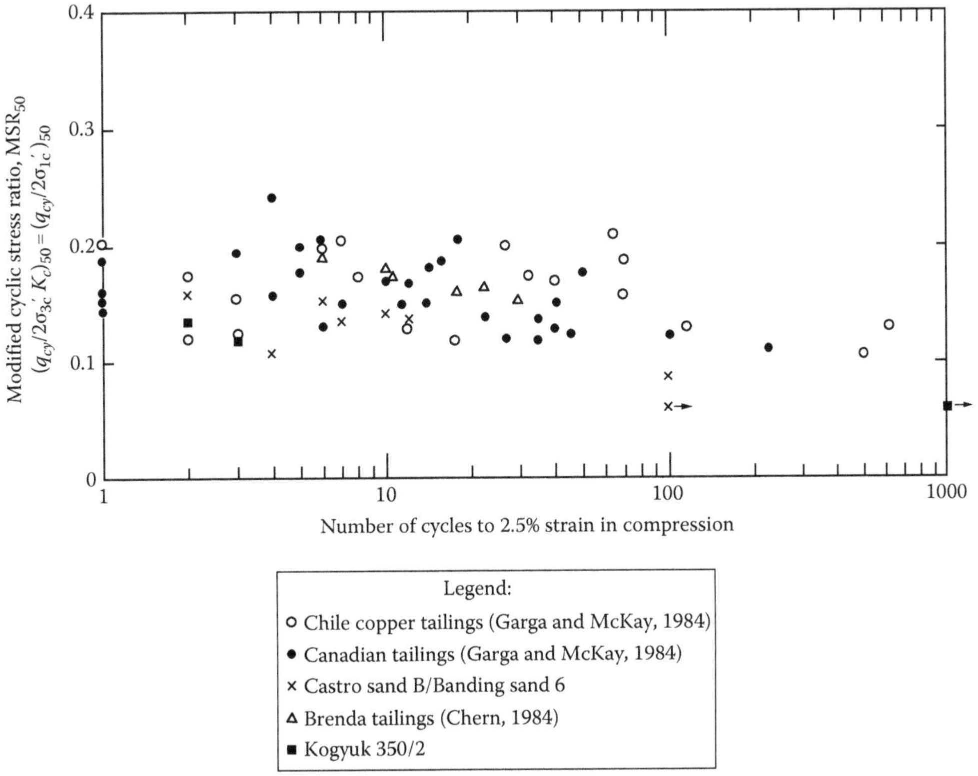

Figure 7.9 Cyclic triaxial data normalized to consolidation stress ratio, K_c, illustrating absence of trend as a function of K_c.

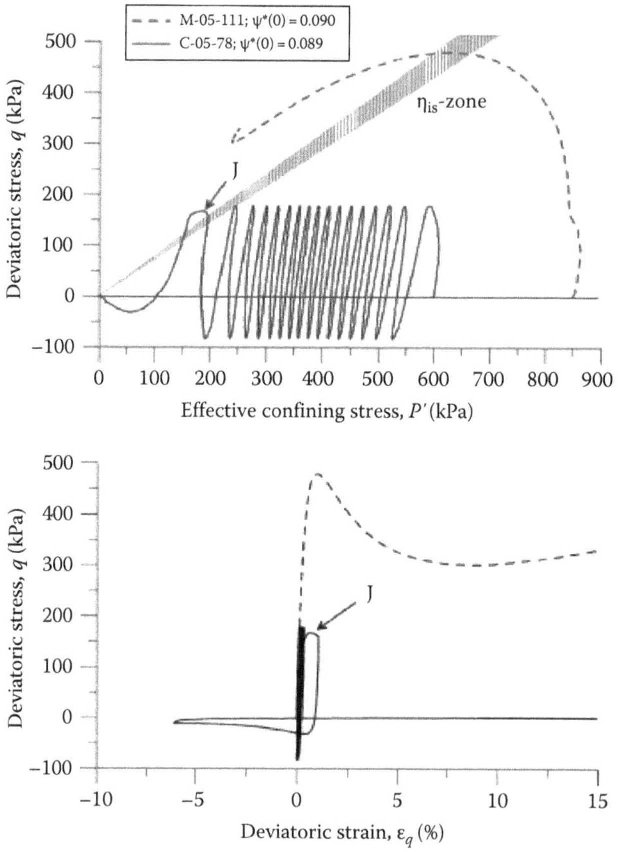

Figure 7.10 Cyclic failure in compression as stress state reaches the instability stress ratio from monotonic test at same state parameter. (From Baki, Md.A.L. et al., *Can. Geotech. J.*, 49(8), 891, 2012. With permission from the NRC, Ottawa, Ontario, Canada.)

from a corresponding monotonic loading test. (They use the word zone because definition of a stress ratio from a single monotonic test is ill-conditioned.) Of course, this corresponds to when the stress path gets close to the phase transformation or transient zero dilatancy condition and when plastic static strains increase. In symmetrical two-way loading, the first indication of large strains occurs in extension since M_{te} is less than M_{tc}. In cyclic loading with a static bias, whether failure occurs in extension or compression depends on which side of the loading cycle approaches the critical stress ratio first, Figure 7.10. A small static bias will increase the cyclic strength by optimizing the stress path so that both the extension and compression stress ratios approach phase transformation at the same time, while a larger static bias will decrease the cyclic strength by bringing in the compression failure mode more quickly.

How does a critical state or state parameter approach apply to the understanding of CRR_{15}? For every group of tests in Figure 7.8, at approximately the same value of ψ, it is possible to determine CRR_{15}. This value of CRR_{15} has been plotted against ψ in Figure 7.11. A reasonable trend is evident, but there is scatter and the question then becomes whether the relationship between CRR_{15} and ψ is unique. At its simplest, there should not be a unique relationship. The effect of a cyclic stress should be related to the critical friction ratio, M. Critical state theory also leads us to expect undrained behaviour to be related to the current ratio of elastic to plastic modulus, and although this ratio seems a function of

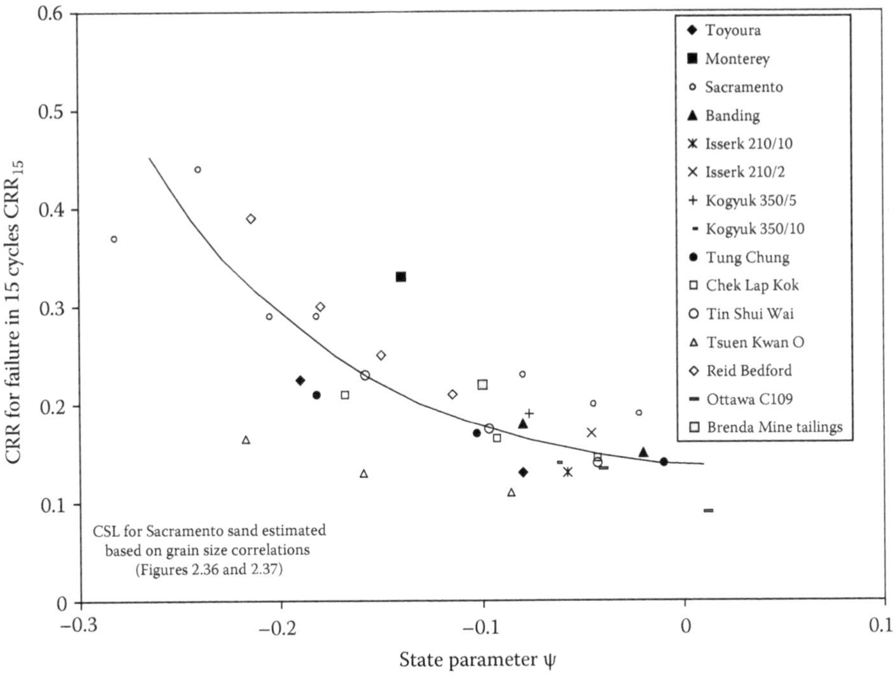

Figure 7.11 CRR at 15 cycles (CRR_{15}) as a function of state parameter for 13 sands.

ψ in sands (Jefferies and Been, 2000), the evidence is that it is different for all soils. Most fundamentally, however, realizing the importance of principal stress direction and fabric as the fundamental driver of cyclic mobility immediately makes it clear that grain contact arrangements are going to be the key. Fabric is not captured with ψ and initial fabric effects should be expected in Figure 7.11.

At a laboratory level, fabric effects will show up as an effect of sample preparation procedure, and Figure 7.12 shows an example of published data on the effect of sample preparation on cyclic strength of sands. This is primarily what was normalized out by taking CRR_{15} in the first place, and it should come as no surprise for it to resurface, as shown in Figure 7.11. Looking at Figure 7.12, it can be seen that the two fabrics tested give a shift in the CRR of about ±0.1; this is exactly the bandwidth around the basic state parameter dominated trend plotted in Figure 7.11.

7.2.3 Cyclic behaviour of silts

Cohesionless silts show very similar behaviour to sands in cyclic triaxial and simple shear tests, as they do for static tests. Figure 7.13 shows a cyclic triaxial test on Bonnie silt, carried out as part of the VELACS project from Arulmoli et al. (1992). Bonny silt contains about 85% silt-sized particles, with 8% sand and 7% clay. In the Unified Soil Classification System, Bonny silt classifies as a CL because it has a liquid limit of 29 and a plasticity index of 15. It is close to the boundary of what might be classified as a silt.

This cyclic triaxial test on Bonnie silt of Figure 7.13 can be compared with a similar test on Nevada sand that is shown in Figure 7.5. The silt displays a similar ramping up of excess pore pressure and a similar reduction in the sample stiffness. Trends in the silt are arguably a little smoother than the sand, which might reasonably be caused by the lesser stiffness of

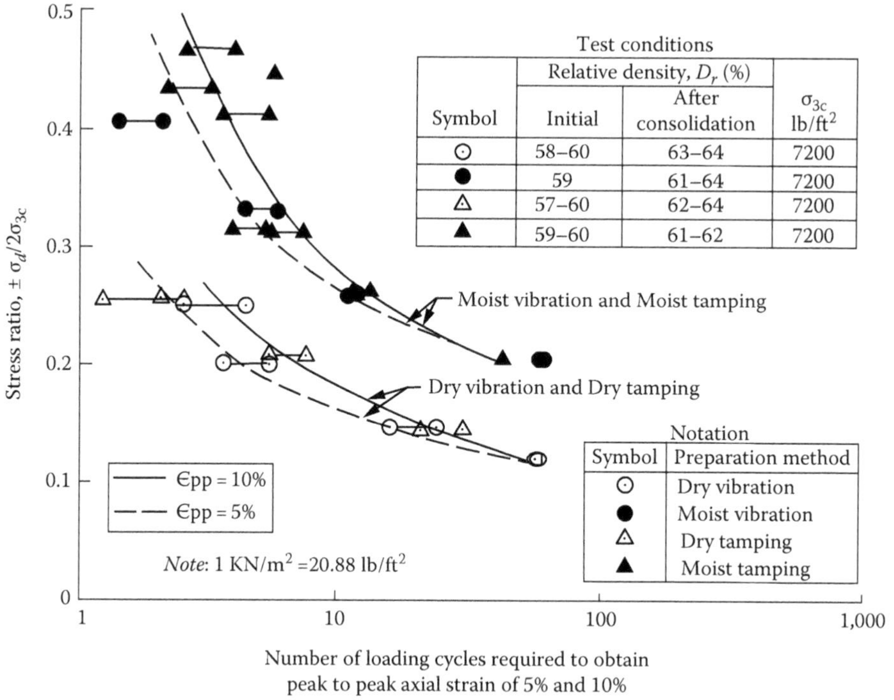

Figure 7.12 Example of effect of fabric (sample preparation method) on cyclic strength of sands. (From Ladd, R.S., *Geotech. Test. J.*, 1(1), 1, 1978. With permission from the ASCE, Reston, VA.)

the silt that makes the test easier to control. This similarity should be no surprise, however, as silts are particulate and without any bonds (true cohesion), just like sands. Silts also exist over a range of densities and are equally well characterized with the state parameter.

Compressibility is important in undrained shear because of dilatancy. Recall that in undrained (constant volume) behaviour the tendency to dilate or contract is offset by pore pressure changes, and the magnitude of pore pressure change is such that the elastic volumetric strain balances the plastic strain due to shear, Equations 6.6 and 6.7. In the critical state world, compressibility is embodied in the slopes of the critical state and the rebound lines, λ and κ. Elastic compressibility is represented by κ and therefore for a given amount of dilatancy, the pore pressure response will depend on κ. Smaller κ values, or stiffer soil, will result in larger pore pressure changes to give the same elastic volume change.

Sangrey et al. (1978) and Egan and Sangrey (1978) in companion papers describe a simple critical state model for pore pressure generation under cyclic loading which gives additional insight into how compressibility affects liquefaction. The normalized CRR curves in Figure 7.8 show a trend that decreases with logarithm of the number of cycles and hint at a level of cyclic loading that would never cause liquefaction no matter how many load cycles occurred. This stress ratio that sets a lower limit to any possible onset of liquefaction is known as the critical level of repeated loading (CLRL) by Sangrey et al. and is illustrated in Figure 7.14. Figure 7.14 also illustrates that when the stress is cycled below the CLRL, pore pressures are generated although an equilibrium condition is reached.

Sangrey et al. (1978) looked at cyclic test data on a number of materials for which critical state parameters were also known and showed that the normalized CLRL varies systematically with compressibility of the material expressed as $\kappa/(1+e)$. Egan and Sangrey (1978)

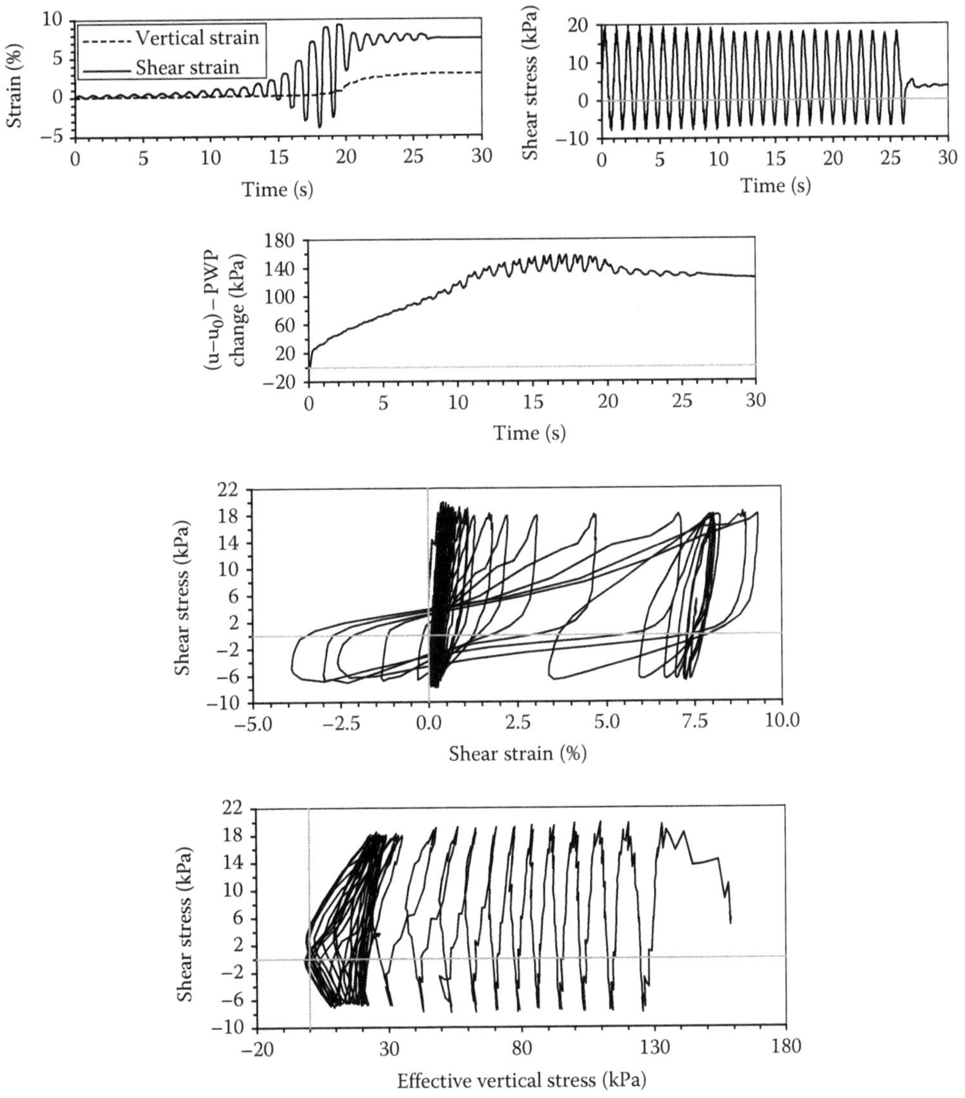

Figure 7.13 Cyclic triaxial test on Bonnie silt. (Velacs project: From Arulmoli, K. et al., VELACS Laboratory Testing Program, Soil Data Report, The Earth Technology Corporation, Irvine, CA, Report to the National Science Foundation, Washington, DC, 1992.)

describe the related critical state model for excess pore pressure potential. These relationships of cyclic strength and pore pressure to the elastic bulk compressibility $\kappa/(1+e)$ are shown in Figure 7.15. Clear trends are evident over the full range of soil types. More compressible soils (clays and silts) are more resistant to liquefaction than sands, and sands show a much higher excess pore pressure potential.

Silt content can now be included in a liquefaction assessment through the soil properties rather than the over simplistic measure of percentage of material passing an arbitrary-sized sieve. The key property is compressibility, which must therefore be measured. Both λ and κ are important; λ captures how plastic behaviour varies with stress level, while κ captures the pore pressure generated for a given plastic volumetric change.

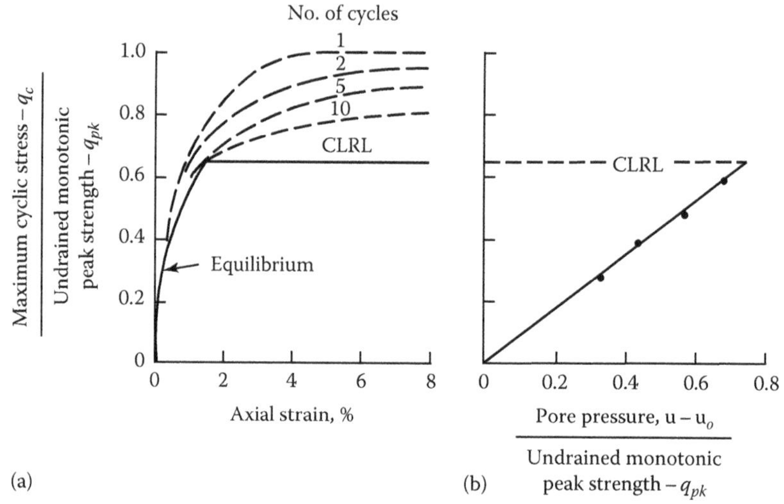

Figure 7.14 Illustration of the critical level of repeated loading. (a) Normalized stress versus strain and (b) normalized pore water pressure change. (From Sangrey, D.A. et al., Cyclic loading of sands, silts and clays, *Proceedings of ASCE Specialty Conference on Earthquake Engineering and Soil Dynamics*, Pasadena, CA, Vol. 2, pp. 836–851, 1978. With permission from the ASCE, Reston, VA.)

7.2.4 Cyclic rotation of principal stress

The laboratory testing of cyclic loading discussed so far was carried out in the context of earthquakes and triaxial loading and largely dominated by intellectual direction from Berkeley. However, the oil price shock of the early 1970s created a large increase in offshore oil production and in particular with the North Sea fields. One consequence of expanding offshore production was platforms being placed in ever more exposed environments and with large cyclic loadings from storm waves acting on the platforms. Storm loadings may be sustained for vastly longer periods than earthquakes. It was realized that understanding how soils responded to principal stress rotation was of fundamental interest to engineering offshore structures for oil production in exposed offshore environments.

Arguably, it was Arthur and co-workers (Arthur et al., 1979, 1980; Wong and Arthur, 1986) who first appreciated the importance of principal stress rotation to soil behaviour and who showed its importance experimentally. Arthur's group at University College London worked from a perspective of sands being anisotropic, from which it is almost self-evident that changing the direction of loading will cause changes in soil behaviour. They developed a directional shear cell (DSC), illustrated in Figure 7.16a, which is somewhat like a simple shear test except that all principal stresses are measured. Using the DSC, for the study reported in 1980, the principal stress was smoothly varied in a sine wave at near-constant stress ratio (Figure 7.16b). The soil tested was Leighton Buzzard sand, and it was placed dense at approximately $D_r = 90\%$. When this dense sand was subjected to principal stress rotation, at near-constant mobilized shear stress, strains accumulated readily with each loading cycle. The greater the principal stress rotation, the greater the plastic straining induced.

A more detailed study was reported by Wong and Arthur (1986), still using the DSC apparatus. This 1986 study used the same Leighton Buzzard sand, but now looked at two sand densities: $D_r = 20\%$ and $D_r = 90\%$. Because the DSC could only impose rather low confining stress (maximum $\sigma_3' = 20$ kPa), when sheared monotonically in plane strain with fixed

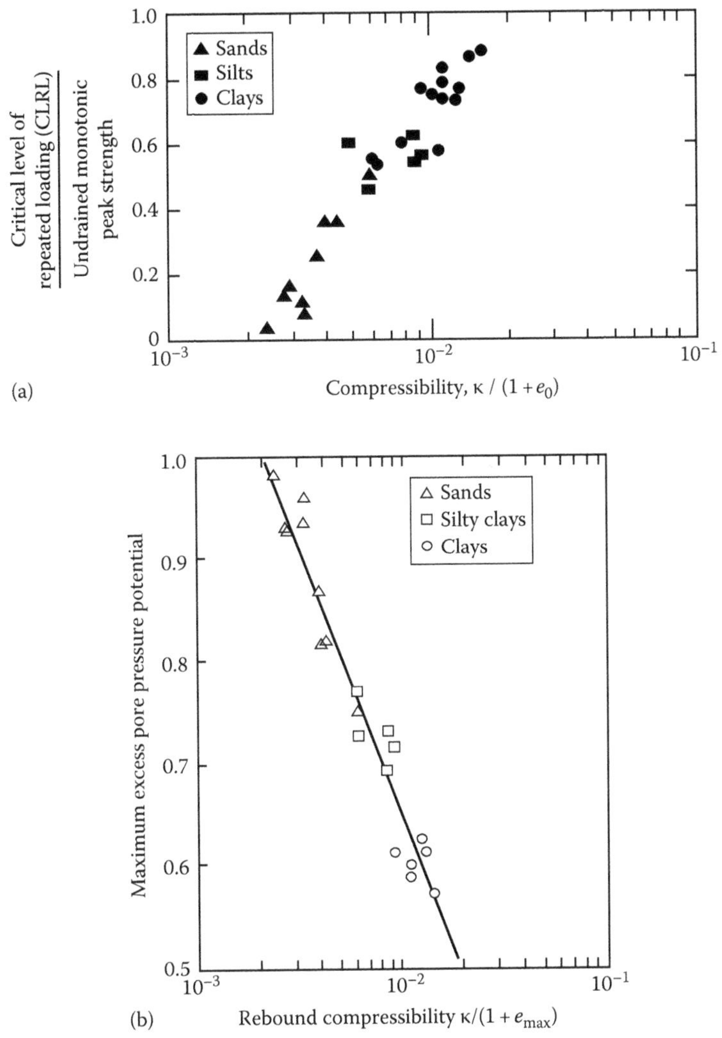

Figure 7.15 Critical state model for liquefaction of sands, silts and clays. (a) Critical level of repeated loading as fraction of monotonic shear strength. (From Sangrey, D.A. et al., Cyclic loading of sands, silts and clays, *Proceedings of ASCE Specialty Conference on Earthquake Engineering and Soil Dynamics*, Pasadena, CA, Vol. 2, pp. 836–851, 1978. With permission from the ASCE, Reston, VA.) and (b) excess pore pressure potential. (From Egan, J.A. and Sangrey, D.A., Critical state model for cyclic load pore pressure, *Proceedings of ASCE Specialty Conference on Earthquake Engineering and Soil Dynamics*, Pasadena, CA, Vol. 1, pp. 410–424, 1978. With permission from the ASCE, Reston, VA.)

principal stress direction, the loose sample ($e_o \approx 0.73$) gave a peak friction angle of 40° and 5° dilation. The dense sample ($e_o \approx 0.52$) gave a peak strength of 49° and 18° dilation. Figure 7.17 plots the relationship between volumetric and shear strain for differing levels of mobilized friction φ_m (i.e. the imposed ratio σ_1'/σ_3') and different imposed rotations of principal stress. These data are for the loose sand only, with no result being reported for the tests on dense sand. Because of the low confining stress, this loosest sand behaviour actually corresponds to initial state parameter of $\psi_o = -0.08$. It is apparent from even casual inspection

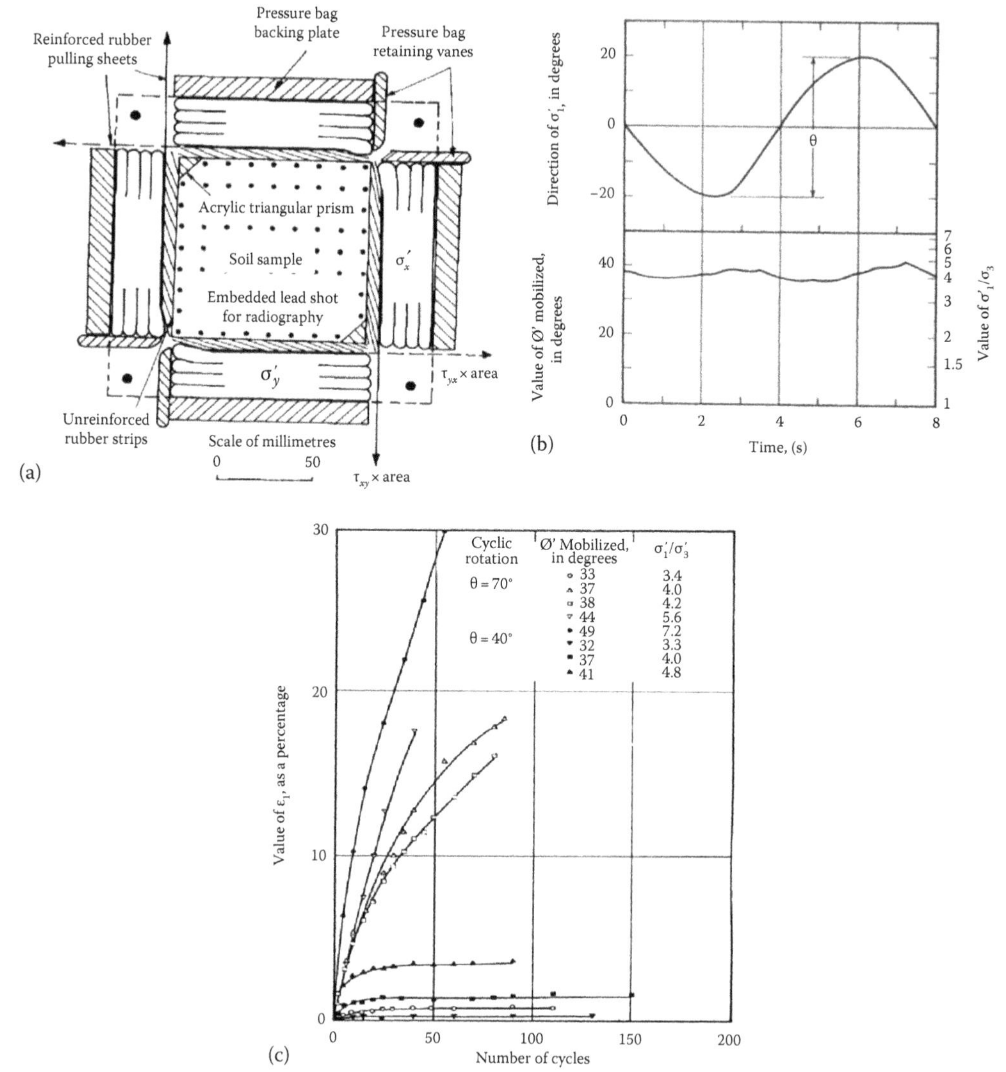

Figure 7.16 Demonstration of the importance of principal stress rotation on behaviour of dense sand by Arthur et al. (a) Directional shear cell, (b) typical loading as a function of time and (c) major principal strain development. (From Arthur, J.R.F. et al., *J. Geotech. Eng., ASCE*, 106(GT 4), 419, 1980. With permission from the ASCE, Reston, VA.)

that principal stress rotation has dominated the sand's behaviour and that even rather small stress rotation can suppress dilation. In their conclusion, Wong and Arthur (1986) state 'Cyclic rotation of principal stress directions in sand which causes strain radically alters the behaviour of the material from that seen in shear under constant directions of principal stress'. This is by no means an overstatement of the situation and something that is essential to recognize in regard to liquefaction.

The DSC equipment never became widely used, possibly because it was never able to test at stress levels of more usual practical importance. The Japanese work never suffered in this regard because the hollow cylinder was adopted from the outset, Ishihara and Towhata (1983) showing a pure cyclic rotation of principal stresses in drained tests

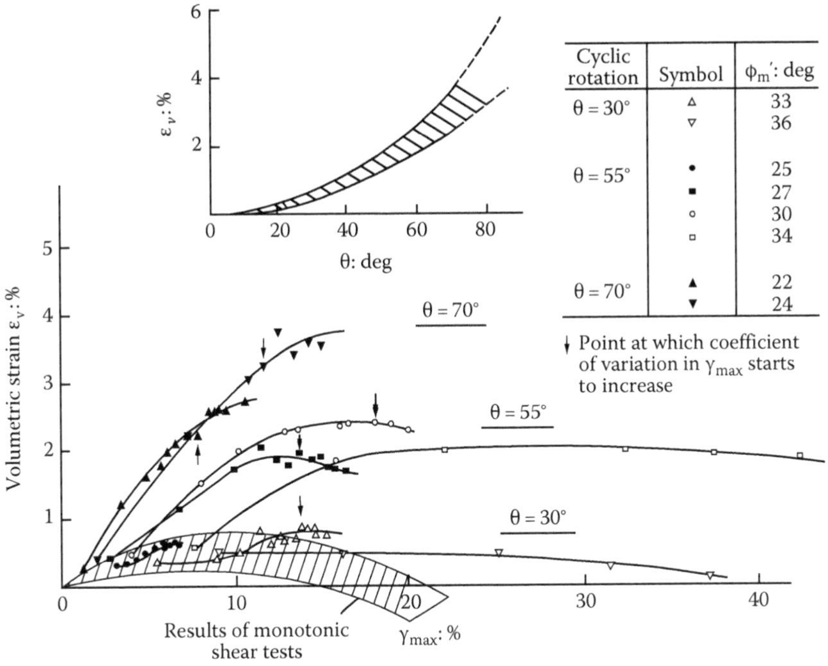

Figure 7.17 Cumulative volumetric strain in lightly dilatant Leighton Buzzard sand caused by principal stress rotation. (From Wong, R.K.S. and Arthur, J.R.F., *Géotechnique*, 36(2), 215, 1986. With permission from the Institution of Civil Engineers, London, U.K.)

on loose Toyoura sand at a mean effective stress of nearly 300 kPa. The direction of the principal stresses was rotated continuously from 0° to ± 45° following a semicircular stress path as shown in Figure 7.18a. This principal stress rotation caused an irrecoverable volumetric strain but its increment gradually decreased as the number of cycles increased, Figure 7.18b.

The trend of decreasing volumetric strain increment from cycle to cycle is a hardening response. What is seen with smooth principal stress rotation is not dissimilar to the approximate log(N) trends seen in cyclic triaxial testing (Figure 7.1), but it is now understood that the mechanism is change in principal stress direction, not cyclic variation in shear stress itself.

These early studies of principal stress rotation were directed at illustrating the importance of this rotation, at a fundamental level, to sand behaviour (and for that matter soil in general). There can be no doubt that principal stress rotation is fundamental from these experimental data, since if stress invariants alone are sufficient, then any constitutive model based solely on invariants will predict essentially no strains in the experiments of either Arthur et al. (1980) or Ishihara and Towhata (1983). This is contrary to what is observed. Further experiments involving principal stress rotation to begin quantifying the effect of rotations in a range of situations have been carried out. For example, Tatsuoka et al. (1986) present a comprehensive set of data on Toyoura, Fuji and Sengenyama sands in which they compare results of hollow cylinder with cyclic triaxial tests. Sample preparation methods were also varied and indicated the substantial effect of preparation method (i.e. initial soil fabric) on the results.

Chapter 1 referred to localized cyclic-induced liquefaction encountered with the offshore platform Molikpaq on April 12, 1986, and when it was subject to some 900 cycles of uniform

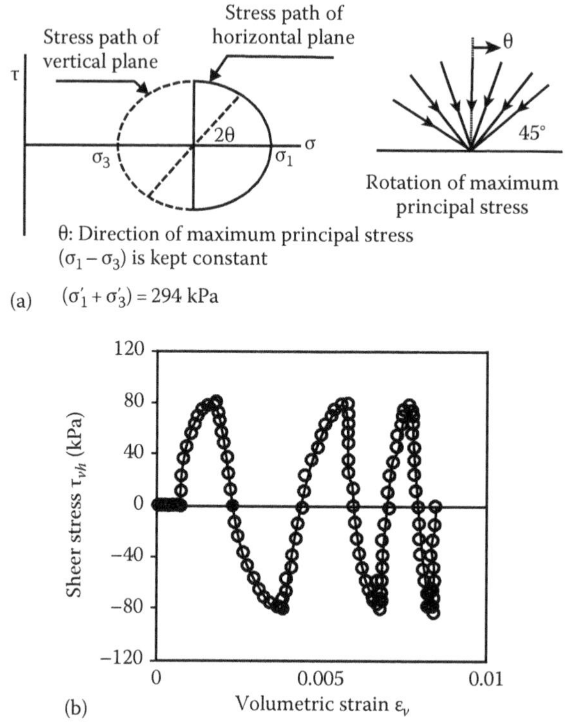

Figure 7.18 Result of a pure principal stress rotation test on loose Toyoura sand. (a) Applied stress path and direction and (b) resulting shear stress and volumetric strain. (From Ishihara, K. and Towhata, I., *Soils Found.*, 23(4), 11, 1983. With permission from the Japanese Geotechnical Society, Tokyo, Japan.)

sawtooth-like load at about 1 Hz frequency. As part of the investigation into the platform response during this ice loading, six hollow cylinder tests were carried out on Erksak sand (the fill used for the core and berm of the platform) at the UBC. In five of the tests, the major principal stress was rotated while keeping the stress invariants σ_q and σ_m constant, essentially duplicating the study of Arthur et al. (1980) to demonstrate that Erksak sand responded to principal stress rotation alone. This was indeed found to be the case, with even the dense samples showing contractive behaviour under principal stress rotation. The sixth sample was loaded along the estimated path followed by sand at mid-side of the loaded caisson face (the maximum excess pore pressure location) during the ice loading event. The principal stress and rotation path were estimated by finite element analysis. The sand was placed by pluviation in the hollow cylinder to match the estimated in-situ characteristic state (i.e. the 80 percentile value of ψ) of the core before the ice loading. The state in the hollow cylinder tests after establishing the initial confining stress was $\psi_o \approx -0.05$. The behaviour of Erksak sand in this simulation of the estimated stress path is shown in Figure 7.19.

At the start of the hollow cylinder test, the sand contracted as the stress state changed from isotropic (at σ'_m = 50 kPa) after saturation to the estimated in-situ state prior to ice loading (σ'_m = 250 kPa, Lode angle $\theta = 0°$ and $\sigma'_1/\sigma'_3 = 2.3$). This is normal stress–dilatancy behaviour for the stress ratio η changing from zero to a maximum $\eta/M_i = 0.63$ (as in this test). Cyclic loading was then applied, varying between a principal stress direction of $\alpha = 35°$ at $\sigma'_1/\sigma'_3 = 1.8$ (the trough of the ice load) to $\alpha = 45°$ at σ'_1/σ'_3 = 2.8 (peak ice load). Drained conditions were

Initial state	HCT test #	Sample preparation		Initial phase R	Consolidation phase	Rotation phases			
		e_i	B			e_o	Ph. RA	e_o	Ph. RAU
Medium dense $\psi = -0.04$	EM 8	0.720	0.98	R = 1.0 to 1.8	σ_m = 50 to 250 kPa b = 0.0 to 0.5 α = 0 to 35°	0.717	8 cycles R=1.8 to 2.8 to 1.8 α = 35 to 45 to 35°	0.716	2 cycles Total stress per Ph. RA

Symbols

- e_i Void ratio before initial phase (σ_m = 50 kPa)
- e_o Void ratio before rotation phase
- B Skempton's pore pressure parameter
- b $(\sigma'_2 - \sigma'_3)/(\sigma'_1 - \sigma'_3)$
- σ_m $(\sigma'_1 + \sigma'_2 + \sigma'_3)/3$
- R σ'_1/σ'_3
- α Angle between σ'_1 and vertical

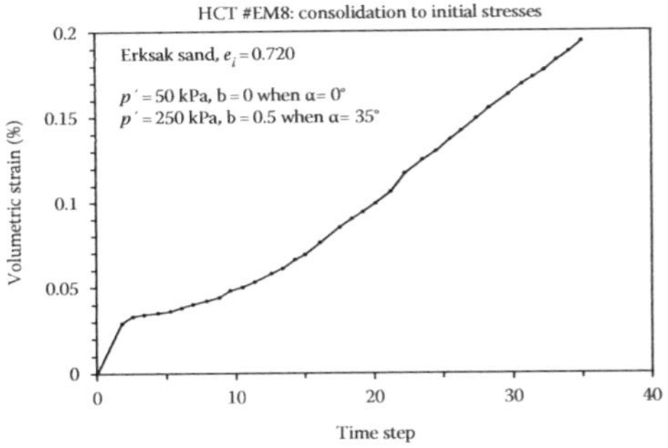

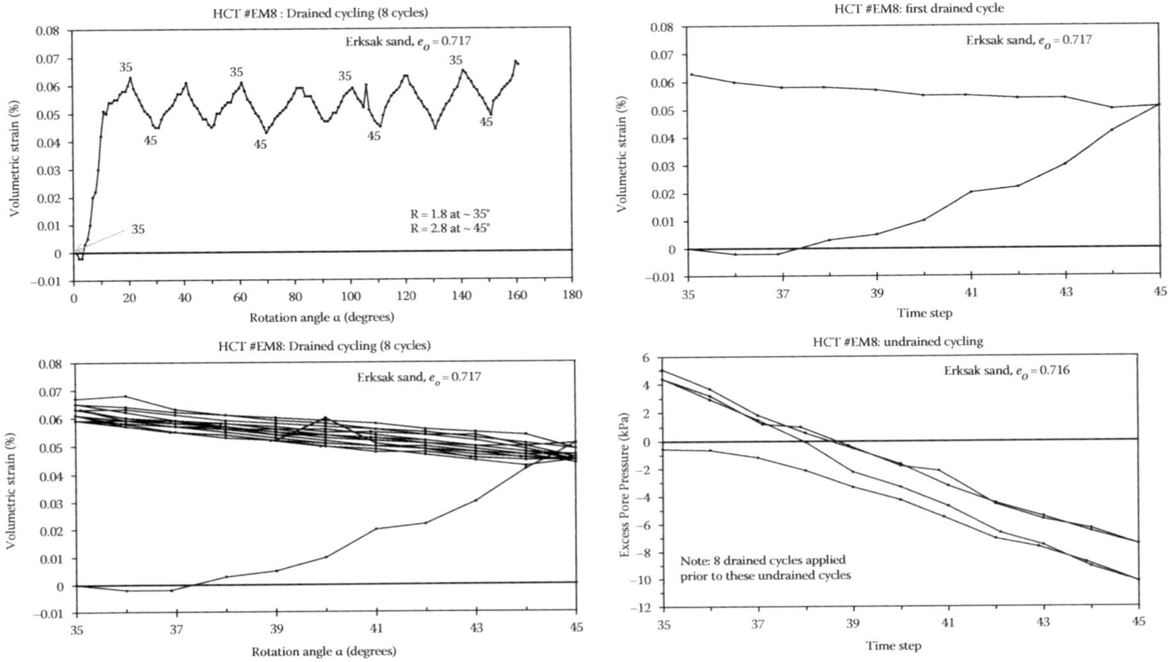

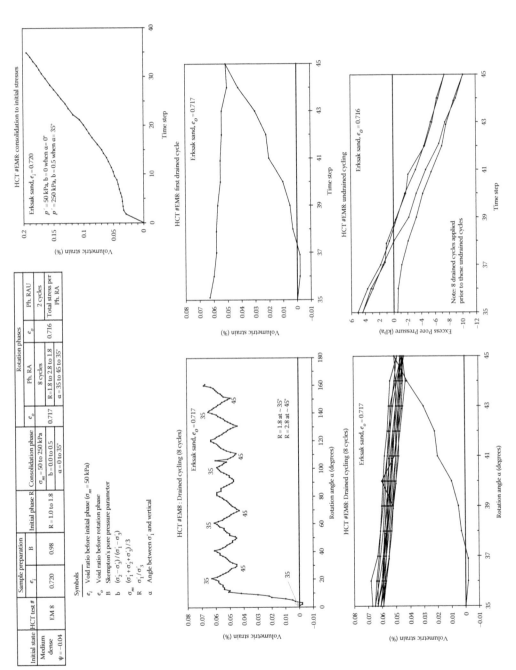

Figure 7.19 Behaviour of Erksak sand in hollow cylinder test simulating principal stress history for Molikpaq piezometer E1 during April 12, 1986, ice loading event.

first simulated, and there was little volumetric strain for the first five cycles before a trend to densification started. When densification started, the boundary conditions were changed to undrained and a gentle accumulation of residual excess pore pressure at about 1.5 kPa per cycle occurred. Unfortunately, this was early in the development of the UBC equipment and it was not possible to continue to several hundred cycles at that time.

The rate of excess pore pressure in the laboratory hollow cylinder can be compared with the rate during the field liquefaction event that averaged 0.8 kPa/cycle (Figure 1.21) and peaked at 1.8 kPa/cycle (Figure 1.22). Although only a few cycles were possible in the laboratory, there is a remarkable correspondence between Erksak sand behaviour in the laboratory modelling of the best-estimated principal stress history and the measured response in-situ.

A difficulty with studying principal stress rotation is that so far the only truly satisfactory test apparatus is the large diameter hollow cylinder equipment such as that described by Hight et al. (1983) and Vaid et al. (1990b). This equipment is rather challenging to use and the role of such hollow cylinder tests appears limited to providing verification data for theoretical constitutive models. (The Molikpaq study had a budget outside normal engineering practice.) However, the cyclic simple shear test has moved into normal consulting practice and, despite its limitations, does offer insight into the effects of principal stress rotation. Cyclic simple shear is the next topic.

7.3 TRENDS IN CYCLIC SIMPLE SHEAR BEHAVIOUR

The cyclic triaxial data shown in the previous section are compromised because the cyclic triaxial test poorly approximates insitu conditions. Nevertheless, such cyclic triaxial data were behind most views about liquefaction until about ~2000. Simple shear tests, on the other hand, are plane strain tests and include the same small and gentle principal stress rotations that arise in cyclically loaded foundations for structures. Simple shear tests are an excellent analogue for vertically propagating seismic waves during an earthquake.

Cyclic simple shear tests are straightforward enough to carry out a reasonable number of tests on a single soil to discern the effects of stress level, void ratio, initial stress state and cyclic stresses on soil behaviour. A feature of laboratory testing is that workers tend to have their favourite equipment, so there are many sets of cyclic simple shear tests without triaxial tests and the opposite. This matters, as measurements made in a simple shear test are not enough to determine soil properties – triaxial tests are needed as well. This situation is compounded by rather few fundamental investigations using cyclic simple shear and a reasonable range of initial conditions. One exception to this situation is the testing of Fraser River sand (FRS), and this testing allows a complete understanding of sand behaviour in cyclic simple shear.

We will use the cyclic simple shear behaviour of FRS as a calibration of *NorSand*, just like we used the Erksak sand database for static *NorSand* in Chapters 2 and 3.

7.3.1 Fraser River sand

FRS is an alluvial deposit widespread in the Fraser River Delta of the Lower Mainland of British Columbia, Canada. The area includes the city of Vancouver and is of considerable economic importance. Lying on the west coast of North America, the area is vulnerable to earthquakes and the FRS deposits are known to have liquefied in past earthquakes (relic sand boils have been excavated from the City of Richmond near the Vancouver airport). Given the wide distribution of FRS in this highly populated and seismically active area, FRS has been extensively tested (both commercially and for research purposes) including a large

Figure 7.20 Microphotograph of FRS grains.

body of work from the UBC (e.g. Shozen, 1991; Vaid and Sivathalayan, 1996, 1998; Vaid et al., 2001; Sriskandakumar, 2004), testing by the University of Alberta (e.g. Chillarige et al., 1997a,b) including for the Canadian Liquefaction Experiment, CANLEX, (summarized in Wride et al., 2000 and Robertson et al., 2000), as well as by geotechnical consulting companies. However, the variability of this natural deposit makes it difficult to generalize properties for FRS, with different authors testing different gradations of this sand and then finding different properties.

A particularly interesting set of cyclic simple shear tests on samples of FRS were undertaken at the UBC between 2002 and 2003 as part of the *Earthquake-induced damage mitigation from soil liquefaction* initiative, described in detail in Sriskandakumar (2004). Both loose and dense FRSs were tested over a range of CSR and various static biases. These data have the added attraction of being available as digital files, making further processing straightforward and providing records for calibration of numerical models. But a deficiency of this UBC work is that it focuses on cyclic simple shear alone, with no determination of the basic properties of the sand tested. The deficiency was remedied by obtaining a sample from the stockpile of FRS at UBC and testing it at Golder Associates laboratory (Ghafghazi and Shuttle, 2010).

The FRS gradation tested by Golder Associates contains around 0.8% fines content and has D_{50} and D_{10} of 0.271 and 0.161 mm, respectively. FRS is a uniform, angular to sub-angular with low to medium sphericity, medium grained clean sand (Figure 7.20) with $e_{min}=0.627$, $e_{max}=0.989$ and $G_s=2.72$. The average mineral composition based on a petrographic examination is 25% quartz, 19% feldspar, 35% metamorphic rocks, 16% granites and 5% miscellaneous detritus. By way of comparison, Sriskandakumar (2004) reports slightly different reference void ratios with $e_{min}=0.62$, $e_{max}=0.94$ (the differing e_{max} is not unusual between differing testing laboratories and is not thought significant, although there is the residual possibility of gradation variation within this stockpile of standard sand used for both testing campaigns).

7.3.2 Triaxial testing programme

The testing programme comprised nine drained and seven undrained triaxial compression tests prepared using the moist tamping technique. Monotonic loading to failure was used in seven of the drained tests with the remaining two tests having repeated unload–reload loops

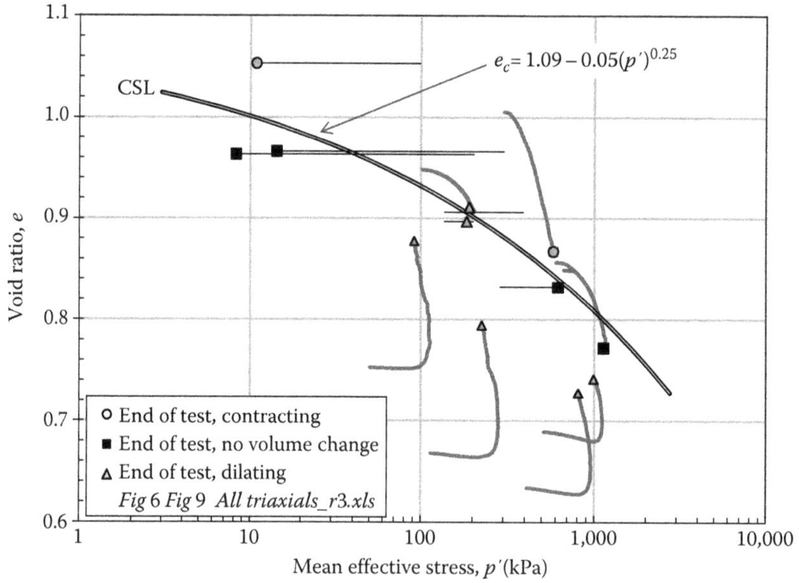

Figure 7.21 State diagram and CSL for triaxial tests on FRS.

to determine yield in unloading. Bender element tests were also carried out during isotropic compression to determine the dependence of shear modulus G_{max} on both void ratio and stress. All tests were conducted on samples that were 142 mm in height and 71 mm in diameter and followed the detailed procedures set out in Appendix B. The freezing method was used for void ratio determination with a repeatability of 0.01 or better being obtained for three pairs of tests that were targeted to start from identical conditions. Corrections were applied for membrane penetration (Vaid and Negussey, 1984) and membrane force (Kuerbis and Vaid, 1988).

The test programme is summarized on a state diagram, Figure 7.21, showing the state paths followed by each test and also indicating whether the sample was dilating, contracting or undergoing negligible volume change at the end of the test. Data from all these tests can be downloaded.

7.3.2.1 Critical state parameters

The critical state locus was developed in the usual way by looking at the end of test conditions in the state diagram. Since the stress range of interest in the cyclic tests is 40 kPa $< p' <$ 200 kPa, fitting of the CSL to data has been optimized over this stress range. A *best fit* conventional semi-log idealization over this stress range is described by the properties $\Gamma = 1.22$, $\lambda_{10} = 0.138$.

This slope of the CSL as measured by λ_{10} is three to four times larger than generally reported for standard laboratory quartz reference sands (see Chapter 2), indicating that FRS is markedly more compressible than standard experience – not entirely surprising given its mineralogy (only 25% quartz). It is also quite clear that a power law CSL would be more accurate for this compressible sand for a range of stress from about 10 to 1000 kPa. The power law fit shown in Figure 7.21 has the following properties:

$$e_c = 1.09 - 0.05(p')^{0.25} \quad (7.1)$$

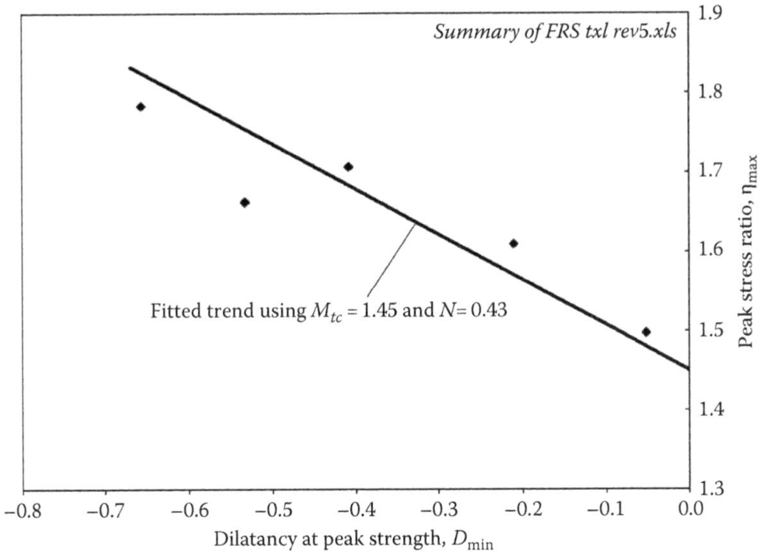

Figure 7.22 Stress–dilatancy of FRS.

The critical state friction ratio, M_{tc}, was obtained by plotting η_{max} against the corresponding dilatancy D_{min} in the usual way. The *Bishop* stress–dilatancy plot for FRS is shown in Figure 7.22 and gives $M_{tc} = 1.45$. Only dilatant samples are plotted, and there is rather more scatter in the data than seen with quartz sands, possibly caused by grain crushing in the higher confining stress tests and which is a further work dissipation mechanism.

7.3.2.2 *Plasticity parameters*

The slope of the stress–dilatancy plot is also used to determine the volumetric coupling parameter, N. The slope of the line in Figure 7.22 gives value of $N = 0.43$ for FRS; again noticeably greater than usual for quartz sands and subject to the same comments about the possible effect of grain crushing as M_{tc}.

The parameter χ_{tc} is the slope of the trend line for peak dilatancy ($= D_{min}$) versus the state parameter at peak, and the FRS results shown in Figure 7.23 give $\chi_{tc} = 3.2$ as a best fit. Like the other properties, this value is different from quartz sands.

7.3.2.3 *Elasticity*

Elasticity during shearing was measured using bender elements. The data were modelled using the equation discussed in Chapter 4:

$$\frac{G}{p_{ref}} = \frac{A}{(e - e_{min})}\left(\frac{p}{p_{ref}}\right)^{b} \tag{7.2}$$

A good fit of this elasticity idealization to the measured data was obtained using $A = 375$, $e_{min} = 0.344$ and $b = 0.466$, illustrated in Figure 7.24 which plots G_{max} derived from this

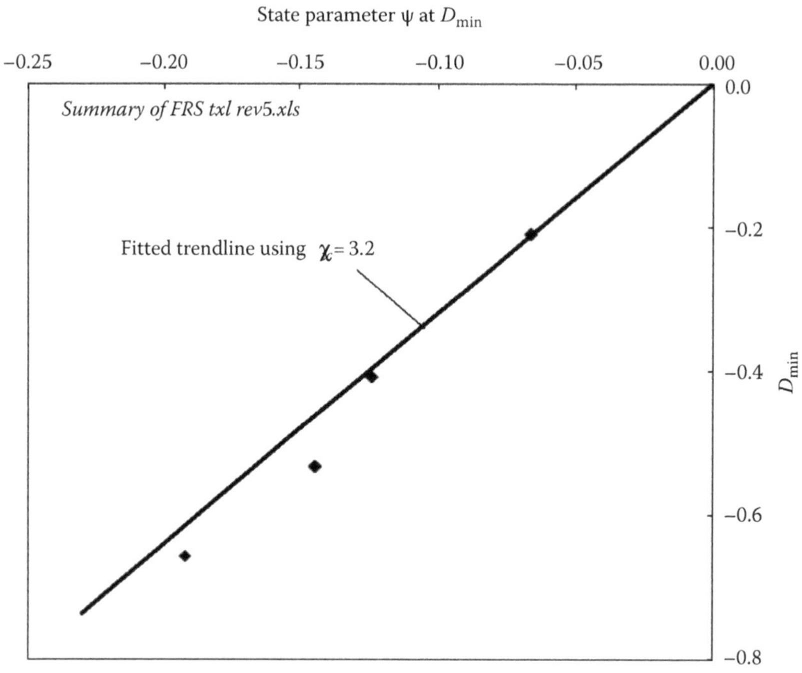

Figure 7.23 State-dilatancy (maximum dilation) of FRS.

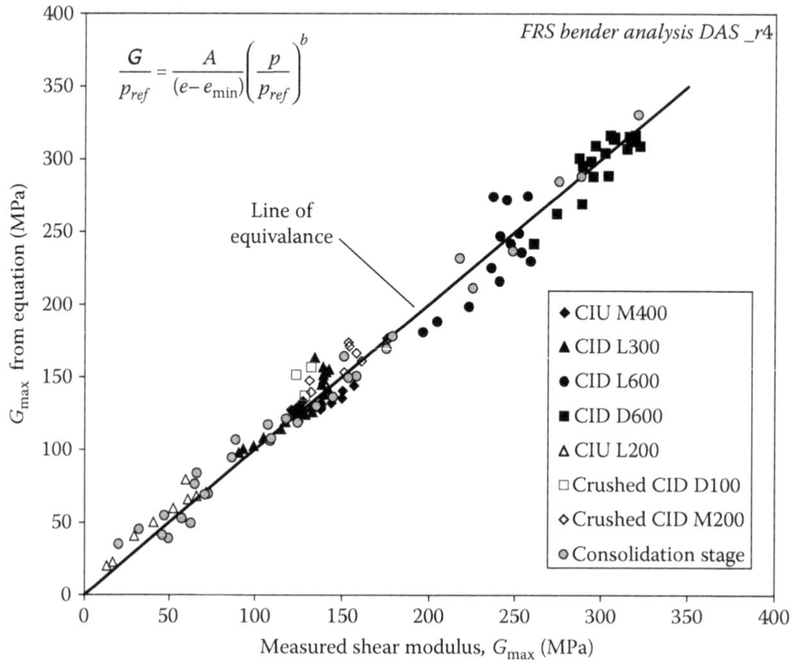

Figure 7.24 Validation of elastic model for FRS.

Table 7.1 Summary of properties of Fraser River sand from *NorSand* calibration

Property	*Value*	*Remark*
CSL		
a	0.947	*CSL is a power law function* of the form
b	0.0006	$e_c = a - b\,(p')^c$
c	0.813	
Plasticity		
M_{tc}	1.45	Critical friction ratio
N	0.43	Volumetric coupling coefficient
H	60–280ψ minimum of 40	Plastic hardening modulus for loading
χ_{tc}	3.2	Relates minimum dilatancy to ψ
Z_r	Equation 7.7	Calibrated using cyclic simple shear
Elasticity		
I_r	$\frac{G}{p_{ref}} = \frac{A}{(e - e_{min})}\left(\frac{p}{p_{ref}}\right)^b$	$A = 375$, $e_{min} = 0.344$, $b = 0.466$
ν	0.1	Poisson's ratio, commonly adopted value

calibrated elastic model with the actual measured G_{max} values. The calibrated model is unbiased and lies within ±5% of the data.

7.3.2.4 *Validation of FRS properties*

The calibrated parameter set for this FRS gradation is provided in Table 7.1. These soil properties were validated by comparing a numerical model of the test, using these properties and the reported initial conditions, with the measured data. In doing this there is one undetermined property, the plastic hardening modulus which was varied to best fit the initial stiffness measured in the tests. Figure 7.25 shows an example validation for a dense sample. Reasonable fits were obtained with the constant soil properties determined earlier despite the concern about values of these properties being outside the normal range caused by angular and crushable particles within FRS.

7.3.3 Cyclic simple shear tests on FRS

7.3.3.1 *Testing programme*

The UBC simple shear apparatus is of the NGI type (Bjerrum and Landva, 1966) and tests a cylindrical sample 70 mm in diameter and about 20 mm in height. The stress controlled cyclic tests were undertaken by enforcing a constant volume boundary condition with the stresses and applying strain rates of 10% or 20% strain per hour. Two relative densities were tested: about 40% and about 80%. All samples were air pluviated to about 40% relative density. The denser samples were then manually tamped prior to confinement being applied. The as-tested void ratio appears not to have been measured, with the work relying

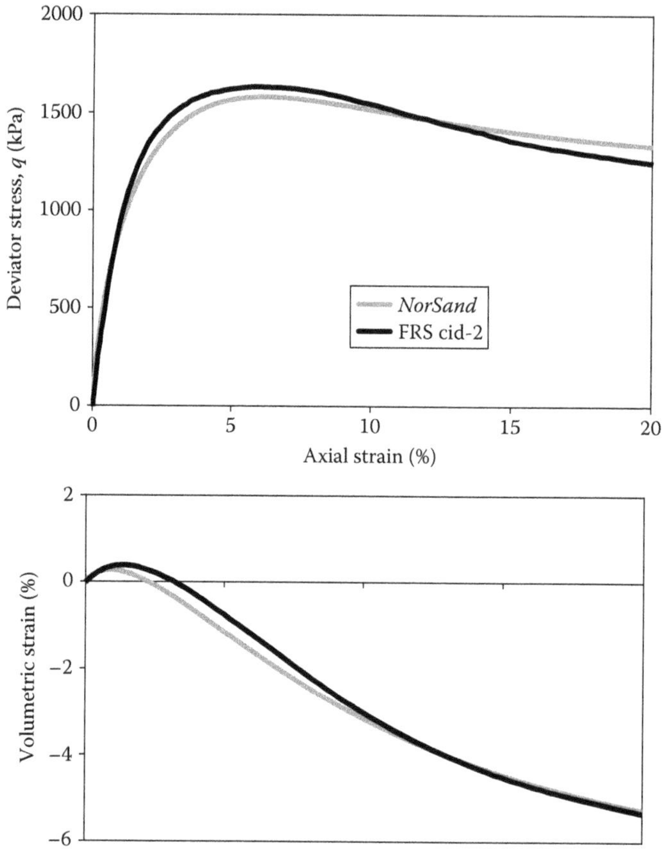

Figure 7.25 Validation of FRS properties using *NorSand*.

on carefully controlled sand placement always producing the same density in the specimen to be tested. However, the consequent accuracy in knowledge of as-tested void ratio stemming from this approach was investigated with Sriskandakumar reporting a bandwidth of as-placed relative densities that was about five percentage points of relative density either side of the target value. Invoking an estimated normal distribution of this scatter, the inferred accuracy in experimental void ratios is about $\Delta e = \pm\ 0.01$, not as good as can be achieved using freezing of saturated samples in triaxial testing but reasonable for modelling these data.

Test conditions are summarized in Table 7.2 and include the file name of the digital data that have been compiled into the downloadable spreadsheet *NorSandPSR_FRS.xls*. The spreadsheet includes a convenient routine to toggle through the data and quickly view the measured soil behaviour.

Figure 7.26 shows the initial conditions of each group of cyclic tests on state diagram together with the CSL for this sand. Despite the term *loose*, all these cyclic tests were substantially dense of the critical state and the *dense* tests were very dense. A quick inspection of the state diagram indicates none of these tests would have been prone to static liquefaction induced by shear, and indeed that is what is found in the data.

Table 7.2 Summary of cyclic simple shear tests on Fraser River sand

No.	*Test*	σ'_{vo} *(kPa)*	*CSR*	D_{rc} *(%)*	*SBR*	e_o	K_o	p'_o *(kPa)*	ψ_o	$N_{3.75}$
1	DSS38-50-0.08	50	0.08	38	–	0.818	0.8	43.3	−0.143	12
2	DSS38-50-0.1	50	0.10	38	–	0.818	0.8	43.3	−0.143	2.5
3	DSS38-50-0.12	50	0.12	38	–	0.818	0.8	43.3	−0.143	1.6
4	DSS40-100-0.08	100	0.08	40	–	0.812	0.8	86.7	−0.125	17.1
5	DSS40-100-0.1	100	0.10	40	–	0.812	0.8	86.7	−0.125	6.2
6	DSS40-100-0.12	100	0.12	40	–	0.812	0.8	86.7	−0.125	3.1
7	DSS44-200-0.08	200	0.08	44	–	0.799	0.8	173.3	−0.109	33.6
8	DSS44-200-0.1	200	0.10	44	–	0.799	0.8	173.3	−0.109	7.2
9	DSS44-200-0.12	200	0.12	44	–	0.799	0.8	173.3	−0.109	3.6
10	DSS44-200-0.15	200	0.15	44	–	0.799	0.8	173.3	−0.109	1
11	DSS80-100-0.25	100	0.25	80	–	0.684	0.8	86.7	−0.253	46
12	DSS80-100-0.30	100	0.30	80	–	0.684	0.8	86.7	−0.253	19.7
13	DSS80-100-0.35	100	0.35	80	–	0.684	0.8	86.7	−0.253	7.7
14	DSS81-200-0.2	200	0.20	81	–	0.681	0.8	173.3	−0.228	90
15	DSS81-200-0.25	200	0.25	81	–	0.681	0.8	173.3	−0.228	20
16	DSS80-200-0.3	200	0.30	80	–	0.684	0.8	173.3	−0.225	5.7
17	DSS40-st0.1-100-0.08	100	0.08	40	0.10	0.812	0.8	86.7	−0.125	–
18	DSS40-st0.1-100-0.065	100	0.07	40	0.10	0.812	0.8	86.7	−0.125	–
19	DSS40-st0.05-100-0.1	100	0.10	40	0.05	0.812	0.8	86.7	−0.125	–
20	DSS40-st0.1-100-0.1	100	0.10	40	0.10	0.812	0.8	86.7	−0.125	–
21	DSS44_st0.05-200-0.1	200	0.01	44	0.05	0.799	0.8	173.3	−0.109	–
22	DSS44_st0.1-200-0.06	200	0.10	44	0.06	0.799	0.8	173.3	−0.109	–
23	DSS44_st0.1-200-0.08	200	0.10	44	0.08	0.799	0.8	173.3	−0.109	–
24	DSS44-st0.1-200-0.1	200	0.10	44	0.10	0.799	0.8	173.3	−0.109	–
25	DSS80-st0.1-100-0.35	100	0.35	80	0.10	0.684	0.8	86.7	−0.253	–
26	DSS80-st0.1-100-0.40	100	0.40	80	0.10	0.684	0.8	86.7	−0.253	–
27	DSS80-st0.1-100-0.45	100	0.45	80	0.10	0.684	0.8	86.7	−0.253	–

$N_{3.75}$ is the number of cycles to *failure* defined as 3.75% double amplitude shear strain; SBR is the static bias ratio; Tests 10, 20 and 23 failed in first cycle because of too large SBR.

7.3.3.2 Loading conditions (static bias)

Cyclic simple shear tests include the concept of *static bias*, which is a drained shear stress applied to the sample before commencing undrained cyclic shear. Thus, the horizontal shear stress imposed on the soil at any instant of time t is

$$\tau = \tau_{st} + \tau_{cyc}\sin(\omega t) \tag{7.3}$$

where ω is the angular frequency of loading (commonly 1 Hz). Clearly, if the combination of static bias and cyclic shear is too large, the sample will fail in monotonic shear in the first half of the loading cycle; this happened in some of the FRS tests (see notes to Table 7.2).

It is conventional to normalize both the static bias stress τ_{st} and the cyclic stress τ_{cyc} by the normal vertical effective stress before the undrained loading σ'_{vo}, leading to the terms static bias ratio (SBR) and CSR, respectively. These ratios are quoted for the FRS tests and are listed in Table 7.2.

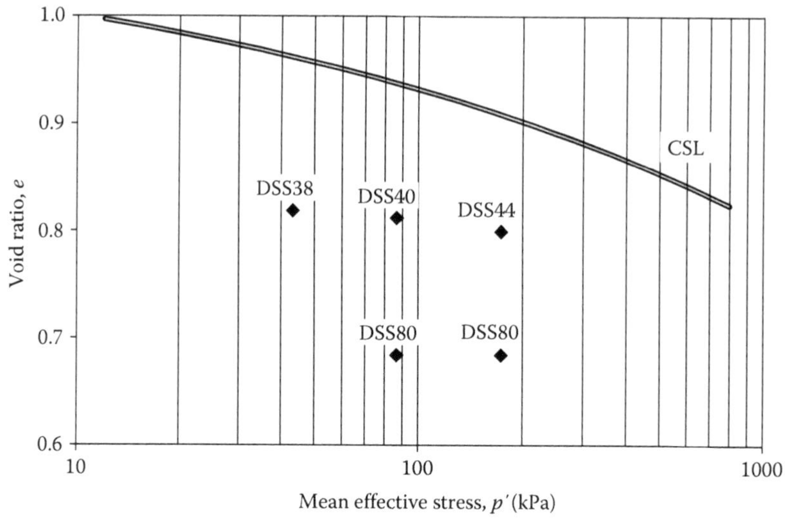

Figure 7.26 State diagram for initial conditions of FRS cyclic simple shear tests.

The literature of test data in cyclic simple shear generally views the CSR as the *loading* and the SBR as a factor that changes the nature of the response, with the SBR being the input to the K_α correction factor of the NCEER method (see Section 7.4.3). This view of the role of SBR and CSR is fundamentally incorrect. Sand behaviour, in all loading paths, depends on the ratio η/M_i. When $\eta/M_i < 1$ then stress–dilatancy forces wholly contractive strains regardless of sand density; conversely, for $\eta/M_i > 1$ sand response will be dominated by dilation and, in the case of cyclic loading, the interplay between dilation caused by accumulating plastic shear strain versus contractive strains caused by principal stress rotation. These basic behaviours need to be kept in mind when assessing data from cyclic simple shear tests.

A limitation of the cyclic simple shear test is that the measured data cannot be simply processed in terms of basic soil behaviour. Only two (τ_{vh}, σ_v) of the four stresses (τ_{vh}, σ_v, σ_h, σ_z) in the complete plane-strain stress tensor are measured in current test equipment. Thus, it is not possible to combine SBR and CSR to determine η, and without the complete stress tensor we can only guess at M_i. However, although a full calculation is not possible we can approximate the situation using monotonic undrained simple shear tests and the assumption that if the sample preparation procedures are comparable, what is measured in monotonic tests allows estimation of an equivalent of the operating M_i under all other simple shear loadings. Figure 7.27 shows this approach applied to the monotonic simple shear of FRS. The normal 'S'-shaped stress path of samples denser than critical was measured, with a trend line drawn through the image condition (phase transition) points. This trend line is the image stress ratio ($ISR \sim M_i/2$) expressed in terms of the stresses measured in simple shear. There is an implicit linkage to sample preparation methods that establish the K_o condition, but this ISR is broadly consistent with M_i in plane strain for FRS based on the triaxial calibration ($M_{tc} = 1.49 \rightarrow M_i \sim 1.05$ at a Lode angle $\theta \sim 17°$ typically developed at image conditions in plane strain shear; see Chapter 2).

The estimation of an equivalent of η is more complicated, as η represents current conditions while the SBR and CSR are both expressed in terms of initial conditions. The initial and current conditions differ by the excess pore pressure, which depends on the loading (and takes us back to the earlier discussion in Chapter 6 that s_u/p' is a better measure of

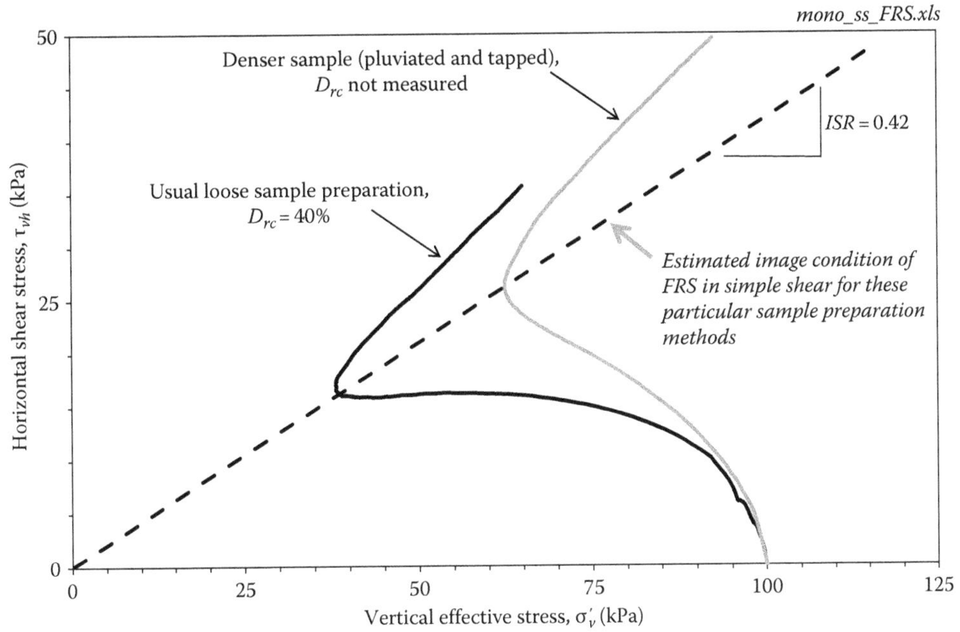

Figure 7.27 Image stress ratio of FRS in monotonic simple shear.

undrained strength than the *instability locus*). A pragmatic way forward is to define an initial *loading* stress ratio LSR_o where the subscript 'o' has been added to emphasize that it represents initial conditions in the test. The ratio LSR_o/ISR would then be anticipated to classify cyclic simple shear behaviour. Let us now consider the FRS data in this framework.

7.3.3.3 *Sand response for* LSR$_o$ < ISR/2

Test 4 (DSS40-100-0.08) of the data set has no static bias with the applied CSR equating to $LSR_o/ISR = 0.19$. Test 18 (DSS40-st0.1-100-0.065) has both static bias and cyclic stress equating to $LSR_o/ISR = 0.39$. The behaviour of these two tests is presented and compared in Figure 7.28. There is considerable similarity in the response.

In terms of the stress paths followed, there is a difference in the first quarter loading cycle, which is a loading that is the same as a normal monotonic test. The difference is simply because Test 18 is taken to a higher shear stress than Test 4, so there is a little excess pore pressure but exactly as would be expected from static loading. During the load cycling that then follows, both tests show a slow increase in excess pore pressure with very little pressure change during the cycle. This is a *fatigue*-like loading. And notice that there is essentially no loss in soil stiffness while the excess pore pressure accumulates. What stiffness change is seen appears largely as a consequence of decreasing effective confining stress causing a slight decrease in G_{max} (as expected from the calibrated elasticity, Equation 7.2).

This situation of fatigue-like generation of excess pore pressure changes dramatically, in both tests, when the stress path realizes a current stress ratio $\tau_{vh}/\sigma'_v \sim 0.95 ISR$; that is, at a mobilized stress ratio slightly less than the *ISR*. Then, strains increase dramatically while the excess pore pressure stabilizes into the familiar *butterfly* stress paths reported by

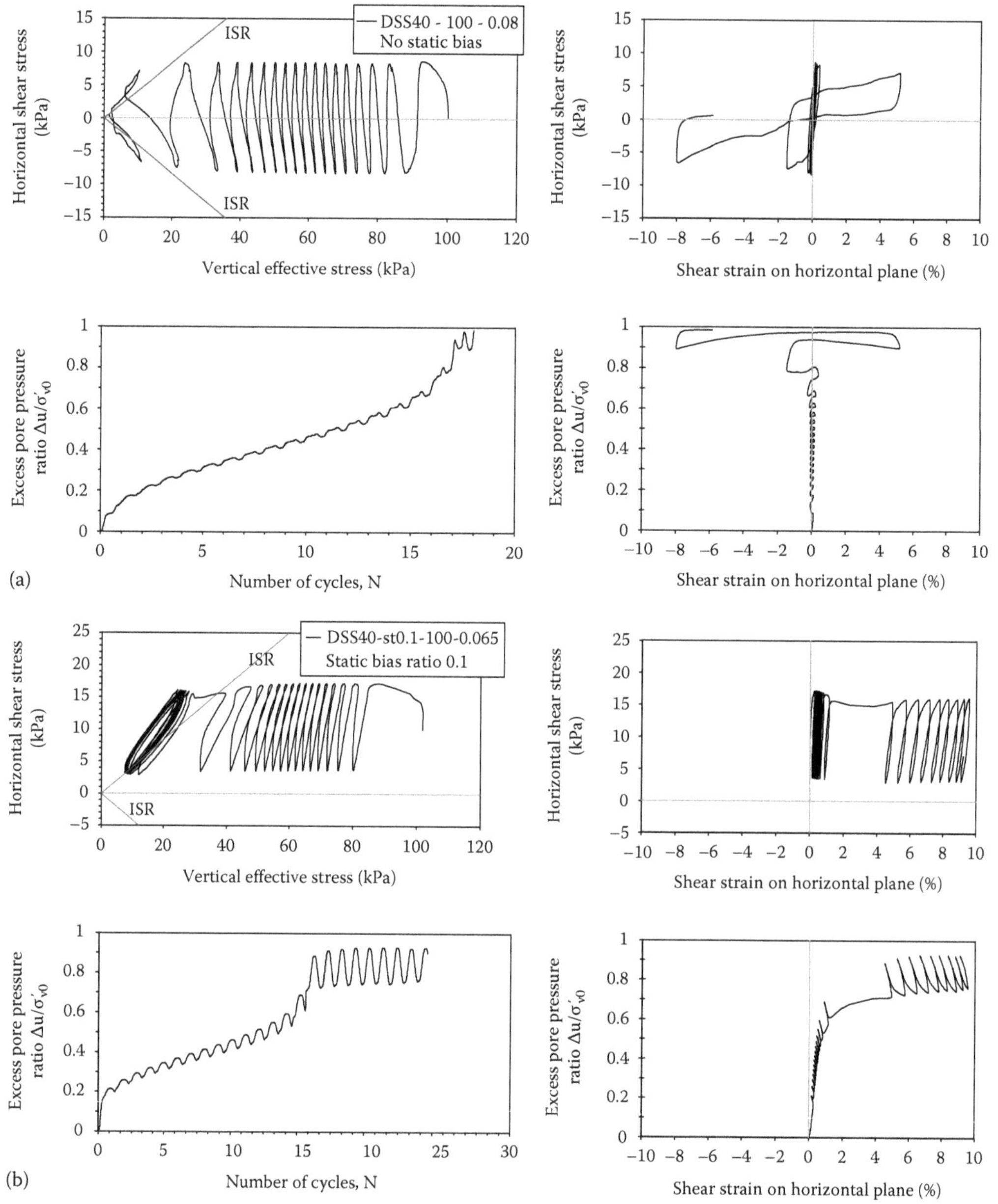

Figure 7.28 FRS in cyclic simple shear with $LSR_0 < ISR/2$. (a) No applied static bias, $\psi_o = -0.125$, CSR = 0.08 and (b) static bias ratio 0.1, $\psi_o = -0.125$, CSR = 0.08.

many investigators. This is a situation of no strength loss – as the sample continues to support the peak imposed load – but severe loss of stiffness. Although the samples were both described as loose, that description is inaccurate as they were both at $\psi_o \sim -0.13$, a state that is substantially dilatant and also denser than the Shuttle and Cunning (2008) criterion separating undrained small strain and flow slide behaviour discussed in Chapter 6.

Notice that there is no great effect of *static bias* in itself during the fatigue-like stage, which is between the end of the first quarter cycle and when the stress path approaches the ISR. During this stage, the rate of increase in excess pore pressure is similar between the two tests despite the rather large difference in the static bias. The rate of increase is also approximately linear with load cycles. However, static bias causes a ratcheting and progressive increase in shear once the stress paths cross the ISR rather than the soil wobbling around a central position (unsurprisingly).

7.3.3.4 Sand response for $LSR_o \approx ISR$

Test 13 (DSS80-100-0.35) of the data set has no static bias but has a rather large CSR that equates to $LSR_o/ISR = 0.83$. Test 25 (DSS80-st0.1-100-0.35) has both static bias and the same large CSR equating to $LSR_o/ISR = 1.07$. The behaviour of these two tests is presented and compared in Figure 7.29. Both samples were similarly dense, at about $\psi_o \sim -0.25$, and there is considerable similarity in their responses.

As expected at $LSR_o/ISR \approx 1$, these samples were loaded to near the image state in the first quarter cycle and the onset to dilation can be seen in the stress paths of both in that first quarter cycle. The only effect of static bias is to push the sample in Test 25 into dilation about half a loading cycle earlier than the sample without static bias.

Once in the dilating zone, pore pressures do indeed accumulate rapidly and the samples liquefy with transient states of near zero effective confining stress and the familiar *butterfly* stress paths. Now consider the stress–strain behaviour. There is only a gradual loss of stiffness, and the occurrence of *initial liquefaction* is not a traumatic event at all. This aspect is further emphasized by Figure 7.30 which shows the shear strain accumulation with loading cycle measured in Test 13. There is a near-linear increase in maximum shear strain with each cycle and nothing that could reasonably called failure. The best analogy is *fatigue softening.* The horizontal lines show a standard *initial liquefaction* criterion (3.75%) that gives $N_{liq} \sim 8$ cycles and the first occurrence of the transient zero effective stress as at $N_{liq} \sim 11$ cycles. Neither event changed the fatigue-like accumulation of strain. Patently, these limits are arbitrary and capture nothing insightful about the soil behaviour at all. (Note that a shear strain limit of 3.75% is taken as equivalent to an axial strain of 5% in a triaxial compression test.)

7.3.4 Nature of liquefaction in simple shear

In terms of general understanding, it is readily apparent in Figures 7.28 and 7.29 that the soil stiffness simply responds to the stress path and the image stress ratio. Whatever the combination of cyclic and static stress ratios, it is the proximity of the peak shear stress to the image stress ratio that controls stiffness. As such, the trend in the literature to talk of a *static bias effect* is at best unhelpful and at worst misleading. Soil behaviour in cyclic loading remains controlled by the critical friction ratio as during static loading. It is the ratio η/M that is fundamental, and all the conflicting accounts of the role of *static bias* are immediately clarified once the stress path is properly expressed using standard stress and strain measures that we use for other aspects of soil behaviour.

Broadly, any combination of static and cyclic shear stress that corresponds to the soil approaching its current image stress ratio (which depends on the soil's properties) will result in rapid transition to dilation-controlled behaviour for soils with $\psi < 0$ and simple immediate static liquefaction for looser soils. In the case of the denser soils, concepts of *initial liquefaction* are unhelpful. The initial response is the same as loading the soil monotonically and this is then followed by a fatigue-like accumulation of strain. In short, for $\psi < 0$ we have loss of stiffness, not loss of strength. The behaviour mentioned earlier is similar to that reported by Baki et al. (2012) for cyclic triaxial tests, except that they considered the instability *zone* measured in a matched monotonic test rather than the image stress ratio. Of course,

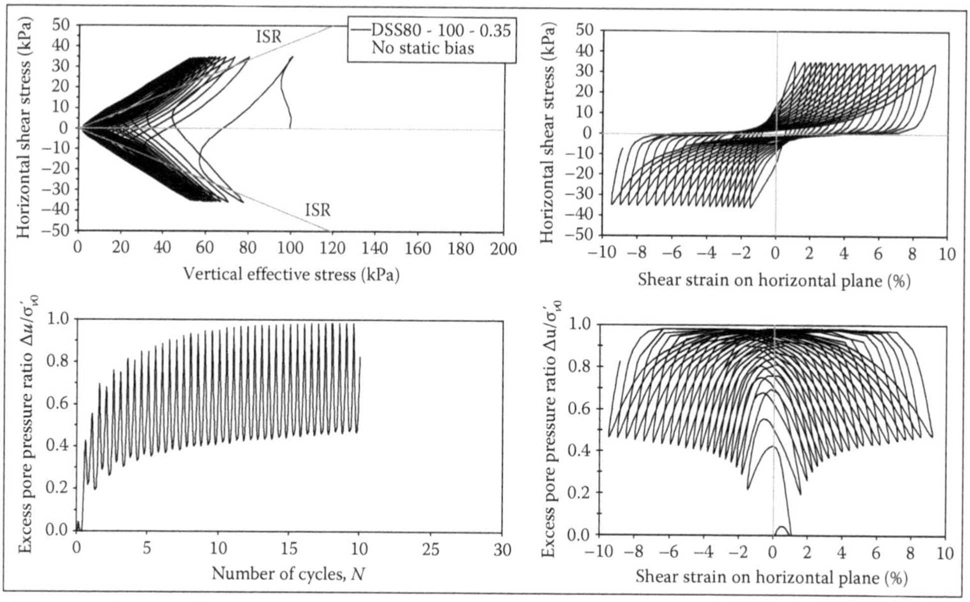

(a) Dense cyclic test without static bias

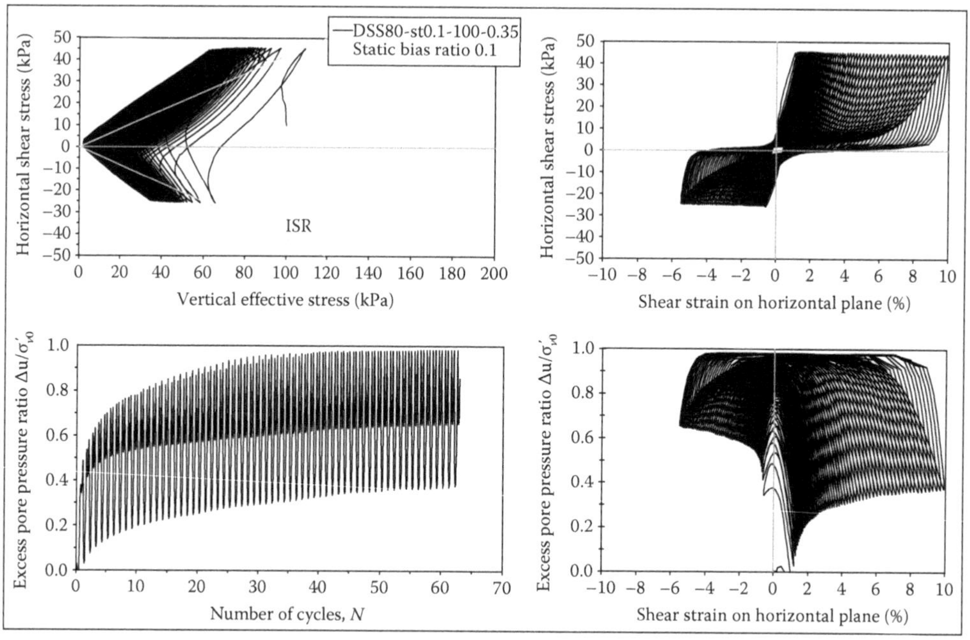

(b) Dense cyclic test with moderate static bias (SBR = 0.1)

Figure 7.29 FRS in cyclic simple shear with $LSR_0 \approx ISR$. (a) and (b) refer to the blocks of the four little plots.

there remains an issue as to whether $\psi \sim 0$ is the exact demarcation between these differing behaviours as the existing experimental data do not cover a sufficient range of initial states.

When using laboratory *element* test data, such as those mentioned earlier, there is always an uncertainty about the extent to which the laboratory reflects full-scale soil behaviour in engineering works. Before moving on to the extensive earthquake case history database in

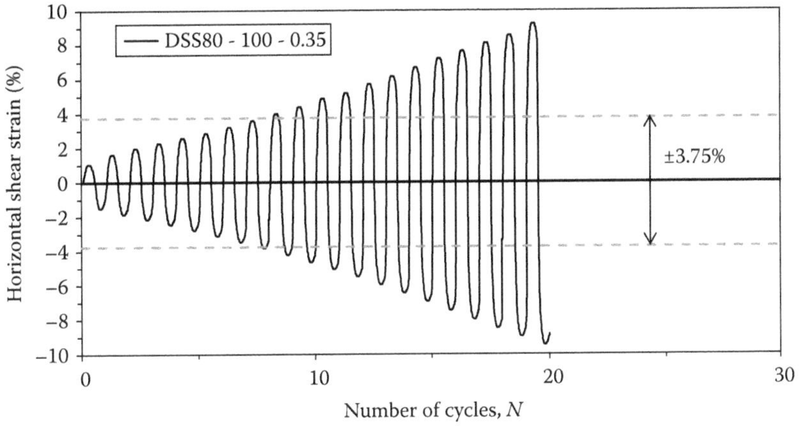

Figure 7.30 Evolution of shear strain in cyclic loading for stress states $\eta_{max} > M_i$.

the next section, consider again the Molikpaq. Compare the pore pressure build-up in Figure 7.28b with the full-scale pore pressure data from the Molikpaq presented in Figure 1.22. Near identical trends for excess pore pressure development with loading cycles are seen in both the laboratory element test and the in-situ piezometers. The laboratory cyclic simple shear appears to be a good representation for the field behaviour in this instance, but is also much simpler than the hollow cylinder method presented in the previous section.

7.4 BERKELEY SCHOOL APPROACH

7.4.1 Background

There is a long tradition in civil engineering of building on precedent and this leads to an emphasis on case history information. Such an approach really comes about because all calculations or methods of analysis involve idealizations and these idealizations lead to a mismatch between calculations and what actually happens. Physical models are no panacea either, as physical models have their own errors, idealizations and scale effects. It is very difficult to do pre-production tests in civil engineering to resolve the consequences of model uncertainty (unlike other branches of engineering), as the prototype is the end result in most cases. Because civil engineering projects have very low probabilities of failure (typically around a 10^{-6} per annum), engineers are beholden to learn the maximum from those failures or performance histories that become available. Liquefaction is no different, and this respect for precedent experience has strongly influenced how the subject has developed.

The disastrous earthquakes in Alaska and Niigata in 1964 produced a realization that the then used pseudo-static method of analysis did not give reliable predictions for stability during earthquakes. Sustained research at Berkeley by the late Professor Seed and his colleagues followed and continues today. Strength reduction during earthquake (cyclic) loading was identified as one of the key issues.

Initial research used the cyclic triaxial test, and indeed some of the data presented earlier were from this work. However, in the mid-1960s when this testing started, there was no constitutive theory capable of modelling the nuances of excess pore pressure generation. Correspondingly, how to translate what was measured in the cyclic triaxial test to in-situ response during an earthquake was uncertain. There was also the important fact that undisturbed samples cannot be obtained for most liquefiable soils, so how were the laboratory conditions using reconstituted samples to be related to conditions in-situ? These factors caused field case history data to be viewed as the principal resource to understand vulnerability to earthquake-induced liquefaction. The concept was to observe where liquefaction occurred in an earthquake and where it did not. These observations were then related to the estimated cyclic shear stress experienced by that ground and the prior state of that ground. This is a geological classification approach, rather than one based in mechanics, and an essentially empirically based protocol was developed for liquefaction assessment. We have used the terminology *Berkeley School* for this liquefaction protocol to indicate that there were many involved other than the late Professor Seed, but that the intellectual direction really remained influenced by the ideas originating at Berkeley (and which continues, today, both at Berkeley and at Davis).

This Berkeley School has become the dominant approach for assessing earthquake-induced cyclic mobility, with many contributors providing further case history information and refinement of some of the inferred trends. Substantial impetus to the formalization of the approach developed in the United States through several workshops, in which invited contributors discussed the state of knowledge in a collegial setting over several days. The National Science Foundation (NSF) with their earthquake hazard reduction programme was a principal instigator in getting these workshops held, and the profession owes a debt to Cliff Astill who was the programme manager at the NSF. The first of these workshops was held in Dedham during March 1985, under the co-sponsorship of the Nuclear Regulatory Commission, with contributions from 28 researchers in the field of liquefaction. Most of these researchers were from the United States but there were also contributions from Japan and the United Kingdom. This workshop produced a paperback book that became a widely used reference at the time, published under the auspices of the National Research Council (NRC, 1985). A decade later, a further workshop was held during January 1996 in Salt Lake City, co-sponsored this time by National Centre for Earthquake Engineering Research (NCEER, now MCEER and where M is for multidisciplinary). This second workshop was less wide ranging than the 1985 workshop, being specifically directed at updating the Berkeley School approach with contributions from 20 researchers. The results of the workshop appeared both as a workshop proceedings (http://mceer.buffalo.edu) (MCEER, 2015) and as a summary in the ASCE *Journal of Geotechnical and Geoenvironmental Engineering* (Youd et al., 2001). Some of the 1996 proceedings were published elsewhere as journal papers. What follows on this section is a description of this updated Berkeley School approach, with the individual contributions cited as they arise. The shortfalls of the Berkeley empirical approach will then be discussed before returning to the critical state approach for guidance.

7.4.2 Liquefaction assessment chart

The key idea in the classification approach of the Berkeley School is to identify cases of liquefaction from no liquefaction. These cases are assessed in terms of the initial state of the soil (characterized by a normalized penetration resistance) and a measure of the earthquake severity (characterized as a CSR normalized for earthquake magnitude). The results are presented on a plot of characteristic state against representative CSR, which divides into two areas: cases of liquefaction and cases on no liquefaction. The bounding line between these two areas is referred to as the *CRR*. This type of plot is sometimes referred to as a *soil liquefaction assessment chart*, and an early example was shown in Chapter 1 (Figure 1.12). Seed et al. (1983) describe the genesis of this plot, starting with Japanese engineers after the Niigata earthquake followed by a collation of data from many locations by Seed and Peacock (1971). More data were added after earthquakes in China, Guatemala, Argentina and Japan in the mid-1970s. Nevertheless, a lot of the case history data remained inaccessible in the usual literature until Fear (1996) provided a comprehensive review of the case histories in what is now termed the Berkeley Catalogue. Subsequent to that, Moss (2003) included the catalogue in his PhD thesis and researchers at Berkeley and Davis are adding to the catalogue and opening it up to be more accessible.

Field evidence of liquefaction for the various case histories generally consisted of surficial sand boils, ground fissures or lateral spreads. Data were collected mostly from flat or gently sloping sites which where underlain by Holocene alluvial or fluvial sediments no deeper than about 15 m. The Wildlife Site data recorded during the Superstition Hills earthquake, magnitude 6.6, in 1987 are unique as a case history with measured excess pore pressures and strong ground shaking (Chapter 1).

The adopted measure of cyclic stress imposed by the earthquake in a particular case history was the dimensionless ratio τ_{cyc}/σ'_{vo}, called the *CSR*, and which we encountered earlier in the context of cyclic simple shear tests. τ_{cyc} is the equivalent uniform cyclic shear stress causing liquefaction and σ'_{vo} is the initial vertical effective stress before the earthquake at the same depth.

Cyclic liquefaction depends on the number of cycles every bit as much as the cyclic shear stress. In the case of earthquakes the number of cycles depends on the source mechanism, but there is an empirical relationship between the number of significant cycles and the source magnitude.

Larger earthquakes tend to longer duration but comparable frequency content to smaller earthquakes, as illustrated in Figure 7.31. Because an earthquake time history contains many smaller cycles as well as a few very large ones, a classification approach must necessarily reduce this variable stress–time record to some standard single number. The concept of significant cycles is this: if the very variable time history of the earthquake shaking is approximated as a single frequency sine wave of constant amplitude (taken as two-thirds the maximum value imposed by the largest spike in the stress time history), then the number of uniform cycles causing equal damage (e.g. excess pore water pressure) as the actual irregular waveform is the number of significant cycles. Figure 7.32 shows an early relationship between the cyclic shear stress and the number of significant cycles to cause liquefaction in laboratory cyclic triaxial tests together with the estimated earthquake magnitude equivalence. The graph is normalized to the shear stress causing liquefaction in 15 cycles and $M = 7.5$ as the reference condition. This early relationship was the basis for a *magnitude scaling factor* in the approach. In the last decade or so there has been much research related to the magnitude scaling factor, because the actually observed range appears to be quite large.

The adopted in-situ state measure was the SPT resistance, primarily because of its prevalence worldwide in the available case history records in the 1960s when this approach to liquefaction started. It was recognized from the outset that the SPT was much influenced by test procedures and in particular the energy delivered by the hammer system. A convention adopted was to modify the measured blow count by the ratio of 60% of the theoretical free fall energy to the (usually estimated) energy actually delivered. This standard of 60% energy efficiency was adopted to approximate much of the data gathered with the old cathead system for the SPT (Chapter 4). The modified penetration resistance is referred to as N_{60}.

It was also recognized that SPT blow counts are much affected by vertical stress at the test depth. The convention adopted was to adjust data to that equivalent to a stress level of $\sigma'_{vo} = 100$ kPa (or approximately 1 tsf), commonly using the square root factor discussed in Section 4.5.2 (Equation 4.9). The energy and stress level–adjusted penetration resistance is then referred to as $(N_1)_{60}$. There are other adjustments for borehole diameter, rod length and the presence of liners in the SPT sampler, but these are secondary and need not be considered for reasons that will become apparent shortly.

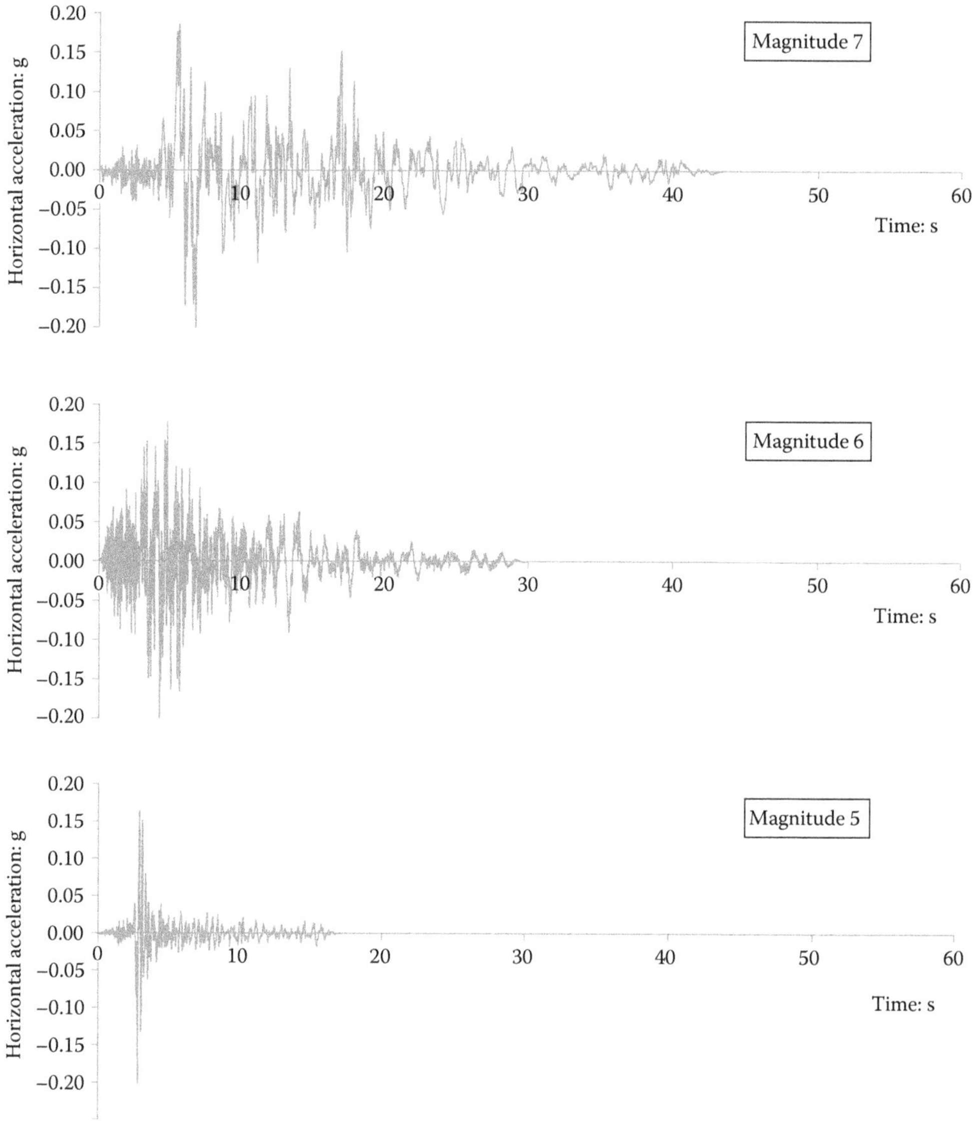

Figure 7.31 Effect of earthquake magnitude on duration of ground motion.

Figure 7.33 presents the variant of the field case history record published in 2001 by members of the *1996 NCEER/NSF Workshop on Liquefaction Resistance of Soils*. It shows the estimated characteristic CSR experienced versus the adjusted blow count for the various sites for which data are available. Sites that liquefied are distinguished from sites that did not, and this then leads to a line distinguishing liquefaction from non-liquefaction. It was observed early in the evaluation of field case histories that, unsurprisingly, soil type mattered. Because the earthquake-based evaluation was focused on loose sandy soils, the soil type classification adopted was that of *fines content*, fines being the fraction of the particle size distribution finer than the #200 sieve. Different liquefaction bounding lines were drawn on the liquefaction assessment chart depending on fines content, as shown in Figure 7.33.

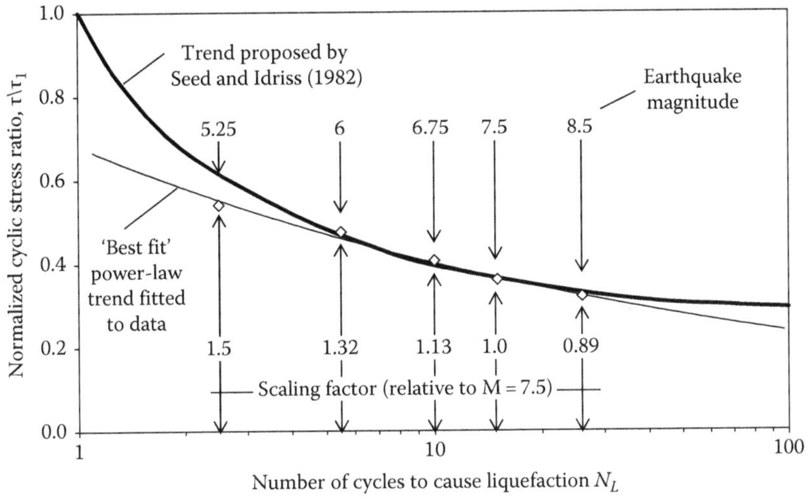

Figure 7.32 Earthquake magnitude scaling factors tabulated and superimposed on cyclic strength curve in Seed and Idriss (1982), with power law trend added. (After Seed, H.B. and Idriss, I.M., *Earthquake Engineering Research Institute Monograph*, Earthquake Engineering Research Institute, Oakland, CA, 1982.)

Although the original work evaluating the case history experience of liquefaction was based on the SPT, the many deficiencies of the SPT are widely known and several workers developed comparable liquefaction assessment charts using the CPT as the input information. This effort started some 30 years ago with Robertson and Campanella (1985) and continued with contributions by Seed and de Alba (1986), Olsen (1988), Olsen and Malone (1988), Shibata and Teparaska (1988), Suzuki et al. (1995), Stark and Olson (1995), Olsen and Koester (1995). This body of experience was summarized by Robertson and Wride (1998) as a contribution to the NCEER workshop on cyclic liquefaction. Figure 7.34 shows the final form of the CPT-based equivalent to Figure 7.33 from that workshop. Conceptually, the chart has the same form with the CPT replacing the SPT, a stress level–adjusted penetration resistance also being used.

Figure 7.34 is preferable to Figure 7.33 because it uses the CPT rather than the poorly repeatable SPT and therefore avoids uncertainty in SPT corrections. However, a missing feature in comparing Figure 7.34 with Figure 7.33 is the soil type. The CPT-based chart is specifically annotated as applying only to sands with less than 5% fines. Although the SPT was originally perceived as essential to use a liquefaction assessment chart, since the original charts invoked fines content to capture the effects of soil type, you do not need soil samples to estimate soil type. Soil type can be evaluated directly from the measured CPT friction and pore pressure data (Section 4.7 and Figure 4.27). However, a more elegant and arguably theoretically sounder approach (because the CPT measures mechanical behaviour, not soil type) is to avoid the intermediate step of estimating a soil type from the CPT and directly express liquefaction resistance in terms of what the CPT measures. This is the basis on which Robertson and Wride (1998) developed an adjustment factor (K_c) from the CPT data itself (soil behaviour–type index I_c) to give CRR directly.

We will return to an updated version of this CPT liquefaction chart, as well as a state parameter version, after first considering a couple of other factors that are needed in the Seed approach.

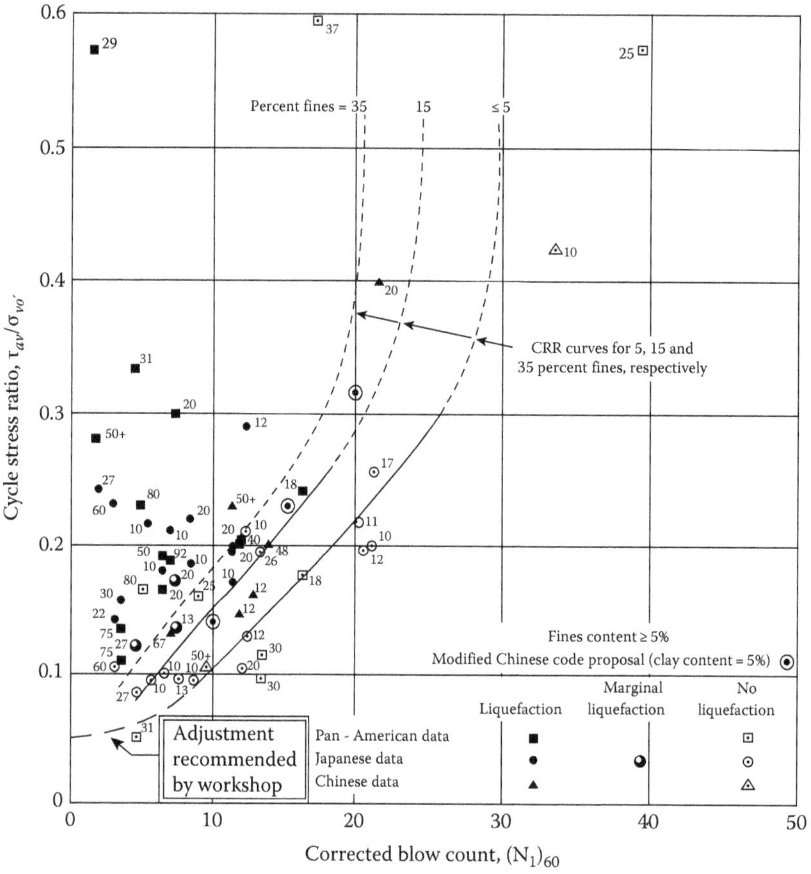

Figure 7.33 NCEER *Soil Liquefaction Assessment Chart* using SPT input. Data point numbers correspond to the case history reference assigned by Fear (1996) based on Ambraseys (1988). (From Youd, T.L. et al., *ASCE J. Geotech. Geoenviron. Eng.*, 127(10), 817, 2001. With permission from the ASCE, Reston, VA.)

7.4.3 CRR adjustment factors

The CRR developed from Figure 7.33 or 7.34 is based on case history data from limited site conditions of shallow liquefaction and near level ground. The line separating liquefaction from non-liquefaction case histories on these figures is for a $M = 7.5$ earthquake and the strength obtained for a given penetration resistance is termed $CRR_{7.5}$. Seed (1983) suggested that this case history data could be used beyond these restricted circumstances by adjusting the $CRR_{7.5}$ to account for actual site conditions. The available CRR for the in-situ conditions and the design (or actual) earthquake is then given by

$$CRR = CRR_{7.5} K_M K_\sigma K_\alpha \tag{7.4}$$

K_M is the earthquake magnitude adjustment factor, and the most commonly used relationship for this was shown earlier in Figure 7.32. Magnitude scaling compensates for the differing number of significant cycles in an earthquake, as larger earthquakes tend to have longer duration of shaking. More recently, Youd et al. (2001) have indicated a range of relationships determined by various workers, summarized in Figure 7.35. Seed and Idriss (1982)

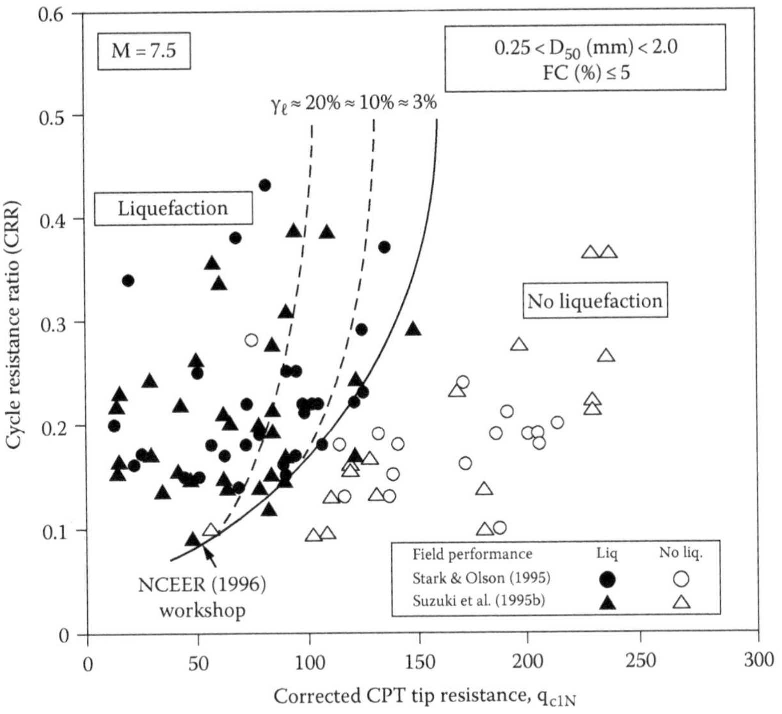

Figure 7.34 NCEER *Soil Liquefaction Assessment Chart* using CPT input. (From Robertson, P.K. and Wride, C.E., *Can. Geotech. J.*, 35(3), 442, 1998. With permission from the NRC, Ottawa, Ontario, Canada.)

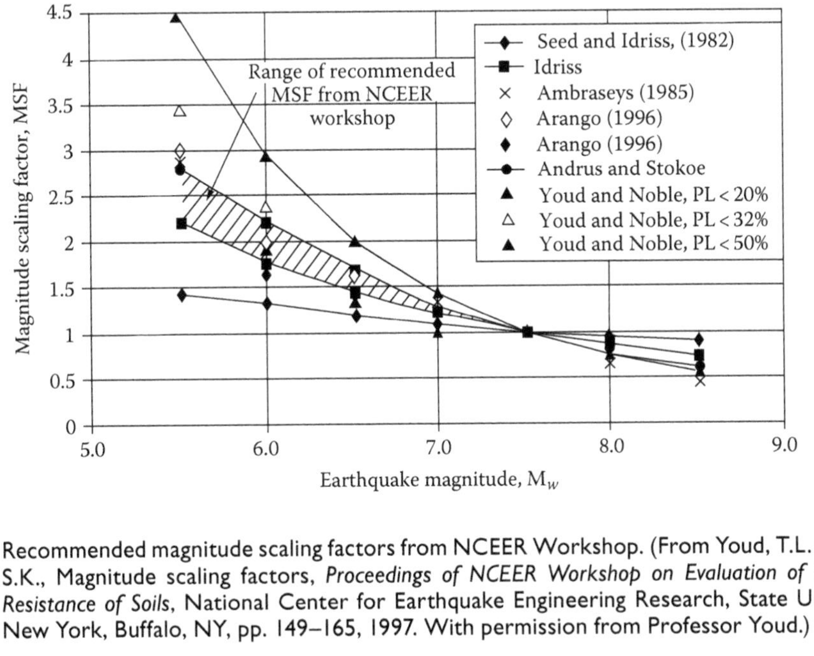

Figure 7.35 Recommended magnitude scaling factors from NCEER Workshop. (From Youd, T.L. and Noble, S.K., Magnitude scaling factors, *Proceedings of NCEER Workshop on Evaluation of Liquefaction Resistance of Soils*, National Center for Earthquake Engineering Research, State University of New York, Buffalo, NY, pp. 149–165, 1997. With permission from Professor Youd.)

essentially normalized the *CRR* by scaling the number of cycles to liquefaction to the 15 loading cycles, which was considered the equivalent to a M7.5 earthquake. Ambraseys (1988), Arango (1996) and Andrus and Stokoe (1997) independently developed slightly different empirical scaling factors depending on their approaches to liquefaction assessment, although all were based largely on the same case history base. Finally, Youd and Noble's recommended scaling factors are based on a probabilistic analysis of the case histories (so that the *PL* values in Figure 7.35 refer to the probability that liquefaction occurred). The original Seed and Idriss factors (Figure 7.32) are still the most widely used in practice, although Youd et al. (2001) recommended taking a slightly more conservative approach for engineering practice identified as *Idriss* in Figure 7.35. (This curve was suggested by Idriss, based on his re-evaluation of the original data, for the NCEER workshop.)

K_σ (*K sigma*) is the stress-level adjustment factor. The database of the liquefaction case histories is dominated by shallow sites, and early cyclic triaxial testing had shown an effect of initial effective confining stress on liquefaction resistance. Seed (1983) therefore introduced K_σ to extrapolate the simplified liquefaction chart to overburden pressures greater than 100 kPa. Isotropically consolidated cyclic triaxial tests on sand samples were used to measure *CRR* for high stress conditions and the correction factor developed by taking the ratio of *CRR* for higher pressures to the *CRR* for approximately 100 kPa (at the same relative density). Other workers added to Seed's data and suggested modifications to K_σ for engineering practice, these suggestions being linked to some further analyses of the actual stress levels of the liquefying ground in the various case histories. Figure 7.36 illustrates the range of values reported by Seed and Harder (1990), using research materials such as Monterey and Reid Bedford sands mentioned in Chapters 2 and 4 as well as sands used for

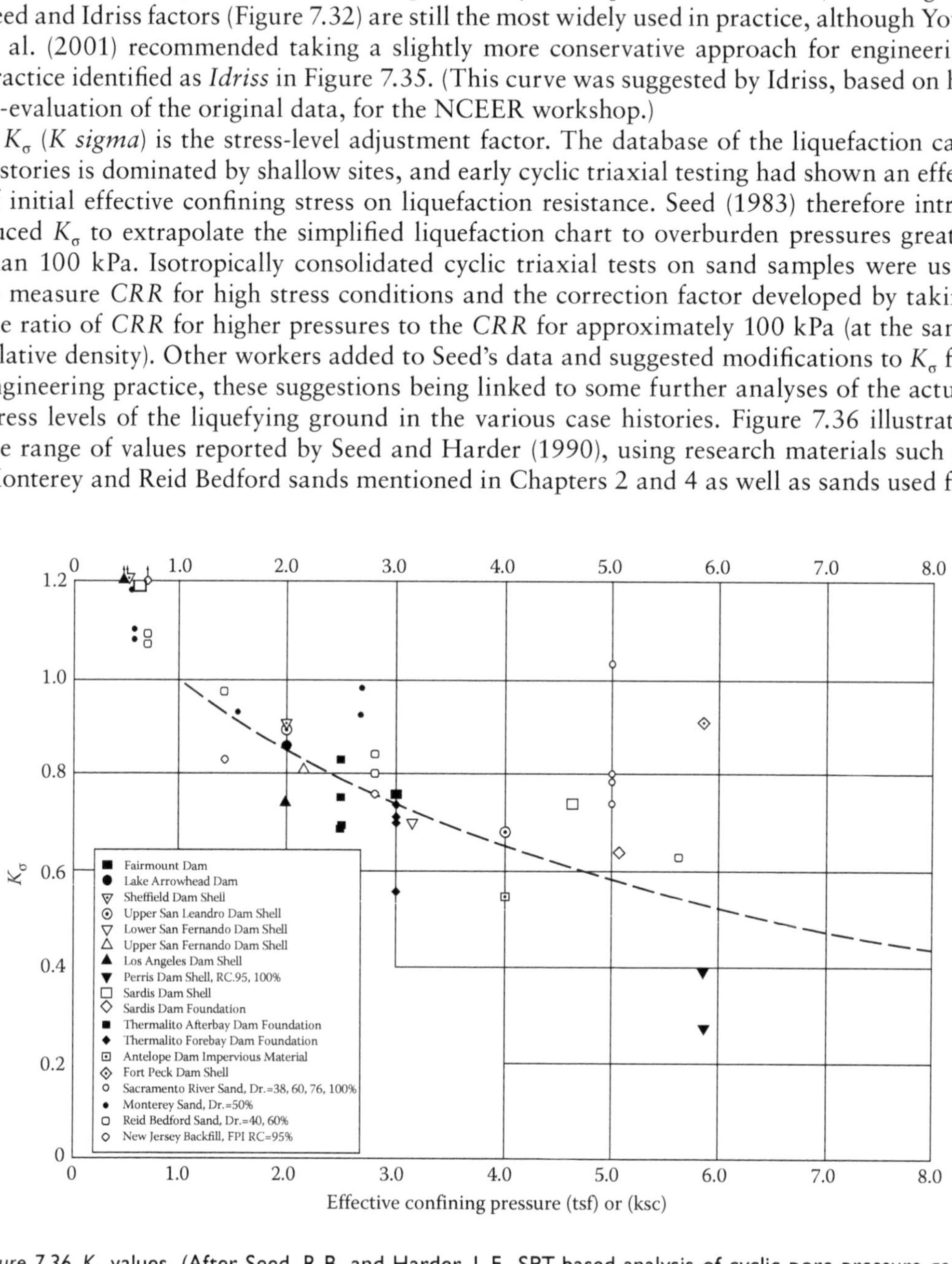

Figure 7.36 K_σ values. (After Seed, R.B. and Harder, L.F., SPT-based analysis of cyclic pore pressure generation and undrained residual strength, *Proceedings of H.B. Seed Memorial Symposium*, Vol. 2, pp. 351–376, 1990. With permission from BiTech, Richmond, British Columbia, Canada.)

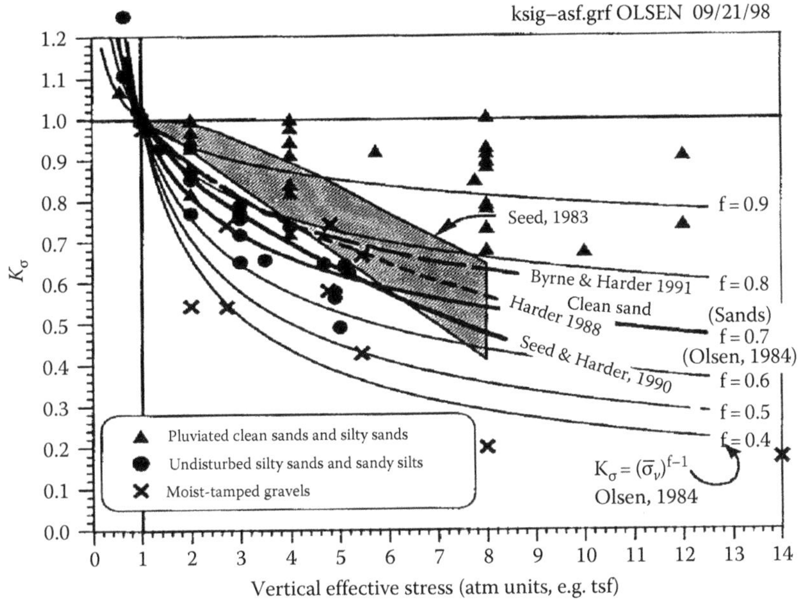

Figure 7.37 K_σ values recommended by Hynes and Olsen. (From Hynes, M.E. and Olsen, R.S., Influence of confining stress on liquefaction resistance, *Proceedings of the International Workshop on Physics and Mechanics of Soil Liquefaction*, Balkema, Rotterdam, the Netherlands, pp. 145–152, 1999. With permission of Taylor & Francis.)

engineering from several dams including Fort Peck Dam. The test data are grouped together as a single trend relating K_σ to initial vertical effective stress and taking 100 kPa (or 1 tsf) as the reference stress level for $K_\sigma = 1$. Note that the *CRR* decreases with increasing confining stress at constant relative density.

Hynes and Olsen (1999) provide a summary of K_σ suggestion that are illustrated in Figure 7.37 and a wide range of behaviour is apparent, with the appropriate K_σ depending on site conditions such as relative density, stress history, ageing and overconsolidation ratio. In some soils (pluviated clean sands), a fivefold increase in initial vertical effective stress results in less than 20% decrease in the *CRR*, while in others (undisturbed silty sands and sandy silts), the *CRR* might reduce by 40%. Hynes and Olsen suggest that the different behaviours are related to initial density of the soil with dense soils having proportionately greater reduction in *CRR* for a given stress increase, despite the data in Figure 7.37 indicating that soil type has a strong influence. Their recommendation is to use the factors $f = 0.8$ for $D_r = 40\%$, $f = 0.7$ for $D_r = 60\%$ and $f = 0.6$ for $D_r = 80\%$ in engineering practice as this provides minimal or conservative estimates for both clean and silty sands and for gravels.

K_α (*K alpha*) is a factor introduced to capture the perceived effect of sloping ground. Recall that the case history record is dominated by near level ground sites. One application of liquefaction analysis is slope failure of dams, and the ground beneath the dam shells has a different in-situ stress state from the level ground case histories. How should this be taken into account? Seed (1983) suggested a further modifier termed K_α. The idea behind K_α comes from cyclic triaxial tests. If cyclic triaxial tests start from an anisotropic stress condition, then a larger *CRR* is obtained for any chosen number of cycles to liquefaction and Figure 7.38 shows some data indicating these experimental trends. Seed (1983) extrapolated from this laboratory result to slope stability by noting that the anisotropic stress conditions in a simple shear test sample could be expressed in terms of the

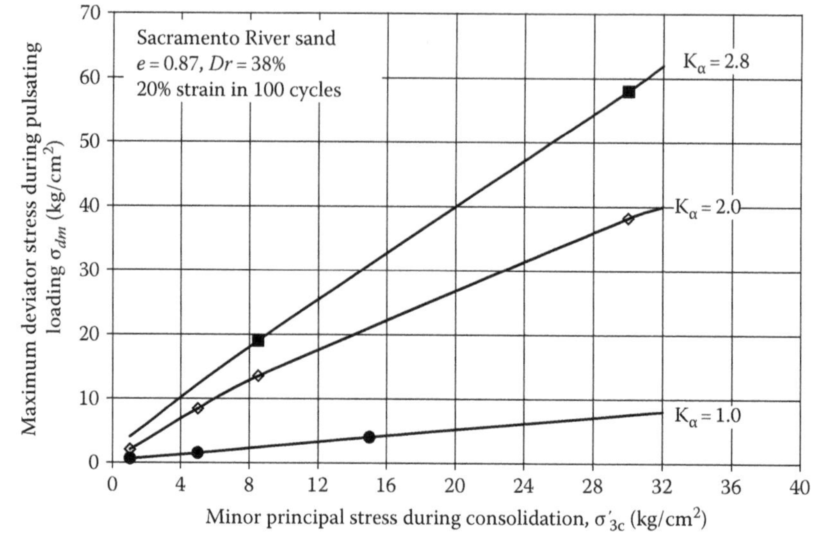

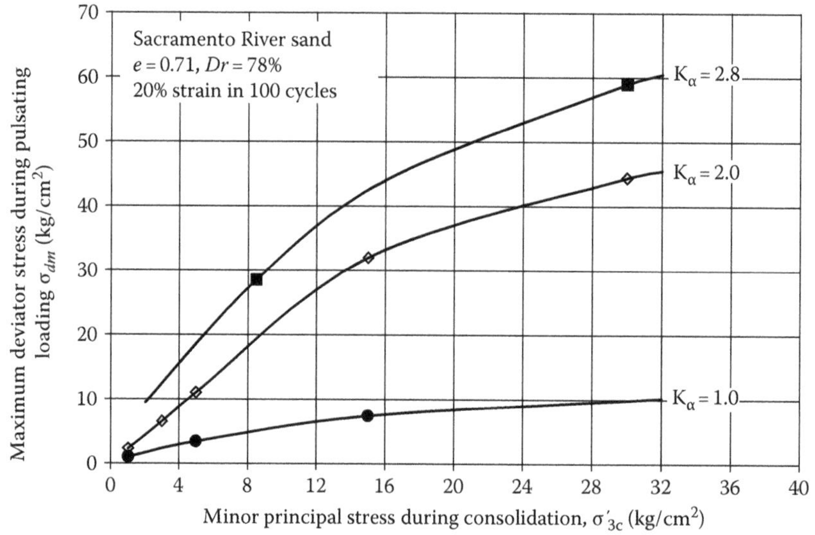

Figure 7.38 Apparent effect of consolidation stress ratio K_c on liquefaction resistance. (Redrawn from Lee and Seed 1967. With permission from ASCE, Reston, VA.)

dimensionless stress ratio $\alpha = \tau_{st}/\sigma_{vo}'$ and that the same ratio could be defined for a layer beneath sloping ground.

Various workers have developed relationships between K_α and α, summarized in Figure 7.39. Conflicting trends are apparent. These conflicts have not been reconciled and present practice is that the K_α correction curves should not be used by 'non specialists' in geotechnical earthquake engineering or in routine engineering practice (Youd et al., 2001) – not a satisfactory situation.

In addition to the stress level and static shear issues, the *CRR* of ground is affected by geologic history (overconsolidation and age). These factors have not been quantified within the Berkeley approach on the basis that these factors directly and similarly affect the measured penetration resistances. Geologic history effects are estimated (i.e. guessed) if necessary.

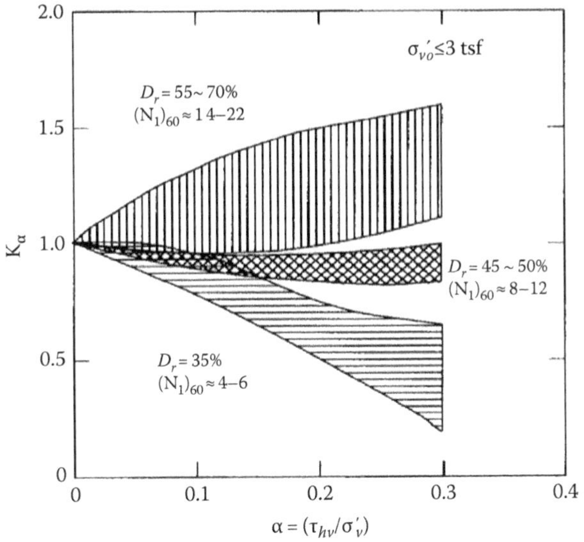

Figure 7.39 Summary of recommended values for K_α. (After Harder, L.F., Jr. and Boulanger, R.W., Application of K_σ and K_α correction factors, *Proceedings of NCEER Workshop on Evaluation of Liquefaction Resistance of Soils*, National Center for Earthquake Engineering Research, State University of New York, Buffalo, NY, pp. 167–190, 1997. With permission from Dr. Harder.)

7.4.4 Deficiencies with the Berkeley School method

There are several basic deficiencies in the Berkeley School approach for assessing the cyclic resistance of soil based on case histories:

- Characteristic penetration resistance is undefined
- Soil properties are neglected
- No mechanistic basis exists for the extrapolations

On the first point, the basic charts used to determine the *CRR*, Figures 7.33 and 7.34, only show a single value for penetration resistance of a stratum under consideration on the x-axis of the plot. Real soils, however, show a spectrum of penetration resistance when tested (illustrated in Chapter 5). What value should be selected from a spectrum of measured data to use these charts? Moreover, what values were used from the case histories in developing the charts? Examination of the records reveals that many case histories are based on a single boring or even a single blow count, and where such limited data might fit in the real spectrum of values at the site is unknown. This leads to considerable uncertainty in the derived trends. Detailed stochastic simulations discussed in Chapter 5 indicate that it is the loosest 5%–15% of the deposit that actually controls liquefaction, but none of that knowledge is present in the case histories used by Seed and co-workers (either in the derivation of the charts or for their subsequent use in design).

On the second point, the charts are anchored to trends derived with example soils (commonly clean sands in laboratory experiments). How should soils at a particular site be related to the charts and how were the case history soils related to the laboratory sands on which stress-level effects and earthquake magnitude factors were estimated? The methodology is based around silt content to relate differences in behaviour to soil type, but where is the basis in mechanics for this and why does silt content matter rather than the whole grain size distribution? Does soil compressibility, which is known to affect penetration resistance, matter? Recall that there was only a very weak relationship between the slope of the CSL λ and fines content in Section 2.6, but that Egan and Sangrey (1978) showed a strong relationship between cyclic loading resistance and compressibility (Section 7.2).

On the third point, equations have been fitted to trends as, for example, the K_σ extrapolation curves suggested by Hynes and Olsen in Figure 7.37, but there is no theoretical basis for these curves. It is one thing to draw a line through data to interpolate, something else entirely to choose a curve to extrapolate beyond the data. In addition, the variable exponent normalization of CPT to the reference stress level is speculative – there is nothing in mechanics to substantiate this framework.

The second and third points really boil down to the absence of a proper constitutive model. Such models explain the measured trends based on fundamental soil properties such as critical friction angle and slope of the critical state locus. Although proper constitutive models for cyclic loading did not exist when Seed started this research at Berkeley in the 1960s, much progress has been made in the last 40 years and there are several good models that are now available. This is evidenced by the *VELACS Conference* in 1993 where some 18 models, implemented in coupled finite element codes, predicted the pattern and details of cyclically induced liquefaction in centrifuge experiments. Generally, predictions were rather good. Why not use this understanding? Why not give the method of liquefaction assessment a basis in mechanics?

7.5 STATE PARAMETER VIEW OF THE BERKELEY APPROACH

7.5.1 State parameter version of the CPT charts

The Berkeley database of seismic liquefaction case histories has been updated since the NCEER workshops and continues to be analyzed by research groups at the California universities. One particularly comprehensive study was that of Moss (2003), who provided us both his thesis and a spreadsheet containing the summary data for his analyses, on which the following is based.

Moss undertook a probabilistic assessment of seismic liquefaction triggering using the CPT database. He looked at more than 600 cases, but reduced that number to 185 records considered to conform to Class A, B and C. Class A consisted of data from ASTM compliant CPT tests in which no correction was required for thin layers and for which there were strong ground motion stations within 100–500 m (expressed as coefficient of variation in CSR $\leq$ 0.2 by Moss). Class B sites required a thin layer correction and the nearest strong ground motion stations were further away, between 500 and 1000 m with the coefficient of variation on CSR between 0.2 and 0.35. Class C data may have used non-standard CPTs without sleeve data and had a coefficient of variation on CSR of 0.35–0.5. We will look only at the Class A and Class B records here, with an emphasis on Class A.

Figure 7.40 shows the CPT liquefaction charts using Moss's Class B data. Chart (a) uses q_{c1} for the penetration resistance and chart (b) uses $q_{c1,\text{mod}}$. Recall from Chapter 4 that q_{c1} is the reference stress–level penetration resistance at 1 atm (~100 kPa). But here the exponent c used in computing q_{c1} is variable, following the logic of Olsen (1988) who suggested that the exponent was soil type dependent varying from 0.5 in sands to 1.0 in clays. However, Moss related c to both the measured q_c and friction ratio R_f (= f_s/q_c) based on further analysis of Olsen's work. The modified form $q_{c1,\text{mod}}$ is the CPT analogue to a fines corrected SPT blow

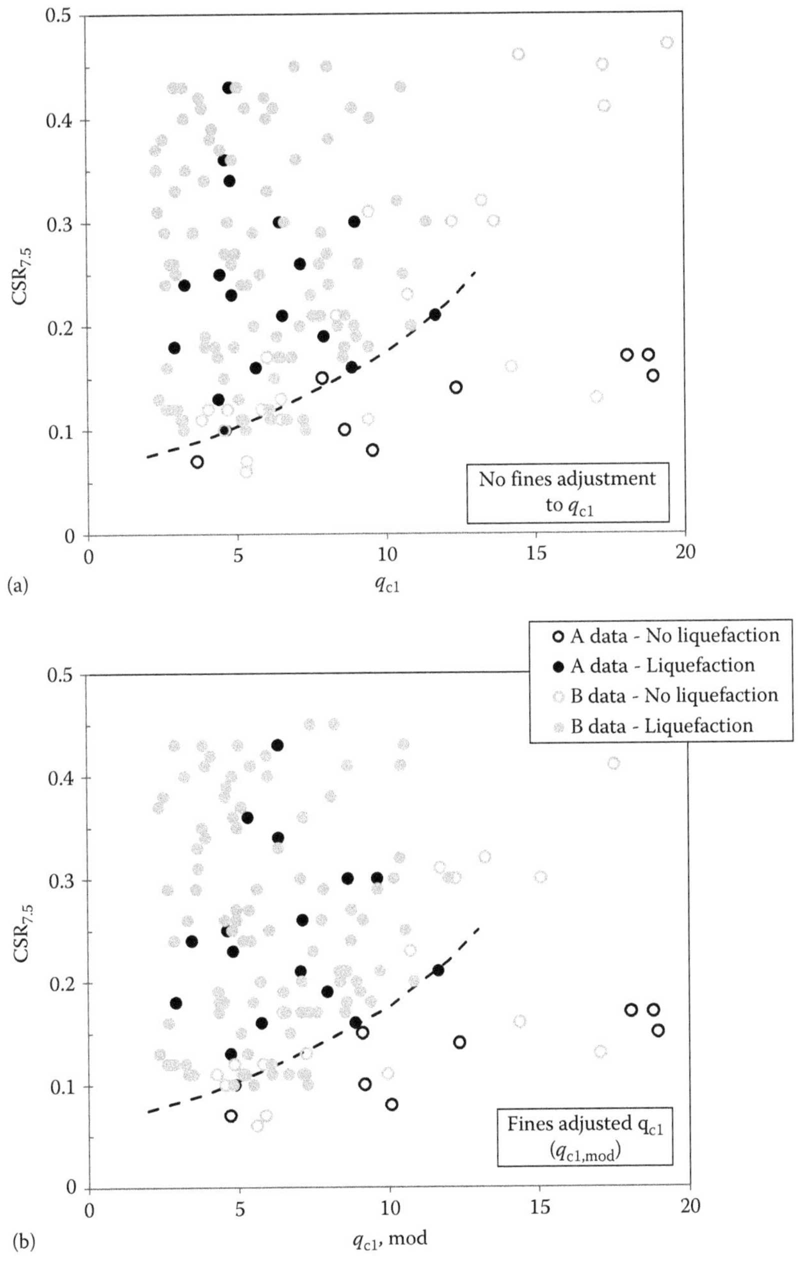

Figure 7.40 Class A and B liquefaction case histories in terms of q_{c1} (a) and $q_{c1,mod}$ (b). (Adapted from Moss, R.E.S., CPT-based probabilistic assessment of seismic soil liquefaction initiation, PhD thesis, University of California, Berkeley, CA, 2003.)

count $N_{160,ecs}$. The modification by Moss is additive to q_{c1} and is a function of both the CSR and CPT friction ratio R_f.

Visually, there is little to choose between Figure 7.40a and b except that a few of the cases with higher friction ratio, that is where the change in q_{c1} to get to $q_{c1,mod}$ is greatest, are graphed at higher values of CPT resistance and as a result there are fewer records that are on the wrong side of a notional division between liquefaction and non-liquefaction case histories. Looking only at the Class A records, there are arguably no cases that are on the wrong side of the dividing line when viewing $q_{c1,mod}$ and only one non-liquefaction case that lies just above the dividing line in terms of q_{c1}.

What happens if we now turn to state parameter for guidance, using the methods developed in Chapter 4 to look at these CPT data? Consider only the Class A records from Moss (2003), which are presented in detail in Appendix G. Table 7.3 is a summary of that data and our state parameter–based processing of it. Figure 7.41 plots the tabulated data as friction ratio versus vertical effective stress in-situ to provide a perspective on the range of soils and stress levels within the database. With one exception, the vertical effective stresses for the Class A case histories are less than 100 kPa. Turning to the friction ratio as an indicator of soil type, the majority of the case histories have $F<0.8\%$, indicative of a clean sand to slightly silty sand. Caution is needed if these data are to be used for projects which involve significantly higher stress conditions or tailings and silts.

The state parameter processing consists of applying the screening method of Plewes et al. (1992) in which λ_{10} is estimated from the normalized friction F and then CPT inversion parameters k, m are calculated from λ_{10} (Section 4.7). Since pore pressure measurements have not been provided with the CPTs, there is a simplification that $B_q=0$ which is accurate for clean sands but may include a slight bias for silty sands and silts where drained penetration may not be a good assumption. In addition, we have had to assume $M=1.25$ across the entire database (again, reasonable for most of the sands in the database) and $K_o=0.7$ (which we consider a reasonable representative value, which cannot be supported or refuted in the absence of measurements).

Recall from Chapters 5 and 6 that it is not the mean state parameter that needs to be used in considering liquefaction. A characteristic value is required and in Chapter 5 our review indicated something like the 80–90 percentile value is appropriate. In Chapter 6, we estimated this percentile value as ψ_k for the analysis of the post-liquefaction undrained shear strength. Since the Moss database includes a mean and standard deviation in the layer of interest for each case, we have used the mean minus one standard deviation value to compute an approximately 85 percentile value for ψ_k (assuming a normal distribution function). Figure 7.42 shows the state parameter version of the CPT liquefaction chart we have estimated using this approach.

We have shown in different shades the case records for which the normalized friction ratio (F) is less than 0.8% or more than 0.8% so as to see whether there might be a bias induced by not having pore pressure data and assuming $B_q = 0$. There is no evidence that the division between liquefied and non-liquefied sites would be affected. Figure 7.42 also shows a simple exponential curve that neatly divides the data for liquefied and non-liquefied sites given by:

$$CRR_{7.5} = 0.06\exp(-9\psi_k) \tag{7.5}$$

Recall how cyclic triaxial test data normalized to a single backbone curve for the effect of CSR, under constant void ratio and initial stress conditions, in Figure 7.8. From this backbone curve, Figure 7.11 showed CRR_{15} as a function of state parameter. Since $M=7.5$ earthquakes implicit in Equation 7.5 are treated as the same as 15 equivalent cycles of loading, we can combine the laboratory and case history trend, Figure 7.43. What jumps out of the page in Figure 7.43 is the correspondence that has been achieved between the state parameter approach based on independent sets of laboratory tests and field CPT tests. The mechanics-based approach works and supports the simplified screening method.

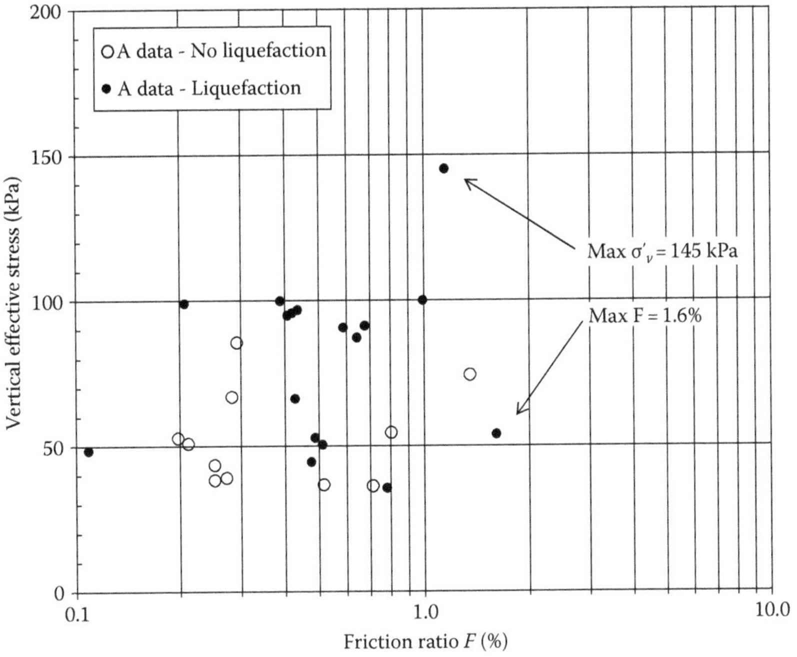

Figure 7.41 Soil types (friction ratio) and stress levels within the Class A liquefaction case history data.

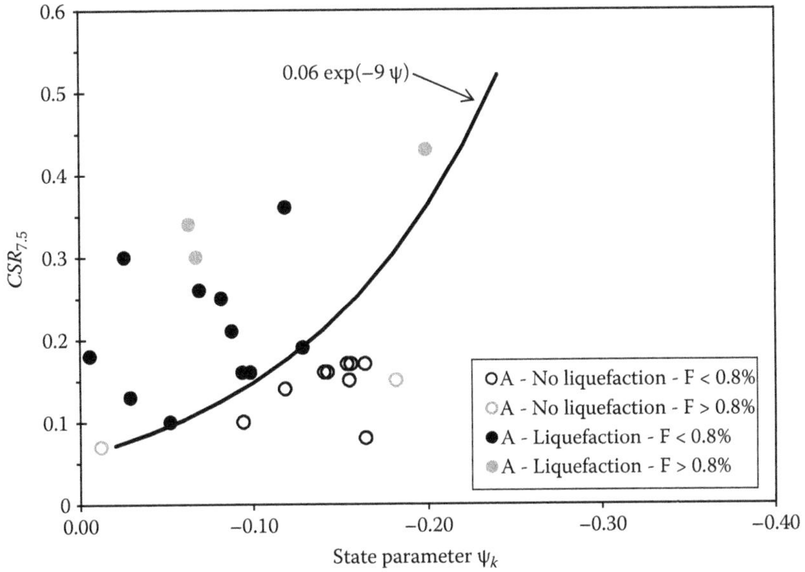

Figure 7.42 Class A liquefaction case history data versus characteristic state parameter.

Table 7.3 Cyclic liquefaction case history data processed in terms of state parameter

#	*Earthquake*	M_w	*Site*	*Liq*	*From (m)*	*To (m)*	*Water table (m)*	σ'_v *(kPa)*	$CSR_{7.5}$	q_c *(MPa)*	$(q_c)_k$ *(MPa)*	*F (%)*	λ_{10}	*k′/M*	*m′*	$Q_p(1-B_q)$	ψ_k
111	1989 Loma Prieta	7	MBAR 13 RC-6	N	3	4.5	2.6	52	0.16	13.4	12.5	0.20	0.020	46.1	11.6	295	−0.140
117	1989 Loma Prieta	7	Moss Landing State Beach 18	N	2.4	3.4	2.4	43.5	0.15	10.4	9.6	0.25	0.025	36.8	11.6	276	−0.155
115	1989 Loma Prieta	7	MBAR 14 CPT-1	N	2.3	3.5	1.9	38.3	0.17	9.6	8.6	0.25	0.025	36.8	11.6	279	−0.156
113	1989 Loma Prieta	7	Sandholdt Rd. UC2	N	3	4.5	2.7	50.9	0.16	16.5	11.6	0.21	0.021	43.2	11.6	283	−0.142
116	1989 Loma Prieta	7	Sandholdt Rd. UC-6	N	6.2	7	2.7	85.6	0.17	18.8	18.2	0.29	0.029	32.1	11.5	264	−0.164
112	1989 Loma Prieta	7	MBAR 13 RC-7	N	4	5	3.7	67.0	0.14	9.3	8.7	0.28	0.028	33.1	11.5	162	−0.118
114	1989 Loma Prieta	7	General Fish CPT-6	N	2.2	3.2	1.7	39.1	0.17	9.4	7.9	0.27	0.027	34.2	11.5	252	−0.154
44	1981 Westmorland	5.9	Radio Tower B2	N	2	3	2.01	36.2	0.08	6.3	3.3	0.71	0.071	14.9	11.0	113	−0.164
34	1979 Imperial Valley	6.5	Radio Tower B2	N	2	3	2.01	36.7	0.1	5.8	2.1	0.52	0.052	19.5	11.2	70	−0.094
110	1989 Loma Prieta	7	Alameda Bay Farm Is.	N	5	6	2.5	74.3	0.15	7.1	4.4	1.35	0.135	9.3	10.1	73	−0.182
32	1979 Imperial Valley	6.5	Kornbloom B	N	2.6	5.2	2.74	54.5	0.07	2.8	0.9	0.80	0.080	13.6	10.8	19	−0.012
88	1989 Loma Prieta	7	Miller Farm CMF8	Y	6.8	8	4.91	99.0	0.23	4.8	3.9	0.21	0.021	44.1	11.6	47	0.013
80	1989 Loma Prieta	7	Sandholdt Rd. UC-4	Y	2.4	4.6	2.7	48.6	0.21	7.6	1.8	0.11	0.011	81.4	11.8	46	0.067
90	1989 Loma Prieta	7	Miller Farm CMF5	Y	5.5	8.5	4.7	99.8	0.26	7.1	5.6	0.39	0.039	24.9	11.4	68	−0.069
78	1989 Loma Prieta	7	Marine Lab. C4	Y	5.2	5.8	2.5	66.3	0.18	2.1	1.7	0.43	0.043	22.9	11.3	30	−0.006
84	1989 Loma Prieta	7	Marine Lab. C4	Y	5.2	5.8	2.5	66.3	0.18	2.1	1.7	0.43	0.043	22.9	11.3	30	−0.006
81	1989 Loma Prieta	7	Moss Landing State Beach 14	Y	2.4	4	2.4	44.6	0.19	4.7	4.0	0.48	0.048	20.9	11.3	111	−0.128
72	1989 Loma Prieta	7	SFOBB-2	Y	6.5	8.5	2.99	96.8	0.16	8.7	6.8	0.44	0.044	22.5	11.3	86	−0.098
71	1989 Loma Prieta	7	SFOBB-1	Y	6.25	7	2.99	90.6	0.16	5.3	4.6	0.59	0.059	17.4	11.1	62	−0.094
87	1989 Loma Prieta	7	Farris Farm Site	Y	6	7	4.5	87.1	0.25	4.0	3.6	0.64	0.064	16.2	11.0	50	−0.082

(Continued)

Table 7.3 (Continued) Cyclic liquefaction case history data processed in terms of state parameter

#	Earthquake	M_w	Site	Liq	From (m)	To (m)	Water table (m)	σ'_v (kPa)	$CSR_{7.5}$	q_c (MPa)	$(q_c)_k$ (MPa)	F (%)	λ_{10}	k'/M	m'	$Q_p (1 - B_q)$	ψ_k
91	1989 Loma Prieta	7	Miller Farm CMF3	Y	5.75	7.5	3	95.7	0.24	3.2	1.8	0.42	0.042	23.2	11.3	22	0.026
42	1981 Westmorland	5.9	Radio Tower B1	Y	3	5.5	2	50.4	0.1	3.2	1.8	0.51	0.051	19.6	11.2	44	−0.052
30	1979 Imperial Valley	6.5	Radio Tower B1	Y	3	5.5	2.01	52.8	0.13	3.1	1.6	0.49	0.049	20.4	11.3	35	−0.029
128	1994 Northridge	6.7	Potrero Canyon Unit C1	Y	6	7	3.3	91.3	0.21	6.2	3.8	0.68	0.068	15.5	11.0	51	−0.088
129	1994 Northridge	6.7	Wynne Ave. UnitC1	Y	5.8	6.5	4.3	94.9	0.3	8.8	3.1	0.41	0.041	23.9	11.4	40	−0.026
31	1979 Imperial Valley	6.5	Mc Kim Ranch A	Y	1.5	4	1.5	35.5	0.36	2.7	1.8	0.78	0.078	13.9	10.9	63	−0.119
130	1994 Northridge	6.7	Rory Lane	Y	3	5	2.7	53.9	0.43	3.6	3.2	1.61	0.161	8.3	9.8	72	−0.199
89	1989 Loma Prieta	7	Miller Farm CMF10	Y	7	9.7	3	99.9	0.34	4.8	2.4	0.99	0.099	11.6	10.6	28	−0.063
126	1994 Northridge	6.7	Balboa Blvd. Unit	Y	8.3	9.8	7.19	145.0	0.3	7.3	3.2	1.15	0.115	10.4	10.4	26	−0.067

Class A sites from Moss (2003).
Note: Column Liq Y/N indicates whether liquefaction was observed (Y) or not (N).

σ'_v is at middle of critical layer.
$K_o = 0.7, M = 1.25, B_q = 0$ assumed.
$(q_c)_k = q_c - 1$ standard deviation.
λ_{10} calculated from Equation 4.16.
k', m' calculated from Equation 4.15.
$Q_p(1 - B_q)$ calculated for characteristic value $(q_c)_k$.
ψ_k calculated using Equation 4.14.

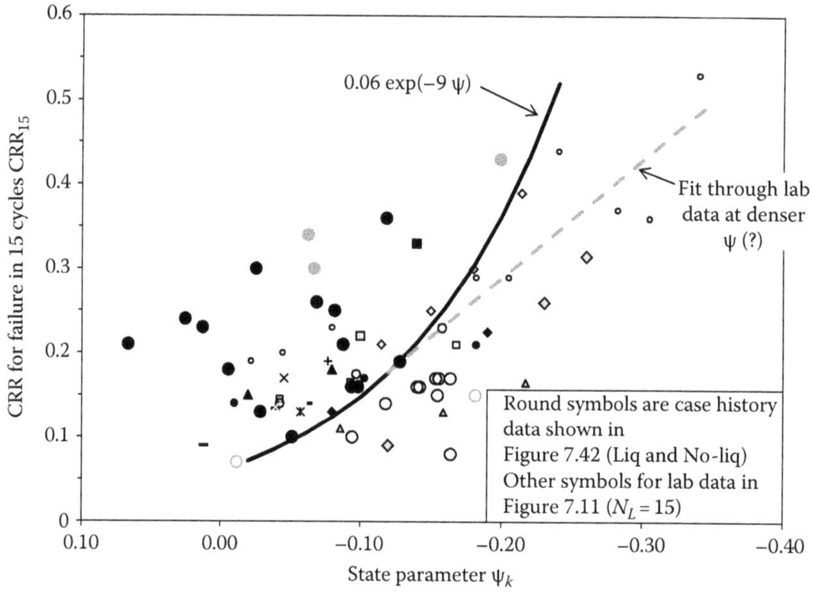

Figure 7.43 Combined laboratory and Class A case history trends of cyclic liquefaction.

Although very encouraging, too much should not be read into Figure 7.43 at this stage because

- The laboratory tests are cyclic triaxial tests, and some adjustment is needed to represent simple shear conditions.
- The field-based line is a limit between liquefied and non-liquefied behaviour considered at the characteristic state, whereas all the laboratory data represent measured liquefaction (the stress ratio at which failure occurs in 15 cycles) at a single sample state.
- The range of soil type and stress levels in the database is rather limited – it is all clean to silty sand and stress levels of less than 150 kPa.

We will return to these points later, once we have looked at simple shear behaviour and have a better theoretical understanding of cyclic loading through *NorSand*.

7.5.2 Nature of K_σ

The effect of the actual soils being at a different stress from the reference condition lead to the development of a stress-level factor, K_σ, in the Berkeley method to modify the assessed CRR for cyclic liquefaction triggering. K_σ is the opposite adjustment for the mapping used in evaluating the soil state from penetration resistance using the reference stress method. The factor K_σ was estimated from soil testing associated with various case histories, and as illustrated in Figures 7.36 and 7.37, there is a wide range of trends as to how K_σ changes with soil conditions. This has led to an equally wide range of recommendations based on opinion not mechanics.

With the understanding from laboratory tests that constant ψ implies constant *CRR* for constant intrinsic soil properties (and in particular fabric), we can calculate just how K_σ should vary. Recall that K_σ was intended to allow for the in-situ stress conditions assuming that soil density remains unchanged. It is now clear that this corresponds to progressively

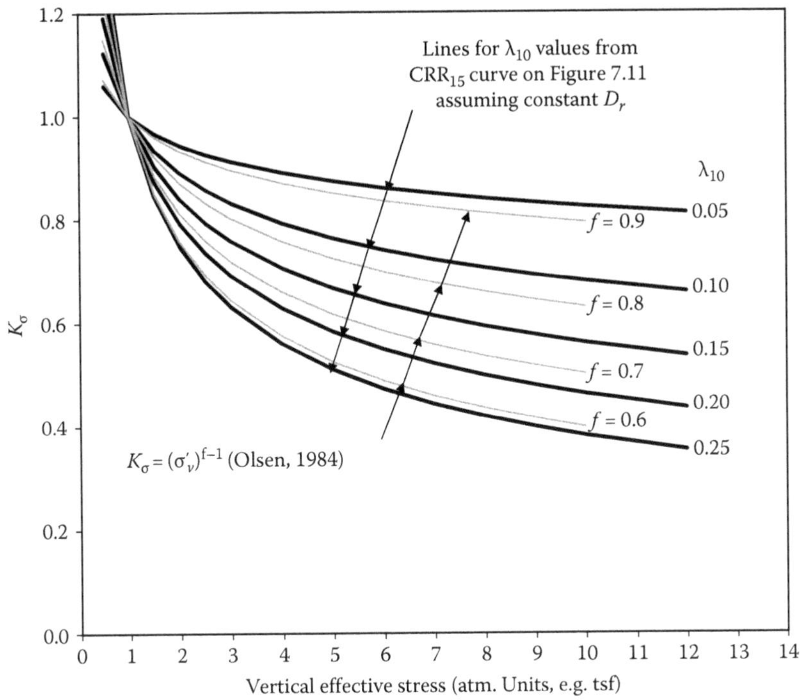

Figure 7.44 Comparison of K_σ recommended by NCEER with K_σ computed from ψ changing with stress level at constant soil density.

changing ψ. Figure 7.11 shows just how the CRR, at which failure occurs in 15 cycles, varies as a function of ψ, and this graph can therefore be used to compute K_σ as a function of stress level and soil properties.

Figure 7.44 shows computed values of K_σ as a function of the slope of the critical state line λ (which is the soil property that determines how ψ varies with stress level for constant density). Also shown in Figure 7.44 is the K_σ as a function of f from Figure 7.37 suggested by Olsen (1984). It is clear from the remarkable correspondence between the two sets of curves that the changing state parameter, because of changing stress level at constant void ratio, explains the stress-level correction represented by K_σ. Conversely, if ψ is the chosen frame of reference then no stress-level correction factor K_σ is required and the liquefaction assessment process becomes a whole lot simpler (as well as being based on measured soil properties rather than geologically based speculation).

7.5.3 Nature of K_α

The factor K_α arose to deal with the observation that sands show a greater resistance to cyclic mobility in anisotropically rather than isotropically consolidated cyclic triaxial tests. However, we showed in Section 7.3.4 that whatever the combination of cyclic and static stress ratios in simple shear, it is the proximity of the peak shear stress to the image stress ratio that controls stiffness. Static bias puts the stress path closer to the image stress ratio. The expectation would therefore be that static bias reduces the cyclic strength. However, this ignores the rate of pore water pressure generation during cyclic loading, which pushes the stress path towards higher stress ratios. The rate of pore pressure generation is more dependent on principal stress rotations than on the CSR, at least below the image stress ratio.

This idea of static bias is often invoked for plane strain slopes; the effect of increased slope is to increase the magnitude of the principal stress difference. This is manifested as a base static shear on which the (usually earthquake caused) cyclic stress is imposed. Fundamentally, this is the same thing as an effect of geostatic stress ratio K_o and it does have an effect on the simple shear stress path and proximity to the image stress ratio, as well as the principal stress rotation. However, there is not a one-to-one mapping because only one of the principal stresses (that extending perpendicular to depth away from the slope surface, generally σ_1) is set by the boundary conditions. The other two principal stresses depend both on K_o and the response of the soil to loading. Correspondingly, the principal stress rotation depends on the imposed cyclic shear stress, the state parameter and the intrinsic properties of the soil. As principal stress rotation is the dominant effect in producing cyclic mobility, the situation becomes implicit and *static bias* cannot give simple trends for a strength adjustment factor like K_α. Rather, formal modelling is needed. This is not difficult, however, and for most purposes cyclic simple shear simulations with *NorSand* will provide the required guidance (covered later in this chapter).

7.5.4 Influence of silt content

The fundamental liquefaction criterion developed by Seed was for *clean* sands, which were taken to be sands with less than 5% fines content. However, many of the case histories involved sands with higher fines contents and these have been assumed to behave in much the same manner as the clean sands. The liquefaction/no-liquefaction lines for higher fines soils have been assumed to be similar to the *clean sand* line such that any sand of arbitrary fines can be mapped onto the clean sand behaviour line by adding a 'correction' to its measured blowcount. The correction is only a function of silt content. The correction factor is referred to as 'ΔN', and the result is denoted as an *equivalent clean sand* value (usually denoted by the subscript *ecs*).

Referring to Figure 7.33 for low cyclic stress events, say $CSR = 0.1$, the 'correction' factor for a sand with 15% silt content compared to less than 5% silt content could amount to the same as the actual penetration resistance. The decision becomes dominated by the correction factor. Further, although the effect of silt content has been drawn as a parallel trend to clean sand, the actual case history data would equally well support a line going through the origin; in effect, making the correction factor a proportion of the measured resistance rather than an additive term. Of course, this may completely change many design decisions.

What insight does the state parameter offer? The first step is to distinguish between soil behaviour and penetration resistance. In Chapter 4, calibration chamber experiments, simple cavity expansion solutions and numerical simulations have all shown that the relationship between state parameter and CPT resistance depends on soil properties. At least part of the silt content 'correction' is to compensate for the fact that the penetration resistance is less in silty sands than clean sands for the same state parameter and stress level. The closed-form solution in particular shows that this correction is in the nature of a multiplier, not an additive term. The silt content 'correction' in the Berkeley School methodology using the SPT has the wrong mathematical form. (However, the Robertson and Wride, 1998, approach to the CPT using an adjustment factor K_c is an appropriate form as it depends on the behaviour of the sand as measured by appropriate CPT dimensionless groups and acts as a multiplier on the resistance.)

Silt content has an important effect on liquefaction behaviour beyond its influence on penetration resistance, although the effect is the result of compressibility rather than silt content. Using the state parameter version of the liquefaction charts recognizes implicitly the effect of compressibility on the penetration resistance. By taking a state parameter view, the effect of material properties on cyclic behaviour (for soils outside the case history database) can also be captured through constitutive modelling.

7.6 THEORETICAL FRAMEWORK FOR CYCLIC LOADING

7.6.1 Alternative modelling approaches for cyclic loading

Elastic–plastic thinking has dominated the development of constitutive laws for liquefaction in the literature. Elastic–plastic models are defined with reference to a stress space associated with a point in the material and are formulated according to classical continuum mechanics: a stress increment is specified by a strain increment (and vice versa) in a tensor relationship. The key step for cyclic loading is how to deal with unloading or load reversals, which relate to the hardening rule rather than the particular form of the yield surface. Three broad classes of hardening law are illustrated in simplified form in Figure 7.45.

The simplest form of work hardening plasticity is isotropic hardening, and such hardening is used in Cam Clay, for example. Isotropic hardening, Figure 7.45a, uniformly expands the yield surface with plastic strain so that, during unloading and reversal of stress, no yield arises until the reversed stress exceeds the yield criterion established in the prior loading direction. Because elastic response does not produce excess pore water pressure, isotropic hardening models cannot simulate the effect of several stress reversals and the gradual build-up of pore pressures that is observed. Multisurface models were developed to avoid this limitation of isotropic hardening by providing several nested yield surfaces of different sizes within stress space, so that a variable rate of plastic straining can be simulated, in particular during stress reversals. Bounding surface models use the concept of an outer limit, or bounding surface, with a single inner surface on which

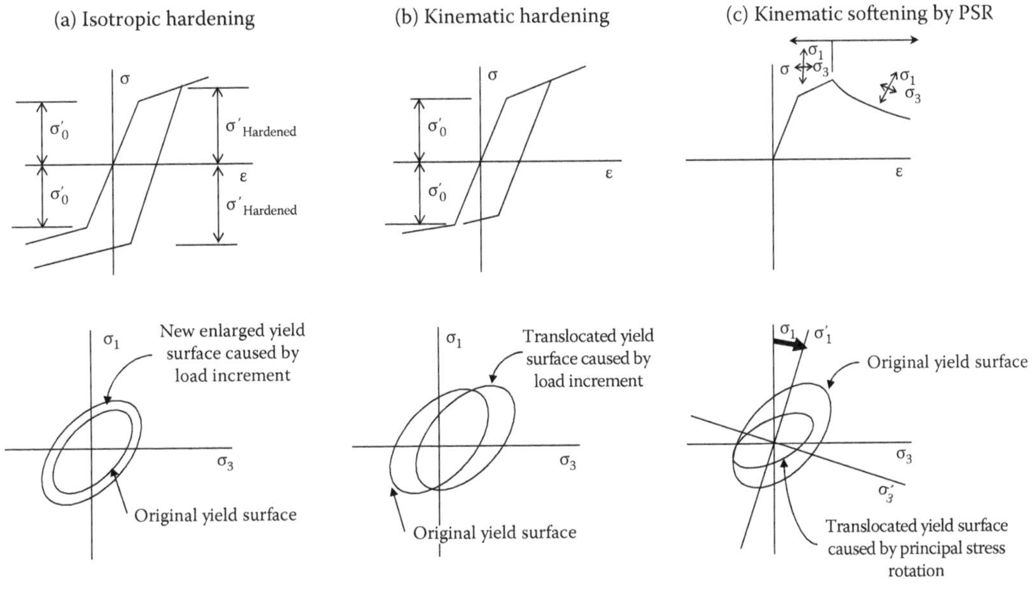

Figure 7.45 Illustration of alternative hardening laws.

yielding is defined with a rate dependent on the distance from the bounding surface. Both multisurface and bounding surface models result in kinematic hardening (Figure 7.45b). Kinematic hardening models can simulate both static liquefaction and cyclic mobility encountered in standard laboratory tests, and these models are relatively popular in the literature. The good models in VELACS were all of this kinematic hardening form and development has continued since then.

Kinematic hardening models do well in single element tests with fixed principal stress direction, however, they are very poor in situations with principal stress rotation. None can simulate the case of constant stress invariants but rotating principal stress direction that was tested in the experiments of Arthur et al. (1980) or Ishihara and Towhata (1983). Clearly, something fundamental is missing.

In kinematic softening (Figure 7.45c), the yield surface is allowed to shrink as a function of principal stress direction changes or stress reversals. Not only does this capture the behaviour in the standard laboratory tests, it also nicely predicts the soil behaviour during principal stress rotation in constant stress invariant tests. There are very few such models. *NorSand* implemented kinematic softening in 1992 and this implementation will be discussed next.

7.6.2 *NorSand* with cyclic loading and principal stress rotation

Kinematic softening is introduced into *NorSand* to account for principal stress rotation during cyclic loading. The idea is that a yield surface reflects the mobilization of particle contacts at a micromechanical level, and therefore, the yield surface has direction in stress space. Since particle contacts have evolved to best carry the imposed stresses, loading from a different direction will always load a less well-configured arrangement of grain contacts. Correspondingly, rotation of principal stress directions is assumed to result in shrinking of the yield surface, as illustrated in Figure 7.45c. The basis of this assumption is illustrated in Figure 7.46 which shows an experimentally determined arrangement of contacts for discs in biaxial compression. In this particular sample, there is a strong preference for vertical contacts over horizontal. Obviously, if the sample is further loaded vertically, the response will be relatively stiff. For horizontal loading, or loading in an intermediate direction, a completely different stress–strain response will arise as the particle arrangement readjusts to establish contacts to carry the forces in the direction of loading. These ideas also illustrate why the principal stress jumps in the cyclic triaxial tests are such a severe loading – the soil is loaded in its weakest direction every half cycle. This idea of principal stress rotation is so important that it overrides, or at least modifies, Second Axiom of critical state theory (that the state migrates to the critical state with shearing, Equation 3.32 and it is regarded as having the status of an axiom in its own right. This gives

- Third Axiom: Principal stress rotation softens (shrinks) the yield surface

Because of the Third Axiom principal stress rotation may or may not move the soil closer to the critical state despite causing plastic strains. Arthur et al. (1980) noted exactly this point. The basic scheme of the Third Axiom softening is illustrated for a plane strain situation in Figure 7.47. Consider a stress point located on a hardened yield surface with ongoing yielding. If the stress point is slightly unloaded to move it just inside the yield surface, plastic yielding will stop. Now, consider that yield surface as fixed in space with a preferred coordinate system (which reflects the mobilized grain contact arrangement). Pure rotation of the principal stresses will not change the stress invariants, but if the stress point is drawn

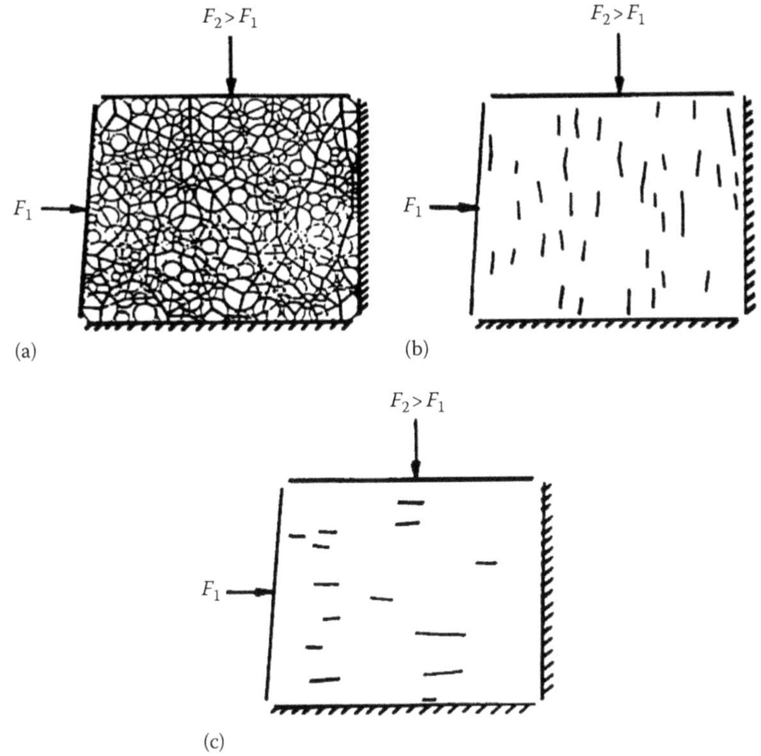

Figure 7.46 Biaxial compression test on an assembly of photo-elastic discs. (a) Full assembly of discs (vertical load F_2 is greater than horizontal load F_1), (b) near vertical contacts and (c) near horizontal contacts. (After de Josselin de Jong, G. and Verruijt, A., *Can. Grpe.Fr. Etude Rheol.*, 2, 73, 1969.)

on the rotated axes it may now lie outside the previous yield surface (which is fixed in space). Obviously, yielding must arise in this situation.

Actually, yield caused by principal stress rotation is not the only possibility. As can be envisaged from Figure 7.47, some rotations could move the stress point further into the yield surface until further yielding would only be triggered on the opposite side of the yield surface. This situation arises every reversal in cyclic simple shear and also causes non-coincidence between directions of principal stress and principal strain increment as the two get out of phase by the size of the elastic zone.

NorSand has two measures of yield surface size, the image stress ratio M_i, which describes the relative size between the deviator and the isotropic directions and the mean stress at the image condition, σ'_{mi}. This image mean stress is the parameter that scales the size of the yield surface and is the object of the hardening law. Principal stress rotation only acts on this image mean stress, and the Third Axiom is implemented as

$$\frac{\dot{\bar{\sigma}}_{mi}}{\bar{\sigma}_{mi}} = -Z_r \frac{\dot{\alpha}}{\pi} \qquad (7.6)$$

The π term in Equation 7.6 is introduced to keep the model dimensionless and arises if α is measured in radians; if α is measured in degrees then the π term becomes 180°. Equation 7.6

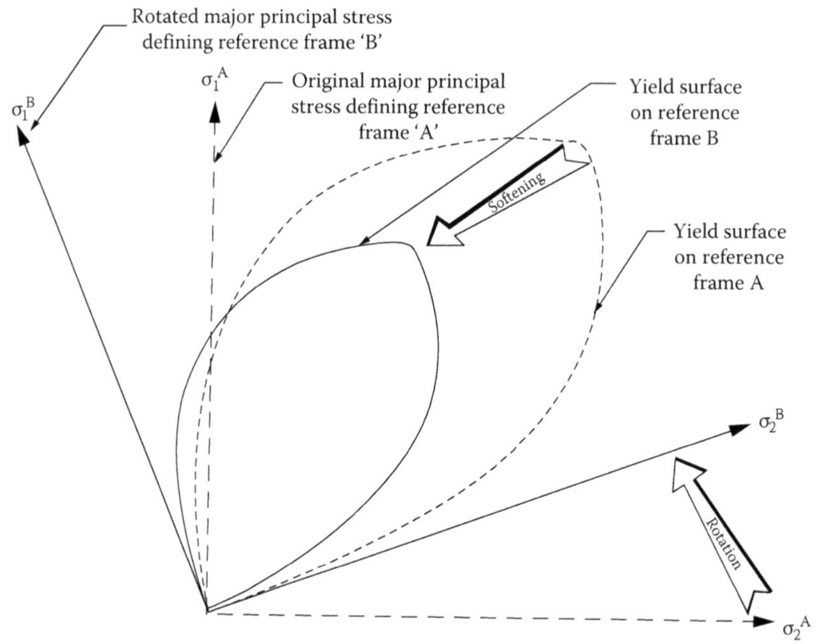

Figure 7.47 Schematic of yield surface softening induced by principal stress rotation. (From Been, K. et al., Class A prediction for model 2, *Proceedings of International Conference on the Verification of Numerical Procedures for the Analysis of Soil Liquefaction Problems* [VELACS], K. Arulanandan, R.F. Scott (eds.), A.A. Balkema, Rotterdam, the Netherlands, 1993. With permission of Taylor & Francis.)

operates as a further step to the hardening law and is simply implemented in any numerical scheme (whether Euler integration of a laboratory element test in a spreadsheet or in finite element analysis). Appendix C includes the details. There are no changes to the stress–dilatancy rule, which continues to operate off the current yield surface. The action of principal stress induced softening aligns the yield surface with the new principal stress directions.

In some respects, the effect of principal stress rotation in *NorSand* is very much like the fatigue model Van Eekelen and Potts (1978) introduced to model Drammen clay under cyclic loading. Their model used a cyclic variation in mobilized stress ratio η to soften the critical mean stress of the yield surface. *NorSand* has exactly the same feature, but largely reacts to changes in principal stress direction as well as yield in unloading.

The present form of *NorSand* does not distinguish, in terms of model soil properties, between principal stress jumps and smooth principal stress rotation. The same value of Z_r is used in both instances. In terms of values for Z_r, the initial estimate was that a 90° principal stress flip should completely soften a yield surface and wipe out all accumulated hardening – implementing hardening as a reflection of grain particle arrangement and taken directly from the micromechanical simulation of particle contacts shown in Figure 7.46. We need to look at some good simple shear data to further examine how the hardening law reflects measured behaviour in sands and silts.

7.6.3 Modelling simple shear with *NorSand*

Earlier in this chapter we illustrated that principal stress rotation was intrinsic to earthquake loading of soils because of the dominant vertical propagation of shear waves from depth to ground surface during an earthquake. The cyclic simple shear test is a direct analogue of an element of soil subject to vertically propagating shear waves (i.e. within the limits of the equipment). But as also pointed out, current versions of the simple shear test are deficient in measuring only two of the four stresses acting on the soil. On the other hand, cyclic simple shear tests are attractive for two reasons: (1) As illustrated by the tests on FRS, these tests do show the key features that we saw in the full-scale extended cyclic loading of the Molikpaq (Chapter 1) and (2) the cyclic simple shear test is the only available test with principal stress rotation that is feasible in current engineering practice. The challenge is to overcome the deficiencies of the experimental data. This can be done by modelling the test and then using the calibrated model to understand what was measured and to transfer the results to engineering design and assessment.

A further wrinkle as to why modelling is needed is that it is exceedingly rare for cyclic simple shear tests to be at the in-situ void ratio. Even with silts, which we can sample and bring to the laboratory, the subsequent extrusion and handling into the test cell causes plastic strains and the as-tested soil is denser, often much denser, than what existed in-situ. In the case of sands, the sample reconstitution procedures seemingly create dense samples. In the case of FRS, even the loose tests had a dilatant initial state and rather more dilatancy than would be encountered with truly loose sands.

The way *NorSand* deals with principal stress rotation was outlined in the previous section with the detailed implementation being provided in Appendix C. The downloadable spreadsheet *NorSandPSR_FRS.xls* has been set up for modelling cyclic simple shear as well as paging through the FRS data. As usual, this principal stress rotation version of *NorSand* is implemented as open-source code in the VBA environment and in the case of undrained cyclic simple shear tests as the routine *NorCSSu* (although the 'housekeeping' in tracking cyclic load makes this code less clear than that for any of the monotonic tests).

Just like the calibration to drained triaxial compression, iterative forward modelling is used in which the computed behaviour in cyclic simple shear is compared to the measured behaviour, over the entire pattern in the test, with model parameters optimized to best align model with data. But it is not possible to use simple shear tests on their own as there are too many unknowns in the test. Thus, the starting point to understand cyclic simple shear is a standard set of triaxial tests to determine the soil properties Γ, λ_{10}, M_{tc}, N, χ_{tc}, H and G_{max}. However, elastic moduli depend on fabric but the shear modulus in any simple shear sample cannot be measured in the present cyclic simple shear equipment using bender elements. Therefore G_{max} is treated as a first estimate and modest changes in elastic modulus are reasonable in fitting cyclic test data. This leaves two parameters for optimizing the fit of *NorSand* to data: the initial geostatic stress ratio K_o and the softening parameter Z_r.

The initial geostatic stress is also not measured in present cyclic simple shear equipment, which makes the initial stress state in the test uncertain and leaves K_o as somewhat of a *free* parameter in the optimization as there is quite a range of plausible values (and depending on the sample preparation method). However, the excess pore pressure induced during liquefaction is strongly related to the mean stress; so, K_o is used to optimize this aspect of the fit. The softening parameter Z_r is a further soil property, akin to the role of H in monotonic tests. Z_r is used to adjust the rate at which excess pore pressure develops in matching the model to data, and optimizing this parameter is straightforward. Figure 7.48 shows fits achieved to tests on both a loose and dense sample of FRS (included in the spreadsheet).

It is unclear whether Z_r is a soil property or something more fundamental related to grain contacts that might be similar for similar roundness of the soil particles. We have carried out detailed calibration of *NorSand* to the cyclic behaviour of Nevada sand and FRS. Despite being rather different soils, the variation of Z_r with ψ appears common to both, Figure 7.49, with a trend that is given by

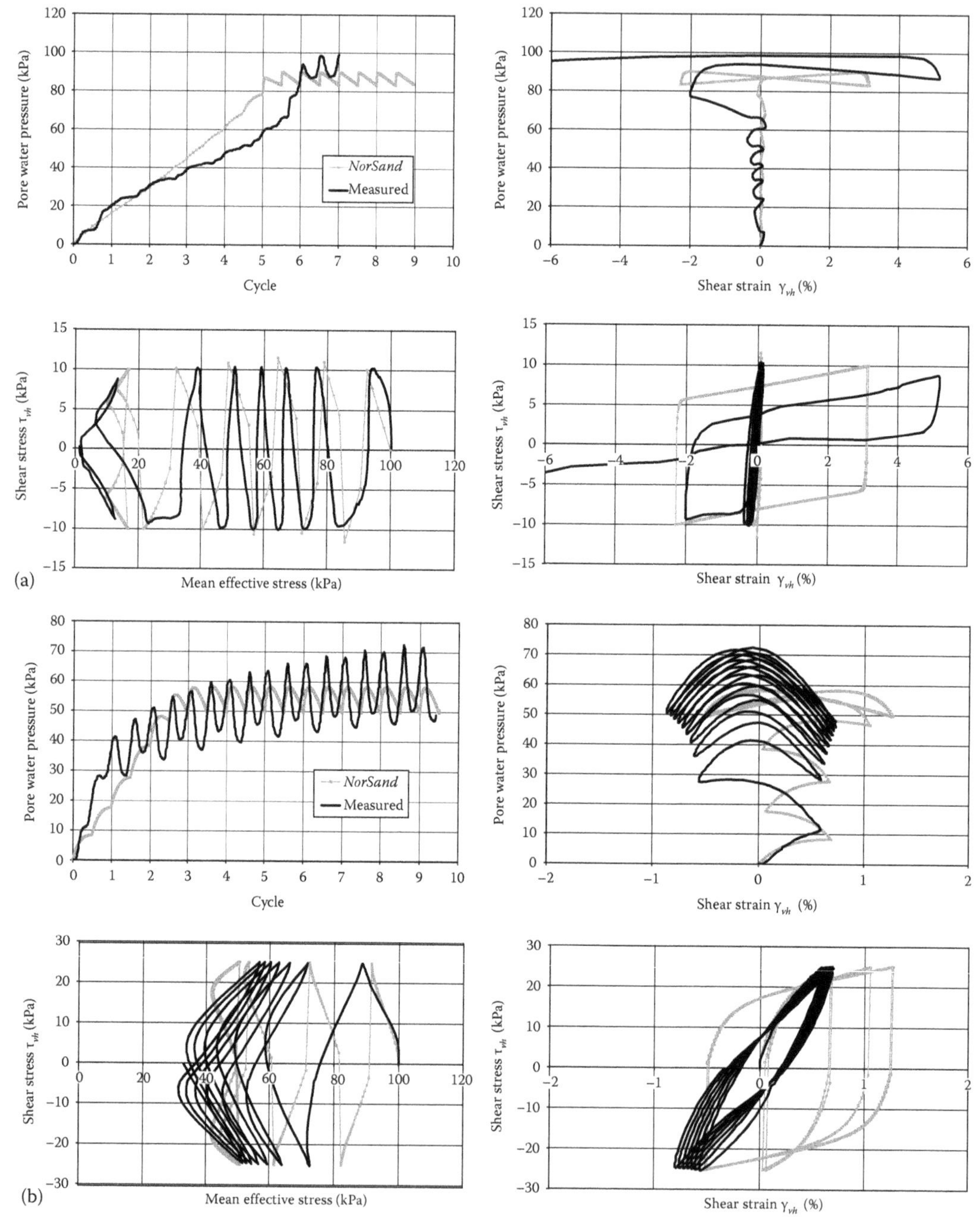

Figure 7.48 Measured FRS behaviour in CSS (black) versus *NorSand* simulations (grey). (a) Test FRS DSS40-100-0p1 (e_o = 0.812) and (b) Test FRS DSS80-100-0p25 (e_o =0.684).

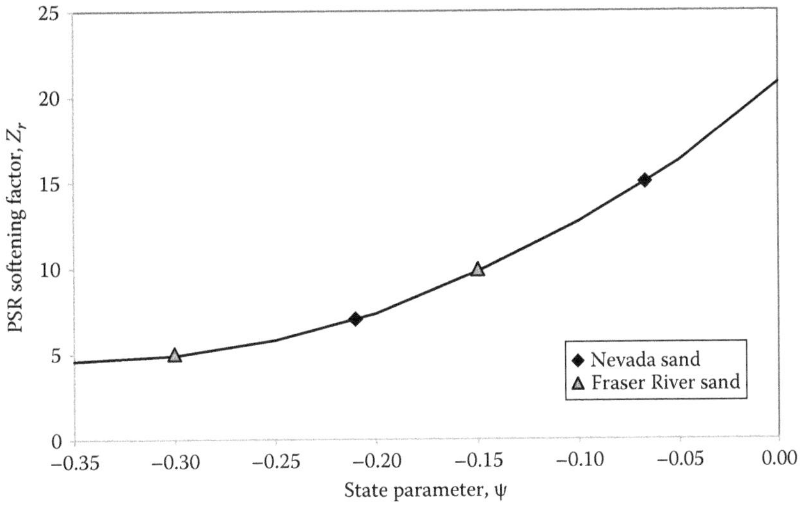

Figure 7.49 Apparent variation in principal stress rotation softening parameter Z_r with soil state parameter ψ.

$$Z_r = 134 + 93\psi + 21\psi^2 \quad (7.7)$$

The downloadable *NorSandPSR_FRS.xls* file is set up for modelling cyclic simple shear and includes the FRS data as well as the table of *NorSand* parameters used to fit all the tests. The fits shown in Figure 7.48 are typical, but do verify this for yourself. And most importantly, this spreadsheet is set up for importing other test data to replace that of FRS. Once a soil has been calibrated, it is then trivial to set the initial state parameter and K_o as to that existing in-situ to develop the expected cyclic behaviour for the real engineering case. In other words, the spreadsheet avoids all issues with 'K_α' as it computes the soil response and you can determine the effects of the in-situ void ratio versus the void ratio in the laboratory test.

As a final comment, the Third Axiom has soil state generally moving away from the CSL during principal stress rotation and which then suggests (7.7) may reflect something else, as ψ is all about moving to the CSL during shear. If the notion is introduced that soil cannot become denser than e_{min} (which may not be the ASTM value, but let us just accept the concept), then it would be plausible that Z_r should scale by the parameter group $e - e_{min}$: Ishihara's *densification potential*. This hypothesis remains to be explored with further testing but would be trivial to incorporate into constitutive models like *NorSand*.

7.7 DEALING WITH SOIL FABRIC IN-SITU

A consequence of principal stress rotation being a dominant driver of cyclic mobility is that the grain particle arrangement (soil fabric) matters. This is not just a question of void ratio, as a whole spectrum of arrangements can arise at one void ratio. The truth of this statement follows from Figure 7.46. In this particular example, there is a strong preference for vertical contacts over horizontal. Obviously, if the sample is further loaded horizontally rather than vertically, a completely different stress–strain response will arise. Void ratio (or state parameter) alone will be insufficient to predict the response of the soil. A description of the anisotropy in the grain contact arrangement is also necessary.

Getting a handle on fabric effects is not just a need to enable confident use of laboratory data for in-situ conditions. The field case histories are also affected, as much as anything because soil fabric is not a dominant factor in CPT response (Been et al., 1987c). The CPT measures the state parameter, but something else is also needed.

Getting to the anisotropy of grain contacts is feasible in the laboratory, albeit requiring some effort. In essence, the sample must be solidified (e.g. by freezing or injection of chemicals) so that thin sections can be cut and the grain contacts examined with a microscope. When looking at the field situation, three methods appear potentially useful: frozen samples, seismic measurements and pressuremeter testing. Freezing is a method to obtain sand samples in a near undisturbed state. Once a frozen sample has been retrieved, similar procedures to those used to get thin sections for laboratory samples could be employed, although nobody appears to have researched this yet. Research is also needed for pressuremeters because, to date, nobody has shown how plastic modulus, which is a result of fabric arrangements, can be retrieved from pressuremeter data. However, it would appear possible in principle because pressuremeter data are rather curved in the early stage of the test and this curvature should reflect the stress–strain behaviour (friction angle and dilation rate are derived from the later stages of the test).

Seismic methods are used to measure shear and compression wave velocities, which give the elastic shear and bulk modulus of the soil. Shear modulus in particular is sensitive to soil fabric (and state parameter) and is relatively easily measured both in the laboratory and in-situ. The evidence suggests that using elastic shear modulus profiles might provide an excellent index of soil fabric at a site and hence a basis to compare various case histories in future.

7.8 SUMMARY

This chapter on cyclic stress–induced liquefaction has

- Discussed cyclic loading conditions, in particular how the direction of the principal stresses changes during different loading conditions
- Examined observed trends in undrained cyclic tests on several sands under different loading and sample preparation conditions
- Confirmed the fact that principal stress rotation is a very important factor in the cyclic degradation of soil strength
- Shown how cyclic resistance curves can be normalized to the resistance at 15 cycles of loading and then how this CRR_{15} is related to the state parameter. (But parameters other than ψ, in particular fabric or sample preparation method, will also have an effect on cyclic behaviour)
- Described the approach to liquefaction assessment based on case histories pioneered by Professor H.B. Seed and researchers at Berkeley
- Used the insight from the state parameter and mechanics framework developed in this book to confirm the basic validity of the Seed chart and at the same time identify some of the deficiencies in the method and explain the basis for the K_σ adjustment and lack of consensus on the K_α factor
- Looked at a theoretical framework for cyclic loading within *NorSand* that captures the key aspects of undrained cyclic loading in tests with principal stress rotations

Although the Seed liquefaction assessment chart has validity in mechanics, there is substantial room for improvement and the state parameter shows how this can be done.

First, it is a big advantage to replace the SPT with the CPT so that proper dimensionless parameter groups can be used to represent the characteristic strength of the ground. Rather than work directly with a penetration resistance *corrected* for stress level and fines content, state parameter estimated from the CPT captures the effect of stress level and soil properties on q_c in a rational and consistent manner.

Second, the cyclic resistance at a given penetration resistance is non-unique and similarly the state parameter does not provide a singular cyclic resistance. The cyclic resistance curve, representative of the predicted or inferred field loading conditions, depends on the soil properties as well as the characteristic state parameter of those soils. The required cyclic resistance can be developed by a suitable combination of laboratory testing and numerical simulations and based on the estimated in-situ state parameter.

The major difference between the empirical Seed chart and the advocated approach based on critical state concepts is that the critical state process is transparent. A critical state approach is not obscured with many dubious *corrections* and adjustments, and how to apply existing knowledge to different situations follows from the laws of mechanics. As with static liquefaction and flow failure, when dealing with cyclic loading, it is essential to consider the range of in-situ conditions, the actual soil properties, the likely effects of drainage and the corresponding material behaviour. Taking a critical state view is by far the simplest and most rational way to do this.

Chapter 8

Finite element modelling of soil liquefaction*

Dawn Shuttle

8.1 INTRODUCTION

The ethos of this book is that liquefaction is simply another computable behaviour of soil, with direct numerical integration of *NorSand* being used to model standard laboratory tests. This is useful in understanding how soils behave, and for obtaining soil properties, but not something that can be used to analyse a dam or foundation. Conversely, the analyses of the various large-scale failures discussed in Chapter 6 and Appendix F used a variety of limit equilibrium methods that had no capability to represent the brittle stress–strain behaviour that is intrinsic to liquefaction. The unsurprising result is that the strengths from these various back-analyses of failures do not match laboratory strengths. In short, we are not using the computable understanding of soil behaviour either to evaluate the case history record (to reduce model uncertainty) or for design. The tools are lacking.

Modern elasto-plastic analysis of geotechnical problems using the finite element (FE) method is one way forward, combining realistic stress–strain models with the ability to evaluate real-world problems, not just laboratory tests. Such FE analyses have been widely accepted in the research arena for many years (some examples of which were shown in Chapter 5), but this type of analysis has not transitioned widely into general engineering practice. This chapter is a contribution to disseminating the research FE capability and presents an implementation of *NorSand* in public domain FE software. It is 'free-to-use' software that can be downloaded from the same site as other data/programs.

There are two types of non-linear behaviour that need to be captured when dealing with soils: material non-linearity, in which the soil stiffness evolves with the deformations, and geometric non-linearity (otherwise known as 'large strain' or 'large displacement') analysis, where higher-order terms of the displacement gradients affect the solution. The FE method can accommodate both types of non-linearity. This chapter considers only the material non-linearity involved in fully capturing the liquefaction stress–strain behaviour. Geometric non-linearity was implemented along with material non-linearity, in a related FE scheme used to evaluate the in-situ state from CPT data (Chapter 4). But geometric non-linearity is omitted for considerations of slopes and foundations to keep things simpler (large strain really becomes important only when there is substantial confinement, such as with the CPT).

Non-linear FE offers real benefits, with the explicit representation of strain softening allowing models to naturally simulate the real failure mechanism. The graphical capabilities of FE programs also allow better understanding of failure mechanisms, simplifying the output to easily understood, and communicated, graphs and pictures.

* Contributed by Dr. Dawn Shuttle

A constraint with FE analysis for engineering practice is the time and effort in setting up models, with further effort in processing the results into design guidance. The man-hour costs can quickly eat into available budget and has resulted in engineering practice being largely limited to geotechnical 'modelling platforms' such as FLAC, PLAXIS or SIGMA/W. But these modelling platforms do not offer 'good' models in their standard menu – Modified Cam Clay is about as good as it gets, which is far from sufficient for looking at liquefaction. And although these modelling platforms do offer the ability to code 'user-defined models' (UDMs), this option comes with limitations. The limitations include tying the numerical implementation to a particular algorithm (e.g. FLAC requires a tangent stiffness formulation) that may not be suited to the loading paths, with the poor 'user' developing models with one hand tied behind their back because there is no access to internal code (e.g. in FLAC, stresses from an individual triangular element with perfect convergence are returned back to the UDM with changed values at the next loadstep).

Open-source FE software is the opposite end of the spectrum to commercial modelling platforms. Open-source code is generally free, and developers have full access to everything. On the other hand, mesh generators are mostly absent (other than for very simple situations), as may be the related visualization modules to create the desired output quickly. In the end, we chose to use free non-proprietary software, primarily because this book is all about encouraging the reader to look at data and make their own decisions – giving the reader access to all of the numerics (nothing hidden!) seemed the right approach. That said, to assist users in implementing *NorSand* into a range of UDMs, Appendix D provides the derivations to implement most solution algorithms.

8.2 OPEN-SOURCE FINITE ELEMENT SOFTWARE

8.2.1 Adopted software

The downloadable FE software for *NorSand* is adapted from the book *Programming the Finite Element Method*, now in the fifth edition (Smith et al., 2013). This text has a programming-oriented style, making it easy to follow, with many example programs. The programs grew out of research and developments in the Department of Civil Engineering at the University of Manchester (and now continuing at the Colorado School of Mines). This work has two particular attributes: (1) the text and programs cover a wide variety of applications with a useful focus on geomechanics and (2) the programs and subroutine libraries are freely available online.

The implementation of *NorSand* presented here is based upon the programs, library functions and subroutines, coded in Fortran90, that were released with the third edition (Smith and Griffiths, 1998).

8.2.2 Prior verification for slope stability analysis

An extensive analysis of slope stability was undertaken by Griffiths and Lane (1999). This paper is among the top five most cited papers in *Géotechnique* from the entire 65-year history of the journal, and is a landmark in applying FE analysis to slope stability.

Griffiths and Lane based their work on Program 6.2 of Smith and Griffiths (1998), extending the program to model more general geometries and soil property variations, including variable water levels and pore pressures. The various slope stability simulations were carried out for plane strain with elastic-perfectly plastic soils and a non-associated Mohr–Coulomb (NAMC) failure criterion utilizing eight-node quadrilateral elements with reduced integration (four Gauss points per element) in the gravity load generation, the stiffness matrix generation and the stress redistribution phases of the algorithm.

Soil is initially elastic, and the model generates normal and shear stresses at all Gauss points within the mesh. These stresses are then compared with the failure criterion. If the stresses at a particular Gauss point lie within the failure envelope, then that location is assumed to remain elastic. If the stresses lie on or outside the failure envelope, then that location is assumed to be yielding. Yielding stresses are redistributed throughout the mesh utilizing the viscoplastic algorithm (Perzyna, 1966; Zienkiewicz and Cormeau, 1974). In this context, viscoplasticity does not refer to any creep behaviour of the soil but instead is a technique for using internal strain increments to redistribute load within the domain proportionally to the amount by which yield has been violated.

There are various ways that slopes can be 'loaded' to failure in numerical analysis. Griffiths and Lane applied gravity in a single increment to an initially stress-free slope. The slope was then brought to 'failure' by reducing soil strengths, exactly analogous to traditional limit equilibrium methods. Defining 'failure' was a little more troublesome.

There are several possible definitions of failure, ranging from rate of change of displacement (say crest settlement) through to the numerical solution failing to converge within an iteration limit. Griffiths and Lane adopted the convergence criterion, arguing that the criterion represented a situation in which no stress distribution could be found that simultaneously satisfied both the constitutive idealization and the global equilibrium. In reality, loss of convergence tended to be where the displacements are rapidly accelerating with strength reduction, which is consistent with this approach to slope stability being 'load controlled'.

Figure 8.1 shows an example of the computed displacement pattern and corresponding deformed mesh at the onset of slope failure of a slope with uniform properties. The relatively coarse mesh used is evident, a modelling choice that prevents clear localization of deformations (adaptive mesh refinement is needed to enable more localized shear bands to form).

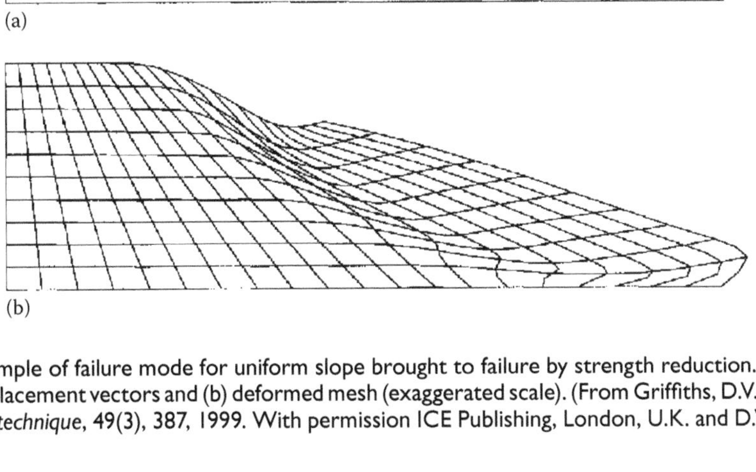

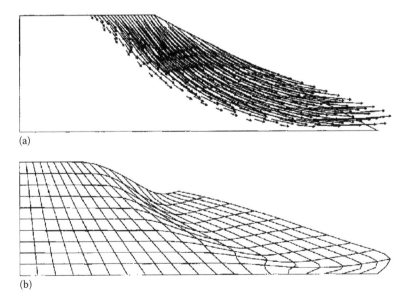

Figure 8.1 Example of failure mode for uniform slope brought to failure by strength reduction. (a) Computed displacement vectors and (b) deformed mesh (exaggerated scale). (From Griffiths, D.V. and Lane, P.A., *Géotechnique*, 49(3), 387, 1999. With permission ICE Publishing, London, U.K. and D.V. Griffiths.)

A focus of Griffith and Lane was the comparison of the FE simulations with conventional limit equilibrium results. An interesting aspect explored was a uniform slope with seepage flowing towards retained water at the toe of the slope, illustrated in Figure 8.2. Various reservoir levels were considered, in what corresponds to 'slow' drawdown with no transient effects on the internal piezometric ('free') surface. The parameter L/H was used to represent a range of analyses where H is the height of the slope and L is the depth to the reservoir measured from the top of the slope ($L/H=0$ is reservoir lapping the top of the slope, $L/H=1$ is no retained water). The results are shown in Figure 8.3. In essence, when the FE method is constrained to match the idealizations inherent in limit equilibrium, identical results are obtained. Of course, real interest lies in taking the FE method and using it to look at mechanisms that result from real geologic situations such as brittle soil behaviour and distributed weak zones.

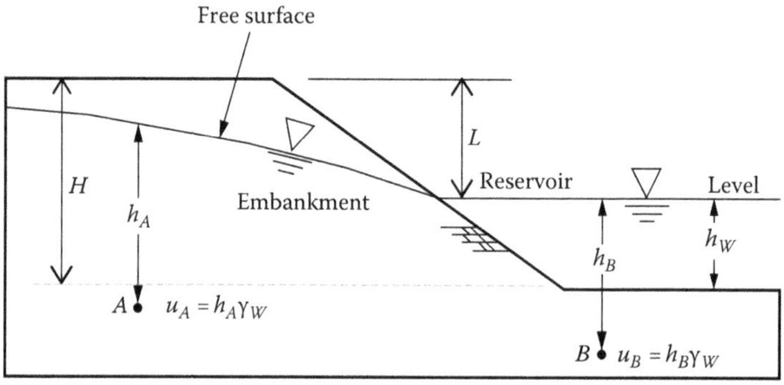

Figure 8.2 Idealized slope used to analyze the effect of reservoir level on stability. (From Griffiths, D.V. and Lane, P.A., *Géotechnique*, 49(3), 387, 1999. With permission ICE Publishing, London, U.K. and D.V. Griffiths.)

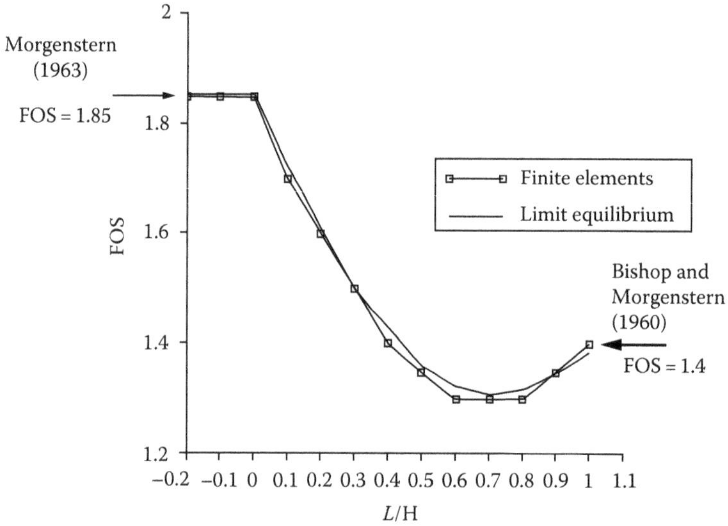

Figure 8.3 Computed factor of safety (FOS) in slow drawdown of idealized slope. (From Griffiths, D.V. and Lane, P.A., *Géotechnique*, 49(3), 387, 1999. With permission ICE Publishing, London, U.K. and D.V. Griffiths.)

8.2.3 *NorSand* implementation

NorSand was implemented in an FE code also adapted from the Program 6.2 software (Smith and Griffiths, 1998). Compared to the work of Griffiths and Lane just discussed, the principal adaptation was to replace the NAMC failure criterion with the general 3D monotonic version of *NorSand* (see Table 3.1). However, as well as changing the failure criterion to *NorSand* there is also a requirement to update the internal model state at each Gauss point as the solution develops because, while NAMC is a fixed yield condition, *NorSand* evolves. The code, called *NorSand* Finite Element Monotonic (*NorSand*FEM), was coded in Fortran90 to match the parent routines. Details of the implementation are presented in Appendix D, and the source code can be downloaded. The simulations shown in this chapter were produced by that code as compiled using Microsoft Powerstation Fortran95.

The downloadable version of *NorSand*FEM is set up to run both axisymmetric and plane strain geometry, with the initial stress condition set as either constant stress with K_o (for verification against laboratory tests) or gravity loading (for field problems). The inputs are simple and are described in detail in Appendix D. They include the basic geometry of the mesh, boundary conditions (i.e. which nodes are fixed in x and/or y) and the *NorSand* material properties.

Throughout the chapter, reference will be made to Appendices and downloads containing the tools to help run the provided core verification suite and example boundary value cases and to extend analyses to general plane strain boundary value problems.

8.2.4 Plotting and visualization

Open-source FE software usually does not include graphical output (in large part because there is no funding for PhD candidates to write the routines). The output of codes like *NorSand*FEM is a tab delimited 'txt' or comma-separated 'csv' file of results that can be imported into other programs for plotting. The results shown in this chapter were plotted using Excel and visualized using Surfer.

A poorly appreciated aspect of interpreting the output from FE codes is that contouring results using kriging does not fully honour the results. Far more is seen, and without funny contours in corners, if the element shape functions are used to map the FE results to the display picture. Golder developed such software two decades ago using VB3, but the demise of support for that programming environment means that all those libraries are no longer functional. This post-processing software is being recovered to modern C++ and will be posted on the download site.

8.3 SOFTWARE VERIFICATION

There are two important reasons to verify analysis software. The first is to ensure that the software is able to provide an accurate solution to the problem you wish to analyse. The second, equally important, use of verification is to ensure that the software is properly compiled on your computer and that you understand the code's inputs.

*NorSand*FEM has been extensively verified against laboratory 'element' tests. These 'element' tests have uniform stress conditions and known loading paths, allowing calculation of the correct solution by direct numerical integration. The verification suite for *NorSand*FEM has been chosen to allow direct comparison to the *NorSandPS.xls* and *NorSandTXL.xls* spreadsheets. These spreadsheets directly integrate the *NorSand* equations using the Euler method, which is about as far different from viscoplasticity as can be found. Comparing the results of the

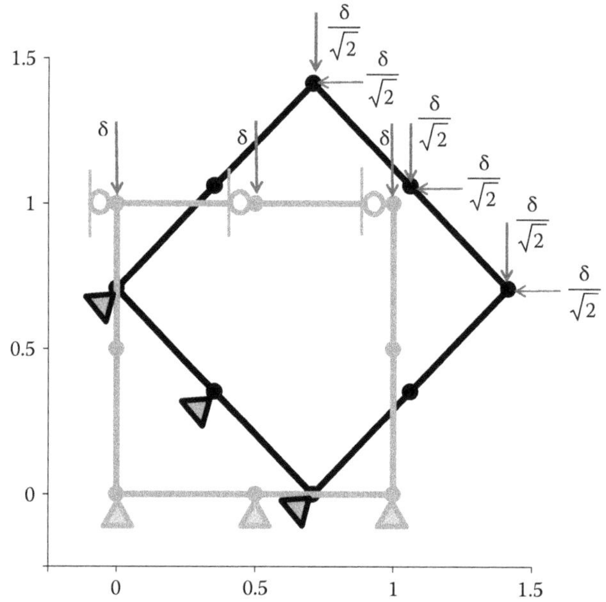

Figure 8.4 In-plane and 45° rotated 'rough platen' mesh.

two calculation methods provides an independent check on the mathematics of both. The full verification suite includes plane strain and triaxial element tests, under drained and undrained conditions, for soils with loose through dense initial states. All the input files are downloadable for you to run yourself and compare with *NorSandPS.xls* and *NorSandTXL.xls*.

Also included in the verification suite is one additional special case of a plane strain element test with rough platens using both an in-plane and rotated mesh, as shown in Figure 8.4. The loading, geometry and material properties are identical between the two meshes; the only difference is the orientation of the mesh. Rotating the orientation is done because internally the code still has x, y as vertical and horizontal, so rotating the problem checks that the stress invariants, strain invariants, 3D representation of yield surfaces, etc., are properly formulated – confirmed by both meshes giving an identical result. This verification case is very important for any 'good' model, which allows the yield surface and dilation to vary realistically as stress conditions vary from triaxial compression, through plane strain, and onto triaxial extension.

Despite all but one of the verification cases being 'element' tests, which could sensibly be verified using a single element (as the stress within the element does not vary spatially), the verification input files include multiple element meshes. This is not accidental. Experience with both commercial and propriety software is that boundary conditions applied to two sides of a single element can provide increased constraint, sometimes leading to stability for a single element that is not repeated for a larger (and less constrained) mesh. So, for added confidence, it is wise to also run cases of special interest with the bigger mesh.

Comparison of the FE and VBA verification results indicates that the match is extremely close, as can be seen in the example loose undrained and dense drained plane strain verifications from Appendix D (Figures D.8 and D.9). Please do change the properties in the input files and run your own parameter combinations. Both the *NorSand*FEM input (.dat) files and the Excel VBA spreadsheets are simple to edit, and the best way to get a feel for the effect of properties is to play with their values.

8.4 SLOPE LIQUEFACTION

The utility of *NorSand*FEM is the ability to investigate how changes in the distribution of soil state (ψ) and loading affect stability. This section presents an example of doing this, both to illustrate how to use the code and to highlight aspects of slope behaviour with soil density. The focus in this section is undrained loading of slopes comprising loose soil (after all, this is a book about liquefaction). But undrained analysis is simply a boundary condition, and *NorSand*FEM works equally well on drained and dense as well as loose soil.

8.4.1 Scenarios analyzed

For simplicity, one slope geometry has been adopted for all of the examples. The slope is 8 m high, with a 1V in 4H outer face. The foundation below the slope is at substantial depth so that the slope behaviour is not constrained by a strong base. In essence, the scenario shown on Figure 8.5 is what might be looked at for a slightly aggressive raise of a tailings dyke over a loose foundation.

The slope was discretized into a uniform mesh of 480, 2 m square elements, as shown in Figure 8.5. The mesh could be refined to provide greater mesh density in the areas of expected greatest displacement gradients, but here the purpose is to illustrate the effect of different modes of loading, and we do not want the mesh to influence the results. The mesh boundary conditions, unless otherwise stated, are fixed laterally at the mesh sides and fixed both vertically and laterally at the base of the mesh.

All of the following examples use the single set of *NorSand* properties shown in Table 8.1, with the initial state being varied. The properties broadly represent a quartz sand with trace silt, what you might find in a tailings beach where natural segregation has resulted in the sand dropping out of the tailings stream close to the spigot. The simulations adopted a uniform soil unit weight of 22.0 kN/m^3 and a geostatic stress ratio $K_o = 0.7$. The water table is set at ground surface. Water has a unit weight of 9.81 kN/m^3 and a bulk modulus of 2×10^7 kPa.

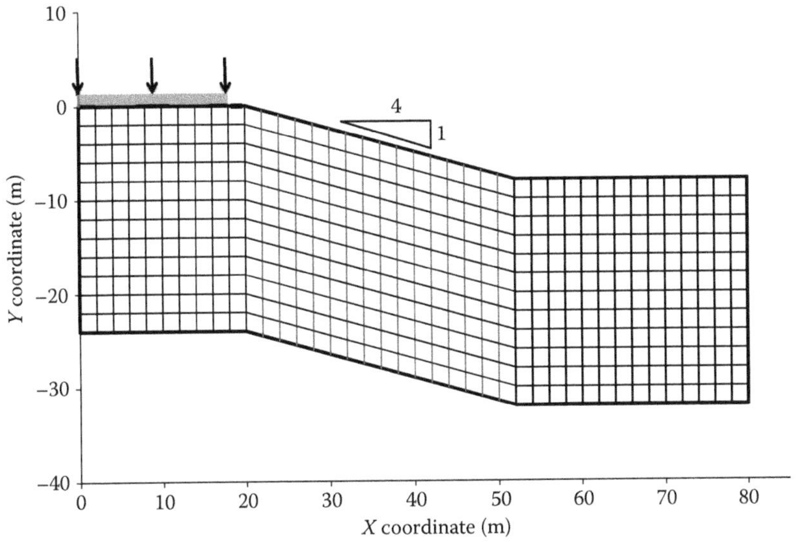

Figure 8.5 Idealized slope used for simulations and showing mesh.

Table 8.1 *NorSand* soil properties used in slope example

Property	*NorSandFEM name*	*Value*
Γ	Gamma	0.875 at $p' = 1$ kPa
λ_e	Lambda	0.03
M_{tc}	Mcrit	1.27
N	Ncrit	0.35
H_0	H0	100
H_ψ	Hy	0
χ_{tc}	Chi	4.0
G_0	G0	30 MPa at $p' = 100$ kPa
G_{power}	Gpower	1.0
ν	Poisson	0.15
R	OCR0	1.001

Elasticity was held constant throughout a simulation, although varied spatially with initial stress conditions using the equation, $G = G_0 * (p/p_{ref})^{Gpower}$, where G_0 and G_{power} are material properties and p_{ref} is the (conventional) reference pressure of 100 kPa to keep dimensions consistent. For simplicity and ease of comparison, the plastic hardening is set constant (i.e. choosing the property $H_\psi = 0$.)

8.4.2 Displacement controlled loading

The analyses of Griffiths and Lane discussed in Section 8.2 used strength reduction throughout the domain to cause failure. This approach is tantamount to load control and means that post-failure modes are never seen as the solution breaks down. In liquefaction case history records, the post-failure deformation is often the primary observation, and so a different approach is warranted.

One possibility would be to change the analysis from static to dynamic and to compute velocities as the slope failed and tracking the progression of failure – but that is certainly complicated, never mind non-standard with all sorts of modelling uncertainty and validation issues. The simpler approach adopted was to load the crest of the slope using displacements and compute the load corresponding to those displacements; in this approach, 'failure' is simply the point of maximum load, but we can track the evolution of stresses and displacements as failure progresses and watch the evolution to the critical state within the domain.

The flip side of the simplicity of displacement control is that the loading is not equivalent to post-earthquake migration of excess pore pressure that seems to have caused many of the actual static liquefaction failures documented in Appendix F. A second displacement control loading scenario of basal yielding was therefore also simulated, which corresponds directly to the triggering mechanism at Nerlerk and Fort Peck, and is 'not unlike' the toe erosion situation triggering the Jamuna large-scale slides.

8.4.3 Surface loading with rough rigid footing

The first loading case, illustrated in Figure 8.5, considers application of a rigid and perfectly rough footing applied 2 m from the crest of the slope. Although applied as a footing, this loading could also be viewed as additional lifts of fill placed on the top of the slope. Initial soil state was taken as uniform throughout the slope and foundation, with four scenarios

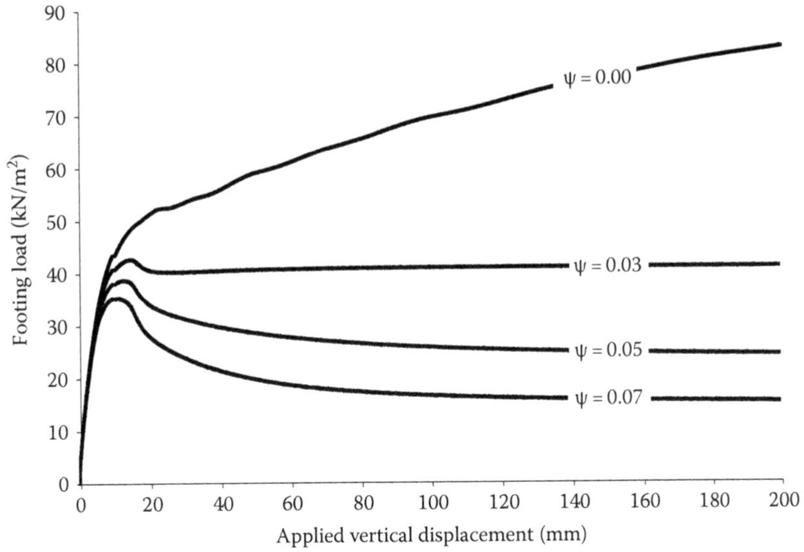

Figure 8.6 Load–displacement response under rough rigid footing for range of ψ_0 from 0.0 to 0.07.

simulated: $\psi_0 = +0.0$, +0.03, +0.05 and +0.07. For the chosen soil properties, these states correspond to, respectively, no strength loss, 'normally consolidated' behaviour, mildly liquefiable and extremely liquefiable. Figure 8.6 shows the resulting load–displacement response, measured at the upper row of stress sampling points.

Initially, the stiffness for all ψ_0 is the same, due simply to both elastic and plastic stiffness being constant for all ψ. As expected, the highest bearing resistance was measured for the densest ($\psi_0 = 0.0$) soil, although more surprisingly, this soil had not reached its maximum bearing capacity within the applied 200 mm of displacement. All of the looser soils reached a peak bearing capacity within 15 mm of displacement, followed by marked reduction in the resistance. Predictably, the largest reduction in load occurred for the loosest ($\psi_0 = 0.07$) soil. But it is noteworthy that a significant proportion of the reduction in load for all three of the reducing load–displacement curves occurred within the first 10 mm of post-peak applied displacement, thus indicating a brittle failure mode.

In terms of liquefaction, the peak load applied varied from 35.4 kPa for $\psi_0 = 0.07$ (equivalent to a slope height rise of just 1.61 m) to 42.8 kPa for $\psi_0 = 0.03$ (equivalent to a height rise of 1.94 m), realistic results for loose soil with high water table. Post liquefaction, both the $\psi_0 = 0.05$ and $\psi_0 = 0.07$ simulations continued to lose resistance with increasing deformation following 'triggering', and at 200 mm displacement, the measured resistances were only 70% and 55% of their peak values (and still continuing to fall). This strength loss indicates that the slope would need to flatten to, respectively, about two-thirds and half the initial slope height before equilibrium stresses matched the post-liquefied strength. These are plausible results based on the case histories discussed in Appendix F.

The low amount of movement to trigger displacement for the looser slopes might seem strange. This is in part an artefact of the idealization of the water table at surface and undrained loading. But equally the small amount of displacement to trigger brittle failure could be seen as a consequence of the mobilized failure surface. Figure 8.7 shows contours of η/M_θ (which is tantamount to the inverse of the local factor of safety [FOS]) at 200 mm footing settlement for the various scenarios of initial state. Figure 8.7a shows that the loosest ($\psi_0 = 0.07$) example develops a very shallow bearing capacity

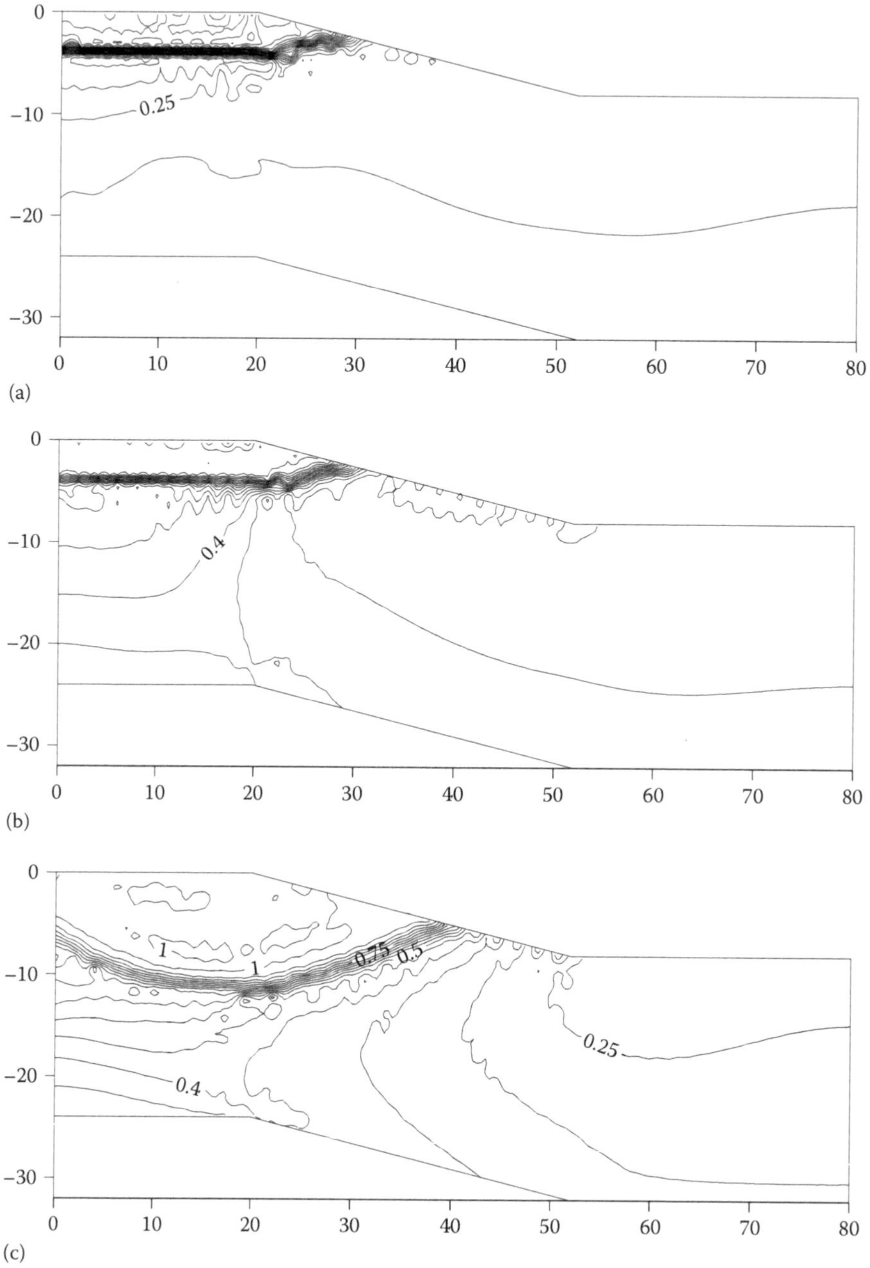

Figure 8.7 Contours of η/M_θ under rough rigid footing at 200 mm vertical displacement. (a) Contractive initial state $\psi_0 = +0.07$, (b) 'normally consolidated' initial state with $\psi_0 = \lambda_e$ (= +0.03) and (c) non-brittle initial state with $\psi_0 = +0.0$.

type of failure. Figure 8.7b shows that making the soil somewhat denser into a 'normally consolidated' state with $\psi_0 = 0.03$ pushes the failure slightly deeper, but it still remains a bearing capacity–type failure. However, as illustrated in Figure 8.7c, setting $\psi_0 = 0.0$ with the removal of brittle failure implied by this choice forces a much deeper, and circular, failure mode.

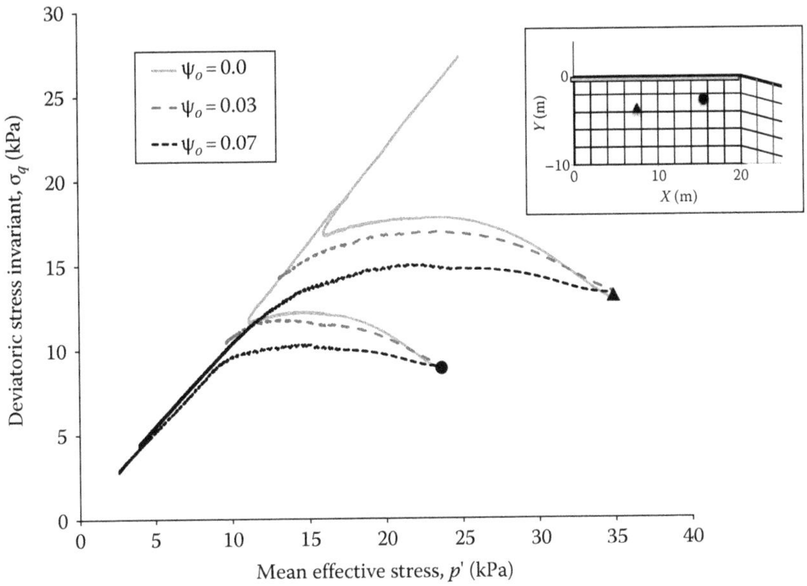

Figure 8.8 Stress paths for Gauss points located below the footing: $\psi_0 = 0.0$, 0.03 and 0.07.

The stress paths for two locations below the footing are shown in Figure 8.8. The locations of the monitoring points are shown on the inset to the figure, with the corresponding symbols marking the initial stress state of the two locations. The very loose ($\psi_0 = 0.07$) stress paths both show negligible, if any, increase in shear stress with decreasing mean stress, prior to the stress paths reducing in shear towards the failure condition. This is very close to the paths found in the plane strain static liquefaction tests of Wanatowski and Chu (2007), which were discussed in Chapter 6 (see Figure 6.9 in particular). The 'normally consolidated' simulation ($\psi_0 = 0.03$) has a stress path initially indicating a small increase in shear stress with decreasing mean stress, followed by a similarly small shear stress reduction. Finally, the $\psi_0 = 0.0$ simulation has the classic 'S' shape of an undrained quasi-steady state before the onset of dilation to reach the critical state.

For this loading scenario, the initial state changes the mode of failure significantly. The movements prior to failure correspondingly differ considerably. In a 'real' situation, a shallow failure mode of the type modelled would provide little, if any, warning of failure, and the situation could go from benign to catastrophic within minutes. This is not a situation suitable for controlling safety by monitoring movements or pore pressures (i.e. the 'observational method').

8.4.4 Crest loading with deep weak zone

Figure 8.7 shows that loading the crest produced bearing capacity–type failures with the loose soil states rather than triggering overall slope movement. This appears a consequence of the soil failing first under the loaded area and then the subsequent loss of strength preventing load being transferred deeper. A further scenario was therefore considered using a 'normally consolidated' soil for most of the slope, which is not prone to much strength loss, with a much looser zone lower in the slope that might attract load, and so cause a deeper failure mode. This is not as artificial an example as it might appear, with the situation of

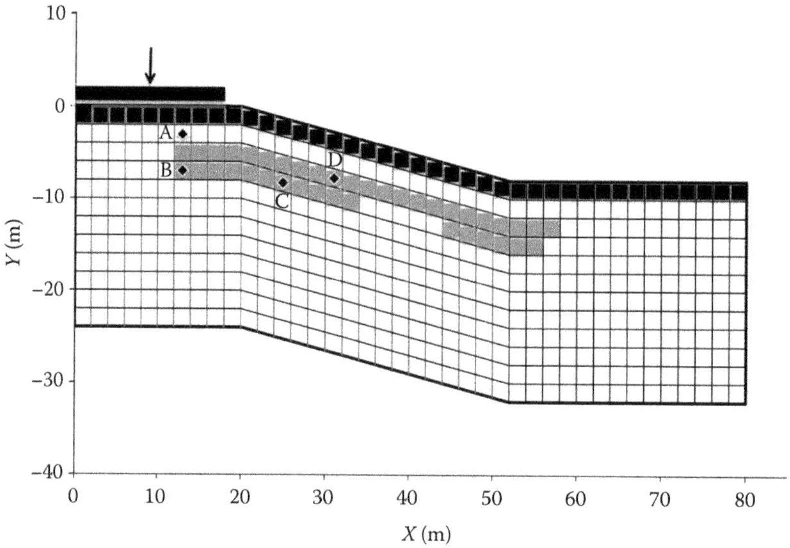

Figure 8.9 Location of weak zone ($\psi_0=0.07$) and surface-compacted layer ($\psi_0=0.0$) with monitoring points.

loose layers developing naturally in real soils (see Chapter 5). Indeed, Hicks and Onisphorou (2005) suggest that extensive weak zones were at least a contributory factor in the Nerlerk failure discussed in Appendix F.

The distribution of the loose zone simulated is illustrated in Figure 8.9. Within a slope with $\psi_0=+0.03$ overall, a slightly denser layer has been applied for the top row of elements much as one might find with a tailings beach exposed to air drying and/or tracking by construction equipment. A rather pervasive loose layer has been simulated running parallel to the slope; this is a little more extreme than inferred to exist at Nerlerk by Hicks and Onisphorou, but interesting as a test of how the failure mode changes. Four monitoring points were used to investigate the stress paths, shown in Figure 8.9 as 'A' to 'D'.

Figure 8.10 shows the computed load–displacement curve for this scenario. The previously presented trends are shown as background. There is a substantial drop in load capacity post-peak, but not to the extent of what developed with the uniformly loose slope. Even though the weak zone is both extensive and located where one might intuitively expect to dominate the slope performance, that turns out not to be the case with the surrounding denser soil also influencing the failure mode.

Figure 8.11 shows both contours of plastic shear strain increment and the matching mobilized stress ratio η/M. The shear strains define a mechanism that is tending to a circular slope failure and does not go down the full length of the idealized weak zone. The stress paths at the four monitored locations are shown in Figure 8.12. Location 'A' within the 'normally consolidated' soil directly below the footing shows a small increase, then reduction in shear stress with strain, reaching a quasi-steady state before dilating towards the critical state. Location 'B', a few metres below in the weak ($\psi_0=0.07$) layer, follows a stress path typical of very loose soil undergoing liquefaction. It experiences a significant loss of strength with increasing strain as it moves towards the critical state. Locations 'C' and 'D' are not directly below the footing, lying in the weak zone at a distance down the slope. Location 'C' indicates predominantly elastic unload/reload behaviour, suggesting that this element is on the edge of the failure zone. The location 'D' stress path indicates an initially elastic

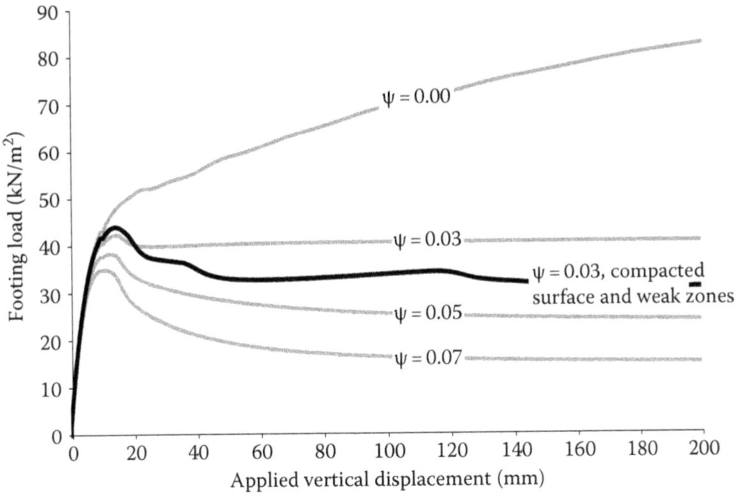

Figure 8.10 Load–displacement response under rough rigid footing $\psi_0 = 0.03$ with compacted surface and weak zone (ψ_0 = from 0.0 to 0.07 with no weak zone shown for comparison).

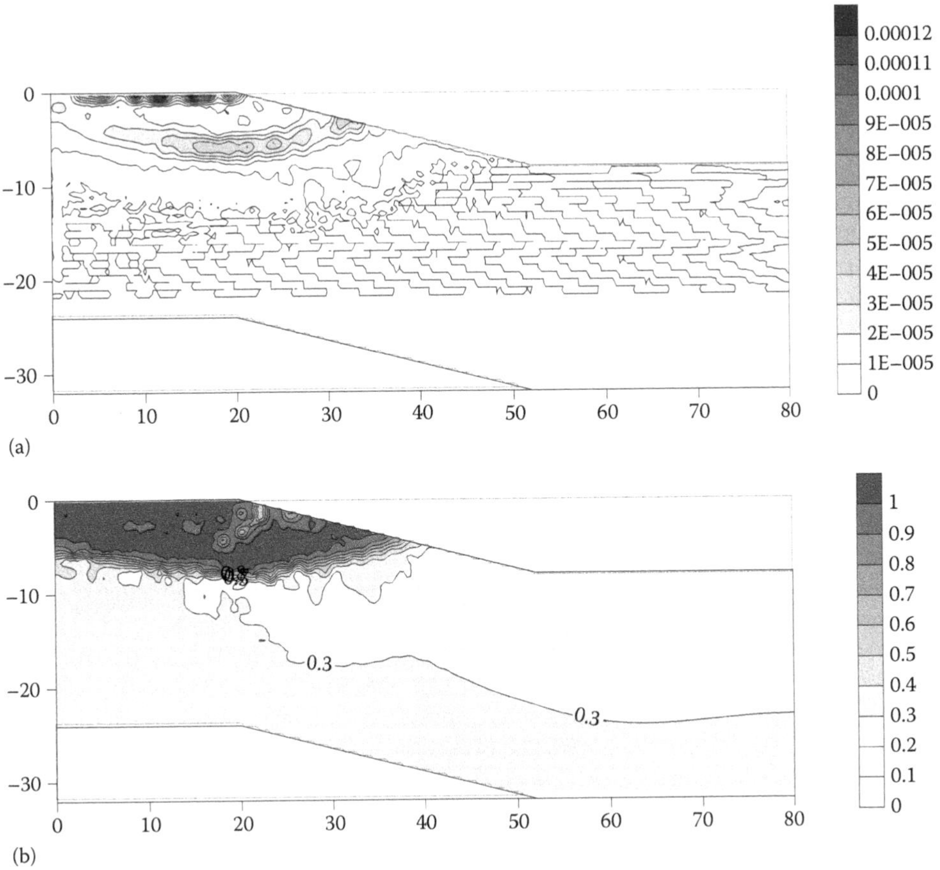

Figure 8.11 Behaviour of 'normally consolidated' slope with loose layer at 200 mm settlement. (a) Contours of incremental plastic shear strain and (b) contours of mobilized stress ratio η/M.

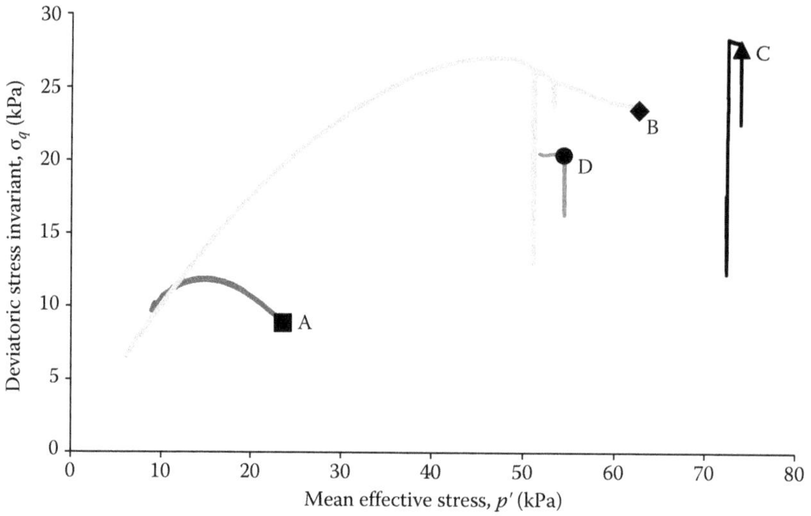

Figure 8.12 Stress paths at four monitoring points in 'normally consolidated' slope with loose layer.

response and then starts to follow a 'liquefaction' stress path, indicating that with increasing strain, this portion of the slope forms part of the slope failure surface (consistent with the location of the plastic shear strain increments in Figure 8.11a).

8.4.5 Movement at depth

An opposite extreme to loading the crest, in terms of slope movement induced, is to simulate foundation creep and, in essence, load the base of the slope. Movement at depth can be the result of construction upon, or excavation into, a local weaker stratum or regional geological feature (such as the Bearpaw Shale in Montana, a credible factor in the Fort Peck failure), amongst other causes.

There are various ways a weak or creeping feature underlying the slope could be modelled. Here, the simplest approach is taken of imposing a 'stretching' base under part of the slope rather than using, say, a viscous layer developing creep with time. Figure 8.13 illustrates the 'loading' that comprised no basal movement for $x \leq 20$ m, a linear increase in lateral basal movement along the length of the slope ($20 < x < 56$), to constant lateral movement for all boundaries with $x \geq 56$. Thus, what has been modelled is a stretch at depth rather than true creep, but equally, the imposed basal displacements are well removed from the slope so that the details at the base of the model should not unduly influence the developing failure mechanism.

An identical range of material properties and groundwater pressure was used as for the crest-loading scenarios, so that the changed slope failure modes can be attributed only to the change in loading mechanism. Two initial states were considered, one very contractive with $\psi_0 = +0.07$ and one with $\psi_0 = +0.0$; these states span the range discussed earlier for crest loading.

Failure zones develop quite quickly, with the failure modes showing their final locations in as little as 0.1 m of base movement. However, it is instructive to allow more creep and see the mechanisms that develop.

Looking to the most contractive $\psi_0 = +0.07$ case first, Figure 8.14 shows both contours of plastic shear strain increment and the corresponding mobilized stress ratio η/M. The shear strains define a localized mechanism that forms a 'classic' two-wedge failure going to full

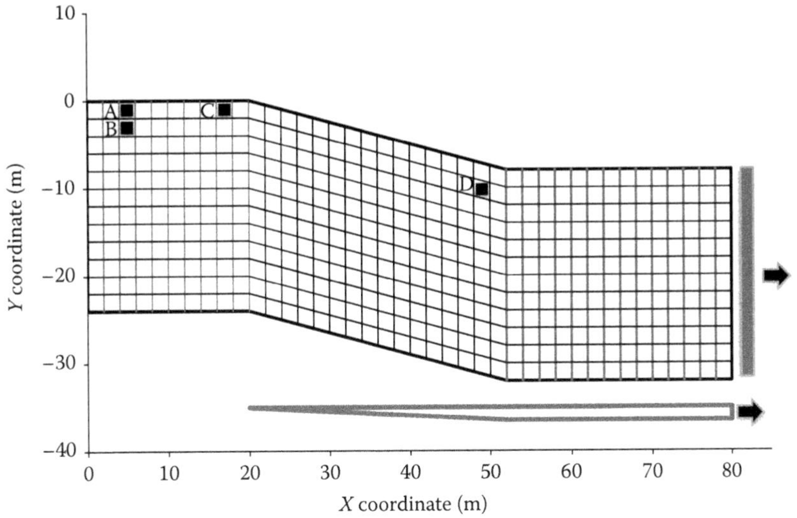

Figure 8.13 Mesh and loading condition used to simulate basal creep (with monitoring points shown).

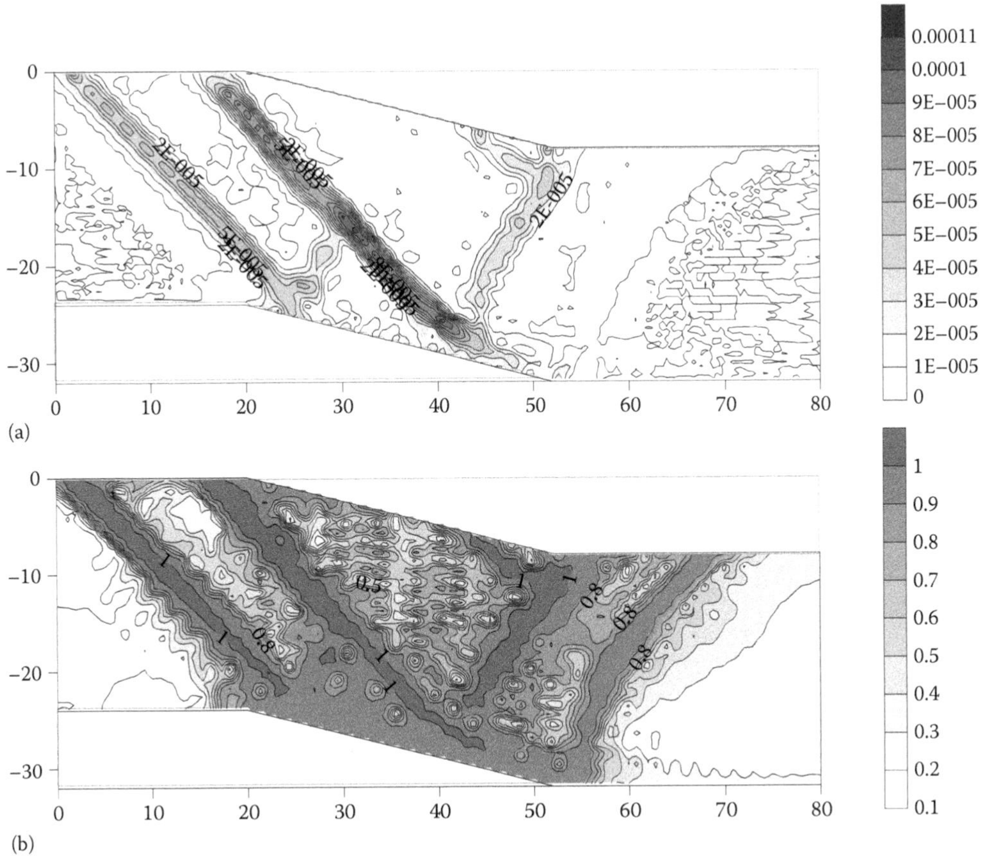

Figure 8.14 Behaviour of loose slope with $\psi_0 = 0.07$ subject to 2 m maximum basal creep. (a) Contours of incremental plastic shear strain and (b) contours of mobilized stress ratio η/M.

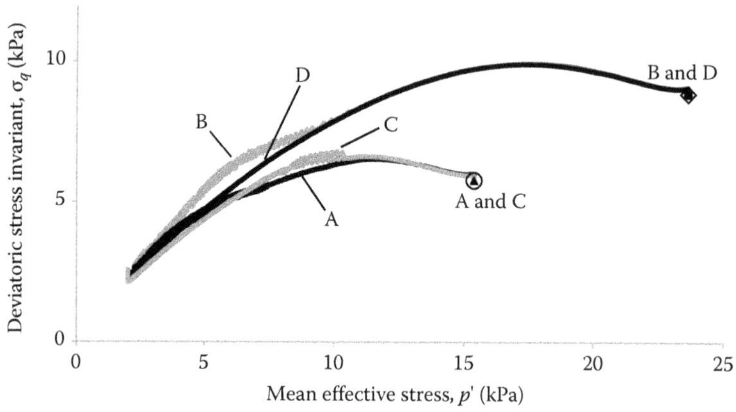

Figure 8.15 Stress paths followed at the four monitoring points during basal creep $\psi_0=0.07$.

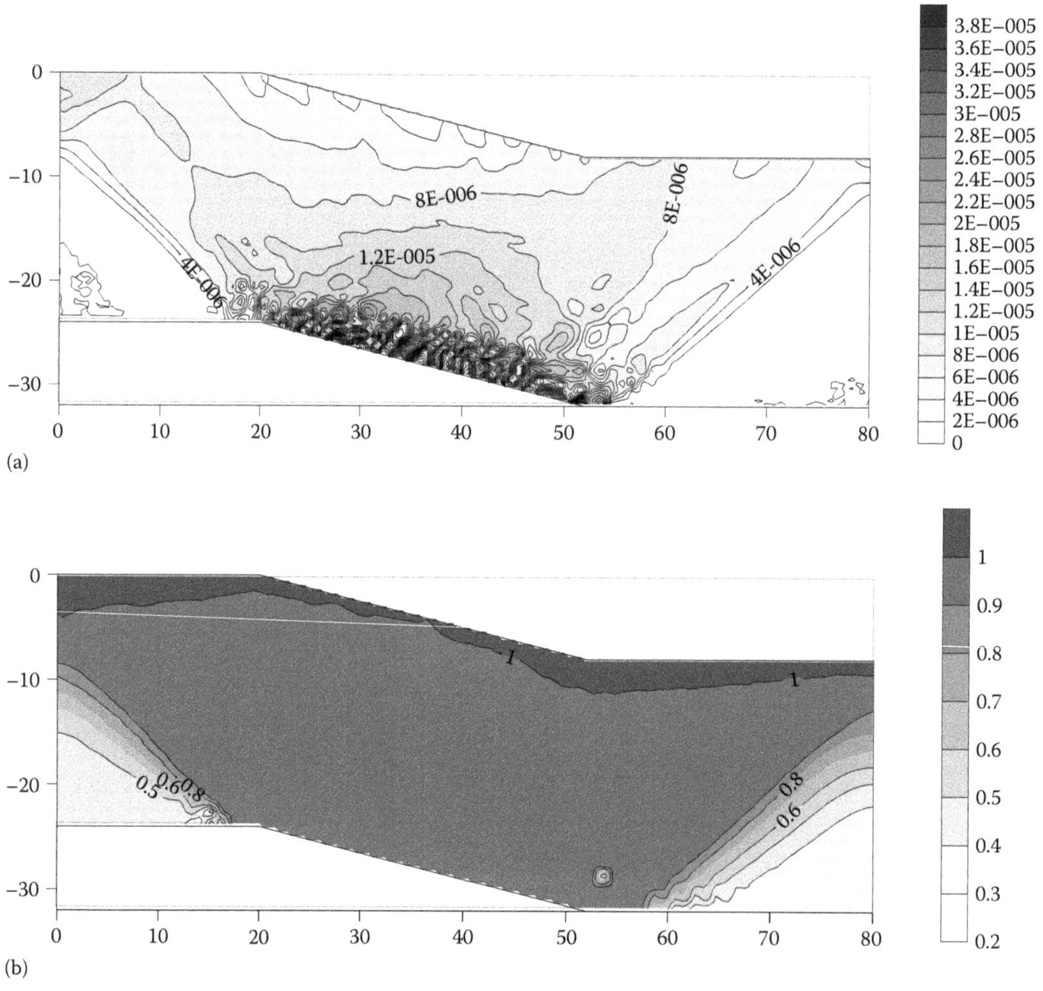

Figure 8.16 Behaviour of loose slope with $\psi_0=0.0$ subject to 2 m maximum basal creep. (a) Contours of incremental plastic shear strain and (b) contours of mobilized stress ratio η/M.

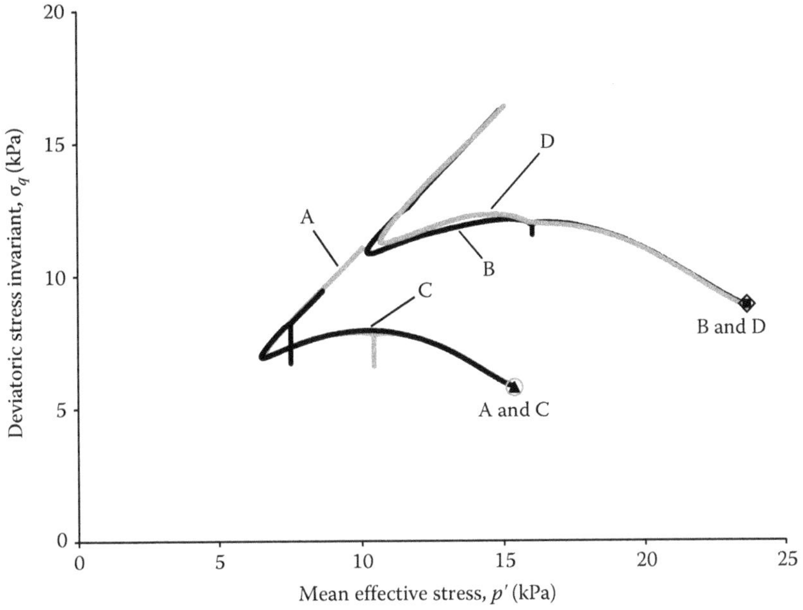

Figure 8.17 Stress paths followed at the four monitoring points during basal creep $\psi_0 = 0.0$.

depth, with hints of a secondary mechanism that might develop into a second back scarp. The stress ratio plots show all the material strength developed in the plastic shearing zones, but with further highly stressed zones that are not yet apparent in the plastic movements – a complete slope-wide instability caused by base creep.

The stress paths followed by the soil at the four monitored points are shown in Figure 8.15. Just as with crest loading, this $\psi_0 = +0.07$ soil is so loose that it is close to its undrained strength under geostatic conditions so that minor perturbation causes loss of strength, with the soil at the monitored points losing about 80% of its strength as it moves to its critical state.

If we now turn to the denser scenario of $\psi_0 = +0.0$, a very different failure mode emerges. Figure 8.16 shows both contours of plastic shear strain increment and the matching mobilized stress ratio η/M. No tendency to form localized shear zones is evident, with the pattern of plastic shear strain increments suggesting something approaching a circular failure mode. However, the mobilized shear stresses are nowhere near the available strength, seen in the stress paths of Figure 8.17, where ongoing dilation is evident following the initial contraction – the 'classic' S-shaped stress path that is consistent with the initial state used. Of course, the interesting question then follows: if pore water migration was allowed locally, what would the mechanism then be?

8.5 COMMENTARY

This chapter has been based on straightforward extensions to a public domain FE program. The underlying code has a long-standing and widely cited capability for slope stability modelling using classic NAMC representation of soil behaviour. The equally long-known deficiencies of the NAMC for undrained soil behaviour (constant dilation prevents realistic excess pore pressure predictions) are overcome by introducing *NorSand*, with its demonstrated accuracy in simulating excess pore pressures caused by yielding. The steps needed to add *NorSand* to this public domain software are in many ways easier than coding a user-defined model in commercial software. And the resulting code accurately verifies across a range of laboratory test stress paths, soil properties and initial conditions.

This chapter has taken the verified FE software and used it to explore static liquefaction of slopes – the situation for the various case histories in Appendix F. Two loading scenarios have been used to trigger static liquefaction, loading of the crest (such as placing another layer of fill) and basal movements (such as occur with creep in a weaker underlying stratum or erosion at the toe of a slope). Neither of these loading conditions at the top or bottom of the slope captures the post-earthquake situation, where failure is triggered by migration of earthquake-induced excess pore water (a far more complicated situation requiring a coupled model). But even within this limited exploration of liquefaction using undrained analysis and external loading, realistic mechanisms naturally emerge from the FE simulations. This ability to predict realistic failure modes is quite an advance for understanding case histories, as these case histories are usually deficient in measured soil properties and in-situ state although there are often reasonable observations of the pattern of failure deformations.

The central issue that should concern geotechnical engineers involved with liquefaction is the mismatch between laboratory strengths and those developed from back-analysis of case histories using limit equilibrium methods. This is a gap in engineering understanding. Although engineering judgement can be invoked to assess a new situation using the case history record as a basis for judgement, it is a long way from showing that the geotechnical profession actually understands the situations they are dealing with. This situation is also not confined to liquefaction – embankments on clay rarely fail at a calculated FS = 1.0 from input soil properties measured during a site investigation. A plausible rational for this mismatch between the measured properties/state of the soil and the actual behaviour of our slopes and structures is the limit equilibrium methodology. Limit equilibrium is far from realistic assumptions of soil behaviour and, correspondingly, cannot capture the internal redistribution of loads.

The hope in providing this chapter is that it will trigger interest in better analysis of the case histories using the FE method. The FE method has seen limited use in this regard to date (e.g. Hicks and Onisphorou, 2005), but such work appears intimidatingly difficult to many geotechnical engineers. We hope we have shown that the FE method is not so difficult, and that it provides a quantum leap in understanding compared to back-analysis using limit equilibrium.

What we have not done in this chapter is to model cyclic mobility. Cyclic mobility is not difficult in terms of the basic soil behaviour (it is easy enough to do for simple shear tests; see *NorSandPSR.xls*), but there is little point in doing that in large-scale problems, or rather in believing in the results of doing that, when the profession has such a mismatch in the static failures. In particular, our understanding–if it can be called that–of the case histories does not include the demonstrable effects of elastic stiffness and plastic compressibility on soil behaviour. As engineers, we are at some distance from removing *model uncertainty* from the practical situations, as we deal with in civil and mining works. We need to improve.

Of course, it is also apparent that situations like Lower San Fernando Dam require coupled analysis with explicit consideration of pore water migration (i.e. a 'Biot' analysis). The public domain software used as the basis of this chapter includes coupled flow, so this is only a small increase in complexity. It may be a stretch today to expect this level of analysis in routine consulting practice (although our own, and a few other, companies do use these methods for some special situations). But, it is certainly reasonable to expect researchers to adopt coupled FE analyses in their investigations of case histories and to disseminate their work into the commercial modelling platforms. Liquefaction assessment where 'fines content' is a major input for strengths used in limit equilibrium analysis ought to be a thing of the past.

Chapter 9

Practical implementation of critical state approach

9.1 OVERVIEW

This chapter collects the various pieces of critical state soil mechanics in the rest of the book together as a *procedure* for engineering of liquefaction problems in sands and silts. The need for this *how to* guide is a response to discussions during projects and following various presentations we have made since the first edition was published.

Let us start by noting that the required procedure is not the result of a choice between laboratory and in-situ, the traditional battleground for engineers looking to determine soil behaviour. Both testing protocols are needed – an intrinsic consequence of being unable to obtain, and then transfer and set up in a laboratory test, undisturbed samples of sands and silts. Since soil response to loading is broadly Response = Properties × State, engineers are put in the position of needing to use in-situ tests to measure State (because you cannot get undisturbed samples) and laboratory tests on reconstituted samples to measure Properties (because in-situ tests cannot distinguish between the effects of state and properties). Accordingly, this chapter gives guidance on

- The scope of field investigations and laboratory testing
- Deriving soil properties from laboratory test data
- Choosing CPT equipment
- Interpretation of in-situ testing data (CPT, shear wave velocity)
- Application of the results to problems in soils ranging from sands to silts

The critical state approach to liquefaction, and for that matter any other aspect of soil behaviour, consists of the following logic:

1. Determine the in-situ state of the material with penetration testing, supplemented by data from other testing and modelling. CPT testing is key, in practice, as the CPT tends to respond to ψ, and this is helpful in dealing with some level of variability of particle size distribution.
2. Determine the properties of the soil in the laboratory, on reconstituted material, over a range of densities and stress levels (i.e. a range of ψ). Some modelling of the soil behaviour (say using *NorSand*) is needed to validate estimated properties.
3. Engineering analyses are based on knowledge of the in-situ state and the soil properties at that state.

It really is that simple and it is not that different from good practice for geotechnical engineering in clay materials which can be sampled nearly undisturbed, but which change state significantly as a result of consolidation under applied loads.

9.2 SCOPE OF FIELD INVESTIGATIONS AND LABORATORY TESTING

The field investigation is about a lot more than liquefaction, as it needs to define thickness of soil units, depth to certain boundaries, groundwater flow conditions and variability of soil conditions over a particular site. Engineering practice will define the frequency and location of probing, as well as broadly what is carried out. When liquefaction is considered to be an issue, and the approach in this book is going to be followed, the investigation needs to include the following:

- CPTs through the material of interest for liquefaction. These need to be the CPTU variety with measurement of the pore pressure in the u2 location.
- Sampling, both to confirm the material behaviour type found in the CPT and to gather material for laboratory testing. Sampling can be whatever method retrieves samples: Mostap (convenient as it works with CPT equipment), driven samplers (possibly with core catchers), Shelby tubes if the material will stay in the tube, California samplers, etc., are all possible.
- Shear wave velocity measurements, in-situ, unless a screening-level assessment is all that is needed. A seismic CPT is the ideal way to collect these data, using vertical seismic profiling as described in Section 4.8. Shear wave velocity will be used for several aspects of the liquefaction analysis. Perhaps the most important need for seismic data is as input to a site-specific seismic response analysis, so that ground motions can be calculated through the soil column. But the same seismic data are also needed input to enhance CPT interpretation beyond a screening-level assessment.
- A carefully planned laboratory testing program to measure the soil properties including those relating to critical state locus, stress–dilatancy, state–dilatancy and plastic shear hardening.
- If dealing with silts or unusual soils (e.g. carbonate sands) under exposure to earthquakes or similar, the laboratory program will need extending to include cyclic simple shear tests.

It is the laboratory program that all too often is neglected in planning a liquefaction study, or may be considered 'too expensive'. But in reality, the laboratory testing is a small cost compared to the fieldwork and is a significant addition of value by removing most of the guesswork and empiricism out of a liquefaction assessment.

In practice, the range of testing specified will depend on the importance of the liquefaction project. For a preliminary screening assessment, the index testing is a starting point and reliance is placed on the screening-level CPT interpretation. However, as soon as you progress beyond this, or have a material that might be 'different', such as a silt, carbonate or highly angular sand, then triaxial compression testing is definitely called for. Triaxial compression testing is the current reference strength test for geotechnical engineering and, with some care to implement the procedures in Appendix B, is quite within the reach of most projects. Bender element, cyclic simple shear and resonant column testing is less common in commercial laboratories, but such advanced tests are only needed for seismic liquefaction studies once you are sure static liquefaction and flow slides have been taken care of. We will show you how to make that determination in Section 9.5 when we are looking at the CPT, after we have derived our soil properties from the triaxial testing and, possibly, cyclic strength trends from simple shear testing.

9.3 DERIVING SOIL PROPERTIES FROM LABORATORY TESTS

The starting point for deriving the critical state line and soil properties is a set of triaxial test data, and it must be triaxial compression data as that is the current reference test for geotechnical engineering. Triaxial testing procedures themselves are set out in Appendix B, but what sort of testing program needs to be specified and how should the resulting data be processed? The triaxial data itself should be available as a digital record for importing into Excel (or comparable spreadsheet), a perfectly reasonable requirement these days and which allows much quicker determination of soil properties. Paper records simply take more time and effort to process while throwing away accuracy.

One concern when viewing the geotechnical literature is that the data are always 'perfect', but that is not what most laboratories supply in their reports and thus what you have to deal with in engineering practice. This chapter is therefore based on the test data for Nerlerk 270/1 sand, a set of 'normal commercial' data of comparable quality as you might encounter in practice. Importantly, what is found is that data do not need to be perfect for a competent engineer to develop reliable estimates of the true soil properties. Of course, if you do have high-quality data, then things get easier but you may still be surprised by the test-to-test variability in results even in the best research environment. The approach set out here will still be relevant.

Most soil properties do not belong to any particular constitutive model beyond the idea that soil is particulate, frictional and compressible. The exception to this is plastic hardening, where no standard idea has emerged for the various good soil models in the literature and despite all those models aiming to mimic exactly the same soil behaviour. We will return to this point later, but for now simply note that much of what is presented in this section is universal. In principle, these properties belong to a laboratory testing report with the data they are derived from. For plastic hardening, we will be using a 'calibration' to derive the values presented based on *NorSand* and for which you can download the spreadsheet *NorTxl_Nerlerk.xls* to follow along. *NorSand* uses the familiar and standard soil properties so to that extent universality is preserved. If you wish to use a different soil model, feel free to do so but the needed validation steps will be exactly the same as presented here using *NorSand*. Validation is a universal aspect of engineering that should be applied in any project, and the steps involved in validation are generic.

9.3.1 Selecting a representative sample

Soil is naturally variable, with differing distributions of particle sizes from place to place even within a single geological unit. If you take 10 samples from a stratum and determine their gradation curves, you will get a bandwidth. This is true even when dealing with man-made soils such as tailings, although the bandwidth with man-made soils will generally be less than that found in natural soils. Figure 9.1 shows an example of grading curves from a single geological unit. Looking at this bandwidth, the first question is: what to test?

There is a body of thought that all samples from a geological stratum are similar, and thus you can combine tests on different samples from that stratum to develop soil properties. For example, this view appears to have underlain the determination of soil properties involved in the Jamuna flowslides (see Appendix F) and, in our experience, is seemingly a common belief within much consulting practice. But the view is misleading and can cause *determination* (if you can call it that) of absurd or misleading soil properties. The problem is that small changes in fines content can cause surprising changes in the CSL (and Γ in particular, see Chapter 2) and that puts you in the position of trying to sort out a soil's properties using (say) four tests on apples, two tests on lemons and one test on a melon. It is nonsense.

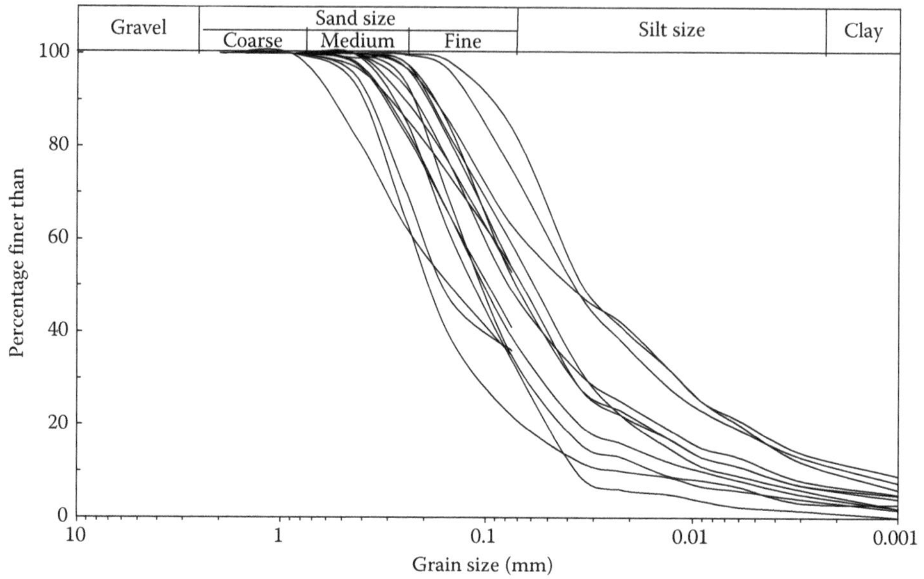

Figure 9.1 Example of gradation bandwidth in a uniform soil stratum.

Step 1, therefore, is to examine the gradation envelope you are dealing with and to pick a single gradation that is representative of the soil stratum (or as representative as you can judge from the available site investigation data). In some cases, you may find that there is more than one material type, such as layered sands and silts, and you might pick two gradations. But for the moment, we will assume that one representative material exists.

Step 2 is to obtain about 10 kg of the selected gradation, if need be by combining samples, sieving, re-blending and mixing. Ten kilograms will be sufficient for about 15 triaxial tests on the chosen representative soil; ten to 12 tests will usually be more than enough in engineering practice. Obviously, the gradation produced for testing must be documented with a particle size distribution test and other index tests for comparison with the database. It is good practice to take a micro-photograph of the sand grains to identify the grain shape and angularity, as well as a qualitative indication of mineralogy.

Step 3 is to decide whether to test additional gradations, looking to both the coarser and finer sides of the in-situ gradation envelope. Whether such step-out testing is done depends on the size of the project and available testing budget. If budget is not a constraint, do a full suite of tests on the step-out gradations. If budget is tight or otherwise constrained, use fewer step-out tests rather than a comprehensive set and then model the measured data to verify the effect of gradation on the soil properties (but note we specify full step-out tests wherever possible).

9.3.2 Minimum test program

The triaxial test program is self-evidently aimed at measuring soil properties, and thus, those properties dictate what sort of test program is needed. Properties fall into five groups:

- Properties describing the soil's CSL (e.g. Γ, λ_{10} for the usual idealization)
- Properties describing the soil's stress–dilatancy (M_{tc}, N)
- A property describing the soil's state–dilatancy (χ_{tc})

- Properties describing the soil's plastic stiffness or compressibility (e.g. H, C_c)
- Elastic properties

In principle, just three triaxial tests are sufficient to determine the five properties Γ, λ_{10}, M_{tc}, N and χ_{tc}. The testing involved is illustrated on the state diagram of Figure 9.2. In reality, we suggest at least 10 tests, but a *minimum three* test suite shown in Figure 9.2 is helpful to understand what is needed and why. While you may be reading this book because of a concern for loose soils, and thus expecting Tests 'A' and 'B', that concern still mandates tests on dense samples because the nature of stress–dilatancy, which applies everywhere, is actually difficult to discern with loose samples. Competing effects make it difficult to distinguish between stress–dilatancy and other aspects of the soil's behaviour in loose samples. Hence, a dense specimen such as Test 'C' must be tested.

All the tests in Figure 9.2 are isotropically consolidated to the initial state prior to shear. It is a good idea to measure the void ratio changes during that consolidation as it is useful data on C_c and available for no extra money (i.e. all it needs is a bit of care on your part while consolidating the sample).

Twenty five years ago, the CSL was defined entirely by undrained tests, but that caused a problem at higher stress levels because the needed initial confining stress would often be outside the limits of the equipment. Further, even with high-pressure equipment, grain crushing often occurred which self-evidently changes the material being tested. Thus, testing practice evolved to a combination of undrained and drained tests on loose sand to determine the CSL, which is the test program shown in Figure 9.2.

Test 'A' is a standard undrained liquefaction test on a loose sample, the sort of test that was extensively discussed in Chapter 2. Its sole purpose is to define the CSL at low confining stress and also not taking that stress so low as to give an issue with transducer resolution. It must be loose enough to give a brittle response and also loose enough that there is not even a whiff of a quasi-steady state. A good initial effective confining stress is 100–200 kPa,

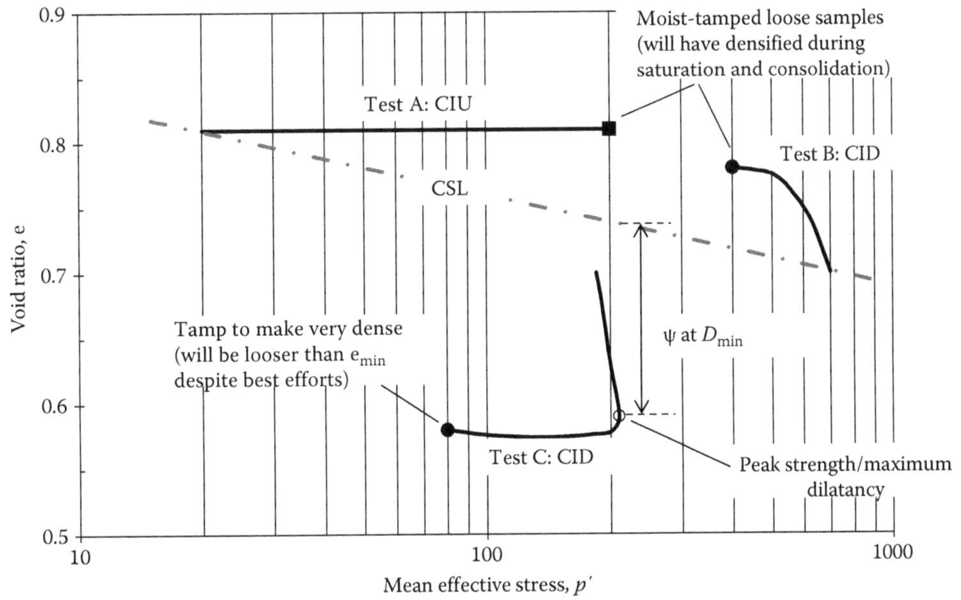

Figure 9.2 Conceptual minimum test program to determine a soil's properties (all tests on reconstituted samples from same blended and homogenous batch of soil).

as that allows a liquefaction event without the end state, which will be on the CSL, getting to such low stresses that the data become compromised by transducer accuracy.

Test 'B' is a similarly loose sample as 'A', but the sample is consolidated to a greater confining stress, about 400 kPa. The sample is then sheared drained to about 20% axial strain. The CSL will be very close to the end state of Test B (either model the test or estimate the true CSL by projecting the measured trend in volumetric strain with a little judgement).

Test 'C' involves a sample as dense as can be prepared. It is then sheared from a low effective confining stress as practically convenient (commonly about 80 kPa) to allow the maximum dilation to develop. This test is directed at the soil properties N and χ. It also provides data to assess the plastic hardening.

Elasticity is normally measured using the *bender element* method to assess G_{max}. This test method was the preserve of universities 10 years ago but is fast becoming standard with available commercial equipment (e.g. http://www.gdsinstruments.com/gds-products/gds-bender-element-system (GDS Instruments, 2014)). Poisson's ratio is generally not measured because it seems to almost invariably be in the range $0.15 < \nu < 0.25$, which is sufficient initial precision for modelling.

9.3.3 Practical test program

Only needing three triaxial tests to determine a suite of soil properties sounds too good to be true, and it is. The problem is that not all tests end up at the desired target initial condition despite best efforts and there is also variability in soil behaviour (two identically prepared and sheared samples will rarely show identical behaviour even in the best research laboratory). You need more than three tests to determine the soil's properties with any reasonable accuracy.

In the case of the CSL, it may show slight curvature plus it is good practice to have redundant data so any engineer can assess the uncertainty in the assessed CSL. In our experience, this leads to at least three *Type A* and two *Type B* tests. It is also prudent to anticipate that getting five tests with the desired range of initial conditions may need more samples to be prepared. *Test C* is aimed at providing data on both stress–dilatancy and state–dilatancy, but this type of test usually runs into the dual difficulty of not being as dense as desired and with questions about whether the single measured data point is representative. Experience suggests that at least three dense drained tests are needed.

Adding these numbers up suggests about 10 tests should be planned for the chosen representative gradation, as indicated in Table 9.1. As clients normally do not want to hear about testing difficulties it has become our practice to quote a fixed price for determination of these soil properties, with that price allowing for the various difficulties that go with this type of testing.

9.3.4 Data handling

9.3.4.1 Data file structure

Triaxial test data should be logged using a computer-controlled data acquisition system and ideally using computer-controlled equipment. There is no excuse for less than this in the modern technological age. Given such digital recording, there is no real limitation on the detail measured, but at the end of the day the data will likely be imported into a spreadsheet for processing and there is no point in having an unwieldy number of points. Something like 4000 scans has been found both convenient and sufficient for tests going to 20% nominal axial strain.

Table 9.1 Laboratory testing program for critical state liquefaction assessment

Test type	*No. of tests (typical)*	*Purpose*
Particle size distribution	20	Define the heterogeneity of the material in-situ by measuring grain size distribution on most samples. Once this is done, one or more representative materials can be identified, or materials considered to bracket the fine and coarse ends of the in-situ spectrum can be defined. Based on this assessment, bulk samples can be blended together for further testing on reconstituted specimens.
Specific gravity	2	Basic property needed for calculation of void ratio from measured water content and density.
Maximum and minimum index density	2	Although not part of the critical state framework, the maximum and minimum index densities are useful for comparison with other materials and to the laboratory technicians in preparation of specimens to target densities.
Triaxial tests, consolidated undrained compression	5	Define the critical state line as well as the peak undrained shear strength and brittleness (see Section 9.3.5).
Triaxial tests, consolidated drained compression	5	Define critical state line and stress–dilatancy parameters. Provide basis for estimating plastic hardening modulus using *NorSand* and iterative forward modelling (see Section 9.3.6).
Bender element tests, with isotropic consolidation	2 sets of about 8	Bender element tests are used to measure shear wave velocity (or small strain shear modulus) in reconstituted samples, which will be compared with in-situ measured shear wave velocity as well as used to develop a shear modulus–stress level–void ratio relationship. Testing is carried out on two specimens, prepared at different initial densities. On each specimen, tests are carried out at about eight different effective stress levels.
Cyclic simple shear tests	8	For seismic liquefaction studies, the cyclic resistance curve is best determined by about four cyclic simple shear tests at a single value of ψ, with cyclic stress ratio varied. Two sets of four should be carried out at different values of ψ.
Resonant column testing	2	While common practice is to use published curves of shear modulus degradation and damping ratio as a function of shear strain, it is always better to check these for soils that may not be *standard*. Testing is carried out on two specimens, prepared at different initial densities.

For archival purposes, the data file should be saved as an ASCII text file using either a fixed format of columns or a *comma-separated variable* (csv) file. Either file type is simply imported into Excel and also has the important attribute of being readable by even the most primitive text editor.

Sufficient *header* information must be included in these ASCII files about the origin of the test and the material tested, so that any engineer looking at the data can see what the test was on from the ASCII file itself. The variable name and units of each column must be given above the column to avoid ambiguity. None of these requirements are in any way special.

Since nobody in future is going to know more about the test than the laboratory making the measurements, we require our own archiving to be in terms of strains, stresses and void ratio. Things like area correction to transform measured loads to stresses are left to the testing laboratory (and documented in the testing report); thus, the laboratory delivers

Test type	: CID
Lab	: GOLDER(C)
Job number	: 882-2086
Test date	: FEB 29/89
Material tested	: NERLERK SAND
Preparation method	: MOIST TAMPED
Specific gravity of solids	: 2.66
Initial void ratio	: 0.604
Post-consolidation void ratio	: 0.572
Initial back pressure (kPa)	: 700

Axial strain (%)	**Vol strain (%)**	**SIGMAA' (kPa)**	**SIGMAR' (kPa)**
0	0	200	200
0.075	0.001	208.39	201.6
0.107	0.002	232.73	201.48
0.135	0.011	249.77	201.26
0.165	0.019	263.61	201.14
0.191	0.028	277.54	200.91
0.221	0.037	291.87	200.91
0.25	0.047	305.31	200.8
0.279	0.056	318.94	200.69

Figure 9.3 Appropriate file structure for reporting and archiving triaxial data.

quality assured data ready for the engineer to work with. The recommended data format is then as illustrated in Figure 9.3. This format works for both drained and undrained tests, since stresses are given as effective (although volumetric strain will of course be zero with an undrained test).

When considering data archiving, keep in mind that some high-value projects (bridges, dams) may quite credibly have an engineer wanting to review or reassess the test data many decades later. We do not jest here, as we have worked on dams that are approaching 100 years old ourselves and have gone back to the original paper records from when they were designed. Hence, our view is that a proper engineering record is something that can be read by any text editor as a rather primitive file. You should not rely on current xls files (or the like) being generally accessible indefinitely. As a profession, our best archive is an ASCII file and that should be stored on non-magnetic media (we have floppy disks from 25 years ago that are no longer readable). Some engineers go as far as to print the ASCII data and bind that into project files on the grounds that it can be scanned should the need arise in the future – that seems primitive, but it might be sensible for high-consequence projects with expected long service lives.

9.3.4.2 Data processing

Data processing will be done in a spreadsheet (assumed here to be Excel), so the first step is to import the quality assured results into a worksheet. It is a matter of style as to whether all tests are put onto a single sheet or whether each test gets its own worksheet, as is the labelling system adopted to show which block of data refers to which test. But do establish a clear style and check if your colleagues can understand it at a glance.

Soil behaviour is controlled by stress and strain invariants, not the individual stresses, and soil properties are framed in terms of these invariants. The test data must be processed to change from measured results to these invariants. This is a straightforward transformation and easily done in the worksheet. Our approach is to put these transformations adjacent to the measured data, with seven additional columns being needed on the worksheet and with a *banner* across the top to indicate which is which (see Figure 9.4).

In terms of strains, the volumetric strain is in itself a strain invariant. This needs to be paired with its matching shear strain ε_q. A little bit of algebra will lead you to the following result:

$$\varepsilon_q = \varepsilon_1 - \frac{\varepsilon_v}{3} \tag{9.1}$$

which is used as the formula to create the column headed ε_q. It operates on every row in the worksheet using the measured data in the first two columns.

Having read this far, stress–dilatancy will be firmly embedded in your soul as fundamental, and thus, the next step is to transform the strain data into dilatancy. This can only be done for drained test data, and there is a a trick to this, with the trick depending on the quality of your laboratory equipment. Dilation is the ratio of strain increments, which is a *differential*. Numerical differentiation can be noisy if there is not good precision in the data being differentiated. One approach is to take the differential over more data points to smooth the differential, but not so many as to smooth too much. We have tended to use a five-interval central difference method, which can be written in terms of the worksheet as

$$D(\text{in row '}n\text{'}) = \frac{\left(\varepsilon_{vn+2} - \varepsilon_{vn-2}\right)}{\left(\varepsilon_{qn+2} - \varepsilon_{qn-2}\right)} \tag{9.2}$$

Clearly, as this expression looks both in front ($n+2$) and behind ($n-2$) the current row (= n), you wind up with an inapplicable expression at the beginning and end of the measured data. We simply leave these rows blank (see Figure 9.4).

Some engineers may prefer to avoid numerical differentiation by using the Excel charting function to graph ε_v versus ε_q and then using the Excel 'fit trend line' function to the data. If a fifth-order polynomial is chosen, with the 'set intercept = 0' option, the displayed equation is often a rather good fit to the data. This displayed equation can then be analytically differentiated to give a noise-free D versus ε_q^p relationship.

However you estimate D is essential to pick out its minimum as that is what will be transferred to a stress–dilatancy plot. This pick is readily achieved by using the Excel 'min' function applied to the column of D values. We place this $D_{\min}$ above the individual D values, with an annotation showing what it is.

The columns for $p = (\sigma_a + 2\sigma_r)/3)$ and $q = (\sigma_a - \sigma_r)$ are the usual transforms of the measured effective stresses, with $\eta = q/p$. The value of particular interest is $\eta_{\max}$ and that is obtained by applying the Excel 'max' function to the data in the 'η' column. Theoretically, stress–dilatancy would have $\eta_{\max}$ and $D_{\min}$ as being coincident, but test data often do not meet this expectation. Our practice is that soil properties are sufficiently well estimated by using the Excel 'min' and 'max' functions as just described and this method is easily cloned across different streams of test data. However, if you wish to force the theoretical view, then identify the row with $D_{\min}$ found by Excel and then pick up the matching value of η as your choice for $\eta_{\max}$. The reason to focus on $D_{\min}$ is that M is varying with fabric and that is what is upsetting the stress–dilatancy theory expectation of $\eta_{\max}$ and $D_{\min}$ being coincident. The value for $\eta_{\max}$ is placed just above the data (see Figure 9.4).

				D_{min} =	−0.194		eta_max	1.540	...@ psi =	−0.089
Measured				Processed						
Axial strain (%)	Vol strain (%)	SIG_Axial' (kPa)	SIG_Radial' (kPa)	epQ (%)	D ...	p (kPa)	q (kPa)	eta ...	e ...	psi ...
0	0	503.48	500.23	0.000		501.3	0.0	0.000	0.619	−0.106
0.028	0.002	526.80	500.00	0.027		508.9	26.8	0.053	0.619	−0.106
0.054	0.006	544.87	500.00	0.052	0.194	515.0	44.9	0.087	0.619	−0.106
0.083	0.013	566.78	500.00	0.079	0.300	522.3	66.8	0.128	0.619	−0.106
0.11	0.02	587.46	500.00	0.103	0.349	529.2	87.5	0.165	0.619	−0.106
0.138	0.032	609.68	500.11	0.127	0.375	536.6	109.6	0.204	0.618	−0.105
0.166	0.041	631.97	500.11	0.152	0.405	544.1	131.9	0.242	0.618	−0.105
0.194	0.05	654.67	500.11	0.177	0.398	551.6	154.6	0.280	0.618	−0.105
0.222	0.06	676.94	500.11	0.202	0.410	559.1	176.8	0.316	0.618	−0.105
0.249	0.071	700.32	500.00	0.225	0.438	566.8	200.3	0.353	0.618	−0.105
0.277	0.081	723.19	500.00	0.250	0.449	574.4	223.2	0.389	0.618	−0.105
0.304	0.092	745.45	500.00	0.273	0.445	581.8	245.5	0.422	0.618	−0.105
0.332	0.103	766.69	500.00	0.298	0.445	588.9	266.7	0.453	0.617	−0.105
0.36	0.114	788.00	499.89	0.322	0.440	595.9	288.1	0.483	0.617	−0.105
0.388	0.124	809.22	499.89	0.347	0.346	603.0	309.3	0.513	0.617	−0.105
0.416	0.135	829.11	499.77	0.371	0.354	609.6	329.3	0.540	0.617	−0.105

Figure 9.4 Layout of worksheet to process triaxial data. Note: Data and processed results extend to end of file.

The void ratio evolution throughout the test is also needed (for drained tests). This information is embedded in the reported volumetric strain, thus we focus on reporting an accurate value of the post-consolidation void ratio at the start of shearing (e_o in the header information in Figure 9.3) with that void ratio generally having been determined by the *freezing method* discussed in Appendix B. We then compute void ratio for use in data processing as

$$e = e_o - (1 + e_o) \times \varepsilon_v \tag{9.3}$$

Equation 9.3 is based on engineering strain, which is the practice of most laboratories. The minus sign is because the compression positive convention of soil mechanics makes positive volumetric strain to be void ratio reduction.

The last thing needed from data processing is to determine the state parameter that corresponds to D_{min}. This is done by calculating a column of ψ values, identifying which row corresponds to D_{min} and selecting that value of ψ to put in the row adjacent to η_{max} (see Figure 9.4). Start by creating a column of ψ to the right of the void ratio column and where

$$\psi = e - (\Gamma - \lambda_{10} \times \ \log p') \tag{9.4}$$

Type Equation 9.4 in the Excel formula exactly as you see it written with Γ being represented by the name 'gamma' and λ_{10} by 'lambda10'. The terms *gamma* and *lamba10* do not exist yet, but they are going to become defined constants within the spreadsheet as your next step.

9.3.5 Evaluation of soil properties

9.3.5.1 Properties worksheet

Soil properties in all *good* models do not depend on the void ratio or the stress level, so the same properties apply to every test in the assembled data of the soil in question. There are different styles as to how this can be reflected in an Excel file, with our preference being to create a worksheet within the spreadsheet whose tab is labelled *properties*. Figure 9.5 is an example.

There are five soil properties that are going to be determined directly from the triaxial data in the properties worksheet, and these properties are both annotated and given initial estimates for their values on the properties worksheet. You will plot various things, but these properties will be used to drive the trend lines with you using your eye to assess the fit of trend line to the data and with you adjusting the properties (*iterating*) until you get a fit you like. This is a process of forward modelling and we prefer it to using regression lines, as forward modelling allows you to weight the assessment to the tests that appear the most reliable. As a further style point, we like to use a bold and blue font in cells that are *user inputs*. Importantly, each of the soil property cells should be *named*. Excel allows users to assign a name to a cell so that instead of using 'C5' you set that cell to 'gamma' (as appropriate) and which then allows you to use the name 'gamma' in your formulas rather than 'C5'. You enter the cell name in the left-most box of the formula bar just above the top of the worksheet. This cell naming is arguably the most important thing you must do in creating a spreadsheet that can be reviewed by another engineer. Thus, name the five boxes containing the soil properties as follows: for the CSL, use gamma and lambda10 (the '10' is there to remind you and others that this particular soil property goes with base 10 logarithms); for the stress–dilatancy behaviour, name the critical friction ratio cell *Mtc* and the volumetric coupling coefficient cell *Ntc*; for the state–dilatancy coefficient, name the cell chi_tc (i.e. χ_{tc}). The 'tc' denotes that these soil properties are all associated with

Nerlerk 270/1 sand from Golder (1989)						
	Index properties			**Soil properties**		
	D50	270 μm		Gamma =	**0.855**	
	fines =	1.90%		lambda10	**0.048**	
	emin =	0.536		Mtc =	**1.27**	
	SG =	2.66		Ntc =	**0.40**	
				chi_tc =	**4.00**	
	see Table 3 of report					
	As tested initial			**At max dilation (= D_{min})**		
Test	p_0	e_0	**psi$_0$**	D_{min}	**eta_max**	**psi**
CID-G151	200.9	0.694	−0.050	−0.135	1.343	−0.035
CID-G152	*No digital data available*					
CID-G154	49.5	0.738	−0.036	−0.095	1.345	−0.023
CID-G155	501.3	0.619	−0.106	−0.282	1.445	−0.068
CID-G156	200.9	0.640	−0.104	−0.352	1.455	−0.085
CID-G157	203.9	0.572	−0.172	−0.561	1.624	−0.143
	As tested initial			**At critical state**		
CIU-G101	500	0.818	0.093	20	0.818	
CIU-G103	500	0.793	0.068	28	0.793	
CIU-G104	700	0.761	0.043	74	0.761	
CIU-G105	500	0.757	0.032	125	0.757	
CIU-G106	500	0.8	0.075	23	0.8	
CIU-G107	700	0.727	0.009	379	0.727	
CIU-G108	500	0.773	0.048	27	0.773	

Figure 9.5 Layout of *properties* worksheet including the summary of test data.

triaxial compression conditions (although N is actually invariant with proportion of intermediate stress, see Appendix C, but naming it N_{tc} in the spreadsheet keeps the emphasis on determining properties under triaxial compression). You will need to annotate what the cell refers to on the sheet as the cell names do not show up when just looking at the worksheet. Figure 9.5 illustrates a style for setting up the soil properties choices. Also note the documentation of the soil index properties and the use of italic text to add comments.

Three plots are going to be used to assess the soil properties, and the trend lines for these plots need to be set up using the soil properties cells just established. Trend line #1 is for the CSL in e-log(p) space; we tend to use $p \sim 10$ kPa and $p = 3000$ kPa as the endpoints for the CSL. Trend line #2 is for the stress–dilatancy plot of η_{max} versus D_{min}. Use $D_{min} = 0$ as one endpoint and $D_{min} = -0.8$ will likely be convenient for the other. The formula used for this trend line is

$$\eta_{max} = M_{tc} - (1 - N_{tc}) \times D_{min} \tag{9.5}$$

where M_{tc} and N_{tc} are the named soil properties set up in the *properties* worksheet. Trend line #3 is for the state–dilatancy plot of D_{min} versus ψ at D_{min} and this trend line must go through $D_{min} = 0$ at $\psi = 0$ (if it does not, there is an error in the estimated CSL). This establishes one point on the trend line, with the other points being chosen consistent with the stress–dilatancy plot since it is the same test data for D_{min} that appears in both plots. The formula used for the trend line for the state–dilatancy plot is

$$D_{min} = \chi_{tc} \times \psi \tag{9.6}$$

where χ_{tc} (chi_tc) is the soil property in the *properties* worksheet.

9.3.5.2 Test summary table

The soil properties are determined from trends in the measured data. This is most easily done if the key values from the various laboratory tests are transferred to a summary table such as on the *properties* worksheet shown in Figure 9.5. Importantly, the values in this table for the drained tests are referenced to the processed data worksheet and not typed in. In the case of the undrained tests, the *at critical state* requires judgement on your part in assessing the data. Copying in the values at greatest strain achieved in the test is a start but that may not be at critical state, especially when the test has gone through a pseudo–steady state and the sample is still dilating at the end of the test. You will need to judge whether the specimen is close to critical state, a long way from critical or whether you can extrapolate just a little to reach a critical state past the end-of-test state. The 'psi' (i.e. ψ) values in this table will automatically update as the soil properties are subsequently adjusted to best fit the data.

9.3.5.3 Critical state line (Γ, λ) from the state plot

The *state* plot is a familiar figure that plots the trajectory of the various tests as void ratio versus the $\log_{10}$ of mean effective stress p. In the case of the drained tests, it is trivial to transfer the $e - p$ values from the processed data worksheet (Figure 9.4) to create a graph in e-log(p) space. In the case of the undrained tests, a further step is needed because the trace of an undrained test is a straight line and the endpoint of the test does not lie on the endpoint of the plotted line if the sample shows dilation – something that often develops if the sample is not truly *loose*. The easiest thing to do is to use the table of critical state conditions (Figure 9.5), which can then be plotted on the e-log(p) graph as discrete points to highlight the end of test conditions within the state path. Figure 9.6 shows the results of plotting the drained and undrained data, from tests on Nerlerk 270/1 sand, in this manner.

The CSL shown in Figure 9.6 is the trend line from the *properties* worksheet; the values of gamma and lambda10 have been adjusted to get the fit shown (gamma moves the trend line up or down, while lambda10 changes its slope). One can argue about the fit a little, but there is not a great deal of wiggle room to meet the critical conditions found in the undrained tests. There are no very loose drained tests in the Nerlerk 270/1 data, leaving the CSL established by only the undrained tests. The drained tests that exist are all dense of the CSL, with most being very dense. These tests cannot be strained enough to attain their critical state with the limits of triaxial equipment. The fitted trend line establishes $\Gamma = 0.855 \pm 0.01$ and $\lambda_{10} = 0.048$ for this sand.

9.3.5.4 Stress–dilatancy plot (M_{tc}, N)

The soil properties M_{tc}, N are found by plotting η_{max} versus D_{min} for the drained tests, a method originated by Bishop (1950). These data pairs are those on the summary table of the *properties* worksheet, developed in the data processing discussed earlier. The result is shown in Figure 9.7 for the Nerlerk 270/1 data set.

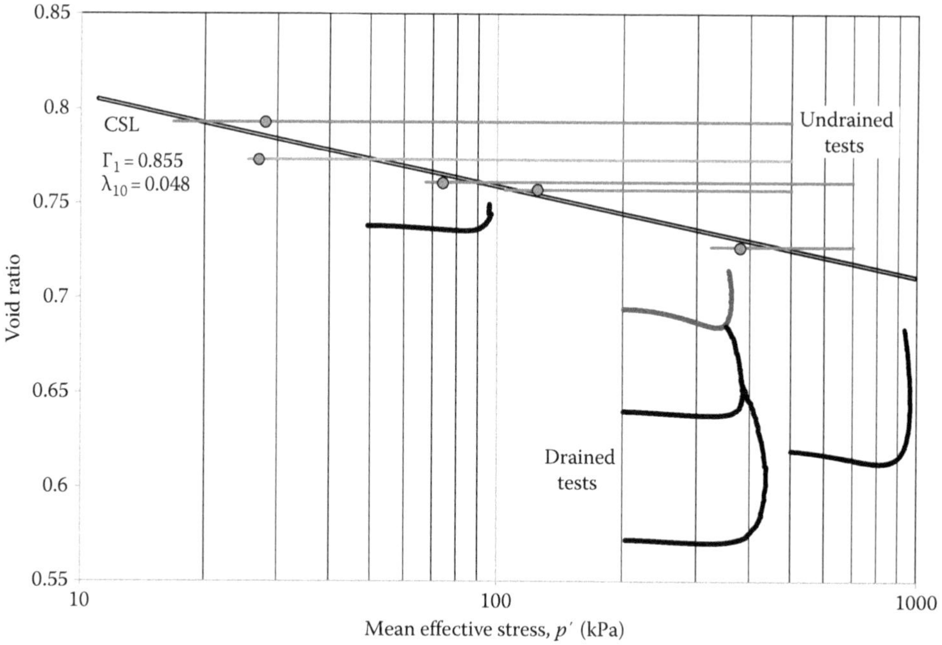

Figure 9.6 State diagram for triaxial tests on Nerlerk 270/1 sand with fitted CSL.

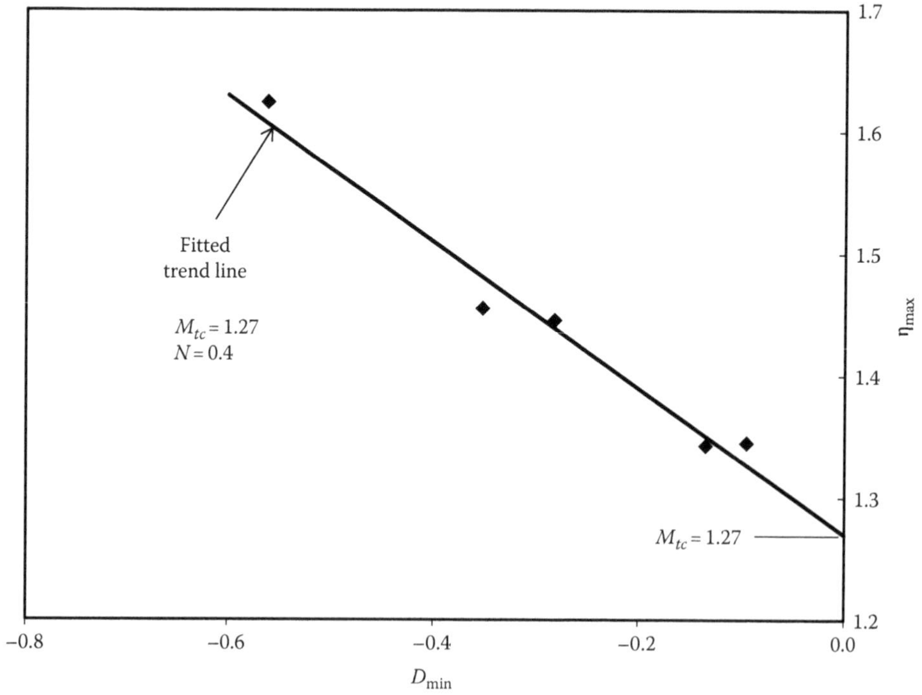

Figure 9.7 Stress–dilatancy in triaxial tests on dense Nerlerk 270/1 sand.

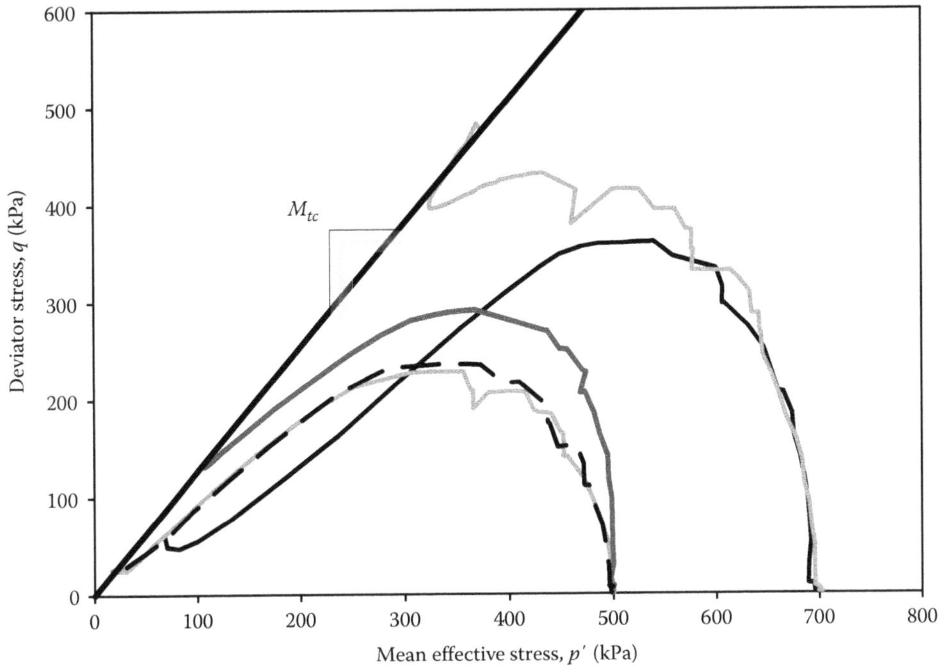

Figure 9.8 Stress paths in undrained triaxial tests on loose Nerlerk 270/1 sand.

The trend line shown in Figure 9.7 is from the *properties* worksheet. The values of M_{tc} and N in the worksheet have been adjusted to get the fit shown (M_{tc} moves the trend line up or down, while N changes its slope). Like the CSL, one can argue about the fit a little, but there is not a great deal of wiggle room. The fitted trend line establishes $M_{tc} = 1.27 \pm 0.01$ and $N = 0.40 \pm 0.02$ for this sand.

As a check, this estimate for M_{tc} can be used to compute a trend line for a stress-path plot and compared to the measured stress paths of the loose undrained tests, Figure 9.8. A very plausible match is found (although this, of course, provides no check on N). Note that one of the tests looks a bit odd, with the suggestion that the pore water pressures were not properly equalized before the start of shear.

9.3.5.5 State–dilatancy plot (χ_{tc})

The final soil property to be determined is that relating the maximum dilation to the state parameter, chi_tc in the *properties* worksheet. When data processing was started, the properties defining the CSL were not known; the state parameter was computed, but it was with initial guesses (or fictional) values for gamma and lambda10. These two guessed properties are now known and if the worksheet has been set up as described the state parameter values for all of the tests imported will have been automatically updated (including the summary of values at D_{min}). Thus, it is now trivial to plot D_{min} versus ψ at D_{min} from the summary table. This plot is shown in Figure 9.9 for the triaxial compression data on dense Nerlerk 270/1 sand. Note that this type of plot only exists for dense samples as loose samples all show $D_{min} \sim 0$ and this is only attained at the end of the test when $\psi \sim 0$ (which makes all the data cluster around the origin).

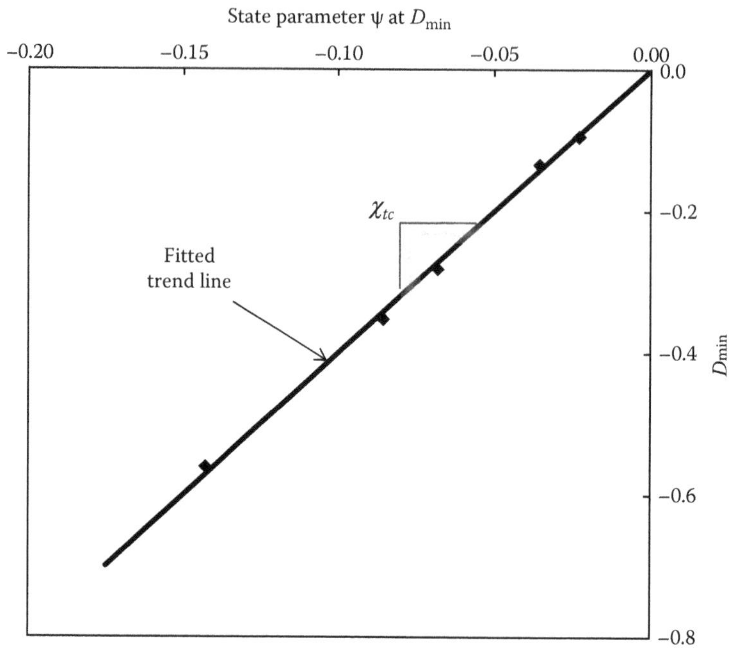

Figure 9.9 State–dilatancy in triaxial tests on dense Nerlerk 270/1 sand.

As with the other property determinations, the appropriate trend line from the *properties* worksheet is pasted into the plot to compare with data and the soil property adjusted to get the best match. For Figure 9.9 the trend line is given by Equation 9.6 and the soil property χ_{tc} is adjusted. An excellent fit is found corresponding to $\chi_{tc} = 4.00 \pm 0.2$. Fits of comparable quality are found quite often, which is thought to be a consequence of D_{min} being a kinematic behaviour while void ratio (on which ψ is based) is a similarly geometric idealization. Certainly, experience is that state–dilatancy shows generally less scatter about the inferred trend than is seen with stress–dilatancy.

9.3.6 Validation of soil properties

The five soil properties Γ, λ_{10}, M_{tc}, N, χ_{tc} have been determined by examining the aspects of measured soil behaviour largely independently and with each aspect being directly associated with one or two particular soil properties. Although a conventional approach, this type of parameter isolation can lead to less than optimum results depending on how test and idealization errors stack up. It is wise to take the estimated soil properties and to put them in a constitutive model to see to what extent the properties lead to reasonable replication of the entire stress–strain behaviour measured: validation, as it is usually called.

Validation necessarily depends on the reasonableness of the soil model selected for the process. All good models (or at least those developed to date) include both a CSL and ψ, so the properties just found are common to nearly all models you might consider. The biggest limitation on model choice is whether the models are conveniently available for use, as most engineers will not want to start from scratch. Many models fall at this hurdle. Here we use *NorSand*, in part because it is simple and in part because it has been implemented in a public domain spreadsheet that you can download. While we encourage you to dig into the details of this model (overviewed in Chapter 3 and detailed in Appendix C), the spreadsheet can be used as a *black box* to see how well the model/properties match the test data. This evaluation of model and properties versus data is done as a simple visual comparison; by all means add a formal *goodness of fit* measure if you wish, but for most engineers a simple visual comparison will be both more satisfactory and more insightful.

The *NorSandTXL* program is an Excel spreadsheet with all coding in the VBA environment (you can access this commented code by pressing the 'Alt' + 'F11' keys). This style of coding is viewed by Excel as a 'macro' so you will have to 'enable macros' when opening the file. This particular spreadsheet simulates drained and undrained triaxial tests (there are other downloadable spreadsheets to model simple shear and other plane strain paths). The spreadsheet computes the drained and undrained behaviour of *NorSand* for the chosen soil properties and initial state, presenting the results on all the plots that you need to see what is going on. It is intended that test data are pasted into the spreadsheet and plotted on the same graphs as the computed results, Figure 9.10 showing such a comparison for a drained triaxial test on the same Nerlerk 270/1 sand just used in discussing how to determine soil properties. There is a very pleasing fit between the data for this particular test and *NorSand*, across all four aspects of the test plotted, suggesting the assessed soil properties are reasonable and representative. How was this fit developed and what are the uncertainties?

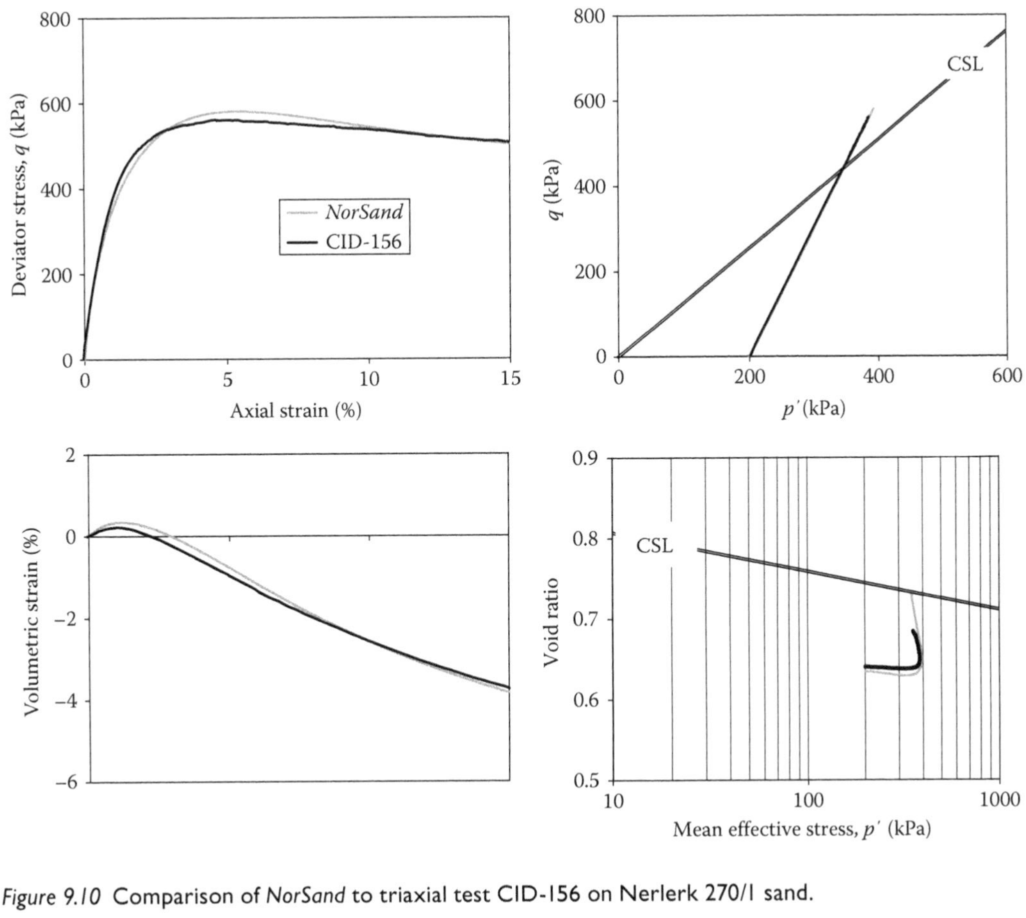

Figure 9.10 Comparison of *NorSand* to triaxial test CID-156 on Nerlerk 270/1 sand.

9.3.6.1 Check plastic properties by simulation of drained tests

After opening *NorSandTXL*, move to the 'Params' & 'Plots' worksheet. This worksheet contains all the inputs and the four graphs of Figure 9.10 (the undrained plots are on a separate worksheet as that is convenient when working with a laptop. By all means move them to the 'Params' & 'Plots' worksheet if you have a large screen). There are two input fields: 'Soil Properties' and 'Initial Soil State' (see Figure 9.11). Most inputs are dimensionless, indicated by ' --- ' in the adjacent units column with the dimensions of the others being indicated. After updating these various inputs, the 'Update Model' button is *clicked* with the mouse to generate the computed soil behaviours for the input parameter set.

Importantly, the drained tests must be fitted before the undrained tests. This is needed because elasticity has only minor effects on the drained behaviour, so drained tests are used to validate the plastic properties of the soil. Undrained tests are then used to check on the elasticity. Let us now consider the various inputs.

Looking at the 'Soil Properties' first, these apply to every test on the soil in question. So these do not change (at least in principle) as you consider Test 3 after Test 2 after Test 1 and so forth, although there are two steps in getting to this level. The upper five soil

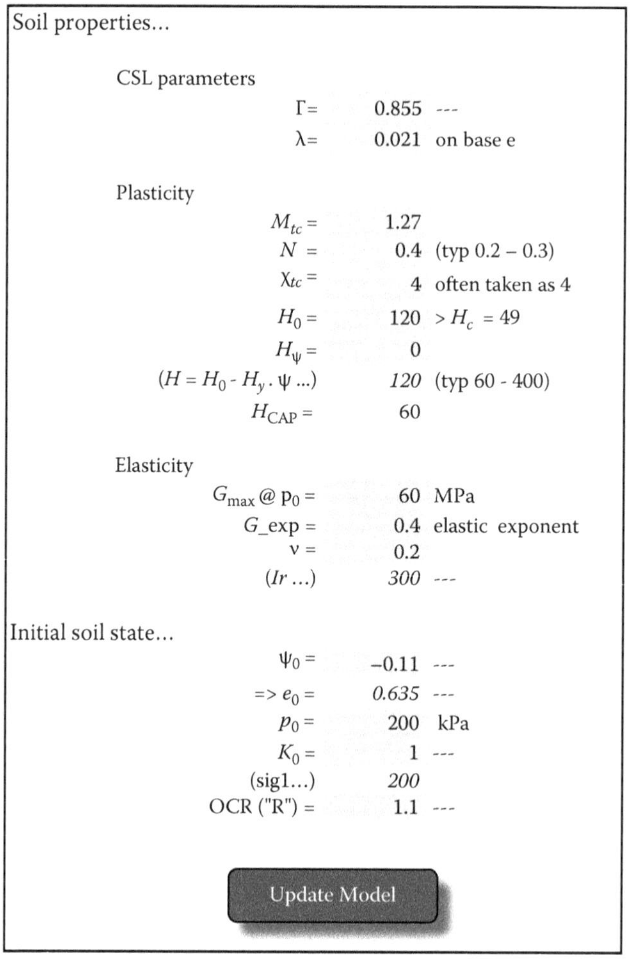

Figure 9.11 Input fields for running the *NorSand* spreadsheet.

properties in Figure 9.11 are exactly the same five properties discussed in Section 9.3. You simply input the values as determined. This will then bring you to H (plastic hardening) and elasticity.

The property H is determined by iterative forward modelling (IFM). In IFM you guess a value, run the model, look at the output, adjust H and rerun the model until you get the best fit. The larger the value of H, the stiffer the stress–strain curve, so it is pretty intuitive to optimize the fit.

What you will find in general is that H varies with the initial state parameter of the test, so that the *do not change* principle means that H is given as

$$H = H_o + H_\Psi \tag{9.7}$$

where H_o, H_ψ are the properties that do not change. This is difficult to optimize at one go, so the procedure is to set $H_\psi = 0$ for a first pass through the data and derive the best values for H_o on a test by test basis. Then plot H_o versus ψ_o, which should largely form a straight trend line, to give the estimated values for H_o and H_ψ. In most cases, this simple procedure and a straight line are sufficient as there will be some variation because of *fabric* effects that are not included in the constitutive model but which are variably present from one test to another even with *identical* sample preparation. Also, H trends may vary between moist-tamped, air-pluviated and water-pluviated samples because of the differing fabrics created by each sample preparation method (discussed in Chapter 3).

But before you do IFM, you need to input elasticity. Elasticity is input as the initial value of G_{max} (in MPa) for the sample as it has been set up. Arguably, this input field belongs to the next section but we have stayed with the convention that elasticity is a soil property. G_{max} usually varies with void ratio, and there are various idealizations for this behaviour, however, we have kept the spreadsheet as simple as possible and programmed

$$G_{max} = G_{ref} \times \left(\frac{p}{p_o} \right)^{G_{exp}} \tag{9.8}$$

This code is in the VBA *Public Function G_max()* so elasticity can readily be changed to a more sophisticated idealization if desired without affecting the rest of the code. G_{ref} is G_{max} at the stress p_o and the property 'G_{exp}' is the dimensionless exponent as to how G_{max} varies with confining stress. The exponent is commonly ~0.5 for sands; the choice $G_{exp} = 0$ sets constant modulus while the choice $G_{exp} = 1$ sets constant rigidity I_r ($I_r = G_{max}/p$), which is the idealization implicit in Cam Clay. If you have bender element data, G_{ref} and G_{exp} will be known from those measurements, but if you do not have bender data, then estimate something reasonable using data on similar soils (Chapter 4 has data if you need to make a first guess). The estimate will be refined later in the calibration using the undrained tests.

Turning from soil properties to the soil-state inputs in the worksheet, the initial mean effective stress is simply chosen to match what was used for the sample being simulated. There is the option of starting from an anisotropic stress state, specified by way of K_o. This option is useful to explore the effect of an initially drained stress path on the subsequent undrained behaviour. The implied major principal stress (σ_1) is calculated by the spreadsheet and shown in italics within the input block. The state of overconsolidation is a further input, with overconsolidation ratio defining an initial elastic zone before the onset of yielding. This makes little practical difference with drained tests but is rather useful in getting better fits to undrained tests as will be seen in the following section.

The final input is to capture the sample density. This spreadsheet is set up so that the state parameter ψ is the input and with the implied initial void ratio of the sample shown underneath in italics. Feel free to reverse the operation as it is a matter of taste whether you want to regard void ratio or ψ as the basic input and thus which is calculated from the other.

The void ratio in the model input *ought* to match the void ratio reported by the testing laboratory at the start of shearing, but this *ought* is affected by the accuracy with which void ratio can be measured and how closely the CSL fits that particular test specimen. So, it is perfectly reasonable to allow a bit of variation between the reported laboratory void ratio and that used in the simulation. In our experience, working within $\Delta e \pm 0.02$ is usually sufficient to line things up rather nicely.

9.3.6.2 Confirm elastic properties by simulation of undrained tests

So far, validation has concentrated on the drained triaxial tests, which is done because it is the drained tests, and the dense drained tests in particular, which give the best validation as to whether the stress and state–dilatancy properties inferred from the trends in the tests are self-consistent. But liquefaction is an undrained behaviour and validation now needs to be extended to show that the estimated soil properties are a reasonable representation for undrained behaviour. Figure 9.12 shows an example of the fit to a liquefaction test on Nerlerk sand, with a good match evident. In principle, it is simply not admissible to have different soil properties undrained to those measured under drained conditions – so what was done to achieve this fit?

The answer to *what was done* is that the input parameters were tweaked a little. Observe that the measured stress path is initially vertical. A vertical path undrained implies no shear-induced excess pore pressure, in turn implying elastic soil response. So, a little overconsolidation was included in the simulation to replicate this effect, with OCR = 1.1 nicely aligning

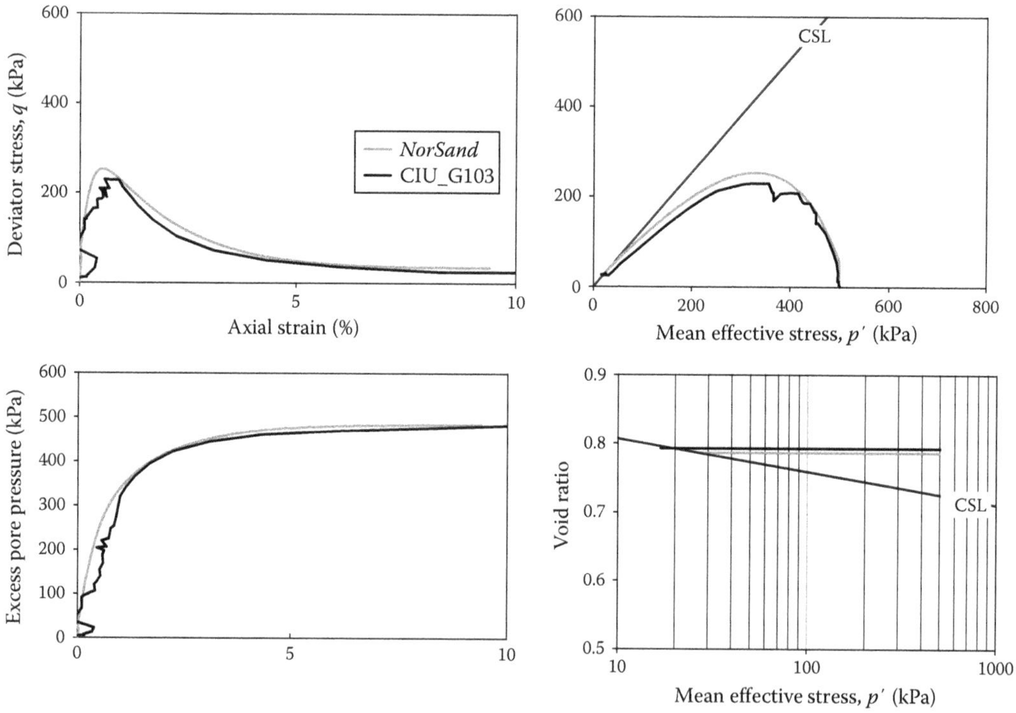

Figure 9.12 Soil properties validated in drained compression applied to undrained test on Nerlerk sand.

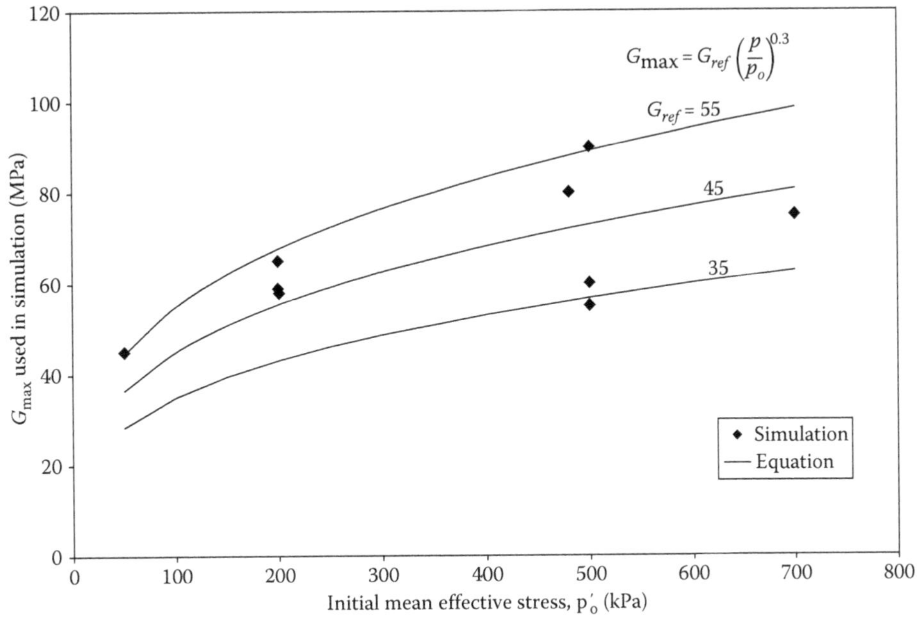

Figure 9.13 Trends in G_{max} found in simulating behaviour of Nerlerk sand.

the measured and simulated stress paths until $q \sim 50$ kPa. Then, G_{max} was refined to align the stress path (recall that G_{max} was estimated but not measured, so here stress-path matching is used to refine the estimate of G_{max}).

When this matching has been carried out systematically across all sample tests, a reasonable trend will emerge with an effect of confining stress and an effect of initial void ratio on G_{max}. Figure 9.13 shows the pattern developed from iteratively simulating the various tests on Nerlerk 270/1 sand.

9.3.7 Document simulation input sets

When starting iterative forward modelling it is awfully easy to lose track of where you are. The way out of this is to use the 'copy' and 'paste' facility of Office. When you have found a fit to a particular test that you find pleasing, group the plots together and use the mouse or menu to create a copy of these Excel plots. Then use 'paste special' in a Word document to make a record of the fit achieved as a picture (enhanced metafile or pdf, your preference). At the same time as doing this, make a copy of the input parameters and then 'paste special' as 'values' into a record of the simulation in an Excel sheet or as a table of values in the Word file.

9.3.8 Reading this section is not enough

Although this section has explained and illustrated both calibration and validation of soil properties, reading it will not be enough to bring yourself *up to speed*. Supporting data and programs are downloadable, and what is needed is for you to pick up the *NorSand_Nerlerk.xls* file and to at least try out for yourself the effect of changing soil properties and initial states on the simulated soil behaviour. Better yet, download the ASCII data files and follow through the calibration yourself, creating your own spreadsheet. If you do this you will realize there is no *magic* or hidden trick to understanding soil behaviour. In short, you will have a *road to Damascus* experience and come away with confidence that there is real substance to critical state soil mechanics.

9.3.9 Reporting soil properties

Although it can be very satisfying to develop soil properties from a set of laboratory tests, it is unlikely that the project owner will share your professional delight. This then leads to considerations on what is an appropriate way to present the information so as to cover both client aspects and looking towards engineering reviews (e.g. a Geotechnical Review Board, or internal senior consultant). In principle, there are three levels of information and associated ways of presenting them:

- Project owners, regulators and similar non-geotechnical stakeholders should be able to readily understand that the testing was relevant to the site conditions and the nature of the resulting soil properties. This is easily accomplished in the engineering report by (1) using a figure showing the gradation envelope of the stratum of interest and the gradation(s) tested within that envelope – a simple visual indicator of relevance – and (2) a table of soil property values (M, N, λ_{10}, etc.) set against the expected range for these properties, establishing whether the project soil is *typical* or *unusual* with some discussion about potential engineering significance of this basic conclusion. This assessment is a *main text* deliverable.
- Review engineers will want to look in more detail at the procedures used and how the reported results were obtained. This means a range of plots, starting with an e-log(p) showing the state path of all the tests and then moving on to the plots discussed earlier and which gave M, N, χ. Some modelling plots are also needed to show that the derived properties are consistent, but not necessarily all the simulations. This is a body of reporting for *technical* review and substantiates the properties that are tabulated in the main report. This is a *report appendix* deliverable, as non-specialists will have little interest in it.
- Fifty years ago, the results of laboratory testing would have been reported as engineering drawings suitable for scaling off values. Today, we do not do that but work with digital data. But the requirement to report basic data has not disappeared. Accordingly, the third level of reporting is the raw information in digital form. Putting such data in company files has not proved a long-term solution (it tends to get lost) so it has become normal to include a CD with the source digital data in the back of all copies of the engineering report, physically linking the data to the engineering. It is helpful to include a table of the tests, with their initial conditions, etc., in a properties appendix (previous bullet point) and to duplicate that table on the CD.

9.4 LABORATORY MEASUREMENT OF CYCLIC STRENGTH

9.4.1 Need for cyclic testing

The case history basis of liquefaction assessment, whether carried forward by the NCEER or state parameter approaches, is biased to soils with less than 15% fines contents and confining stress levels of less than 150 kPa. How can such an approach be used for high tailings dams or deep dam foundations on liquefiable silts? The general answer to these and related questions in current engineering practice is to use the computer-controlled cyclic simple shear test of the soils involved with a project and at representative stress levels. It has become normal to specify quite extensive amounts of CSS testing.

9.4.2 Cyclic strength ratio from simple shear tests

The cyclic resistance curve is derived from a set of cyclic simple shear tests. The tests can be on your best *undisturbed* samples or on reconstituted samples, but you cannot assume that the undisturbed samples are at the same state or void ratio as in-situ. As discussed in Section 10.4, even frozen sampling can result in volume changes as the samples thaw and there is a near guarantee of overestimating the in-situ cyclic strength. Thus, your objective is at least one well-defined curve of cyclic resistance ratio CSR versus N_L (number of cycles to liquefaction or 3.75% double amplitude strain) at a certain value of ψ plus one or two data points each at higher and lower values of ψ. These tests need to cover the in-situ confining stress range for the stratum of interest, so something like four CSS tests and more likely six are needed to allow for some samples not being at the desired void ratio prior to cyclic shearing. Figure 9.14a shows an example.

If you opt to use undisturbed samples, then carefully evaluate the effect of sample to sample gradation differences to make sure that the samples are similar enough to use a single CSL when computing their ψ_o at the start of load cycling. In many instances, there will be so much disturbance in sample recovery, extrusion and consolidation to in-situ stress that it may be simpler to develop a representative gradation and reconstitute the samples directly into the test equipment (much as done for sands as routine in triaxial tests).

The end result of the CSS testing is the graph you really need, which is Figure 9.14b showing CSR as a function of ψ. The number of cycles selected for Figure 9.14b depends on the earthquake magnitude. $N_L = 15$ corresponds to $M = 7.5$, $N_L = 10$ to $M = 7.0$, $N_L = 6$ to $M = 6.0$ as a guideline, but more detail is found in Section 7.4.

9.4.3 Representing trends in cyclic strength ratio

The data trends shown in Figure 9.14 need to be captured as an equation for transfer to the CPT processing, as it is this CPT data that will provide the actual, in-situ, ψ from which the design cyclic strength will come. Rather than use the Excel 'fit trend line' function, we prefer to fit by eye, as that allows judgement about the *best* representation of the test data and to use an equation for CSR versus ψ that crops up in the analysis of general trends. In other words, we tend to view individual testing campaigns as having their own errors with a wider view defining the most likely trend. Of course, you are free to disagree with us although you will then need to update the CPT processing routine to capture your view (easy enough to do in VBA, discussed shortly). Our preference is to use the following:

$$\text{CRR} = a - b \times \psi_o \tag{9.9}$$

where values of a, b are defined from fitting this linear trend line to the tests results that have been reduced to the form of Figure 9.14b. These coefficients a, b will be input in the CPT processing, Section 9.5.

9.4.4 Modelling cyclic simple shear tests

For simplicity, it is generally best to avoid static bias in the CSS campaign, and avoiding static bias certainly reduces the number of tests you need to develop the trend in the soil behaviour. If you need to include bias effects in your assessment of the in-situ situation,

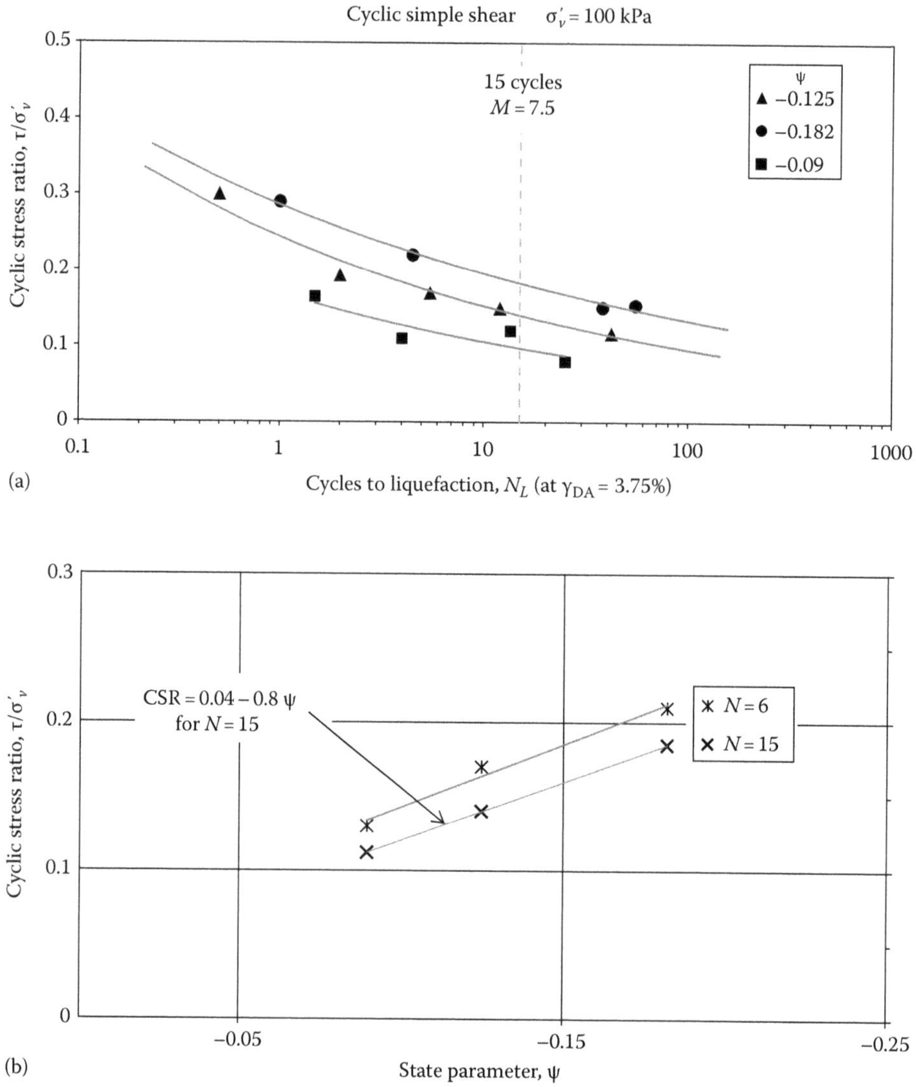

Figure 9.14 Cyclic stress ratio from laboratory cyclic simple shear tests. (a) Number of cycles to liquefaction at different values of state parameter and (b) cyclic stress ratio to give failure in 6 and 15 cycles (derived from [a]).

then calibrate *NorSandPSR.xls* to all your CSS data fitting all aspects including the rate of excess pore pressure development per cycle. With this calibration done, then it is easy enough to vary static bias (and state for that matter) within the spreadsheet and compute additional design trends. You can then update the CSR trend properties a, b for input into the CPT data reduction by fitting the *NorSand* results.

9.4.5 Reporting CSR trends

Review engineers will want to look in detail at the data used to develop the CSR trend(s) that is applied to the processing of the project's CPT data, but the detail to make a reviewer comfortable is beyond what most clients will interest themselves in. So, the CSR data summary and reduction become an appendix with a brief summary in the main text of the report discussing which soil was tested and a brief commentary on the results (e.g. if a silt was tested, does it match other experience with silts and is the trend well defined by the available tests). It has become our practice to include a further appendix with the results sheets from each CSS test but there is an argument that this is too much detail in print and the data on a bound-in CD are sufficient documentation to the reported summary trends.

9.5 DETERMINING SOIL STATE BY CPT SOUNDINGS

Now that you have derived the properties for your soil from laboratory testing, the next step is to characterize the in-situ state of the soil. Chapter 4 provides the background to how this is done using the CPT. Remember that the CPT responds to ψ much better than it does to void ratio or relative density. So minor changes in the particle size distribution, in particular fines content, which can affect the critical state line are less of an issue than you might think using a traditional relative density or void ratio approach. But even though the CPT is the best approach and with wide availability, there is a bit more needed than calling your local testing contractor. Here, we go through the issues from choosing appropriate equipment to processing the data into in-situ state characterization.

A factor in adopting the CPT is that the test was originally devised as a stratigraphic logging tool, whereas its use here, and very widely in contemporary geotechnical engineering, is as a quantitative measure of soil strength or state. This enhanced use of the CPT has implications for the quality of the equipment used and how the testing is done. These aspects are discussed before presenting the data processing procedures.

9.5.1 CPT equipment and procedures

9.5.1.1 Standards and requirements

There are codes/standards for CPT work and with some commonality amongst them worldwide. Examples of standards include the following: ISO 22476-1, ASTM D5778, EN 1997-3. These days a right cylindrical geometry and 60° apex are almost universal for probe geometry, but a piezoecone cannot be assumed. So make sure a CPTU is specified in addition to referring to a standard. In doing this, also specify the 'u2' (shoulder) location for the piezometric element (the location of the sensing element affects the measured excess pore pressure and the standard charts are based on the 'u2' location).

There is a strong element of *consensus* in all the standards, which means the lowest commonly acceptable approach is effectively built into standards – standards are not innately *best practice*. Also, keep in mind that much CPT sounding is done for stratigraphic characterization while interest here centres on quantitative use of CPT data. These are two rather different motivations in terms of equipment needed. The situation is that CPT contractors like robust, reliable equipment, which is fine for stratigraphic logging. But these attributes come at the cost of giving up accuracy. For example, 18-bit data recording is useless for an engineer testing soft silt to assess its state using a standard-compliant CPT rated at 100 MPa tip capacity. Despite 18-bit digital recording giving an apparent resolution of better than 1 kPa, the accuracy in the strain gauge element feeding the digital converter will not support this level of resolution. Equally, when specifying CPT equipment it is essential to acknowledge that a probe configured for soft clay can be damaged (or even destroyed) if pushed in dense sand. What to do?

It is perfectly reasonable to expect 0.5% of full-scale output (FSO) accuracy on tip resistance from competent testing companies with carefully calibrated equipment. Check with the testing company that this quality of equipment can be delivered. Then specify that several probes are sent to site. Use a robust 50 MPa capacity tip for an initial sounding and then change down to a 20 MPa, or even a 10 MPa, probe for the soft and/or loose soils that you are likely interested in.

Transducer accuracy is particularly important when dealing with loose silts compared to sands. Standards-compliant CPT soundings can be woefully inadequate for assessing the liquefaction potential in slurried tailings or paste (non-segregating tailings) in the mining industry.

Of course an issue here is whether you are using CPT equipment as an adjunct to a geotechnical drill or whether a self-contained CPT unit has been mobilized. When using the CPT with a drill rig it is straightforward to drill down through hard/dense layers to provide access for the CPT to underlying loose material. When using a stand-alone CPT system, it may be necessary to compromise accuracy to get any data at all. Do think about the likely site conditions before choosing between a self-contained truck and drill-rig deployment of the CPT, as the choice may affect the quality of the measurements.

In the case of friction transducers, the principal issue tends to be keeping the seals clean so the sleeve does not stick and then being very careful to zero the equipment before sounding and confirm the zero has not shifted at the end of the sounding. This is more about how the equipment is used rather than the equipment itself. There are two internal transducer configurations for measuring sleeve friction: the subtraction type and the independent type (see Figure 1a of the ASTM D5778 Standard). Both configurations are allowed by the CPT standards but the subtraction type is a poorly conditioned measurement. With a subtraction CPT, friction f_s is measured as the difference between two large numbers with minor errors in each of these large numbers producing a much larger error in f_s. Because using CPT data quantitatively compensates for soil compressibility (or soil type/fines content) through the measured f_s (every methodology does this whether the *equivalent clean sand* approach or by estimating λ), it is important to seek out the independent transducer type for quantitative use of CPT data. Specifying a standard is not going to be enough, and you need to check the details of the equipment proposed.

We have found far fewer issues with the accuracy of the u2 measurements, provided the CPTU is operated using silicon oil or glycerine and properly zeroed with each sounding. We do not recommend using water as the operating fluid because of the tip de-saturating above the water table. Cavitation is not an issue and is easily recognized in the data.

Overall, codes of practice are helpful in specifying procedures for saturation, data recording, tolerances on CPT geometry and so forth. ASTM D5778 is a pretty good guide to appropriate procedures for CPT sounding. Thus, specify testing to one of the three standards listed earlier (depending on your geographic location, although the ISO should prove acceptable everywhere) and then add in the specific accuracy requirements for working with soft and/or loose soils.

9.5.1.2 Data recording

Any modern CPT testing will use digital data acquisition. However, there is an issue of scan rate (or depth interval between scans). Data acquisition is normally time based, with depth being logged at the same time as the transducer readings. For a standard 20 mm/s pushing velocity, a scan rate of 1 Hz averages 20 mm between scans. In electronic terms, 1 Hz can hardly be viewed as an onerous rate. However, some contractors argue that a 50 mm interval between scans is sufficient and use that, but there is not much redundant data at a 50 mm scan interval and some of the detail from the u2 sensor is lost. We used 15 mm nominal interval when we first got involved with digital CPT soundings 30 years ago and rather like that rate – it is a nice compromise between conveniently sized files and making sure that the detail of soil bedding and layering revealed by the piezometric element is fully documented.

In reality, the scan rate used may come down to the choice of testing contractor. If it comes to a choice, opt for a more accurate transducer rather than a faster scan rate. However, it really should not become a choice and 0.5% accuracy at 1 Hz scanning is a reasonable expectation for *best-practice* testing. You will need to set this expectation out when calling for quotes from CPT contractors (or purchasing CPT equipment from suppliers) as the issue is poorly covered in any of the existing standards.

9.5.1.3 Data structure

As with laboratory data, you should insist on getting digital data in a usable ASCII or 'csv' form from the CPT contractor. Any good contractor will already do this, and some will also give you the data already imported into an Excel spreadsheet. Figure 9.15 shows an acceptable ASCII data file with the header and first few lines of data from one of many CPT providers in North America.

Also keep in mind the earlier comments on archive quality for triaxial test data. If the CPT testing is for a dam or bridge, it is very likely there will be ongoing safety reviews (typically every 10 years) and those review engineers likely will want to look at the CPT data as changes in understanding soil behaviour evolve. Just consider the evolution of liquefaction assessment over the past 20 years. So well-done CPT soundings will be as valuable in 50 years as today provided that the testing is properly archived and an engineer (who is possibly not even born yet) can use it.

9.5.1.4 Dissipation tests

The standards leave it to the user to specify dissipation tests in terms of number and location within a planned CPT investigation. The implication is that dissipation tests might be carried out to measure the coefficient of consolidation in clays and silts. This misses an important role of dissipation tests.

Determining the piezometric situation is almost a *Number 1 task* for any site investigation, since piezometric pressures are the input to so many calculations (including processing of the CPT data itself); never mind the contribution to general geological understanding of the site. If the site has at least some sand strata, the u2 sensor will measure the current piezometic pressure during the sounding (as the sounding will be drained in sands). However, if the site is largely silty sands to clays, there will be excess pore pressure during the CPT soundings that prevent direct identification of the piezometric regime (we deliberately do not refer to it as hydrostatic as the situation may involve groundwater flow). A way forward is to recognize that the background piezometric pressure is also an output of a dissipation test so, even if minimally interested in consolidation properties, plan a CPT investigation so that the investigation clearly documents the piezometric conditions at the site during the investigation, for example by carrying out a few long-duration dissipation tests at different depths.

9.5.2 Interpretation of CPT data

Procedures for assessing soil state will be illustrated using a layered tailings profile for which we have good laboratory testing–based soil properties as well as cavity expansion modelling of the CPT and in-situ shear wave velocity measurements.

1	Date	:	1-Jan-80
2	Operator	:	John
3	Location	:	BH-01
4	Reference elev	:	2.03 m AMSL
5	Depth to soil (m)	:	0.5 (m)
6	Depth to water	:	2 (m)
7	Zeroing depth	:	0.5 (m)
8	Push start at	:	0.5 (m)
9	Cone tip number	:	306
10	Cone test ID	:	80CPT01
11	Datafile name	:	CPTU_01

1	Tip resistance calibration factor	:	3104
2	Tip resistance range (MPa)	:	50
3	Tip resistance zero offset	:	802
4	Pore pressure calibration factor	:	1425
5	Pore pressure range (MPa)	:	1
6	Pore pressure zero offset	:	790
7	Skin friction calibration factor	:	1600
8	Skin friction range (MPa)	:	0.2
9	Skin friction zero offset	:	1445

Scan	Rod	Cone depth (m)	Tip qt (MPa)	PWP u2 (MPa)	Friction fs (MPa)
1	1	0.03	2.438	0.005	0.006
2	1	0.035	2.438	0.005	0.006
3	1	0.045	2.438	0.007	0.006
4	1	0.065	2.438	0.005	0.006
5	1	0.085	2.478	0.005	0.006
6	1	0.105	2.517	0.005	0.007
7	1	0.125	2.596	0.005	0.006
8	1	0.145	2.596	0.007	0.007
9	1	0.165	2.674	0.005	0.007
10	1	0.18	2.753	0.005	0.006

Figure 9.15 Example of data format for CPT for archiving or general processing.

9.5.2.1 CPT processing software

There are several excellent commercial programs available for CPT processing, and some give you a state parameter option. Some of the commercial software (e.g. RapidCPT, www.dataforensics.net; Datgel, www.datgel.com) is linked in to a database for geotechnical information such as gINT (www.bentley.com/en-US/Products/gINT), and some has been developed by contractors for their own use (e.g. Fugro, Conetec). It does not matter which you use, as long as you know what you have and how it calculates state parameter.

In general, the state parameter offered will be a screening-level approach using the friction ratio to obtain λ_{10} (i.e. the Plewes et al., 1992, approach) or I_c to obtain λ_{10} (after Been and Jefferies, 1992). Our experience in the 20 odd years since 1992 is that the Plewes method works slightly better and has the advantage of being simpler. It is a reasonable first step, but a lot more can, and should, be done with CPT data. It is easiest to demonstrate the wider possibilities and options using public domain software, so we will use the downloadable spreadsheet *CPT_plot*.xls since you can use the VBA macro features of Excel ('alt' plus 'F11') to see how the equations work.

CPT_plot allows you to go beyond a screening-level assessment and to use the soil parameters you have derived from laboratory testing and numerical modelling in the CPT interpretation. *CPT_plot* has three graph templates, corresponding to (1) reporting basic CPT data; (2) producing profiles of derived soil state and strengths with depth and (3) investigating the behaviour type of specific layers chosen by user-selected depth intervals. There is a title block for including corporate logo and project information, so *CPT_plot* can generate the plots needed for reporting an investigation (and has been used for this purpose by several consulting companies to our knowledge as well as Golder Associates). It is easy to use *CPT_plot* in the field to look at and assess data as those data are obtained. It can be used to guide the fieldwork as well as for subsequent reporting of results.

9.5.2.2 Using CPT_plot

Like the *NorSand* spreadsheets, *CPT_plot* is coded in the VBA environment that lies behind Excel. When opening *CPT_plot*, a dialog will pop up and you need to select 'enable macros' for the VBA code to function. The code is open source and written in plain English, so hopefully it is easy to follow; equations are referenced (mostly to Chapter 4). View the code by pressing 'Alt' + 'F11' keys together or via the '/Tools/Macro/Visual Basic Editor' menu. The code is structured to read the CPT data as a block and operate on that within VBA. The results are passed back to Excel for graphing in the usual way. It may be helpful to download *CPT_plot* and have it open while reading this section.

The arrangement of *CPT_plot* is three worksheets for inputs (with the tabs labelled Project Data, CPT Data and Soil Properties), three worksheets with preset plotting formats to view test data and interpreted results in a report-ready form (tabs labelled Report Fig 1, Report Fig 2 and Report Fig 3) and one worksheet (tab labelled Processed Results). There is also a worksheet labelled Notes, which contains the revision history, comments on the program architecture and the statement of the program being released under the GNU V2 license. To avoid inadvertently moving plots around (and with plots then not lining up), the plotting sheets are locked. This locking password is given in the Notes worksheet (this is an open-source program).

The *Project Data worksheet* inputs are just client name, project title, etc. These inputs are simply echoed to the title blocks of the three report plots.

The *CPT Data worksheet* (Figure 9.16) is where the test to be processed is input. There are a lot of CPT data formats, so the approach adopted is to provide space for a simple 'paste' of the data provided by the testing contractor. These pasted data are then copied across to a second area with all the units changed to the standard form used by *CPT_plot* (i.e. m or MPa). What is needed in importing data should be obvious in comparing to the example CPT file given (but do remember to blank out cells before pasting new data so the plot is not a mixture of two CPTs). If the CPT data have not been corrected for unequal end area, you should calculate q_t here by setting the unequal end area factor in this worksheet. Thereafter, *CPT_plot* works by first transferring the measured and now standard CPT data (in columns *E* to *H*) to a global array within the VBA environment and operating on that array, with results then being transferred back to the Processed Results worksheet for plotting. The calculation is activated by 'clicking' on the 'Process Data' button at the top of the worksheet, which will take you to Report Fig 1 to view the results after a second or so.

The *Soil Properties worksheet* (Figure 9.17) is where soil properties are input and choices made on how the data should be processed. The key soil *properties* are the soil bulk unit

Worksheet: CPT data

1. Paste the raw CPT data into the area highlighted in light blue

2. Convert measured data to standard metric form in the brown area
(this is the data read by the VBA code, and unequal area and other corrections should be applied here)

3. Then... [Process CPT data into Results]

4. Move to the 'Soil Properties' sheet to refine processing coefficients

CPTu sounding: CPTU-01

Measured Readings				Standard Form for CPT data in SI Units			
Depth (m)	q_c (MPa)	f_s (MPa)	u_2 (MPa)	Depth (m)	Tip, q_t (MPa)	Friction (MPa)	PWP (MPa)
0.000	0.2578	0.0006	0.2578	0.00	0.258	0.001	0.258
0.020	0.2578	0.0006	0.2578	0.02	0.258	0.001	0.258
0.040	0.2578	0.0006	0.2578	0.04	0.258	0.001	0.258
0.060	0.2583	0.0000	0.2590	0.06	0.258	0.000	0.259
0.080	0.2583	0.0008	0.2590	0.08	0.258	0.001	0.259
0.100	0.2578	0.0006	0.2578	0.10	0.258	0.001	0.258
0.120	0.2370	0.0011	0.2566	0.12	0.237	0.001	0.257
0.140	0.2384	0.0006	0.2602	0.14	0.239	0.001	0.260
0.160	0.2587	0.0014	0.2602	0.16	0.259	0.001	0.260
0.180	0.2786	0.0011	0.2590	0.18	0.279	0.001	0.259

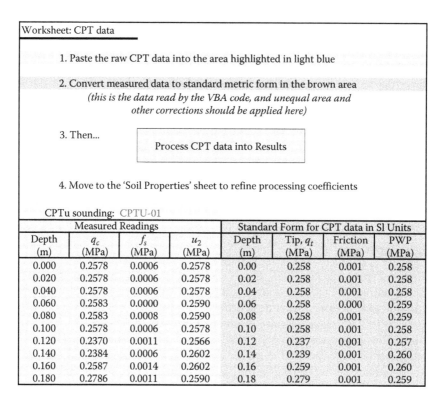

Figure 9.16 First page of CPT_PLOT ('CPT data' worksheet).

weight, the depth to the water table and the unit weight of water. These three inputs are used to compute the vertical effective stress, which is then used for processing the CPT data. At present, *CPT_plot* only has uniform values of bulk unit weight with depth above and below the water table, which is a reasonable approximation (although estimated bulk unit weight can be estimated from soil type and some programs do indeed do this). Underdrainage at a site can be captured by using a reduced bulk unit weight for water. The plot of piezometric pressure on Report Fig 1 also shows the groundwater pressure profile computed from these inputs on top of that measured by the CPT, allowing quick checking that appropriate inputs were made.

The next item on the Soil Properties sheet is to guess at K_o if you have not estimated or measured it some other way. This estimate of K_o is needed because so much of the CPT response is controlled by mean, not vertical, stress. In normally consolidated soils choosing $K_o = 0.7$ is a reasonable start. See Section 4.9 for further guidance.

In terms of liquefaction triggering, *CPT_plot* always computes the cyclic resistance ratio (CRR) using the NCEER method (see Section 7.4) as well as estimates based on the state parameter.

In the case of the state parameter approach, there are two options for computing CRR from ψ: (1) using cyclic simple shear tests in the laboratory with trends for that testing input as values of a, b discussed in Section 9.4 and (2) and the case history–based trends based on Moss (2003) as presented in Section 7.5. The strength trend is specified by inputting a, b values on the *Soil Properties* worksheet (the case history fit values are shown as a prompt).

There are choices on how the state parameter is calculated for input to the selected CRR-ψ trend, input on the *Soil Properties* worksheet as follows:

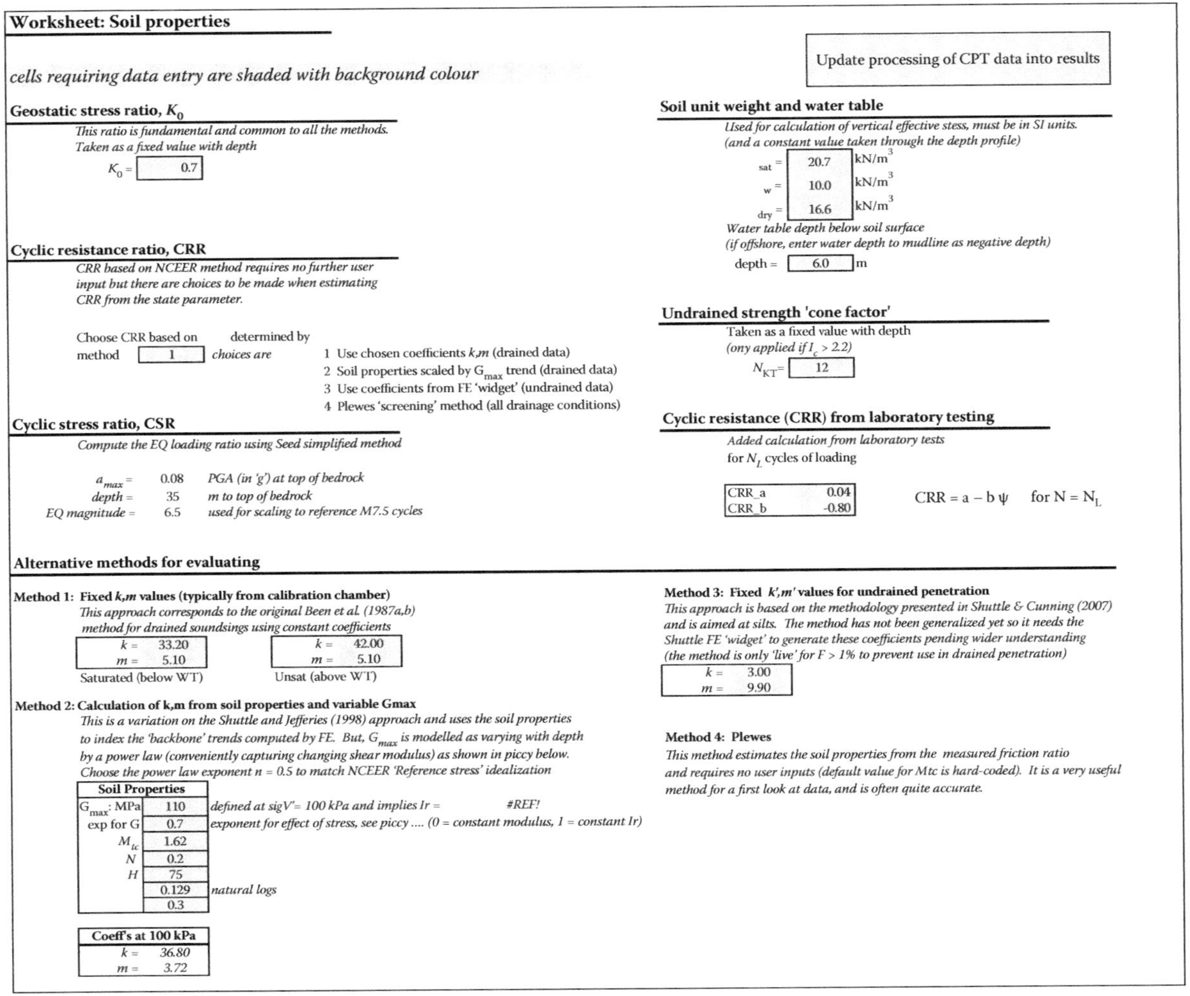
Worksheet: Soil properties
Update processing of CPT data into results
cells requiring data entry are shaded with background colour
Geostatic stress ratio, K0
This ratio is fundamental and common to all the methods.
Taken as a fixed value with depth
K0 = 0.7
Soil unit weight and water table
Used for calculation of vertical effective stess, must be in SI units.
(and a constant value taken through the depth profile)
sat = 20.7 kN/m3
w = 10.0 kN/m3
dry = 16.6 kN/m3
Water table depth below soil surface
(if offshore, enter water depth to mudline as negative depth)
depth = 6.0 m
Cyclic resistance ratio, CRR
CRR based on NCEER method requires no further user input but there are choices to be made when estimating CRR from the state parameter.
Choose CRR based on method 1
determined by choices are
1 Use chosen coefficients k,m (drained data)
2 Soil properties scaled by Gmax trend (drained data)
3 Use coefficients from FE 'widget' (undrained data)
4 Plewes 'screening' method (all drainage conditions)
Undrained strength 'cone factor'
Taken as a fixed value with depth
(ony applied if Ic > 2.2)
NKT = 12
Cyclic stress ratio, CSR
Compute the EQ loading ratio using Seed simplified method
amax = 0.08 PGA (in 'g') at top of bedrock
depth = 35 m to top of bedrock
EQ magnitude = 6.5 used for scaling to reference M7.5 cycles
Cyclic resistance (CRR) from laboratory testing
Added calculation from laboratory tests
for NL cycles of loading
CRR_a 0.04
CRR_b -0.80
CRR = a − b ψ for N = NL
Alternative methods for evaluating
Method 1: Fixed k,m values (typically from calibration chamber)
This approach corresponds to the original Been et al. (1987a,b) method for drained soundsings using constant coefficients
k = 33.20
m = 5.10
Saturated (below WT)
k = 42.00
m = 5.10
Unsat (above WT)
Method 3: Fixed k',m' values for undrained penetration
This approach is based on the methodology presented in Shuttle & Cunning (2007) and is aimed at silts. The method has not been generalized yet so it needs the Shuttle FE 'widget' to generate these coefficients pending wider understanding (the method is only 'live' for F > 1% to prevent use in drained penetration)
k = 3.00
m = 9.90
Method 2: Calculation of k,m from soil properties and variable Gmax
This is a variation on the Shuttle and Jefferies (1998) approach and uses the soil properties to index the 'backbone' trends computed by FE. But, Gmax is modelled as varying with depth by a power law (conveniently capturing changing shear modulus) as shown in piccy below.
Choose the power law exponent n = 0.5 to match NCEER 'Reference stress' idealization
Soil Properties
Gmax: MPa 110 defined at sigV'= 100 kPa and implies Ir = #REF!
exp for G 0.7 exponent for effect of stress, see piccy .... (0 = constant modulus, 1 = constant Ir)
Mtc 1.62
N 0.2
H 75
0.129 natural logs
0.3
Coeff's at 100 kPa
k = 36.80
m = 3.72
Method 4: Plewes
This method estimates the soil properties from the measured friction ratio and requires no user inputs (default value for Mtc is hard-coded). It is a very useful method for a first look at data, and is often quite accurate.

Figure 9.17 Second page of CPT_PLOT ('Soil Properties' worksheet).

- *Method 1*: Specify k, m directly as constants for CPT processing. This is the original method based on calibration chamber testing and Equation 4.8. This method allows users to go back to calibration chamber data to make a judgement as to what is appropriate for the site and then process data based on that judgement. However, filtering is applied so that this method is only used for drained penetration, initially making the processing *live* for the parts of the CPT profile where $F<1.5\%$ and $-0.02<B_q<0.02$ (these limits can be changed; go to the Declarations part of the VBA code).
- *Method 2*: Compute k, m from soil properties. Specify soil properties M, N, H, λ, ν, G_{ref} and n as inputs with *CPT_plot* then computing k, m using the approach of Shuttle and Jefferies (1998) in Equation 4.12 and Table 4.3. This method allows the effects of the soil properties, measured as discussed in Section 9.3, to be used to allow for the effect of soil properties on the CPT response. This is basically a site-specific calibration of the CPT. G varies as indicated in Equation 9.8 and considers any apparent stress level inherent in the calibration chamber test data discussed in Chapter 4 (unless n is set to zero). You can use the value of G_{ref} and n from various test methods but the best is to use a seismic cone and then input the values of G_{ref} and n that best fit the shear wave velocity data for each CPT location.
- *Method 3*: Specify $\bar{k}$, $\bar{m}$ for soundings in silts and sandy silts where $B_q>0$ develops. This method is only live in the spreadsheet for $F>1\%$ to avoid its use for drained penetration (again, the limit can be reset in the Declarations part of the VBA). In this case, $\bar{k}$, $\bar{m}$ are computed using the method of Shuttle and Cunning (2007) and then used in the inverted form of Equation 4.14 to obtain ψ from Q_p and B_q. You need to do the cavity expansion calculations outside the spreadsheet; the *widget* to do this is a downloadable file.
- *Method 4*: Plewes. This is the method that captures the effect of soil compressibility on the CPT by using F to determine λ (Equation 4.16) and $\bar{k}$ and $\bar{m}$ (Equation 4.15). The critical friction ratio is approximated as $M_{tc}=1.25$ for all soils (sands and silts) in this approach. This method is based on Plewes et al. (1992); experience across many sites suggests it is surprisingly accurate. The disadvantage of this method is that it neglects G_{max} and other soil properties. The advantage is that it reflects in-situ point to point variation in fines content and related aspects of the soil.

Now 'click' on the 'Update processing of CPT data into results' button in either the 'CPT data' or 'Soil properties' worksheet for the macro to do the calculations for you. Note that your choice of method for calculating ψ will be displayed above the plot and will print automatically. Also, the output of the Plewes method is always shown. If the state calculation is selected as Methods 1–3, the Plewes results are shown in the background as a light grey line. If the Plewes method is chosen (= Method 4) then it is plotted in the normal way.

9.5.2.3 Viewing CPT results

The results from the calculations will now be in the Processed Results worksheet, but are best viewed graphically in separate worksheets labelled Report Fig 1, Report Fig 2 and Report Fig 3.

The Report Fig 1 worksheet presents basic plots and is just what you would expect from the CPT, Figure 9.18. They include the tip resistance q_t, sleeve friction f_s and pore pressure u measurements followed by the normalized parameters F, B_q and soil behaviour–type index I_c. Everything is aligned so you can scan across the plot to see what response developed during sounding, where you might place the stratigraphic boundaries and what is interbedded or uniform. It includes everything that a 'cone head' needs to develop a first

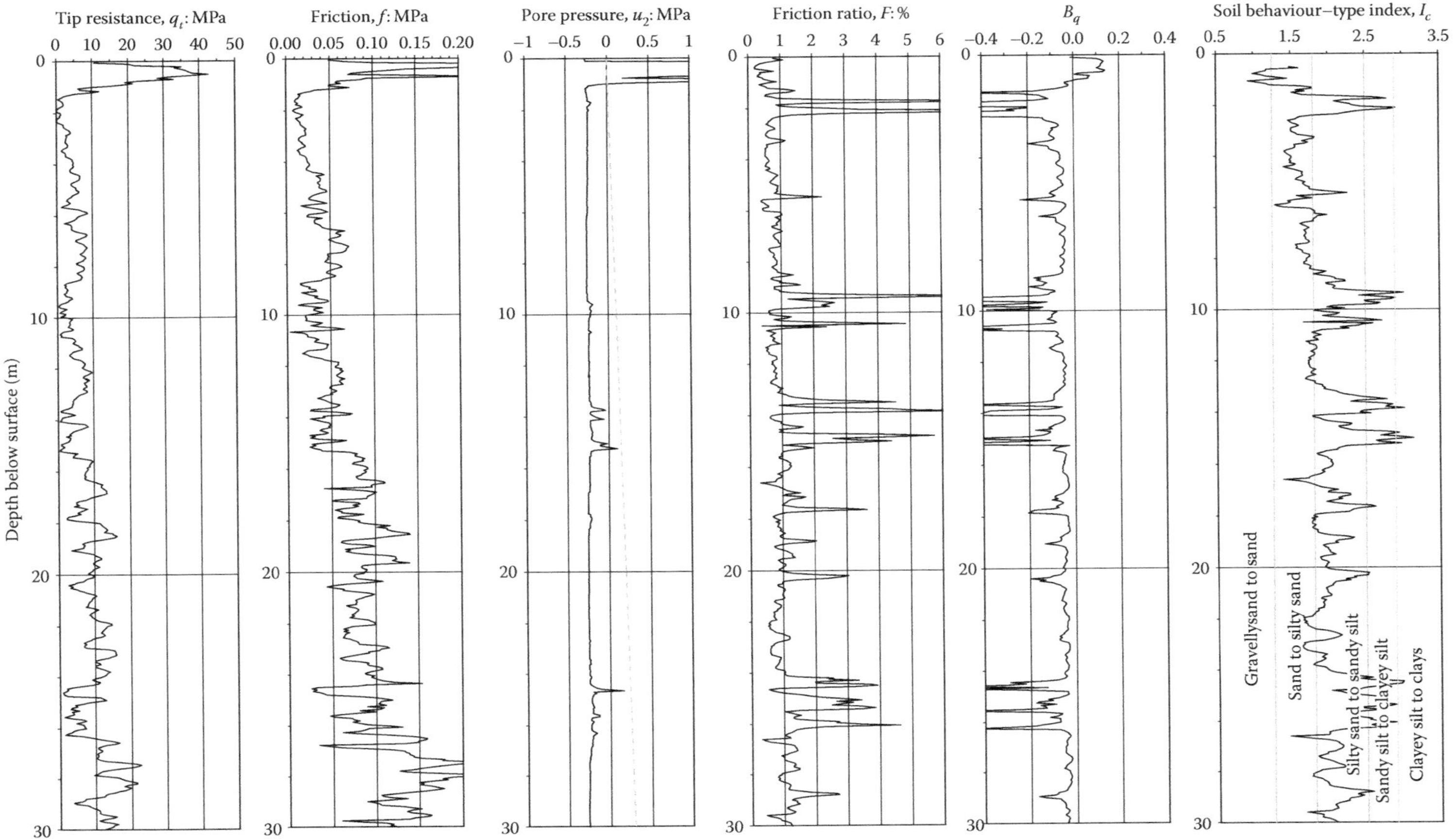

Figure 9.18 Basic processed CPT output ('Report Fig 1 Basic' worksheet).

appreciation of the site. There is space to add an interpreted soil profile on the far right, not shown in Figure 9.18, but this is not done automatically (you can be artistic).

Slightly out of sequence, it may be best to look at the Report Fig 3 next, this figure presenting the details of a chosen layer (or window) of the data on a *soil behaviour–type* plot. In the case of the basic plots just shown, it is evident that there is a relatively uniform material from 15.5 to 23.5 m; this depth range is entered on the Report Fig 3 sheet, the 'update' button clicked, and the individual processed CPT data are then presented for review, Figure 9.19.

This worksheet provides a tool to interrogate the data to form judgements about stratigraphic details as well as a first assessment of soil state. It would be usual to create a 'Fig 3' type plot for every stratigraphic unit identified. The left-hand side of the figure shows the q_t-depth trace and with the windowed zone highlighted so that a casual viewer can quickly identify which part of the CPT profile the soil behaviour–type plot corresponds to. This soil behaviour–type plot was introduced in Chapter 4 as Figure 4.27, with circular arcs representing constant values of soil behaviour–type index I_c and thus *the same soil along an arc*. What is new on this classification diagram is a family of curves of ψ, which have been calculated using cavity expansion along the lines of Shuttle and Cunning (2007) described in Section 4.7, including a line shown as 'demarcation between strain softening and strain hardening behaviour following initial liquefaction' (Shuttle and Cunning, 2008), or roughly $\psi = -0.05$. This is your starting point for a liquefaction assessment.

The two report-type plots just discussed are mostly *geologic* in aim, and it is now time to turn to *engineering*: Report Fig 2 shows example results from the data of Figure 9.18 as interpreted results in Figure 9.20. The left-hand plot is a copy of the q_t-depth to allow a viewer to correlate the derived engineering results with the previous geological-type figures. The figure then presents the following:

- The ψ profile computed by the chosen method. The method choice itself is shown above the plot.
- A profile of the drained friction angle ϕ', using the selected ψ method and the general state parameter – friction angle relationship implied by Figure 2.7, $\phi' = 32 - 50\ \psi$. If your soil differs from this relationship, and you have selected Method 2 for calculation of state using soil-specific soil properties, this general equation will be replaced using the input soil properties. Note that when the material behaviour–type index is greater than $I_c = 2.5$, a total stress of $\phi = 0°$ is plotted.
- If the material behaviour–type index is greater than $I_c = 2.5$, indicating a clay-like material behaviour, then undrained strength s_u is calculated by the conventional undrained total stress method using the N_{kt} factor that appears in the 'Soil Properties' worksheet (which you can change like any of the other parameters on that worksheet, and click 'update' to reprocess with a new value).
- CRR is shown calculated using both the NCEER method (thin grey line) and the chosen ψ-based method and the fitted laboratory Equation 9.9. If you do not have site-/soil-specific parameters a, b for Equation 9.9, the empirical fit to the ψ-based liquefaction case histories in Figure 7.42 would be plotted (or you could consider both). For comparison, the plotted CRR profiles are adjusted for duration of shaking using the chosen earthquake magnitude input in the Soil Properties sheet (and which is echoed back as a banner above the plot). Do consider whether the site conditions (soil type, depth of loose layer) lie within the limits of the NCEER database (see Figure 7.41) before relying on that CRR profile; the state-based profile is more general to depth and soil type.
- Also shown on the CRR graph is a simplified Seed calculation of earthquake loading or cyclic stress ratio (CSR) using the approach documented in the Youd et al. (2001) paper following the NCEER workshops in the late 1990s to agree the approach.

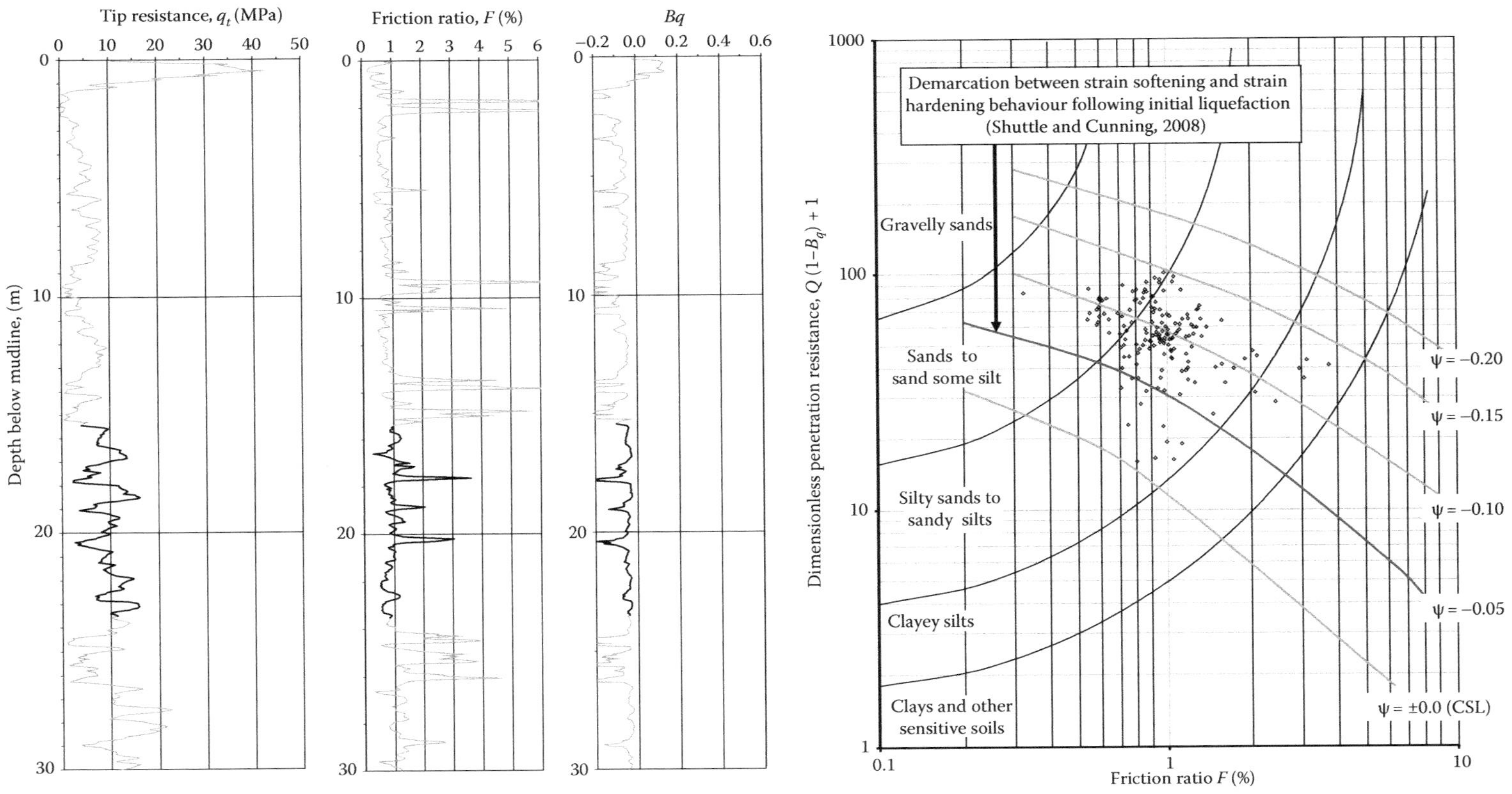

Figure 9.19 Normalized tip resistance and friction ratio classification ('Report Fig 3 Classification' worksheet) showing demarcation between undrained strain softening and hardening materials.

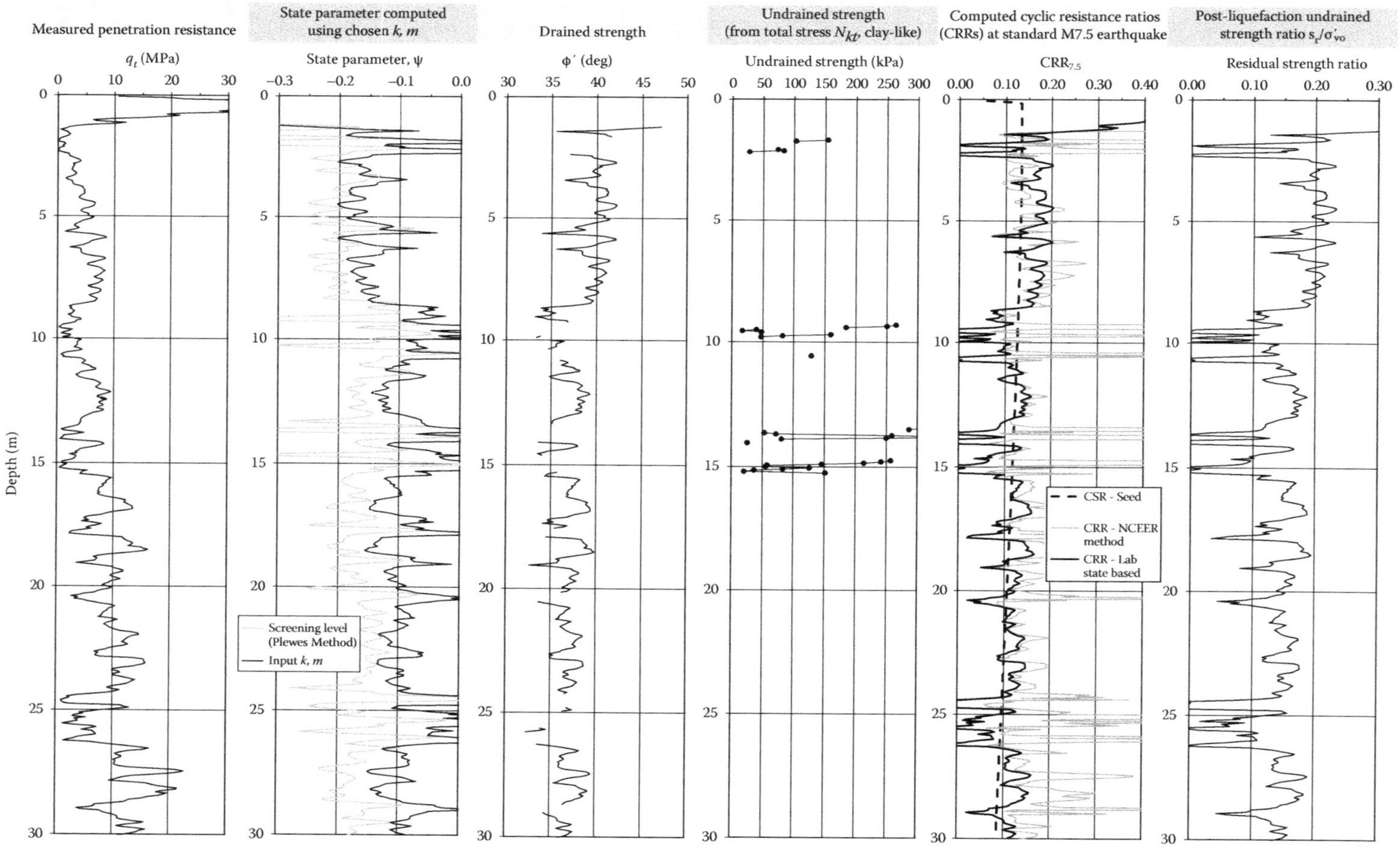

Figure 9.20 Interpreted CPT parameters – ψ, ϕ, CRR, s_u, s_r ('Report Fig 2 Interpreted' worksheet).

The PGA at the ground surface and the earthquake magnitude are needed for this calculation and should also be input to the Soil Properties worksheet. A word of caution is needed here. It is the surface PGA that is needed for that calculation, not the bedrock ground motion more commonly given in a seismic hazard assessment or design codes. You really need to do a site-specific ground response analysis to determine the surface PGA, but in that case you will have calculated the entire CSR profile with depth so that the simplified approach becomes unnecessary.

- The last graph is the residual post-liquefaction undrained shear strength ratio, s_r/σ'_{vo}, approximated depending on whether the soil is stiff, intermediate or compressible based on the λ value (refer to Chapter 6 and Figure 6.47 in particular).

9.5.2.4 Reporting CPT data

The *CPT_plot* program provides three types of plots, but how should these plots be used in engineering practice? Just as there was a reporting protocol for laboratory data, so there is a similar consideration with the results of a CPT investigation to assess soil state and strengths. Our practice, and the reasons for it, is the following:

- Owners and Regulators and non-technical stakeholders will not be interested in the details of any individual CPT, but are often interested in the stratigraphy across the site. It is common to use q_t versus depth data from sounding at the site to create engineering section drawings – standard work for your graphics department. What is then needed is to illustrate the type of behaviour within the key strata identified this way, which comes down to the *classification* plots (i.e. Report Fig 3) with their implications of whether the engineering needs to be directed to run-out distances and consequences, ground improvement or tolerable displacements. This is the high-level conclusion of the investigation and a *main text* item.
- Review engineers will be interested in the robustness of the assessed state as well as the assessed strength profiles (both cyclic and post-liquefaction), as it is these assessments that will be used to drive the engineering. This information is contained in the *interpretation* plots (i.e. Report_Fig 2), and it is usual to provide a plot of this type for each CPT sounding in an appendix. What is also needed is to discuss and justify the assessed groundwater conditions, the choices of methods used in the CPT processing and the effect of variability from one CPT to another – all key considerations, but really rather more than non-technical people want to read. So, use an appendix to present the methodology and results from CPT processing but not necessarily the same one that contains the pages of plots from individual CPTs. The Processed Results worksheet contains the processed results for a CPT sounding and it is often convenient to compile a project workbook containing the various set of processed results so that, for example, composite plots of state profiles from all soundings can be shown as a single figure.
- For data archiving, the raw CPT data need to go onto a CD bound into the report. However, it is often helpful to quickly scan CPT plots so as to assess what sounding is showing what and therefore we normally include a further appendix with the Report Fig 1 plots as a *basic* overview for each sounding.

9.6 APPLICATION TO TYPICAL PROBLEMS IN SANDS AND SILTS

A first decision is whether you are dealing with a strain softening or hardening material (in an undrained condition), or whether your in-situ material is sufficiently dense that strain hardening is assured. The normalized classification graph on Figure 9.18 provides a good indication, on a screening level, of where you stand.

- If the soil is all denser than $\psi < -0.05$, the problem is likely to be one of soil displacement during and after an earthquake. Post-earthquake settlement, dynamic slope stability and lateral spreads need to be considered.
- If enough of the soil is looser than $\psi > -0.05$, undrained behaviour will result in strain softening and potentially large flow slide displacements. In most civil engineering projects, ground treatment of some nature will be called for to increase the in-situ density to an acceptable level. Alternatively, you may need to consider the residual, post-liquefaction undrained shear strength in your stability calculations.

A good liquefaction assessment really is as simple as that.

Chapter 10

Concluding remarks

This book promised a view of liquefaction based on mechanics, while respecting full-scale experience. We sincerely hope that this has been delivered. In concluding this work, we would like to expand on these practical issues, how they affect the current approach to liquefaction and what should be expected for the future.

The decade since the first edition went to the publisher has seen progress in the understanding of liquefaction and a large increase in both experimental data (for silts in particular) and the case history experience (the events in Christchurch being noteworthy). From our perspective, it has been pleasing to see widespread acceptance that the framework of critical state soil mechanics (CSSM) is relevant to all aspects of liquefaction. Here, in this concluding chapter, we comment on aspects where further research and progress are needed.

10.1 MODEL UNCERTAINTY AND SOIL VARIABILITY

An important issue going forward is reconciling theoretical and laboratory-based expectations with full-scale experience: model uncertainty as it is known. Practically, for example, we have post-liquefaction strength trends from case histories, and these are reasonable enough for engineering use, but it is a long way from intellectually satisfactory that we cannot predict why there is a mismatch with what we measure in the laboratory. This is not exactly a new topic for geotechnical engineers, nor something that is restricted to liquefaction.

There are two perceived contributions to model uncertainty: (1) the effect of soil variability and (2) our calculation methods not reflecting how soil behaves. We now consider each of these in turn.

10.1.1 Quantifying soil variability

Geotechnical engineering is quite a way from properly handling real material variability in the field, in terms of both how soil state varies from point to point throughout a geological stratum and how one soil type blends into another. Yet, weak layers can be the trigger for liquefaction of a soil that otherwise looks satisfactory. The Nerlerk berm is a pertinent case history. Recall that Nerlerk was an underwater hydraulic fill structure constructed to support an offshore drilling platform and failed by liquefaction that was caused by yield of the underlying soft clay. Nerlerk was not especially loose, and this caused much confusion between engineers on one side asserting the cone chamber calibration tests were providing a misleading understanding of the in-situ fill state, while engineers on the other side struggling to understand how something that looked vaguely adequate could have failed so catastrophically and repeatedly so (there were actually five slides). It is now clear from the stochastic studies by Hicks and Onisiphorou (2005) that the arrangement of the looser zones within the fill allowed a liquefaction failure mechanism to develop in what was, on average, a lightly dilatant fill.

Chapter 5 considered issues posed by real material variability and how they may be dealt with today. It is difficult to think of another engineering discipline that works with as limited knowledge of material properties and distributions as confronts every geotechnical engineer every day. Liquefaction, with its issues of control by loose layers (and how loose zone effects can propagate through the domain), exposes this weakness. A decade ago, this quantification and simulation of soil variability was a topic dealt with by only a few leading researchers (Manzari and Dafalias, Griffiths and Fenton being noteworthy in addition to Hicks and Onisiphorou), but now there has been a modest uptake of these ideas into high-risk projects by consulting companies (including our own). In part, the limitation from a consulting perspective has been the lack of tools, both to simulate the geology as a stochastic domain and then to compute the response of that stochastic domain. The recent availability of state parameter–based models (including, but by no means limited to, *NorSand*) is a way forward, since all that is needed is to vary the state parameter stochastically as the model looks after the implied distribution of soil behaviour from this simple stochastic input. Stochastic simulations do seem key to closing the gap on idealized predictions and engineering reality.

10.1.2 Analytical methods

One reason for a mismatch between theory and reality is that the calculation method has inappropriate assumptions. It has been known for a long time that limit equilibrium methods, which are the standard way post-liquefaction strengths are estimated, can seriously mislead if brittle failure mechanisms are involved. Static liquefaction is about as brittle as soil behaviour gets. Yet, our profession largely looks at all case histories through the eyes of limit equilibrium.

What we found striking in Chapter 8 was the wide range of behaviour observed in one slope depending on how that slope was brought to failure: failures caused by too much crest load were quite unlike failures caused by basal creep. Toe erosion is another cause of failure, and then there is the post-earthquake situation where failure is caused (presumed) by large-scale migration of pore water as, for example, at Lower San Fernando. The Griffith and Lane (1999) work hinted at this result, and Chapter 8 takes it further. Yet, none of this quite different behaviour appears to have been accounted for in the various back-analyses discussed in Chapter 6. As a profession, we can do far more relevant back-analyses now than has been done so far (and on which we all base our designs/assessments).

Perhaps we are being a little harsh in our views here, as developing software such as used in Chapter 8 is a different endeavour from that for which most geotechnical engineers are trained (or even wish to do). Certainly, the downloadable software that goes with Chapter 8 is there for use, but it is not as user-friendly as needed in a consulting office (nor even for back-analyses of case histories by postgraduate students unless they want to immerse themselves in the numerical details). What our profession needs is for state parameter–based models to be included in the various commercial modelling platforms, and we are only going to get that if we all start asking for it. User-defined models are not enough, and the code verification of appropriate state-based models needs to be put into the lap of the software houses. We are waiving the GNU license restrictions for any software house that wants to incorporate our routines into their platforms (provided that there is an appropriate citation). We also encourage academics to contribute their models to the public domain, whether one of the *NorSand* or Bounding Surface variants. If we do not *disseminate*, then we will continue, as a profession, to do a lesser job than we could.

10.2 STATE AS A GEOLOGICAL PRINCIPLE

When dealing with in-situ soils and their natural variability, an interesting consideration is how the depositional conditions and geological history affect soil state. For example, nobody is surprised if the overconsolidation ratio in a clay stratum represents a near-constant maximum past pressure from eroded overburden or the loading of an ice sheet (for a glaciated deposit). How might similar geological idealizations apply to the state parameter and sands?

Tailings impoundments present an opportunity to examine the idea that the state parameter might reflect geological processes. Tailings are the result of grinding up rock to recover metals, with each ton of ore resulting in almost a ton of sandy silts. Tailings are usually discharged into valleys that have been closed off by dams (full perimeter dams are sometimes used), with the discharge from the slurry pipeline being from one or more spigot points. The tailings segregate with the sand-sized fraction settling out more quickly in a beach and with the remaining soil becoming progressively finer with distance from the spigot point. In effect, a tailings impoundment is similar to a controlled delta depositional environment.

Dams used to retain tailings are becoming some of the largest engineered structures in the world, with heights in excess of 250 m (and in highly seismic regions) now being designed and constructed. A lot of CPT soundings have been carried out in tailings to establish the conditions upstream of these dams. Figure 10.1 shows an example of the measured CPT data, in a soil behaviour–type plot, in three soundings used to explore tailings variability within one impoundment. The tailings grain size ranges from sands with little silt through to fine silts (the CPT data plot as clayey silt, but there was no plasticity or clay-sized particles) and yet plots consistently with $\psi \sim 0$ when using the Plewes et al. liquefaction screening method. We have encountered similar data at other tailings impoundments.

There are intriguing aspects to trends as seen in Figure 10.1. First, if you have any residual desire to use relative density, then Figure 10.1 should put an end to that desire. How can relative density be assessed for such a wide range of soils, and is relative density even measurable with silts? These difficulties do not arise with the state parameter. Second, we do not imply that the natural depositional state of sand and silts is $\psi \sim 0$. If a higher-energy depositional environment occurs, then denser states result. For example, hydraulic fill construction of sands with the bottom discharge method consistently develops $\psi \sim -0.1$ (Jefferies et al., 1988b). Natural beach sands exposed to wave action tend to have ψ denser than -0.2 (in our experience).

The utility that the geological environment, in its widest sense, gives a particular state parameter regardless of soil gradation lies in site characterization. In effect, this concept is a further constraint to reduce the uncertainty when considering natural variability at a site. For example, one might skip thin layer corrections on the grounds that if the thin layer lies within a defined geological unit, it ought to have the same characteristic state as the remainder of the stratum.

Clearly, this linking of geological environment to state parameter is new for sands and silts (although well established for clays). It is based on observations within the deltaic conditions of mine tailings impoundments with supporting data from engineered hydraulic fills, but it certainly has the potential to act as a unifying principle if further data support the framework across a wider range of geologies.

10.3 IN-SITU STATE DETERMINATION

Proper characterization of stochastic soil necessarily means basing engineering on the CPT. There is simply no other test that offers as much data, with such precision, at a reasonable cost.

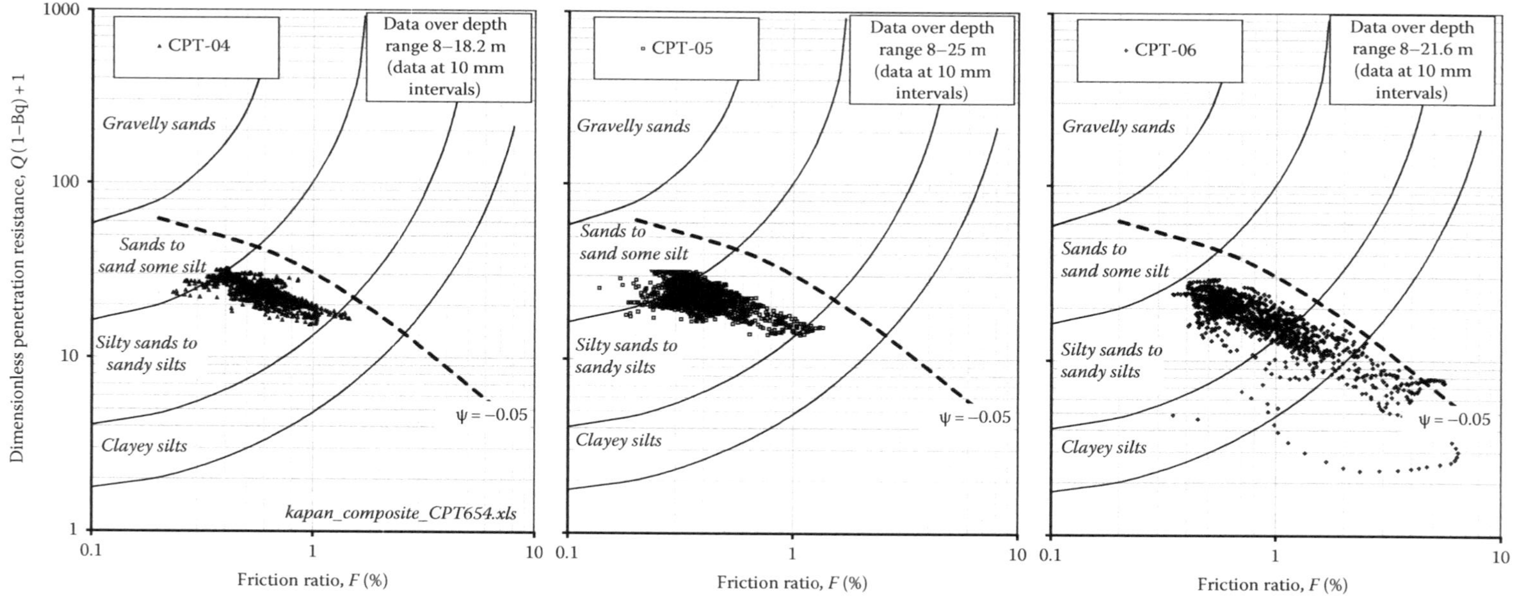

Figure 10.1 CPT screening-level classification of tailings from three test locations, showing remarkable consistency of state within the deposit.

We hope that a convincing case has been made that site characterization for liquefaction should use the CPT. Nothing less than a standard right cylindrical piezocone to international reference configuration standards should be used, but keep in mind the number of soundings required. Two or three CPT soundings for a 100 m × 100 m site are not enough to get an appreciation of the true characteristic values. In the order of 10–15 soundings are required, which is not that onerous given a goodly part of the cost of doing CPTs is mobilization. As experience develops regarding soil property distributions with differing geological environments during soil deposition, there will be prior distributions for the soil properties or state, leading to the possibility that fewer soundings are needed to characterize any particular site.

A difficulty with all in-situ tests, however, is that they are not like laboratory element tests in which there is complete knowledge of all stresses and strains. The CPT is a continuous bearing capacity test, not a continuous test of peak friction angle (because more than friction angle determines bearing capacity). The relationship between measured CPT data and soil state depends on the soil's properties, in particular compressibility (or more accurately, plastic hardening) and critical friction ratio. This has been understood within the literature for a while, but is also almost always neglected in actual liquefaction assessments. Hopefully, what was presented in Chapter 4 has made this point firmly. And the software *widget* needed to create a soil-specific calibration for the CPT is yet another downloadable file.

When dealing with the effect of soil properties on the evaluation of CPT data, two approaches have been presented. One approach uses the additional friction data measured by the CPT to infer soil behaviour type. Soil properties are then estimated based on experience for that soil behaviour type. Plewes et al. (1992) called this a *screening level* assessment, and it is an easy data processing task that delivers a first estimate of the soil state. The methodology was described in Chapter 4, and our experience over the past two decades (the method is widely used) suggests that it is quite a good method. The second approach is to use measured soil properties. Only a limited amount of laboratory testing is needed to determine CSL parameters of the soil, but these are fundamental to correct evaluations of site conditions. CSL testing is not particularly expensive, about $15,000 at the time of writing, and the procedures have been presented in Chapter 9 and Appendix B. If a range of gradations exist, such as found in most tailings impoundments or natural deltaic sands, then more than one soil should be tested and interpolation used for intermediate materials.

The downloadable software *CPT_plot.xls* provides the options for these alternative methods of determining ψ from CPT data. The way it has been programmed in VBA allows adding further methods, for example a blend of the Plewes method with soil-specific calibration might be useful at some sites. This reinforces one aim of this book: we are sharing data and software so that you too can gain a better understanding of soil liquefaction. What we have provided is sound and useful, although you may need to dig into the background and take it forward for your specific situations (and, remember, all the source information is downloadable).

There is a requirement to define the in-situ elastic shear modulus. In part, this is needed for site response calculations, but it is now widely appreciated to be an important input when assessing soil behaviour. If ever there was a research development that transitioned quickly to routine practice in geotechnical engineering, this is it. Possibly, this transition has been helped because it is not an onerous testing requirement, being conveniently and inexpensively done using the seismic CPT.

Two difficulties then stand out: horizontal geostatic stress and plastic hardening. Neither can be viewed as routine engineering. Understanding of both could be enhanced by further research and by engineers in consulting practice keeping their eyes open for trends in different geological environments. This information then also needs to be published.

It is an inconvenient fact of life that soil behaviour is seemingly controlled by the mean effective stress, not the easily estimated vertical effective stress. We have used the self-bored pressuremeter on many projects to measure the horizontal geostatic stress, but obtaining good data in sands with this device is challenging. But does the geostatic ratio vary greatly in deltaic sands? Much of the data presented in Chapter 4 have been for man-made hydraulic fills, and while there are certainly local fluctuations in K_o, there are also some clear trends. What is needed is to expand this database to include tailings, deltaic sands and silty sands and compacted fills. Given time, accumulation of knowledge will make assessing the horizontal stress easier, especially on lower-budget projects that cannot support the costs of pressuremeter tests. For now, the only acceptable way forward is to explicitly consider different K_o scenarios in assessing the characteristic state of a site or earth fill structure. We also like $K_o \approx 0.7$ as a starting point for normally consolidated deposits whether clays, silts or sands based on the data presented in Chapter 4.

A relatively recent development is the use of anisotropy of shear wave velocity measurements to estimate in-situ stresses and void ratio (Fioravante et al., 2013). The idea is that you measure shear and compression wave propagation in a large calibration CPT chamber, but not just in one direction. The measurements are made vertically, along both horizontal axes, and diagonally, across the chamber sample. This allows you to generate a calibration between horizontal and vertical stresses, and void ratio, to the shear and compression wave velocities. A seismic CPT is then used to measure vertical shear and compression wave velocities and state parameter (from the tip resistance) in-situ, from which the in-situ stresses are computed. If this sounds complicated and expensive, it is because it is indeed expensive. However, large and critical projects may well justify this type of approach.

10.4 LABORATORY STRENGTH TESTS ON UNDISTURBED SAMPLES

A basic premise of the approach of this book is that getting undisturbed samples of sands is very difficult, but that has not proved an impediment in some circumstances. Undisturbed sampling with subsequent strength testing is a widely accepted protocol for clays. Silts lie between sands and clays, with samples generally recoverable but likely to be disturbed.

An aspect of the past decade has been a rising awareness of the vulnerability of silts (or *high fines* soils) to earthquake loading coupled with recognition that the empirical underpinning of the National Center for Earthquake Engineering Research (NCEER) approach does not stand too much scrutiny in regard to silt behaviour. One response has been to divide soils into categories of behaviour: *sand-like* for which penetration tests are the accepted basis of engineering, *clay-like* where engineering focuses on recovering high-quality samples that are then tested in the laboratory and *intermediate*, which encompasses everything in between and which was a much discussed topic at recent CPT conferences in 2010 and 2014.

Here, we describe some experience with both sands and clayey silts, illustrating what some might view as state-of-the-art work involving soil sampling and testing rather than reliance on the CPT.

10.4.1 Undisturbed sampling and testing of sands (Duncan Dam)

In reaction to the apparent associated uncertainties in determining ψ from the CPT, research in Canada in the early 1990s focused on avoiding penetration tests and measuring density in-situ. The work was part of a liquefaction assessment for Duncan Dam, with a heroic effort made to test truly undisturbed samples, which resulted in a series of linked papers in the December 1994 issue of the *Canadian Geotechnical Journal* (following a special session on the work at the 1993 Canadian Geotechnical Conference). Imrie (1994) presents an overview of the work at Duncan Dam, and the work is interesting from the perspective of what was tried and what did not work. The Duncan Dam data also underlie the correction for stress-level effects found in the NCEER approach to seismic liquefaction assessment (discussed in Chapter 7).

Duncan Dam, completed in 1967, is a zoned earth fill embankment some 39 m high located on the Duncan River in southeastern British Columbia. Foundation conditions include loose sands and compressible silts that extend to a depth of as much as 100 m below surface. It has been known for a long time that Duncan Dam is vulnerable to earthquake-induced liquefaction, with the adequacy of the dam depending on sufficient post-liquefaction strength in the foundation soils to prevent a flow failure.

A comprehensive field and laboratory investigation of Duncan Dam foundation was undertaken between 1988 and 1992. There were three broad thrusts to the investigation: (1) standard penetration tests (SPTs) analyzed using the reference stress approach, (2) ground freezing to allow undisturbed sampling by coring with subsequent testing in the laboratory (predominantly monotonic and cyclic simple shear tests) and (3) direct measurement of in-situ void ratio by geophysical techniques. Three CPT soundings were put down, but they were apparently only used to identify soil stratigraphy and CPT/SPT correlations (these CPT data appear to have been lost as they could not be recovered from the owner's and testing company's files for a dam safety review in 2010).

Freezing the ground so that undisturbed samples can be obtained is an interesting development from Duncan Dam, as it allows loose sands to be notionally tested as *undisturbed* samples, much as one might expect to test undisturbed samples of clay. Sego et al. (1994) describe in detail the ground freezing and sampling at the Duncan Dam, and it could be regarded as a reasonably *doable* approach for any large dam foundation. Freezing and undisturbed sampling were also carried out at several of the CANLEX sites (Wride et al., 2000) with indicative costs at the time of about $50,000 per site. The void ratio of the frozen samples showed good agreement with the in-situ void ratio measured by gamma–gamma density logging in the same borehole (Plewes et al., 1994). However, obtaining undisturbed frozen samples only solves half the problem; one still has to thaw the sample without imposing plastic strains. Also, how does one know what in-situ stress conditions to impose on the sample as it thaws? Patently, if the wrong stress state (mean and deviator) is applied on the sample during thawing in the laboratory, then plastic strains will occur, and the sample is no longer representative of in-situ soils.

Data from the laboratory testing programme on the Duncan Dam samples obtained by ground freezing are provided in Pillai and Stewart (1994). Specimens for monotonic and cyclic simple shear testing were machined to the required size using a precision lathe in a cold room at −15°C and then set up in the frozen state in the testing device and allowed to thaw. After thawing, samples were consolidated to vertical effective stresses ranging from 200 to 981 kPa. Given the sampling depth of 13.7–16.3 m, only data for specimens consolidated to 200 kPa are considered representative of in-situ conditions. Figure 10.2 shows the initial void ratio of the samples plotted against the void ratio after consolidating back to approximately the in-situ vertical effective stress, that is 200 kPa. There is a densification of approximately $\Delta e = 0.05$ that occurs during the thawing and consolidation, even in the easiest case of simple shear samples (for triaxial testing, guessing the in-situ horizontal stress to apply in the laboratory adds additional complexity).

What is the consequence of this sample densification of the undisturbed frozen samples as they thaw? Olson (2006) evaluated the residual strength at Duncan Dam using the SPT data and found strengths about half to two-thirds of those from the laboratory tests on

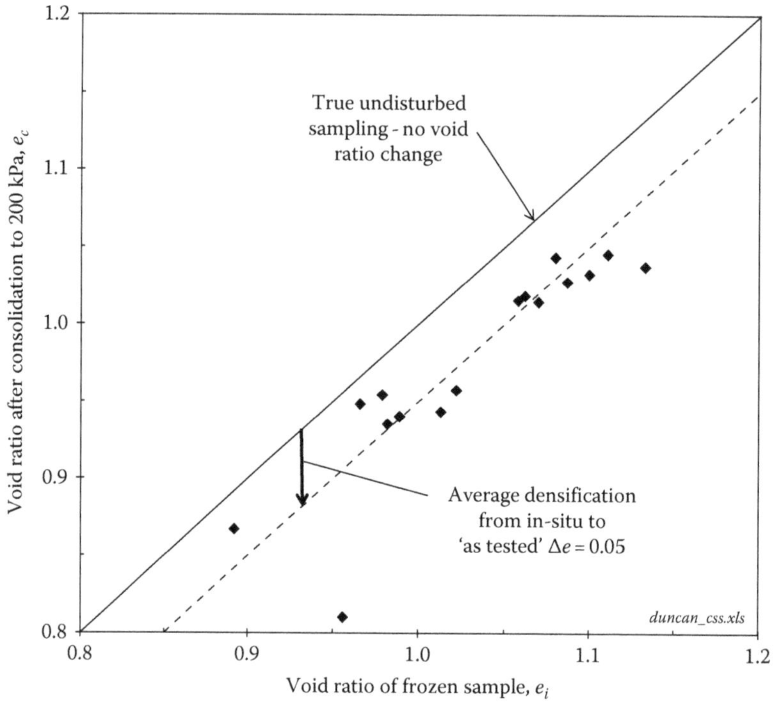

Figure 10.2 Comparison of void ratio of frozen samples before and after thawing and consolidation to approximately in-situ stress conditions in simple shear device. (Data from Pillai, V.S. and Stewart, R.A., *Can. Geotech. J.*, 31(6), 951, 1994.)

previously frozen samples. The apparent conflict between the 2006 and the 1988–1992 work disappears if the reported sample densification during thawing is considered. Konrad et al. (1997) determined the CSL for the Duncan Dam foundation sand and reported that $\lambda_{10} \sim 0.15$. For such a λ_{10}, the average sample densification of 0.05 in the laboratory would be expected to cause an overestimation in the available strength by more than a factor of two. Since the reduced residual strength profile shown in Olsen (2006) is about 60% of that obtained from the laboratory tests on thawed samples reported by Pillai and Salgado (1994), the difference is entirely explained by the sample densification.

In summary, frozen samples are interesting, but it is essential to track the sample densification from in-situ to the as-tested conditions and then to correct the measured data for that densification.

10.4.2 Undisturbed sampling and testing of clay-like soils

Besides the Duncan Dam, British Columbia has several large hydropower projects in high-seismicity areas. One of these has a horizontal peak ground acceleration assessed to reach 0.48*g*, with a requirement that the structure must return to service after 72 h following such an earthquake event. It has potential liquefaction aspects to part of the facility and was subject to an intensive and detailed site investigation. Aspects of that investigation were reported by Mohajeri and Ghafghazi (2012), with about half the laboratory testing carried out by Golder Associates.

Sediments at the project site were deposited during the Fraser glaciation and postglacial periods between 30,000 and 13,000 years ago. The glacial deposits generally consist of a very dense layer of sand, known as Quadra sand, overlain by a glacial till blanket over bedrock; these glacial units were not a concern. Interest here is in two clay-like postglacial deposits: (1) the lower clay, about 20 m of grey low-plasticity clay deposited in relatively deep water on the coastal lowlands with occasional seashells which attest to its marine origin, and (2) the interbedded clay and sand, which were deposited in a complex sedimentary sequence (because of changing sea levels amongst other factors), which is generally clay dominant ranging between thinly laminated beds to massive layers up to 10 m in thickness.

The site investigation programme used both CPT and undisturbed sampling, with the CPT data being used to target the sampling intervals. Sampling was carried out in a cased drill hole which was advanced carefully with tricone and drilling mud to the sample target depth. The casing closely followed the hole advance to assure easy cleaning of the bottom of the hole without causing additional disturbance. Sampling used a 76 mm diameter fixed piston sampler. The samples were kept in a temperature- and moisture-controlled environment for 2 weeks during the fieldwork and then shipped to the testing laboratories in isolated containers, with the 250 km distance between the project site and the laboratories equally split between slow driving and a ferry ride. The maximum accelerations applied to the sample boxes were recorded.

When the samples arrived in the testing laboratory, they were first visually assessed for damage and then gamma ray imaging was used to identify any potential defects inside the tubes. After carefully selecting the exact testing locations, the tubes were cut using manual rotary tube cutters by applying very mild pressures. The soil was then extruded using a hydraulic piston extrusion device at maximum 15 cm long sections and trimmed to the appropriate lengths and diameters using a wire cutter.

Much of the testing was carried out in the simple shear equipment. Figure 10.3 shows the as-tested void ratio after the samples were brought to the in-situ vertical effective stress and compares the void ratio to the as-recovered value, that is the value determined from the saturated water content measured in the field. Despite as good as could be reasonably expected sampling and handling protocol, substantial densification of the samples occurred, averaging about

$$\Delta e = -0.1$$

And, these samples showing substantial densification by disturbance were indeed clay like, with plasticity indices mostly in the range $10 < PI < 20$. Gradationally, these soils were about 50% silt sized and finer with little clay-sized fraction (it appears to be rock-flour-derived soil, not untypical for western Canada).

10.4.3 Correcting for sampling disturbance (void ratio matters)

The dominant thinking, generally tacit but occasionally explicit, is that if sufficiently high-quality samples are obtained, then what is measured by the laboratory strength test is reliable. It is accepted that sample quality has an effect, one commonly cited study being that by Lunne et al. (2006), who showed a strength reduction of about a third in sensitive Onsoy clay when going from Sherbrooke block samples to 54 mm diameter piston samples (and with arguably larger changes in plastic stiffness). Various correction schemes then develop, of which the stress history and normalized engineering parameters (SHANSER) approach (Ladd and Foott, 1974) is popular. The SHANSEP approach does not admit a role for void

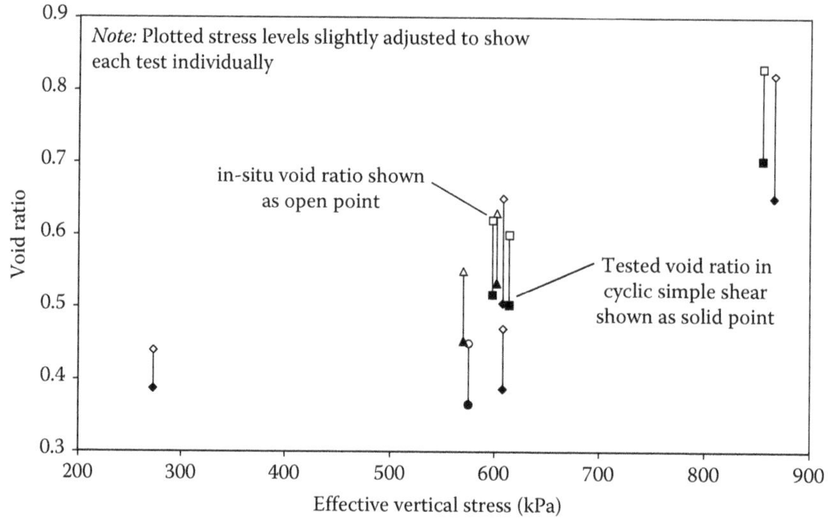

Figure 10.3 Void ratio changes between field sampling and cyclic simple shear testing of low-plasticity clay-like soil in a dam foundation.

ratio, with everything being normalized by overconsolidation ratio. Although Wroth (1975) demonstrated that the SHANSEP framework was entirely derivable from Cam Clay, this does not answer the question about how sample densification actually affects the measured strength. SHANSEP is focused on high-plasticity clays with rather high λ (or equivalently C_c) values, and Cam Clay has no concept of state other than overconsolidation, which is why it does not work for sands as discussed in Chapter 3. The following three-step approach is suitable for sands:

- Step 1 in dealing with *undisturbed* samples is to measure their in-situ void ratio (equivalent to water content assuming they come from below the water table and are saturated). This is not difficult, requiring only that the water content be measured on the trimmings from the piston or Shelby sample before sealing it for transport. Routine water content logging in the field should be standard practice and costs very little to do.
- Step 2 is to measure the void ratio as tested. By all means, test at slightly greater effective confining stress than in the field to remove some aspects of disturbance. Ideally, freeze the sample after testing (see Appendix B) for greatest accuracy in void ratio measurement.
- Step 3 is to correct for the disturbance, considering the change in state parameter. This can be empirical, from laboratory testing at a range of ψ that gives you strength change as ψ varies, or it can be modelled and computed. Figure 10.4 shows an undrained triaxial test on an *undisturbed* silt sample which has been fitted with the *NorSand* model in the downloadable spreadsheet. A good fit of model to data is evident in all aspects. Then, the void ratio (and stress) in-situ (known from Step 2) is used in *NorSand* rather than the as-tested value while changing nothing else in the model calibration (since all the properties are independent of void ratio). The in-situ strength is markedly less than the laboratory-measured strength.

Modelling laboratory tests was discussed in several chapters (Chapters 3, 6, 7 and 9), but for now, note that it is not a question of whether the material is sand, silt or clay like.

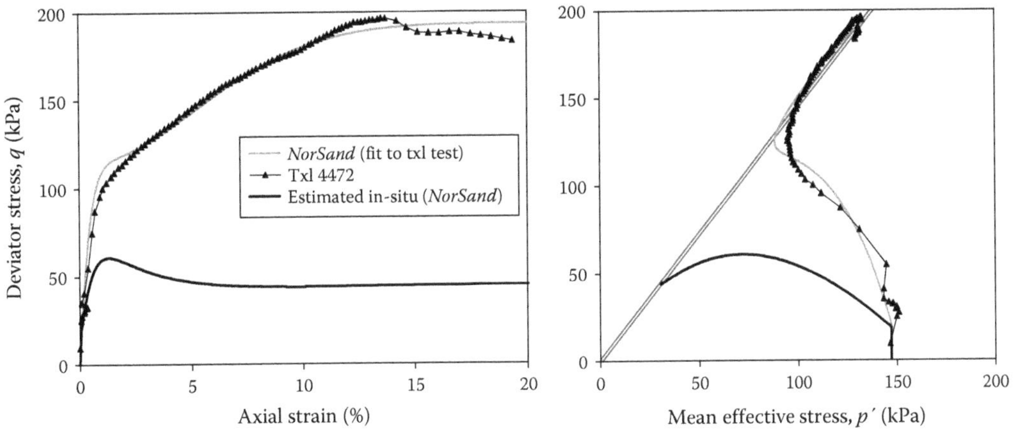

Figure 10.4 Example of *NorSand* modelling of triaxial testing of *undisturbed* silt (clay-like soil) sample to correct void ratio changes.

The soil behaviour is *automatically* simulated as a consequence of the best fit *NorSand* parameters, no matter what its grain size or other characteristics. All you need to do when you are sampling is to make sure you measure void ratio changes between as-recovered and as-tested conditions and then model the effect of that change.

Formally considering the effect of sample disturbance will show in-situ strengths markedly less than reported by the laboratory, as was observed at Duncan Dam. However, this may still be an improvement on using the cyclic strength of reconstituted bulk laboratory samples, because hopefully, your careful sampling and handling procedures will have preserved most of the in-situ fabric of the soil.

10.5 SOIL PLASTICITY AND FABRIC

Plastic strains are a very basic soil response to load, and the magnitude of these strains is controlled by work hardening. Consequently, any reasonable model for soil behaviour will have one plastic work hardening modulus to describe this aspect of the behaviour (or sometimes several moduli). This book has been based in part on the *NorSand* model in which the dimensionless plastic modulus H determines the magnitude of plastic strains for a given stress increment. Undrained soil strength is very much affected by H, as is the CPT resistance. It looks very much like H influences resistance to cyclic mobility. But how can H be measured? It is easy enough in laboratory tests as the stress–strain curve can be fitted in one test and then used to predict another. However, it is very quickly learnt that H depends on the sample preparation method, or more correctly on the particular pattern of soil particle contacts established by that sample preparation method. This approach to determining H has been used throughout the book when fitting the various laboratory tests. What about determining H in-situ?

What is becoming a conventional view in the literature is that laboratory water-pluviated samples are a reasonable model for deltaic deposits. Such laboratory-reconstituted samples could then be used to calibrate soil properties for design, and plastic hardening in particular. Although this does seem a reasonable suggestion, it is not an established fact, even though repeated citing of the suggestion has resulted in many seeing it as such. There are two obvious problems. First, one cannot presently form uniform water-pluviated samples in the laboratory with sand and silt mixtures (e.g. 80% sand, 20% silt), despite the abundance of such soils in nature. Second, water-pluviated samples in the laboratory are not especially loose, but there is no trouble forming a very loose fill by hydraulic dredging and filling. There appears to an issue of scale affecting what is achievable in the laboratory versus in the field. An alternative to laboratory tests is therefore to use an in-situ test to determine H, and both downhole plate bearing tests and the self-bored pressuremeter have been used. However, evaluation of the data has not been straightforward because both ψ and H have similar effects in terms of computed response to load, and the effects of each are not easily separated. One thought is that if the ratio of elastic and plastic moduli, H/I_r, could be linked or considered sensibly constant in a given soil, then the in-situ measurement of shear wave velocities could be a great help. This is a difficult topic, and it is presently being researched. Practically, for the moment, it seems sensible to follow the suggestion to test water-pluviated samples in the laboratory and allow some further stiffening from ageing and other geological processes in the estimation of H in-situ. Of course, one way to estimate this stiffening is to use the undisturbed sampling techniques discussed in Section 10.4, as long as proper correction is made for state parameter changes.

10.6 RELATIONSHIP TO CURRENT PRACTICE

The approach advocated for the CPT data reduction to infer in-situ state of the soil differs in almost every aspect from what is the practice within North America today. What has been presented has been firmly anchored in mechanics, especially plasticity theory, with much effort going into verification of numerical procedures and validation of the models. None of the advocated framework is outside good engineering (e.g. the framework for the evaluation of the CPT is not much different from pile bearing capacity estimation); however, it is outside usual practice for liquefaction. How has this situation come about, and does it matter?

In the Introduction, it was noted that widespread recognition of the importance of liquefaction followed the major earthquake damage from two earthquakes in 1964, one in the United States and one in Japan. Although there was awareness of the issues of liquefaction amongst some experienced engineers three decades earlier (Casagrande and others in the Corps of Engineers in particular), the intellectual direction to the work starting in the 1960s came from what has been termed the Berkeley School in this book. This school focused on a geological classification approach, in essence cataloguing what happened in certain circumstances. This classification approach was not unreasonable given the rather limited understanding of soil plasticity at the time, remembering that in 1960, stress–dilatancy theory had just emerged and critical state soil mechanics (CSSM) was at a very early stage of development. The early studies resulted in a simple design chart that expressed available cyclic strength in terms of a stress-normalized penetration resistance. With time, and further studies, additional factors were identified as affecting the assessed strength, and these factors were referred to as *corrections*. These *corrections* have remained based on inferred trends from comparing different case histories, and there has never been an underlying framework based on mechanics for any of this.

A notable, and very recent, suggestion arises from Moss (2014). He identified three sites in the Berkeley liquefaction database where the natural sand was essentially Monterey research sand, for which the CSL is known. There is also a substantial cyclic triaxial test database for Monterey sand. Figure 10.5 shows his assessment of these sites, in the context of using the state parameter as an approach to link the field case histories to laboratory testing and

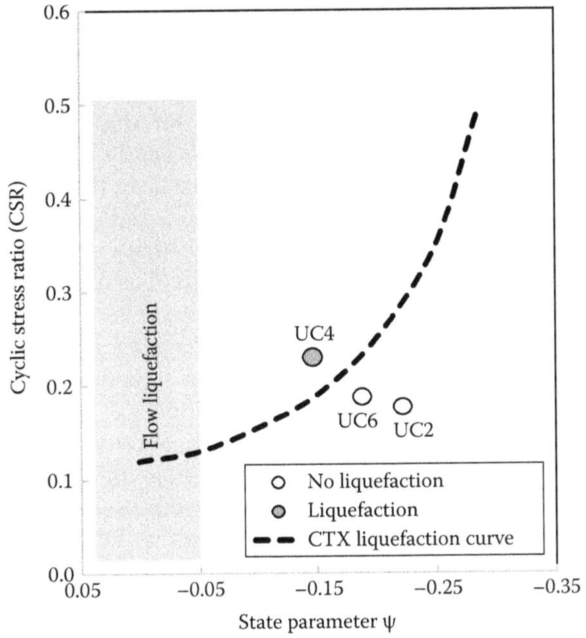

Figure 10.5 Loma Prieta case histories involving Monterey sand presented in state parameter format. (From Moss, R.E.S., A critical state framework for seismic soil liquefaction triggering using *CPT, Proceedings of the Third International Symposium on Cone Penetration Testing*, CPT14, Las Vegas, NV, pp. 477–486, May 2014. With permission.)

(by inference) the underlying mechanics. A rather fruitful avenue of research would be to measure the CSL and cyclic strength of the better documented case history soils, in the same way as Moss did for three Loma Prieta sites.

Of course, it could be argued that the Berkeley School is based on actual occurrences of liquefaction and that it is perfectly reasonable to work within that experience, but such an argument neglects that the data have rather limited range for soil type and depth of liquefaction. Working within the databases might be argued, but extrapolating outside it to large dams and silty tailings (which is what you find in North America) is simply inappropriate. Extrapolations must be anchored to trends derived from mechanics if they are to have any reliability for engineering. It does not reflect well on geotechnical engineering that we are in the situation of having to defend mechanics from geologically based extrapolation, and the approach suggested earlier by Moss (2014) is a step in the right direction.

In contrast to the Berkeley School, the state parameter is anchored in CSSM theory. CSSM is still the only complete framework for representing soil behaviour, and it is not some wacky idea put forward by one university. As we illustrated in the history of the subject in Chapter 3, CSSM is a thread in understanding how and why soil behaves as it does that has developed over 125 years and with as many important contributions from the United States (Harvard, MIT, Brown, Corps of Engineers) as the United Kingdom (Manchester, Imperial, Cambridge).

It was noted at the outset that liquefaction assessments tend to fall into two schools: (1) the situation is pretty much satisfactory, and there is as much uncertainty in the loading (earthquake, storm or ice) as in the assessed response; (2) the soil is not strong enough, and ground treatment is called for. Both of these are dominated by the question 'How dense is dense enough?'

In case (1), no cyclic model used in a soil–structure interaction analysis with a criterion of, say, 3% strain can be relied on for design. Reflection on the uncertainties involved will convince any engineer of this reality. Design proceeds by requiring no liquefaction under credible service-level loads and no catastrophic failure under extreme conditions. In case (2), most money is spent on mobilization and set up for the treatment works, with little distinguishable cost between compaction to, say, $Q_k > 70$ and $Q_k > 100$ (Q is the normalized CPT resistance, and the subscript k indicates *characteristic* in the sense of limit state codes). To some extent, the cyclic behaviour examined and discussed in Chapter 7 is an interesting aspect of soil behaviour but is not something that should drive design or assessment. Rather, it is Chapter 6 and the potential for large displacements that matters, and that is driven by the *how dense* question.

This concern for how dense is dense enough has overtones of the steady-state approach, but with two key differences. First, assessment of site or structural adequacy is based on the CPT (although requiring some laboratory testing in support of the data evaluation) with no attempt whatsoever at retrieving undisturbed samples. Second, it is not enough to be just denser than the CSL. Going through soil behaviour using constitutive modelling as guide, considering localization effects and considering the various uncertainties that must be faced in-situ (primarily K_o and plastic hardening), gives the rather easy-to-remember criterion for satisfactory behaviour: $\psi_k < -0.1$. The subscript k denotes characteristic, about the 80th–90th percentile from present knowledge. This criterion is independent of soil gradation, fabric, stress path, compressibility, etc. The identification and determination of this value of ψ in-situ have dominated this book, in particular Chapters 4 through 6.

Under circumstances of no untoward layering in the ground, reasonably careful evaluation of soil properties and the subsequent evaluation of the CPT data and consideration of possible sensitivity to plastic hardening and fabric effects, the criterion might be relaxed to $\psi_k < -0.05$. Anything looser than $\psi_k = -0.05$ implies potential flowslides. Flowslides would be regarded as unacceptable in most circumstances (e.g. unless the material is somehow contained within an impoundment).

How does the state parameter approach stack up against current engineering practice? A characteristic state in the range $-0.05 < \psi_k < -0.1$ gives an undrained phase transition deviator stress approximately equal to the drained strength. This is the same idea, although argued for different reasons, that has cropped up in the literature over the past two decades as the quasi-steady-state strength or the strength for limited liquefaction. Examples of engineers arguing for this behaviour criterion include Ishihara's Rankine lecture and the various publications by Konrad and by Vaid et al.

Finally, it seems that current practice is to steer away from finite element modelling and to rely on empirical methods and limit equilibrium stability calculations. This is a satisfactory approach to calculate embankment or foundation stability only after you have shown that liquefaction (whether triggered by seismic or static loads) is not a concern. But as soon as there may be some softer zones in the ground, or the possibility of spreading of pore pressures from loose zones, or concerns about displacements, you need to take it a step further and start to use a numerical model with implementation of a good constitutive model. We hope we have shown that the FE method is not so difficult and that it provides a quantum leap in understanding compared to back-analysis using limit equilibrium. While we would not expect this level of analysis in routine consulting practice, we encourage more geotechnical engineers to adopt coupled finite element analyses and researchers and teachers to disseminate their work into the commercial modelling platforms to make it possible. Our industry needs a mechanics-based approach to liquefaction assessment so that we can stop using fines content as a major input to analyses.

10.7 WHAT NEXT?

Looking to the future, areas where further research could contribute usefully to liquefaction assessments have been identified. Consulting engineers have much to offer, and contributions to understanding should not remain the domain of universities alone. Because soils are variable, a lot of examples are needed so that trends can stand out from variability. Obtaining lots of test data assembled into geologically understood contexts is something that is difficult for university researchers to do because they tend to have access to a few research sites and to test few samples in great detail. The consulting community, on the other hand, sees many sites and much test data in the normal course of business. That is how the utility of the state parameter became obvious and lead to the 1985 paper introducing state parameter. Perhaps the best way forward is more joint industry–university research.

10.8 DO DOWNLOAD!

What has been presented in this book is soundly based on constitutive modelling using ideas in applied mechanics that extend back more than a century, and with as many substantive contributions from the United States as from the United Kingdom. But that brings us back to the Russian Proverb "trust is wonderful, distrust is better": We have provided a large amount of data and open-source code as a downloadable companion to this book so that you do not have to trust us. Do download!

A further reason to download, flippancy aside, is that a lot of the plots are in colour,and many are dynamic so the effects of changing properties can be seen. These downloads are more than a 'resource' and will greatly help you to the, road to Damascus, moment that generalized CSSM via ψ is the answer to geotechnical questions (and in contrast to engineering geology, aka 'fines content').

Finally, we hope that this book has been an enjoyable read as well as providing all that is needed to apply the state parameter approach in engineering practice.

Appendix A: Stress and strain measures

This appendix overviews the stress and strain measures used throughout this book. Much of what follows is standard, but there are a few things that may be unfamiliar and which are important.

Because the approach to liquefaction looks towards numerical modelling, and indeed numerical analyses are used for some aspects, it is helpful to get away from friction angles in favour of stress invariants. Most ideas in soil mechanics have developed in the context of the triaxial test, and for this test, the usual stress invariants are

$$\bar{p} = \frac{(\bar{\sigma}_1 + \bar{\sigma}_2 + \bar{\sigma}_3)}{3} \tag{A.1}$$

$$q = (\bar{\sigma}_1 - \bar{\sigma}_3) \tag{A.2}$$

The two invariants are commonly referred to as the *mean stress* ($\bar{p}$ or p') and the *deviator stress* (q). The bar superscript on stresses denotes effective where $\bar{\sigma}_1 = \sigma_1 - u$ and so forth (u being the pore water pressure). The corresponding strain invariants associated with these stress measures are

$$\dot{\varepsilon}_v = \dot{\varepsilon}_1 + 2\dot{\varepsilon}_3 \tag{A.3}$$

$$\dot{\varepsilon}_q = \frac{2}{3}(\dot{\varepsilon}_1 - \dot{\varepsilon}_3) \tag{A.4}$$

The strain definitions are associated with the convention that length reduction is a positive strain, so that positive volumetric strain ε_v is associated with void ratio reduction. The dot superscript on strains denotes increment.

These stress and strain measures are *work conjugate*, which is no more than a fancy term for the equivalence:

$$q\dot{\varepsilon}_q + \bar{p}\dot{\varepsilon}_v = \bar{\sigma}_1\dot{\varepsilon}_1 + \bar{\sigma}_2\dot{\varepsilon}_2 + \bar{\sigma}_3\dot{\varepsilon}_3 \tag{A.5}$$

When looking at soil behaviour, it is usually the relative amount of deviator to mean stress that matters, and it is usual to adopt the stress ratio:

$$\eta = \frac{q}{\bar{p}} \tag{A.6}$$

The strain measure matching the stress ratio η is the *dilatancy* or *dilation rate*, defined as

$$D = \frac{\dot{\varepsilon}_v}{\dot{\varepsilon}_q} \tag{A.7}$$

Dilation D is expressed in terms of strain increments, not strains themselves. It is not usual to use a dot superscript on D, even though D is a ratio of strain increments. Further, because of the compression positive convention, the usual phraseology that dense soils 'dilate' actually corresponds to negative values of D.

As useful as the triaxial test is (and it has provided the basis for understanding soil), more general stress and strain measures must be introduced for useful engineering. This generalization of stress and strain measures is essential to encompass plane strain, which is by far the most common practical situation, and for the implementation of useful models in finite element codes. It is helpful to work in terms of the usual invariants of plasticity theory. These invariants trace back to Lode (1926), but were brought to the attention of the English-speaking world by Nayak and Zienkiewicz (1972). Following Zienkiewicz and Naylor (1971), the deviatoric stress q is generalized as the invariant $\bar{\sigma}_q$, where

$$\bar{\sigma}_q = \left[\frac{1}{2} s_{ij} s_{ij}\right]^{1/2} \quad \text{with } s_{ij} = \sigma_{ij} - \delta_{ij}\bar{\sigma}_m \tag{A.8}$$

and

$$\bar{\sigma}_m = \frac{(\bar{\sigma}_1 + \bar{\sigma}_2 + \bar{\sigma}_3)}{3} \tag{A.9}$$

These stress invariants are ubiquitous in modern numerical approaches to modelling soils (e.g. Smith and Griffiths, 1988). The mean effective stress is only a change of notation (i.e. $\bar{\sigma}_m \equiv \bar{p}$). The deviatoric invariant can be written in terms of the principal effective stresses as

$$\bar{\sigma}_q = \left[\frac{1}{2}(\bar{\sigma}_1 - \bar{\sigma}_2)^2 + \frac{1}{2}(\bar{\sigma}_2 - \bar{\sigma}_3)^2 + \frac{1}{2}(\bar{\sigma}_3 - \bar{\sigma}_1)^2\right]^{1/2} \tag{A.10}$$

The familiar triaxial stress invariant q is given by (A.10) under triaxial conditions.

Of course, there are three principal stresses, so these stresses cannot be reduced to two invariants without losing information. A third invariant is needed, conveniently taken as the Lode angle, θ:

$$\theta = \frac{1}{3}\arcsin\left(\frac{13.5 s_1 s_2 s_3}{\bar{\sigma}_q^{\,3}}\right) \quad \text{with } s_1 = \frac{(2\bar{\sigma}_1 - \bar{\sigma}_2 - \bar{\sigma}_3)}{3}, \text{etc.} \tag{A.11}$$

Triaxial compression conditions correspond to θ = 30° with triaxial extension being θ = −30°. Plane strain depends on the intermediate principal stress developed during straining and typically lies in the range 15° < θ < 20°. Plane strain is not a fixed stress condition like the triaxial test.

There is an information loss in reducing σ_1, σ_2, σ_3 to invariants as a direction is associated with each of the principal stresses. In general, it will be desirable to include measures of the direction of the three principal stresses to the chosen coordinate frame of reference (typically, a rectangular Cartesian frame denoted as *x*, *y*, *z*). Of course, σ_1, σ_2, σ_3 can readily be recovered from the stress invariants:

$$\bar{\sigma}_1 = \bar{\sigma}_m - \frac{2}{3}\bar{\sigma}_q \sin(\theta - 120) \tag{A.12a}$$

$$\bar{\sigma}_2 = \bar{\sigma}_m - \frac{2}{3}\bar{\sigma}_q \sin(\theta) \tag{A.12b}$$

$$\bar{\sigma}_3 = \bar{\sigma}_m - \frac{2}{3}\bar{\sigma}_q \sin(\theta + 120) \tag{A.12c}$$

The familiar stress ratio η continues to be defined as

$$\eta = \frac{\bar{\sigma}_q}{\bar{\sigma}_m} \tag{A.13}$$

Turning to generalized strain measures, critical state models are based on postulates as how the work done by stresses on an element of soil is stored or dissipated. With such a work-based fundamental approach, it is obviously necessary that stresses and strain must be expressed in work conjugate measures (the strain rate invariants used by Naylor and Zienkiewicz are not work conjugate), as otherwise the theory becomes inconsistent and so causing ambiguities in dilation rates, stress dilatancy, etc. (see Jefferies and Shuttle, 2002). The issue, then, becomes that of defining the work conjugate shear strain rate measure $\dot{\gamma}_q$ (the triaxial $\dot{\varepsilon}_v$ is already a proper and general strain rate invariant). Stating the equivalence from work conjugacy that

$$\bar{\sigma}_q\dot{\gamma}_q + \bar{\sigma}_m\dot{\varepsilon}_v = \bar{\sigma}_1\dot{\varepsilon}_1 + \bar{\sigma}_2\dot{\varepsilon}_2 + \bar{\sigma}_3\dot{\varepsilon}_3 \tag{A.14}$$

gives, on rearranging, the appropriate work conjugate shear strain measure:

$$\dot{\gamma}_q = \frac{s_1\dot{\varepsilon}_1 + s_2\dot{\varepsilon}_2 + s_3\dot{\varepsilon}_3}{\bar{\sigma}_q} \tag{A.15a}$$

On substituting the principal stresses, Equations A.11, A.12 and A.15a may be written as

$$\dot{\gamma}_q = \frac{1}{3}\left(\left(\sin\theta + \sqrt{3}\cos\theta\right)\dot{\varepsilon}_1 - 2\sin\theta\dot{\varepsilon}_2 + \left(\sin\theta - \sqrt{3}\cos\theta\right)\dot{\varepsilon}_3\right) \tag{A.15b}$$

The strain measure $\dot{\gamma}_q$ reduces to the triaxial variable $\dot{\varepsilon}_q$ under triaxial conditions. This strain measure was originally introduced by Resende and Martin (1985).

As (A.15b) is linear, the usual elastic plastic decomposition of strain $\dot{\gamma}_q = \dot{\gamma}_q^e + \dot{\gamma}_q^p$ can be used. Critical state models may then be generalized using the chosen stress invariants, while preserving the postulated work dissipation basis, provided that plastic dilatancy is defined as

$$D^p = \frac{\dot{\varepsilon}_v^p}{\dot{\gamma}_q^p} \tag{A.16}$$

Just as with η, the general measure of dilation from (A.16) reduces to the familiar triaxial variable under triaxial conditions.

Appendix B: Laboratory testing to determine the critical state of sands

Ken Been and Roberto Olivera

B.1 OVERVIEW

The critical state line (CSL) is commonly determined in the laboratory by means of triaxial tests. Paths followed during conventional isotropically consolidated undrained (CIU) and isotropically consolidated drained (CID) testing are shown on a state plot in Figure B.1 for both initially loose and dense conditions.

The loose specimen shown in Figure B.1 with a solid circle has an initial positive state parameter. When shear undrained, the tendency of the sample to contract will result in positive pore pressures and a decrease in mean effective stress. Undrained conditions restrict volume change resulting in a horizontal shift of the state plot towards the CSL (path A in Figure B.1). When shear drained, the specimen will tend to decrease its volume resulting in a reduction in void ratio accompanied with an increase in the mean effective stress with increasing deviatoric load, resulting in a shift towards the bottom right on a state plot (path B in Figure B.1). The behaviour of a dense specimen with an initial negative state parameter is also illustrated in Figure B.1 for common CID and CIU tests with paths C and D, respectively.

The ideal laboratory test program will be targeted at determining behaviour for a range of initial positive and negative state parameters; however, there are practical difficulties associated with testing a range of initial conditions. As will be shown shortly (Section B.6), achieving a desired density after consolidation could prove challenging for certain soils when specimens are reconstituted to a loose state. Also, when shear drained, dense specimens may develop shear bands with deformation and volume changes occurring along complex localization zones that depend on testing conditions (Desrues et al., 1996). Experimental evidence presented by Desrues et al. (1996) using computer tomography indicates that the void ratio within the shear zone (termed the local void ratio by Desrues et al.) tends towards the critical state. This local void ratio is greater than the global average measured conventionally in the laboratory using the volume of the test specimen. The practical consequence of this behaviour is illustrated in Figure B.1, with path C, which plots a global average void ratio that indicates that the CSL determined using loose specimens is not reached with dense soils.

Loose samples do not form shear planes and do not have the tendency to localization that is normal in dense (dilatant) sands. Originally, the standard protocol followed Castro and concentrated on undrained tests. Undrained tests are more convenient and should always be the starting point for the practical reason that the strains required to reach the critical state are well within the limits of triaxial equipment for loose samples. Small strains result

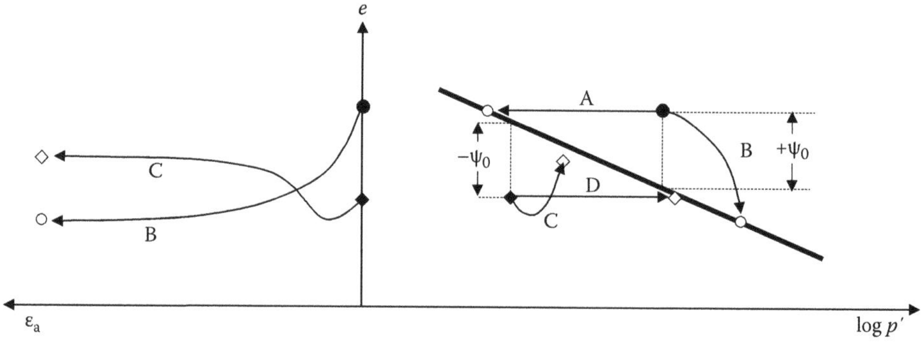

Figure B.1 Stress–strain–void ratio paths for conventional CID and CIU tests.

in large pore pressure changes, and therefore undrained samples can change state (i.e. move to the critical state) relatively quickly.

However, it turns out that it is difficult to obtain data on the CSL above about $p' = 400$ kPa with undrained tests, as it is necessary to consolidate the sample to $p' = 2$ MPa or more prior to shearing. Such high pressures are both inconvenient for most commercial triaxial equipment and often involve grain crushing effects. Drained tests are therefore used as well as undrained. In drained tests on loose samples, the sample moves to the critical state at a much slower rate, and displacements to the limits of the triaxial equipment are required.

The preferred method of determining the CSL is a series of triaxial compression tests on loose samples, generally markedly looser than the critical state. The number of tests needed depends on various factors including the amount of material available and more generally on budgetary and schedule restrictions. The range of confining pressures will also depend on the problem at hand. For liquefaction studies, interest is generally concentrated on shallow depths and hence relatively low confining pressures; however, other projects may be concerned with determining behaviour at high stresses like pressures below a dam.

The range of testing stresses and number of tests should be evaluated on a case-by-case basis. However, a good starting point and one commonly used is to run three loose tests under undrained conditions (CIU) at initial confining pressures of 100, 200 and 400 kPa, and one loose drained test (CID) at an initial pressure of about 400 kPa (see Figure B.2). This selection of initial test conditions will typically provide a definition of the CSL in the 10–700 kPa range, which is common in many geotechnical engineering problems. Additional tests may be carried out to refine the CSL, expand the range of initial state parameters or expand the stress range as required.

Successful CSL testing is dependent on getting certain details of the triaxial testing correct:

- Uniform samples must be prepared in a suitably loose state at a predetermined void ratio (the operator must be able to achieve a desired void ratio).
- When reconstituting samples, the soil must be thoroughly mixed at the predetermined water content.
- Samples must be fully saturated.
- The void ratio must be known accurately (to within about ±0.003).
- The measurement system must be capable of measuring low stresses and pore pressures at a high rate with very little system compliance (a 'liquefied' sample may be at a mean effective stress of ≈1 kPa, derived as the difference between a measured total stress of 300 kPa and pore pressure of 299 kPa).

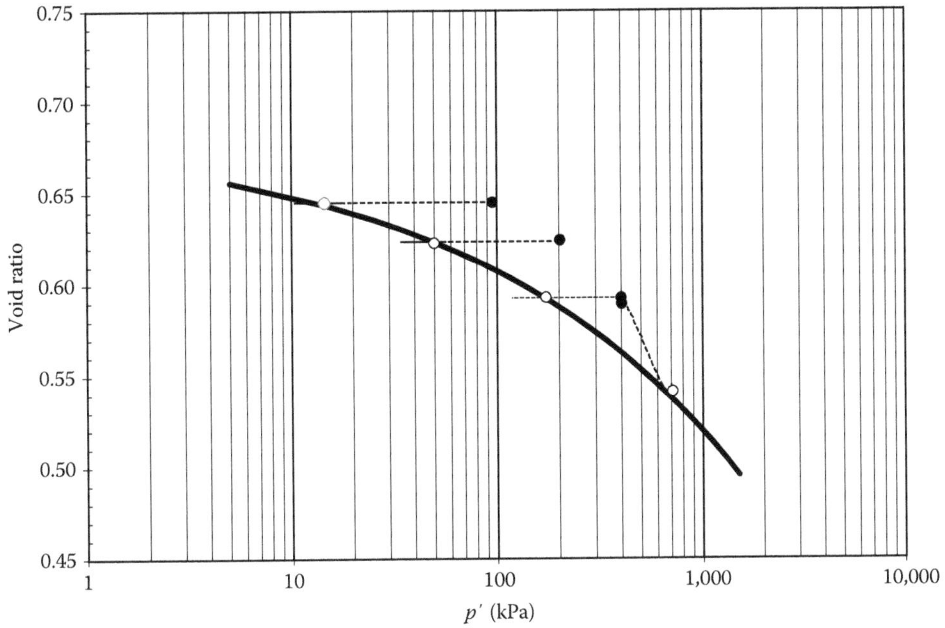

Figure B.2 State diagram for initial tests to determine the CSL using four specimens.

Each of these aspects is covered in some detail in the following sections. However, a familiarity with standard laboratory practice is assumed, in particular a good understanding of triaxial testing methods and equipment. There are good textbooks and papers that address the laboratory testing techniques that are discussed in this chapter. Bishop and Henkel (1962) is a geotechnical classic on the triaxial test, while the *ASTM Symposium on Advanced Triaxial Testing of Soil and Rock* in 1986 (Donaghe et al., 1988) contains many excellent papers. A significant step in computer-controlled testing was the development of the hydraulic cell for controlled stress path testing by Bishop and Wesley (1975). What we present in this appendix are the additional aspects that you need to be aware of, and implement, to measure the critical state reliably.

B.2 EQUIPMENT

A detailed description about the laboratory testing equipment and requirements is found in testing standards including ASTM D4767 for Consolidated Undrained Triaxial Compression tests and D7181 for Consolidated Drained Triaxial Compression Tests. Other standards (e.g. British, Australian and Norwegian) may be consulted as well. What follows are recommended modifications to the standard equipment and specific equipment modifications for critical state testing.

B.2.1 Computer control

The use of computer-controlled volume/flow pumps and load frame is highly recommended for critical state testing. Computer-controlled pumps provide added accuracy in the measurement of volumes and pressures over the classical pressure control panel, where water volumes are measured from burettes. The use of this equipment used to be limited to research institutions, but nowadays it can be found in most commercial laboratories. The initial capital investment is higher; however, testing is more efficient, and with practice, testing time of a single specimen can be reduced to a couple of days, as the test and measurements are continuous (24 h a day) throughout the test. With time, improved accuracy, quality and efficiency can offset the initial investment.

B.2.2 Platens

For sand testing, lubricated end platens are essential to reduce the influence of platen restraint on stresses in the sample and on non-uniformity of strains (Rowe and Barden, 1964). The use of enlarged platens is also recommended, as dilative samples may expand their volume radially beyond the initial sample diameter.

A simple system is illustrated in Figure B.3. The lubricated end consists of two discs of standard triaxial latex membrane, with a thin layer of silicone grease sandwiched between them. The platens should ideally be some 5 mm larger in diameter than the sample to allow uniform radial strains at the ends of the sample. Naturally, the lubricating discs mean that a full-sized porous stone cannot be used, but this is not a problem as sands are relatively permeable. A 30 mm diameter porous stone in the centre of a 71 mm sample is quite adequate and must be inset into the platen as illustrated. A disadvantage with the use of a central porous disc is that properties measured during the consolidation stage (i.e. coefficient of consolidation and permeability) cannot be determined as the flow boundary is modified.

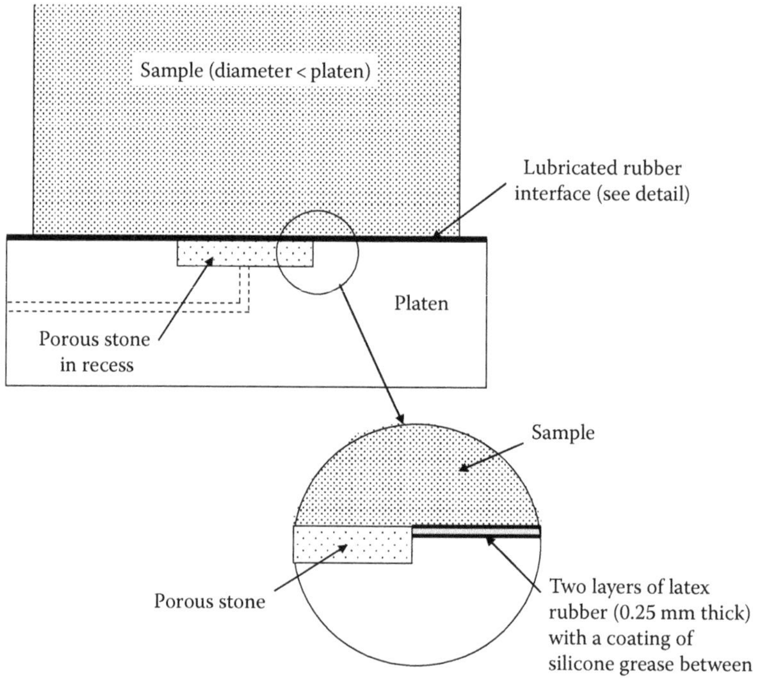

Figure B.3 Lubricated end platen for triaxial testing of sands.

B.2.3 Axial load measurement

Load measurement should use an internal load cell to avoid friction in the piston bearings affecting the results. This is especially important for loose sand samples where the deviator stress at the critical state may be in the order of 1–5 kPa. Another tactic we have employed, which avoids the issue of waterproofing load cells and electrical connections through a pressurized triaxial cell, is to measure loads at the bottom of the sample. This requires a piston below the lower platen, through the cell base, resting on a load cell. This piston does not move (other than for compliance of the load cell itself, which is minimal), and therefore friction is minimized, as illustrated in Figure B.4.

B.2.4 Compaction mould

Sample preparation requires the use of a non-standard compaction mould with modifications to accommodate enlarged platens and allow reconstitution. A schematic of such a modified preparation mould is presented in Figure B.5, which can be used for moist tamping and dry pluviation. This mould can also be used for wet pluviation and slurry deposition with minor modifications. The internal surface of the mould can be provided with a grid of small groves that will help distributing the vacuum and allow the triaxial membrane to attach to the mould when reconstituting the specimen.

B.2.5 Tamper

Although tampers available in most laboratories are standard, the schematic (Figure B.6) shows a modification that allows controlling the layer height when reconstituting samples. The rod is set up with a couple of clamps that allow fixing the drop height of the tamper inside the mould to a predetermined level providing the desired layer thickness. This tamper eliminates the need to draw marks on the membrane to indicate the height of individual layers and has been found by the authors to accelerate the sample reconstitution process. This type of tamper is useful for implementing the sample reconstitution process presented in Section B.3.5 and can be used with the compaction mould shown in Figure B.5.

Figure B.4 Load cell underneath triaxial chamber to minimize piston friction effects (note bowl 478 for observing CO_2 bubbling through sample).

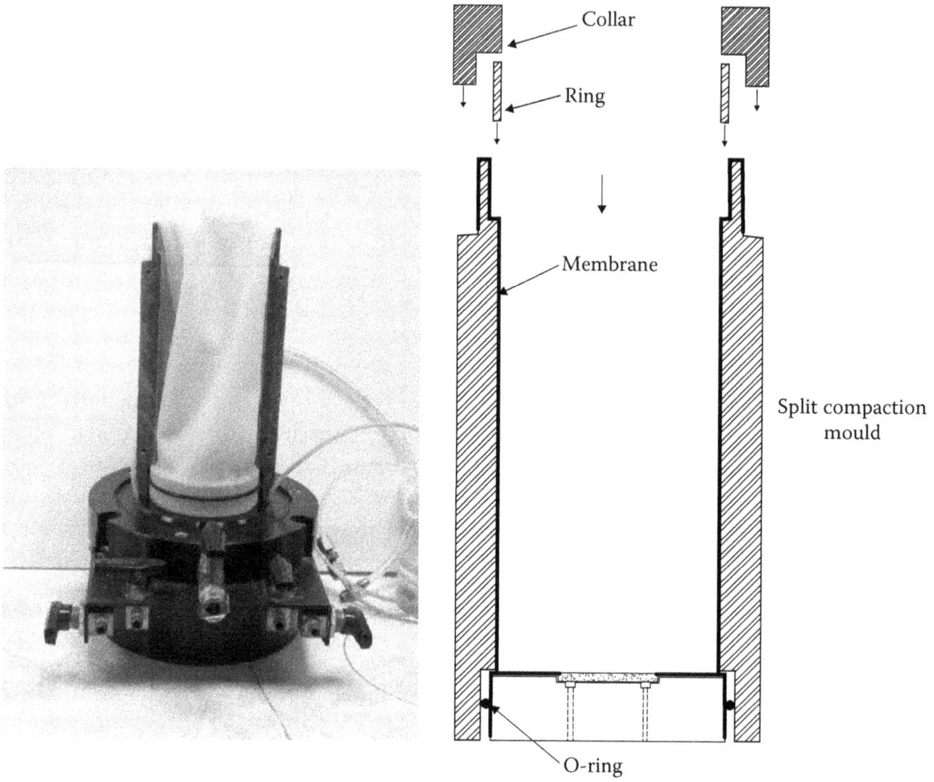

Figure B.5 Sample preparation mould to accommodate enlarged platens (note ring, with same diameter as top platen, between upper collar and mould to hold membrane in place).

B.3 SAMPLE PREPARATION

Sample preparation has much concerned laboratory testers of sands for many years. In particular, studies in the mid-1970s showed how specimen preparation markedly affected the cyclic strength of sands. There are now many different methods of sample preparation, with minor differences in detail between laboratories.

Much of the concern with sample preparation has been its major effect on the behaviour of the sample during the test. Ladd (1977), for example, made this point very clearly for cyclic resistance in triaxial tests (Figure B.7). Since then, many other workers, in particular Vaid and Thomas (1995), have shown that liquefaction under monotonic loading is markedly affected by soil fabric and stress path to failure. This is because the method of specimen preparation determines the structure, or fabric, of the sand. These aspects have been tested in a thorough program of experimentation on Erksak 330/0.7 sand, reported in Been et al. (1991), and are discussed later. However, for the CSL determination, this is not a major concern. The critical state is reached only after the initial structure has been destroyed, and the sample reaches a very different particle arrangement at large strains. The main concern for specimen preparation in the CSL testing is therefore that uniform samples are obtained at predetermined void ratios.

Moist tamping is the easiest method of preparation to achieve a full range of densities and is therefore described in detail. Wet and dry pluviation techniques are also described briefly, as they are useful techniques for preparing samples for other testing to determine

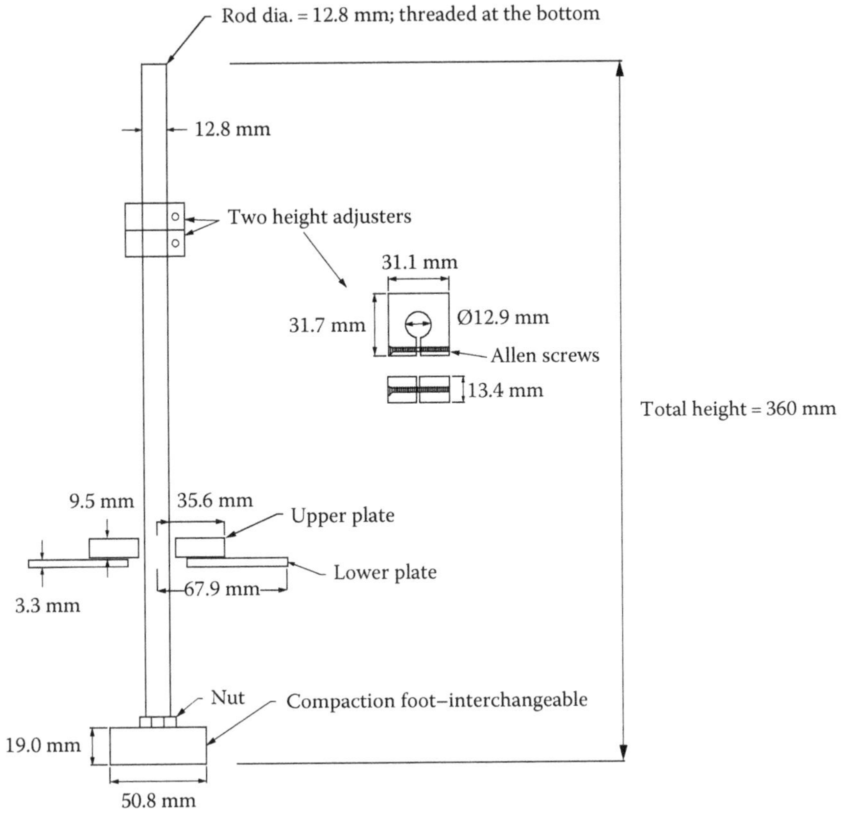

Figure B.6 Compaction hammer to control lift height during sample compaction.

design parameters. Figure B.8 illustrates these methods schematically. A variation on wet pluviation, called the slurry deposition method by Kuerbis and Vaid (1988), is also described for completeness.

It is useful before dealing with specimen preparation to repeat here the sequence of preparation for the testing of samples in the triaxial test:

- Prepare platens by placing double-lubricated rubber membranes.
- Assemble base of triaxial cell with lower platen.
- Place membrane and split mould over lower platen.
- Draw membrane to mould with vacuum.
- Deposit or tamp sample into mould.
- Place upper platen on formed sample and attach membrane to platen.
- Apply partial vacuum (negative pore pressure) to sample to keep its shape.
- Remove mould.
- Measure sample height and diameter (to determine density).
- Assemble remainder of triaxial cell.
- Saturate sample (always keeping a positive effective stress).
- Check saturation with a B-value measurement.
- Consolidate sample to the desired stress level.
- Run test (extension/compression/cyclic/etc.).

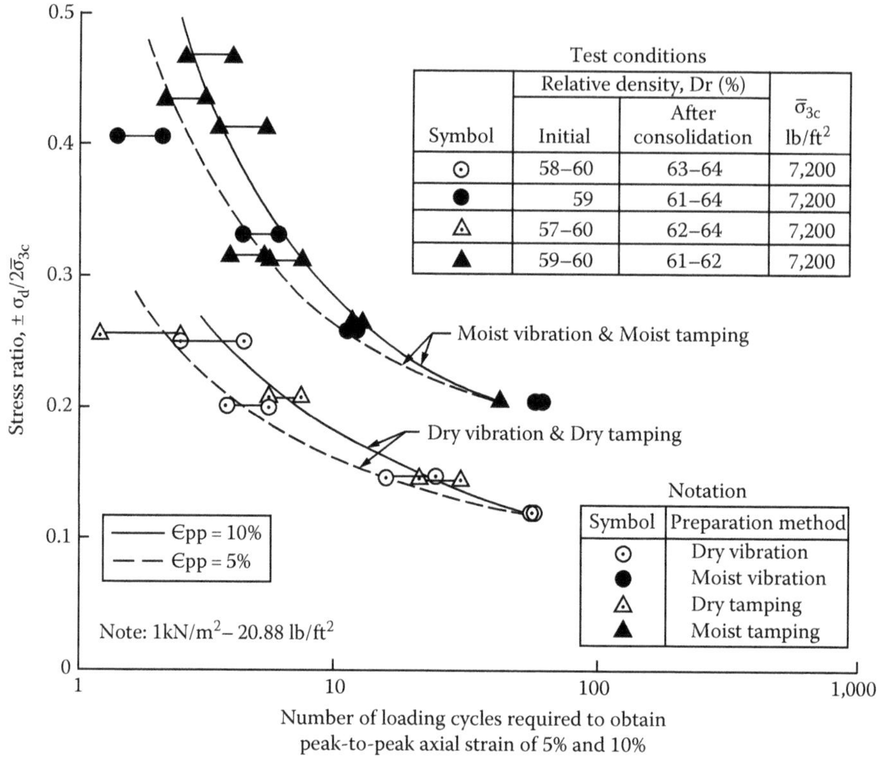

Figure B.7 Effect of sample preparation on the cyclic resistance of sand samples. (From Ladd, R.S., *J. Geotech. Eng. Div., ASCE*, 103, 535, 1977. With permission from ASCE.)

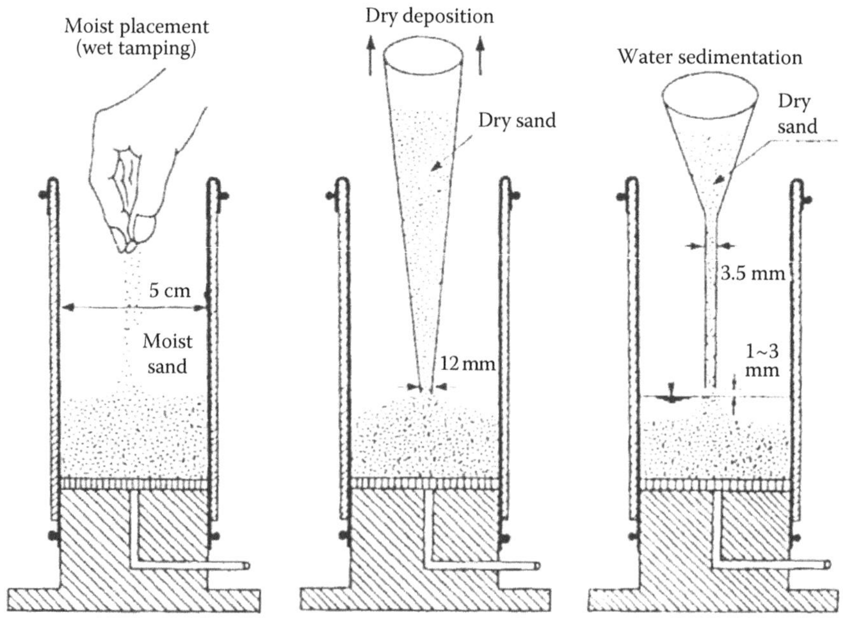

Figure B.8 Illustration of sample preparation methods for clean sands. (From Ishihara, K., *Géotechnique*, 43(3), 349, 1993. With permission from Institution of Civil Engineers and Prof. Ishihara.)

At the end of the test, the sample density is once again checked preferably using the freezing technique as described later. This post-test measurement is usually considered the most accurate density determination because the process of saturation during sample preparation leads to a volume change of the sample that is difficult to measure in the laboratory. Ignoring this volume change, especially for loose samples, can lead to a systematic error in the determination of the CSL, although the error is not as big as suggested by Sladen and Handford (1987).

B.3.1 Moist tamping method

Moist tamping uses a moisture content of about 5%, which results in capillary forces between the sand grains and allows bulking of the sand to low densities not achievable with wet or dry samples. The effective stress induced in the sample by the capillary forces also helps to keep the sample shape once the split mould is removed. The sample is prepared in six (or more) equal layers of equal density.

The first step in specimen preparation is to calculate the target void ratio of the test and then work back to the target preparation void ratio based on estimates of volume changes during saturation and consolidation. From the target void ratio and the size of the specimen mould, the dry density and dry weight of sand for the sample is calculated. The most useful equation needed here is the relationship between void ratio and dry density:

$$1+e=G_s\left(\frac{\gamma_w}{\gamma_d}\right) \tag{B.1}$$

and therefore if γ_d is expressed in units of water density (1000 kg/m^3), the equations are

$$e=\frac{G_s}{\gamma_d}-1 \quad \text{and} \quad \gamma_d=\frac{G_s}{(1+e)} \tag{B.2}$$

Once the target dry weight of the sample has been computed, the specimen is prepared as follows:

- **Preweigh** six equal portions of oven-dried sand material (i.e. one-sixth of the calculated total dry sample weight) into six preparation dishes.
- **Mix** the material in each dish with distilled water to give a moisture content of about 5% and allow to cure for a minimum of 16 h (overnight, in a humid room or with a lid on the dish is ideal to allow the fines in the sample to become properly 'wet').
- **Tamp** the first layer into the membrane-lined mould. A technique is needed to ensure that the layer is not over compacted; it should end up being exactly one-sixth of the sample height. A tamper with an adjustable stop is illustrated in Figure B.6. Before tamping, the stop is adjusted so that the tamping foot stops at exactly the top of the layer. The entire sixth portion of the sample is placed in the mould, and the sample layer is gently tamped until the tamper hits the stop at every stroke.
- **Scarify** the top of the tamped layer gently to avoid a smooth planar surface between layers.
- **Repeat** the layer deposition and tamping process until all six layers are formed.

B.3.2 Wet pluviation

Many workers have pointed out that moist tamping results in a specimen fabric or structure that is dissimilar to that which will be obtained in nature, and therefore pluviation techniques of sample preparation are preferable. It is no doubt true that moist tamping is not representative of natural sand deposition, but this raises the question of whether any laboratory sample preparation method is representative. Pluviation in the quiescent laboratory conditions is unlikely to be similar to underwater deposition in rivers and seabeds where strong currents are usually active at the time of deposition. The assertion that wet pluviation matches in-situ conditions is presently only speculation. Nevertheless, wet pluviation is a useful sample preparation technique when samples without any pre-consolidation due to capillary tension, or samples with a different fabric from moist tamping, are required. It is, however, difficult to control the ultimate void ratio of a pluviated sample. The steps for sample preparation are as follows:

- Calculate the total dry weight of sand for the target void ratio.
- Weigh out a single oven-dried sample of the correct amount.
- Place dry sample in a long-necked flask.
- Add de-aired water to fill the flask.
- Apply a vacuum to the top of the flask to ensure saturation of the sample.
- Leave sample to cure for several hours.
- Fill the membrane-lined sample mould with de-aired water.
- With a thumb over the neck of the flask, invert the flask and insert the neck into the water in the mould to approximately 25 mm above the bottom of the mould.
- Remove thumb (if it has managed to stay in place with insertion that deep into the mould!). The sand will now gradually flow out of the flask under gravity, and excess water will flow up into the flask to replace the sand.
- Allow the sand to pluviate like this while moving the neck of the flask slowly and continuously in a circular motion. The neck should be kept at a constant height of about 25 mm above the top of the forming sample.
- Remove flask when all the sand has pluviated out.

At this stage, the top of the sample will hopefully be above the final target height, and gentle tapping of mould will densify the sand to the correct height. (If the top of the sample is below the target sample height, there is not much one can do other than to accept a denser sample or start preparation of the sample from scratch.)

Some minor additions to the preparation equipment are useful. In particular, it is useful to have an insert to extend the height of the forming mould. Initial deposition looser than the target density will then not result in overflowing and loss of sand. Once the sample has been tapped down to the correct height, the excess water at the surface has to be removed. An ear (as in medical) syringe is ideal for this purpose.

It is important to note that wet pluviation does result in some fines loss from the sand, roughly 50%. Thus, a sample that starts with 2% fines may end up with only 1% fines after pluviation into the mould. This should be accounted for in the density and dry weight calculations, and it is also advisable to check the final fines content of the sample after the test. (The fines is usually neatly collected in the flask during pluviation and can also be weighed or 'recycled' with future samples.)

B.3.3 Slurry deposition

The slurry deposition method was developed by Kuerbis and Vaid (1988) mainly to overcome the problem of particle segregation in poorly graded or silty sand samples. First, the silt or clay fines must be separated from the sand. The coarse and fine fractions are then mixed with water and boiled to de-air the mixtures.

The sample is initially prepared in a mixing tube with a slightly smaller diameter than the final sample. The fines mixture is poured into the mixing tube, and the fines is allowed to settle before pluviating the sand mixture into the tube as described earlier. The tube, which is now full of water, needs to be sealed closed. The bottom of the tube should initially have been sealed with a rubber stopper, while the top is covered with a porous stone (de-aired), a thin metal plate and a stretched rubber membrane. The porous stone will ultimately be placed on the bottom platen of the triaxial cell. To maintain saturation, Kuerbis and Vaid recommend that sealing of the mixing tube be carried out in a water bath. The sample is then mixed by vigorously rotating the mixing tube for about 20 min until a completely homogenous sample is obtained.

Next, the mixed slurry must be transferred to the triaxial cell. The mixing tube is placed on the lower platen, which must be in a water bath so that the stretched membrane can be rolled back and the steel plate removed leaving the porous stone held in place by water tension. Once the triaxial testing membrane has been stretched over this assembly, the assembly can be removed from the water bath and a split mould assembled around the mixing tube. Vacuum is applied to stretch the membrane to the sides of the mould, and water is added to the gap between the mould and the mixing tube. The rubber stopper on the top of the mixing tube is then removed to release the water tension, and the mixing tube is withdrawn slowly leaving a uniform very loose sand slurry in the mould.

The sample can be densified by gentle vibration or tapping as described earlier for pluvation, and the cell assembly completed as usual.

B.3.4 Dry pluviation

Dry pluviation is mentioned here as it is a commonly used and reliable method to achieve a uniform density in clean sands. By close control on the rate of deposition and the drop height of the sand, a range of densities (a range of density index of about 30%–70%) can be achieved with the technique. An assessment of the technique and factors affecting the density is provided by Rad and Tumay (1987). This sample preparation method needs more sophisticated equipment than moist tamping or wet pluviation, and there are therefore a number of variants of the method.

The principle of dry pluviation is that the correct dry weight of sand is contained in a hopper of the same diameter as the sample mould. This hopper is placed directly above the sample mould. Sand is then allowed to pluviate through a diffuser, for example, a coarse mesh sieve, into the mould. Drop height is controlled by ensuring the diffuser is at a constant height above the sample surface. Pluviation rate is controlled by the size and the number of holes in the bottom of the hopper.

While dry pluviation results in the most uniform sample compared to wet pluviation and moist tamping, its application is limited. Sands with plastic fines cannot be prepared this way as the drying process coagulates the fines. It is also difficult to prepare very loose samples with this technique. Finally, as with moist tamping, unmeasurable volume changes may occur during saturation of the samples.

B.3.5 Recommended sample reconstitution procedure

The recommended procedure for sample preparation is presented in this section. It follows the moist tamping technique previously described adapted for the use of six (or more) layers of equal volume.

The sand specimen is prepared at a moisture content of about 5%, although this initial content may vary and could be brought closer to optimum in the case of dense specimens. The sample is thoroughly mixed and is stored in a covered container for a minimum of 16 h prior to compaction.

An important aspect of sample reconstitution involves achieving a uniform sample density throughout the specimen. When sand is compacted in layers, the compaction of each succeeding layer can further densify the sand below it. These aspects have been identified by Ladd (1978), who proposed the under-compaction method consisting of compacting the bottom layers to a lower density than the final desired value by a predetermined amount defined as percent under-compaction, U_n. The value of U_n in each layer is linearly varied from the bottom to the top layer, with the bottom first layer having the maximum U_n value (Ladd, 1978).

Under-compaction becomes important when reconstituting loose specimens, and as previously described, reconstituting specimens to densities looser than critical is recommended for defining the CSL. Determining the percent under-compaction is an empirical procedure, but after a couple of trials, and with practice, the laboratory operator will develop a sense as to what this number should be. It may be necessary to reconstitute one or two trial samples before the test program is initiated.

The procedure presented by Ladd has been modified to consider n layers of constant height. First, the total weight of moist sample (W_{mt}) required to achieve the desired density or void ratio is calculated using the known specimen (mould) volume. Then, the weight per layer is calculated using the following equation:

$$\frac{W_{mt}}{n}\left[1+\frac{U_n}{100}*\left(\frac{2n_i-n+1}{n-1}\right)\right] \quad \text{for } i=1\ldots n \text{ and with bottom layer } i=1 \qquad \text{(B.3)}$$

Figure B.9 shows a schematic with the sample reconstitution process summarized as follows:

- Prepare top and bottom platens. Clean the surfaces and place a thin coat of vacuum grease on the surface avoiding touching and clogging the porous stones. Place the first latex membrane on the prepared surface and place a second coat of vacuum grease on the latex membrane. Place the second latex membrane on top of the first membrane. The vacuum grease should be sticky enough to allow the two latex membranes in place throughout the reconstitution process.
- Mount the compaction mould and triaxial membrane around the bottom platen and apply vacuum to stretch the triaxial membrane against the walls of the mould. Some moulds are provided with a grid of small groves that will help distributing the vacuum throughout the internal face of the mould.
- Set the base of the tamper to a predefined height using one or two lightweight (acrylic or aluminium) spacers.
- Pour the first portion of wet soil into the membrane-lined mould calculated with Equation B.3. Distribute the soil around the base of the mould with the aid of a long wire or needle. It is important to thoroughly mix the soil to ensure that the particle gradation in each layer is as uniform as possible and to avoid segregation.
- Tamp the first layer into the membrane-lined mould.
- Scarify the top of the tamped layer gently to avoid a smooth planar surface between layers.
- Repeat the layer deposition and tamping process until all layers are formed.
- Remove the mould by applying vacuum to the sample through the drainage lines.
- Measure the dimensions of the specimen using a caliper with a minimum of three height measurements (120° apart) and at least three diameter measurements at the

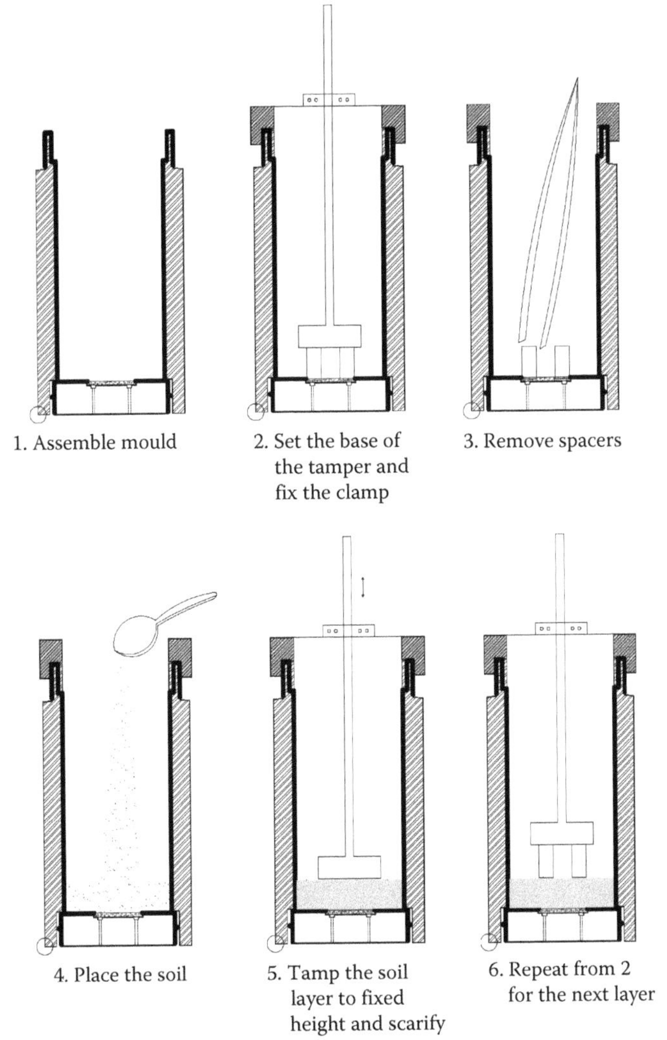
1. Assemble mould
2. Set the base of the tamper and fix the clamp
3. Remove spacers
4. Place the soil
5. Tamp the soil layer to fixed height and scarify
6. Repeat from 2 for the next layer

Figure B.9 Illustration of recommended sample preparation method (moist tamping and under-compaction of lower layers).

quarter points of the height. A pi tape is recommended to measure the circumference rather than a caliper. The standard here is to have individual measurements of height or diameter not varying from average by more than 5%. The effectiveness of the under-compaction method and the selected degree of under-compaction can be verified at this stage with the variation in the circumference measurements.

B.4 SAMPLE SATURATION

There are a number of techniques to aid saturation of sand samples prepared dry or moist, some of which are described later. Saturation is important in that constant-volume (or undrained) conditions cannot be assumed unless the sample is fully saturated.

The basic saturation process for specimens prepared dry or moist is that de-aired water is flushed through the sample (always from bottom to top) to displace the air. This process does not result in full saturation. The water pressure is then increased gradually, which results in both a reduction of the volume of air due to compression and increased dissolution. The degree of saturation is checked by carrying out a 'B-test' in which a step increment in total cell pressure (σ_3) is applied with the sample undrained, and the corresponding increment in pore pressure (u) is measured. Skempton's B value is then determined as

$$B = \frac{\Delta u}{\Delta \sigma_3} \tag{B.4}$$

In a fully saturated sample where the water is incompressible compared to the soil skeleton, B should be 1. In practice, there is some compliance in the test apparatus, and sand samples are not as compressible relative to water as clay samples. Therefore, a B of about 0.97 is achievable and recommended as a target to indicate full saturation.

In general, the larger the grain size of a sand and the less fines it contains, the easier it is to saturate. Sands with a D_{50} of less than 0.200 mm and a fines content of 5% or more can be difficult to saturate, and back pressures of 400 kPa or more may be required to achieve a B value of 0.97.

B.4.1 Carbon dioxide treatment

Familiarity with the safe use and handling of CO_2 is assumed. Carbon dioxide is many times more soluble in water than air is soluble in water. One method to reduce the time and back pressure for saturation is to bubble CO_2 through the sample prior to saturation. A low-volume and low-pressure CO_2 source, controlled through a needle valve from a regular gas bottle and regulator, is connected to the lower platen water line. The CO_2 is bubbled through the sample after the top platen and membrane have been assembled, the sample mould removed and a nominal confining stress applied to the sample. The CO_2 is vented through a thin tube from the top platen, which is best left with its open end under water to observe the bubbles. A bubble rate of one to five bubbles per second is about right, with the process lasting 1–2 h.

Failure to vent the CO_2 will result in a pore pressure build-up and collapse of the sample. Too large a flux of CO_2 will result in the CO_2 piping and flowing up preferential pathways through the sand, rather than displacing the air. (It is, of course, important to note that CO_2 is denser than air and therefore bubbling from the bottom is effective.)

It is helpful to start the CO_2 bubbling process during specimen preparation. The sand is tamped, or pluviated, into a sample mould that essentially contains CO_2 rather than air.

Some laboratories are opposed to the use of CO_2, sometimes because of practical reasons (i.e. mobile laboratories with little space or temporal laboratories set in remote locations). In some cases, and depending on the gradation (permeability) of the specimen, it is still possible to achieve a good saturation without the use of CO_2, but in general, higher back pressures would be required. In some other cases, however (i.e. samples with low permeabilities and/or high densities), saturation will not be possible without the use of CO_2, and flushing times in excess of 2 h (and up to 24 h) may be required to limit back pressures to a practical level.

B.4.2 Saturation under vacuum

A more complicated, but nevertheless effective, sample saturation technique is to conduct the flushing process under a vacuum. Figure B.10 illustrates the apparatus required for this technique. The sample and two de-aired water containers are all attached to the same

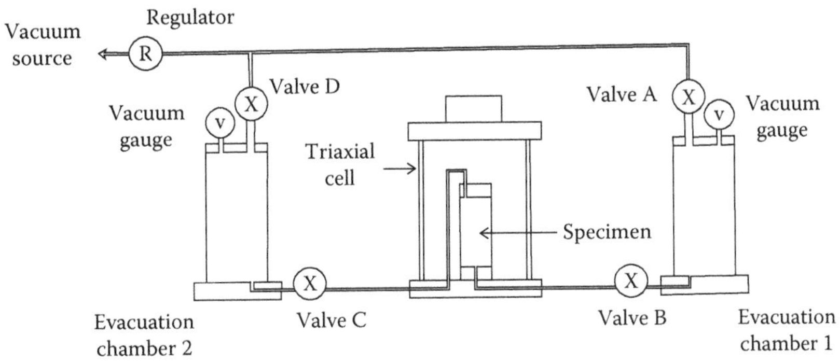

Figure B.10 Illustration of vacuum saturation apparatus for triaxial sample preparation. (Reproduced from Shen, C.K. and Lee, K.-M., A study of hydraulic fill performance in Hong Kong, GEO Report No. 40, Report to Geotechnical Engineering Office of the Hong Kong Government, Hong Kong University of Science and Technology, Hong Kong, 1995. With permission Head of Geotechnical Engineering Office and Director of Civil Engineering Department, Hong Kong SAR Government.)

vacuum line. Water then flows under gravity from the container connected to the lower platen to the upper platen (by placing the containers at different heights if necessary). An enhancement to this scheme may be to use graduated burettes for the water so that accurate measurements of volumes of water in and out of the sample are obtained. In addition, a differential pressure regulator between the source and waste containers could be used to provide a greater driving pressure across the sample.

A disadvantage of this system is that the sample is in effect over-consolidated by the effective stresses induced by the applied vacuum to the sample. This is not a major problem as test consolidation pressures are usually well above the maximum vacuum-induced stress (theoretically about 100 kPa).

B.5 VOID RATIO DETERMINATION

Measurement of void ratio of sand samples in the triaxial test can be subject to potentially large errors, especially for loose samples. Some of these errors and suggested methods to circumvent poor resolution in measurements are presented by Vaid and Sivathalayan (1996b). While it is a relatively simple matter to determine initial sample dimensions and the dry weight of the sample, it is the volume changes during sample saturation and consolidation (membrane penetration effects) that can lead to large errors if they are ignored.

The final volume (V_f) is used to calculate the void ratio at the critical state and is obtained from the initial volume after sample reconstitution (V_o). The volume changes during the lifetime of a triaxial sample are

$$V_f = V_o + \Delta V_T \tag{B.5}$$

where ΔV_T is an incremental volume change which combines different changes during saturation, consolidation and shear as follows:

$$\Delta V_T = \Delta V_{sat} + (\Delta V_c + \Delta V_m) + \Delta V_s \tag{B.6}$$

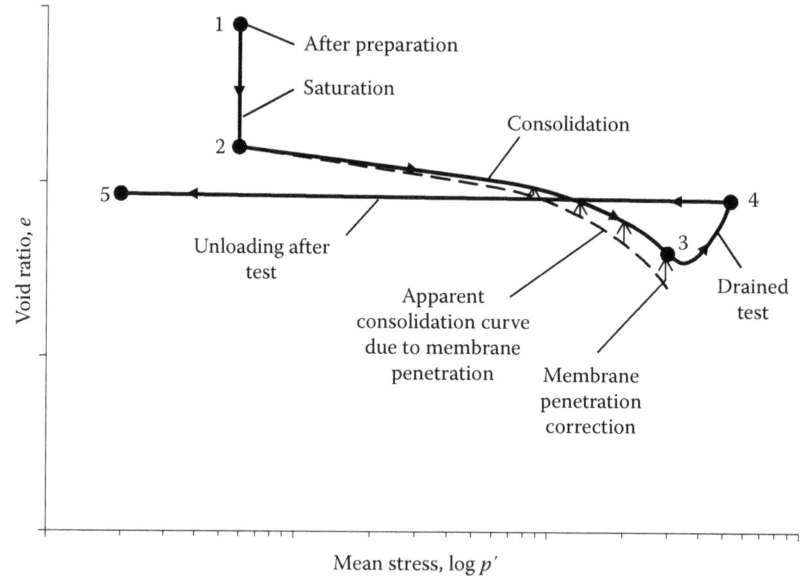

Figure B.11 Volume changes during triaxial sample lifetime (for a drained test on a dilatant sample).

where ΔV_{sat} is the incremental volume change during saturation including sample flushing and back pressurization, the quantity in parenthesis is the change during consolidation (including ΔV_c, the incremental volume change caused by soil deformation, and ΔV_m caused by membrane penetration effects), and ΔV_s is the incremental volume change during shear (note that for undrained tests, ΔV_s would be zero).

Combining (B.5) and (B.6) gives the final volume used for determining the void ratio at the end of the test:

$$V_f = V_o + \Delta V_{sat} + (\Delta V_c + \Delta V_m) + \Delta V_s \tag{B.7}$$

The changes are illustrated schematically in Figure B.11. High-resolution measurements of ΔV_{sat} are difficult to obtain as the sample is not yet saturated and hence measurements of pore water leaving or entering the sample are not available at this stage. All other quantities in the right-hand side of (B.7) are measured except for ΔV_m, which can be estimated using some of the procedures described in Section B.5.2.

B.5.1 Volume changes during saturation (ΔV_{sat})

Samples undergo strains during saturation as a result of the changes in effective stresses. Effective stress changes are induced by changes in the external applied stresses and by the release of surface tension effects in moist sands. Volume changes during saturation are particularly difficult to measure. Sladen and Hanford (1987) illustrate how significant errors may be if volume change during saturation is ignored (Figure B.12). The filled circles represent the CSL using void ratio determined on initial sample dimensions, without accounting for any volume change during saturation. Open circles represent the CSL using the void ratio after testing determined by freezing the samples (as described later in this section).

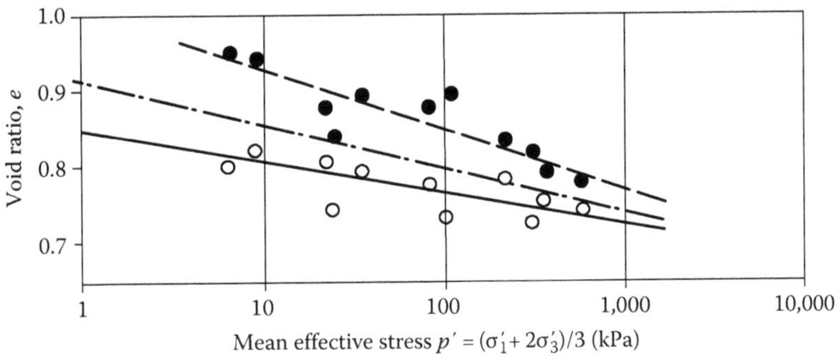

Figure B.12 Potential error in void ratio if volume changes during saturation are not considered. (From Sladen, J.A. and Handford, G., *Can. Geotech. J.*, 24(3), 462, 1987. With permission from NRC of Canada). Filled circles – no volume change considered; open circles – void ratio measured on frozen samples after test; intermediate line – assuming isotropic strains and measured axial strain.

Typical practice at the time was to estimate the volume change by measuring the axial strain, which would give the intermediate broken line in Figure B.12. An error in terms of void ratio of about 0.04 is quite possible, and this represents an error in relative density of about 10% for the Syncrude sand tested by Sladen and Handford.

The water flow from the saturation burette or pressure pump into the sample (minus the flow out of the sample) is not equal to the void volume of the sample as air in the water lines and porous stones is also displaced by water. Another common technique is to measure the axial shortening of the sample during saturation, which is easily done by the piston displacement when the sample is in a triaxial cell, and then to assume that the volumetric strain is isotropic. The total volumetric strain is thus three times the axial strain. However, experience has shown that this assumption is not at all valid. Prepared sand specimens are generally anisotropic, and this anisotropy has a strong influence at the low effective stresses associated with sample saturation.

A solution to the problem of sample volume changes during saturation is to measure the sample volume directly after saturation, using a pi-tape for diameter and vernier calipers for height. This is possible only if the sample is not yet in a triaxial cell, whereas the triaxial cell is needed to apply the necessary confining stress during saturation. The double vacuum container saturation method described earlier in this chapter does however open the possibility of saturating the sample outside the triaxial cell, measuring the volume directly after saturation and then assembling the cell before releasing the vacuum. The only assumption in this method is that the volume change which takes place as the back pressure is increased is negligible. The assumption is probably reasonable – most of the volume change occurs during CO_2 and water flushing.

Another approximate method is to measure the sample volume change by measuring the change in the triaxial cell volume. This requires de-aired water in the chamber and flushing of the lines to minimize bubbles. The cell has to be calibrated so that corrections can be made to the measured cell volume for creep/expansion of the outer cell jacket and piston movements. The 'corrected' cell volume change is then assumed to be equal to the sample volume change during saturation. Experience is that sample volume changes during saturation calculated using this method can be in good agreement with those calculated using the freezing method presented later, generally within ±2 cm^3, which translate in a void ratio variation of ±0.01 between the two methods.

One of the best ways to deal with volume change during saturation is to measure the sample volume at the end of the test (V_f in Equation B.7) and then calculate ΔV_{sat}. By the end of the tests, the sample is quite distorted and irregular in shape. Sladen and Handford (1987) suggest that the sample is frozen, undrained, in the cell after the test, removed while frozen and the water content then determined. This method is accurate. Some modifications to the triaxial cell are required for the method to be used without damage to the equipment due to expansion of the water on freezing. This method cannot be used if the platens are set up with electronic equipment like bender elements, as freezing will damage these components.

The procedure to determine sample void ratio using the freezing method is as follows:

Immediately on completion of the test, isolate the sample drainage by closing valves as close to the top and bottom platens as possible. (In general, these valves would need to be installed especially for this purpose.)

Reduce the cell pressure gradually. Because the sample is saturated, a negative pore pressure is developed in the sample.

Drain the cell.

Disassemble as much of the cell as possible without breaking the 'seal' on the sample and dry the outside membrane to remove any free water.

Place the sample and associated cell parts in a freezer. A cell design that has the water line from both the top and bottom platens feeding into the cell base is ideal, as then only the base plate, sample and platens need to be frozen. A typical sample after freezing is shown in Figure B.13.

Allow the sample to freeze. This will need some trials when starting out, as the aim is to reduce the sample temperature to −3°C which is sufficient to immobilize the pore

Figure B.13 Triaxial specimen after compression test and freezing.

water but avoids getting the sample so cold that it becomes difficult to remove the platens (if you use a domestic chest freezer, which is our laboratory practice, and leave the sample overnight, it will freeze to –18°C, which is far too cold).

Once the sample is frozen, it can be removed intact, with the same water content as at the end of the test. Note that the volumetric expansion of water on freezing is not an issue, as only the water content is used, not the sample volume.

Compute the sample end-of-test void ratio from the water content, assuming 100% saturation and the measured specific gravity of the solids, G_s.

B.5.2 Membrane penetration correction

When effective confining pressure is applied to a sample of sand through a rubber membrane, the membrane deforms and is pushed into the pore spaces between the grains. This results in expulsion of some pore water from the sample, without a change in void ratio of the sample. Thus, the measured volume change during consolidation must be corrected for membrane penetration when void ratio is calculated.

There are a number of theoretical studies (Molenkamp and Luger, 1981; Baldi and Nova, 1984; Kramer et al., 1990) summarized by Ali et al. (1995), suggesting the form of the equation for membrane penetration. For practical purposes, membrane penetration can be quantified in terms of a normalized membrane penetration:

$$\varepsilon_m = \frac{\Delta V_m}{A_s \log(p'_1/p'_2)} \tag{B.8}$$

where

ε_m is the normalized membrane penetration
ΔV_m is the volume change due to membrane penetration
A_s is the sample area covered by the membrane ($2\pi rh$ for a cylindrical sample)
p'_1, p'_2 are net pressure acting across the membrane before and after the volume change

For sands, ε_m is primarily dependent on grain size, assuming other factors such as membrane thickness and modulus are constant. Figure B.14 summarizes data for ε_m for a range of sands. Using the appropriate value of ε_m, the volume change associated with membrane penetration can be calculated for a given sample area and change in net pressure. The void ratio 'correction' for membrane penetration is thus given by

$$\Delta e = (1+e_o)\frac{\Delta V_m}{V_o} = (1+e_o)\frac{\varepsilon_m A_s \log(p'_1/p'_2)}{V_o} \tag{B.9}$$

Because the membrane penetration correction is so dependent on sand type and testing equipment, it is advisable in each laboratory test program to measure membrane penetration directly. There are at least three methods to do this relatively simply:

- Carry out an isotropic consolidation and rebound test. Vaid and Negussey (1982) have shown that strains are generally isotropic during the rebound part of the test, and thus the volume change due to membrane penetration can be calculated approximately from measurements of axial strain and total volume change made during the rebound portion of the test.

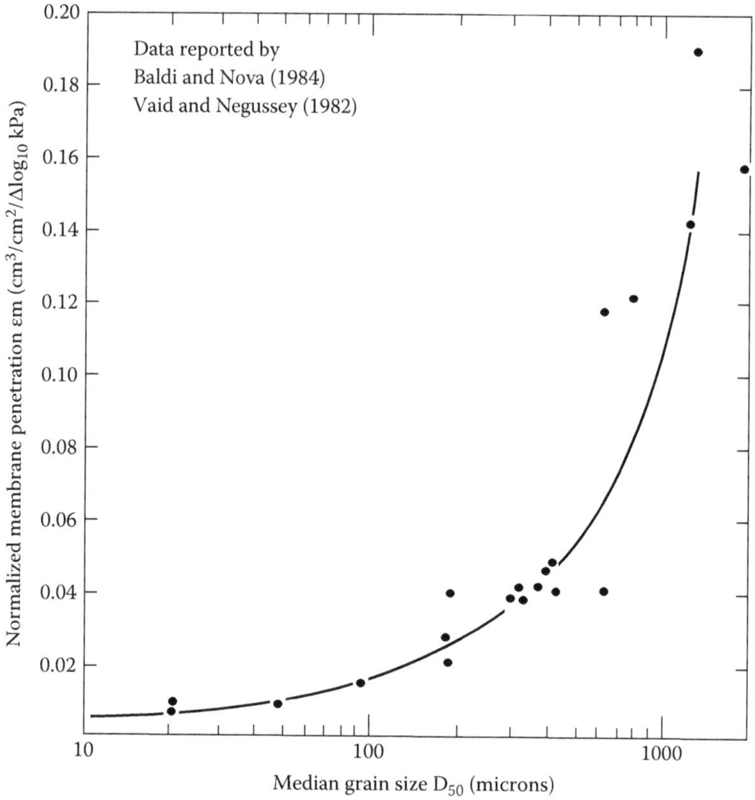

Figure B.14 Normalized membrane penetration coefficient as a function of median grain size.

- Prepare a cylindrical sand sample around a steel insert with a diameter of about 6 mm less than the full sample. The total volume of this thin (3 mm thick) cylindrical sample of sand is small, and therefore the volume change on unloading and reloading can be assumed to be small, compared to membrane penetration. This special sample is then 'consolidated' in the normal way in the triaxial cell, but the measured volume change is due entirely to membrane penetration effects.
- Prepare a sample in the normal way, but replace the fine material in the sample structure by cement (Ali et al., 1995). The surface texture of this sample will represent the real sample, but the volume change with applied pressure should be negligible because of cementation. As with the steel insert method, the volume change during a consolidation test is then attributed solely to membrane penetration.

B.6 DATA REDUCTION

The principal responsibility of the testing laboratory is to deliver quality-assured data, which, for a triaxial test, will comprise a series of measurements of axial strain, volumetric strain and associated stresses. A printed record is a minimum requirement, but most testing is best reported as text files with the data in columns.

The reported stresses must be corrected for sample cross-section evolution as the cylindrical sample is compressed ('area correction'). Critical state testing generally goes to larger displacements than standard for the triaxial test, with large axial strains (about 20%) commonly being encountered, and the sample cross-section changes to match. Computed stress must allow for the changing area over which load is applied. This is one reason that lubricated end platens are necessary, as they assist in the sample maintaining a 'cylindrical' geometry throughout the test and which leads to the deviator stress being computed as

$$q = \frac{\text{Deviator load}}{A_o} \times \frac{(1-\varepsilon_v)}{(1-\varepsilon_a)} \tag{B.10}$$

where A_o is the initial cross-section area of the sample at the start of shear, and for undrained tests, $\varepsilon_v = 0$. When the sample shape is not cylindrical, other corrections may be applied, but this is not common practice. The simplest alternative correction is for a 'bulging' sample, in which case a factor, typically 1.5, is applied to the ε_a term, so that

$$q = \frac{\text{Deviator load}}{A_o} \times \frac{(1-\varepsilon_v)}{(1-1.5\varepsilon_a)} \tag{B.11}$$

Parabolic equations for the area correction exist, but such sophistication is hardly justified when the deviator load is measured only at the end of the sample.

For very soft clays, a membrane stiffness correction may also be applied, and the same is true for sand samples which reach the critical state at stresses of only a few kPa. However, when a thin membrane is used on a 75 mm diameter sample, the correction is generally negligible. A recommended approach, given the other uncertainties in the membrane correction, is to calculate the correction and to apply it only if it is greater than 5% of the deviator stress. For convenience, the membrane correction is given by

$$\Delta q = \frac{4E_m t_m \varepsilon_a}{D} \tag{B.12}$$

In (B.12), E_m and t_m are the membrane modulus and thickness, respectively, and D is the sample diameter. For reference, 1400 kPa is a reasonable modulus for latex, but you should always measure it directly on each batch of membranes.

Returning to the data reported by the laboratory, some laboratories prefer to report effective axial and radial stresses accompanying the measured strains, while others standardize on deviator and effective radial stress. It does not matter which convention is used, but that convention must be indicated in the data file by naming each column of data as well as indicating the units of measurement. It is good practice to make the first column of data the elapsed time in the test. Generally, our laboratories have not reported void ratio as that is implicit in the volumetric strain and the measured void ratio at the start of shearing. But it is certainly acceptable to add void ratio as a further data column.

Each data file must record basic 'housekeeping' items such as project number, sample number, soil gradation, G_s and date tested. ASTM Standard D7181 gives a good guide to these items. We also like to add a note with a reference to the area correction equation used.

The related task of taking a set of test data and developing the soil properties, 'interpretation', is normally viewed as involving engineering judgement and carried out by the geotechnical engineer using the data, not by the test laboratory. Chapter 9 provides guidance on the 'interpretation' through a worked example.

Appendix C: *NorSand* derivations

PREAMBLE

This appendix derives the various equations that comprise *NorSand*, and substantiates the equations found in the various VBA subroutines of the downloadable spreadsheets. The *NorSandTxl.xls* spreadsheet implements *NorSand* drained and undrained triaxial tests, and is set up to model laboratory data from such tests – it is the starting point to look at the model and see how it works. Of course, triaxial tests have particular symmetry and are not general for engineering; so, there is a second spreadsheet, *NorSandM.xls*, which is set up for more general monotonic stress paths (hence the 'M' in the file name) including drained and undrained simple shear as well as the corresponding plane strain paths in Cornforth's apparatus. This second spreadsheet is there for interest, but it is not a 'production' modelling tool. Its main use has been to generate the verification cases for the finite element implementations of *NorSand*. Finally, there is a third spreadsheet *NorSandPSR.xls*, which is set up to simulate and model the now popular cyclic simple shear test based on adding in principal stress rotation (hence the 'PSR') to the framework.

Of course, a reasonable question would be: why use spreadsheets? This comes down to the fact that all proper plasticity models give a current soil stiffness that varies along the loading path, not a stress–strain curve per se. Producing stress–strain curves to compare with laboratory test results requires that the stiffness be integrated along the imposed stress path, and that means numerical integration. As numerical integration is reasonably easy to do in the Excel VBA environment, VBA routines become a core component to understanding soil behaviour. The routines that go with this appendix are well commented and the code itself is written in a 'plain English' style. It may be helpful to view the VBA code as you read (work through…) this appendix.

In what follows, the superscripts e, p refer to elastic and plastic strains, respectively, and where the usual strain decomposition $\varepsilon = \varepsilon^e + \varepsilon^p$ is used. The 'dot' notation is used over a variable to denote increments of that variable. The remainder of the notation used is for general 3D stress conditions, rather than the particular conditions of the triaxial test (see Appendix A).

All *NorSand* properties are defined under triaxial compression conditions, emphasized using the subscript 'tc', with the internal physical idealizations in *NorSand* generalizing these properties for 3D stress states. Nearly all *NorSand* properties are dimensionless numbers (i.e. with no physical units); the exception is the elastic shear modulus G_{max}, where it seemed better to stick with the familiar engineering usage and express this modulus in the units of stress (i.e. as MPa).

C.1 EVOLUTION OF *NorSand*

NorSand has evolved since the initial triaxial variant (Jefferies, 1993). Apart from the generalization from triaxial to 3D (Jefferies and Shuttle, 2002), the evolution has included the following: the introduction of the soil state–dilatancy property χ; a simplification in the representation of the critical friction ratio M; and changes in the treatment of the volumetric coupling property N. A slightly subtle modification has been the introduction of engineering strain as the default output for the spreadsheets. These issues are briefly discussed here. What then follows is an exposition of the current development of *NorSand* without looking back to how it got to this stage in any detail.

C.1.1 State–dilatancy (χ_{tc})

NorSand grew out of experience with large-scale hydraulic fill construction in the Canadian offshore. Extensive testing of various sand gradations was used to support the offshore construction, and this resulted in the original proposition of Been and Jefferies (1985), which asserted that maximum dilation rate was independent of sand gradation and depended only on the initial value of the state parameter (i.e. ψ_o). This proposition was embedded in the original *NorSand* (Jefferies, 1993).

Although the Beaufort Sea sands, and the 'standard' laboratory sands also tested at that time, were thought to have a reasonable gradation range, this was in the context of hydraulic sand fills. Such construction sands are rather uniformly graded with differences between them mostly lying in the silt content (ranging from 0% to about 15%). In 1997, *NorSand* was used in the remediation of Bennett Dam, and that involved testing well-graded silty sands. Testing these silty sands rapidly showed that the uniformity of a soil changed the effect of void ratio on maximum dilation rate – not entirely surprising in that if there is less void space occurs with well-graded soils, then the effect of void ratio change (dilation) is amplified. But linear trends were still found and resulted in the slope of the D_{min}-ψ trend being introduced as a new soil property χ for the general 3D version of *NorSand* (Jefferies and Shuttle, 2002). There was a subtlety here too though.

In *NorSand*, χ is used to control the hardening limit. Any yield surface has a single void ratio associated with it, but mean stress changes as you move around the yield surface, which has the effect of making ψ also change. Since the concept of a hardening limit is that it is fixed for the current soil state, a particular choice of ψ is needed to define a unique yield surface for the current soil state. As *NorSand* is a critical state model, the obvious choice was the state parameter at the current 'image' condition on the yield surface: ψ_i (see Section C.3). Thus, the 3D development of *NorSand* invoked χ_i as the soil property where

$$D_{min} = D^p_{min} = \chi_i \psi_i \tag{C.1}$$

Defining the state–dilatancy property as χ_i is elegant but causes two difficulties. First, state–dilatancy is a general soil behaviour regardless of whether you choose to represent that behaviour with *NorSand* or an alternative model; but, χ_i is *NorSand* specific since it assumes a particular yield surface shape (the isotropic and Cam Clay like 'bullet'). In principle, true soil properties ought to be widely accepted and as general as possible – χ_i does not meet such a test. Second, it was found that many laboratory technicians had difficulty using the formula for computing ψ_i as it involved more processing of data than you could do with a simple calculator. Engineering practice demanded something simpler than χ_i. These dual demands resulted in the state–dilatancy property χ_{tc} being defined as the slope of a trend line through drained triaxial compression data when plotted as D_{min} versus the concurrent

ψ at D_{min}; this is the form used in Chapter 2 (see Figure 2.13) and where χ_{tc} is now the soil property. A D_{min} versus ψ at D_{min} plot is universal since it does not assume any shape of the yield surface or CSL (although you must know what the CSL is from testing), and this form is not specific to any constitutive model. Further, computing ψ at D_{min} is straightforward as all that is needed from the laboratory technician is to present void ratio evolution with axial strain alongside the volumetric strain data.

Although one could use χ_{tc} directly in *NorSand* (by working out the value of ψ at the current cap), this is not particularly elegant as everything else is centred on the image condition. A better approach is to use a mapping to convert the input soil property χ_{tc} to the internal soil property χ_i; this is done as follows.

Since D_{min} is the same in either approach, we have (for triaxial compression):

$$\chi_i \psi_i = \chi_{tc} \psi_{D\min} \tag{C.2}$$

For a semi-log CSL of slope λ_e,

$$\psi_i = \psi_{D\min} - \lambda_e \ln\left[\left(\frac{p_i}{p}\right)_{\max}\right]$$

Introducing the limiting dilation into the *NorSand* hardening limit (Equation C.26),

$$\psi_i = \psi_{D\min} - \frac{\lambda_e \chi_{tc} \psi_{D\min}}{M_{itc}}$$

which gives after substituting in (C.1) and rearranging

$$\chi_i = \frac{\chi_{tc}}{(1 - \lambda_e \chi_{tc} / M_{itc})} \tag{C.3a}$$

Equation C.3a shows that for typical values of soil properties, $\chi_i \approx 1.1\chi_{tc}$. This is a small shift in the soil property value between the two definitions, but one might as well be accurate. However, there is a small catch as M_{itc} itself depends on χ_i (Section C.1.3) so that, strictly, you need the bisection algorithm to solve for χ_i. Given the accuracy to which soil properties are determined (and which will be validated anyway as part of model calibration), a simpler approximation is sufficient:

$$\chi_i = \frac{\chi_{tc}}{(1 - \lambda_e \chi_{tc} / M_{tc})} \tag{C.3b}$$

The various *NorSand* spreadsheets embed (C.3b) in the *CheckInputParameters* subroutine, leaving χ_{tc} as the input soil property for the user consistent with the testing report received from the laboratory. This is a matter of taste, as the conversion could be 'external' and the model input taken as χ_i.

In terms of backward compatibility, the original state parameter paper (Been and Jefferies, 1985) and the subsequent initial triaxial version of *NorSand* (Jefferies, 1993) correspond to an implicit $\chi_{tc} \sim 4$ across all the sands in those papers (approximate because of the shift from ψ_o to ψ at D_{min}).

C.1.2 Critical friction ratio (*M*)

Most investigations into the relationship between stresses at the critical void ratio have been for triaxial compression, but include a rather wide range of stress. The data are largely for sands as this is experimentally convenient, but this is no restriction as in critical state theory all soils are regarded as particulate without any true cohesion. Although there are alternative approaches for determining M (see Chapter 2), high-precision work usually follows Bishop (1971) and plots data from several tests in the stress–dilatancy form η_{max} versus D_{min}. M is then taken as the value of η_{max} corresponding to the intersection of the trend line through the data with the axis $D_{min} = 0$. Figure C.1 illustrates this procedure for triaxial test data on Erksak sand extending to a mean stress as great as 4.4 MPa (the data are from Vaid and Sasitharan, 1992). A linear trend is a good fit to the data and gives $M_{tc} = 1.26$ (equivalent to $\phi_c = 31.4°$) in triaxial compression. Based on results like these, it is uncontroversial to take

$$\bar{\sigma}_q = M\bar{\sigma}_m \tag{C.4}$$

where M is independent of void ratio. This independence of M from void ratio may not be actually true for very loose soils. Although the data have wide error bars, there is a trend suggesting reduced M for very high void ratio. This is not altogether surprising, but something that is not included in current models.

The lack of controversy over (C.4) did not extend to the effect of Lode angle on M. Early critical state models idealized M as constant, which was mathematically elegant but conflicted with experimental work at Imperial College (most importantly, the experiments of Cornforth). Famously, Bishop used part of his Rankine Lecture (Bishop, 1966) to 'trash' the idea of constant M. The concern is easy to see even with only triaxial data, with Figure C.1

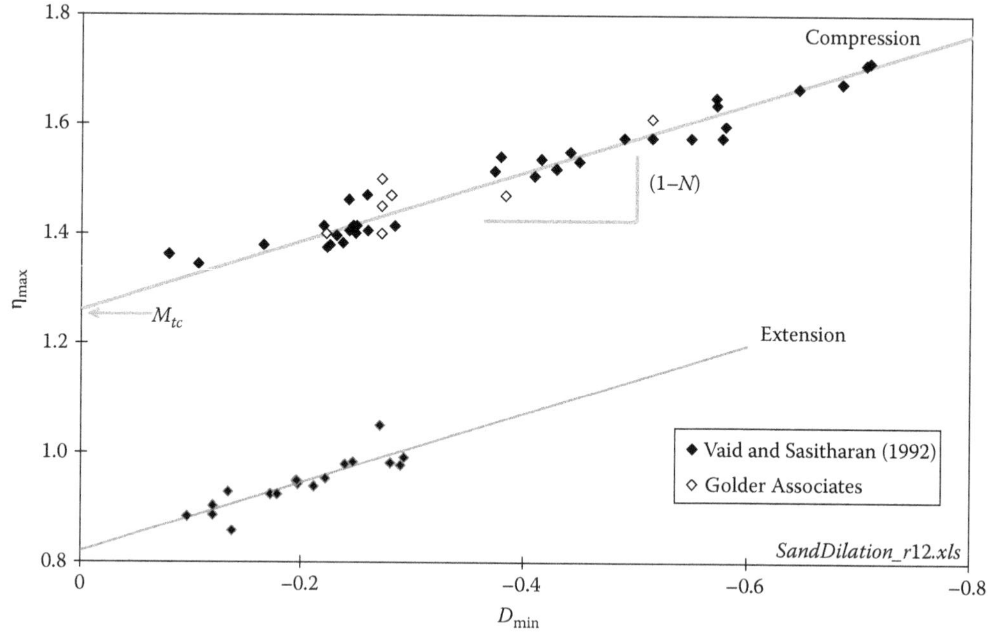

Figure C.1 Erksak sand stress–dilatancy in triaxial compression and extension. (After Jefferies, M.G. and Shuttle, D.A., *Géotechnique*, 52(9), 625, 2002. With permission from Institution of Civil Engineers.)

also showing triaxial extension data on Erksak sand. Although there is a smaller range of dilation to define the trend, a trend is nevertheless evident and indicates $M_{te}=0.82$, where the subscript *te* denotes triaxial extension. This variation of *M* with Lode angle corresponds to a reduction in friction angle (to about $\phi_c=28.3°$) in triaxial extension compared to compression. Where Bishop erred was in implying that constant *M* was in some way necessary to theoretical plasticity for soils – it is not.

The effect of intermediate principal stress on the failure criteria of sand was actively researched throughout the 1960s and early 1970s (e.g. Cornforth, 1964; Bishop, 1966; Green and Bishop, 1969; Green, 1971; Reades, 1971; Lade and Duncan, 1974). This interest covered a wide range of sand densities, but was directed at peak strength with substantial dilatancy. This body of work does not provide adequate guidance for critical state models, which require the friction ratio at D=0: the mechanism that dissipates plastic work. The dilatant strength component merely transfers work between the principal directions.

Available data on *M* (or the alternative identity ϕ_c) as a function of Lode angle are sparse, and the results obtained by Cornforth (1964) on Brasted sand are the dominant data set. Wanatowski and Chu (2007) have provided a smaller set of results on Changi sand using a modern version of Cornforth's equipment.

One common error is to treat plane strain as an alternative situation to (say) triaxial compression. The error in doing this is that the stress state in plane strain varies from one plane strain state to another because the stress state develops to accommodate the imposed strain condition; this stress state is usually denoted by Bishop's parameter '*b*' or the Lode angle θ. We use the Lode angle because that measure is more common in the finite element literature. Figure C.2 shows the variation in Lode angle at peak strength with the dilation rate for that peak strength; all the data on this figure are plane strain.

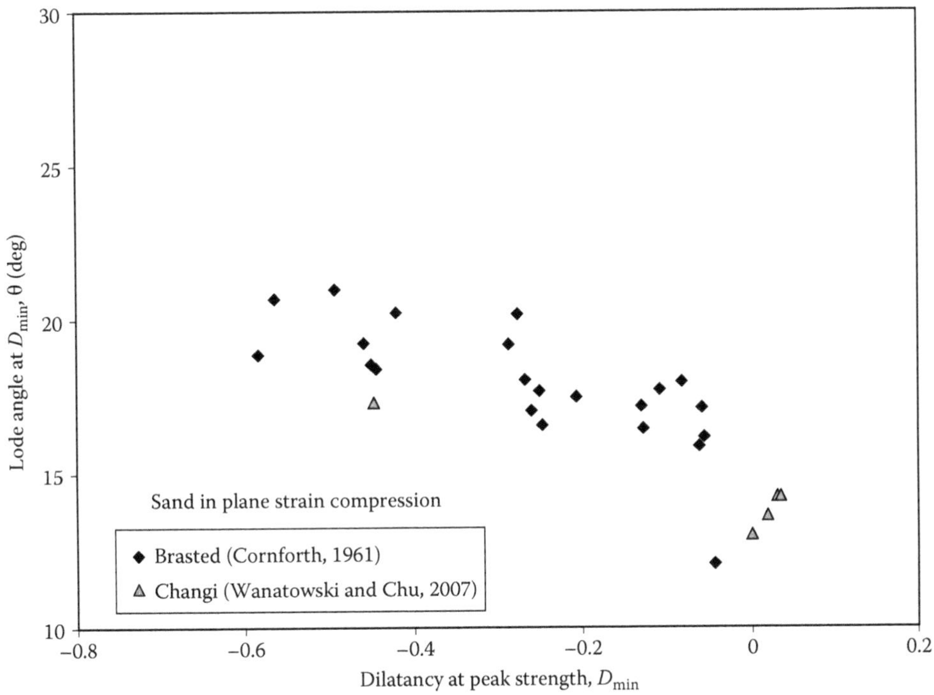

Figure C.2 Lode angle of sand at peak strength in plane strain.

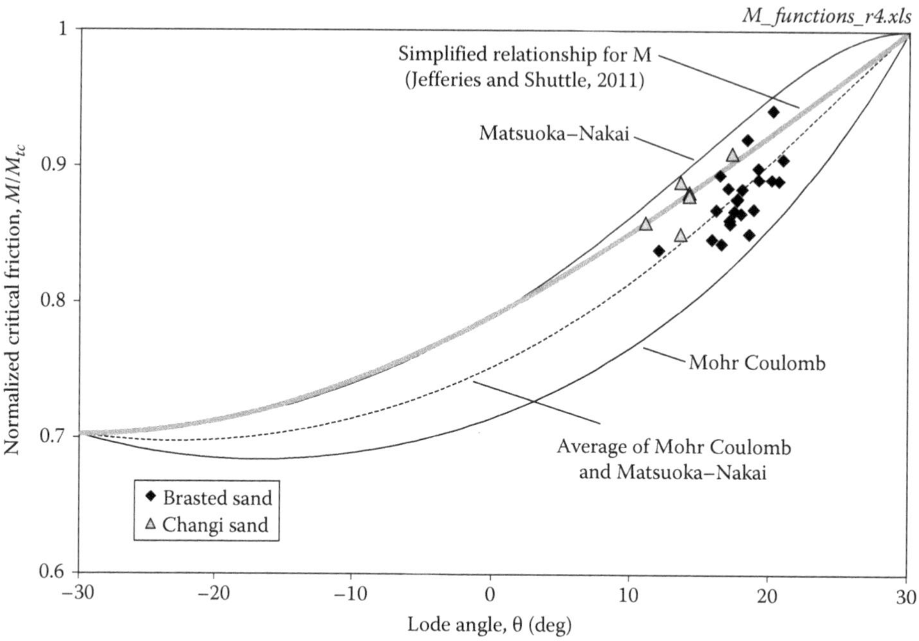

Figure C.3 Comparison of functions for $M(\theta)$ with plane strain data on Brasted and Changi sands. The test data for Brasted and Changi have been normalized by M_{tc} to bring both data sets to a common basis for comparison.

Cornforth's tests on Brasted sand covered a range of soil densities, and each test needs to be assessed using the stress–dilatancy flow rule to extract the operating value of M. This processing is straightforward (for details, see Jefferies and Shuttle, 2002), with the resulting ratio M/Mtc plotted versus the Lode angle of each test (Figure C.3).

Two standard soil strength models are also plotted in Figure C.3, both taking triaxial compression as the reference condition. The Mohr–Coulomb criterion with a constant critical friction angle is a widely held idealization for the critical state and gives

$$M = \frac{(3\sqrt{3})}{\left(\cos\theta(1+6/M_{tc}) - \sqrt{3}\sin\theta\right)} \tag{C.5}$$

The less familiar Matsuoka–Nakai (1974) idealization is based on the physically appealing concept of spatially mobilized planes but has M as implicit:

$$27 - 3M^2 = A\left[3 - M^2 + \frac{8}{9}M^3\sin(\theta)\left(\frac{3}{4} - \sin^2(\theta)\right)\right] \tag{C.6a}$$

where the coefficient A is written in terms of property M_{tc} as

$$A = \frac{27 - 3M_{tc}^2}{3 - M_{tc}^2 + 2/9M_{tc}^2} \tag{C.6b}$$

There is no analytical solution of (C.6) for M, and the bisection algorithm is used to find M for θ of interest such that $M_{te} \le M \le M_{tc}$.

The original generalization of *NorSand* to 3D (Jefferies and Shuttle, 2002) adopted an average of the Mohr–Coulomb and Matsuoka–Nakai idealizations as a reasonable model, something that is self-evident from Figure C.3. This was programmed as the VBA functions *Mpsi_v1*, *MatNak* and *MohrColomb*, which can be found in the *NorSandTxl.xls* spreadsheet. Similar functions were used in finite element implementations of *NorSand*.

A consequence of M evolving as stress conditions change is that M must be computed at every step in a numerical model. This is not too onerous for laboratory element tests, but once analysis moves to boundary value problems, the required number of calculations becomes an impediment to a responsive and interactive analysis. Thus, Jefferies and Shuttle (2011) proposed that an operationally adequate, and computationally, efficient approach would be to use

$$M(\theta) = M_{tc} - \frac{M_{tc}^2}{3+M_{tc}}\cos\left(\frac{3\theta}{2}+\frac{\pi}{4}\right) \tag{C.7}$$

where θ is measured in radians. This function is also shown in Figure C.3 and is a plausible fit to the plain strain test data (which is the practical interest). The notation $M(\theta)$ has been invoked in (C.7) to emphasize that M varies with Lode angle, while M_{tc} is a constant (i.e. a property) for a soil.

Obviously, the 'average' and 'simplified' idealizations for M differ moving away from plane strain towards triaxial extension, with the 'simplified' idealization following the Matsuoka–Nakai trend. Which to choose? The difficulty is that there are no data at present on critical stress ratios other than under triaxial compression, plane strain and triaxial extension on which to base a selection. Equation C.7 was proposed as having an elegant mathematical form that matches the flow rule used in *NorSand*. Its similarity to Matsuoka–Nakai was a pleasing bonus.

C.1.3 Volumetric coupling in stress–dilatancy (*N*)

When data on soil strength are viewed in stress–dilatancy space, such as Figure C.1, it is a universally acknowledged truth that

$$D^p_{\min} = \frac{(M-\eta_{\max})}{(1-N)} \tag{C.8}$$

which is usually referred to as Nova's flow rule (after Nova, 1982). N is a soil property that pairs with the critical friction ratio M. Although N represents the slope of a trend through data, as per Figure C.1, if the work flow in the soil is considered, one finds that N represents a volumetric coupling between mean and distortional strains (Jefferies, 1997). The Original Cam Clay (OCC) flow rule, which is based on a postulated/idealized work dissipation, is simply (C.8) with the property $N=0$. Many natural sands tend to show $N \approx 0.3$ (see Chapter 2).

Soil models that follow the framework of Drucker et al. (1957) invoke normality, with the yield surface shape derived by integrating the direction perpendicular to the flow rule (see Chapter 3). The original version of *NorSand* (Jefferies, 1993) followed that approach and derived the yield surfaces from (C.8) with a family of shapes depending on the value of N. Then, Dafalias and co-workers made a particularly insightful contribution that provided both simplicity and better representation of soil behaviour. Their innovation was promptly adopted by *NorSand*.

Recall from Chapter 2 that when the first 'micromechanical' view of stress–dilatancy was derived (Rowe, 1962), it was found that constant ϕ_c was not particularly accurate in fitting the theory to measured soil behaviour. Rowe suggested that the operating friction (ϕ_f) should lie between that corresponding to slip of soil particles against each other (i.e. mineral to mineral friction), ϕ_μ, and that of the critical state, ϕ_c. This idea of an operating friction ratio less than critical appears to have slipped from collective memory for several decades after Rowe's findings until it was resurrected by Dafalias and co-workers (Manzari and Dafalias, 1997; Li and Dafalias, 2000). Accepting that ϕ_f varies turns out to be a great idea, giving better results with a simpler model. In the present context, the aim is to clearly associate the operating critical friction with the current yield surface, in particular at the *image* condition (defined shortly in Sections C.2 and C.3). So, we use the subscript '*i*', and thus M_i, rather than M_f, which would be implied if Rowe's notation was used.

What controls how M_i evolves throughout a test? Strain is, in itself, not an admissible input to M_i as, even with a perfectly sampled 'element' of ground, there is no test you can make to determine the reference configuration from which strain is measured. The insight of Dafalias and co-workers was that M_i must satisfy the following condition:

$$M_i \Rightarrow M \text{ as } \varepsilon_q \Rightarrow \infty \tag{C.9a}$$

which is naturally expressed in terms of the state parameter:

$$M_i \Rightarrow M \text{ as } \psi \Rightarrow 0 \tag{C.9b}$$

Equation C.9 is invoked alongside the general flow rule:

$$D^p = M_i - \eta \tag{C.10}$$

Equation C.10 is a generalization of a flow rule derived from a simple idealization of the dissipation of work by plastic strains (Chapter 3), reflecting what has been known, but neglected, about soil behaviour for some 50 years.

One consequence of adopting a work-based idealization of soil behaviour is that the deviatoric shear strain invariant ε_q (Appendix A) is also a function of Lode angle, which means that the limiting dilation established under triaxial compression must be generalized. This is simply done (Jefferies and Shuttle, 2002) using the expression

$$D^p_{\min} = \frac{M}{M_{tc}} \chi_i \psi_i \tag{C.11}$$

which is the general version of (C.1) and where we explicitly associate χ_i with triaxial compression conditions as implied by (C.3).

Substituting (C.8) for η in (C.10) and then further substituting (C.11) gives an expression consistent with the desired framework and written entirely in terms of our familiar soil properties:

$$M_i = M(\theta)\left(1 + \frac{N\chi_i}{M_{tc}} \psi_i\right) \tag{C.12}$$

Other workers have proposed different expressions to (C.12) for M_i, but these do not have such easily recognizable soil properties as (C.12). In all fairness, the best representation of M_i could be viewed as 'work in progress' with the addition of some measure of soil fabric an obvious missing aspect. There is also uncertainty over loose soils. Equation C.11 applies to dense (dilatant) samples, since it is dilating samples that give us the soil property χ. The situation for loose samples is less clear. Recognizing that M_i represents plastic work dissipation within the soil, it seems strange that loose soil could have a greater work dissipation potential than dense soil; one alternative would be to take $M_i = M(\theta)$ for $\psi > 0$ where the idea of M_i then applies only to dense soil. But a better fit to data seems to be a symmetric version of (C.12) (Been and Jefferies, 2004), which is

$$M_i = M(\theta)\left(1 - \frac{N\chi_i}{M_{tc}}|\psi_i|\right) \tag{C.13}$$

Turning back to the development history of *NorSand*, (C.13) has been used in the various programs since about 2000. Between then and today, N went away and then returned. In many soils, the parameter group $N\chi_{tc}$ is close to unity, which suggests that it might be a compensating factor with unity being a 'not unreasonable' value for the parameter pair (Jefferies and Shuttle, 2005), and could be viewed as a useful simplification by removing one soil property from the calibration. However, calibrations to various tailings sands and tailings 'paste' (a non-segregating sandy silt) over the last few years have indicated that N and χ are not complimentary and do capture differing aspects of soil behaviour. Thus, N is now reinstated to the *NorSand* parameter list.

Operationally, the pair of Equations C.7 and C.13 give a good fit to the available plane strain data (within experimental precision) as illustrated in Figure C.4. These equations are

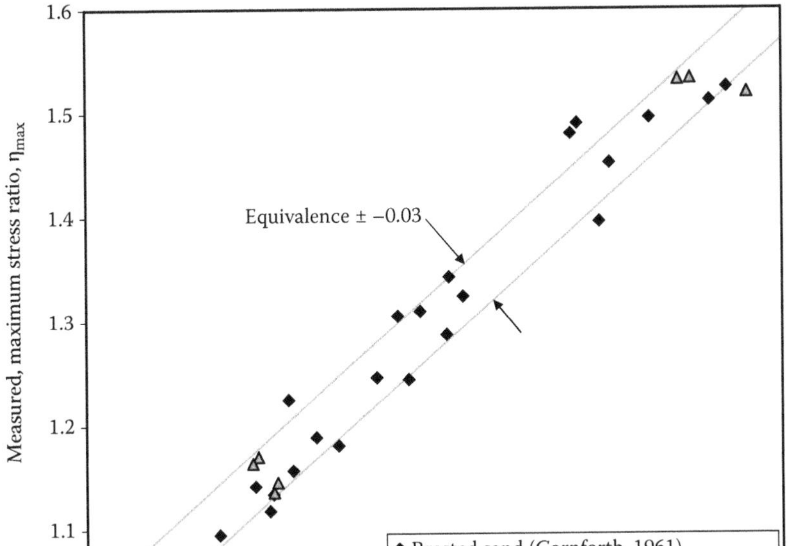

Figure C.4 Performance of simplified idealization for M_i. (After Jefferies, M.G. and Shuttle, D.A., *Géotechnique*, 61(8), 709, 2011.)

embedded in the spreadsheets as the *Mpsi_v3* function and supersede the *Mpsi_v1* function, which is no longer called (the earlier function is left in the code for historical interest).

C.1.4 Engineering strain

Historically, engineering theories of material behaviour started with the *Theory of Elasticity* (e.g. the classic Timoshenko and Goodier text). As elastic strains are small, say in the order of 1% for practical safety factors, second-order terms could be neglected to simplify the framework from the consideration of finite deformations experienced by a material. These ideas from elasticity continue to dominate the geotechnical literature, and indeed civil engineering education, but may cause errors when dealing with constitutive models for soils.

NorSand, along with all variants of Cam Clay, is an intrinsically 'large strain' theory because it is defined on the basis of stress and void ratio, that is, an *equation of state* or

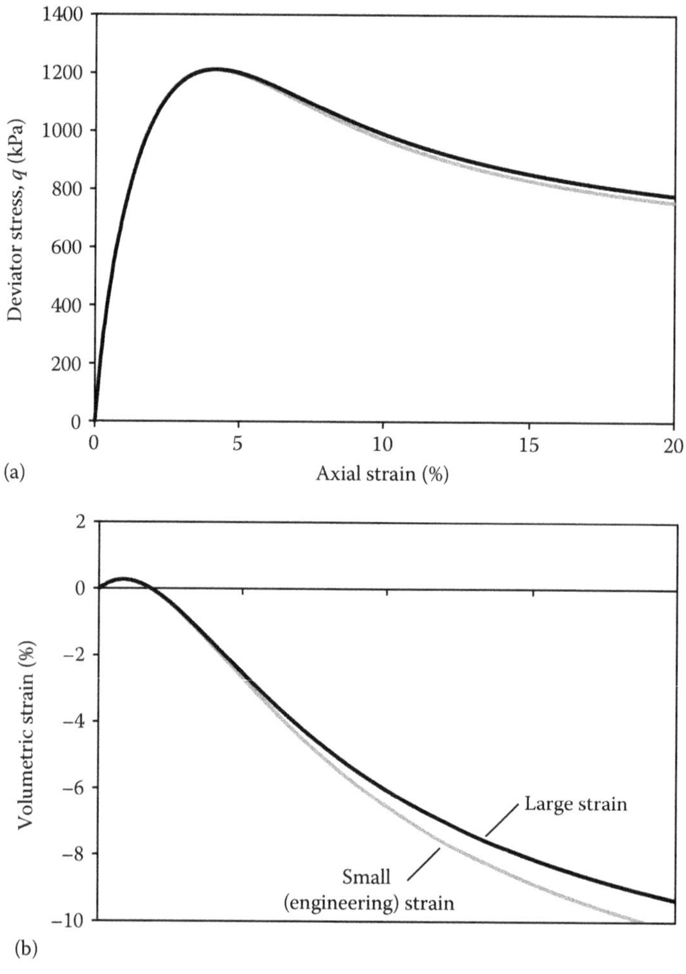

Figure C.5 Effect of using large and small strain when plotting sample deformation. The soil behaviour is identical between the two trends plotted, with the apparent differing volumetric strain caused only by the changed strain measure. Critical state models are intrinsically large strain, but most laboratories report small strains. (a) deviator stress versus axial strain and (b) volumetric strain versus axial strain.

finite deformation view. But most testing laboratories report test data in terms of the initial sample configuration with, for example, axial strain defined as the ratio of height change to initial height. This might seem a small detail, but it affects the parameter estimates obtained by fitting constitutive models to data. Figure C.5 shows two simulations using *NorSand* with identical soil properties and identical soil behaviour/deformations; the difference arises because one set of output is the natural 'large strain' results of *NorSand*, while the other is the computed results transformed into familiar 'engineering' strain. The implications for fitting a particular laboratory test are obvious.

To get over this possible issue in fitting test data, all the downloadable *NorSand* VBA code now includes a toggle *StrainMode* in the *Declarations* part of the code. When this toggle is set to the defined constant 'small' (see the VBA code), the computed output is transformed to familiar engineering strain; otherwise, natural strains are output. The toggle is placed in the declarations part of the code as most users will make the choice only once after checking the data reporting format used by their testing laboratory. Do make it a user input from the modelling window if you want to switch it in and out routinely.

C.2 YIELD SURFACE

The yield surface specifies the size of the elastic zone and is the locus of stress states leading to plastic strains. The derivation of the yield surface equation depends on just two assumptions: normality and the stress–dilatancy relationship. From the definition of the stress ratio η, take the differential to express the change in shear stress as

$$\dot{\bar{\sigma}}_q = \bar{\sigma}_m \dot{\eta} + \eta \dot{\bar{\sigma}}_m \tag{C.14}$$

Treating soil as a work-hardening plastic material, and following Drucker (1951), normality specifies that the plastic strain rates are perpendicular to the yield surface, so

$$\frac{\dot{\bar{\sigma}}_q}{\dot{\bar{\sigma}}_m} = \frac{-1}{\dot{\varepsilon}_q^p / \dot{\varepsilon}_v^p} = -D^p \Rightarrow \dot{\bar{\sigma}}_q = -D^p \dot{\bar{\sigma}}_m \tag{C.15}$$

Substituting (C.15) in (C.14) gives

$$-D^p \dot{\bar{\sigma}}_m = \bar{\sigma}_m \dot{\eta} + \eta \dot{\bar{\sigma}}_m \Rightarrow \frac{\dot{\bar{\sigma}}_m}{\bar{\sigma}_m} + \frac{\dot{\eta}}{D^p + \eta} = 0 \tag{C.16}$$

Equation C.16 is an identity of the normality condition and is true regardless of the soil's internal mechanisms for the dissipation of plastic work as long as stable yielding prevails (work-hardening or perfectly plastic conditions). This condition for stable yielding can be expressed in terms of the *stress increments* as

$$\dot{\bar{\sigma}}_q \dot{\varepsilon}_q + \dot{\bar{\sigma}}_m \dot{\varepsilon}_m > 0 \Rightarrow \text{stable yielding} \tag{C.17}$$

When (C.17) is violated, the soil becomes progressively easier to deform with further straining as the rate of plastic working decreases. This allows strains to localize and form shear bands (with adjacent soil unloading elastically and dumping their stored energy into the shear band) – an important consideration for post-liquefaction strengths. For the present, we consider the evolution of soil behaviour only during stable yielding.

As (C.16) is separated, it can be integrated directly with the details of the integration depending on the stress–dilatancy relation between D^p and η. Accepting the concept of an operating critical friction ratio M_i, Equation C.10 is substituted in C.16 giving

$$\frac{\dot{\bar{\sigma}}_m}{\bar{\sigma}_m} + \frac{\dot{\eta}}{M_i} = 0 \tag{C.18}$$

Because M_i is not a function of η (M_i is a function of the Lode angle, and possibly the state parameter, neither of which is part of the integral and so can be treated as constant during the integration), (C.18) readily integrates as

$$\ln(\bar{\sigma}_m) + \frac{\eta}{M_i} = C \tag{C.19}$$

where C is the coefficient of integration. This coefficient is chosen as the mean stress when $\eta = M_i$, which is referred to as the *image* condition and denoted by the subscript '*i*'. Making the substitution for the integration coefficient C gives the equation of the yield surface:

$$\frac{\eta}{M_i} = 1 - \ln\left(\frac{\bar{\sigma}_m}{\bar{\sigma}_{mi}}\right) \tag{C.20}$$

where the image stress is denoted by $\bar{\sigma}_{mi}$ and $\eta = M_i \Leftrightarrow \bar{\sigma}_m = \bar{\sigma}_{mi}$. The image condition is the point on a yield surface where the plastic strain rates give $D^p = 0$, which is one of the two conditions for the critical state (hence the name *image*, because it is not the critical state in general). Equation C.20 describes a 'bullet'-shaped yield surface exactly as found in OCC but with the subtlety that M_i now varies with Lode angle and state parameter – the idealized symmetry of OCC is lost when we add in real soil behaviour trends although the framework remains familiar.

The image stress is one of the internal variables of the model. Work hardening (softening) operates by changing $\bar{\sigma}_{mi}$. The evolution of yield surface size as the soil strains depends only on the hardening law, which operates directly on $\bar{\sigma}_{mi}$. However, several aspects must be developed before introducing the hardening law.

C.3 IMAGE STATE PARAMETER

The idea of the image condition was an obvious choice for the integration constant in derivation of the yield surface. There is now a consequence for the definition of the state parameter. Recall that the state parameter is defined looking at the familiar e-log(p) plot and treating that plot as a 'state diagram' much as you might do in thermodynamics; this gives $\psi = e - e_c$. The issue then becomes, because mean stress varies around a yield surface, so does ψ. It is not possible to associate a unique value of ψ with a fixed yield surface. This variation of ψ certainly introduces complexity; a more strict view might be that a state measure varying around a yield surface is theoretical nonsense. The simplest way to incorporate a correct 'state view' into the derived yield surface is to define an internal variable ψ_i, which is the state parameter at the same image condition as used to define the size of that yield surface; this parameter ψ_i is defined as

$$\psi_i = e - e_i \tag{C.21}$$

where e_i is the critical state void ratio at the image mean stress $\bar{\sigma}_{mi}$. Since there is a unique image stress state for any yield surface, this also makes ψ_i unique. Equation C.21 is quite general with no implication for any particular idealization of the CSL. Noting that $\eta = M_i \Leftrightarrow \bar{\sigma}_m = \bar{\sigma}_{mi}$, it immediately follows from (C.21) that the condition $\psi_i = 0$ uniquely defines yield surfaces that intersect the critical state.

If we give up some generality and work with a conventional semi-log idealization of the CSL, the ψ_i and ψ are simply related by

$$\psi_i = \psi + \lambda \ln\left(\frac{\bar{\sigma}_{mi}}{\bar{\sigma}_m}\right) \quad \text{(C.22)}$$

Curved CSL are most elegantly implemented by direct evaluation, so that (C.21) becomes the embedded form for numerical implementations.

C.4 HARDENING LIMIT AND INTERNAL YIELDING

A key feature of dense soils, whether sands or clays, is that dilatancy is limited to a maximum value for any specific soil state. Conventionally, this is represented by invoking a non-associated flow rule with appropriate choice of dilation angle. However, such a conventional approach is not acceptable for models based on Drucker's stability postulate (Drucker, 1959) that requires normality (or an *associated* flow rule). Normality was used earlier in deriving the yield surface equation for *NorSand*, just as done for the variants of Cam Clay. One of the kernel ideas in *NorSand* is to limit maximum dilation to replicate dense soil behaviour without resorting to non-associated flow rules. Realistic maximum dilatancy is obtained, despite normality, by controlling $\bar{\sigma}_{mi}$. Substitution of the flow rule (C.10) in the yield surface (C.20) gives

$$D^p = M_i \ln\left(\frac{\bar{\sigma}_m}{\bar{\sigma}_{mi}}\right) = -M_i \ln\left(\frac{\bar{\sigma}_{mi}}{\bar{\sigma}_m}\right) \quad \text{(C.23)}$$

Inversion of (C.23) allows a current maximum yield surface size (hardness) to be determined from the minimum dilation rate (minimum because of the compression positive sign convention):

$$(\bar{\sigma}_{mi})_{\max} = \bar{\sigma}_m \exp\left(\frac{-D^p_{\min}}{M_i}\right) \quad \text{(C.24)}$$

As discussed in Chapter 2, a large body of test data on the maximum dilatancy of soils in triaxial compression supports the simple first-order rate equation relationship:

$$D^p_{\min,tc} = \chi_i \psi_i \quad \text{(C.25)}$$

where χ_i is a *NorSand* defined under triaxial compression. Substitution of (C.25) into (C.24) leads to an evolving hardening limit for the yield surface:

$$(\bar{\sigma}_{mi})_{\max} = \bar{\sigma}_m \exp\left(\frac{-\chi_i \psi_i}{M_{i,tc}}\right) \quad \text{(C.26a)}$$

To make the notation less klutzy, we rename the variable $(\bar{\sigma}_{mi})_{\max}$ as simply $\bar{\sigma}_{mx}$ as in

$$\bar{\sigma}_{mx} = \bar{\sigma}_m \exp\left(\frac{-\chi_i \psi_i}{M_{i,tc}}\right) \tag{C.26b}$$

Equation C.26 is no more than a strict implementation of the ideas of Drucker et al. (1957), who first clarified that the Mohr Coulomb criterion was a locus of failure states, not a yield surface. If the limiting hardness were plotted as $\eta_{\max}$ in a q-p plot, (C.26) forms the Hvorslev surface (see Figure 3.10).

It is also helpful to understand that (C.26) implies an internal cap to the yield surface from consideration of self-consistency. Figure C.6a sketches a yield surface with the dilation limit operating to stop the yield surface expanding. Now consider an arbitrary stress path involving elastic unloading, illustrated as path 1 in Figure C.6a. If subsequent reloading follows path 2, then a dilatancy greater than the supposed maximum could be realized. This inconsistency is removed by requiring that stress states traversing from within the yield surface to the shaded zone in Figure C.6a contract (isotropically soften) the yield surface, as illustrated in Figure C.6b; yield arises in 'unloading' paths as a natural consequence of invoking state–dilatancy while requiring normality. To implement plastic softening, we must have an internal cap to the yield surface. Although a range of shapes are possible for the internal cap, an acceptable and simple shape is to take the cap as a plane perpendicular

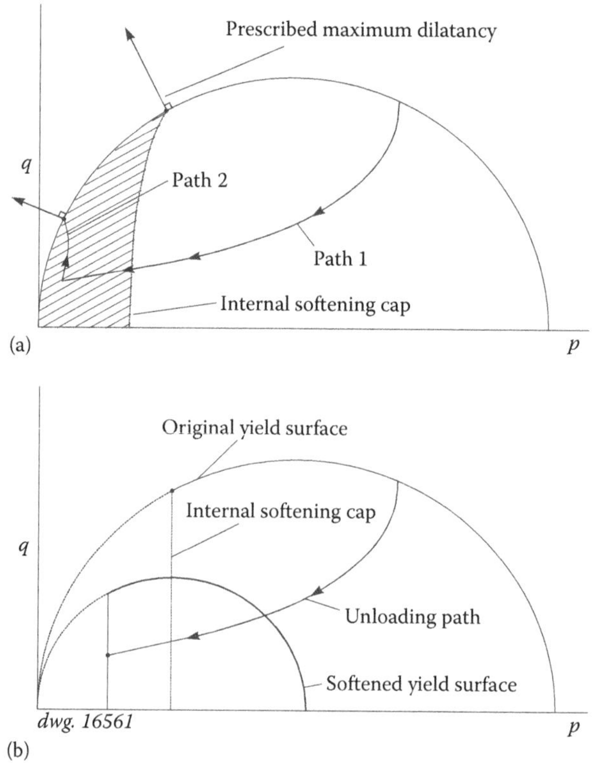

Figure C.6 Schematic illustration of self-consistency requirement for internal cap to yield surface. (a) possibility of inconsistent dilation and (b) isotropic softening on unloading. (From Jefferies, M.G., *Géotechnique*, 47, 1037, 1997. With permission from Institution of Civil Engineers.)

to the mean stress axis, as illustrated in Figure C.6b. This planar internal cap idealization is embedded in all variants of *NorSand*.

C.5 HARDENING RULE

C.5.1 Outer yield surface hardening with fixed principal directions

The hardening law describes how the yield surface evolves with plastic strain, which is the nature of all plastic work-hardening models. The natural form for a hardening law that complies with the Second Axiom, while respecting the constraint on maximum allowable dilatancy, is a simple difference equation between the current hardness (i.e. size) and the *current* maximum allowable value of hardness [from (C.26b)], which gives the *rate* of change in yield surface size with incremental plastic shear strain:

$$\dot{\bar{\sigma}}_{mi} = H(\bar{\sigma}_{mx} - \bar{\sigma}_{mi})\dot{\varepsilon}_q^p \tag{C.27}$$

In (C.27), H is a hardening modulus (which could be a function of soil fabric and ψ). Shear strain must be used as the strain measure because, at the image condition, $\eta = M_i \Rightarrow \dot{\varepsilon}_v = 0$, which means that purely volumetric strain-based hardening will not get past the image condition during shear. Equation C.27 is akin to a radioactive decay equation but with the additional feature that the target is evolving as well as the state variable.

To make the mechanics independent of units, the hardening law is better expressed in dimensionless form by dividing through by current image mean stress and further normalizing through by current mean stress to make the role of (C.26) explicit. Thus, (C.27) is restated as

$$\frac{\dot{\bar{\sigma}}_{mi}}{\bar{\sigma}_{mi}} = H\left(\frac{\bar{\sigma}_m}{\bar{\sigma}_{mi}}\right)\left[\frac{\bar{\sigma}_{mx}}{\bar{\sigma}_m} - \frac{\bar{\sigma}_{mi}}{\bar{\sigma}_m}\right]\dot{\varepsilon}_q^p \tag{C.28}$$

An issue with general 3D conditions and the hardening modulus H must now be addressed. The Lode angle is the third stress invariant in *NorSand*, but the Lode angle is undefined when the stress state is isotropic. So H cannot be a function of Lode angle; otherwise, it will be undefined also. Further, isotropic compression is a perfectly reasonable loading, and the hardening law must allow for such a stress path.

Introducing the ratio $M_i/M_{i,tc}$ as a modifier on hardening in (C.28),

$$\frac{\dot{\bar{\sigma}}_{mi}}{\bar{\sigma}_{mi}} = H\frac{M_i}{M_{i,tc}}\left(\frac{\bar{\sigma}_m}{\bar{\sigma}_{mi}}\right)\left[\frac{\bar{\sigma}_{mx}}{\bar{\sigma}_m} - \frac{\bar{\sigma}_{mi}}{\bar{\sigma}_m}\right]\dot{\varepsilon}_q^p \tag{C.29}$$

Now, by definition, $\dot{\varepsilon}_v^p = D^p\dot{\varepsilon}_q^p$, while from (C.10), $D^p \Rightarrow M_i$ as $\eta \Rightarrow 0$

$$\therefore \quad \dot{\varepsilon}_v^p \Rightarrow M_i\dot{\varepsilon}_q^p \text{ as } \eta \Rightarrow 0 \tag{C.30}$$

Substituting the limit condition of (C.30) into (C.29) shows that

$$\frac{\dot{\bar{\sigma}}_{mi}}{\bar{\sigma}_{mi}} \Rightarrow \frac{H}{M_{i,tc}}\left(\frac{\bar{\sigma}_m}{\bar{\sigma}_{mi}}\right)\left[\frac{\bar{\sigma}_{mx}}{\bar{\sigma}_m} - \frac{\bar{\sigma}_{mi}}{\bar{\sigma}_m}\right]\dot{\varepsilon}_v^p \text{ as } \eta \Rightarrow 0 \tag{C.31}$$

There are no terms involving the Lode angle on the right-hand side of (C.30), and therefore hardening is fully and uniquely defined under isotropic stress conditions despite the Lode angle itself becoming indeterminate. The term $M_{i,tc}$ is introduced in (C.29) for backward compatibility with previously published values for H, but equally, it would be perfectly reasonable to drop $M_{i,tc}$ from the hardening law and simply redefine H.

The hardening law was developed from the physical consideration of treating soil behaviour as a rate process. However, when such a law is fitted to sand stress–strain data, it is not a perfect fit. Hardening also depends on the shear stress level, not a new observation given the preponderance of hyperbolic stiffness models in the geotechnical literature and certainly perfectly acceptable within a critical state framework. Such a dependence on mobilized η is readily introduced by multiplying through by the ratio $\bar{\sigma}_m/\bar{\sigma}_{mi}$, which has the effect of about a threefold change in the hardening as η changes from zero to its typical maximum. The form of the hardening law is then

$$\frac{\dot{\bar{\sigma}}_{mi}}{\bar{\sigma}_{mi}} = H\frac{M_i}{M_{i,tc}}\left(\frac{\bar{\sigma}_m}{\bar{\sigma}_{mi}}\right)^2\left[\frac{\bar{\sigma}_{mx}}{\bar{\sigma}_m} - \frac{\bar{\sigma}_{mi}}{\bar{\sigma}_m}\right]\dot{\varepsilon}_q^p \tag{C.32a}$$

or, on explicitly introducing the state parameter using (C.26) to make the state dependence and role of model properties clear,

$$\frac{\dot{\bar{\sigma}}_{mi}}{\bar{\sigma}_{mi}} = H\frac{M_i}{M_{i,tc}}\left(\frac{\bar{\sigma}_m}{\bar{\sigma}_{mi}}\right)^2\left[\exp\left(\frac{\chi_i\psi_i}{M_{i,tc}}\right) - \frac{\bar{\sigma}_{m,i}}{\bar{\sigma}_m}\right]\dot{\varepsilon}_q^p \tag{C.32b}$$

It is a matter of programming elegance as to which version of (C.32) is appropriate when writing numerical code to implement *NorSand*. Functionally, they are simply alternative versions of the same hardening rule.

C.5.2 Additional softening

The hardening law (C.32) works superbly for drained tests, but shows insufficient control of hardening during loose undrained tests. The problem is that σ_{mx} changes with the mean effective stress (see Equation C.26), and for loose soils with $\psi_i > 0$, σ_{mx} can change more rapidly in an undrained test than a pure 'rate equation' law (C.32) can follow. The hardening law (C.32) gets there in the end, but the rate of strength loss with strain is less rapid than that found in laboratory tests. This too slow response can be improved by 'rolling in' an additional term dealing with the change in the hardening 'target' (i.e. σ_{mx}) during a strain step:

$$\frac{\dot{\bar{\sigma}}_{mi}}{\bar{\sigma}_{mi}} = H\frac{M_i}{M_{i,tc}}\left(\frac{\bar{\sigma}_m}{\bar{\sigma}_{mi}}\right)^2\left[\frac{\bar{\sigma}_{mx}}{\bar{\sigma}_m} - \frac{\bar{\sigma}_{mi}}{\bar{\sigma}_m}\right]\dot{\varepsilon}_q^p + \frac{\eta}{\eta_L}\frac{\dot{\bar{\sigma}}_{mx}}{\bar{\sigma}_{mi}} \tag{C.33}$$

where the ratio η/η_L is a linear 'rolling in' so that the additional term on the RHS of the hardening law becomes progressively more important as the stress ratio η approaches its current limiting value η_L (= M_i (1 − $D_{\min}/M_{i,tc}$)).

This additional term on the RHS of (C.33) was referred to as 'cap softening' in the first edition. In some ways this cap softening can be viewed as a smooth transition to the current limiting hardness (or internal cap), at which point the yield surface changes from evolving with plastic shear strain to now only depending on the evolution of ψ. Mathematically, the requirement for an additional softening term traces back to the form of rate model used and where we have two things evolving, not one. Practically, (C.33) provides excellent matches to loose soil behaviour in general stress paths; numerically, it nicely controls $\bar{\sigma}_{mi}/\bar{\sigma}_m$ from exceeding its postulated limiting ratio (Equation C.26). The nature of cap softening is derived by taking the differential of Equation C.26:

$$\dot{\bar{\sigma}}_{mx} = \dot{\bar{\sigma}}_m \exp\left(-\frac{\chi_i}{M_{itc}}\psi_i\right) + \bar{\sigma}_m \exp\left(-\frac{\chi_i}{M_{itc}}\psi_i\right)\left(-\frac{\chi_i}{M_{itc}}\dot{\psi}_i + \frac{\chi_i}{M_{itc}^2}\psi_i \dot{M}_{itc}\right)$$

$$= \bar{\sigma}_{mx}\left(\frac{\dot{\bar{\sigma}}_m}{\bar{\sigma}_m} + \frac{\chi_i}{M_{itc}}\left(\frac{\psi_i \dot{M}_{itc}}{M_{itc}} - \dot{\psi}_i\right)\right)$$

After dividing through by the image stress,

$$\frac{\dot{\bar{\sigma}}_{mx}}{\bar{\sigma}_{mi}} = \frac{\bar{\sigma}_{mx}}{\bar{\sigma}_{mi}}\left(\frac{\dot{\bar{\sigma}}_m}{\bar{\sigma}_m} + \frac{\chi_i}{M_{itc}}\left(\frac{\psi_i \dot{M}_{itc}}{M_{itc}} - \dot{\psi}_i\right)\right) \qquad \text{(C.34)}$$

Although (C.34) is explicit, numerical implementation is actually simpler if σ_{mx} is tracked as an internal variable and the differential computed from a simple backward difference (trailing) estimate – far fewer calculations than implementing (C.34). And remember that (C.34) is applied only for loose states where the yield surface is contracting to the critical void ratio; for dense states, no cap softening is invoked as such soils are dilating to the critical void ratio.

C.5.3 Softening of outer yield surface by principal stress rotation

Principal stress rotation is fundamentally important to soil liquefaction under earthquake or other cyclic loading. These loadings may have cyclic variation in the magnitude of deviator stress, but in almost all cases, they vary the principal stress directions cyclically. These aspects are discussed at some length in Chapter 7.

As plastic hardening is a macro-scale abstraction of the underlying micro-scale reality of grain contact arrangements developing to carry the imposed loads, and these grain contacts are orientated, changing the principal stress direction loads an existing arrangement of soil particles suboptimally (see the experimental data in Chapter 7). *NorSand* captures this behaviour by principal stress rotations always softening (shrinking) the yield surface, as illustrated in Figure C.7, since plastic hardening expresses the effect of particle contact 'chains' developing to carry the imposed load.

Implementation of principal stress rotation is straightforward. An increment of plastic strain is imposed to harden the yield surface under fixed principal stress direction using (C.33). Then, the computed hardening is reduced proportionally to the amount of principal stress rotation. A new soil property, a second plastic modulus Z_r, is introduced as the coefficient of proportionality. Correspondingly, the hardening law is modified with a term of the form

$$\frac{\dot{\bar{\sigma}}_{mi}}{\bar{\sigma}_{mi}} = -Z_r \frac{\dot{\alpha}}{\pi} \qquad \text{(C.35)}$$

The value π appears in the right-hand side of (C.35) as a reminder that the change of principal stress direction is measured in radians. Z_r is then a dimensionless softening modulus.

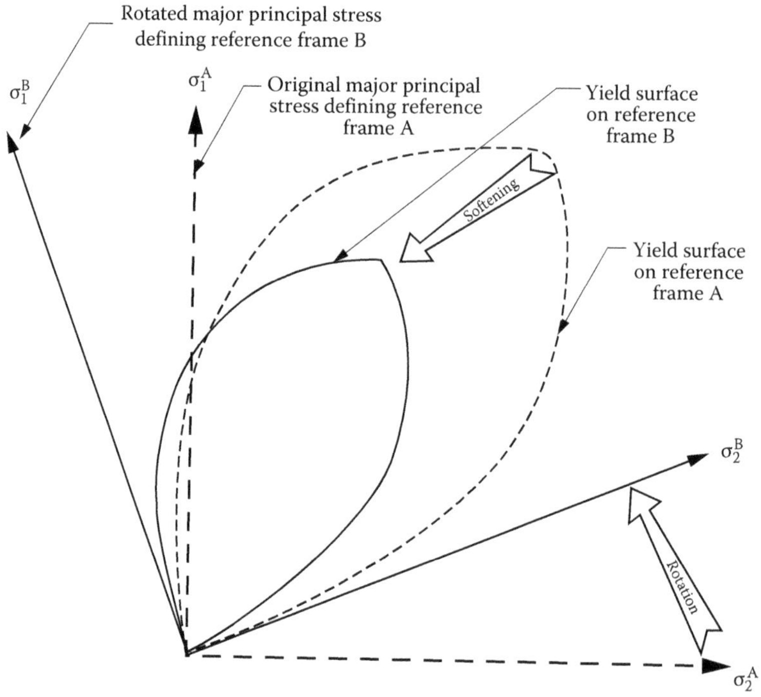

Figure C.7 Yield surface softening induced by principal stress rotation. (From Been, K. et al., Class A prediction for model 2, in *Proceedings of the International Conference on the Verification of Numerical Procedures for the Analysis of Soil Liquefaction Problems [VELACS]*, eds. K. Arulanandan and R.F. Scott, A.A. Balkema, Rotterdam, the Netherlands, 1993. With permission from Taylor & Francis.)

The form of a softening law (C.35) is a first-order approach. The effect of principal stress rotation will often move the soil state away from critical, at least initially, in violation of the Second Axiom. Such 'violation' is not fundamental, as the initial response involves small shear strain. It is only when the soil approaches $\eta > M_i$ that shear strains develop, and indeed under this condition, the Second Axiom asserts control, and we get the familiar 'butterfly' stress paths seen in cyclic simple shear of dense soils. A more significant limit is that softening from principal stress rotation can exceed the fixed-direction hardening, but that softening cannot go past the equivalent normally consolidated state. There is also the principle that there is a lower limit to void ratio, certainly $e_{min} > 0$. A reasonable expectation is that Z_r will be scaled by $(e-e_{min})$ so that there is no softening once a soil gets to its minimum void ratio. These 'issues of principle' require further work, and are not just for *NorSand* – there is a dearth of understanding about constitutive implications of principal stress rotation in general. As that understanding emerges, it is trivial to update the *NorSand* hardening law. For the moment, (C.35) is adopted as it replicates cyclic simple shear rather well, and that is sufficient for the present purpose.

C.5.4 Softening of inner yield surface

Liquefaction can arise during reduction of mean effective stress at more or less constant load, the situation being caused by increasing pore pressure. The Aberfan disaster is an example of this situation. In the context of *NorSand*, this loading comprises plastic yield on the internal cap. Such yield always causes softening of the yield surface, and again the Second Axiom is neglected since loading is directed inwards and away from the CSL. The hardening law for the internal yield surface is

$$\frac{\dot{\bar{\sigma}}_{m,i}}{\bar{\sigma}_{m,i}} = -\frac{H}{2}\frac{M_i}{M_{i,tc}}\left|\dot{\varepsilon}_q^p\right| \tag{C.36}$$

The cap always contracts, softening the yield surface, when loaded. There is a case for introducing a further plastic modulus to describe the cap behaviour, but available tests to date suggest that about half the loading modulus seems to fit the data. Hence, the empirical factor of two in (C.36).

It should be recognized that tests to determine the behaviour of the inner cap are rare, although constant-shear-drained tests have become a recent research interest (see Chapter 6). Loading under decreasing shear as well as decreasing mean stress does not move the soil towards critical (this is a stress path when reducing axial load in a triaxial cell after the soil has reached its maximum dilation rate). And, theoretically, there are issues with the recovery of internal non-elastic stored energy which affect the stress–dilatancy rule (e.g. Jefferies, 1997; Collins and Muhunthan, 2003). Further developments to (C.36) should be expected as these issues are explored and reconciled.

C.5.5 Constraint on hardening modulus

Although the essence of *NorSand* is to decouple hardening from the CSL, self-consistency requires that positive ψ must be accessible as that is a fundamental premise of the model and further supported by the data showing an infinity of NCL. This means that there is a limiting lower value of the plastic hardening H that is related to the hardening of the NCL. While stiffer behaviour (greater H) is acceptable, H is constrained at the lower end by the slope of the CSL. All NCL must be able to cross the CSL.

Assume the usual semi-log form of the CSL (what follows can be derived for different CSL idealizations, which changes the numerical value of the constraint but not the principle of it), and further assume that volumetric elasticity can be represented by constant rigidity (i.e. the standard κ model). Because H is a plastic parameter, the equivalent plastic compliance of interest from the CSL is given by $\lambda - \kappa$. For isotropic conditions, the plastic volumetric stiffness of the CSL is given by

$$K_c^p = \frac{1+e}{\lambda-\kappa}\bar{\sigma}_m \tag{C.37}$$

Self-consistency of the model with the postulates on which it is based requires that

$$K^p > K_c^p \ \forall \ \psi_i < 0 \wedge \eta = 0 \tag{C.38}$$

where the condition $\eta = 0$ is invoked because the restriction can be imposed only under isotropic conditions. As the shear stress increases and the soil moves to the critical state, then very different behaviours come into play because of Axiom 2. Although the restriction is written for all negative states, it is the limiting condition on the CSL that matters as, if this requirement is met, it is always true from experiments (denser states are always stiffer than loose ones for the same mean stress).

The hardening rule under isotropic conditions (C.31) can be further simplified using the *spacing ratio* '*r*' as

$$\frac{\dot{\bar{\sigma}}_{mi}}{\bar{\sigma}_m} = \left\{ \frac{H}{M_{tc}} r^2 \left[\exp\left(\frac{\chi \psi_i}{M_{tc}} \right) - \frac{1}{r} \right] \right\} \dot{\varepsilon}_v^p \tag{C.39}$$

Expressing (C.39) as a plastic stiffness gives directly, for $\psi_i = 0$:

$$K^p = \left\{ \frac{H}{M_{tc}} r \left[r \exp(0) - 1 \right] \right\} \bar{\sigma}_m \tag{C.40}$$

Invoking the limit (C.38) and using (C.40) with (C.37) gives

$$\left\{ \frac{H}{M_{tc}} r \left[r \exp(0) - 1 \right] \right\} \bar{\sigma}_m > \frac{1+e}{\lambda - \kappa} \bar{\sigma}_m$$

On rearranging,

$$H > \frac{M_{tc}}{r} \frac{1}{r-1} \frac{1+e}{\lambda + \kappa} \tag{C.41}$$

Putting in typical values that $M_{tc} \cong 1.25$, $r = 2.718$, $e \cong 0.7$ gives the simple and approximate limit on H:

$$2H > \frac{1}{\lambda - \kappa} \forall \psi_i < 0 \tag{C.42}$$

This theoretical self-consistency constraint turns out to be very much a lower limit of experience. This right-hand side of (C.42) is recognized as the Cam Clay hardening.

C.6 OVERCONSOLIDATION

Overconsolidation actually involves two concepts. On one hand, overconsolidation is defined in, say, an oedometer test as the ratio of current vertical stress to the vertical pre-consolidation stress determined from the measured soil behaviour. In effect, measuring the relationship of current stress to where yielding recommences. On the other hand, there is the geologic definition that overconsolidation relates the current vertical stress to its maximum past value. The two concepts are not the same because yield in unloading shrinks the yield surface as stress levels are reduced. *NorSand* recognizes and uses both concepts of overconsolidation ratio:

$$\text{Yield definition: } R = \frac{\bar{\sigma}_{mi}}{\bar{\sigma}_{me}} \tag{C.43}$$

$$\text{Stress history definition: } P = \frac{\bar{\sigma}_{1,\max}}{\bar{\sigma}_1} \tag{C.44}$$

These two definitions of overconsolidation are not intellectual 'niceties' – real soils show both effects, discussed shortly after dealing with the definitions themselves.

Taking the mechanical sense of overconsolidation first, what is the stress σ_{me} in the measure R? The familiar 'taught' idea of yield overconsolidation, such as you find in texts

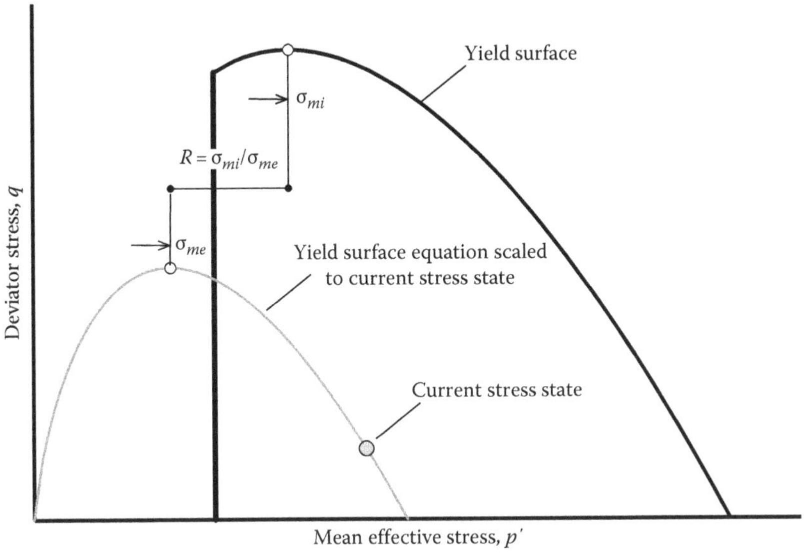

Figure C.8 Definition of overconsolidation used in *NorSand*.

discussing compression of clays, is to look along the loaded stress axis (i.e. mean stress for an isotropic consolidation or the vertical stress axis for an oedometer) and define the overconsolidation ratio on that axis. This concept is not really helpful for general stress states, and there is a better way that still conforms to the familiar usage of 'overconsolidation'. Consider a yield surface as shown in Figure C.8 and with a current stress point lying inside that yield surface. As illustrated in the figure, define a second pseudo-yield surface going through that stress state using the same equation as the yield surface itself, but now with a reduced scaling stress σ_{me}:

$$\bar{\sigma}_{me} = \bar{\sigma}_m \exp\left(\frac{\eta}{M_i} - 1\right) \tag{C.45}$$

Using (C.45), the 'proximity' of the elastic state to yielding defined by (C.43) automatically accommodates the effect of geostatic shear stress on the proximity to the yield surface. The *NorSand.xls* spreadsheets allow input of R to set an elastic range. This elastic range pushes the yield surface out from the geostatic stress state (i.e. K_o) input as the starting point for the simulation. The related finite element software (Appendix D) implements the same concept of overconsolidation.

The yield concept of overconsolidation is used to define the onset of plastic yielding, while the stress history (or geological) concept of overconsolidation affects the dilatancy and plastic hardening. There are three possible combinations of R and P on the main yield surface (i.e. excluding yield on the inner cap):

- Normally consolidated behaviour $R=1$ and $P=1$
- Reloading plastic behaviour $R=1$ and $P>1$
- Elastic behaviour $R>1$ and $P>1$

Within these three possible combinations of overconsolidation, there are two restrictions/conditions on R.

The first restriction is that $R \leq P$. One might find this restriction violated if true bonds have developed between the particles (say by cementation), but if that is the case, then true cohesion (or similar) needs introducing to the constitutive framework. This restriction applies for 'particulate' idealizations of soil.

The second restriction is that the elastic stress state must lie within the existing yield surface. If the 'overconsolidated' stress state moves past the internal cap, then yield in unloading develops, and the yield surface shrinks to keep the cap on the current stress state (see Section C.7.2). The maximum yield overconsolidation ratio depends on the deviator stress, with an absolute maximum for R corresponding to the condition $\sigma_q = 0$, which is

$$R < r\left(\frac{\bar{\sigma}_{mi}}{\bar{\sigma}_m}\right)_{\max} \tag{C.46}$$

If some common soil properties and geostatic stress states are invoked, the practical result is that a reasonable expectation is $R < 3$, with yield in unloading, and the subsequent effect of that on reloading, being credibly normal in many natural 'overconsolidated' soils.

C.6.1 Effect of reloading

A corollary to the limiting yield overconsolidation ratio is that the yield surface softening so induced leaves a memory within the soil that shows up during reloading. Reloading was investigated in the triaxial compression of Erksak sand, with unload–reload cycles from both pre- and post-maximum strengths. The data can be downloaded from the website and comprises the tests CID-G860 to G874 inclusive. The measured behaviour of test CID-G687 is shown in Figure C.9 together with the *NorSand* simulation. This simulation used $H_{reload} = 4H$ for $R > 1$ and $P < 1$. This simulation used Nova's flow rule (Equation C.8) modified so that αN was used rather than N. The reloading was best fitted with $\alpha = 2$ as can be seen from Figure C.9. Reloading is obviously intimately linked to yield in unloading, and further developments should be expected in how reloading is handled. Note that, regrettably, α is used here as a scaling of stress-dilatancy for consistency with the original reference; α is not the direction of σ_1 from the vertical direction in this context.

C.7 CONSISTENCY CONDITION

Plastic strain of the soil causes the yield surface to change size (harden or soften). Since the yield surface is expressed in terms of the dimensionless stress ratio $\bar{\sigma}_{mi}/\bar{\sigma}_m$, this gives the relationship between the current values at step 'j' and those sought at the end of a strain step (i.e. at 'j+1') as

$$\left[\frac{\bar{\sigma}_{mi}}{\bar{\sigma}_m}\right]_{j+1} = \left[\frac{\bar{\sigma}_{mi}}{\bar{\sigma}_m}\left(1 + \frac{\dot{\bar{\sigma}}_{mi}}{\bar{\sigma}_{mi}} - \frac{\dot{\bar{\sigma}}_m}{\bar{\sigma}_m}\right)\right]_j \tag{C.47}$$

Equation C.47 is fine for advancing the integration provided the mean stress increment is known. But this is generally not the case, with often only the ratio of mean to shear stress increment being known (as, e.g. in a drained triaxial test). The *consistency condition* is used in such situations. The consistency condition is simply that, as plastic yield occurs and changes the size of the yield surface, the stress state must remain on the yield surface. The consistency condition is conventionally expressed by the notation that the yield surface

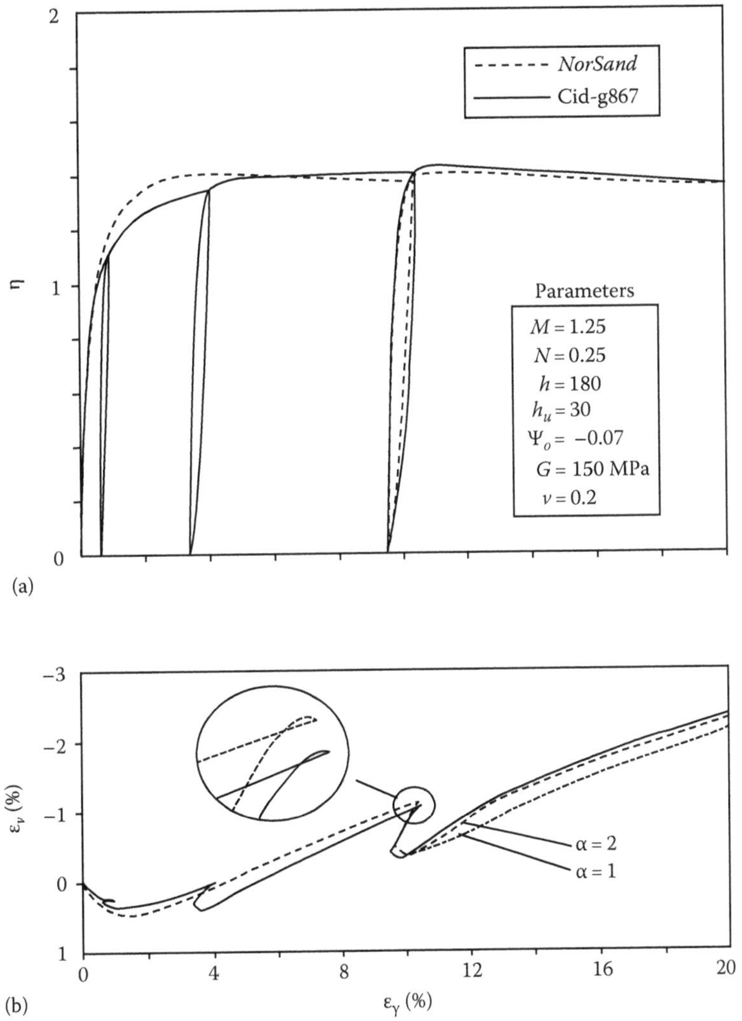

Figure C.9 Fit of *NorSand* to data with modified stiffness and dilatancy for reloading. (a) stress ratio η versus shear strain with $H_{reload} = 4H$ and (b) volumetric strain versus shear strain with N scaled by α. (From Jefferies, M.G., *Géotechnique*, 47, 1037, 1997. With permission from Institution of Civil Engineers.)

corresponds to the equation $F=0$, and thus, the consistency condition is simply $dF=0$. Applying $dF=0$ provides an additional equation to compute the effect of plastic yield.

C.7.1 Consistency case 1: on outer yield surface

The *NorSand* yield surface is expressed in the standard form $F=0$ as

$$F = \eta - M_i + M_i \ln\left(\frac{\bar{\sigma}_m}{\bar{\sigma}_{mi}}\right) \tag{C.48}$$

Noting that (C.48) is a function of three variables (M_i, η, and the stress ratio $\bar{\sigma}_{mi}/\bar{\sigma}_m$; θ influences the yield surface only through affecting M_i); taking differentials allows the consistency condition to be expressed as

$$dF = \frac{\partial F}{\partial M_i} dM_i + \frac{\partial F}{\partial \eta} d\eta + \frac{\partial F}{\partial(\bar{\sigma}_m/\bar{\sigma}_{mi})} d\left(\frac{\bar{\sigma}_m}{\bar{\sigma}_{mi}}\right) = 0 \tag{C.49}$$

On taking the required partial differentials of (C.48) and substituting in (C.49),

$$dF = d\eta - dM_i + \ln\left(\frac{\bar{\sigma}_m}{\bar{\sigma}_{mi}}\right) dM_i + M_i \left(\frac{\bar{\sigma}_m}{\bar{\sigma}_{mi}}\right)^{-1} d\left(\frac{\bar{\sigma}_m}{\bar{\sigma}_{mi}}\right) \tag{C.50}$$

Applying the consistency condition that $dF = 0$ (and changing to dot notation),

$$\dot{\eta} = \left(1 - \ln\left(\frac{\bar{\sigma}_m}{\bar{\sigma}_{mi}}\right)\right) \dot{M}_i - M_i \left(\frac{\bar{\sigma}_m}{\bar{\sigma}_{mi}}\right)^{-1} d\left(\frac{\bar{\sigma}_m}{\bar{\sigma}_{mi}}\right)$$

$$\Rightarrow \dot{\eta} = \frac{\eta}{M_i} \dot{M}_i - M_i \left(\frac{\bar{\sigma}_m}{\bar{\sigma}_{mi}}\right)^{-1} d\left(\frac{\bar{\sigma}_m}{\bar{\sigma}_{mi}}\right)$$

$$\Rightarrow \dot{\eta} = \frac{\eta}{M_i} \dot{M}_i - M_i \left(\frac{\bar{\sigma}_m}{\bar{\sigma}_{mi}}\right)^{-1} \left(\frac{\bar{\sigma}_{mi}\dot{\bar{\sigma}}_m - \bar{\sigma}_m\dot{\bar{\sigma}}_{mi}}{\bar{\sigma}_{mi}^2}\right) \tag{C.51}$$

$$\Rightarrow \dot{\eta} = \frac{\eta}{M_i} \dot{M}_i - M_i \left(\frac{\bar{\sigma}_{mi}\dot{\bar{\sigma}}_m - \bar{\sigma}_m\dot{\bar{\sigma}}_{mi}}{\bar{\sigma}_m\bar{\sigma}_{mi}}\right)$$

$$\Rightarrow \dot{\eta} = \frac{\eta}{M_i} \dot{M}_i - M_i \left(\frac{\dot{\bar{\sigma}}_{mi}}{\bar{\sigma}_{mi}} - \frac{\dot{\bar{\sigma}}_m}{\bar{\sigma}_m}\right)$$

Equation C.51 gives the change in the shear stress ratio η given a dimensionless change in the image stress σ_{mi} (a term given directly by the hardening law), the dimensionless change in the mean stress σ_m (which depends on the stress path and loading conditions), and a dimensionless change in the image stress ratio M_i (which will actually be small). It may generally be convenient to use (C.49) restated in terms of the stress invariants themselves, which is done as follows. From the definition of η,

$$\dot{\eta} = d\left(\frac{\bar{\sigma}_q}{\bar{\sigma}_m}\right) = \frac{\dot{\bar{\sigma}}_q}{\bar{\sigma}_m} - \eta \frac{\dot{\bar{\sigma}}_m}{\bar{\sigma}_m} \tag{C.52}$$

On substituting (C.52) in the consistency condition (C.51),

$$\frac{\dot{\bar{\sigma}}_q}{\bar{\sigma}_m} - \eta \frac{\dot{\bar{\sigma}}_m}{\bar{\sigma}_m} = \eta \frac{\dot{M}_i}{M_i} + M_i \left(\frac{\dot{\bar{\sigma}}_{mi}}{\bar{\sigma}_{mi}} - \frac{\dot{\bar{\sigma}}_m}{\bar{\sigma}_m}\right)$$

$$\Rightarrow \frac{\dot{\bar{\sigma}}_q}{\bar{\sigma}_m} = \eta \frac{\dot{M}_i}{M_i} + M_i \frac{\dot{\bar{\sigma}}_{mi}}{\bar{\sigma}_{mi}} + (\eta - M_i) \frac{\dot{\bar{\sigma}}_m}{\bar{\sigma}_m} \tag{C.53}$$

As $\eta - M_i = -D^p$, substituting this identity in (C.53),

$$\Rightarrow \frac{\dot{\bar{\sigma}}_q}{\bar{\sigma}_m} = \eta \frac{\dot{M}_i}{M_i} + M_i \frac{\dot{\bar{\sigma}}_{mi}}{\bar{\sigma}_{mi}} - D^p \frac{\dot{\bar{\sigma}}_m}{\bar{\sigma}_m} \tag{C.54}$$

Equations C.54 and C.51 are alternative forms of the consistency condition. Either can be used as most convenient, with the choice depending on the form of numerical integration.

C.7.2 Consistency case 2: on inner cap

On the inner cap, the yield surface is a flat plane which has a yield function written in the standard form $F=0$ as

$$F = \frac{\bar{\sigma}_{mi}}{\bar{\sigma}_m} - \exp\left(-\frac{\chi_{tc}\psi_i}{M_{tc}}\right) \tag{C.55}$$

Recalling that χ_{tc} and M_{tc} are material properties (i.e. constants), differentiation gives

$$dF = \frac{\dot{\bar{\sigma}}_{mi}}{\bar{\sigma}_m} - \frac{\bar{\sigma}_{mi}\,\dot{\bar{\sigma}}_m}{\bar{\sigma}_{mi}\,\bar{\sigma}_m} + \exp\left(-\frac{\chi_{tc}\psi_i}{M_{tc}}\right)\frac{\chi_{tc}}{M_{tc}}\dot{\psi}_i \tag{C.56}$$

On multiplying through by the ratio of mean to image stress, and setting $dF=0$ to enforce consistency,

$$\frac{\dot{\bar{\sigma}}_{mi}}{\bar{\sigma}_{mi}} = \frac{\dot{\bar{\sigma}}_m}{\bar{\sigma}_m} + \frac{\bar{\sigma}_m}{\bar{\sigma}_{mi}}\exp\left(-\frac{\chi_{tc}\psi_i}{M_{tc}}\right)\frac{\chi_{tc}}{M_{tc}}\dot{\psi}_i \tag{C.57}$$

If we use the approximation that $\chi_{tc}/M_{tc} \approx \chi_i/M_{i,tc}$, then the equation may be simplified by substituting (C.26) in (C.57) for the exponential term. In so doing, the mean and image stress terms also then cancel (because $\sigma_{mi}=\sigma_{mx}$ when the stress state is on the cap). Thus, (C.57) becomes

$$\frac{\dot{\bar{\sigma}}_{mi}}{\bar{\sigma}_{mi}} = \frac{\dot{\bar{\sigma}}_m}{\bar{\sigma}_m} + \frac{\chi_{tc}}{M_{tc}}\dot{\psi}_i \tag{C.58}$$

Equation C.58 is the basic form of the consistency condition for yield in unloading. During an increment of yielding on the inner cap, any change in the state parameter will tend to be small so that the earlier equation is almost a statement that the geometry of the yield surface does not change much. However, it is sometimes useful to have the change in state term explicit. Differentiating the image state parameter,

$$\dot{\psi}_i = \dot{e} - \dot{e}_{ci} \tag{C.59}$$

The first term in (C.59) is simply related to the total volumetric strain increment, since by definition, $\dot{\varepsilon}_v = -\dot{e}/(1+e)$. For the second term, if the semi-logarithmic form of the CSL is adopted (and any 'curved' CSL can be reduced to an equivalent locally semi-log idealization),

$$\dot{e}_{ci} = -\lambda\frac{\dot{\bar{\sigma}}_{mi}}{\bar{\sigma}_{mi}} \tag{C.60}$$

Putting (C.60) and (C.59) in (C.58) gives, after rearranging,

$$\frac{\dot{\bar{\sigma}}_{mi}}{\bar{\sigma}_{mi}}\left(1 - \frac{\chi_{tc}}{M_{tc}}\lambda\right) = \frac{\dot{\bar{\sigma}}_m}{\bar{\sigma}_m} - (1+e)\dot{\varepsilon}_v\frac{\chi_{tc}}{M_{tc}} \tag{C.61}$$

If the *k* model for volumetric elasticity is introduced,

$$\frac{\dot{\bar{\sigma}}_{mi}}{\bar{\sigma}_{mi}}\left(1-\frac{\chi_{tc}}{M_{tc}}\lambda\right)=\frac{\dot{\bar{\sigma}}_m}{\bar{\sigma}_m}\left(1-\frac{\chi_{tc}}{M_{tc}}\kappa\right)-(1+e)\dot{\varepsilon}_v^p\frac{\chi_{tc}}{M_{tc}} \tag{C.62}$$

For the undrained case, which is of particular interest during simulation of static liquefaction, (C.61) reduces to

$$\frac{\dot{\bar{\sigma}}_{mi}}{\bar{\sigma}_{mi}}\left(1-\frac{\chi_{tc}}{M_{tc}}\lambda\right)=\frac{\dot{\bar{\sigma}}_m}{\bar{\sigma}_m} \quad \text{iff} \quad \dot{\varepsilon}_v=0\wedge\dot{\bar{\sigma}}_m<0 \tag{C.63}$$

where the condition $\dot{\bar{\sigma}}_m<0$ clarifies that the loading vector is directed inwards.

C.8 STRESS DIFFERENTIALS

An essential step in numerical implementations of work-hardening plasticity is to determine the differentials of the stress invariants in terms of differentials of the three principal stresses. These derivatives are often quoted in finite element texts, but it is useful to see their origin. The derivations that follow will be given in terms of σ_1, with the derivatives in terms of σ_2 and σ_3 following by cyclic substitution of principal stresses. Recall that the stress invariants in terms of the three principal stresses are

$$\bar{\sigma}_m=\frac{(\bar{\sigma}_1+\bar{\sigma}_2+\bar{\sigma}_3)}{3} \tag{C.64}$$

$$\bar{\sigma}_q=\left[\frac{1}{2}(\bar{\sigma}_1-\bar{\sigma}_2)^2+\frac{1}{2}(\bar{\sigma}_2-\bar{\sigma}_3)^2+\frac{1}{2}(\bar{\sigma}_3-\bar{\sigma}_1)^2\right]^{\frac{1}{2}} \tag{C.65}$$

$$\sin(3\theta)=13.5\frac{s_1s_2s_3}{\bar{\sigma}_q^3} \quad \text{with } s_1=\frac{(2\sigma_1-\sigma_2-\sigma_3)}{3} \text{ etc.} \tag{C.66}$$

From (C.64), it immediately follows that

$$\frac{\partial\bar{\sigma}_m}{\partial\bar{\sigma}_1}=\frac{\partial\bar{\sigma}_m}{\partial\bar{\sigma}_2}=\frac{\partial\bar{\sigma}_m}{\partial\bar{\sigma}_3}=\frac{1}{3} \tag{C.67}$$

Turning to the deviatoric stress, from (C.65),

$$\frac{\partial\bar{\sigma}_q}{\partial\bar{\sigma}_1}=\frac{1}{2}\left[\frac{1}{2}(\bar{\sigma}_1-\bar{\sigma}_2)^2+\frac{1}{2}(\bar{\sigma}_2-\bar{\sigma}_3)^2+\frac{1}{2}(\bar{\sigma}_3-\bar{\sigma}_1)^2\right]^{\frac{-1}{2}}$$

$$d\left(\frac{1}{2}(\bar{\sigma}_1-\bar{\sigma}_2)^2+\frac{1}{2}(\bar{\sigma}_2-\bar{\sigma}_3)^2+\frac{1}{2}(\bar{\sigma}_3-\bar{\sigma}_1)^2\right)$$

$$\Rightarrow\frac{\partial\bar{\sigma}_q}{\partial\bar{\sigma}_1}=\frac{1}{2\bar{\sigma}_q}d\left(\tfrac{1}{2}2(\bar{\sigma}_1-\bar{\sigma}_2)1+0+\tfrac{1}{2}2(\bar{\sigma}_3-\bar{\sigma}_1)(-1)\right)$$

$$\Rightarrow\frac{\partial\bar{\sigma}_q}{\partial\bar{\sigma}_1}=\frac{1}{2\bar{\sigma}_q}(2\bar{\sigma}_1-\bar{\sigma}_2-\bar{\sigma}_3) \tag{C.68}$$

The other required partial differentials are found by cyclic substitution in (C.68). The derivatives with respect to θ are tedious and best approached by going through the invariant J_3:

$$\frac{\partial\theta}{\partial\bar{\sigma}_1} = \frac{\partial\theta}{\partial J_3}\frac{\partial J_3}{\partial\bar{\sigma}_1} \quad \text{where } J_3 = s_1 s_2 s_3 \tag{C.69}$$

with s_1, s_2, s_3 being given in (C.66). Rewriting (C.66) in terms of J_3 gives

$$\sin(3\theta) = 13.5\frac{J_3}{\bar{\sigma}_q^3} \tag{C.70}$$

On taking the partial derivative of (C.70) with respect to J_3,

$$3\cos(3\theta)\frac{\partial\theta}{\partial J_3} = \frac{13.5}{\bar{\sigma}_q^3} \Rightarrow \frac{\partial\theta}{\partial J_3} = \frac{4.5}{\cos(3\theta)\bar{\sigma}_q^3} \tag{C.71}$$

Moving to the definition of J_3, (C.69), on differentiating with respect to σ_1,

$$\frac{\partial J_3}{\partial\bar{\sigma}_1} = \frac{\partial(s_1 s_2 s_3)}{\partial\bar{\sigma}_1} = \frac{\partial s_1}{\partial\bar{\sigma}_1}(s_2 s_3) + \frac{\partial(s_2 s_3)}{\partial\bar{\sigma}_1}s_1$$

$$\Rightarrow \frac{\partial J_3}{\partial\bar{\sigma}_1} = \frac{\partial s_1}{\partial\bar{\sigma}_1}(s_2 s_3) + \frac{\partial s_2}{\partial\bar{\sigma}_1}(s_3 s_1) + \frac{\partial s_3}{\partial\bar{\sigma}_1}(s_2 s_1)$$

and as (see C.66) $\frac{\partial s_1}{\partial\bar{\sigma}_1} = \frac{2}{3}$ with $\frac{\partial s_2}{\partial\bar{\sigma}_1} - \frac{1}{3} = \frac{\partial s_3}{\partial\bar{\sigma}_1}$

It then follows that

$$\frac{\partial J_3}{\partial\bar{\sigma}_1} = \frac{2}{3}(s_2 s_3) - \frac{1}{3}(s_3 s_1) - \frac{1}{3}(s_2 s_1) \tag{C.72}$$

So, finally, on substituting (C.72) with (C.71) into (C.69),

$$\frac{\partial\theta}{\partial\bar{\sigma}_1} = \frac{3(2s_2 s_3 - s_3 s_1 - s_2 s_1)}{2\cos(3\theta)\bar{\sigma}_q^3} \tag{C.73}$$

The other required partial differentials are found by cyclic substitution in (C.73).

C.9 DIRECT NUMERICAL INTEGRATION FOR ELEMENT TESTS

Models such as *NorSand* are used in finite element analyses, direct numerical integration of is also useful to get model predictions of standard laboratory tests for a couple of reasons. Although some parameters can be determined by regression of data points abstracted from tests or from identification of a particular aspect of the test (e.g. M_{tc}), the goal of a model is to represent the entire behaviour. This goal is best achieved by simulating the entire test and iterating on the inferred soil properties to get the best overall fit of the model to the data. Models also have much to offer in their own right as predictive tools as in, for example, understanding how the post-liquefaction strength is affected by initial conditions.

Plasticity models are defined in terms of differentials or incremental strains, with often delightfully simple form. But, a stress–strain curve is needed to compare with the laboratory test. This means models have to be integrated over the stress or stain path of the laboratory test, which is straightforward as the tests have uniform stress conditions with known stress or strain paths (or some combination of the two). These known stress and strain conditions (often known as ratios, but it depends on the particular test) are substituted in the consistency condition to give the relationship between hardening and change in yield surface size for the particular test. In this way, the yield surface and hardening equations are integrated over the known stress–strain path to get the prediction of the test behaviour.

NorSand does not have any closed-form solution, so numerical integration is needed. This is not difficult and may be done in a standard Euler manner of working out the current conditions and incremental gradients at that current condition (say loading step j), applying an increment of plastic shear strain, and then computing the corresponding values at the end of the strain step (= loadstep $j+1$) by applying the computed gradients. For example, the shear stress ratio is simply $\eta_{j+1} = \eta_j + \delta\eta$ where $\delta\eta$ is the increment of the shear stress ratio that develops in the loading step.

Direct numerical integration is easily and conveniently implemented within VBA subroutines of an Excel spreadsheet, and these are provided in the various *NorSand.xls* spreadsheets. They can be accessed using the Visual Basic editor (open the subroutines using the menu/*Tools/Macro/Visual Basic Editor* or just press 'Alt+F11'). The equations used in the various subroutines are derived in the following sections. Several subroutines are used because the details of the integration depend on the test and whether it is drained or undrained, most conveniently implemented as a separate subroutine for each. Using separate subroutines also speeds up the solution by avoiding 'if' statements in the integration loop. The common integration used in all subroutines is as follows:

Loop over …
Apply plastic shear strain increment
Recover all plastic strain increments from stress–dilatancy
Use hardening rule to get increment of image stress
Apply consistency condition to determine new stress state
Add in elastic strains from stress changes in loadstep
Update strains, void ratio, state parameter

The key is how the consistency condition is used on a test-by-test basis. Everything else is common amongst all tests. The reason that the use of the consistency condition changes is that each element test has its own stress and strain paths, which means the relative amounts of the stress increments change from one test to another (and even during the test). This use of the consistency condition is now derived in detail for various laboratory tests. Note that the derivations are all for monotonic (continuous) loading from a normally consolidated initial state. Of course, unloading/reloading and overconsolidation are important and, indeed, are crucial to cyclic loading. However, these aspects are conceptually straightforward but tedious in code (they are coded with annotations in the *NorSand.xls*, so refer to the VBA code if interested).

C.9.1 Undrained triaxial tests

This is by far the simplest case, and there is little difference between triaxial extension and compression. The Euler rule for the integration is

$$\begin{bmatrix} \bar{\sigma}_{mi} \\ \bar{\sigma}_m \end{bmatrix}_{j+1} = \begin{bmatrix} \bar{\sigma}_{mi} \\ \bar{\sigma}_m \end{bmatrix} \left(1 + \frac{\dot{\bar{\sigma}}_{mi}}{\bar{\sigma}_{mi}} - \frac{\dot{\bar{\sigma}}_m}{\bar{\sigma}_m} \right) \Bigg]_i \qquad \text{(C.47 bis)}$$

There are two terms between the brackets on the right-hand side. The first of these is given by the hardening law. The second term follows from the undrained condition:

$$\dot{\varepsilon}_v = 0 \Leftrightarrow \dot{\varepsilon}_v^p = -\dot{\varepsilon}_v^e \quad \text{from which } \dot{\bar{\sigma}}_m = -\dot{\varepsilon}_v^p K \tag{C.74}$$

where K is the elastic bulk modulus. The plastic volumetric strain is known from the stress–dilatancy rule, (C.10). The consistency condition is essentially trivial as we immediately know the new mean stress without any further algebraic manipulation. Hence, using (C.47), the yield surface hardness is updated, and then the deviator stress recovered as the mean stress and current value for M_i are both known. It is then trivial to calculate the elastic shear strain increment:

$$\dot{\varepsilon}_q^e = \frac{\dot{\bar{\sigma}}_q}{3G} \tag{C.75}$$

from which all the strain increments are now fully defined.

Notice that the effective stresses in the undrained triaxial test respond only to the shear component of load. An external load increment that increases the total mean stress on the sample creates an equal response in the pore water pressure (i.e. it is assumed that there is full sample saturation and so $B = 1$). Partial saturation effects can be simulated easily enough by adding B to the parameter list, but doing so adds no insight for the present application and is therefore neglected.

Principal stresses are readily recovered from the stress invariants and are (for both compression and extension)

$$\bar{\sigma}_1 = \tfrac{2}{3}\bar{\sigma}_q + \bar{\sigma}_m \tag{C.76a}$$

$$\bar{\sigma}_3 = \bar{\sigma}_m - \tfrac{1}{3}\bar{\sigma}_q \tag{C.76b}$$

There are only two differences between triaxial compression and triaxial extension. First, a much reduced M will be used in extension (see Chapter 3). Second, the recovery of the principal strain increments from the strain increment invariants differs because of the differing symmetry between extension and compression.

For triaxial *compression*, $\dot{\varepsilon}_2 = \dot{\varepsilon}_3 \Rightarrow \dot{\varepsilon}_v = \dot{\varepsilon}_1 + 2\dot{\varepsilon}_3$ and $\dot{\varepsilon}_q = \tfrac{2}{3}(\dot{\varepsilon}_1 - \dot{\varepsilon}_3)$

as $\dot{\varepsilon}_v = 0$ (the undrained condition)

$$\Rightarrow \dot{\varepsilon}_1 = \dot{\varepsilon}_q \tag{C.77}$$

For triaxial extension: $\dot{\varepsilon}_2 = \dot{\varepsilon}_1 \Rightarrow \dot{\varepsilon}_v = 2\dot{\varepsilon}_1 + \dot{\varepsilon}_3$ (C.78)

and $\dot{\varepsilon}_q$ is unchanged, similarly for triaxial extension

$$\Rightarrow \dot{\varepsilon}_1 = \tfrac{1}{2}\dot{\varepsilon}_q \tag{C.79}$$

Equations C.77 and C.78 illustrate one source of confusion in comparison of triaxial compression and extension test data. Such comparison is often on the basis of q versus axial strain. And what is now clear is that in such a comparison, apples are being compared with oranges (or half an apple). It is then unsurprising that 'different' stress–strain behaviours are reported by experimenters.

C.9.2 Drained triaxial compression

Drained triaxial compression differs from the undrained in that the mean stress increment is not immediately known (unless we have the particular case of constant mean effective stress tests when it is, of course, zero). Rather, from the configuration of the test, we know the ratio of increase in shear stress to increase in mean stress. This ratio is 1/3 for a standard triaxial compression test and –1/3 for a triaxial extension tests. Other values can be programmed into controlled stress path tests. For the derivation, the value does not matter, as the load direction parameter, L, is used, and which is known from the test conditions:

$$L = \frac{\Delta\bar{\sigma}_q}{\Delta\bar{\sigma}_m} \tag{C.80}$$

The notation Δ is used to indicate that there are finite changes over which this ratio is expected to be constant. The definition of η is now used. On differentiating η,

$$\dot{\eta} = \left(\frac{\dot{\bar{\sigma}}_q}{\dot{\bar{\sigma}}_m} - \eta\right)\frac{\dot{\bar{\sigma}}_m}{\bar{\sigma}_m} \Rightarrow \frac{\dot{\bar{\sigma}}_m}{\bar{\sigma}_m} = \frac{\dot{\eta}}{L-\eta} \tag{C.81}$$

Substituting (C.81) in the consistency condition (C.51) gives

$$\dot{\eta} = \eta\frac{\dot{M}_i}{M_i} + M_i\left(\frac{\dot{\bar{\sigma}}_{mi}}{\bar{\sigma}_{mi}} - \frac{\dot{\eta}}{L-\eta}\right)$$

$$\Rightarrow \dot{\eta} = \left(\eta\frac{\dot{M}_i}{M_i} + M_i\frac{\dot{\bar{\sigma}}_{mi}}{\bar{\sigma}_{mi}}\right)\Big/\left(1 + \frac{M_i}{L-\eta}\right) \tag{C.82}$$

Principal strain increments follow as the undrained case, except that there is now non-zero volumetric strain.

C.9.3 Drained plane strain: Cornforth's apparatus

Cornforth (1961, 1964) was the first to investigate plane strain behaviour systematically and in comparison to triaxial compression and extension conditions. As explained in Chapter 3, this is an important body of data even though it is 40 years old. The apparatus Cornforth used was much like a square triaxial test with one axis being restrained to force plane strain and the other imposing constant σ_3 (see Figure 2.43).

With Cornforth's apparatus, the combination of strain constraint and constant stress, and the loss of symmetry, requires a somewhat more elaborate treatment of how the consistency condition is used. The consistency condition in terms of stress invariants is

$$\frac{\dot{\bar{\sigma}}_q}{\bar{\sigma}_m} = \eta\frac{\dot{M}_i}{M_i} + M_i\frac{\dot{\bar{\sigma}}_{mi}}{\bar{\sigma}_{mi}} - D^p\frac{\dot{\bar{\sigma}}_m}{\bar{\sigma}_m} \tag{C.54 bis}$$

Progress requires (C.54) be recast in terms of the increment in σ_1. Remembering that these tests were constant σ_3; the differentials of the stress invariants are

$$\dot{\bar{\sigma}}_m = \frac{\partial\bar{\sigma}_m}{\partial\bar{\sigma}_1}\dot{\bar{\sigma}}_1 + \frac{\partial\bar{\sigma}_m}{\partial\bar{\sigma}_2}\dot{\bar{\sigma}}_2 \Rightarrow \dot{\bar{\sigma}}_m = \frac{1}{3}\dot{\bar{\sigma}}_1 + \frac{1}{3}\dot{\bar{\sigma}}_2 \tag{C.83}$$

$$\dot{\bar{\sigma}}_q = \frac{\partial \bar{\sigma}_q}{\partial \bar{\sigma}_1}\dot{\bar{\sigma}}_1 + \frac{\partial \bar{\sigma}_q}{\partial \bar{\sigma}_2}\dot{\bar{\sigma}}_2 \tag{C.84}$$

In (C.84), recall that the partial differential terms were previously derived as straightforward equations in terms of the current stress state in Section C.8 (Equations C.67 and C.68). On substituting (C.83) and (C.84) in (C.54),

$$\frac{\partial \bar{\sigma}_q}{\partial \bar{\sigma}_1}\frac{\dot{\bar{\sigma}}_1}{\bar{\sigma}_m} + \frac{\partial \bar{\sigma}_q}{\partial \bar{\sigma}_2}\frac{\dot{\bar{\sigma}}_2}{\bar{\sigma}_m} = \eta\frac{\dot{M}_i}{M_i} + M_i\frac{\dot{\bar{\sigma}}_{mi}}{\bar{\sigma}_{mi}} - D^p\left(\frac{\dot{\bar{\sigma}}_1}{3\bar{\sigma}_m} + \frac{\dot{\bar{\sigma}}_2}{3\bar{\sigma}_m}\right) \tag{C.85}$$

The plane strain condition gives (continuing to remember that $\dot{\bar{\sigma}}_3 = 0$ for these tests)

$$-\dot{\varepsilon}_2^p = \dot{\varepsilon}_2^e = \frac{1}{E}(\dot{\bar{\sigma}}_2 - \nu\dot{\bar{\sigma}}_1) \Rightarrow \dot{\bar{\sigma}}_2 = \nu\dot{\bar{\sigma}}_1 - E\dot{\varepsilon}_2^p \tag{C.86}$$

And, finally, on substituting (C.86) in (C.85),

$$\frac{\partial \bar{\sigma}_q}{\partial \bar{\sigma}_1}\frac{\dot{\bar{\sigma}}_1}{\bar{\sigma}_m} + \frac{\partial \bar{\sigma}_q}{\partial \bar{\sigma}_2}\frac{\nu\dot{\bar{\sigma}}_1 - E\dot{\varepsilon}_2^p}{\bar{\sigma}_m} = \eta\frac{\dot{M}_i}{M_i} + M_i\frac{\dot{\bar{\sigma}}_{mi}}{\bar{\sigma}_{mi}} - D^p\left(\frac{\dot{\bar{\sigma}}_1}{3\bar{\sigma}_m} + \frac{\nu\dot{\bar{\sigma}}_1 - E\dot{\varepsilon}_2^p}{3\bar{\sigma}_m}\right)$$

$$\Rightarrow \frac{\dot{\bar{\sigma}}_1}{\bar{\sigma}_m}\left(\frac{\partial \bar{\sigma}_q}{\partial \bar{\sigma}_1} + \nu\frac{\partial \bar{\sigma}_q}{\partial \bar{\sigma}_2} + \frac{(1+\nu)D^p}{3}\right) = \eta\frac{\dot{M}_i}{M_i} + M_i\frac{\dot{\bar{\sigma}}_{mi}}{\bar{\sigma}_{mi}} + \frac{E\dot{\varepsilon}_2^p}{\bar{\sigma}_m}\left(\frac{D^p}{3} + \frac{\partial \bar{\sigma}_q}{\partial \bar{\sigma}_2}\right) \tag{C.87}$$

Equation C.87 gives the major principal stress increment in terms of known variables in the loadstep, which is what is sought to progress the integration loop. The intermediate principal stress is then recovered through (C.86).

C.9.4 Undrained simple shear tests

The simple shear test aims to approximate the situation in which soil is sheared in plane strain under constant mean normal total stress, the situation illustrated in Figure C.10. Simple shear is an approximation to the situation prevailing during slope stability, and hence this test ought to be of considerable influence. The difficulty is that the test apparatus approximates only the idealized conditions of the in-situ situation, and the horizontal stress is commonly unmeasured, which leaves an ambiguity when modelling test data. There is an interesting 'Symposium in Print' on the simple shear test in the March 1987 issue of *Geotechnique*, which discusses some of these issues. But simple shear is becoming an increasingly popular test, and we can make progress theoretically.

The experimental difficulties of getting to simple shear do not constrain the numerical modelling, as we start from known (or assumed) conditions and integrate over the idealized situation of interest. This integration is one step more complicated than that for Cornforth's tests, as there are even more mixed stress and strain boundary conditions combined with the rotation of the principal directions. The following boundary conditions apply to the idealization shown in Figure C.10:

$\dot{\varepsilon}_x = 0$, applying the infinite extent in the horizontal direction (C.88)

$\dot{\varepsilon}_2 = 0$, applying the plane strain condition (C.89)

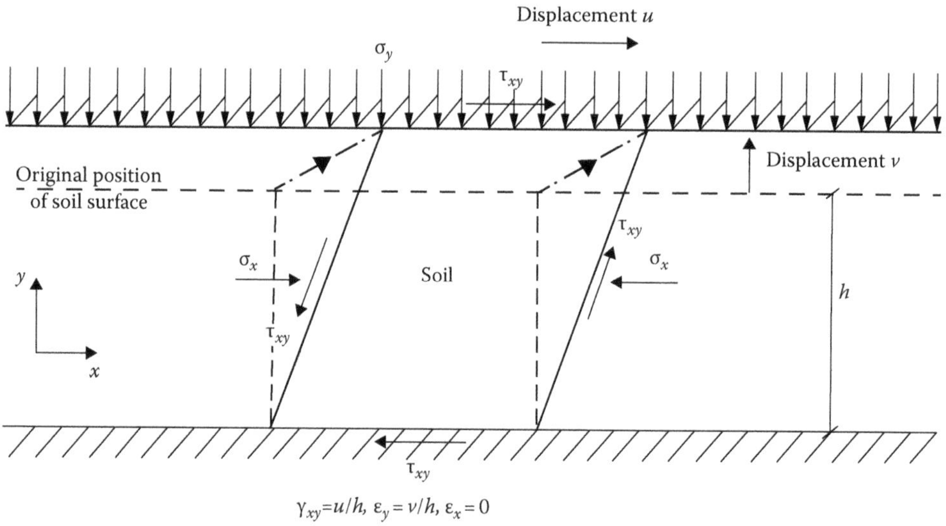

Figure C.10 Simple shear conditions. (From Potts, D.M. et al., *Géotechnique*, 37(1), 11, 1987. With permission from Institution of Civil Engineers.)

$$\dot{\varepsilon}_v = 0, \text{ the undrained condition ("}v\text{" is volumetric, not vertical)} \tag{C.90}$$

$$\dot{\sigma}_y = 0, \text{ the loading condition of simple shear (note total stress)} \tag{C.91}$$

The start of the integration is the general consistency condition expressed in terms of increments in the stress invariants:

$$\frac{\dot{\bar{\sigma}}_q}{\bar{\sigma}_m} = \eta\frac{\dot{M}_i}{M_i} + M_i\frac{\dot{\bar{\sigma}}_{mi}}{\bar{\sigma}_{mi}} - D^p\frac{\dot{\bar{\sigma}}_m}{\bar{\sigma}_m} \tag{C.54 bis}$$

This equation can be simplified in the undrained case as, on application of the shear strain increment, all terms on the RHS are known either from the hardening law or from the undrained condition. Write this simplified equation as

$$\frac{\dot{\bar{\sigma}}_q}{\bar{\sigma}_m} = Y \tag{C.92}$$

Taking the differential of the shear stress invariant (and recall that the partial differentials were previously derived as simple functions of the stress state (C.68)),

$$\dot{\bar{\sigma}}_q = \frac{\partial\bar{\sigma}_q}{\partial\bar{\sigma}_1}\dot{\bar{\sigma}}_1 + \frac{\partial\bar{\sigma}_q}{\partial\bar{\sigma}_2}\dot{\bar{\sigma}}_2 + \frac{\partial\bar{\sigma}_q}{\partial\bar{\sigma}_3}\dot{\bar{\sigma}}_3 \tag{C.93}$$

Invoking the elastic–plastic strain decomposition, the plane strain condition in simple shear gives

$$\dot{\varepsilon}_2 = 0 \Rightarrow -\dot{\varepsilon}_2^p = \dot{\varepsilon}_2^e = \frac{1}{E}\left(\dot{\bar{\sigma}}_2 - \nu\dot{\bar{\sigma}}_1 - \nu\dot{\bar{\sigma}}_3\right)$$

$$\Rightarrow \dot{\bar{\sigma}}_2 = \nu\dot{\bar{\sigma}}_1 + \nu\dot{\bar{\sigma}}_3 - E\dot{\varepsilon}_2^p \tag{C.94}$$

Eliminating σ_2 by substituting (C.94) in (C.93)

$$\dot{\bar{\sigma}}_q = \left(\frac{\partial\bar{\sigma}_q}{\partial\bar{\sigma}_1} + \nu\frac{\partial\bar{\sigma}_q}{\partial\bar{\sigma}_2}\right)\dot{\bar{\sigma}}_1 + \left(\frac{\partial\bar{\sigma}_q}{\partial\bar{\sigma}_3} + \nu\frac{\partial\bar{\sigma}_q}{\partial\bar{\sigma}_2}\right)\dot{\bar{\sigma}}_3 - E\dot{\varepsilon}_p^2\frac{\partial\bar{\sigma}_p}{\partial\bar{\sigma}_2} \tag{C.95}$$

On substituting (C.92) in (C.95)

$$\left(\frac{\partial\bar{\sigma}_q}{\partial\bar{\sigma}_1} + \nu\frac{\partial\bar{\sigma}_q}{\partial\bar{\sigma}_2}\right)\frac{\dot{\bar{\sigma}}_1}{\bar{\sigma}_m} + \left(\frac{\partial\bar{\sigma}_q}{\partial\bar{\sigma}_3} + \nu\frac{\partial\bar{\sigma}_q}{\partial\bar{\sigma}_2}\right)\frac{\dot{\bar{\sigma}}_3}{\bar{\sigma}_m} - \frac{E\dot{\varepsilon}_p^2}{\bar{\sigma}_m}\frac{\partial\bar{\sigma}_p}{\partial\bar{\sigma}_2} = Y \tag{C.96}$$

Equation C.96 is better written in terms of local variables a, b, c for further manipulation; thus

$$a\frac{\dot{\bar{\sigma}}_1}{\bar{\sigma}_m} + b\frac{\dot{\bar{\sigma}}_3}{\bar{\sigma}_m} = c \tag{C.96a}$$

where

$$a = \left(\frac{\partial\bar{\sigma}_q}{\partial\bar{\sigma}_1} + \nu\frac{\partial\bar{\sigma}_q}{\partial\bar{\sigma}_2}\right) \tag{C.96b}$$

$$b = \left(\frac{\partial\bar{\sigma}_q}{\partial\bar{\sigma}_3} + \nu\frac{\partial\bar{\sigma}_q}{\partial\bar{\sigma}_2}\right) \tag{C.96c}$$

$$c = Y + \frac{E\dot{\varepsilon}_p^2}{\bar{\sigma}_m}\frac{\partial\bar{\sigma}_q}{\partial\bar{\sigma}_2} \tag{C.96d}$$

All of a, b, c are explicitly known for the loadstep from the stress state at the start of the step, the flow rule and the result of the strain increment operating on the hardening law. This leaves two unknowns, σ_1 and σ_3. Returning to the condition of no volumetric strain (recall this is for undrained loading), the principal strains are split into elastic and plastic components to get (noting the plane strain condition of $\dot{\varepsilon}_2 = 0$)

$$\dot{\varepsilon}_v = 0 = \dot{\varepsilon}_1 + \dot{\varepsilon}_3 \Rightarrow \dot{\varepsilon}_1^e + \dot{\varepsilon}_3^e = -\left(\dot{\varepsilon}_1^p + \dot{\varepsilon}_3^p\right)$$

$$\Rightarrow \dot{\bar{\sigma}}_1 - \nu\dot{\bar{\sigma}}_2 - \nu\dot{\bar{\sigma}}_3 + \dot{\bar{\sigma}}_3 - \nu\dot{\bar{\sigma}}_2 - \nu\dot{\bar{\sigma}}_1 = -E\left(\dot{\varepsilon}_1^p + \dot{\varepsilon}_3^p\right) \tag{C.97}$$

Again using (C.94) to eliminate σ_2 in (C.97),

$$\Rightarrow \dot{\bar{\sigma}}_1 - \nu\left(\nu\dot{\bar{\sigma}}_1 + \nu\dot{\bar{\sigma}}_3 - E\dot{\varepsilon}_2^p\right) - \nu\dot{\bar{\sigma}}_3 + \dot{\bar{\sigma}}_3 - \nu\left(\nu\dot{\bar{\sigma}}_1 + \nu\dot{\bar{\sigma}}_3 - E\dot{\varepsilon}_2^p\right) - \nu\dot{\bar{\sigma}}_1 = -E\left(\dot{\varepsilon}_1^p + \dot{\varepsilon}_3^p\right)$$

$$\Rightarrow \dot{\bar{\sigma}}_1(1 - \nu - \nu^2) + \dot{\bar{\sigma}}_3\left(1 - \nu - \nu^2\right) = -E\left(\dot{\varepsilon}_1^p + \dot{\varepsilon}_3^p + 2\nu\dot{\varepsilon}_2^p\right) \tag{C.98}$$

or in terms of additional local variables d, f,

$$\dot{\bar{\sigma}}_1 d + \dot{\bar{\sigma}}_3 d = -f$$

$$\Rightarrow \dot{\bar{\sigma}}_3 = -\left(\frac{f}{d + \dot{\bar{\sigma}}_1}\right) \tag{C.99a}$$

where

$$d = 1 - \nu - 2\nu^2 \tag{C.99b}$$

$$f = E\left(\dot{\varepsilon}_1^p + \dot{\varepsilon}_3^p + 2\nu\dot{\varepsilon}_2^p\right) \tag{C.99c}$$

On eliminating σ_3 between (C.96) and (C.99),

$$(a - b)\dot{\bar{\sigma}}_1 = c\bar{\sigma}_m + \frac{bf}{d} \tag{C.100}$$

Equation C.100 gives the increment in major principal stress in terms of variables all of which are known in the loadstep, and the minor principal stress increment is then recovered through (C.99). The principal stresses are then updated. However, these principal stresses are not what is measured during the simple shear test. We proceed by applying the normal stress boundary condition of simple shear and recalling that the undrained condition gives the pore pressure change:

$$\dot{\sigma}_v = 0 \Rightarrow \dot{\bar{\sigma}}_v = -K\dot{\varepsilon}_v^p \tag{C.101}$$

from which we update the vertical effective stress. The relationship between vertical (y-direction) stress and the principal stresses in the plane of loading is

$$\bar{\sigma}_v = \bar{\sigma}_1 \cos^2\alpha + \bar{\sigma}_3 \sin^2\alpha \tag{C.102}$$

where α is the angle between the '1' direction and the vertical. On rearranging (C.102) using $\cos^2 x = 1 - \sin^2 x$ and inverting

$$\sin^2\alpha = \frac{(\bar{\sigma}_1 - \bar{\sigma}_y)}{(\bar{\sigma}_1 - \bar{\sigma}_3)} \tag{C.103}$$

The measured shear stress in the simple shear test τ_{xy} then follows immediately as

$$\tau_{xy} = \frac{1}{2}(\bar{\sigma}_1 - \bar{\sigma}_3)\sin 2\alpha \tag{C.104}$$

All that remains to be determined is the shear strain. Because stress and strain increments are coaxial, the shear strain increment is recovered most simply from the updated principal strain increments:

$$\dot{\gamma}_{xy} = (\dot{\varepsilon}_1 - \dot{\varepsilon}_3)\sin 2\alpha \tag{C.105}$$

Strictly, modelling of simple shear should use the hardening law with the principal stress rotation–induced softening. However, there is so much uncertainty in fitting simple shear test data (because the test measures only two of the three stresses needed to specify the stress state) that it is simply not worth dealing with an additional plastic modulus. The simpler form of hardening is used as adequate given the other uncertainties with simple shear tests.

C.9.5 Drained simple shear tests

It may sometimes be interesting to evaluate drained simple shear tests. The derivation proceeds much like the undrained case except that the conditions in the test change to

$$\dot{\varepsilon}_x = 0 \text{ applying the infinite extent in the horizontal direction} \tag{C.106}$$

$$\dot{\varepsilon}_2 = 0 \text{ applying the plane strain condition} \tag{C.107}$$

$$\dot{\bar{\sigma}}_y = 0 \text{ the loading condition of drained simple shear} \tag{C.108}$$

The start of the integration is again the general consistency condition expressed in terms of the increments in the stress invariants:

$$\frac{\dot{\bar{\sigma}}_q}{\bar{\sigma}_m} = \eta\frac{\dot{M}_i}{M_i} - M_i\frac{\dot{\bar{\sigma}}_{mi}}{\bar{\sigma}_{mi}} - D^p\frac{\dot{\bar{\sigma}}_m}{\bar{\sigma}_m} \tag{C.54 bis}$$

It is convenient to rewrite this consistency condition distinguishing between what is known in the loadstep (Y') and what is being solved for:

$$\frac{\dot{\bar{\sigma}}_q}{\bar{\sigma}_m} = Y' - D^p\frac{\dot{\bar{\sigma}}_m}{\bar{\sigma}_m} \quad \text{where: } Y' = \eta\frac{\dot{M}_i}{M_i} - M_i\frac{\dot{\bar{\sigma}}_{mi}}{\bar{\sigma}_{mi}} \tag{C.109}$$

Then, exactly as just done for the undrained case, by taking the differential of the deviator stress invariant and eliminating the intermediate principal stress by invoking the plane strain condition,

$$\dot{\bar{\sigma}}_q = \left(\frac{\partial\bar{\sigma}_q}{\partial\bar{\sigma}_1} + \nu\frac{\partial\bar{\sigma}_q}{\partial\bar{\sigma}_2}\right)\dot{\bar{\sigma}}_1 + \left(\frac{\partial\bar{\sigma}_q}{\partial\bar{\sigma}_3} + \nu\frac{\partial\bar{\sigma}_q}{\partial\bar{\sigma}_2}\right)\dot{\bar{\sigma}}_3 - E\dot{\varepsilon}_p^2\frac{\partial\bar{\sigma}_p}{\partial\bar{\sigma}_2} \tag{C.95 bis}$$

Similarly, a relationship is derived for the mean stress increment:

$$\dot{\bar{\sigma}}_m = \left(\frac{\partial\bar{\sigma}_m}{\partial\bar{\sigma}_1} + \nu\frac{\partial\bar{\sigma}_m}{\partial\bar{\sigma}_2}\right)\dot{\bar{\sigma}}_1 + \left(\frac{\partial\bar{\sigma}_m}{\partial\bar{\sigma}_3} + \nu\frac{\partial\bar{\sigma}_m}{\partial\bar{\sigma}_2}\right)\dot{\bar{\sigma}}_3 - E\dot{\varepsilon}_p^2\frac{\partial\bar{\sigma}_m}{\partial\bar{\sigma}_2}$$

$$\Rightarrow \dot{\bar{\sigma}}_m = \frac{1}{3}(1+\nu)\left(\dot{\bar{\sigma}}_1 + \dot{\bar{\sigma}}_3\right) - \frac{E}{3}\dot{\varepsilon}_2^p \tag{C.110}$$

On substituting (C.95) and (C.110) in (C.109) and introducing dimensionless local variables a', b', c',

$$a'\frac{\dot{\bar{\sigma}}_1}{\bar{\sigma}_m} + b'\frac{\dot{\bar{\sigma}}_3}{\bar{\sigma}_m} = c' \tag{C.111a}$$

where

$$a'\left(\frac{\partial\bar{\sigma}_q}{\partial\bar{\sigma}_1} + \nu\frac{\partial\bar{\sigma}_q}{\partial\bar{\sigma}_2}\right) + D^p\frac{1+\nu}{3} \tag{C.111b}$$

$$b'\left(\frac{\partial\bar{\sigma}_q}{\partial\bar{\sigma}_3} + \nu\frac{\partial\bar{\sigma}_q}{\partial\bar{\sigma}_2}\right) + D^p\frac{1+\nu}{3} \tag{C.111c}$$

$$c' = Y' + \frac{E}{\bar{\sigma}_m}\left(\frac{\partial\bar{\sigma}_q}{\partial\bar{\sigma}_2} - \frac{D^p}{3}\right)\dot{\varepsilon}_2^p \tag{C.111d}$$

All of the variables needed to evaluate a', b', c' are known at the start of the loadstep. The volumetric strain is expressed in terms of the principal strains (split into elastic and plastic components), noting the plane strain condition of $\varepsilon_2 = 0$, as

$$\dot{\varepsilon}_v = D^p \dot{\varepsilon}_q + \frac{\dot{\bar{\sigma}}_m}{K} = \dot{\varepsilon}_1 + \dot{\varepsilon}_3 = \dot{\varepsilon}_1^e + \dot{\varepsilon}_1^p + \dot{\varepsilon}_3^e + \dot{\varepsilon}_3^p \tag{C.112}$$

Using (C.110) to eliminate $\dot{\bar{\sigma}}_m$ in (C.112),

$$D^p \dot{\varepsilon}_q + \frac{1+\nu}{3K}\left(\dot{\bar{\sigma}}_1 + \dot{\bar{\sigma}}_3\right) - \frac{E}{3K}\dot{\varepsilon}_2^p = \dot{\varepsilon}_1^e + \dot{\varepsilon}_1^p + \dot{\varepsilon}_3^e + \dot{\varepsilon}_3^p$$

and then writing the elastic stains in terms of stress increments,

$$D^p \dot{\varepsilon}_q + \frac{1+\nu}{3K}\left(\dot{\bar{\sigma}}_1 + \dot{\bar{\sigma}}_3\right) - \frac{E}{3K}\dot{\varepsilon}_2^p = \frac{\left(\dot{\bar{\sigma}}_1 - \nu\dot{\bar{\sigma}}_2 - \nu\dot{\bar{\sigma}}_3\right)}{E} + \dot{\varepsilon}_1^p + \frac{\left(\dot{\bar{\sigma}}_3 - \nu\dot{\bar{\sigma}}_2 - \nu\dot{\bar{\sigma}}_1\right)}{E} + \dot{\varepsilon}_3^p \tag{C.113}$$

Eliminating σ_2 using the plane strain condition in simple shear (C.94),

$$D^p \dot{\varepsilon}_q + \frac{1+\nu}{3K}\left(\dot{\bar{\sigma}}_1 + \dot{\bar{\sigma}}_3\right) - \frac{E}{3K}\dot{\varepsilon}_2^p = \frac{\dot{\bar{\sigma}}_1(1-\nu-2\nu^2)}{E} + \dot{\varepsilon}_1^p + \frac{\dot{\bar{\sigma}}_3(1-\nu-2\nu^2)}{E} + \dot{\varepsilon}_3^p + 2\nu\dot{\varepsilon}_2^p$$

$$\Rightarrow \dot{\bar{\sigma}}_1\left(1-\nu-2\nu^2 - \frac{E(1+\nu)}{3K}\right) + \dot{\bar{\sigma}}_3\left(1-\nu-2\nu^2 - \frac{E(1+\nu)}{3K}\right) = E\left(D^p\dot{\varepsilon}_q - \dot{\varepsilon}_1^p - \left(2\nu + \frac{E}{3K}\right)\dot{\varepsilon}_2^p - \dot{\varepsilon}_3^p\right) \tag{C.114}$$

or in terms of additional local variables d', f',

$$\dot{\bar{\sigma}}_1 d' + \dot{\bar{\sigma}}_3 d' = f' \tag{C.114a}$$

where

$$d' = 1-\nu-2\nu^2 - \frac{E(1+\nu)}{3K} \tag{C.114b}$$

$$f' = E\left(D^p\dot{\varepsilon}_q - \dot{\varepsilon}_1^p - \left(2\nu + \frac{E}{3K}\right)\dot{\varepsilon}_2^p - \dot{\varepsilon}_3^p\right) \tag{C.114c}$$

Finally, on eliminating σ_3 between (C.111) and (C.114),

$$\frac{\dot{\bar{\sigma}}_1}{\bar{\sigma}_m}(a' - b') = c' - \frac{b'f'}{d'\bar{\sigma}_m} \tag{C.115}$$

Equation C.115 gives the increment in major principal stress in terms of variables all of which are known in the loadstep, and the minor principal stress is then recovered through (C.111). However, just like undrained simple shear, neither of these stresses is what is measured during the test. These principal stress increments must be expressed in terms of the horizontal shear stress applied to the sample and compared to the shear strain experienced by the sample. A boundary condition of drained simple shear (C.108) is that the applied vertical stress is constant; thus, we immediately recover the principal stress directions as

$$\sin^2 \alpha = \frac{(\bar{\sigma}_1 - \bar{\sigma}_y)}{(\bar{\sigma}_1 - \bar{\sigma}_3)} \qquad \text{(C.103 bis)}$$

where α is the angle between the '1' direction and the vertical. Likewise, with α now known, the applied shear stress τ_{xy} then follows immediately as

$$\tau_{xy} = \frac{1}{2}(\bar{\sigma}_1 - \bar{\sigma}_3)\sin 2\alpha \qquad \text{(C.104 bis)}$$

Finally, stress and strain increments are coaxial so the shear strain increment is recovered most simply from the updated principal strain increments:

$$\dot{\gamma}_{xy} = (\dot{\varepsilon}_1 - \dot{\varepsilon}_3)\sin 2\alpha \qquad \text{(C.105 bis)}$$

Appendix D: Numerical implementation of *NorSand*[*]

PREAMBLE

This appendix derives the equations used in implementing the monotonic version of *NorSand* (*NorSand*-M) within general numerical models, with a focus on static liquefaction. Two numerical strategies are explicitly considered: (1) *viscoplasticity* and (2) *elastic predictor–plastic corrector* (EP–PC). The viscoplastic solution method was used in the downloadable *NorSand*-M finite element code, *NorSandFEM*, to produce the examples discussed in Chapter 8. Verification of the *NorSandFEM* code is also discussed, and the downloadable verification cases are listed and described. The EP–PC approach is also presented in detail. EP–PC is adopted in the user-defined models of some commercial software packages (e.g. FLAC), as well as it is used as a precursor to the full tangent stiffness solution technique used in many other software codes.

The derivations given here assume plane stress conditions apply, the most widely adopted idealization for commercial geotechnical modelling. Adaptation to triaxial conditions is simple and already implemented within *NorSandFEM* as an option.

An inelegant wrinkle is that finite element codes (including those in the Smith and Griffiths book and which are the basis of this appendix) commonly have a tension positive convention, while soil mechanics uses the opposite of compression positive. This means that one either has finite element routines that do not obviously correspond to the familiar soil mechanics derivations, or one has to swap sign convention to give the elastic–plastic matrix used in solving for the stresses and strains. The *NorSandFEM* code adopted the second choice, so the embedded *NorSand* equations can be read in the source code much as derived. But, this comes with the penalty of an inelegant handover from the main code to the *NorSand* subroutines to deal with the changed stress convention.

D.1 PRINCIPAL VERSUS CARTESIAN

A Cartesian frame of reference is adopted: x = horizontal, y = vertical and z = out of plane. Positive x is to the right, and positive y is upward. The stresses and strains for this coordinate system are σ_{xx}, ε_{xx}, σ_{yy}, ε_{yy}, σ_{xy}, $\varepsilon_{xy,}$ σ_{zz}, ε_{zz}. In general, the principal directions (by convention, denoted using numbers rather than letters … σ_1 as opposed to σ_{xx}, etc.) will be at an angle to the x and y directions – and that angle will be different at every point throughout the domain being analysed. The out-of-plane direction (z-axis) is always principal.

* Contributed by Dawn Shuttle.

NorSand has implicit direction and aligns with principal stress and principal strain increment space. Computed stress increments for the imposed stress state must be mapped back into the Cartesian frame as the final step. This mapping is based on the angle between the '1' direction and the 'y' (vertical) direction, denoted as α:

$$\alpha = \tfrac{1}{2}\arctan\left(\frac{2\sigma_{xy}}{\sigma_y - \sigma_x}\right) \quad \text{If } \sigma_x < \sigma_y \text{ then } \alpha = \alpha + \pi/2 \tag{D.1}$$

This convention is chosen because the situation of $K_0 < 1$ is most frequent as a 'greenfield' condition, and this conveniently corresponds to $\alpha = 0$ as the starting point for the analysis.

Also note that although α is used only for changing between frames of reference in the monotonic version of *NorSand*, in the more general version, *NorSand-PSR* α is a further 'state measure' and used to drive plastic softening caused by principal stress rotation.

D.2 VISCOPLASTICITY

Despite its name, *viscoplasticity* does not refer to any creep behaviour of the soil but is a technique for using internal strain increments to redistribute load within the domain proportionally to the amount by which yield has been violated. Historically, this solution method has mainly been used with extremely simple soil plasticity models, such as Von-Mises and Mohr–Coulomb. In part, this is understandable as the viscoplastic solution method typically requires more iterations to converge than tangent stiffness approaches, so viscoplasticity is most numerically efficient if the global stiffness matrix is constructed only once – and which requires the element elastic stiffness to remain constant during a simulation. But more importantly, researchers have tended to believe that viscoplasticity is numerically unstable for strain-hardening constitutive models. This is untrue; implementation with *NorSand* has shown viscoplasticity to be a simple and surprisingly stable solution technique, well worth the slight loss of efficiency to accommodate stress-dependent elasticity by reforming the global stiffness matrix.

The viscoplastic method was first applied to soil mechanics by Olszak and Perzyna (1964) and later expanded to model creep, together with strain hardening and softening, using the finite element method by Zienkiewicz and Cormeau (1974) and more generally by Zienkiewicz et al. (1975). The viscoplastic technique uses only the elastic stiffness matrix. The strains are divided into two components, an elastic component $\dot{\varepsilon}^e$ and a viscoplastic component $\dot{\varepsilon}^{vp}$. The rate of movement of the viscous 'dashpot' is a function of the magnitude of the yield violation:

$$d\dot{\varepsilon}^{vp} = \gamma f\left[\frac{F}{F_0}\right]\frac{\partial Q}{\partial \sigma} \tag{D.2}$$

where

γ is the viscous fluidity parameter (controlling the rate of convergence)
F is the yield surface function
F_0 is the stress scalar to non-dimensionalize F (Zienkiewicz and Cormeau, 1974)
Q is the potential surface function (equal to the yield surface for *NorSand*'s associated flow rule)

The parameter grouping $\gamma f(F/F_0)$ is used only as a computational factor and has no connotation of true soil creep behaviour. Therefore, generally (D.2) is implemented in the following form:

$$d\dot{\varepsilon}^{vp} = F\frac{\partial Q}{\partial \sigma} \tag{D.3}$$

The violation (or overshoot) of yield, in terms of ε^{vp}, causes the elasto-plastic forces to be overestimated, and so these must be balanced by equivalent nodal loads called 'bodyloads'. In order to compute the term ε^{vp}, simple Euler integration in time is usually used, such that

$$d\dot{\varepsilon}^{vp} = \int d\dot{\varepsilon}^{vp}dt \cong \Delta t\, d\dot{\varepsilon}^{vp} \tag{D.4}$$

and for each iteration j,

$$(d\varepsilon^{vp})^{j} = (d\varepsilon^{vp})^{j-1} + (\Delta t\, d\dot{\varepsilon}^{vp})^{j} \tag{D.5}$$

For numerical stability, a maximum 'critical' value of Δt should not be exceeded, which has been derived for Von-Mises and Mohr–Coulomb materials by Cormeau (1975). For Mohr–Coulomb, this critical timestep is given by

$$\Delta t_{mc} = \frac{4(1+\nu)(1-2\nu)}{E(1-2\nu+\sin^2\phi)} \tag{D.6}$$

where
- E is Young's elastic modulus
- v is Poisson's ratio
- ϕ is the friction angle

For more complex constitutive models, this critical timestep is not known, and the value of Δt must be estimated (discussed later in the context of *NorSand*).

The stresses computed to be outside of the current yield surface, termed 'illegal stresses', are given by

$$\Delta\sigma^{p} = D^{e}\Delta\varepsilon^{vp} \tag{D.7}$$

giving the stress increment

$$\Delta\sigma = D^{e}(\Delta\varepsilon - \Delta\varepsilon^{vp}) \tag{D.8}$$

In the viscoplastic method, the viscoplastic strain component, $\Delta\varepsilon^{vp}$, is incremented each iteration until convergence is achieved. Concurrently, the excess 'illegal' stresses are balanced by bodyloads, R. In standard finite element notation, these bodyloads are given by

$$\Delta R = \int^{\text{volume}} B^{T}D^{e}\Delta\varepsilon^{vp}\,dV \tag{D.9}$$

or

$$\Delta R^{j} = \Delta R^{j-1} + \sum_{\text{elements}}^{\text{all}} \int B^T D^e (\Delta \dot{\varepsilon}^{vp})^j \, d(\text{element}) \tag{D.10}$$

where B is a matrix containing the differentials of the shape functions, N (in ε_{xx}, ε_{yy}, ε_{xy}, ε_{zz} order), typically represented as $B = AN$, where

$$A = \begin{bmatrix} \partial/\partial x & 0 \\ 0 & \partial/\partial y \\ \partial/\partial y & \partial/\partial x \\ 0 & 0 \end{bmatrix} \tag{D.11}$$

and for the eight-node quadrilateral elements used for examples in Chapter 8,

$$N = \begin{bmatrix} N_1 & N_2 & N_3 & N_4 & N_5 & N_6 & N_7 & N_8 & 0 & 0 & 0 & 0 & 0 & 0 & 0 & 0 \\ 0 & 0 & 0 & 0 & 0 & 0 & 0 & 0 & N_1 & N_2 & N_3 & N_4 & N_5 & N_6 & N_7 & N_8 \end{bmatrix} \tag{D.12}$$

where

N_1–N_8 are the shape functions of each node (with the two rows referring to the x and y directions, respectively)

The plane strain elastic constitutive matrix D^e is:

$$D^e = \frac{E(1-v)}{(1+v)(1-2v)} \begin{bmatrix} 1 & \frac{v}{1-v} & 0 & \frac{v}{1-v} \\ \frac{v}{1-v} & 1 & 0 & \frac{v}{1-v} \\ 0 & 0 & \frac{1-2v}{2(1-v)} & 0 \\ \frac{v}{1-v} & \frac{v}{1-v} & 0 & 1 \end{bmatrix} \tag{D.13}$$

In the B and D^e matrices, the terms corresponding to the z direction are included, because although there is no net displacement in the z plane, plastic yield can occur in this out-of-plane direction resulting in the generation of equal and opposite elastic and plastic strains, $\varepsilon_z^p = -\varepsilon_z^e$.

This viscoplastic solution algorithm has been adopted for the *NorSandFEM* downloadable software.

D.3 *NorSandFEM* VISCOPLASTICITY PROGRAM

NorSand Finite Element Monotonic, *NorSandFEM*, is the core component of the downloadable FE software for *NorSand*. *NorSandFEM* is adapted from the material non-linearity section of the book, *Programming the Finite Element Method*, now in the fifth edition (Smith et al., 2013). This text has a programming-oriented style, making it easy to follow and with many example programs. The programs grew out of research and developments in the Department of Civil Engineering at the University of Manchester (with work now continuing at the University of Manchester and Colorado School of Mines). This work has two particular attributes: (1) the text and programs cover a wide variety of finite element capabilities and with a useful focus on geomechanics and (2) the programs and subroutine libraries are freely available online.

The implementation of *NorSand* presented here is based upon the programs, library functions and subroutines, coded in Fortran90, which were released with the third edition (Smith and Griffiths, 1998). The *NorSandFEM* code, and associated library routines, is compatible with most Fortran90 or later Fortran compilers; the executable code was compiled using Microsoft Powerstation Fortran95.

D.3.1 *NorSandFEM* conventions

The *NorSandFEM* code adopts eight-node quadrilateral elements, which have a node at each of the element's four corners and at the middle of each of the four sides. Globally, the elements and nodes in the mesh are numbered vertically (although it is a simple task to swap *NorSandFEM* to horizontal numbering). On a local (or individual element) level, each element is numbered clockwise from the bottom left-hand corner. This convention is embedded within the Smith and Griffiths (1998) library routines and used for all of the 2D codes in their library routines.

D.3.2 *NorSandFEM* freedom numbering

As a 2D code, *NorSandFEM* code allows for movement in two directions: x and y. The finite element method tracks the movements at the nodes, with the element shape functions (N) operating on these nodal displacements to control the distribution of displacement over the entire element. For the eight-node elements used here, there are potentially 16 movements being computed per element. These x and y movements appear in a single displacement array (as required by the finite element library of solution subroutines).

Within *NorSandFEM*, it is necessary to map each of the individual nodal x and y displacements, called *freedoms* as they are 'free' to move as the finite element solution is computed, back to their nodes and elements. This is done using an integer array of numbers, each number corresponding to a computed nodal displacement. At some nodes, one or both of the node movements are constrained to enforce a boundary condition and thus are no longer free – such freedoms are removed from the assembled matrices.

Consistent with the global node numbering, *NorSandFEM freedoms* are numbered vertically from the top left-hand corner of the mesh. Every node has two associated freedoms corresponding to the x and y movements, respectively. If the node is restrained (i.e. prevented from moving) in x and/or y, the movement is assigned a freedom of '0'. All unrestrained nodes are numbered sequentially (e.g. see Figure D2 on page 530).

D.4 VISCOPLASTICITY IN *NorSandFEM*

The appeal of viscoplasticity is its simplicity, as illustrated by the flow chart of Figure D.1. *NorSandFEM* adopts a Forward Euler approach. In essence, the algorithm is looping through each Gauss (or sampling) point in the mesh to check whether the current stress

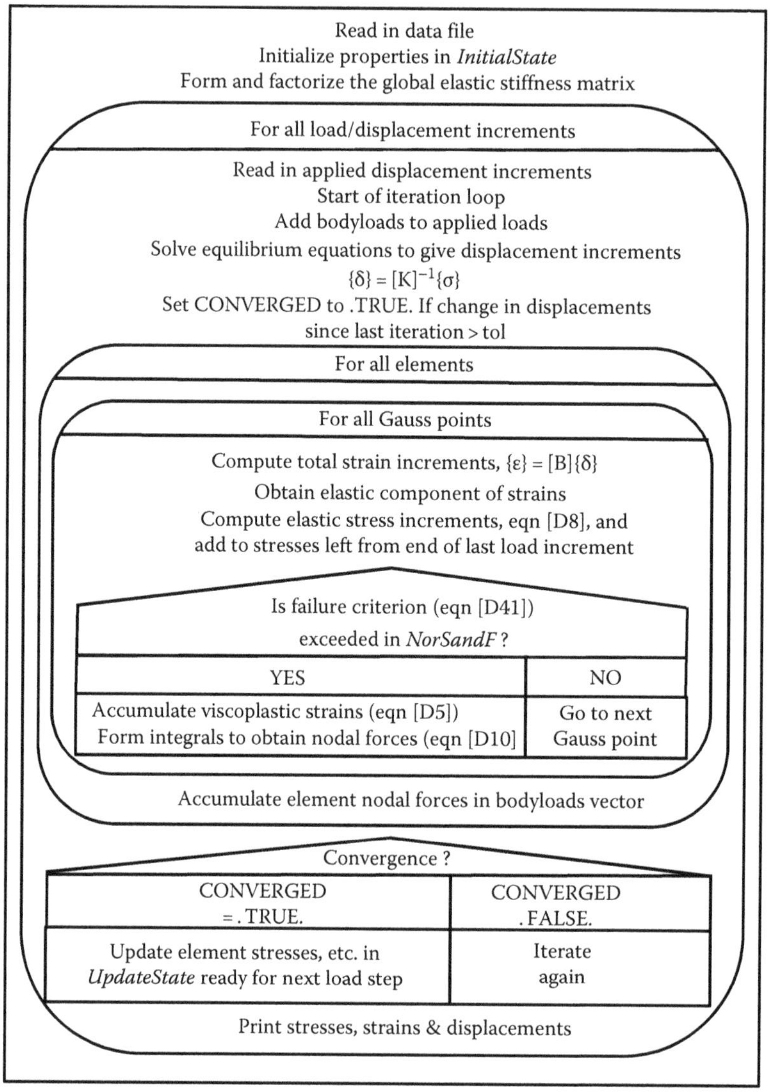

Figure D.1 Structure chart for the *NorSandFEM* viscoplasticity algorithm. (Adapted from Smith, I.M. and Griffiths, D.V., *Programming the Finite Element Method*, 2nd ed, Wiley, Chichester, U.K., 3rd ed, Wiley, New York, 1998.)

state exceeds yield. At every Gauss point location exceeding yield, viscoplastic strains are incremented (see [D.4]), and the incremented viscoplastic strains used to accumulate bodyloads (see [D.9]) to balance the excess load. This continues until the change in updated global displacements is below some predefined limit – and convergence has been achieved.

The modifications required to add *NorSand-M* to the basic code (which is described in detail in Smith and Griffiths, 1998) are quite limited and comprise just four steps: (1) initialize *NorSand* properties, (2) determine yield and update the image mean stress, (3) compute the viscoplastic strain increments for *NorSand-M* and (4) update *NorSand* properties. Of these, (1) and (4) are required for any new constitutive model, however simple, and require only minimal description.

The *NorSand* material properties (M_{tc}, N_{tc}, Γ, λ_e, etc.) and initial state measures (ψ_0, R_0) are read into the code, and derivative properties (such as void ratio, e, and image mean stress, $\bar{\sigma}_{m,i}$) are computed in the *InitialState* subroutine prior to assembling the global elastic stiffness matrix.

Following convergence at the end of each load increment, the state measures and derivative properties are updated (apart from $\bar{\sigma}_{m,i}$, which is incrementally updated every iteration, as well as at the end of the step). This update is done within the *UpdateState* subroutine.

Steps (2) and (3) are central to this viscoplastic implementation for *NorSandM* and are discussed later. In the following the superscript 'O' refers to the 'old' stress state from the previous load step, the superscript 'T' refers to a value in a current iteration (T = transient), and the superscript 'C' is the converged stress state for that loadstep.

D.4.1 Viscoplastic yield routine

The *NorSandFEM* viscoplastic yield subroutine, named *NorSandF*, fulfils two main functions. First, the subroutine determines whether yielding is occurring at any of the Gauss points this iteration. Second, if the current Gauss point is yielding, the value of $\bar{\sigma}_{m,i}$ is appropriately hardened/softened. How this is achieved is most easily explained by following the main equations in the subroutine.

The yield surface, F (termed *NorBullet* in *NorSandF* to indicate the bullet yield surface), is determined by writing the standard *NorSand* bullet-shaped yield surface in terms of the stresses (see (D.14)). All of the values in (D.14) are current estimates of the 'new' stress state. A value of $F=0$ lies on the yield surface, $F > 0$ is yielding and $F < 0$ is elastic.

$$F = \bar{\sigma}_m^E \exp\left(\frac{\bar{\sigma}_q^E}{\bar{\sigma}_m^E M_i^E} - 1\right) - \bar{\sigma}_{m,i}^T \tag{D.14}$$

where

$\bar{\sigma}_m^E$ is the 'elastic' mean effective stress
$\bar{\sigma}_q^E$ is the 'elastic' deviatoric stress invariant
$\bar{\sigma}_{m,i}^T$ is the value of image mean stress of this iteration
M_i^E is the value of M_i computed using the 'elastic' guess

The term 'elastic' is within parentheses in (D.14) because the stresses are calculated using (D.8), and although the stresses are computed from the elastic stress–strain matrix, D^e, the strains are the current elasto-plastic guess. At convergence, these 'elastic' stresses will be equal to the true converged elasto-plastic solution.

The M_i^E value is computed within a function named *M_psii*, again using the current elastic guess at the true converged solution. The inputs into *M_psii* are the constant material properties, M_{tc}, N_{tc}, χ_i, and the evolving Lode angle, θ^E, and image state, ψ_i^E. The latter requires a current estimate of volumetric strain increment, ε_v^E, to calculate the current estimate of void ratio, e^E, which is used with the current estimate of critical void ratio, e_c^E, in the current state estimate, ψ^E, and finally in ψ_i^E.

D.4.2 Update image mean stress: *NorSand* hardening

Hardening of the bullet is separated into two distinct parts. The first is a 'standard' component based on the increment of plastic strain, which can both harden and soften the yield surface. The second 'softening' portion of the hardening is triggered only for loose image states ($\psi_i^O > 0.0$), softens only the yield surface (hence the name adopted) and is a function of the magnitude of the change in mean effective stress. In this implementation of *NorSand*, the two types of hardening are implemented in separate portions of the code.

D.4.3 Standard bullet hardening

Standard bullet hardening is implemented within the *NorSand*F subroutine. If yielding is occurring, the current estimate of the converged state does not yet correspond with the yield surface, and the value of $\bar{\sigma}_{m,i}^T$ requires further hardening. The $\bar{\sigma}_{m,i}^T$ at each Gauss point hardens every iteration that yield is being violated, as indicated by the '*T*' superscript in (D.14). Consistent with the forward Euler algorithm, all properties used in the calculation to increment $\bar{\sigma}_{m,i}^T$ are values related to the converged stress state at the previous loadstep, annotated by the superscript '*O*'.

The standard bullet hardening term, X_H, is given by

$$X_H = H_\theta \frac{\bar{\sigma}_m^O}{\bar{\sigma}_{m,i}^O}\left(\bar{\sigma}_{mx}^O - \bar{\sigma}_{m,i}^O\right) \tag{D.15}$$

where

$$M_i^O = M\_psii\left(M_{tc}, N_{tc}, \chi_i, \psi_i^O, \theta^O\right) \tag{D.16a}$$

$$M_{i,tc}^O = M\_psii\left(M_{tc}, N_{tc}, \chi_i, \psi_i^O, 30°\right) \tag{D.16b}$$

$$H_{tc} = set\_H\left(H_0, H_\psi, \psi^O\right) \tag{D.16c}$$

$$H_\theta = H_{tc}\frac{M_i^O}{M_{i,tc}^O} \tag{D.16d}$$

The increment to the image mean stress, $\dot{\bar{\sigma}}_{m,i}$, is equal to the X_H multiplied by the increment of plastic shear strain occurring in this iteration, $\dot{\varepsilon}_q^p$:

$$\dot{\bar{\sigma}}_{m,i} = X_H \dot{\varepsilon}_q^p \tag{D.17}$$

giving the current 'best estimate' of $\bar{\sigma}_{m,i}^T$ in this iteration to be

$$\bar{\sigma}_{m,i}^T = \bar{\sigma}_{m,i}^T + \dot{\bar{\sigma}}_{m,i} \tag{D.18}$$

The value of $\bar{\sigma}_{m,i}^T$ continues to increment while the Gauss point exceeds yield; hence, on the final iteration of the timestep, the value of $\bar{\sigma}_{m,i}^T$ equals $\bar{\sigma}_{m,i}^C$ and is consistent with the 'converged' stress state.

D.4.4 Additional softening term

Unlike standard hardening, the 'softening term' is implemented only if $\psi^O > 0$. For numerical efficiency, this term is also implemented as a single calculation after convergence has been achieved, in the *UpdateState* subroutine. We tolerate imperfect convergence, which is then carried forward into the next loadstep.

The limiting dilation D_{min} changes during the loadstep within the following range:

$$D_{min}^{O} = \chi_i \psi_i^{O} \quad \text{to} \quad D_{min}^{C} = \chi_i \psi_i^{C} \tag{D.19}$$

The softening term is based on the change in $\bar{\sigma}_{m,i\max}^{O}$ occurring during the loadstep:

$$\bar{\sigma}_{mx}^{O} = \exp\left(-\frac{D_{min}^{O}}{M_{i,tc}^{O}}\right)\bar{\sigma}_{m}^{O} \tag{D.20}$$

$$\bar{\sigma}_{mx}^{C} = \exp\left(-\frac{D_{min}^{C}}{M_{i,tc}^{C}}\right)\bar{\sigma}_{m}^{C} \tag{D.21}$$

with $\bar{\sigma}_{mx}$ replacing $\bar{\sigma}_{m,i\max}$ for notational brevity.

(where (D.20) was calculated at the end of the previous loadstep and (D.21) is calculated as part of the property update).

If $\bar{\sigma}_{m,i}$ has changed during this loadstep (i.e. yielding has occurred) and $\psi_i^{O} > 0.0$, then the additional softening term (D.22) is implemented:

$$\dot{\bar{\sigma}}_{m,i} = \left(\bar{\sigma}_{mx}^{C} - \bar{\sigma}_{mx}^{O}\right)\left(\frac{\bar{\sigma}_{q}^{O}}{\bar{\sigma}_{m}^{O}}\right)\frac{1}{\eta_L} \tag{D.22}$$

where

$$\eta_L = M_i^{O}\left(1 - \frac{D_{min}^{O}}{M_{i,tc}^{O}}\right) \tag{D.23}$$

Giving the converged value of $\bar{\sigma}_{m,i}^{C}$:

$$\bar{\sigma}_{m,i}^{C} = \bar{\sigma}_{m,i}^{T} + \dot{\bar{\sigma}}_{m,i} \tag{D.24}$$

While discussing the $\bar{\sigma}_{m,i}$ update, it seems most appropriate to mention here a modification to the standard *NorSand*-M that is included in both the *NorSandF* and *UpdateState* subroutines. To improve numerical stability, a limit has been applied to the value of $\bar{\sigma}_{m,i}$:

$$\bar{\sigma}_{m,i} \geq \bar{\sigma}_{m,i}^{\text{LIMIT}} \tag{D.25}$$

The reason for this limit is that when modelling boundary value problems (e.g. slopes), the near-surface in-situ gravity stresses are typically low, and during loading, the stress state may transiently move into tensile mean effective stress – a condition inconsistent with the *NorSand* constitutive model. To minimize the occurrence of this condition, a lower limit is placed on $\bar{\sigma}_{m,i}$, presently hard-coded at 2 kPa; essentially this sets *NorSand* to a minimum shear strength $s \sim 1$ kPa.

D.4.5 Viscoplastic strain increments

In (D.3), the viscoplastic strain increments are a function of $\partial Q/\partial\sigma$, and codes from Smith and Griffiths (1998) compute $\partial Q/\partial\sigma$ numerically. This is not possible for *NorSand* because, although *NorSand* is an associated flow model in terms of $\bar{\sigma}_m$ and $\bar{\sigma}_q$, the fact that no work is involved with the third stress invariant θ means that *NorSand* is not fully associated (Jefferies and Shuttle, 2002). Instead, $(d\dot{\varepsilon}^{vp})^j$ in (D.5) are computed for the principal strain directions directly from dilatancy, D^p, and then transposed to x, y, z coordinates.

The subroutine *GetStrainRateRatios* uses the dilatancy D^p at the start of the loadstep to calculate the ratios $\varepsilon_2^p/\varepsilon_1^p$ (termed *ep21* in *NorSandFEM*) and $\varepsilon_3^p/\varepsilon_1^p$ (termed *ep31* in *NorSandFEM*). These principal strain rate ratios are computed by taking a cosine interpolation (similar to *NorSand's* Mi function in the π-plane [Jefferies and Shuttle, 2011]).

The calculations follow the following process.

First, recover the dilatancy for the current mobilized stress ratio at triaxial compression and extension, using the fact that η/M_i is invariant with θ:

$$D_{tc}^P = \frac{M_{i,tc}}{M_i} D^P$$

$$D_{te}^P = \frac{M_{i,te}}{M_i} D^P \tag{D.26}$$

Then, calculate the ratio $\varepsilon_3^P/\varepsilon_1^P$ for triaxial compression ($\varepsilon_2 = \varepsilon_3$) and triaxial extension ($\varepsilon_2 = \varepsilon_1$) as

$$Z_{tc} = \frac{2D_{tc}^P - 3}{6 + 2D_{tc}^P}$$

$$Z_{te} = \frac{2D_{te}^P - 6}{3 + 2D_{te}^P} \tag{D.27}$$

Now interpolate for the value of $\varepsilon_3^P/\varepsilon_1^P$ in Lode angle space. The interpolation rule runs from 0 to –1, hence signs

$$\frac{\varepsilon_3^p}{\varepsilon_1^p} = Z_{tc} + (Z_{te} - Z_{tc})\cos\left(0.5\left(3\theta + \frac{\pi}{2}\right)\right) \tag{D.28}$$

Finally, recover the value of $\varepsilon_2^P/\varepsilon_1^P$ by interpolating on θ:

$$a = \frac{\sin\theta + \sqrt{3}\cos\theta}{3} \tag{D.29a}$$

$$b = \frac{-2\sin\theta}{3} \tag{D.29b}$$

$$c = \frac{\sin\theta - \sqrt{3}\cos\theta}{3} \tag{D.29c}$$

$$\frac{\varepsilon_2^p}{\varepsilon_1^p} = \frac{a D^p - 1 + \frac{\varepsilon_3^p}{\varepsilon_1^p}(c D^p - 1)}{1 - b D^p} \tag{D.30}$$

Having found the strain ratios in the principal directions, and having previously determined whether the z (out-of-plane) direction contains the major, intermediate or minor principal stress, the subroutine *compute_dilation* uses the principal direction angle, α, to convert $\varepsilon_2^p/\varepsilon_1^p$, $\varepsilon_3^p/\varepsilon_1^p$ into $\varepsilon_x^p/\varepsilon_1^p$, $\varepsilon_y^p/\varepsilon_1^p$ and $\varepsilon_z^p/\varepsilon_1^p$. Equation D.3, giving $d\dot{\varepsilon}^{vp}$, then continues as

$$d\dot{\varepsilon}_x^{vp} = F\frac{\varepsilon_x^p}{\varepsilon_1^p} \tag{D.31}$$

and similarly for the y and z directions.

Equation D.31 is then multiplied by the viscoplastic timestep to give increment of viscoplastic strain, $d\varepsilon^{vp}$. As the critical timestep is unknown for *NorSand*, the *NorSandFEM* code uses a scaled version of the Mohr–Coulomb critical timestep Δt_{mc} (D.6):

$$\Delta t = tolfac\,\Delta t_{mc} \tag{D.32}$$

where *tolfac* is a user-defined input and Δt_{mc} is calculated assuming that the friction angle is 30° and Young's modulus is computed at the initial stress state (or 200 kPa for varying initial stresses). It has been found that the viscoplastic implementation of *NorSand* is typically stable with *tolfac* ~ 0.25 (but values as large as 0.5 have been adequate with some parameter combinations).

D.5 INPUTS TO *NorSandFEM*

The downloadable version of *NorSandFEM* is set up to run both axisymmetric and plane strain geometry, with either the initial stress condition set as constant stress with K_0 (for verification against laboratory tests) or gravity loading with K_0 (for field problems). The inputs are simple and include the basic geometry of the mesh, boundary conditions (i.e. which nodes are fixed in x and/or y) and the *NorSand* material properties. Note that *NorSandFEM* is based on effective stresses and that the computed pore pressure values do not affect the results of the analysis; total stresses are computed by adding the pore pressure, but are not used within the analysis. For simplicity, *NorSandFEM* reads only meshes with a constant number of elements in the 'x' and 'y' directions.

Displacement loading has been adopted. This is a requirement for liquefaction analyses if the post-triggering stresses and displacements are to be followed (in addition to being more numerically efficient with the viscoplastic solution algorithm).

There are two variants of the input file, corresponding to 'constant stress' or 'gravity loading', which are described later. Both variants are adapted from the input file format used in Smith and Griffiths (1998). Each row of numerical data in *NorSandFEM* input files is preceded by a 'character string input', used to annotate the inputs (and hopefully make the files easier to use).

In the following descriptions, focus is placed on explaining the numerical inputs (e.g. mesh generation and numerical tolerances); the *NorSand* input properties are described in detail in the main text of this book. Both file formats are explained by reference to two simple examples: (1) a plane strain two-by-two element test and (2) a slope.

D.5.1 Constant stress input file format

The mesh used for the element test is shown in Figure D.2, with the corresponding *NorSandFEM* data file shown in Figure D.3 (with the implemented data annotated by line number for clarity). As stated earlier, all of the text lines in the data file are for readability only (they are read by *NorSandFEM* but not used).

A description of the inputs in the Figure D.3 input file, by reference line number, is given as follows:

Line 1: igeom – flag to toggle between axisymmetry (= 0) and plane strain (= 1)
Line 2: Basic mesh geometry
  nels – number of elements in the mesh
  nxe – number of elements in the 'x' (horizontal direction)
  nye – number of elements in the 'y' (vertical direction)
  nn – number of nodes (= nxe*(nye + 1) + (nxe + 1)*(2*nye + 1))
Line 3: *NorSand* soil (material) properties
  Mcrit – M_{tc}
  Gamma – Γ at 1 kPa
  Pref – equal to 1 kPa in the units used for the analysis
  Lambda – λ_e
  Ncrit – N_{tc}
  Chi – χ_{tc}
  H0 – H_0, used to compute hardening H according to the equation $H = H_0 - H_\psi \psi$
  Hy – H_ψ
Line 4: *NorSand* state properties
  Psi0 – ψ_0
  R0 – measure of the overconsolidation ratio, R_0

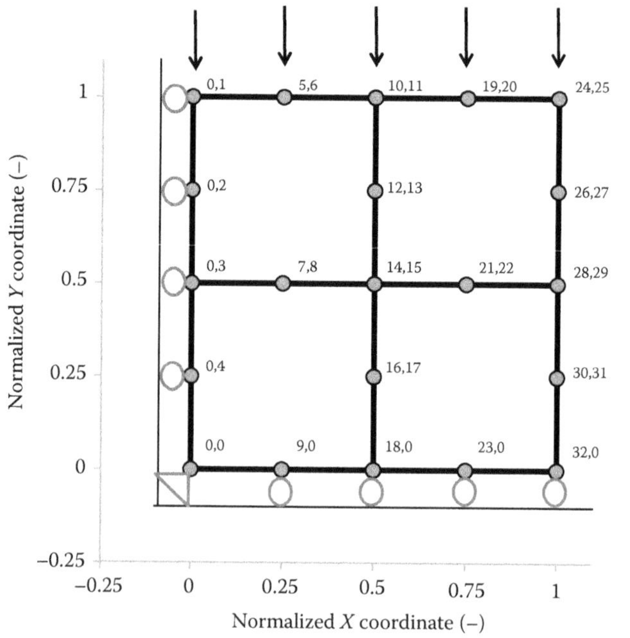

Figure D.2 Mesh used for the element test example with initial 'constant stress' conditions.

```
igeom - 0=axisymmetric 1=plane strain
1                                                          1
nels nxe nye nn nip
4   2   2  21 4                                            2
Mcrit, Gamma, Pref, Lambda, Ncrit Chi H0  Hy
1.25   0.875   1.0   0.04   0.35   4.0  100. 1000.         3
psi0  OCR
0.05  1.001                                                4
G0,  Gpower  v  (elasticity)
50.e3  0.0     0.15                                        5
istress - 1=constrant stress 2=gravity loading
1                                                          6
bulkW  sigmaV0     k0     pore0
2.e7      202.  0.985148515  0.0                           7
nstep   tol     tol2    tolfac limit  MeshFlag
4001  2.0e-5  2.5e-2  0.05  2500        1                  8
nload
5                                                          9
freedom presc
1         -1.e-4                                           10
6         -1.e-4                                           11
11        -1.e-4                                           12
20        -1.e-4                                           13
25        -1.e-4                                           14
loaded elements range
1 2                                                        15
geometry
.0  .50  1.0                                               16
.0  .50  -1.0                                              17
fixity
9                                                          18
1       0      1                                           19
2       0      1                                           20
3       0      1                                           21
4       0      1                                           22
5       0      0                                           23
8       1      0                                           24
13      1      0                                           25
16      1      0                                           26
21      1      0                                           27
Frequency of plotting output, numout (steps per output)
2000                                                       28
elements for stress-path plotting
1 2 3 4                                                    29
Number of elements for which psi is reset
0                                                          30
Reset psi - element numbers and psi values
```

Figure D.3 Example of constant stress data file.

Line 5: Elasticity

G_0, G_{power} – shear modulus $G = G_0\left(\dfrac{\bar{\sigma}_m}{\bar{\sigma}_{m,\mathrm{ref}}}\right)^{G_{power}}$ where $\bar{\sigma}_{m,ref} = 100$ kPa

v – Poisson's ratio

Line 6: Flag to define initial stress state – istress

1 = constant stress

2 = gravity loading

Line 7: Stress state properties (for istress = 1)

BulkW – bulk modulus of water in adopted units (use 0.0 for drained analysis)

sigmaV0 – vertical effective stress

K0 – lateral earth pressure coefficient

Pore0 – initial pore pressure

Line 8: Numerical control properties

nstep – number of loadsteps

tol – numerical convergence tolerance on fractional change in displacement

tol2 – numerical convergence tolerance on maximum overshoot of yield

tolfac – multiplier on the Mohr–Coulomb critical timestep (see (D.32))

limit – iteration limit for the loadstep before continuing

MeshFlag – geometry flag where 1 = uniform mesh, 2 = deformed 'slope' mesh

Line 9: Number of loads

nload – the number of displacements applied

Line 10–14: Applied displacement loading

nload pairs of numbers containing the freedom number (see D.3.2) and magnitude of the applied displacement

(Figure D.2 indicates the freedom numbers for the vertical applied load)

Line 15: Range of surface elements (measured in x) to use in computing net load per m

Line 16: Mesh geometry x direction

Coordinates of the element corners in the x direction (m)

Line 17: Mesh geometry y direction

Coordinates of the element corners in the y direction (m)

Line 18: Fixity

Number of restraints (NR) – number of nodes fixed in the x and/or y directions

Line 19–27: Fixity

For each restrained node

Node number

Fixed in x (0 = yes, 1 = no)

Fixed in y (0 = yes, 1 = no)

Line 28: Frequency of plotting output, numout steps per output

(used to create output files containing measures of stress and strain for post-processing and subsequent plotting)

Line 29: Elements for stress path plotting

Values of $\bar{\sigma}_m, \bar{\sigma}_q, \bar{\sigma}_{m,i}, \psi, e$, etc., at the Gauss points of four elements are output to file every loadstep. The four numbers select the elements for output. Note that different Gauss points are used for each file – so in a boundary value problem, four outputs from the same element will likely differ.

Line 30: Number of elements for which ψ is reset (numPsiSet in *NorSandFEM*)

Used to assign specific elements a different initial ψ to model weaker or denser zones.

Line 31: Element number and value of reset psi (for numPsiSet > 0)

Listing of element number and assigned ψ for weak/dense zones

D.5.2 Gravity loading input file format

An example slope is used to represent the gravity loading input file format. The slope mesh is shown in Figure D.4, with the corresponding *NorSandFEM* data file shown in Figure D.5. For readability, this data file has some portions hidden (indicated by '....'). A quick perusal of Figure D.5 will show that the file is near identical to that for 'constant stress', so only the differences in the file are annotated with a line number and discussed in the following:

Line 1: Flag to define initial stress state – istress
 1 = constant stress
 2 = gravity loading
Line 2: Stress state properties (for istress = 2)
 BulkW – bulk modulus of water in adopted units (use 0.0 for drained analysis)
 unitWt – unit weight of soil in adopted units
 H2OWt – unit weight of water in adopted units (use 0.0 for drained analysis)
 K0 – lateral earth pressure coefficient
 WaterTable – depth of the water table below the ground surface (m)
Line 3: Numerical control properties
 nstep – number of loadsteps
 tol – numerical convergence tolerance on fractional change in displacement
 tol2 – numerical convergence tolerance on maximum overshoot of yield
 tolfac – multiplier on the Mohr–Coulomb critical timestep (see (D.32))
 limit – iteration limit for the step before continuing
 MeshFlag – geometry flag where 1 = uniform mesh, 2 = deformed 'slope' mesh
 If MeshFlag = 2, the mesh is deformed in the *Y* direction
Line 4: Mesh deformation information (used if MeshFlag = 2) in *X* direction
 X0 – First reference *X* value
 X1 – Second reference *X* value
 X2 – Third reference *X* value
 X3 – Final reference *X* value

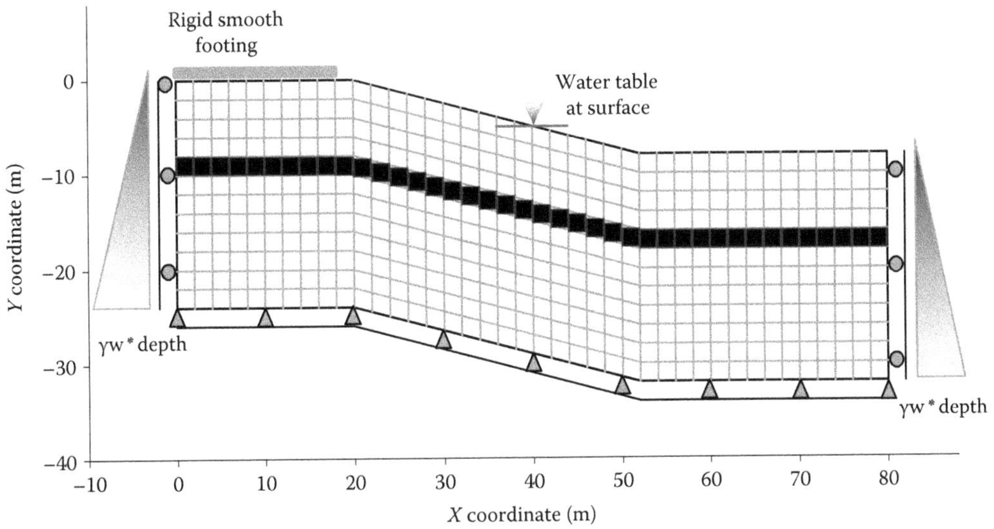

Figure D.4 Mesh used for the slope example incorporating gravity stresses.

```
igeom - 0=axisymmetric 1=plane strain
1
nels  nxe  nye  nn  nip
480   40   12  1545  4
Mcrit, Gamma, Pref, gLambda, Ncrit  Chi  H0  Hy
1.2     0.875     1.0    0.03      0.35    4.0  100.  0.
psi0  OCR
0.03  1.001
G0,  Gpower  poisson
30.e3  1.0  .    15
istress - 1=constrant stress 2=gravity loading
2                                                          1
bulkW,   unitWt,   H2OWt   k0   WaterTable
2.e7       22.        9.81     0.7        0.0              2
nstep  tol   tol2  dt_scale  tolfac  limit  MeshFlag
4002  4.e-3  0.5     0.25      0.4     1000      2         3
X0   X1   X2   X3
0.0   20.   52.   80.                                      4
Y0   Y1   Y2   Y3
0.0   0.0   -8.0   -8.0                                    5
nload
19
freedom presc
1         -5.00E-05
26        -5.00E-05
...
626       -5.00E-05
loaded elements range
1 9
geometry
.0 2. 4. 6. 8. 10. 12. 14. 16. 18. 20. 22. 24. 26. 28. 30. 32. 34. 36. 38. 40.
42. 44. 46. 48. 50. 52. 54. 56. 58. 60. 62. 64. 66. 68. 70. 72. 74. 76. 78. 80.
.0 -2. -4. -6. -8. -10. -12. -14. -16. -18. -20. -22. -24.
fixity
129
1       0     1
2       0     1
...
1544    0     1
1545    0     0
numout (steps per output)
800
elements for P-Q tracking
49 86 73 38
Number of elements for which psi is reset
40                                                         6
psi reset element numbers and psi values
  5  0.07                                                  7
 17  0.07                                                  8
...                                                        ..
476  0.07                                                  9
```

Figure D.5 Example of gravity loading data file.

Line 5: Mesh deformation information (used if MeshFlag = 2) in *Y* direction

Y0 – First reference *Y* value

Y1 – Second reference *Y* value

Y2 – Third reference *Y* value

Y3 – Final reference *Y* value

Relative to the original mesh, the mesh is moved vertically by Y0 at X0 and Y1 at X1, with linear interpolation of *Y* between X0 and X1. Similarly, the mesh is moved vertically by Y2 at X2, with linear interpolation of *Y* between X1 and X2, and the mesh is moved vertically by Y3 at X3, with linear interpolation of *Y* between X2 and X3. Figure D.4 shows the slope geometry developed with this command using the properties in Figure D.5.

Line 6: Number of elements for which psi is reset

Used to assign specific elements a different initial ψ to model weaker or denser zones

Line 7–9: Element number and value of reset psi

Listing of element numbers and assigned ψ for weak/dense zones (Figure D.4 illustrates the weaker zone developed with this command using the Figure D.5 input file)

D.6 VERIFICATION AND EXAMPLES

There are two important reasons to verify analysis software. The first is to ensure that the software provides an accurate solution to the problem you wish to analyse. The second, equally important, use of verification is to ensure that the software is properly compiled on your computer and that you understand the code's inputs – so, while you can read about the verification cases here, you also need to download the input files and run them yourself.

NorSandFEM has been extensively verified against laboratory 'element' tests. These 'element' tests have uniform stress conditions and known loading paths, allowing calculation of the correct solution by direct numerical integration. The verification suite for *NorSandFEM* has been chosen to enable the outputs to be directly compared against the *NorSandPS.xls* and *NorSandTXL.xls* spreadsheets. This spreadsheet directly integrates the *NorSand* equations using the Euler method, which is about as far different from viscoplasticity as can be found. Comparing the results of the two calculation methods provides an independent check on the mathematics of both. The full verification suite includes plane strain and triaxial element tests, under drained and undrained conditions, and for soils with loose through dense initial states. The number of elements is also varied. Table D.1 summarizes the verification scenarios adopted. All the input files for these scenarios are downloadable for you to compare with *NorSandPS.xls* and *NorSandTXL.xls*.

A schematic representation of the boundary conditions used for verification in the plane strain and triaxial element tests is shown in Figure D.6. Plane strain compression is the test developed at Imperial College (see Figure 2.43) and has a fixed loading direction (σ_1) and constant stress normal to that direction (σ_3); it is not a constant intermediate stress σ_2, as σ_2 evolves from the specified starting value (usually the same as the σ_3) throughout the test. Triaxial compression is similar, but with symmetry in the two horizontal stresses.

Also included as the final two files in the verification suite in Table D.1 is an additional special case of a plane strain element test with rough platens using both an in-plane and a rotated mesh, as shown in Figure D.7. The loading, geometry and material properties are identical between the two meshes; the only difference is the orientation of the mesh. Rotating the orientation is done because internally the code still has x, y as vertical and

Table D.1 Summary of *NorSandFEM* verification suite

File name	*Geometry*	*Drained/undrained*	*State*	*Elements*
PS1_D_D.dat	PS	D	Dense	1 by 1
PS1_UD_L.dat	PS	UD	Loose	1 by 1
PS4_D_L.dat	PS	D	Loose	2 by 2
PS4_D_D.dat	PS	D	Dense	2 by 2
PS4_UD_L.dat	PS	UD	Loose	2 by 2
PS4_UD_D.dat	PS	UD	Dense	2 by 2
Txl4_D_L.dat	Triaxial	D	Loose	2 by 2
Txl4_D_D.dat	Triaxial	D	Dense	2 by 2
Txl4_UD_L.dat	Triaxial	UD	Loose	2 by 2
Txl4_UD_D.dat	Triaxial	UD	Dense	2 by 2
PSR1_D_D1.dat	PS – rough – unrotated	D	Dense	1 by 1
PSR1_D_D2.dat	PS – rough – rotated	D	Dense	1 by 1

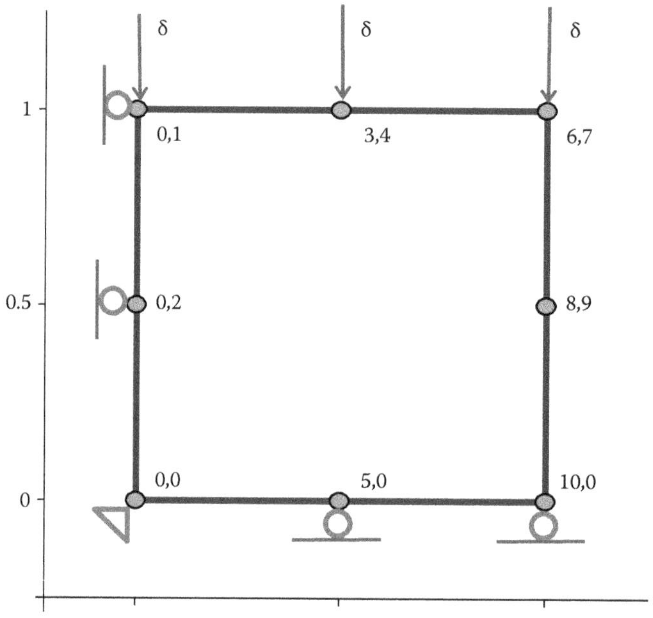

Figure D.6 Geometry, boundary and loading conditions for vertical loading verifications.

horizontal, so rotating the problem checks that the stress invariants, strain invariants, 3D representation of yield surfaces, etc., are properly formulated – confirmed by both meshes giving an identical result. This verification case is very important for any 'good' model, which allows the yield surface and dilation to vary realistically as stress conditions vary from triaxial compression, through plane strain and onto triaxial extension.

Despite all but one of the verification cases being 'element' tests, and which could sensibly be verified using a single element (as the stress within the element is constant), the verification input files include multiple element meshes. This is not accidental. Experience with both

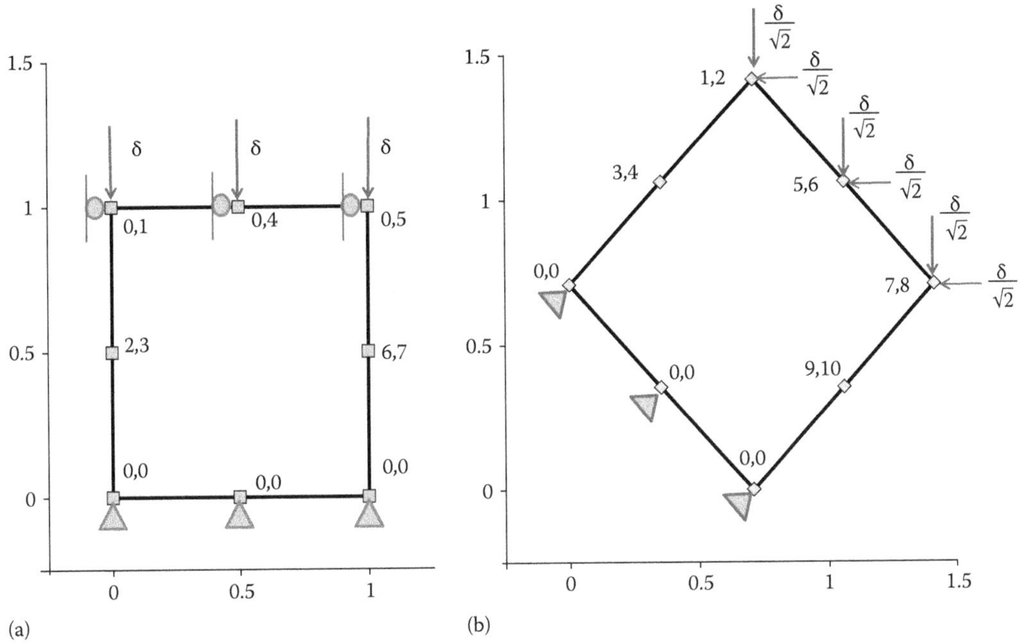

Figure D.7 Geometry, boundary and loading conditions for rotated verification. (a) In-plane 'rough' mesh and (b) rotated 'rough' mesh.

commercial and propriety software is that boundary conditions applied to two sides of a single element can provide increased constraint, sometimes leading to stability for a single element that is not repeated for a larger (and less constrained) mesh. So for added confidence, it is wise to also run cases of special interest with the bigger mesh.

The close correspondence between the VBA and *NorSandFEM* verification results is illustrated for an undrained plane strain loose element test (Figure D.8) and a drained plane strain dense element test (Figure D.9). In both cases, the VBA solution is plotted as a solid black line, while the *NorSandFEM* solution is plotted in grey. These verification results range from close to essentially coincident over the full modelled strain range, which is also found in the other verification scenarios of Table D.1. Do change the properties in the input files and run your own parameter combinations. Both the *NorSandFEM* input (.dat) files and the Excel VBA spreadsheets are simple to edit, and the best way to get a feel for the effect of properties is to play with their values.

D.7 DOWNLOAD NOTES FOR *NorSandFEM*

Although the basics of *NorSandFEM* are described in this appendix, a full appreciation of the capabilities (and limitations) of the code really requires downloading the software and then running the code. Apart from the code (Fortran source and executable) and data files (to run the verifications and examples), the download also contains additional information to assist in using the software. This includes instructions on how to quickly run the verification cases, a detailed description of the required format of the input file (each input files is also annotated with the name of the variable being read).

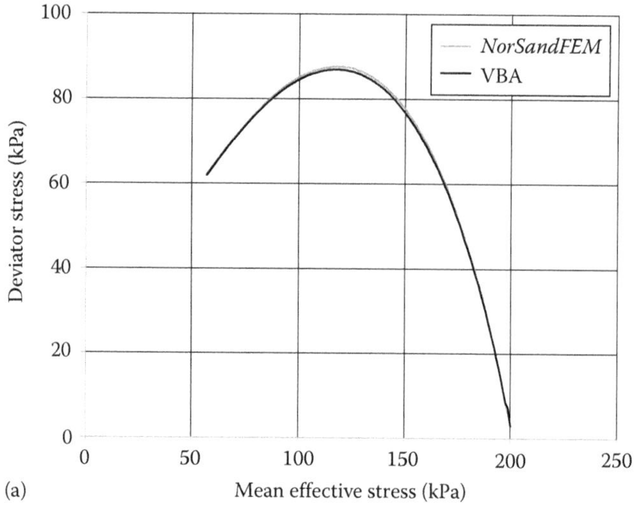

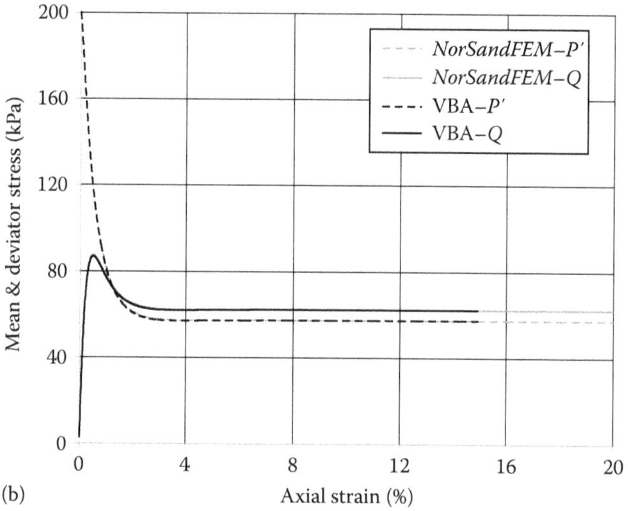

Figure D.8 Verification for loose undrained soil loaded in plane strain compression. (a) Stress path and (b) stress–strain.

There is an issue with compilers. *NorSandFEM* is written in Fortran 90 for compatibility with the adopted library of numerical routines. Although Fortran was developed for scientific computing, and so numerically efficient for this type of application, in recent years the popularity of Fortran has declined with commercial software producers. Inexpensive, easy to use, 'Windows-based' compilers that include good debugging capabilities are becoming harder to find – Fortran compilers are no longer found in all (or even most) geotechnical consulting companies or even university engineering departments. Therefore, it is anticipated that *NorSandFEM* will be ported to C++ in the near future. Although C++ is not ideal for scientific programming (it does not even have inbuilt multidimensional arrays), both Microsoft™ and Apple™ currently provide easy to use, and free, compilers. This makes C++ a good platform for interested readers to use to 'test the waters'. So do check the download site periodically to obtain this version.

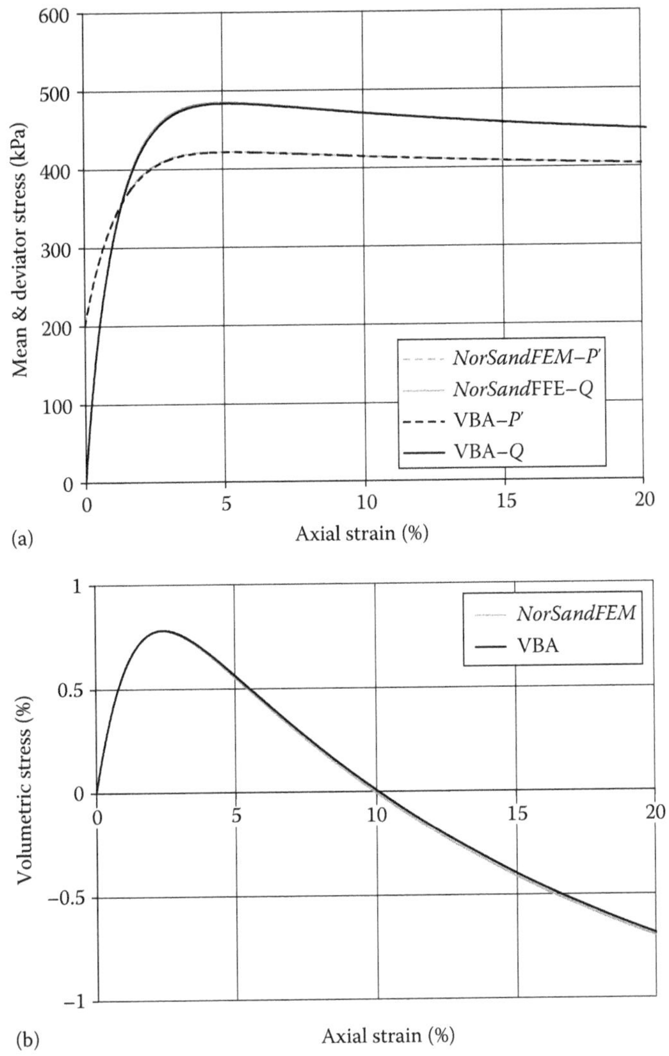

Figure D.9 Verification for dense drained soil loaded in plane strain compression. (a) Stress–strain and (b) volumetric strain.

D.8 ELASTIC PREDICTOR–PLASTIC CORRECTOR

The EP–PC method is conceptually very similar to the widely used *initial stress* solution technique (e.g. Zienkiewicz et al., 1969). The EP–PC method is differentiated from visco-plasticity by the way the load (or displacement) is applied. Whereas viscoplasticity applies an increment of elastic stress and iterates to the true elasto-plastic solution, the EP–PC approach explicitly separates the load into its elastic and plastic components in one step:

$$\dot{\bar{\sigma}} = D(\dot{\varepsilon} - \dot{\varepsilon}^p) \tag{D.33}$$

This is graphically represented in Figure D.10. In its simplest form, and using very small steps to maintain accuracy, EP–PC is equivalent to a simple explicit Euler approach. EP–PC

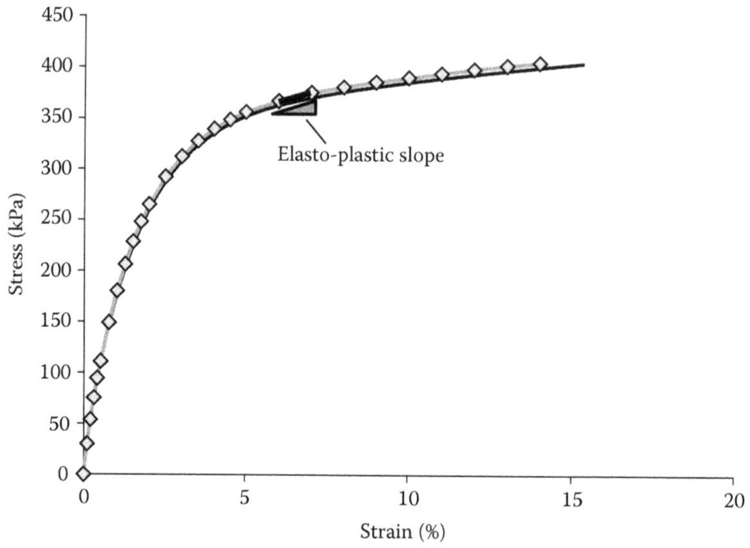

Figure D.10 Schematic representation of constant stiffness solution method.

is particularly advantageous for load-controlled problems as the limit load is approached because 'elastic guess' methods (such as viscoplasticity) tend to be very slow to converge in this situation.

In practice, iterations are typically used to refine the solution in more advanced algorithms (enabling larger loadsteps to be used), but this is a modification on the basic scheme.

D.8.1 Plasticity in EP–PC

A forward Euler integration scheme is described here, where the rate equation is integrated at the start of the step (or when the yield surface is crossed). The analysis steps forward in time, with the new converged stress state $\bar{\sigma}^C$ related to the converged stress state at the previous loadstep ($\bar{\sigma}^O$) by

$$\bar{\sigma}^C = \bar{\sigma}^O + \dot{\bar{\sigma}} \tag{D.34}$$

$$\dot{\bar{\sigma}} = D\left(\dot{\varepsilon} - \lambda^s \frac{\partial Q}{\partial \sigma}\right) \tag{D.35}$$

equivalent in (D.33) to

$$\dot{\varepsilon}^p = \lambda^s \frac{\partial Q}{\partial \sigma} \tag{D.36}$$

where

Q is the plastic potential
λ^s is the plastic multiplier, discussed later

Importantly, note that λ^s is not the slope of the CSL with its usual semi-log idealization. Although both uses of λ are long-standing conventions, the same Greek letter (confusingly) stands for two very different things in numerical implementations of constitutive models with a CSL. We follow the convention that superscript 's' applied to λ denotes the plastic multiplier while the subscript 'e' denotes the soil property.

The change in stress in (D.34) can be decomposed into two parts (see (D.35)), comprising an EP and a PC in which an elastic stress increment is added to the converged stress state at the previous 'old' loadstep ($\bar{\sigma}^0$) to give the initial 'elastic' estimate of the new stress state ($\bar{\sigma}^E$). This approach is also referred to in the finite element literature as the 'initial stress' method, but in this implementation of EP–PC, (D.36) is used directly, and not used to form the elasto-plastic D matrix explicitly (as is usual with initial stress). The elastic estimate is then corrected for the plastic strains in the current loadstep to give the new, 'converged', stress state $\bar{\sigma}^C$. Importantly, note that this algorithm assumes that stress increments are coaxial with the stress state at the start of the loadstep. This means that α is used to map stress increments, with α then being updated for the consequent changes in the stress state at the next loadstep (i.e. principal stress rotation lags stress changes by one step).

The *NorSand* equations are most simply derived using the principal coordinate frame. Conversely, finite element codes working with general stresses require the results in a fixed x,y,z coordinate frame. Assuming that σ_z is the intermediate principal stress (and σ_2 transposes with σ_3 if not), the converged stress state in terms of the principal stresses increments is

$$\bar{\sigma}_x^C = \bar{\sigma}_x^O + \left(\Delta\dot{\bar{\sigma}}_1 \cos^2(90+\alpha) + \Delta\dot{\bar{\sigma}}_3 \sin^2(90+\alpha)\right) \tag{D.37a}$$

$$\bar{\sigma}_y^C = \bar{\sigma}_y^O + \left(\Delta\dot{\bar{\sigma}}_1 \cos^2\alpha + \Delta\dot{\bar{\sigma}}_3 \sin^2\alpha\right) \tag{D.37b}$$

$$\bar{\sigma}_z^C = \bar{\sigma}_z^O + \Delta\dot{\bar{\sigma}}_2 \tag{D.37c}$$

$$\bar{\sigma}_{xy}^C = \bar{\sigma}_{xy}^O + 0.5(\Delta\dot{\bar{\sigma}}_1 - \Delta\dot{\bar{\sigma}}_3)\sin 2\alpha \tag{D.37d}$$

Writing the elastic strain increments in terms of total and plastic increments, using the standard plastic strain decomposition $\varepsilon^p = \varepsilon - \varepsilon^e$, the increments of principal stresses are given in terms of strain increments by

$$\dot{\bar{\sigma}}_1 = A\left(\dot{\varepsilon}_1 - \dot{\varepsilon}_1^p\right) + B\left(\dot{\varepsilon}_2 - \dot{\varepsilon}_2^p\right) + B\left(\dot{\varepsilon}_3 - \dot{\varepsilon}_3^p\right)$$

$$\Rightarrow \dot{\bar{\sigma}}_1 = A\dot{\varepsilon}_1 + B(\dot{\varepsilon}_2 + \dot{\varepsilon}_3) - \left(A\dot{\varepsilon}_1^p + B\dot{\varepsilon}_2^p + B\dot{\varepsilon}_3^p\right) \tag{D.38}$$

where A, B are two elastic coefficients:

$$A = K + 4G/3 \tag{D.39a}$$

$$B = K - 2G/3 \tag{D.39b}$$

The other two principal stress increments follow similarly. Now, defining the ratios of plastic principal strain rates as z_2, z_3 (which are unique functions of the converged stress state $\bar{\sigma}^0$ at the previous loadstep),

$$z_2 = \dot{\varepsilon}_2^p \,/\, \dot{\varepsilon}_1^p \tag{D.40a}$$

$$z_3 = \dot{\varepsilon}_3^p \,/\, \dot{\varepsilon}_1^p \tag{D.40b}$$

Using the z_2, z_3, rewrite (D.38) as

$$\dot{\bar{\sigma}}_1 = A\dot{\varepsilon}_1 + B(\dot{\varepsilon}_2 + \dot{\varepsilon}_3) - \dot{\varepsilon}_1^p(A + Bz_2 + Bz_3) \tag{D.41a}$$

$$\dot{\bar{\sigma}}_2 = A\dot{\varepsilon}_2 + B(\dot{\varepsilon}_3 + \dot{\varepsilon}_1) - \dot{\varepsilon}_1^p(Az_2 + Bz_3 + B) \tag{D.41b}$$

$$\dot{\bar{\sigma}}_3 = A\dot{\varepsilon}_3 + B(\dot{\varepsilon}_1 + \dot{\varepsilon}_2) - \dot{\varepsilon}_1^p(Az_3 + B + Bz_2) \tag{D.41c}$$

It is helpful to work in terms of the shear strain invariant to implement *NorSand*, rather than ε_1, and this is readily done by noting

$$\dot{\varepsilon}_v^p = \dot{\varepsilon}_1^p + \dot{\varepsilon}_2^p + \dot{\varepsilon}_3^p \Rightarrow \dot{\varepsilon}_1^p = \frac{\dot{\varepsilon}_v^p}{1 + z_2 + z_3} \Rightarrow \dot{\varepsilon}_1^p = \frac{D^p \dot{\varepsilon}_q^p}{1 + z_2 + z_3} \tag{D.42}$$

On substituting (D.42) in (D.41), the principal stress increments become

$$\dot{\bar{\sigma}}_1 = A\dot{\varepsilon}_1 + B(\dot{\varepsilon}_2 + \dot{\varepsilon}_3) - \left(\frac{A + Bz_2 + Bz_3}{1 + z_2 + z_3}\right) D^p \dot{\varepsilon}_q^p \tag{D.43a}$$

$$\dot{\bar{\sigma}}_2 = A\dot{\varepsilon}_2 + B(\dot{\varepsilon}_3 + \dot{\varepsilon}_1) - \left(\frac{Az_2 + Bz_3 + B}{1 + z_2 + z_3}\right) D^p \dot{\varepsilon}_q^p \tag{D.43b}$$

$$\dot{\bar{\sigma}}_3 = A\dot{\varepsilon}_3 + B(\dot{\varepsilon}_1 + \dot{\varepsilon}_2) - \left(\frac{Az_3 + B + Bz_2}{1 + z_2 + z_3}\right) D^p \dot{\varepsilon}_q^p \tag{D.43c}$$

Notice in (D.43) that A, B are elastic constants (for any location in the domain of interest and at any converged stress state, they can vary with stress level when stepping the solution forward) and there is now a single unknown, the plastic shear strain increment that drives the hardening law, to be solved for at the integration point in the mesh. As usual, this unknown is approached by the plastic multiplier λ^s. With that strain solved, substituting (D.43) in (D.37) immediately gives the new, converged stress state.

D.8.2 Plastic multiplier in EP–PC

Plastic strain increments are normal to the 'plastic potential' (denoted as Q), and their magnitude is an unknown to be found as part of the solution. However, the plastic strain increments are proportional to each other. Thus, the standard plastic 'flowrule' is

$$\dot{\varepsilon}^p = \lambda^s \frac{\partial Q}{\partial \sigma} \tag{D.36 bis}$$

where λ^s is an unknown scalar. *NorSand* is an 'associated' model for the bullet-shaped part of the yield surface with the plastic potential function Q identical to the yield surface function F under triaxial conditions. *NorSand* has an ambiguity in the π-plane as no work is associated with movements around this plane (Jefferies and Shuttle, 2002). This will be addressed later; for now, we will adopt $F = Q$.

It is convenient to rewrite the *NorSand* yield surface from dimensionless ratios into standard '$F = 0$' form with units of stress:

$$\frac{\eta}{M_i} = 1 - \ln\left(\frac{\bar{\sigma}_m}{\bar{\sigma}_{m,i}}\right) \Leftrightarrow F = \bar{\sigma}_q - M_i\bar{\sigma}_m + M_i\bar{\sigma}_m \ln(\bar{\sigma}_m) - M_i\bar{\sigma}_m \ln(\bar{\sigma}_{m,i}) \tag{D.44}$$

with $F = 0$ indicating plastic yield and $F < 0$ elastic states.

Derivation of plastic strains has simple form if written using the invariants underlying *NorSand*, that is,

$$\dot{\varepsilon}_v^p = \lambda^s \frac{\partial F}{\partial \bar{\sigma}_m} \tag{D.45a}$$

$$\dot{\varepsilon}_q^p = \lambda^s \frac{\partial F}{\partial \bar{\sigma}_q} \tag{D.45b}$$

Taking the partial differential of (D.44) with respect to mean stress,

$$\frac{\partial F}{\partial \bar{\sigma}_m} = -M_i + M_i(\ln(\bar{\sigma}_m) - \ln(\bar{\sigma}_{m,i})) + M_i\bar{\sigma}_m \frac{1}{\bar{\sigma}_m}$$

$$\Rightarrow \frac{\partial F}{\partial \bar{\sigma}_m} = M_i(\ln(\bar{\sigma}_m) - \ln(\bar{\sigma}_{m,i}))$$

and on substitution of (D.44) to eliminate the log terms,

$$\frac{\partial F}{\partial \bar{\sigma}_m} = M_i - \eta = D^p \tag{D.46}$$

where dilatancy, D^p, comes from the stress-dilatancy 'work' equation (see 3.4.1).

Next, taking the partial differential of (D.44) with respect to $\bar{\sigma}_q$, on inspection,

$$\frac{\partial F}{\partial q} = 1 \tag{D.47}$$

Substituting (D.46) and (D.47) in (D.45) gives the pleasingly simple expressions for the *NorSand* plastic strain rates:

$$\dot{\varepsilon}_v^p = \lambda^s D^p \tag{D.48a}$$

$$\dot{\varepsilon}_q^p = \lambda^s \tag{D.48b}$$

The next step is to write the stress increments in terms of the unknown plastic multiplier λ^s. Because the shear strain measure ε_q is linear, as is ε_v, the strain invariants can be written as direct expressions of the plastic strain decomposition:

$$\dot{\varepsilon}_q^e = \dot{\varepsilon}_q - \dot{\varepsilon}_q^p \tag{D.49a}$$

$$\dot{\varepsilon}_v^e = \dot{\varepsilon}_v - \dot{\varepsilon}_v^p \tag{D.49b}$$

Because the stress increment corresponds to the elastic strain increment through the elastic modulus, on introduction of the shear and bulk stiffness into (D.49),

$$\dot{\bar{\sigma}}_q = 3G\dot{\varepsilon}_q^e = 3G\dot{\varepsilon}_q - 3G\dot{\varepsilon}_q^p \tag{D.50a}$$

$$\dot{\bar{\sigma}}_m = K\dot{\varepsilon}_v^e = K\dot{\varepsilon}_v - K\dot{\varepsilon}_v^p \tag{D.50b}$$

Combining (D.50) with the plastic strain rates through normality, (D.48), the increments of the stress invariants are

$$\dot{\bar{\sigma}}_q = 3G\dot{\varepsilon}_q - 3G\lambda^s = 3G\left(\dot{\varepsilon}_q - \lambda^s\right) \tag{D.51a}$$

$$\dot{\bar{\sigma}}_m = K\dot{\varepsilon}_v - K\lambda^s D^p = K\left(\dot{\varepsilon}_v - \lambda^s D^p\right) \tag{D.51b}$$

Notice that in the critical state when shearing continues at constant stress state, then it follows that $\lambda^s = \dot{\varepsilon}_q$ for these particular circumstances.

D.8.3 Consistency condition

The plastic multiplier λ^s is found through the consistency condition that plastic strains developing during loading must leave the stress state still on the yield surface: $\dot{F} = 0$. Thus, the new stress state depends on the combination of how the yield surface changes size during the load increment (the hardening law) and how the stress state moves across the yield surface ('neutral' loading). The equations implementing the consistency condition are different for outward loading of the bullet-like yield surface to the inward loading on the cap.

In the case of the bullet-like yield surface, taking the total differential of the yield surface (D.44), and setting to zero, gives the following consistency condition:

$$\dot{F} = 0 = \frac{\partial F}{\partial \bar{\sigma}_m}\dot{\bar{\sigma}}_m + \frac{\partial F}{\partial \bar{\sigma}_q}\dot{\bar{\sigma}}_q + \frac{\partial F}{\partial \bar{\sigma}_{m,i}}\dot{\bar{\sigma}}_{m,i} + \frac{\partial F}{\partial M_i}\dot{M}_i \tag{D.52}$$

The partial differentials with respect to stress were established in the previous section:

$$\frac{\partial F}{\partial \bar{\sigma}_m} = D^p \tag{D.46 bis}$$

$$\frac{\partial F}{\partial q} = 1 \tag{D.47 bis}$$

Taking the partial differential of (D.44) with respect to the image mean stress,

$$\frac{\partial F}{\partial \bar{\sigma}_{m,i}} = -M_i \frac{\bar{\sigma}_m}{\bar{\sigma}_{m,i}} \tag{D.53}$$

Finally, taking the partial differential of (D.44) with respect to the operating critical friction ratio M_i,

$$\frac{\partial F}{\partial M_i} = -\bar{\sigma}_m + \bar{\sigma}_m(\ln(\bar{\sigma}_m) - \ln(\bar{\sigma}_{m,i}))$$

and again using (D.44) to eliminate the log terms,

$$\Rightarrow \frac{\partial F}{\partial M_i} = -\bar{\sigma}_m + \bar{\sigma}_m\left(1 - \frac{\eta}{M_i}\right) = -\frac{\bar{\sigma}_q}{M_i} \tag{D.54}$$

On collecting (D.46), (D.47), (D.53) and (D.54), the consistency condition (D.52) can be written as

$$\dot{F} = 0 = D^p\dot{\bar{\sigma}}_m + \dot{\bar{\sigma}}_q - M_i \frac{\bar{\sigma}_m}{\bar{\sigma}_{m,i}}\dot{\bar{\sigma}}_{m,i} - \frac{\bar{\sigma}_q}{M_i}\dot{M}_i \tag{D.55}$$

The last term in (D.55) involving M_i is a tad tedious and also loses generality as it depends on the chosen idealization for the CSL. The M_i term also varies in the π-plane, and as this variation of M_i does not involve work, it is this component that prevents *NorSand* from being truly associated. In numerical implementations, it is both conceptually and practically simplest to use a trailing measure (backward difference) for M_i as it changes quite slowly.

The *NorSand* hardening rule is (see Equation C.33 from the model definition)

$$\dot{\bar{\sigma}}_{m,i} = H\frac{M_i}{M_{i,tc}}\left(\frac{\bar{\sigma}_m}{\bar{\sigma}_{m,i}}\right)\left[\bar{\sigma}_m \exp\left(-\frac{\chi_i\psi_i}{M_{i,tc}}\right) - \bar{\sigma}_{m,i}\right]\dot{\varepsilon}_q^p + Z\frac{\eta}{\eta_L}\dot{\bar{\sigma}}_{m,i\max}$$

where Z is a flag to toggle the softening on/off, and $\eta_L = M_i\left(1 - \frac{\chi_i\psi_i}{M_{i.tc}}\right)$ and $\dot{\bar{\sigma}}_{mx} = \bar{\sigma}_{mx}\left(\frac{\dot{\bar{\sigma}}_m}{\bar{\sigma}_m} + \frac{\chi_i}{M_{i,tc}}\left(\frac{\psi_i\dot{M}_{i,tc}}{M_{i,tc}} - \dot{\psi}_i\right)\right)$ from Equation C.34 with $\dot{\bar{\sigma}}_{mx}$ replacing $\dot{\bar{\sigma}}_{m,i\max}$ for notational brevity.

The hardening rule can be viewed as comprising four terms: the first two (X_H and X_C) are multipliers of the current unknown plastic strain increment as $\dot{\varepsilon}_q^p$ and $\dot{\bar{\sigma}}_m$ are related to λ^s through (D.51). The remaining two (X_D and X_E) take values estimated from the previous loadstep:

$$\dot{\bar{\sigma}}_{m,i} = X_H\dot{\varepsilon}_q^p + X_C\dot{\bar{\sigma}}_m + X_D\dot{M}_{i.tc} - X_E\dot{\psi}_i = X_H\lambda^s + X_C\dot{\bar{\sigma}}_m + X_D\dot{M}_{i,tc} - X_E\dot{\psi}_i \tag{D.56}$$

where

$$X_H = H\frac{M_i}{M_{i,tc}}\left(\frac{\bar{\sigma}_m}{\bar{\sigma}_{m,i}}\right)\left[\bar{\sigma}_m \exp\left(-\frac{\chi_i\psi_i}{M_{i,tc}}\right)-\bar{\sigma}_{m,i}\right] \tag{D.57a}$$

$$X_C = Z\frac{\bar{\sigma}_{mx}}{\bar{\sigma}_m}\frac{\eta}{\eta_L} \tag{D.57b}$$

$$X_D = Z\bar{\sigma}_{mx}\frac{\eta}{\eta_L}\frac{\psi_i\chi_i}{M_{i,tc}^2} \tag{D.57c}$$

$$X_E = Z\bar{\sigma}_{mx}\frac{\eta}{\eta_L}\frac{\chi_i}{M_{i,tc}} \tag{D.57d}$$

Using the new parameters X_H, X_C, X_D and X_E, the consistency condition (D.52) becomes

$$\begin{aligned}\dot{F} = 0 = D^p\dot{\bar{\sigma}}_m + \dot{\bar{\sigma}}_q - M_i\frac{\bar{\sigma}_m}{\bar{\sigma}_{m,i}}X_H\lambda^s - M_i\frac{\bar{\sigma}_m}{\bar{\sigma}_{m,i}}X_C\dot{\bar{\sigma}}_m \\ - M_i\frac{\bar{\sigma}_m}{\bar{\sigma}_{m,i}}X_D\dot{M}_{i,tc} + M_i\frac{\bar{\sigma}_m}{\bar{\sigma}_{m,i}}X_E\dot{\psi}_i - \frac{\bar{\sigma}_q}{M_i}\dot{M}_i\end{aligned} \tag{D.58}$$

Substituting the stress increments in terms of the unknown plastic multiplier λ^s, derived earlier as (D.51), allows the consistency condition (D.52) to be written as

$$\begin{aligned}D^pK\left(\dot{\varepsilon}_v - \lambda^s D^p\right) + 3G\left(\dot{\varepsilon}_q - \lambda^s\right) - M_i\frac{\bar{\sigma}_m}{\bar{\sigma}_{m,i}}X_H\lambda^s - M_i\frac{\bar{\sigma}_m}{\bar{\sigma}_{m,i}}X_C\dot{\bar{\sigma}}_m \\ - M_i\frac{\bar{\sigma}_m}{\bar{\sigma}_{m,i}}X_D\dot{M}_{i,tc} + M_i\frac{\bar{\sigma}_m}{\bar{\sigma}_{m,i}}X_E\dot{\psi}_i - \frac{\bar{\sigma}_q}{M_i}\dot{M}_i = 0\end{aligned} \tag{D.59}$$

On collecting terms,

$$\left(3G + D^pKD^p + M_i\frac{\bar{\sigma}_m}{\bar{\sigma}_{m,i}}X_H\right)\lambda^s =$$

$$3G\dot{\varepsilon}_q + D^pK\dot{\varepsilon}_v - \frac{\bar{\sigma}_q}{M_i}\dot{M}_i - M_i\frac{\bar{\sigma}_m}{\bar{\sigma}_{m,i}}X_C\dot{\bar{\sigma}}_m - M_i\frac{\bar{\sigma}_m}{\bar{\sigma}_{m,i}}X_D\dot{M}_{i,tc} + M_i\frac{\bar{\sigma}_m}{\bar{\sigma}_{m,i}}X_E\dot{\psi}_i$$

or

$$\lambda^s = \frac{3G\dot{\varepsilon}_q + D^pK\dot{\varepsilon}_v - \frac{\bar{\sigma}_q}{M_i}\dot{M}_i - M_i\frac{\bar{\sigma}_m}{\bar{\sigma}_{m,i}}X_C\dot{\bar{\sigma}}_m - M_i\frac{\bar{\sigma}_m}{\bar{\sigma}_{m,i}}X_D\dot{M}_{i,tc} + M_i\frac{\bar{\sigma}_m}{\bar{\sigma}_{m,i}}X_E\dot{\psi}_i}{3G + D^pKD^p + M_i\frac{\bar{\sigma}_m}{\bar{\sigma}_{m,i}}X_H} \tag{D.60}$$

Now setting

$$X_m = M_i \frac{\bar{\sigma}_m}{\bar{\sigma}_{m,i}} \tag{D.61}$$

we obtain the simplest explicit form for the plastic multiplier:

$$\lambda^s = \frac{3G\dot{\varepsilon}_q + D^p K\dot{\varepsilon}_v - \frac{\bar{\sigma}_q}{M_i}\dot{M}_i + X_m\left(X_E\dot{\psi}_i - X_C\dot{\bar{\sigma}}_m - X_D\dot{M}_{i,tc}\right)}{3G + K\left(D^p\right)^2 + X_m X_H} \tag{D.62}$$

D.9 CONCLUDING COMMENTS

The information provided in this appendix relates to the monotonic version of *NorSand*, and therefore principal stress rotation is not included in the derivations. The internal cap has also been neglected, both for simplicity and because the cap is of lesser importance under monotonic loading conditions.

Appendix E: Calibration chamber test data

This appendix contains tabulated CPT calibration chamber test data on which the inversion methods in Chapter 4 are based. Symbols in the table titles and headings are as follows:

γ_d	Dry density of sample
D_r	Relative density/density index $(e_{max} - e)/(e_{max} - e_{min})$
e	Void ratio, calculated from relative density or dry density
σ'_v	Effective vertical stress on sample during test
σ'_h	Effective horizontal stress on sample during test
K_o	$\frac{\sigma'_h}{\sigma'_v}$
OCR	Overconsolidation ratio
BC	Boundary condition code (see Table E.1)
p'	Mean effective stress on sample during test
p	Mean total stress on sample during test
ψ	State parameter
q_c	Measured cone penetration resistance (corrected for unequal end area)
$Corr$	Correction to normalized resistance for chamber boundary conditions (after Been et al., 1986, unless indicated otherwise in notes for each sand)
$\frac{q_c - p}{p'}$	Normalized cone penetration resistance (corrected)
σ_{hc}	Horizontal stress measured on sleeve of horizontal stress cone
u	Pore pressure in saturated samples
$\frac{\sigma_{hc} - u}{\sigma'_h}$	Horizontal stress amplification factor
V_s	Shear wave velocity (measured for Chek Lap Kok sand only)
Dia	Cone diameter (assumed to be 3.57 cm unless otherwise noted)
Γ_1	Critical state line parameter (e_c at 1 kPa)
λ_{10}	Critical state line slope (assumed log linear CSL, to log base 10)
e_{max}	Maximum void ratio
e_{min}	Minimum void ratio

Table E.1 Boundary condition codes, after Parkin et al. (1980)

Boundary condition	*Side restraint*	*Base restraint*
BC 1	Constant stress	Constant stress
BC 2	Constant volume	Constant volume
BC 3	Constant volume	Constant stress
BC 4	Constant stress	Constant volume

Chek Lap Kok sand (Lee, 2001)

$\Gamma_l = 0.905$ $\lambda_{10} = 0.130$ $e_{max} = 0.68$ $e_{min} = 0.41$

Test	D_r (%)	e	σ'_v (kPa)	σ'_h (kPa)	K_o	*OCR*	*BC*	p' (kPa)	p (kPa)	ψ	q_c (MPa)	*Corr*	$(q_c - p)/p'$	V_s (m/s)
1	26.2	0.610	53	34	0.64	1	1	40	40	−0.086	2.31	1.048	59	154
2	63.0	0.510	151	97	0.64	1	1	115	115	−0.037	8.91	1.051	81	206
3	35.2	0.586	100	59	0.58	1	1	72	72.4	−0.077	3.10	1.052	44	181
4	37.6	0.579	200	117	0.58	1	1	145	145	−0.045	4.65	1.052	33	220
5	42.8	0.565	101	62	0.61	1	1	75	75	−0.097	3.39	1.053	47	183
6	45.1	0.559	201	122	0.61	1	1	149	149	−0.063	6.20	1.044	43	221
7	52.7	0.538	51	33	0.64	1	1	39	39	−0.160	3.85	1.049	103	158
8	55.0	0.532	151	97	0.64	1	1	115	115	−0.105	6.95	1.055	63	210
9	80.4	0.463	53	31	0.58	1	1	38	38	−0.236	8.90	1.145	267	186
10	81.4	0.460	152	88	0.58	1	1	109	109	−0.180	15.78	1.081	156	243

All tests carried out using 20 mm diameter cone. Correction factors from Salgado et al. (1997) as reported by Lee (2001).

Test	γ_d (t/m³)	e	σ'_v (kPa)	σ'_h (kPa)	K_o	OCR	BC	p' (kPa)	p (kPa)	ψ	q_c (MPa)	Corr	$(q_c - p)/p'$	σ_{hc} (kPa)	$(\sigma_{hc} - u)/\sigma'_h$	u (kPa)
Erksak sand (Been et al., 1987b)																
$\Gamma_1 = 0.845$	$\lambda_{10} = 0.054$	$e_{max} = 0.96$	$e_{min} = 0.53$													
3A	1.626	0.630	100	100	1.00	1	4	100	400	−0.105	6.7	1.00	63	425	1.25	300
3B	1.677	0.580	100	100	1.00	1	4	100	400	−0.155	14.1	1.05	144			300
05	1.732	0.530	127	89	0.70	1	4	102	502	−0.205	26.2	1.15	290	550	1.94	400
6A	1.596	0.660	306	214	0.70	1	4	245	398	−0.054	12.4	1.00	49			153
6B	1.636	0.620	307	214	0.70	1	4	245	399	−0.094	18.6	1.00	74	570	1.94	154
6C	1.677	0.580	309	214	0.69	1	4	246	402	−0.134	30.4	1.00	122	625	2.19	156
07	1.699	0.560	307	214	0.70	1	4	245	399	−0.154	31.5	1.05	133			154
08	1.656	0.600	374	266	0.71	1	4	302	453	−0.109	29.7	1.00	97	490	1.27	151
09	1.699	0.560	63	44	0.70	1	4	50	200	−0.191	10.5	1.11	229	280	3.28	150
10	1.688	0.570	188	131	0.70	1	4	150	300	−0.155	27.8	1.05	193	360	1.68	150
11	1.732	0.530	180.0	126.0	0.70	1	4	144	294	−0.196	31.2	1.12	240	330	1.60	150
12	1.656	0.600	180.0	126.0	0.70	1	4	144	294	−0.126	12.9	1.00	88	470	2.54	150
18	1.616	0.640	30.0	22.0	0.73	1	4	25	175	−0.128	1.9	1.00	65	181	1.41	150
19	1.732	0.530	63.0	45.0	0.71	1	4	51	451	−0.221	11.5	1.20	260	620	5.87	400

Hilton mines tailings (Harmon, 1976)

$\Gamma_1 = 1.315$ $\lambda_{10} = 0.17$ $e_{max} = 1.05$ $e_{min} = 0.63$

Test	γ_d *(t/m³)*	*e*	σ'_v *(kPa)*	σ'_h *(kPa)*	K_o	*OCR*	*BC*	*p′ (kPa)*	*p (kPa)*	ψ	q_c *(kg/cm²)*	*Corr*	$(q_c - p)/p'$
27	1.380	0.920	61	28	0.46	1	1	39.0	39.0	−0.125	24.2	1.02	62
28	1.393	0.902	51	23	0.45	1	1	32.3	32.3	−0.156	16.4	1.08	54
29	1.381	0.919	61	26	0.43	1	2	37.7	37.7	−0.128	19.1	1.00	50
30	1.542	0.718	60	25	0.42	1	1	36.7	36.7	−0.331	94.5	1.95	502
31	1.528	0.734	60	26	0.43	1	2	37.3	37.3	−0.314	104.5	1.20	335
32	1.464	0.810	60	22	0.37	1	1	34.7	34.7	−0.243	65.3	1.37	257
33	1.476	0.796	60	24	0.40	1	2	36.0	36.0	−0.254	60.0	1.10	182
34	1.476	0.796	52	21	0.40	1	1	31.3	31.3	−0.265	36.0	1.48	169
35	1.545	0.715	51	25	0.49	1	1	33.7	33.7	−0.340	70.6	2.00	418
40	1.371	0.933	61	29	0.48	1	1	39.7	39.7	−0.110	21.7	1.01	54
41	1.377	0.925	61	27	0.44	1	2	38.3	38.3	−0.121	18.1	1.00	46
42	1.382	0.918	61	31	0.51	1	1	41.0	41.0	−0.123	17.6	1.02	43
49	1.422	0.863	271	117	0.43	1	1	168.3	168.3	−0.074	87.2	1.00	51
50	1.488	0.781	271	112	0.41	1	1	165.0	165.0	−0.157	189.0	1.08	123
51	1.573	0.685	270	114	0.42	1	1	166.0	166.0	−0.253	235	1.42	200
52	1.575	0.683	271	121	0.45	1	2	171.0	171.0	−0.252	244	1.08	153
55	1.406	0.885	272	122	0.45	1	1	172.0	172.0	−0.050	69.9	1	40
56	1.395	0.900	272	121	0.44	1	2	171.3	171.3	−0.035	67.6	1	38
57	1.410	0.880	272	121	0.44	1	2	171.3	171.3	−0.055	80.6	1	46
58	1.495	0.773	271	105	0.39	1	2	160.3	160.3	−0.167	207.6	1.02	131

Hokksund sand (Baldi et al., 1986; Lunne, 1986)

$\Gamma_I = 0.934$ $\lambda_{10} = 0.054$ $e_{max} = 0.91$ $e_{min} = 0.55$

Test	γ_d (t/m^3)	e	σ'_v (kPa)	σ'_h (kPa)	K_o	OCR	BC	p' (kPa)	p (kPa)	ψ	q_c (MPa)	Corr	$(q_c - p)/p'$	Cone Dia. (cm)
E174	1.692	0.596	120.7	54.4	0.45	1	1	76.5	76.5	−0.237	21.7	1.27	359	3.57
E175	1.690	0.598	68.7	29.9	0.44	1	1	42.9	42.9	−0.248	15.0	1.31	457	3.57
E177	1.692	0.596	266.8	117.1	0.44	1	1	167.0	167.0	−0.218	34.7	1.20	248	3.57
E178	1.742	0.550	68.7	30.1	0.44	1	1	42.9	42.9	−0.296	18.7	1.54	669	3.57
E179	1.742	0.550	68.7	30.4	0.44	1	1	43.1	43.1	−0.296	23.0	1.00	532	2.54
E180	1.742	0.550	70.6	29.9	0.42	1	1	43.5	43.5	−0.296	26.0	1.00	596	2.00
E184	1.533	0.761	116.7	57.1	0.49	1	1	77.0	77.0	−0.071	2.8	1.00	35	3.57
E185	1.528	0.767	62.8	30.1	0.48	1	1	41.0	41.0	−0.080	1.5	1.00	35	3.57
E186	1.525	0.770	312.0	157.9	0.51	1	1	209.2	209.2	−0.038	5.9	1.00	27	3.57
N001	1.734	0.557	60.8	28.0	0.46	1	1	38.9	38.9	−0.291	16.5	1.56	658	3.57
N002	1.721	0.569	60.8	21.3	0.35	1	3	34.5	34.5	−0.282	20.5	1.12	665	3.57
N005	1.510	0.788	58.9	24.1	0.41	1	1	35.7	35.7	−0.062	2.2	1.00	61	3.57
N006	1.514	0.783	58.9	21.8	0.37	1	3	34.1	34.1	−0.068	2.9	1.00	83	3.57
N009	1.725	0.565	60.8	21.3	0.35	1	1	34.5	34.5	−0.286	15.0	1.54	669	3.57
N013	1.710	0.579	60.8	23.7	0.39	1	1	36.1	36.1	−0.271	18.2	1.00	503	2.54
N015	1.706	0.583	60.8	19.5	0.32	1	1	33.2	33.2	−0.269	22.6	1.00	679	2.54
N018	1.495	0.806	58.9	23.5	0.40	1	1	35.3	35.3	−0.044	3.6	1.00	100	2.54
N019	1.499	0.801	58.9	22.4	0.38	1	3	34.5	34.5	−0.050	1.2	1.00	33	2.54
N022	1.709	0.580	60.8	21.9	0.36	1	1	34.9	34.9	−0.271	21.4	1.00	612	2.54
N023	1.714	0.575	60.8	21.3	0.35	1	1	34.5	34.5	−0.276	18.4	1.00	532	2.54
N024	1.533	0.761	58.9	23.5	0.40	1	3	35.3	35.3	−0.089	6.0	1.00	168	2.54
S002	1.715	0.574	58.9	21.8	0.37	1	1	34.1	34.1	−0.277	17.3	1.00	506	2.54
S003	1.717	0.573	58.9	20.0	0.34	1	1	33.0	33.0	−0.280	13.0	1.52	598	3.57
S004	1.727	0.563	107.9	34.5	0.32	1	1	59.0	59.0	−0.275	20.0	1.50	507	3.57
S005	1.730	0.561	206.0	65.9	0.32	1	1	112.6	112.6	−0.263	27.0	1.44	344	3.57
S006	1.720	0.570	402.2	124.7	0.31	1	1	217.2	217.2	−0.238	37.0	1.33	225	3.57
S008	1.711	0.578	402.2	124.7	0.31	1	3	217.2	217.2	−0.230	41.0	1.06	199	3.57

(Continued)

Hokksund sand (Baldi et al., 1986; Lunne, 1986)														
$\Gamma_l = 0.934$		$\lambda_{10} = 0.054$		$e_{max} = 0.91$		$e_{min} = 0.55$								
Test	γ_d (t/m^3)	e	σ'_v (kPa)	σ'_h (kPa)	K_o	OCR	BC	p' (kPa)	p (kPa)	ψ	q_c (MPa)	Corr	$(q_c - p)/p'$	Cone Dia. (cm)
S009	1.740	0.552	206.0	65.9	0.32	1	3	112.6	112.6	−0.271	33.0	1.10	321	3.57
S010	1.720	0.570	107.9	35.6	0.33	1	3	59.7	59.7	−0.268	22.0	1.10	404	3.57
S011	1.706	0.583	58.9	23.5	0.40	1	3	35.3	35.3	−0.268	16.0	1.10	497	3.57
S022	1.733	0.558	107.9	39.9	0.37	1	1	62.6	62.6	−0.279	26.0	1.00	414	2.54
S023	1.706	0.583	206.0	76.2	0.37	1	1	119.5	119.5	−0.239	36.0	1.00	300	2.54
S024	1.700	0.588	304.1	115.6	0.38	1	1	178.4	178.4	−0.224	40.0	1.00	223	2.54
S025	1.713	0.576	402.2	148.8	0.37	1	1	233.3	233.3	−0.230	49.0	1.00	209	2.54
S026	1.731	0.560	58.9	21.2	0.36	1	3	33.7	33.7	−0.292	19.0	1.00	562	2.54
S027	1.720	0.570	107.9	38.8	0.36	1	3	61.9	61.9	−0.267	26.0	1.00	419	2.54
S028	1.710	0.579	206.0	74.2	0.36	1	3	118.1	118.1	−0.243	40.0	1.00	338	2.54
S029	1.720	0.570	402.2	148.8	0.37	1	3	233.3	233.3	−0.236	53.0	1.00	226	2.54
S032	1.720	0.570	58.9	22.6	0.38	1	1	34.7	34.7	−0.281	7.3	1.00	210	2.54
S033	1.53	0.765	58.7	21.6	0.37	1	1	33.9	33.9	−0.087	9.6	1.12	316	3.57
S034	1.628	0.658	58.7	19.6	0.33	1	3	32.6	32.6	−0.194	9.3	1.02	290	3.57
S035	1.627	0.659	58.7	20.1	0.34	1	1	33.0	33.0	−0.193	8.2	1.12	278	3.57
S036	1.61	0.677	206.0	76.5	0.37	1	3	119.7	119.7	−0.145	3.3	1.00	27	3.57
S037	1.62	0.667	206.0	72.6	0.35	1	1	117.1	117.1	−0.156	2.4	1.06	21	3.57
S038	1.57	0.720	57.9	22.1	0.38	1	3	34.0	34.0	−0.132	5.6	1.00	164	3.57
S039	1.55	0.742	205.0	81.4	0.39	1	3	122.6	122.6	−0.079	1	1.00	7	3.57
S040	1.52	0.776	106.9	42.2	0.394	1	3	63.8	63.8	−0.060	7.5	1.00	117	3.57
S041	1.55	0.742	401.2	159.9	0.398	1	3	240.3	240.3	−0.063	0.5	1.00	1	3.57
S042	1.47	0.837	56.9	26.5	0.47	1	3	36.6	36.6	−0.013	2	1.00	54	3.57
S043	1.48	0.824	204.0	95.2	0.46	1	3	131.5	131.5	0.005	5.6	1.00	42	3.57
S044	1.46	0.849	400.2	177.6	0.46	1	3	251.8	251.8	0.045	1.9	1.00	7	3.57
E181	1.744	0.548	62.8	79.9	1.27	14.6	1	74.2	74.2	−0.285	20.3	1.49	406	3.57
E182	1.745	0.547	62.8	79.4	1.27	14.5	1	73.9	73.9	−0.286	34.7	1.00	469	2.54
E183	1.744	0.548	63.8	79.9	1.25	14.4	1	74.5	74.5	−0.285	38.2	1.00	511	2.00

(Continued)

Test	γ_d (t/m^3)	e	σ'_v (kPa)	σ'_h (kPa)	K_o	OCR	BC	p' (kPa)	p (kPa)	ψ	q_c (MPa)	Corr	$(q_c - p)/p'$	Cone Dia. (cm)
Hokksund sand (Baldi et al., 1986; Lunne, 1986)														
				$\Gamma_l = 0.934$		$\lambda_{10} = 0.054$		$e_{max} = 0.91$		$e_{min} = 0.55$				
E187	1.528	0.767	111.8	97.3	0.87	7.3	1	102.1	102.1	−0.058	4.5	1.00	43	3.57
N003	1.727	0.563	60.8	60.8	1.00	8	1	60.8	60.8	−0.274	22.5	1.48	547	3.57
N004	1.723	0.567	60.8	60.2	0.99	8	3	60.4	60.4	−0.271	22.8	1.10	414	3.57
N005	1.510	0.788	58.9	53.6	0.91	8	1	55.3	55.3	−0.052	4.9	1.00	88	3.57
N008	1.489	0.813	58.9	44.1	0.75	8	3	49.1	49.1	−0.029	5.1	1.00	103	3.57
N010	1.718	0.572	60.8	51.1	0.84	8	1	54.3	54.3	−0.269	15.8	1.46	423	3.57
N011	1.510	0.788	58.9	53.0	0.90	8	1	54.9	54.9	−0.052	4.1	1.00	74	3.57
N012	1.502	0.798	58.9	54.7	0.93	8	3	56.1	56.1	−0.042	3.9	1.00	69	3.57
N014	1.719	0.571	60.8	54.1	0.89	8	1	56.4	56.4	−0.269	29.4	1.00	520	2.54
N016	1.710	0.579	60.8	54.7	0.90	8	3	56.8	56.8	−0.260	29.4	1.00	518	2.54
N017	1.713	0.576	60.8	59.6	0.98	8	1	60.0	60.0	−0.262	30.8	1.00	512	2.54
N020	1.487	0.816	58.9	46.5	0.79	8	1	50.6	50.6	−0.026	2.9	1.00	56	2.54
N021	1.480	0.824	58.9	51.2	0.87	8	3	53.8	53.8	−0.016	2.6	1.00	47	2.54
N025	1.496	0.805	58.9	43.6	0.74	8	1	48.7	48.7	−0.038	4.8	1.00	98	2.54
N026	1.720	0.570	60.8	55.3	0.91	8	1	57.2	57.2	−0.269	21.1	1.00	368	2.54
S016	1.719	0.571	107.9	52.9	0.49	2	1	71.2	71.2	−0.263	24.0	1.44	484	3.57
S017	1.680	0.607	107.9	63.7	0.59	4	1	78.4	78.4	−0.225	29.0	1.25	461	3.57
S018	1.718	0.572	58.9	27.7	0.47	2	1	38.1	38.1	−0.277	16.0	1.50	629	3.57
S019	1.730	0.561	58.9	37.1	0.63	4	1	44.3	44.3	−0.284	18.0	1.55	628	3.57
S020	1.733	0.558	58.9	48.9	0.83	8	1	52.2	52.2	−0.283	20.0	1.55	593	3.57
S021	1.701	0.587	206.0	82.4	0.40	2	1	123.6	123.6	−0.234	34.0	1.30	356	3.57
S030	1.730	0.561	107.9	52.9	0.49	2	1	71.2	71.2	−0.273	30.0	1.00	420	2.54
S031	1.730	0.561	107.9	72.3	0.67	4	1	84.2	84.2	−0.269	36.0	1.00	427	2.54
S045	1.476	0.829	57.8	31.0	0.54	2	3	39.9	39.9	−0.018	3.8	1.00	93	3.57
S046	1.481	0.823	57.8	36.3	0.63	4	3	43.5	43.5	−0.022	3.5	1.00	79	3.57
S048	1.469	0.838	57.7	32.1	0.56	2	1	40.6	40.6	−0.009	3.0	1.00	72	3.57
S049	1.473	0.833	106.7	59.9	0.56	2	1	75.5	75.5	0.000	5.2	1.00	67	3.57

(Continued)

Hokksund sand (Baldi et al., 1986; Lunne, 1986)														
$\Gamma_I = 0.934$ $\lambda_{10} = 0.054$ $e_{max} = 0.91$ $e_{min} = 0.55$														
Test	γ_d (t/m^3)	e	σ'_v (kPa)	σ'_h (kPa)	K_o	OCR	BC	p' (kPa)	p (kPa)	ψ	q_c (MPa)	Corr	$(q_c - p)/p'$	Cone Dia. (cm)
S051	1.478	0.827	106.8	56.6	0.53	2	1	73.4	73.4	−0.006	5.9	1.00	79	3.57
S052	1.470	0.837	57.7	39.9	0.69	8	1	45.8	45.8	−0.008	6.7	1.00	145	3.57
S054	1.480	0.824	106.8	70.2	0.66	4	3	82.4	82.4	−0.006	6.5	1.00	78	3.57
S055	1.472	0.834	204.8	115.9	0.57	2	3	145.6	145.6	0.017	0.1	1.00		3.57
S056	1.475	0.831	106.8	68.3	0.64	4	1	81.2	81.2	0.000	6.2	1.00	75	3.57
S057	1.481	0.823	204.8	114.7	0.56	2	1	144.7	144.7	0.006	2.5	1.00	16	3.57
S058	1.489	0.813	57.7	35.2	0.61	8	3	42.7	42.7	−0.033	5.5	1.00	127	3.57

Monterey sand (Huntsman, 1985)																
$\Gamma_I = 0.875$ $\lambda_{10} = 0.029$ $e_{max} = 0.83$ $e_{min} = 0.54$																
Test	*Dr %*	e	σ'_v *(kPa)*	σ'_h *(kPa)*	K_o	*OCR*	*BC*	p' *(kPa)*	p *(kPa)*	ψ	q_c *(MPa)*	*Corr*	$(q_c - p)/p'$	σ_{hc} *(kPa)*	$(\sigma_{hc} - u)/\sigma'_h$	*u*
B6	39	0.713	212	212	1.00	1	1	212.0	212.0	−0.095	25.0	1.02	119	210	1.01	0
B7	59	0.654	204	64	0.31	1	1	110.7	110.7	−0.162	18.0	1.32	213	90	1.86	0
B8	30	0.740	198	64	0.32	1	1	108.7	108.7	−0.076	10.0	1.00	91	30	0.47	0
B9	27	0.749	205	106	0.52	1	1	139.0	139.0	−0.064	15.0	1.00	107	50	0.47	0
B10	33	0.731	302	158	0.52	1	1	206.0	206.0	−0.077	20.0	1.00	96	90	0.57	0
B11	72	0.617	199	106	0.53	1	1	137.0	137.0	−0.196	28.0	1.60	325	220	3.32	0
B12	69	0.627	212	212	1.00	1	1	212.0	212.0	−0.181	25.0	1.47	172	220	1.53	0
B13	43	0.702	212	212	1.00	1	1	212.0	212.0	−0.106	25.0	1.04	122	210	1.03	0
B14	69	0.627	201	64	0.32	1	1	109.7	109.7	−0.189	23.0	1.54	321	140	3.37	0
B15	42	0.704	219	106	0.48	1	1	143.7	143.7	−0.108	19.0	1.05	138	150	1.49	0
B16	27	0.748	321	101	0.31	1	1	174.3	174.3	−0.062	15.0	1.00	85	100	0.99	0
B17	46	0.692	315	315	1.00	1	1	315.0	315.0	−0.111	32.0	1.06	107	205	0.69	0
B18	31	0.737	315	315	1.00	1	1	315.0	315.0	−0.066	23.0	1.00	72	185	0.59	0
B19	53	0.673	319	95	0.30	1	1	169.7	169.7	−0.137	24.0	1.17	164	140	1.72	0
B20	54	0.668	318	158	0.50	1	1	211.3	211.3	−0.140	34.0	1.18	189	180	1.34	0
B21	70	0.627	322	95	0.30	1	1	170.7	170.7	−0.183	33.0	1.48	285	160	2.49	0
B22	70	0.624	201	64	0.32	1	1	109.7	109.7	−0.192	21.0	1.56	297	180	4.39	0
B23	52	0.674	199	106	0.53	1	1	137.0	137.0	−0.139	20.0	1.18	171	120	1.34	0
B24	34	0.728	212	212	1.00	1	1	212.0	212.0	−0.080	19.0	1.00	89	150	0.71	0
B27	63	0.643	212	212	1.00	1	1	212.0	212.0	−0.165	27.0	1.34	169	140	0.88	0
B29	33	0.729	212	212	1.00	1	1	212.0	212.0	−0.079	16.0	1.00	74	100	0.47	0
B31	69	0.625	212	212	1.00	1	1	212.0	212.0	−0.183	29.0	1.48	201	175	1.22	0

Monterey sand (Tringale, 1983)													
				$\Gamma_I = 0.875$		$\lambda_{10} = 0.029$		$e_{max} = 0.83$		$e_{min} = 0.54$			
Test	*Dr (%)*	e	σ'_v *(kPa)*	σ'_h *(kPa)*	K_o	*OCR*	*BC*	p' *(kPa)*	p *(kPa)*	ψ	q_c *(MPa)*	*Corr*	$(q_c - p)/p'$
TB1	58	0.658	280	135	0.48	1	1	183.2	183.2	−0.152	32.0	1.25	217
TB3	71	0.622	278	133	0.48	1	1	181.2	181.2	−0.188	31.8	1.52	265
TB4	58	0.658	149	70	0.47	1	1	96.4	96.4	−0.160	17.4	1.30	233
TB5	62	0.648	88	45	0.51	1	1	59.5	59.5	−0.175	11.6	1.42	276
TB6	67	0.634	285	139	0.49	1	1	187.5	187.5	−0.175	33.4	1.42	252
TB7	61	0.651	200	100	0.50	1	1	133.4	133.4	−0.162	23.4	1.32	230
TB10	32	0.734	241	119	0.49	1	1	159.6	159.6	−0.077	15.3	1.00	95
TB11	27	0.749	141	70	0.49	1	1	93.4	93.4	−0.069	8.5	1.00	90
TB12	74	0.614	225	110	0.49	1	1	148.4	148.4	−0.198	30.3	1.62	329

Ottawa sand (Harman, 1976)

$\Gamma_I = 0.754$ $\lambda_{10} = 0.028$ $e_{max} = 0.79$ $e_{min} = 0.49$

Test	γ_d (t/m^3)	e	σ'_v (kPa)	σ'_h (kPa)	K_o	OCR	BC	p' (kPa)	p (kPa)	ψ	q_c (MPa)	Corr	$(q_c - p)/p'$
14	1.576	0.681	51	24	0.47	1	1	33.0	33.0	−0.030	1.8	1.00	54
15	1.572	0.686	59	28	0.47	1	1	38.3	38.3	−0.024	2.05	1.00	52
16	1.581	0.676	272	132	0.49	1	1	178.7	178.7	−0.015	7.42	1.00	41
17	1.583	0.674	263	131	0.50	1	1	175.0	175.0	−0.017	7.4	1.00	41
18	1.574	0.684	61	27	0.44	1	1	38.3	38.3	−0.026	2.0	1.00	50
19	1.539	0.722	61	29	0.48	1	1	39.7	39.7	0.013	1.5	1.00	36
20	1.534	0.728	61	28	0.46	1	1	39.0	39.0	0.019	1.5	1.00	36
21	1.534	0.728	53	24	0.45	1	1	33.7	33.7	0.017	1.4	1.00	40
22	1.722	0.539	60	26	0.43	1	1	37.3	37.3	−0.171	10.0	1.11	296
23	1.709	0.551	52	18	0.35	1	1	29.3	29.3	−0.162	8.3	1.09	309
25	1.698	0.561	60	24	0.40	1	1	36.0	36.0	−0.149	10.0	1.06	292
26	1.706	0.553	52	22	0.42	1	1	32.0	32.0	−0.159	8.5	1.08	287
36	1.701	0.558	60	23	0.38	1	1	35.3	35.3	−0.153	14.7	1.01	419
37	1.577	0.680	61	30	0.49	1	2	40.3	40.3	−0.029	2.3	1.00	56
38	1.533	0.729	61	30	0.49	1	2	40.3	40.3	0.020	1.53	1	37
43	1.716	0.544	271	117	0.43	1	2	168.3	168.3	−0.148	40.9	1	242
44	1.702	0.557	271	111	0.41	1	2	164.3	164.3	−0.135	27.46	1.03	171
45	1.652	0.604	60	25	0.42	1	1	36.7	36.7	−0.106	6.18	1.01	169
46	1.664	0.593	271	122	0.45	1	1	171.7	171.7	−0.098	23.14	1	134
47	1.553	0.706	272	130	0.48	1	1	177.3	177.3	0.015	5.67	1	31
48	1.554	0.705	263	125	0.48	1	1	171.0	171.0	0.014	5.58	1	32
59	1.633	0.623	60	24	0.40	1	2	36.0	36.0	−0.087	7.01	1	194
60	1.610	0.646	271	114	0.42	1	2	166.3	166.3	−0.046	17.96	1	107
61	1.641	0.615	58	27	0.47	1	1	37.3	37.3	−0.095	3.83	1	102
63	1.562	0.697	272	128	0.47	1	2	176.0	176.0	0.006	6.67	1	37
64	1.543	0.717	272	134	0.49	1	2	180.0	180.0	0.026	5.62	1	30
75	1.532	0.73	64	28	0.44	1	1	40.0	40.0	0.021	1.34	1	33
76	1.563	0.695	294	126	0.43	1	1	182.0	182.0	0.004	5.5	1	29
79	1.712	0.548	284	100	0.35	1	1	161.3	161.3	−0.144	29.81	1.04	191
80	1.713	0.547	291	99	0.34	1	1	163.0	163.0	−0.145	29.42	1.05	189

Reid-Bedford sand (Lhuer, 1976)

$\Gamma_I = 1.014$ $\lambda_{10} = 0.065$ $e_{max} = 0.87$ $e_{min} = 0.55$

Test	γ_d (t/m^3)	e	σ'_v (kPa)	σ'_h (kPa)	K_o	OCR	BC	p' (kPa)	p (kPa)	ψ	q_c (MPa)	Corr	$(q_c - p)/p'$
84	1.494	0.781	60	20	0.33	1	1	33.1	33.1	−0.134	1.57	1.04	48
	1.499	0.775	267	118	0.44	1	1	167.9	167.9	−0.094	6.38	1	37
104	1.485	0.791	60	29	0.49	1	1	39.4	39.4	−0.119	1.85	1.02	47
	1.494	0.781	266	120	0.45	1	1	168.8	168.8	−0.088	6.82	1	39
87	1.488	0.788	267	113	0.42	1	1	164.1	164.1	−0.082	3.98	1	23
85	1.648	0.614	59	28	0.47	1	1	38.2	38.2	−0.297	7.75	1.67	337
	1.654	0.608	266	107	0.40	1	1	160.1	160.1	−0.263	23.8	1.45	214
103	1.648	0.614	59	26	0.44	1	1	36.8	36.8	−0.298	7.69	1.67	348
	1.654	0.608	266	111	0.42	1	1	163.0	163.0	−0.262	25.3	1.45	224
88	1.494	0.781	66	24	0.37	1	1	37.9	37.9	−0.130	1.7	1.03	45
95	1.496	0.778	66	27	0.41	1	1	39.6	39.6	−0.132	1.95	1.04	50
92	1.501	0.772	259	117	0.45	1	1	164.2	164.2	−0.098	5.99	1	35
97	1.499	0.775	259	103	0.40	1	1	155.0	155.0	−0.097	6.19	1	39
91	1.654	0.608	70	20	0.28	1	1	36.4	36.4	−0.305	7.5	1.72	353
96	1.648	0.614	65	29	0.45	1	1	41.0	41.0	−0.295	9.64	1.65	386
94	1.656	0.606	258	116	0.45	1	1	163.6	163.6	−0.264	25	1.45	220
100	1.654	0.608	258	99	0.38	1	1	151.9	151.9	−0.264	22.2	1.46	212

Sydney sand (Pournaghiazar et al., 2011)

$\Gamma_I = 1.037$ $\lambda_{10} = 0.066$ $e_{max} = 0.92$ $e_{min} = 0.60$

Test	D_r (%)	e	σ'_v (kPa)	σ'_h (kPa)	K_o	OCR	BC	p' (kPa)	p (kPa)	ψ	q_c (MPa)	Corr	$(q_c - p)/p'$
1	33	0.814	25	25	1	1		24	24	−0.132	2.0	1.047	86
2	33	0.814	50	50	1	1		49	49	−0.112	3.4	1.037	72
3	33	0.814	50	50	1	1		49	49	−0.112	3.3	1.037	69
4	33	0.814	100	100	1	1		98	98	−0.092	5.8	1.029	60
5	61	0.725	30	30	1	1		29	29	−0.217	3.1	1.065	114
6	61	0.725	50	50	1	1		48	48	−0.202	5.0	1.058	109
7	61	0.725	100	100	1	1		96	96	−0.183	11.0	1.054	120
8	61	0.725	150	150	1	1		143	143	−0.171	18.0	1.052	132

For test boundary conditions and chamber size correction factors, see M. Pournaghiazar et al. (2012).

Syncrude oilsands tailings (Golder Associates project files)

$\Gamma_1 = 0.860$ $\lambda_{10} = 0.065$ $e_{max} = 0.90$ $e_{min} = 0.54$

Test	γ_d (t/m^3)	e	σ'_v (kPa)	σ'_h (kPa)	K_o	OCR	BC	p' (kPa)	p (kPa)	ψ	q_c (MPa)	Corr	$(qc - p)/p'$
CC 101	1.553	0.699	600	300	0.50	1	4	400.0	400.0	0.008	14	1	34
CC 102	1.583	0.666	150	75	0.50	1	4	100.0	100.0	−0.064	5.5	1	54
CC 103	1.687	0.564	50	25	0.50	3	4	33.3	33.3	−0.197	26.5	1.07	850
CC 104	1.606	0.643	300	150	0.50	1	4	200.0	200.0	−0.067	14.4	1	71
CC 105	1.614	0.634	75	38	0.51	1	4	50.3	50.3	−0.115	14	1.03	285
CC 106	1.580	0.670	400	200	0.50	1	4	266.7	266.7	−0.032	12.8	1	47
CC 107	1.592	0.657	150	75	0.50	1	4	100.0	100.0	−0.073	15.6	1	155
CC 108	1.588	0.661	75	38	0.51	1	4	50.3	50.3	−0.088	3.3	1	65

Ticino sand (Baldi et al., 1982, 1986)															
$\Gamma_I = 0.975$				$\lambda_{10} = 0.056$				$e_{max} = 0.89$				$e_{min} = 0.60$			
Test	*Sand*	γ_d *(t/m³)*	e	σ'_v *(kPa)*	σ'_h *(kPa)*	K_o	*OCR*	*BC*	p' *(kPa)*	p *(kPa)*	ψ	q_c *(MPa)*	*Corr*	$(q_c - p)/p'$	*Dia. (cm)*
E019	TS 1	1.677	0.598	515.0	217.9	0.42	1	3	316.9	316.9	−0.237	46.5	0.95	138	3.57
E020	TS 1	1.679	0.596	313.9	128.4	0.41	1	3	190.2	190.2	−0.251	39.1	0.94	192	3.57
E021	TS 1	1.679	0.596	115.8	45.1	0.39	1	3	68.7	68.7	−0.276	23.9	0.93	323	3.57
E022	TS 1	1.615	0.659	311.0	131.5	0.42	1	3	191.4	191.4	−0.188	26.1	0.97	131	3.57
E023	TS 1	1.610	0.665	113.8	47.3	0.42	1	3	69.5	69.5	−0.207	15.6	0.97	217	3.57
E024	TS 1	1.615	0.659	514.0	224.1	0.44	1	3	320.8	320.8	−0.175	34.4	0.98	104	3.57
E025	TS 1	1.616	0.658	716.1	316.5	0.44	1	3	449.7	449.7	−0.168	40.7	0.98	88	3.57
E028	TS 1	1.679	0.596	312.9	126.7	0.41	1	3	188.8	188.8	−0.251	36.2	0.94	179	3.57
E030	TS 1	1.560	0.718	311.0	135.3	0.44	1	3	193.8	193.8	−0.129	13.4	0.99	67	3.57
E031	TS 1	1.573	0.704	513.1	227.3	0.44	1	3	322.5	322.5	−0.131	20.1	0.99	61	3.57
E032	TS 1	1.578	0.698	712.2	318.4	0.45	1	3	449.6	449.6	−0.128	25.0	0.99	54	3.57
E033	TS 1	1.573	0.704	113.8	49.4	0.43	1	3	70.9	70.9	−0.168	9.1	0.98	125	3.57
E034	TS 1	1.675	0.600	65.7	27.7	0.42	1	3	40.4	40.4	−0.285	18.4	0.92	420	3.57
E035	TS 1	1.619	0.655	65.7	27.0	0.41	1	3	39.9	39.9	−0.230	10.9	0.95	257	3.57
E036	TS 1	1.578	0.698	63.8	27.1	0.43	1	3	39.3	39.3	−0.187	5.6	0.97	138	3.57
E050	TS 1	1.624	0.650	115.8	47.7	0.41	1	1	70.4	70.4	−0.221	13.6	1.03	197	3.57
E059	TS 2	1.548	0.731	115.8	50.5	0.44	1	1	72.2	72.2	−0.140	7.0	1.01	97	3.57
E060	TS 2	1.554	0.725	510.1	230.1	0.45	1	1	323.4	323.4	−0.110	15.5	1.00	47	3.57
E061	TS 2	1.678	0.597	512.1	218.7	0.43	1	1	316.5	316.5	−0.238	43.7	1.03	141	3.57
E062	TS 2	1.679	0.596	121.6	49.1	0.40	1	1	73.3	73.3	−0.274	20.9	1.06	302	3.57
E063	TS 2	1.619	0.655	114.8	47.4	0.41	1	1	69.9	69.9	−0.216	12.1	1.03	177	3.57
E065	TS 2	1.622	0.652	313.9	135.6	0.43	1	1	195.0	195.0	−0.194	22.1	1.02	115	3.57
E070	TS 2	1.636	0.638	507.2	225.7	0.45	1	1	319.5	319.5	−0.197	31.6	1.02	100	3.57
E074	TS 2	1.547	0.732	64.7	26.9	0.42	1	1	39.5	39.5	−0.153	4.5	1.01	115	3.57
E075	TS 2	1.556	0.722	715.1	329.0	0.46	1	1	457.7	457.7	−0.104	19.9	1.00	42	3.57
E076	TS 2	1.675	0.600	68.7	25.4	0.37	1	1	39.8	39.8	−0.285	12.0	1.06	317	3.57

(Continued)

Ticino sand (Baldi et al., 1982, 1986)															
$\Gamma_I = 0.975$ $\lambda_{10} = 0.056$ $e_{max} = 0.89$ $e_{min} = 0.60$															
Test	Sand	γ_d (t/m^3)	e	σ'_v (kPa)	σ'_h (kPa)	K_o	OCR	BC	p' (kPa)	p (kPa)	ψ	q_c (MPa)	Corr	$(q_c - p)/p'$	Dia. (cm)
E081	TS 3	1.676	0.599	312.0	138.2	0.44	1	1	196.1	196.1	−0.248	33.6	1.04	177	3.57
E082	TS 3	1.618	0.656	713.2	324.5	0.46	1	1	454.1	454.1	−0.170	32.1	1.01	70	3.57
E083	TS 2	1.623	0.651	67.7	27.5	0.41	1	1	40.9	40.9	−0.233	8.8	1.03	220	3.57
E084	TS 2	1.637	0.637	715.1	306.1	0.43	1	1	442.4	442.4	−0.190	37.2	1.02	85	3.57
E113	TS 2	1.675	0.600	119.7	47.3	0.40	1	1	71.4	71.4	−0.271	20.7	1.05	303	3.57
E114	TS 2	1.540	0.740	115.8	54.3	0.47	1	1	74.8	74.8	−0.130	7.3	1.00	97	3.57
E115	TS 2	1.626	0.648	117.7	48.3	0.41	1	1	71.4	71.4	−0.223	16.0	1.02	227	3.57
E121	TS 2	1.674	0.601	313.9	126.8	0.40	1	1	189.2	189.2	−0.247	40.2	1.04	220	3.57
E123	TS 2	1.681	0.594	314.9	130.1	0.41	1	1	191.7	191.7	−0.253	36.4	1.04	196	3.57
E132	TS 4	1.474	0.818	114.8	56.2	0.49	1	1	75.8	75.8	−0.052	2.8	1.00	37	3.57
E136	TS 4	1.679	0.596	121.6	50.7	0.42	1	1	74.4	74.4	−0.274	18.8	1.06	267	3.57
E138	TS 4	1.498	0.789	116.7	62.1	0.53	1	1	80.3	80.3	−0.079	2.4	1.00	29	3.57
E139	TS 4	1.615	0.659	115.8	51.7	0.45	1	1	73.1	73.1	−0.211	6.6	1.03	92	3.57
E140	TS 4	1.681	0.594	121.6	54.1	0.45	1	1	76.6	76.6	−0.275	18.7	1.06	257	3.57
E141	TS 4	1.680	0.595	122.6	53.3	0.44	1	3	76.4	76.4	−0.274	22.6	0.93	274	3.57
E143	TS 4	1.632	0.642	117.7	52.7	0.45	1	1	74.4	74.4	−0.228	11.7	1.03	160	3.57
E167	TS 4	1.638	0.636	118.7	48.7	0.41	1	3	72.0	72.0	−0.235	18.1	0.95	238	3.57
E168	TS 4	1.642	0.632	308.0	135.5	0.44	1	3	193.0	193.0	−0.215	22.9	0.96	113	3.57
E170	TS 4	1.643	0.631	308.0	133.7	0.43	1	3	191.8	191.8	−0.216	27.0	0.96	134	3.57
E172	TS 4	1.520	0.763	507.2	249.5	0.49	1	3	335.4	335.4	−0.070	19.9	1.00	58	3.57
E173	TS 4	1.521	0.762	508.2	250.5	0.49	1	3	336.4	336.4	−0.071	13.3	1.00	38	3.57
I009	TS 4	1.667	0.608	319.8	137.5	0.43	1	1	198.3	198.3	−0.239	31.0	1.04	162	3.57
I010	TS 4	1.663	0.612	113.8	46.3	0.41	1	1	68.8	68.8	−0.261	19.6	1.05	298	3.57
I011	TS 4	1.657	0.617	63.8	26.4	0.41	1	1	38.9	38.9	−0.269	14.7	1.06	400	3.57
I015	TS 4	1.598	0.677	114.8	50.4	0.44	1	1	71.9	71.9	−0.194	13.6	1.02	193	3.57
I016	TS 4	1.600	0.675	114.8	49.0	0.43	1	1	70.9	70.9	−0.196	12.5	1.02	179	3.57

(Continued)

Ticino sand (Baldi et al., 1982, 1986)

$\Gamma_I = 0.975$ $\lambda_{10} = 0.056$ $e_{max} = 0.89$ $e_{min} = 0.60$

Test	Sand	γ_d (t/m³)	e	σ'_v (kPa)	σ'_h (kPa)	K_o	OCR	BC	p' (kPa)	p (kPa)	ψ	q_c (MPa)	Corr	$(q_c - p)/p'$	Dia. (cm)
1018	TS 4	1.604	0.671	315.9	139.9	0.44	1	1	198.6	198.6	−0.175	24.8	1.01	125	3.57
1019	TS 4	1.592	0.683	62.8	25.8	0.41	1	1	38.1	38.1	−0.203	9.6	1.02	255	3.57
1020	TS 4	1.577	0.699	113.8	51.2	0.45	1	1	72.1	72.1	−0.172	12.7	1.01	177	3.57
1021	TS 4	1.606	0.669	315.9	141.8	0.45	1	1	199.8	199.8	−0.177	24.2	1.02	122	3.57
1023	TS 4	1.598	0.677	113.8	50.5	0.44	1	1	71.6	71.6	−0.194	10.8	1.02	153	3.57
1024	TS 4	1.603	0.672	315.9	144.7	0.46	1	1	201.7	201.7	−0.174	20.9	1.01	104	3.57
1028	TS 4	1.501	0.785	112.8	56.7	0.50	1	1	75.4	75.4	−0.084	6.5	1.00	85	3.57
1029	TS 4	1.601	0.674	113.8	50.4	0.44	1	3	71.5	71.5	−0.197	15.6	0.97	210	3.57
1031	TS 4	1.439	0.862	61.8	39.2	0.64	1	1	46.8	46.8	−0.019	2.8	1.00	60	3.57
1032	TS 4	1.447	0.852	314.9	198.7	0.63	1	1	237.4	237.4	0.010	7.8	1.00	32	3.57
1033	TS 4	1.437	0.865	61.8	39.4	0.64	1	1	46.9	46.9	−0.016	1.8	1.00	37	3.57
1035	TS 4	1.475	0.817	112.8	64.5	0.57	1	1	80.6	80.6	−0.051	3.4	1.00	41	3.57
1037	TS 4	1.465	0.829	61.8	35.8	0.58	1	1	44.5	44.5	−0.053	2.2	1.00	48	3.57
1038	TS 4	1.482	0.808	515.0	293.0	0.57	1	1	367.0	367.0	−0.023	12.2	1.00	32	3.57
1040	TS 4	1.526	0.756	61.8	28.6	0.46	1	1	39.7	39.7	−0.129	6.4	1.01	163	3.57
1043	TS 4	1.532	0.749	112.8	50.8	0.45	1	1	71.4	71.4	−0.122	8.9	1.00	124	3.57
1045	TS 4	1.664	0.611	43.2	17.0	0.39	1	1	25.7	25.7	−0.285	11.5	1.07	478	3.57
1046	TS 4	1.604	0.671	47.1	19.1	0.41	1	1	28.4	28.4	−0.223	7.6	1.03	276	3.57
1047	TS 4	1.539	0.741	42.2	19.4	0.46	1	1	27.0	27.0	−0.153	3.6	1.01	133	3.57
1048	TS 4	1.541	0.739	42.2	19.1	0.45	1	3	26.8	26.8	−0.156	3.9	1.01	146	3.57
1049	TS 4	1.601	0.674	62.8	26.1	0.42	1	1	38.3	38.3	−0.212	9.8	1.03	262	3.57
1050	TS 4	1.597	0.672	60.8	25.7	0.42	1	3	37.4	37.4	−0.215	10.9	0.96	278	3.57
1051	TS 4	1.536	0.738	61.8	31.7	0.51	1	1	41.7	41.7	−0.146	6.5	1.00	155	3.57
1052	TS 4	1.488	0.794	310.0	178.3	0.58	1	1	222.2	222.2	−0.049	8.9	1.00	39	3.57
1053	TS 4	1.482	0.802	110.8	63.9	0.58	1	1	79.6	79.6	−0.067	4.3	1.00	53	3.57
1054	TS 4	1.593	0.676	41.2	17.6	0.43	1	3	25.4	25.4	−0.220	8.5	0.96	318	3.57

(Continued)

Ticino sand (Baldi et al., 1982, 1986)

$\Gamma_I = 0.975$ $\lambda_{10} = 0.056$ $e_{max} = 0.89$ $e_{min} = 0.60$

Test	Sand	γ_d (t/m^3)	e	σ'_v (kPa)	σ'_h (kPa)	K_o	OCR	BC	p' (kPa)	p (kPa)	ψ	q_c (MPa)	Corr	$(q_c - p)/p'$	Dia. (cm)
1055	TS 4	1.602	0.667	63.7	27.8	0.44	1	1	39.7	39.7	−0.219	15.6	1.03	402	3.57
1056	TS 4	1.603	0.666	63.7	28.1	0.44	1	1	40.0	40.0	−0.220	15.5	1.03	398	3.57
1057	TS 4	1.544	0.729	61.8	28.8	0.47	1	1	39.8	39.8	−0.156	7.7	1.01	196	3.57
1058	TS 4	1.485	0.798	112.8	58.1	0.52	1	1	76.3	76.3	−0.072	7.3	1.00	94	3.57
1059	TS 4	1.487	0.796	316.8	165.4	0.52	1	1	215.8	215.8	−0.049	14.0	1.00	64	3.57
1060	TS 4	1.433	0.863	62.8	31.8	0.51	1	1	42.1	42.1	−0.021	2.1	1.00	49	3.57
1061	TS 4	1.436	0.859	60.8	30.7	0.51	1	1	40.7	40.7	−0.026	2.1	1.00	51	3.57
1062	TS 4	1.545	0.728	112.8	51.1	0.45	1	1	71.7	71.7	−0.143	5.3	1.00	72	3.57
1063	TS 4	1.535	0.739	63.7	29.0	0.46	1	1	40.6	40.6	−0.146	3.9	1.00	95	3.57
1064	TS 4	1.532	0.743	317.8	149.4	0.47	1	1	205.5	205.5	−0.103	13.5	1.00	65	3.57
1065	TS 4	1.669	0.600	317.8	134.1	0.42	1	1	195.3	195.3	−0.247	32.0	1.03	168	3.57
1066	TS 4	1.468	0.819	113.8	67.6	0.59	1	1	83.0	83.0	−0.049	4.1	1.00	49	3.57
1072	TS 4	1.480	0.804	314.9	153.0	0.49	1	3	207.0	207.0	−0.041	13.9	1.00	66	3.57
1073	TS 4	1.484	0.799	112.8	55.0	0.49	1	3	74.3	74.3	−0.071	6.2	1.00	82	3.57
1075	TS 4	1.472	0.814	42.1	18.9	0.45	1	1	26.6	26.6	−0.081	4.0	1.00	150	3.57
1076	TS 4	1.667	0.608	114.8	47.1	0.41	1	3	69.6	69.6	−0.264	24.1	0.93	321	3.57
1080	TS 4	1.606	0.669	113.8	48.1	0.42	1	3	70.0	70.0	−0.203	14.7	0.96	200	3.57
1085	TS 4	1.518	0.765	112.8	52.5	0.47	1	3	72.6	72.6	−0.105	7.5	0.99	101	3.57
1113	TS 4	1.561	0.717	111.5	49.1	0.44	1	1	69.9	69.9	−0.155	7.7	1.00	109	3.57
1161	TS 4	1.683	0.592	209.9	87.3	0.42	1	1	128.2	128.2	−0.265	26.4	1.05	215	3.57
1162	TS 4	1.685	0.591	212.6	89.3	0.42	1	3	130.4	130.4	−0.266	29.3	0.93	208	3.57
1163	TS 4	1.685	0.591	312.3	132.4	0.42	1	1	192.4	192.4	−0.257	32.3	1.05	175	3.57
1164	TS 4	1.684	0.591	313.8	132.4	0.42	1	3	192.9	192.9	−0.256	34.6	0.94	168	3.57
1168	TS 4	1.610	0.665	112.7	46.1	0.41	1	1	68.3	68.3	−0.208	13.8	1.02	206	3.57
1169	TS 4	1.614	0.660	111.8	47.1	0.42	1	3	68.7	68.7	−0.212	15.7	0.96	218	3.57
E037	TS 1	1.622	0.652	112.8	79.2	0.70	2.8	3	90.4	90.4	−0.213	16.4	0.96	174	3.57

(Continued)

Ticino sand (Baldi et al., 1982, 1986)

$\Gamma_I = 0.975$ $\lambda_{10} = 0.056$ $e_{max} = 0.89$ $e_{min} = 0.60$

Test	Sand	γ_d (t/m³)	e	σ'_v (kPa)	σ'_h (kPa)	K_o	OCR	BC	p' (kPa)	p (kPa)	ψ	q_c (MPa)	Corr	$(q_c - p)/p'$	Dia. (cm)
E038	TS 1	1.620	0.654	113.8	68.8	0.61	2.8	3	83.8	83.8	−0.213	18.0	0.96	206	3.57
E039	TS 1	1.627	0.647	112.8	103.7	0.92	5.5	3	106.7	106.7	−0.214	21.4	0.96	191	3.57
E040	TS 1	1.626	0.648	111.8	117.1	1.05	6.6	3	115.3	115.3	−0.211	22.6	1.10	214	3.57
E051	TS 2	1.555	0.723	113.8	98.0	0.86	5.4	3	103.3	103.3	−0.139	6.6	0.99	63	3.57
E052	TS 2	1.563	0.715	210.9	149.5	0.71	2.9	3	170.0	170.0	−0.135	9.7	0.99	56	3.57
E053	TS 2	1.561	0.717	65.7	65.9	1.00	9.3	3	65.8	65.8	−0.156	6.6	1.00	99	3.57
E054	TS 2	1.563	0.715	61.8	65.3	1.06	9.4	1	64.1	64.1	−0.159	5.3	1.01	82	3.57
E056	TS 2	1.550	0.729	210.9	154.4	0.73	2.9	1	173.2	173.2	−0.121	14.7	1.00	84	3.57
E058	TS 2	1.556	0.722	112.8	99.8	0.89	5.5	1	104.2	104.2	−0.140	10.3	1.01	98	3.57
E066	TS 2	1.639	0.635	113.8	114.7	1.01	6.3	1	114.4	114.4	−0.225	20.8	1.03	186	3.57
E067	TS 2	1.632	0.642	112.8	128.3	1.14	8.1	1	123.1	123.1	−0.216	20.6	1.10	183	3.57
E068	TS 2	1.632	0.642	109.9	113.4	1.03	5.6	1	112.2	112.2	−0.218	23.6	1.10	230	3.57
E069	TS 2	1.628	0.646	111.8	85.4	0.76	2.8	1	94.2	94.2	−0.218	18.2	1.03	198	3.57
E071	TS 2	1.542	0.738	108.9	94.5	0.87	5.6	1	99.3	99.3	−0.125	6.9	1.00	69	3.57
E072	TS 2	1.548	0.731	211.9	151.9	0.72	2.9	1	171.9	171.9	−0.119	10.3	1.00	59	3.57
E086	TS 2	1.625	0.649	117.7	126.7	1.08	7.8	1	123.7	123.7	−0.209	25.7	1.10	228	3.57
E087	TS 2	1.627	0.647	115.8	109.6	0.95	5.4	1	111.7	111.7	−0.213	21.7	1.02	197	3.57
E088	TS 2	1.629	0.645	114.8	92.2	0.80	3.6	1	99.7	99.7	−0.218	23.7	1.03	244	3.57
E089	TS 2	1.629	0.645	114.8	91.7	0.80	3.6	1	99.4	99.4	−0.218	22.4	1.03	231	3.57
E090	TS 2	1.628	0.646	112.8	80.5	0.71	2.8	1	91.3	91.3	−0.219	19.2	1.03	216	3.57
E094	TS 2	1.678	0.597	112.8	102.2	0.91	4.5	1	105.7	105.7	−0.265	30.1	1.05	298	3.57
E099	TS 2	1.676	0.599	113.8	66.9	0.59	1.9	1	82.5	82.5	−0.269	26.2	1.05	332	3.57
E100	TS 2	1.680	0.595	110.9	119.1	1.07	7.2	1	116.3	116.3	−0.264	32.5	1.15	321	3.57
E101	TS 2	1.675	0.600	110.9	90.6	0.82	3.7	1	97.3	97.3	−0.264	29.3	1.05	315	3.57
E102	TS 2	1.539	0.741	109.9	77.2	0.70	2.8	1	88.1	88.1	−0.125	6.6	1.00	74	3.57
E103	TS 2	1.545	0.735	109.9	107.9	0.98	7.3	1	108.6	108.6	−0.126	9.0	1.00	82	3.57

(Continued)

Ticino sand (Baldi et al., 1982, 1986)

$\Gamma_I = 0.975$ $\lambda_{10} = 0.056$ $e_{max} = 0.89$ $e_{min} = 0.60$

Test	Sand	γ_d (t/m^3)	e	σ'_v (kPa)	σ'_h (kPa)	K_o	OCR	BC	p' (kPa)	p (kPa)	ψ	q_c (MPa)	Corr	$(q_c - p)/p'$	Dia. (cm)
E104	TS 2	1.541	0.739	109.9	85.9	0.78	3.7	1	93.9	93.9	−0.125	8.6	1.00	91	3.57
E105	TS 2	1.522	0.761	108.9	94.2	0.87	4.7	1	99.1	99.1	−0.102	4.7	1.00	47	3.57
E106	TS 2	1.548	0.731	113.8	69.2	0.61	1.9	1	84.1	84.1	−0.136	7.2	1.01	85	3.57
E108	TS 2	1.542	0.738	105.0	93.6	0.89	4.9	1	97.4	97.4	−0.126	11.6	1.00	118	3.57
E109	TS 2	1.544	0.736	111.8	105.6	0.94	6.4	1	107.7	107.7	−0.125	10.0	1.00	92	3.57
E110	TS 2	1.548	0.731	112.8	101.8	0.90	5.4	1	105.4	105.4	−0.130	11.1	1.00	104	3.57
E112	TS 2	1.679	0.596	114.8	78.4	0.68	2.8	1	90.5	90.5	−0.269	26.5	1.05	307	3.57
E124	TS 4	1.503	0.783	306.1	236.9	0.77	2.9	1	260.0	260.0	−0.057	21.9	1.00	83	3.57
E125	TS 4	1.673	0.602	189.3	174.8	0.92	4.74	1	179.6	179.6	−0.247	38.1	1.04	220	3.57
E131	TS 4	1.683	0.592	62.8	87.2	1.39	14.6	1	79.1	79.1	−0.276	20.4	1.06	273	3.57
E133	TS 4	1.508	0.777	113.8	73.6	0.65	1.9	1	87.0	87.0	−0.089	3.1	1.00	35	3.57
E135	TS 4	1.526	0.756	111.8	106.6	0.95	6.4	1	108.3	108.3	−0.105	7.0	1.00	63	3.57
E147	TS 4	1.635	0.639	62.8	81.4	1.30	4.7	1	75.2	75.2	−0.231	14.7	1.15	224	3.57
E149	TS 4	1.684	0.591	62.8	85.1	1.36	4.4	3	77.7	77.7	−0.278	18.3	1.20	282	3.57
I014	TS 4	1.665	0.610	108.9	98.0	0.90	4.8	1	101.6	101.6	−0.253	25.9	1.05	267	3.57
I022	TS 4	1.664	0.611	63.8	36.7	0.58	2.6	1	45.7	45.7	−0.271	16.6	1.06	384	3.57
I026	TS 4	1.598	0.677	61.8	44.3	0.72	2.6	1	50.1	50.1	−0.203	12.5	1.02	254	3.57
I027	TS 4	1.597	0.678	62.8	55.4	0.88	4.3	1	57.9	57.9	−0.198	15.2	1.02	266	3.57
I030	TS 4	1.602	0.673	113.8	99.0	0.87	4.5	3	103.9	103.9	−0.189	22.6	0.97	210	3.57
I077	TS 4	1.667	0.608	113.8	83.5	0.73	2.8	3	93.6	93.6	−0.257	26.2	0.94	262	3.57
I078	TS 4	1.668	0.607	112.8	115.3	1.02	5.5	3	114.5	114.5	−0.253	28.1	1.10	269	3.57
I079	TS 4	1.668	0.607	112.8	112.1	0.99	5.4	4	112.4	112.4	−0.253	31.0	1.04	286	3.57
I081	TS 4	1.604	0.671	113.8	83.2	0.73	2.8	3	93.4	93.4	−0.194	16.6	0.97	171	3.57
I082	TS 4	1.606	0.669	113.8	108.4	0.95	5.5	3	110.2	110.2	−0.192	18.2	0.97	159	3.57
I083	TS 4	1.612	0.663	116.8	113.5	0.972	5.4	4	114.6	114.6	−0.197	23.98	1.02	212	3.57

(Continued)

Test	Sand	γ_d (t/m^3)	e	σ'_v (kPa)	σ'_h (kPa)	K_o	OCR	BC	p' (kPa)	p (kPa)	ψ	q_c (MPa)	Corr	$(q_c - p)/p'$	Dia. (cm)
Ticino sand (Baldi et al., 1982, 1986)															
$\Gamma_l = 0.975$ $\lambda_{10} = 0.056$ $e_{max} = 0.89$ $e_{min} = 0.60$															
1086	TS 4	1.528	0.754	111.8	80.7	0.72	2.8	3	91.1	91.1	−0.111	13.4	0.99	144	3.57
1088	TS 4	1.531	0.750	113.8	101.5	0.89	5.5	4	105.6	105.6	−0.111	12.9	1.00	122	3.57
1094	TS 4	1.536	0.745	112.8	119.7	1.06	8.2	4	117.4	117.4	−0.114	16.2	1.00	137	3.57
1095	TS 4	1.683	0.592	111.8	136.3	1.22	8.3	4	128.2	128.2	−0.265	33.2	1.00	258	3.57
1096	TS 4	1.634	0.640	111.8	129.3	1.16	8.3	4	123.5	123.5	−0.218	22.7	1.00	183	3.57
1148	TS 4	1.678	0.597	60.8	46.2	0.76	3.6	3	51.0	51.0	−0.282	16.7	0.93	303	3.57
1149	TS 4	1.684	0.591	61.8	63.9	1.03	6.8	4	63.2	63.2	−0.283	23.0	1.00	362	3.57
1159	TS 4	1.684	0.591	61.8	63.0	1.02	6.7	1	62.6	62.6	−0.283	18.4	1.20	352	3.57
1160	TS 4	1.682	0.593	61.8	45.4	0.73	3.4	1	50.8	50.8	−0.286	16.7	1.06	347	3.57
1173	TS 4	1.605	0.670	61.8	72.9	1.18	14.7	1	69.2	69.2	−0.202	16.5	1.10	261	3.57
1174	TS 4	1.608	0.667	61.8	71.4	1.16	14.9	3	68.2	68.2	−0.206	15.9	1.10	255	3.57
1175	TS 4	1.511	0.774	61.8	55.5	0.90	14.6	1	57.6	57.6	−0.103	8.7	1.00	151	3.57
1178	TS 4	1.605	0.670	61.8	66.9	1.08	14.6	4	65.2	65.2	−0.204	19.4	1.00	297	3.57
E127	TS 4	1.679	0.596	191.3	174.7	0.91	4.7	1	180.2	180.2	−0.252	41.0	1.00	226	2.54
E128	TS 4	1.682	0.593	191.3	177.7	0.93	4.7	1	182.2	182.2	−0.255	40.6	1.00	222	2
E129	TS 4	1.678	0.597	62.8	86.1	1.37	14.6	1	78.3	78.3	−0.272	26.6	1.00	339	2
E130	TS 4	1.680	0.595	62.8	86.1	1.37	14.6	1	78.3	78.3	−0.274	25.9	1.00	329	2.54
E137	TS 4	1.680	0.595	135.4	56.6	0.42	1.0	1	82.9	82.9	−0.272	26.1	1.00	314	2.54
E144	TS 4	1.687	0.589	125.6	52.2	0.42	1.0	1	76.7	76.7	−0.281	19.9	1.00	258	2.54
E145	TS 4	1.676	0.599	129.5	57.1	0.44	1.0	2	81.2	81.2	−0.269	23.2	1.00	284	2
E146	TS 4	1.638	0.636	62.8	82.7	1.32	14.7	1	76.1	76.1	−0.234	16.4	1.00	215	2.54
E151	TS 4	1.680	0.595	62.8	83.9	1.34	14.7	3	76.8	76.8	−0.274	23.8	1.00	308	2.54
E152	TS 4	1.681	0.594	62.8	83.9	1.34	14.5	2	76.9	76.9	−0.275	27.3	1.00	354	2.54
E153	TS 4	1.678	0.597	63.8	83.4	1.31	14.4	1	76.9	76.9	−0.272	21.9	1.00	284	2.54
E154	TS 4	1.681	0.594	62.8	82.8	1.32	14.6	4	76.1	76.1	−0.275	26.2	1.00	343	2.54
E155	TS 4	1.676	0.599	62.8	82.0	1.31	14.5	2	75.6	75.6	−0.271	26.4	1.00	348	2

(Continued)

Ticino sand (Baldi et al., 1982, 1986)

$\Gamma_I = 0.975$ $\lambda_{10} = 0.056$ $e_{max} = 0.89$ $e_{min} = 0.60$

Test	Sand	γ_d (t/m³)	e	σ'_v (kPa)	σ'_h (kPa)	K_o	OCR	BC	p' (kPa)	p (kPa)	ψ	q_c (MPa)	Corr	$(q_c - p)/p'$	Dia. (cm)
E156	TS 4	1.678	0.597	62.8	83.2	1.33	14.5	3	76.4	76.4	−0.272	23.6	1.00	308	2
E157	TS 4	1.678	0.597	62.8	83.9	1.34	14.5	4	76.8	76.8	−0.272	25.6	1.00	332	2
E158	TS 4	1.676	0.599	62.8	83.2	1.33	14.6	1	76.4	76.4	−0.271	23.1	1.00	301	2
E159	TS 4	1.676	0.599	120.7	51.2	0.42	1.0	4	74.3	74.3	−0.271	20.3	1.00	273	2.54
E160	TS 4	1.672	0.603	120.7	53.1	0.44	1.0	3	75.6	75.6	−0.267	22.8	1.00	300	2.54
E161	TS 4	1.679	0.596	62.8	22.8	0.36	14.6	1	36.1	36.1	−0.292	25.3	1.00	699	2.54
E162	TS 4	1.676	0.599	120.7	48.3	0.40	1.0	1	72.4	72.4	−0.272	26.5	1.00	365	2
E163	TS 4	1.675	0.600	120.7	50.9	0.42	1.0	3	74.2	74.2	−0.270	22.4	1.00	301	2
E164	TS 4	1.673	0.602	121.6	51.0	0.42	1.0	4	74.5	74.5	−0.268	21.3	1.00	285	2
I090	TS 4	1.530	0.752	111.8	83.5	0.75	2.8	1	93.0	93.0	−0.113	12.4	1.00	133	2
I091	TS 4	1.512	0.772	111.8	86.3	0.77	2.8	1	94.8	94.8	−0.092	14.5	1.00	151	2
I092	TS 4	1.632	0.642	113.8	135.1	1.19	8.1	1	128.0	128.0	−0.215	25.5	1.00	198	2
I093	TS 4	1.677	0.598	114.8	142.3	1.24	8.1	1	133.1	133.1	−0.258	32.9	1.00	246	2
I117	TS 4	1.556	0.722	112.8	46.6	0.41	1	1	68.7	68.7	−0.150	9.0	1.00	130	2
I118	TS 4	1.628	0.646	112.8	46.1	0.41	1	1	68.4	68.4	−0.226	13.8	1.00	201	2
I119	TS 4	1.624	0.650	113.8	45.7	0.40	1	3	68.4	68.4	−0.222	14.1	1.00	205	2
I120	TS 4	1.689	0.587	212.9	95.6	0.45	1	1	134.7	134.7	−0.269	32.5	1.00	240	2
I121	TS 4	1.673	0.602	213.9	89.2	0.42	1	3	130.7	130.7	−0.255	29.6	1.00	226	2
I122	TS 4	1.692	0.584	312.9	130.2	0.42	1	1	191.1	191.1	−0.263	34.8	1.00	181	2
I124	TS 4	1.676	0.599	313.9	137.2	0.44	1	3	196.1	196.1	−0.248	34.4	1.00	175	2
I125	TS 4	1.674	0.601	62.8	51.0	0.81	3.4	1	54.9	54.9	−0.277	19.7	1.00	358	2
I126	TS 4	1.674	0.601	62.8	56.1	0.89	6.6	1	58.3	58.3	−0.275	21.2	1.00	362	2
I150	TS 4	1.679	0.596	61.8	47.8	0.77	3.5	3	52.5	52.5	−0.283	21.3	1.00	405	2
I151	TS 4	1.682	0.593	62.8	62.8	1.00	6.7	4	62.8	62.8	−0.281	24.6	1.00	391	2
I152	TS 4	1.681	0.594	62.8	47.8	0.76	3.5	3	52.8	52.8	−0.284	20.3	1.00	384	2.54
I153	TS 4	1.684	0.591	61.8	60.6	0.98	6.7	4	61.0	61.0	−0.284	24.7	1.00	404	2.54

(Continued)

Ticino sand (Baldi et al., 1982, 1986)

$\Gamma_l = 0.975$ $\lambda_{10} = 0.056$ $e_{max} = 0.89$ $e_{min} = 0.60$

Test	Sand	γ_d (t/m³)	e	σ'_v (kPa)	σ'_h (kPa)	K_o	OCR	BC	p' (kPa)	p (kPa)	ψ	q_c (MPa)	Corr	$(q_c - p)/p'$	Dia. (cm)
1154	TS 4	1.684	0.591	60.8	61.1	1.01	6.8	1	61.0	61.0	−0.284	22.3	1.00	364	2.54
1155	TS 4	1.683	0.592	61.8	47.7	0.77	3.5	1	52.4	52.4	−0.286	20.1	1.00	383	2.54
1156	TS 4	1.684	0.591	211.9	89.0	0.42	1	1	130.0	130.0	−0.265	28.2	1.00	216	2.54
1157	TS 4	1.684	0.591	212.9	88.6	0.42	1	3	130.0	130.0	−0.265	30.2	1.00	231	2.54
1158	TS 4	1.686	0.590	312.0	133.2	0.43	1	1	192.8	192.8	−0.257	33.8	1.00	175	2.54
1165	TS 4	1.661	0.613	312.9	128.3	0.41	1	3	189.8	189.8	−0.234	35.5	1.00	186	2.54
1166	TS 4	1.554	0.725	111.8	47.6	0.43	1	3	69.0	69.0	−0.147	10.1	1.00	145	2.54
1167	TS 4	1.609	0.666	111.8	47.2	0.42	1	1	68.7	68.7	−0.206	13.5	1.00	196	2.54
1170	TS 4	1.521	0.762	111.8	47.2	0.42	1	1	68.7	68.7	−0.110	6.9	1.00	99	2.54
1171	TS 4	1.556	0.722	60.8	65.4	1.08	14.8	1	63.9	63.9	−0.152	11.7	1.00	182	2.54
1172	TS 4	1.606	0.669	61.8	71.6	1.16	14.8	3	68.4	68.4	−0.204	18.9	1.00	276	2.54
1176	TS 4	1.515	0.769	61.8	55.5	0.90	14.8	1	57.6	57.6	−0.107	10.8	1.00	186	2
1177	TS 4	1.604	0.671	60.8	69.0	1.14	15.0	1	66.3	66.3	−0.202	15.4	1.00	231	2
1179	TS 4	1.605	0.670	60.8	65.2	1.07	14.9	4	63.7	63.7	−0.204	21.5	1.00	336	2
1180	TS 4	1.606	0.669	60.8	65.1	1.07	14.9	3	63.7	63.7	−0.205	18.8	1.00	295	2
1181	TS 4	1.604	0.671	61.8	66.3	1.07	14.6	4	64.8	64.8	−0.203	21.8	1.00	335	2.54
1182	TS 4	1.510	0.775	59.8	53.6	0.90	15.1	1	55.7	55.7	−0.102	11.8	1.00	210	2.54

Ticino 9 sand (Golder Associates project files)

$\Gamma_I = 0.975$ $\lambda_{10} = 0.056$ $e_{max} = 89$ $e_{min} = 60$

Test	γ_d (t/m³)	e	σ'_v (kPa)	σ'_h (kPa)	K_o	OCR	BC	p' (kPa)	p (kPa)	ψ	q_c (MPa)	Corr	$(q_c - p)/p'$
CC 01	1.610	0.658	150	75	0.50	1	4	100.0	100.0	−0.205	23	1	229
CC 02	1.636	0.632	100	100	1.00	1	4	100.0	100.0	−0.231	23	1	229
CC 03	1.616	0.652	100	100	1.00	6	4	100.0	100.0	−0.211	23	1	229
CC 04	1.560	0.712	450	250	0.56	1	4	316.7	316.7	−0.123	28	1	87
CC 05	1.615	0.653	45	23	0.51	1	4	30.3	30.3	−0.239	9.3	1	306
CC 06	1.419	0.882	150	75	0.50	1	4	100.0	100.0	0.019	4.2	1	41
CC 07	1.427	0.871	200	200	1.00	1	4	200.0	200.0	0.025	4.1	1	20
CC 10	1.498	0.782	150	75	0.50	1	4	100.0	100.0	−0.081	6.4	1	63
CC 08	1.554	0.718	75	38	0.51	1	4	50.3	100.3	−0.162	10.5	1	207
CC 09	1.428	0.870	50	50	1.00	1	4	50.0	50.0	−0.010	0.7	1	13

Toyoura 160/0 sand (Fioravante et al., 1991)

$\Gamma_I = 1.043$ $\lambda_{10} = 0.085$ $e_{max} = 0.977$ $e_{min} = 0.605$

Test	γ_d (t/m^3)	e	σ'_v (kPa)	σ'_h (kPa)	K_o	OCR	BC	p' (kPa)	p (kPa)	ψ	q_c (MPa)	Corr	$(q_c - p)/p'$	Dia. (cm)
311	15.61	0.659	107	51	0.48	1	1	70	70	−0.227	18.3	1.27	332.6	3.57
312	15.56	0.664	115	78	0.68	1	3	90	90	−0.213	22.5	1.04	258.0	3.57
313	15.58	0.662	144	91	0.63	1	2	109	109	−0.208	24.9	1.03	235.0	3.57
319	15.21	0.702	115	71	0.62	1	1	86	86	−0.177	19.2	1.13	252.5	3.57
320	15.21	0.702	111	53	0.48	1	1	72	72	−0.183	16.2	1.13	252.1	3.57
321	15.23	0.700	131	78	0.60	1	2	96	96	−0.175	20.0	1.02	212.2	3.57
323	14.58	0.777	113	62	0.55	1	3	79	79	−0.105	11.3	1.0	142.0	3.57
360	14.85	0.743	110	51	0.47	1	1	71	71	−0.143	12.9	1.05	190.7	3.57
362	14.83	0.746	122	69	0.57	1	2	87	87	−0.132	15.2	1.0	173.9	3.57
363	14.66	0.766	69	35	0.51	1	1	47	47	−0.135	7.5	1.04	166.3	3.57
314	15.59	0.662	110	50	0.45	1	1	70	70	−0.224	23.8	1.0	339.0	2.0
316	15.55	0.666	125	68	0.54	1	3	87	87	−0.212	26.5	1.0	303.6	2.0
340	15.64	0.656	120	60	0.50	1	2	80	80	−0.225	27.4	1.0	340.5	2.0
342	14.66	0.766	110	51	0.46	1	1	71	71	−0.120	13.0	1.0	182.8	2.0
346	15.66	0.654	120	60	0.50	1	2	80	80	−0.227	27.4	1.0	340.5	2.0
358	14.80	0.749	113	56	0.50	1	3	75	75	−0.135	15.2	1.0	202.5	2.0
359	14.86	0.742	114	56	0.50	1	2	75	75	−0.141	16.3	1.0	215.3	2.0
365	14.66	0.766	62	31	0.50	1	3	41	41	−0.140	9.5	1.0	229.0	2.0
366	14.65	0.767	61	28	0.46	1	1	39	39	−0.141	8.6	1.0	221.2	2.0
367	14.22	0.821	61	29	0.48	1	1	40	40	−0.086	4.3	1.0	106.8	2.0
368	14.20	0.823	110	54	0.49	1	1	73	73	−0.062	6.0	1.0	81.2	2.0
364	14.67	0.765	61	29	0.47	1	1	39	39	−0.143	8.2	1.0	207.8	1.0
381	15.75	0.645	62	29	0.46	1	1	40	40	−0.262	21.2	1.0	529.9	1.0
318	15.57	0.664	116	100	0.86	7.2	3	105	105	−0.207	27.1	1.05	269.1	3.57
322	15.64	0.656	118	112	0.95	7.2	1	114	114	−0.212	29.0	1.22	309.4	3.57
341	15.81	0.638	113	111	0.98	7.3	3	112	112	−0.231	40.4	1.0	360.5	2.0
317	15.60	0.660	113	113	1.00	7.3	1	113	113	−0.208	38.9	1.0	343.2	2.0
361	14.87	0.741	113	96	0.85	7.3	3	101	101	−0.131	19.4	1.03	196.0	3.57

West Kowloon sand (Shen and Lee, 1995)

$\Gamma_I = 0.710$ $\lambda_{10} = 0.080$ $e_{max} = 0.69$ $e_{min} = 0.44$

Test	γ_d (t/m^3)	e	σ'_v (kPa)	σ'_h (kPa)	K_o	OCR	BC	p' (kPa)	p (kPa)	ψ	q_c (MPa)	Corr	$(q_c - p)/p'$
437ia	1.719	0.544	106	66	0.62	1	1	79.1	79.1	−0.015	6.1	1.00	76
437ib	1.723	0.540	204	135	0.66	1	1	158.0	158.0	0.006	10.5	1.00	65
438ia	1.720	0.543	36	20	0.56	1	1	25.4	25.4	−0.055	2.95	1.00	115
438ib	1.724	0.539	71	39	0.55	1	1	49.5	49.5	−0.035	4.9	1.00	98
440ia	1.717	0.545	37	21	0.56	1	1	25.9	25.9	−0.051	2.3	1.00	88
440ib	1.721	0.542	70	39	0.56	1	1	49.5	49.5	−0.033	3.5	1.00	69
442ia	1.709	0.553	206	119	0.58	1	1	147.9	147.9	0.016	6.9	1.00	45
442ib	1.712	0.550	301	172	0.57	1	1	215.0	215.0	0.027	9.9	1.00	45
443ia	1.650	0.608	52	34	0.65	1	1	39.6	39.6	0.026	1.3	1.00	33
443ib	1.655	0.604	101	66	0.65	1	1	77.6	77.6	0.045	2.7	1.00	33
444ia	1.652	0.606	199	113	0.57	1	1	141.6	141.6	0.069	5.4	1.00	37
444ib	1.657	0.602	301	172	0.57	1	1	214.8	214.8	0.078	7.7	1.00	35
453ia	1.777	0.494	101	63	0.63	1	1	75.7	75.7	−0.066	7.7	1.00	100
453ib	1.782	0.490	301	185	0.61	1	2	223.5	223.5	−0.032	16.9	1.00	75
455ia	1.700	0.561	51	33	0.65	1	2	39.0	39.0	−0.022	1.77	1.00	44
455ib	1.704	0.557	101	62	0.61	1	2	74.8	74.8	−0.003	3.71	1.00	49
456ia	1.744	0.522	99	62	0.63	1	2	74.5	74.5	−0.038	4.85	1.00	64
456ib	1.745	0.521	300	186	0.62	1	1	224.1	224.1	−0.001	12.07	1.00	53

All tests carried out using 20 mm diameter cone.

Yatesville silty sand (Brandon et al., 1990)													
			$\Gamma_1 = 0.791$		$\lambda_{10} = 0.164$		e_{max} = n/a		e_{min} = n/a				
Test	γ_d (t/m^3)	e	σ'_v (kPa)	σ'_h (kPa)	K_o	OCR	BC	p' (kPa)	p (kPa)	ψ	q_c (MPa)	Corr	$(q_c - p)/p'$
1	1.834	0.445	140.0	56.0	0.40	1	1	84.0	84.0	−0.030	1.1	1.00	12
2	1.839	0.441	100.0	40.0	0.40	1	1	60.0	60.0	−0.058	1.4	1.00	22
3	1.757	0.508	70.0	28.0	0.40	1	1	42.0	42.0	−0.017	0.4	1.00	9
4a	1.870	0.417	280.0	112.0	0.40	1	1	168.0	168.0	−0.009	2.6	1.00	14
4b	1.858	0.426	140.0	56.0	0.40	2	1	84.0	84.0	−0.049	2.2	1.00	25

Appendix F: Some case histories involving liquefaction flow failure

F.1 NINETEENTH- AND TWENTIETH-CENTURY ZEELAND COASTAL SLIDES (THE NETHERLANDS)

The coast of the Netherlands comprises young alluvial sediments with ongoing active geologic processes. In particular, the deposition and erosion of river channels has caused many flowslides over the centuries. A well-known report on the situation is a paper by Koppejan et al. (1948), and their descriptions are used in what follows.

A total of 229 slides were registered for the period 1881–1946, ranging from very small slumps to large flowslides involving three million m^3 of moving soil, the general location of these slides being indicated in Figure F.1.

Koppejan et al. distinguish between slope failures cause by toe erosion and true flowslides, both occurring in Zeeland. Flowslides were noted as causing unexpected sliding of a large portion of the foreshore and sometimes taking part of the flood prevention dyke with it. Zeeland flowslides are somewhat gradual with soil masses sliding downward and out at intervals of a few minutes, although these observations are possible only once the slide is well established with the scarp visible above water level. The rearward regression rate is typically about 50 m/h with the slide taking as much as a day from start to completion. An example of a flowslide geometry is shown in Figure F.2, this being from an 1889 slide at Vlietpolder involving nearly one million m^3 of soil. The steepest slope prior to failure was 27°, while the post-failure slope was about 4°.

The soils involved in the flowslides are predominantly fine uniform sand of the Older Holocene formation and with 90% of the gradation within the particle size range 70–200 μm. The in-situ state can be judged from four cone penetration test (CPT) soundings presented by Koppejan et al., which are reproduced in Figure F.3. The usual variability of tip resistance in sand is apparent as is the clear trend for increased penetration resistance with depth (which is arguably linear). Adopting the criterion that about the 80–90 percentile strength of soil controls its characteristic behaviour, the range of normalized penetration resistance that has been involved in flowslides is $30 < Q_k < 50$. These characteristic Q_k values are seemingly constant with depth.

Most interestingly, critical density tests were carried out on the Older Holocene sands. A critical porosity of 47.5% was quoted as differentiating between flowslides and non-flow failures. These critical density tests determine the volumetric strain caused by shear alone which is not the same as the modern critical sate in the U.S. usage (Chapter 2), with the Netherlands critical density being typically at around $\psi \approx -0.05$. It is therefore consistent that Koppejan et al. report a peak friction angle of $\phi = 37°$. Since the flowslides initiated in the foreshore, and the soil profile of apparently loose sand extends to substantial depth, the residual strength is estimated using infinite slope analysis. This type of analysis is documented in standard texts (e.g. Lambe and Whitman, 1968) and gives, assuming the water

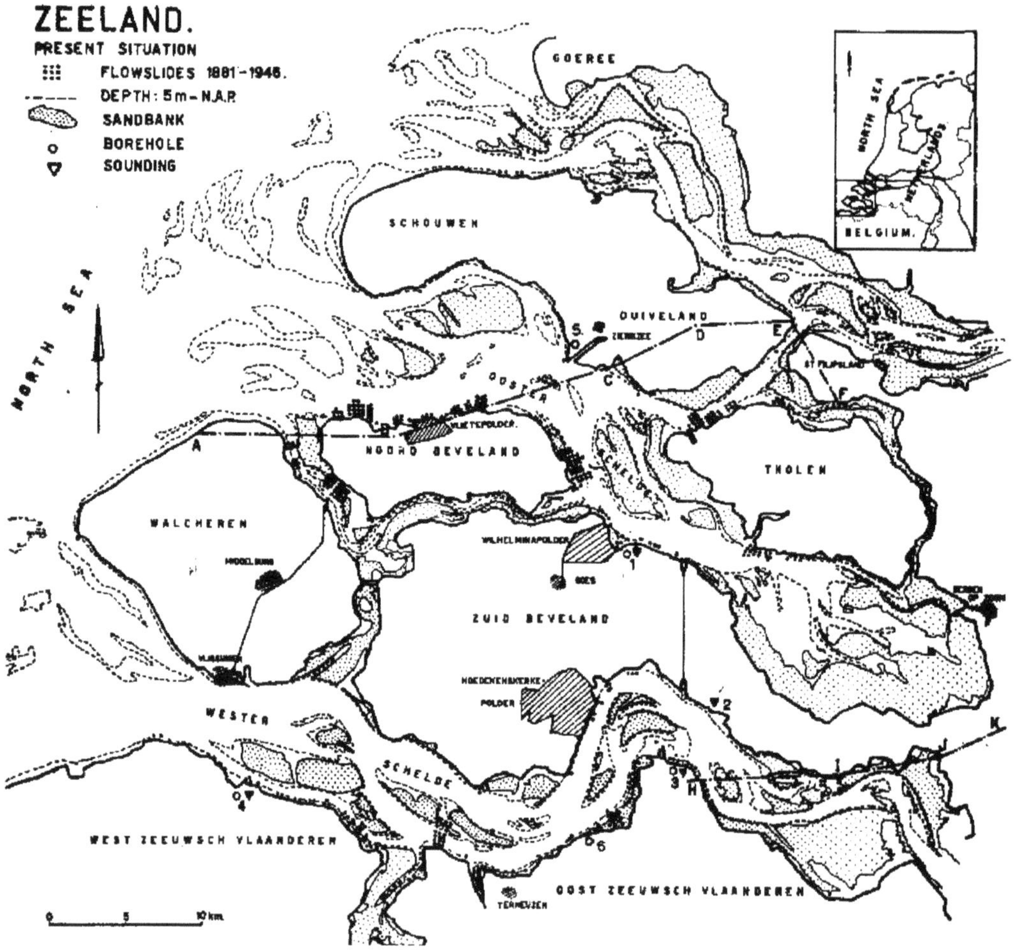

Figure F.1 Location of flowslide on the coast of Zeeland from 1881 to 1946. (From Koppejan, A.W. et al., Coastal flow slides in the Dutch province of Zeeland, in: *Proceedings of the Second International Conference on Soil Mechanics and Foundation Engineering*, Rotterdam, the Netherlands, 1948, Vol. V, pp. 89–96.)

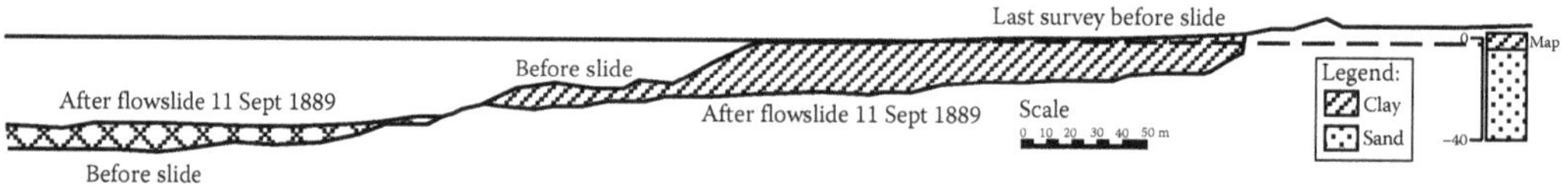

Figure F.2 Vlietpolder flowslide geometry. (From Koppejan, A.W. et al., Coastal flow slides in the Dutch province of Zeeland, in: *Proceedings of the Second International Conference on Soil Mechanics and Foundation Engineering*, Rotterdam, the Netherlands, 1948, Vol. V, pp. 89–96.)

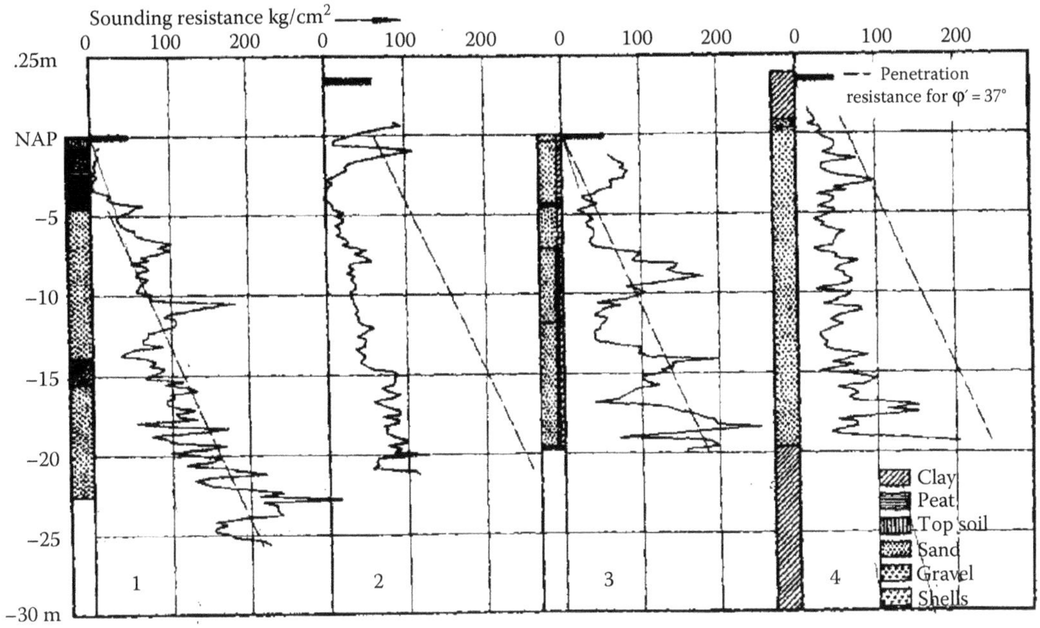

Figure F.3 Typical CPT soundings in flowslide material. (From Koppejan, A.W. et al., Coastal flow slides in the Dutch province of Zeeland, in: *Proceedings of the Second International Conference on Soil Mechanics and Foundation Engineering*, Rotterdam, the Netherlands, 1948, Vol. V, pp. 89–96.)

table is at ground surface (a reasonable assumption given was that it was a coastal foreshore that failed at low water),

$$\frac{s}{\sigma'_{vo}} = \frac{\gamma_t}{\gamma'}\cos\theta\sin\theta \qquad \text{(F.1)}$$

Applying this equation and using the reported final slope of 4° gives an estimated strength ratio $s_r/\sigma'_{vo} \approx 0.13$ assuming no model uncertainty in the rigid-plastic infinite slope model (i.e. taking factor of safety = 1.0 as corresponding to the instant of failure).

The last aspect to consider is the estimated CSL parameters and the in-situ state. The sands involved in the slides are medium to fine sand and with only traces of silt; this is not dissimilar to some of the dredged sands used in the Beaufort island construction whose data have been presented at some length in Chapter 2. Looking at these data, a typical slope to the CSL would be about $\lambda_{10} \approx 0.06$ and $M_{tc} \approx 1.25$ (which is also consistent with the reported $\phi = 37°$ because of the difference between the Dutch critical density and the true critical state). The corresponding ranges for the CPT coefficients over the depth range of 5–20 m below ground surface are about $25 < k < 35$ and $m \approx 6.5$. This range of CPT coefficients has been calculated using calibration chamber test data (Figure 4.19) at the low end and at the upper end using Equation 4.12 with an estimated hardening parameter $H \approx 150$ and elastic rigidity of $I_r \approx 600$. Neither H nor I_r values are controversial for a clean sand with only traces of silt (see the calibrations presented in Chapter 3). Adopting $0.7 < K_o < 0.9$ (this was normally consolidated but aged natural ground), the estimated characteristic in-situ state range is about $-0.09 < \psi_k < -0.02$. The CPT data shown in Figure F.3 indicate that some of the zones at the looser end of this spectrum are rather extensive.

F.2 1907: WACHUSETT DAM, NORTH DYKE (MASSACHUSETTS)

Wachusett Dam is some 48 km from Boston and retains about 240 million m³ of water. Even today, it remains an important water supply reservoir for the city, and it was the most important reservoir before 1939. However, an adjacent saddle dam, referred to as the North Dyke, failed with an upstream flowslide during first filling of the reservoir in 1907. What follows is based on Olson et al. (2000), this paper being one of the most comprehensive evaluations of residual strength back-analysed from a static flowslide that is found in a journal. As such, it is a good example of how such case histories should be analysed, the only aspects of regret being the absence of CPT data and rather too little laboratory testing to establish basic properties of the soils involved.

The main dam is a masonry structure, constructed during 1898–1907, and of no interest here. Two dykes were constructed in low-lying parts of the reservoir rim. The North Dyke, some 3200 m long by maximum 25 m high, was a zoned earth fill dam comprising a sandy silt core with mainly fine sand shells that traversed a relict glacial lake. Longitudinal and transverse sections of the North Dyke are shown in Figure F.4.

The North Dyke was constructed using compacted fill for the trench cut-off and the core, with uncontrolled fill placement in the shells. The core was taken from the reservoir area and comprised sandy silt to silty sand. It was placed in 150 mm lifts and compacted using horse-drawn carts. The downstream shell comprises silty sand to sand, reportedly placed in about 2 m lifts and compacted by flooding (which induced 150–300 mm of settlement by saturation). The upstream shell comprised the same material as the downstream shell, was placed in the same way, but was neither compacted nor flooded. Construction of the North Dyke was completed in 1904, 3 years before the slope failure.

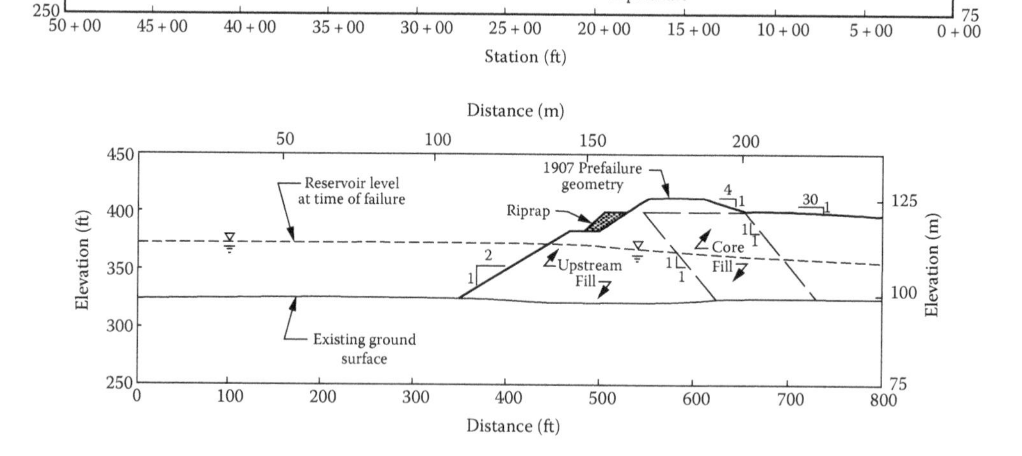

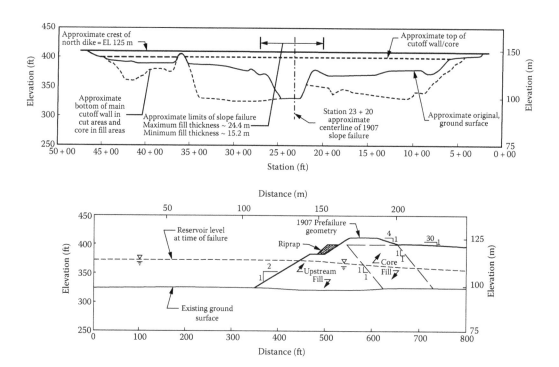

Figure F.4 Longitudinal and transverse sections of North Dyke of Wachusett Dam. (After Olson, S.M. et al., *J. Geotech. Geoenviron. Eng.*, 26(GE12), 1184, 2000. With permission ASCE.)

On April 11, 1907, about 46,000 m³ of upstream fill slid as much as 100 m upstream during the initial impounding of the reservoir and when the reservoir was at about half pool. The sliding mass extended for 213 m along the crest of the dyke and was centred on a former river channel as indicated in Figure F.4. The slide is assumed to have been a static liquefaction as there were no triggers other than the rising reservoir and because the soil moved so far under its own weight. The dyke was reconstructed using compacted fill during 1907, with the dam being finally brought into service later that year. It has performed adequately since.

Site investigations were carried out at the North Dyke in 1984 and 1991, the latter as part of an earthquake vulnerability assessment. A reasonably extensive set of borings exists, with rather frequent standard penetration tests (SPTs). These borings indicated that the liquefying soil had a D_{50} of about 420 µm and 5%–10% fines content. Olson et al. related these SPTs to the estimated sliding surface (Figure F.5). Thirteen of the SPTs were close to the 1907 liquefaction zone, with representative resistances being in the range $6 < (N_1)_{60} < 7$. Densification post-slide and densification during reconstruction of the dam were considered, with Olson et al. concluding that there was no rational means to allow for these effects on the penetration resistances measured nearly a century after the actual failure. There is also the issue that Olson et al. focus on average SPTs, whereas it is something in the 80–90 percentile range that governs, and there are certainly several very low penetration resistances in borings WND-105 and WND-2. In the present circumstances, about the best that can be estimated is that the dimensionless characteristic CPT resistance for the liquefying material was in the range $10 < Q_k < 30$. In suggesting this range, the q_c/N conversion factors discussed in Chapter 4 have been used together with the view that the characteristic resistance in 1907 pre-slide can hardly have been greater than the lower end of the measured average of 1991 and equally plausibly might actually correspond to the average of the three very low resistances measured in 1991. And although the equivalent clean sand fiction has been used in reporting the SPT data, given the range of data and the large effect of using an 80–90 percentile rather than the average, this does not seem a dominant issue in assessing the likely range of Q_k.

A striking feature of the Olson et al. paper is the effort put into estimating the initial effective stress conditions and their distribution along the liquefied zone, with a weighted average lying in the range 142–151 kPa (depending on the assumptions made). Back-analysis for mobilized strength was even more comprehensive, with extensive consideration of the

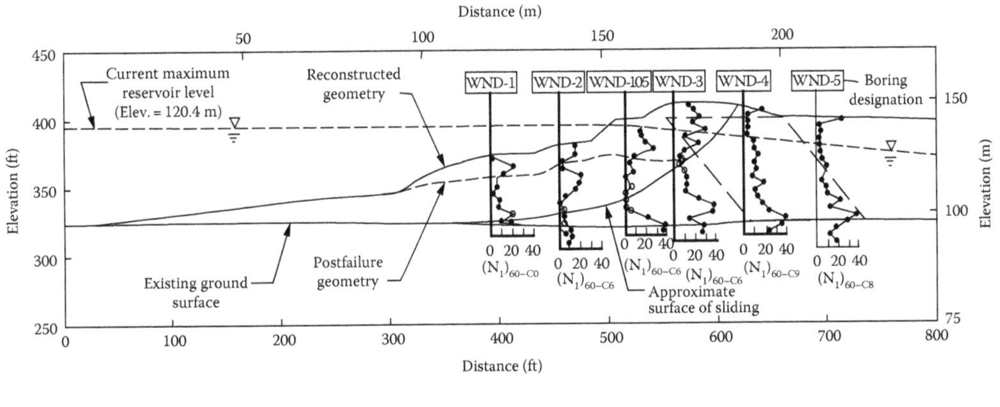

Figure F.5 Cross section of Wachusett Dam failure with 1991 investigation results. (After Olson, S.M. et al., *J. Geotech. Geoenviron. Eng.*, 26(GE12), 1184, 2000. With permission ASCE.)

Table F.1 Summary of strengths and strength ratios determined by Olson et al. (2000)

Strength	*Low bound*	*Best estimate*	*Upper bound*
Peak, s_u	37.6 kPa	Not given	41.9 kPa
Liquefied, s_r	10.4 kPa	16.0 kPa	19.1 kPa
Peak, s_u/σ'_{vo}	0.26	Not given	0.30
Liquefied, s_u/σ'_{vo}	0.07	0.11	0.13

acceleration and deceleration of the sliding mass. Other factors included were the potential for the toe of the slide to either entrain water (with dramatic strength reductions) and proportion of soil not liquefying (because it was not saturated). Both s_r and s_u were estimated. Table F.1 summarizes the results.

CSL parameters and the in-situ state for this case history are estimated as follows. The sands involved in the slides are medium sand and a silt content on the high end of the sand data presented in Chapter 2. Looking at these data, a typical slope to the CSL would be about $0.06 < \lambda_{10} < 0.10$. There are no data on the critical friction ratio or angle, but taking $M_{tc} \approx 1.25$ would appear uncontroversial. The corresponding range for the CPT coefficient k is about 18–30 based in Figure 4.19 (calibration chamber data) and Equation 4.12 over the depth range of 20–40 m below ground surface while $m \approx 5.0$. These CPT coefficients have been calculated using an estimated plastic hardening $H \approx 100$ and an elasticity range of $250 < I_r < 400$. Assuming $K_o = 0.7$ (an average of measurements in other hydraulic fills, see Chapter 4), the estimated characteristic in-situ state range is about $-0.05 < \psi_k < +0.07$. The looser zones, and the SPT data shown in Figure F.5, indicate that some of these loose zones are rather extensive and were statically liquefiable with clear potential for flowslides. It is unsurprising that this fill liquefied during reservoir impoundment, and it could credibly have failed during construction like some other hydraulic fill dams of the era.

F.3 1918: CALAVERAS DAM (CALIFORNIA)

This case history is described in two articles in Engineering New Record (Hazen, 1918; Hazen and Metcalf, 1918) and a paper in the Transactions of the ASCE (Hazen, 1920). The discussion that accompanies the transaction paper is illuminating, and it appears that several other dams failed similarly to Calaveras. Calaveras Dam was completed as a 64 m (210 ft) high earth fill dam. It suffered a flow failure near the end of its construction which led to a redesign. The original dam that failed was of uncompacted fill shells ('steam shovel fill') which were used to contain additional hydraulic filling. The hydraulic fill was placed at the outside limits of the shell, so that soil settled out preferentially leaving relatively sandy shells and a very soft silt core. This core consolidated under its own weight, but a slower rate than that of further fill placement.

This scheme of construction was not uncommon at the end of the nineteenth and early twentieth centuries, and is illustrated schematically in Figure F.6. Even though a clear distinction is shown between toe (shell in modern parlance) and core, in reality, this was somewhat gradational. The fill was primarily material taken from the surrounding hillsides. The steam shovel fill was a broken-up soft sandstone, although Hazen (1918) noted that it was not true sandstone as the broken rock decomposed into particles 'almost as fine in grain size as clay'. The steam shovel fill was not compacted other than from traffic moving around its surface; an average in place bulk unit weight of 18.8 kN/m^3 (120 pcf) is quoted.

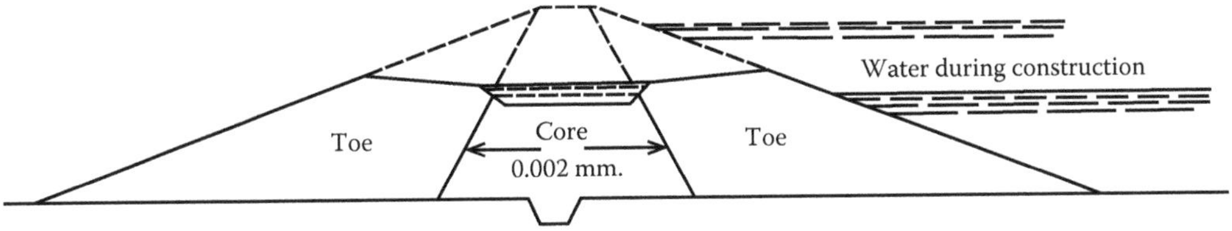

Figure F.6 Typical section of hydraulic fill dam during construction. (From Hazen, A., *Eng. News Rec.*, 81(26), 1158, 1918.)

The hydraulic fill comprised both surface soil and disintegrated soft rock, and was not compacted. The specific gravity was noted as being unusually light with $G_s \approx 2.3$ rather than 2.6–2.7 associated with siliceous materials.

On March 28, 1918, about 610,000 m^3 (800,000 yd^3) moved 90 m (300 ft) upstream while dropping 30 m (100 ft) in elevation as it did so, within a period of about 5 min. The whole failing mass first moved forward as a unit, afterwards separating into blocks. This mechanism suggests failure on fill lower down in the dam, and is not dissimilar in description to the failure at Fort Peck (Chapter 1). Hazen noted that the material that flowed had a porosity of 65%, while that remaining in place had a porosity of no more than 50%. The post-failure configuration is illustrated in Figure F.7 and shows the top surface slope of the failed mass as 1V:10H. If the failure line is defined to the toe of the moved mass, a somewhat steeper slope of about 1V:7.5H is found.

The mobilized residual strength can be estimated from the post-failure configuration if the inertial forces during deceleration of the slide as it came to a halt are neglected. For the range of final slopes, Taylor's stability charts give $0.05 < s_r/\gamma_s H < 0.06$ (the dam was founded on rock). There is a possible range for H depending on the interpretation placed on Hazen's sketch, but a reasonable range is 43 m $< H <$ 52 m (note the sketch is in ft, i.e. 140 ft $< H <$ 170 ft). An average bulk unit weight of the hydraulically placed sandy silts and the confining steam shovel fill would have been about $\gamma_s \approx 18$ kN/m^3 from the data quoted by Hazen and allowing for a higher void ratio in the sandy silts. Putting these values together gives an estimated residual strength in the range 38 kPa $< s_r <$ 56 kPa.

Seed (1987) quoted a residual strength of 35 kPa (750 psf) from his back-analysis of the Calaveras failure, which is the least that would follow from the Taylor's chart-based

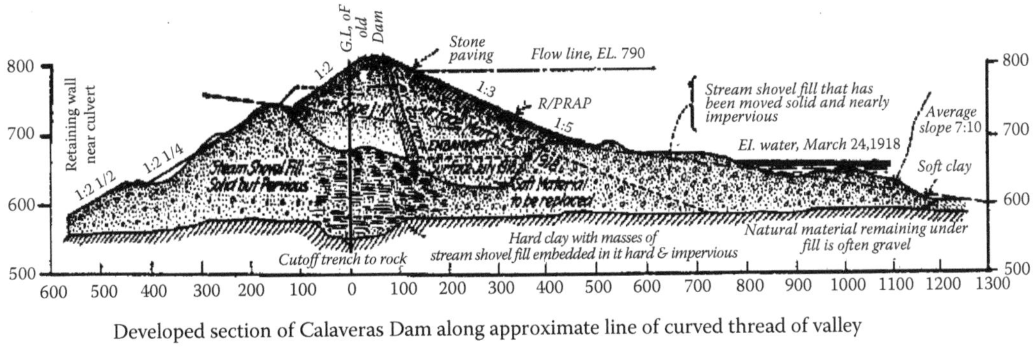

Developed section of Calaveras Dam along approximate line of curved thread of valley

Showing surface before and after the slip and the proposed outline for reconstruction

Figure F.7 Sketch of Calaveras Dam failure showing surface before and after slip. (From Hazen, A., *Eng. News Rec.*, 81(26), 1158, 1918.)

evaluation as given earlier. Poulos (1988) estimated the range of 29 kPa $< s_r <$ 53 kPa (600 psf $< s_r <$ 1100 psf), which agrees rather well with the earlier estimated range. Seed and Harder (1990) quote a narrow range of $s_r = 34 \pm 3$ kPa (650 ± 50 psf), which is a much exaggerated and illusory precision for such limited historic information. Subsequent workers (e.g. Stark and Mesri, 1992; Wride et al., 1999) appear to have followed Seed and Harder without further analysis.

The average initial mean effective stress in the failure zone is less widely estimated. Stark and Mesri (1992) report $\sigma'_{vo} = 137$ kPa (2855 psf). Given the construction method, the piezometric conditions in the shell correspond to about hydrostatic, but there is a possibility of excess pressure in the core given Hazen's description of the lack of consolidation during construction. There is also the depth to the average failure zone, which can hardly be more than about 52/2 = 26 m (85 ft) on one hand, while on the other hand, it can hardly be less than about 15 m (50 ft) (Figure F.7). A corresponding range in effective stress is about 110 kPa $< \sigma'_{vo} <$ 180 kPa, given the unit weights quoted earlier. It would be perverse, although strictly possible, to combine uncertainties in developing the credible range of residual stress ratio. The estimated mobilized strength ratio is therefore constrained by taking the highest strength as corresponding to the highest initial effective stress and so forth. This gives the stress ratio range $0.31 < s_r/\sigma'_{vo} < 0.35$, which can be compared with the ratio $s_r/\sigma'_{vo} \approx 0.23$ estimated by Stark and Mesri.

Regarding the initial in-situ state, Seed (1987) states that the 'tests performed in recent years show that the SPT $(N_1)_{60} = 12$' for the liquefying sand. Seed's assessment is difficult to comprehend in several ways. First, as is clear from Hazen's description of construction, the material that liquefied was predominantly silt, not sand. Second, this is an unreasonably large penetration resistance for sandy silt with a reported in-situ porosity of between 50% and 65%. Seed and Harder assert that the estimated penetration resistance should be thought of in terms of the equivalent clean sand fiction and that an actual $(N_1)_{60} = 7$ corresponds to $(N_1)_{60,ecs} = 12$. This does not add up either, as it is still too great a penetration resistance for the soil conditions described by Hazen. Further, no data on the supposed 'recent tests' in the fill have ever come to light. Poulos (1988) estimates that, given the soil conditions, $(N_1)_{60} = 2$.

To modern eyes, Calaveras looks similar to an upstream-constructed tailings dam, and there are examples of CPTs in such sandy silts. In our experience, the range for undried and unconsolidated true sandy silts is a normalized characteristic penetration resistance of about $20 < Q_k < 30$, although the deposits generally divide into layers of sands and silts. This is where the absence of real CPT data impedes understanding, as it is unclear that the grading given as a global average of the various layers is actually of any relevance. Continuing with the hydraulically placed tailings analogy, the predominantly silt-sized soils (slimes) would usually have the more positive state parameter. Assuming that it was such soils that actually controlled the Calaveras failure, then the relevant characteristic penetration resistance would be about $4 < Q_k < 8$. This latter estimate is consistent with the penetration resistance estimated by Poulos (1988).

Consider first the scenario that the sandy silt was the cause of liquefaction. For a sandy silt, there is considerable uncertainty over the CSL. Although λ tends to increase with fines content as fines are progressively added to clean sand, once the fines content approaches 30%, the fines appear to fully fill the void space between the sand particles and correspondingly then start reducing the λ values. One of the soils in the CSL database is a well-graded material with 35% fines, and it has a λ that would usually be associated with a clean quartz sand. Hazen's description of the fill, however, suggests the possibility of crushable soil particles. The credible range is therefore about $0.07 < \lambda_{10} < 0.15$. There are no data on the critical friction ratio or angle, but again based on Hazen's description of crushable particles, a slightly lower than normal value may be appropriate, say $M_{tc} \approx 1.20$. The corresponding range for the CPT coefficients at the average depth of interest (20–25 m below ground surface) then is about $15 < k < 25$ and $m \approx 4.5$ for $M_{tc} \approx 1.20$, $H \approx 75$–150 and $I_r \approx 200$–300. Adopting $K_o = 0.7$ since it was a hydraulic fill, the estimated characteristic in-situ state range is about $-0.1 < \psi_k < -0.05$.

Now consider an alternative scenario with the failure dominated by the siltier soils. In this case, it is credible that $\lambda_{10} \approx 0.15$, but the critical friction ratio should be more usual. The corresponding range for the CPT coefficients at the average depth of interest (20–25 m below ground surface) then is about $k \approx 10$–15 and $m \approx 4.0$ using $M_{tc} \approx 1.25$, $H \approx 50$ and $I_r \approx 300$. For this silt-dominated scenario, again assuming $K_o = 0.7$, the estimated characteristic in-situ state range is about $+0.11 < \psi_k < +0.14$. The silt appears much more likely to be the material that caused the liquefaction failure.

F.4 1925: SHEFFIELD DAM (CALIFORNIA)

Sheffield Dam was constructed in 1917 to the north of Santa Barbara. It is largely a homogeneous section dam, maximum height 7.6 m by 220 m crest length, with an upstream clay lining that was protected in turn by a concrete facing. The dam failed during an M6.3 earthquake on 29 June 1925. The epicentre was about 7 miles from the dam. At the time of the earthquake, the reservoir was only partly full, but nevertheless some 12,000 m^3 of water was released and flooded part of the city (O'Shaughnessy, 1925). The failure comprised about the 90 m central part of the dam, which slid downstream some 30 m to release the reservoir.

The dam and its foundation were investigated by the Corp of Engineers in 1949 (U.S. Army Corp of Engineers, 1949), while further laboratory strength tests (cyclic and static) were reported by Seed et al. (1969). Arguably, the Seed et al. (1969) paper ended the acceptability of pseudo–static methods for assessing dam adequacy during earthquakes.

Sheffield Dam was largely constructed of undifferentiated fill taken from what became the reservoir upstream of the dam. This fill was a silty sand to sandy silt (with some cobbles and boulders) and appears to have been compacted through construction traffic across it but without a formal compaction protocol. Rock was at shallow (<3 m) depth beneath the dam, overlain by the silty sand to sandy silt. Seed et al. note that 'it has been fairly well established' that there was no stripping of the foundation soils prior to placement of dam fill, and this is not an unreasonable or unusual situation for dams of that era. The 1949 investigation by the Corps indicated that the upper foot or so of the foundation was looser than the remainder, with $\gamma_d \approx 14.2$ kN/m^3 for the loose zone versus $\gamma_d \approx 18.5$ kN/m^3 for the foundation in general. Although the upstream clay blanket was effective, seepage of reservoir water occurred. Seed et al. suggested a piezometric surface somewhat above the foundation as illustrated on the maximum height cross section through the dam (Figure F.8). The suggested phreatic surface is uncontroversial.

The soils involved in the failure were about 50% silt sized and finer. Seed et al. (1969) reported that triaxial testing of samples from immediately downstream of the dam, and with a gradation similar to that reported by the Corps, gave a peak drained strength of $c = 0$ and $\phi = 34.5°$ at a reconstituted $\gamma_d \approx 14.5$ kN/m^3.

The 1925 earthquake caused a peak ground acceleration at the dam site of about 0.15 g with the ground shaking lasting perhaps about 18 s. The dam failed. Willis (1925) reported that 'the rise of water as the ground was shaken formed a liquid layer of mud under the dam, on which it floated out...'. However, there were actually no eyewitnesses to the failure, and the description is based on post-failure morphology. Seed et al. (1969)

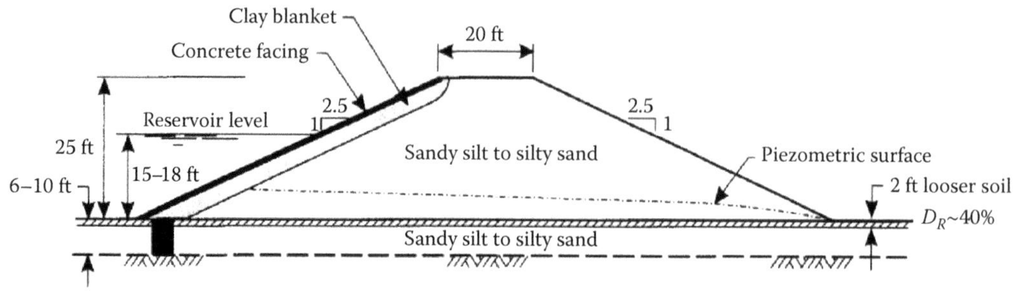

Figure F.8 Sheffield Dam, based on Seed et al. (1969), modified to show the liquefying layer.

analysed the dam's cyclic response, based on cyclic strength tests on reconstituted samples, and computed liquefaction in the confined loose foundation layer (i.e. the 2 ft thick $D_r \approx 40\%$ stratum).

Seed (1987) quotes a post-liquefaction strength for the liquefied zone of $s_r = 2.4$ kPa (50 psf). This strength appears based on the thrust of the water alone and also requires assuming that the retained water was 4.6 m (rather than the range of 4.6–5.5 m found in earlier studies). Both Seed and Harder (1990) and Stark and Mesri (1992) quote the somewhat higher strength estimate of $s_r = 3.6 \pm 1.2$ kPa. A difficulty with the Sheffield Dam failure is the large displacement on what is reasonably only a thin layer (most of the dam fill was plausibly dry and could not have liquefied). However, Seed's calculation was for level ground, and even a minor ground slope increases the strength estimate. The dam was built in a valley, and self-evidently, the reservoir was on the upslope side. If the ground was sloping at say 1° to the downstream, then immediately the strength estimate rises to $s_r = 3.8$ kPa, and if the minimal slope was actually 2°, then $s_r = 4.8$ kPa. Resolving this issue requires detailed evaluation of preconstruction survey drawings if they still exist, but it can certainly be noted that the Seed (1987) estimate is biased on the low side. Seed and Harder (1990) appears more reasonable, but the post-liquefaction strength could be yet larger.

The average vertical effective stress on the failure zone is straightforward. The average depth of the failure zone below the dam surface is about 4.3 m (slightly weighted above half dam height because of the proportion under the crest), and the average piezometric head under normal reservoir operation appears to have been only a few feet in this layer. Treating the dam compaction as giving about the same in-place density as the non-loosened foundation (i.e. using $\gamma_s \approx 18.5$ kN/m^3 as an average fill density), the estimated initial stress is $\sigma'_{vo} \approx 70$ kPa. For comparison, Stark and Mesri (1992) estimate $\sigma'_{vo} = 95$ kPa.

Combining the initial stress and liquefied strength estimate leads to a credible range $0.04 < s_r/\sigma'_{vo} < 0.07$, which can be compared with the best estimate ratio $s_r/\sigma'_{vo} = 0.04$ by Stark and Mesri (1992). Our higher estimates stems from allowing for the possible slope of the ground and a lower estimate of average initial vertical effective stress.

Turning to the in-situ condition, the data from the Corps give a probable initial density, but the representation of this as a relative density involves a judgement. First, there is the question as to whether representative maximum and minimum densities can be measured as about half the soil is silt sized and finer. Soils like this tested in Golder Associates laboratories indicate that minimum density is almost impossible to determine reliably. Second, no maximum and minimum densities have been reported. Seed et al. (1969) estimated a relative density for the liquefied zone of 35% < D_r < 40%; Seed (1987) raised the estimate to 40% < D_r < 50%. The corresponding estimated penetration resistance using the equivalent clean sand adjustment was $6 < (N_1)_{60,ecs} < 8$.

Seed's estimate for penetration resistance can be deconstructed somewhat as his 1987 paper also deals with the ΔN adjustment for the effect of fines content on this resistance. From Table 1 of that paper, an adjustment of $\Delta N_1 = 4$ could be inferred for the Sheffield Dam. Seed seems to have thought in terms of an $(N_1)_{60}$ in the range of say 2–4. In terms of a normalized penetration resistance, and avoiding any soil-type 'corrections', this would correspond to about $6 < Q_k < 12$ using an SPT–CPT conversion factor of 3 for silty sand to sandy silt (Figure 4.3).

CSL parameters and the in-situ state for this case history are estimated as follows. The soils involved in the slides are sandy silt, and a typical slope to the CSL would be about $0.10 < \lambda_{10} < 0.15$. There are no direct data on the critical friction ratio or angle, but taking $M_{tc} \approx 1.25$ would appear uncontroversial given the measured $\phi = 34.5°$ at a density that might be near the critical state (in effect, assuming the tested samples were slightly denser than critical). The range for the CPT coefficients, at an average depth of 4.3 m below ground surface, is about $12 < k < 22$ and $m \approx 4.5$ (from Figure 4.19 and calculated using $70 < H < 100$ and $600 < I_r < 700$ in Equation 4.12). Adopting $0.7 < K_o < 0.9$ (this was not a hydraulic fill, but is natural ground and subjected to construction traffic), the estimated characteristic in-situ state range is about $+0.04 < \psi_k < +0.15$. This is a rather large range for ψ, a consequence of the factor of two ranges in the estimated penetration resistance and considerable uncertainty about the basic soil properties.

F.5 1938: FORT PECK (MONTANA)

The Fort Peck slide is one of the largest liquefaction failures, and aspects of this failure were presented in Chapter 1. The information on Fort Peck here is a summary of the measurements of density and strengths from the post-failure investigation (based on Middlebrooks, 1940). The estimates of the in-situ state prior to the failure are then discussed, as are the calculated strengths mobilized during the slide.

To recap, the Fort Peck dam was a hydraulic fill structure that failed because the hydraulically placed sand fill in the upstream shell was brought to its peak undrained strength by movements in the underlying shale foundation. Although the literature gives the impression that the entire upstream shell failed, this was not the case. Only a small part near the right abutment failed as illustrated by the aerial photographs in Figure F.9.

Void ratio data for the shell were measured after the failure in several test pits put down into undisturbed shell. Critical density tests were carried out on representative samples with various silt contents in the fill, the results being shown in Figure F.10. Although there is quite a wide range for the measured critical states, a consequence of the differing silt contents from one sample to another, there is a clear pattern of behaviour measured. In modern parlance, the CSL parameters are in the range $0.84 < \Gamma < 1.04$ and $\lambda_{10} \approx 0.19$. The range for Γ is unremarkable for a sand fill with some silt, although λ indicates perhaps a little more compressibility than would have been expected from the fill gradations.

The measured in-situ void ratios are also shown in Figure F.10, and these are denser than the measured critical states. How much denser depends on the silt content of the individual samples, but it has not been possible to ascertain such details from available records. The measured void ratios lie just below the band of measured critical states so that the loosest $\psi \approx -0.01$. Assuming a median for the critical state measurements gives $\psi \approx -0.05$

(a)

(b)

Figure F.9 Aerial photographs of Fort Peck Dam failure. (a) View of slide from the left bank. (From U.S. Army Corps of Engineers, Washington, DC, 1939.) (b) Vertical view of failure. (From Sigmundstad, R., http://www.fortpeckdam.com, accessed March 15, 2015.)

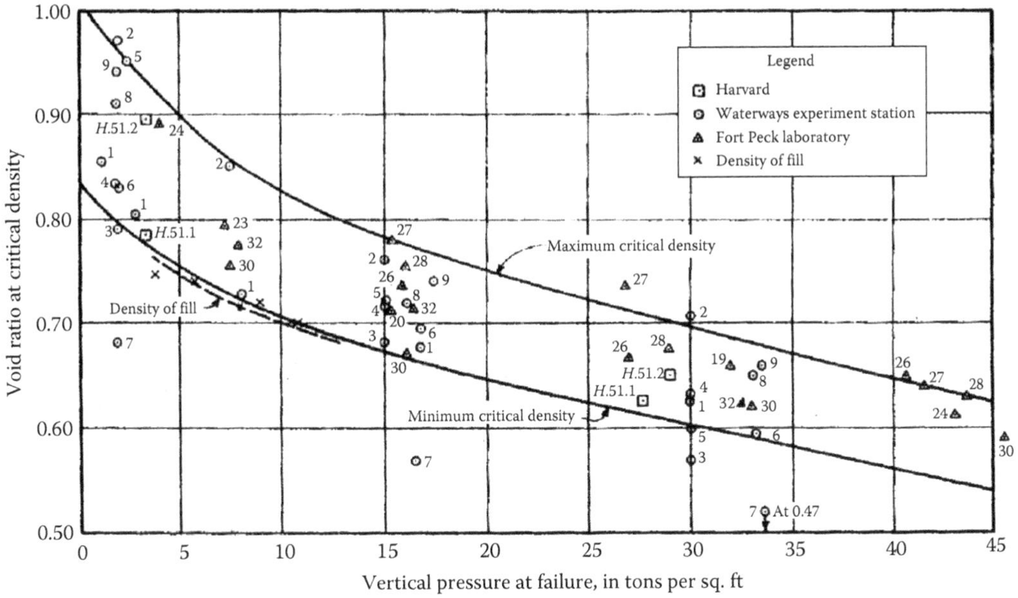

Figure F.10 Critical state summary for Fort Peck Dam shell material. (From Middlebrooks, T.A., *Trans. ASCE*, 107[Paper 2144], 723, 1940.)

and the upper bound $\psi \approx -0.10$. A reasonable range for the characteristic state parameter of the failed fill is the lower half of this range, say $-0.05 < \psi < -0.01$. There are no known penetration tests from the time of the failure, even in that part of the upstream shell that did not fail.

The mobilized residual strength can be estimated from the post-failure configuration if the inertial forces during deceleration of the slide as it came to a halt are neglected. However, there are uncertainties about just what is regarded as the post-failure slope. The widely cited cross section (from Casagrande, 1975) through the failed part of the dam is shown in Figure F.11. If the gross final configuration of the dam section is used so that distances are measured from toe to back-scarp, then post-failure height is about 49 m with about a third of this height counterbalanced by retained water in the reservoir. The corresponding horizontal distance from toe to scarp is about 730 m giving an average slope of 1V:15H. If attention is limited to the upstream shell material, then a range of 24 m $< H <$ 30 m is inferred with effectively all of this counterbalanced by retained water; the corresponding horizontal distance over which this slope existed was in the range 300–450 m, depending on where the 'crest' is denoted. The range of slopes for this view on the failure is from about 1V:12H to 1V:15H.

Taking the full-height view first, an extrapolation of Taylor's stability charts to the 1V:15H slope gives $s_r/\gamma H < 0.04$ (limiting the slide mechanism to the fill). Allowing for the lower third of the slope being submerged, an average bulk unit weight is $\gamma \approx 15$ kN/m^3. Using this average unit weight in the ratio from the stability chart gives $s_r \approx 30$ kPa. The alternative view that only the upstream shell post-failure configuration is relevant, where the submerged unit weight is used because of the retained water, leads to $s_r \approx 10$ kPa. These strengths might be increased a little to capture the inertial effects of the slide coming to a halt, but there are marked 3D influences as well (Figure F.9) which would have acted in the

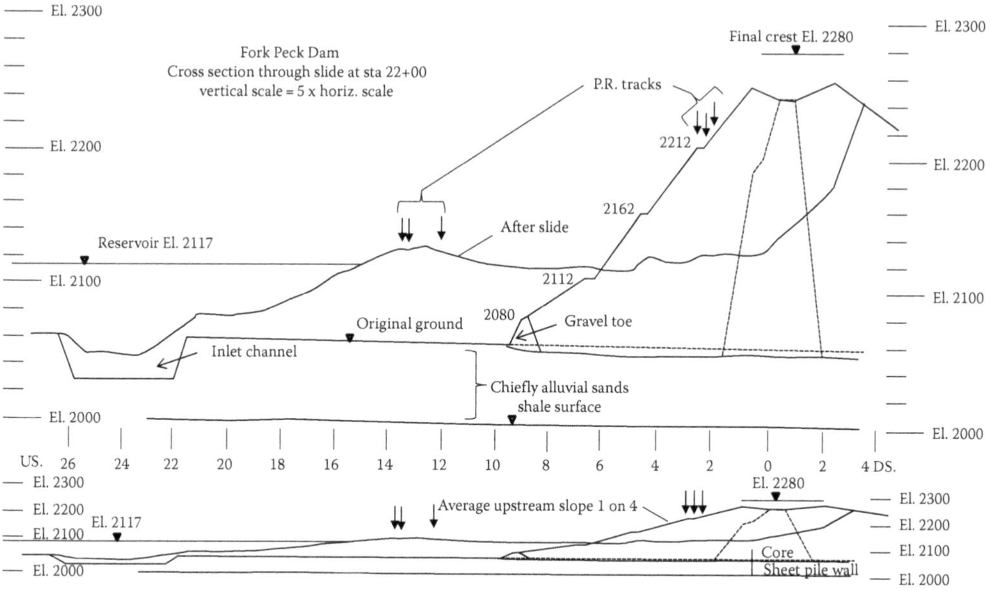

Figure F.11 Section through Fort Peck Dam failure. (From Casagrande, A., Liquefaction and cyclic deformation of sands: A critical review, in *Proceedings of the Fifth Pan-American Conference on Soil Mechanics and Foundation Engineering*, Buenos Aires, Argentina, 1975, Vol. 5, pp. 79–133.)

opposite sense. On balance, it seems better to accept the uncertainties in this case history and adopt 10 kPa < s_r < 30 kPa for the average post-liquefaction strength that was mobilized during the slide.

By comparison to the strength range just discussed, Seed (1987) estimated $s_r \approx 33$ kPa (700 psf), while Davis et al. (1988) estimated 24 kPa (500 psf) < s_r < 53 kPa (1100 psf). Subsequently, Seed and Harder (1990) and Stark and Mesri (1992) both used the strength $s_r = 16 \pm 5$ kPa (350 ± 100 psf).

The uncertainties in the details of the failure mechanism are reflected in uncertainties in the initial effective stress on the liquefying layer. If it is assumed that it was the lower hydraulic fill below the retained reservoir that liquefied, then this lies at average depths between 30 and 36 m below the dam slope. Further assuming that the reservoir had fully saturated this upstream fill, then the credible range of average stress conditions prior to failure are 400 kPa < σ'_{vo} < 530 kPa. Stark and Mesri (1992) quote $\sigma'_{vo} \approx 530$ kPa as representative of average pre-failure conditions in the fill that liquefied.

Combining the range in the estimate of strength with initial effective stress, it appears that the mobilized liquefied strength ratio reasonably lies in the range $0.04 < s_r/\sigma'_{vo} < 0.06$ (rounded to reflect underlying imprecision).

F.6 1968: HOKKAIDO TAILINGS DAM (JAPAN)

Details of this case history are from Ishihara et al. (1990). This tailings retention dam was breached during an earthquake-induced liquefaction of the retained silty sand tailings in 1968. The section for this dam appears homogeneous (Figure F.12), which is unusual,

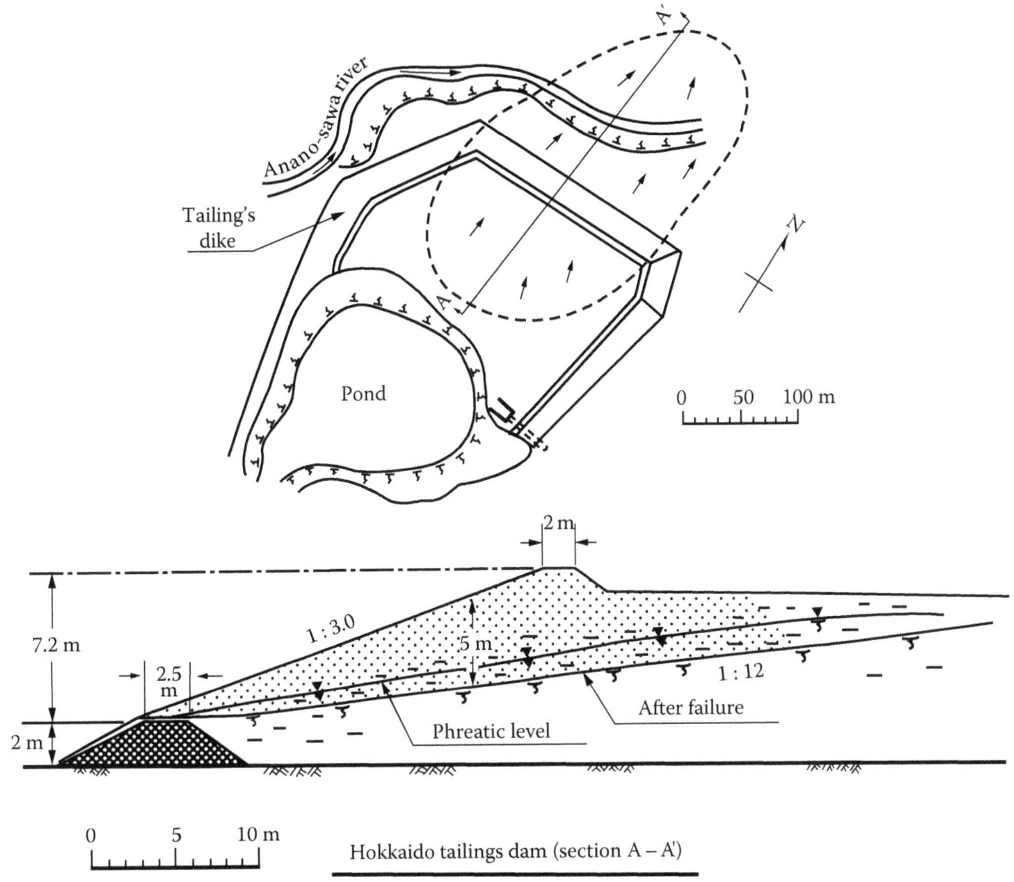

Figure F.12 Plan and section of Hokkaido Tailings Dam. (From Ishihara, K. et al., *Soils Found.*, 30(3), 69, 1990. With permission of Japanese Geotechnical Society.)

as upstream-constructed dams such as this would usually have been raised with preferentially reclaimed sands for the outer berms, which are apparent at depth in the profile on the CPT soundings. The pre-failure slope is 1V:3H, and this reduced to about 1V:12H after the failure. The tailings were tested after failure (presumably in a related but unaffected area of the impoundment) using a Dutch cone CPT, and the two reported penetration resistance profiles are presented in Figure F.13. The CPT profiles are quite similar, although the resolution for the data of interest is low at the published plot scales. These CPT resistances show what looks close to slimes beneath the tailings surface, as often arises with upstream-constructed tailings dams, with sands deeper than about 6 m. Apart from the description that the liquefied soil was a silty sand tailings, no gradation data or other soil properties are reported.

Liquefaction is assumed to have arisen in the 3–4 m thick zone of very loose soil overlying the denser deposits that remained in place after the flowslide, with Ishihara et al. suggesting that the liquefied zone ('sliding surface' in their description) lay at a depth of 2–5 m. The quoted penetration resistance is the range 0.2 MPa $< q_c <$ 0.3 MPa at depths corresponding to the estimated liquefaction zone. However, inspection of Figure F.13 shows a linear

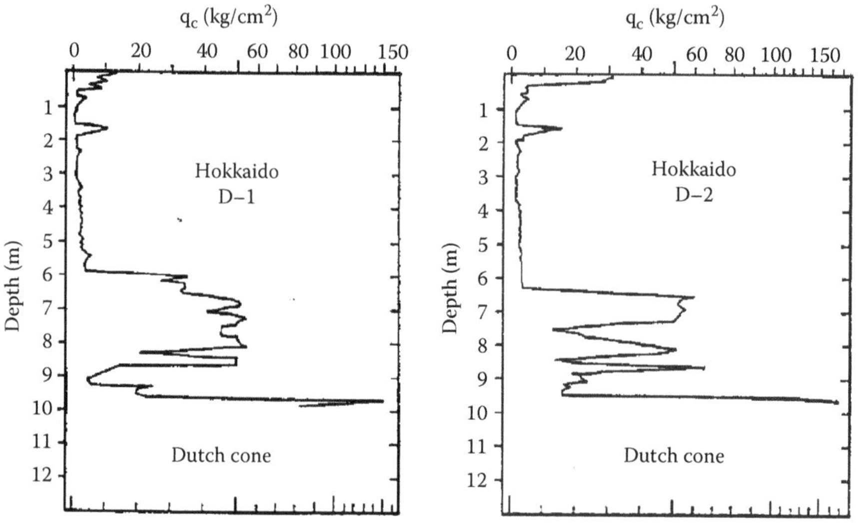

Figure F.13 CPTs from Hokkaido Tailings Dam. (From Ishihara, K. et al., *Soils Found.*, 30(3), 69, 1990. With permission of Japanese Geotechnical Society.)

increase in q_c with depth giving a sensibly constant value of the dimensionless resistance Q. The characteristic value is $5 < Q_k < 7$ in the liquefying zone.

Ishihara et al. estimated the post-liquefaction residual strength using the total stress infinite slope approach. For a sliding surface at depth z below the slope,

$$s = \gamma_t z \cos\theta \sin\theta \tag{F.2}$$

A bulk unit weight of $\gamma = 18$ kN/m^3 was used in (F.2) giving a strength of $s_r \approx 7$ kPa for sliding on a plane at an assumed depth of 4 m at the final slope of 1V:12H. This might be better expressed in terms of a strength ratio since the actual depth of sliding is unknown, which is $0.08 < s_r/\sigma'_{vo} < 0.12$, for final slopes between 1V:12H and 1V:8H.

CSL parameters and the in-situ state for this case history are estimated as follows. The soils involved in the slides are quoted as being silty sand, and the CPT penetration resistances substantiate a loose but surprisingly uniform deposit. However, it is difficult to credit that such loose soils were really a silty sand because the reported penetration resistance is so low. In our experience, the reported resistances are much closer to what would be expected in the hydraulically separated silt-sized fraction (i.e. slimes) even though the plan shows the pond at the opposite end of the impoundment to the dam (see Figure F.12). There is some uncertainty over a typical slope to the CSL, but a range of about $0.1 < \lambda_{10} < 0.2$ would be credible. There are no direct data on the critical friction ratio or angle, and $M_{tc} \approx 1.25$ is adopted for lack of other evidence (generally, M_{tc} should always be measured with tailings as experience indicates tailings can be quite different in their properties from natural sands).

The CPT coefficients, at an average depth of 4 m below ground surface, are estimated to be about $k \approx 13$ and $m \approx 3.5$ (calculated using $H = 50$, $I_r = 300$). Adopting $K_o = 0.7$ (assuming spigotting is equivalent to a hydraulic fill), the estimated characteristic in-situ state range is about $+0.07 < \psi_k < +0.12$. This was an extremely loose deposit.

F.7 1978: MOCHIKOSHI TAILINGS DAMS NO. 1 AND NO. 2 (JAPAN)

Details of this case history are from Ishihara et al. (1990). These were a pair of tailings dam failures triggered by an earthquake in 1978. Dam No. 1 failed during or shortly after the main shock. Dam No. 2 failed 4 h after an aftershock that occurred a day later.

The dams were constructed using the upstream construction method on silty sand (about 50% fines) tailings, see cross sections in Figure F.14 (Dam No. 1 is at the top of Figure F.14). In reality, it was the tailings that failed rather than the outer containment bunds used during construction. The phreatic surface is not quoted in the records, but might be taken as near ground surface given the high silt content of the tailings.

Non-standard mechanical CPT soundings were carried out on intact tailings adjacent to the slide material, although Ishihara et al. note that this non-standard cone provides similar resistance to the standard CPT in cross-calibration checks. While substantial scatter was found, the minimum penetration resistance increased linearly with depth (Figure F.15). Because the penetration resistance increases linearly with depth, a characteristic penetration resistance ratio Q is simpler to deal with than a range of q_c values. The quoted bulk unit weight is 18 kN/m^3 and, assuming that the groundwater table was at ground surface, gives $3 < Q_k < 5$ (depending on whether a low bound is taken through the penetration test data or an estimated 80 percentile).

The simplified infinite slope approach was used to estimate strength in the back-analysis, Ishihara et al. determining $s_r = 15$ kPa for Slide 1 and $s_r = 18$ kPa for Slide 2. These values were based on taking the height of the failing soil mass as the average depth to the failure plane (6 m for both slides) and further assuming that the post-failure slope represented the residual condition. An alternative approach to calculating strengths is Taylor's stability charts. For the pre-failure slope of 1V:3H (the same for both dams), these charts give $0.1 < s_u/\gamma_s H < 0.15$ (depending on the assessed depth factor). This then gives a peak undrained strengths range

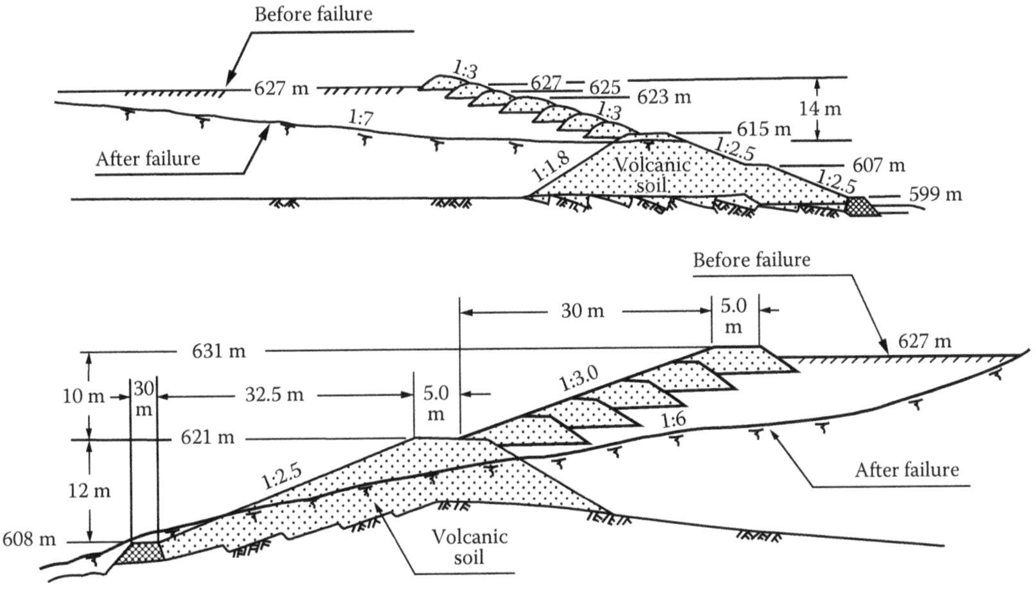

Figure F.14 Cross section of Mochikoshi Tailings Dams. (Dam No. 1 is top, and Dam No. 2 is the bottom.) (From Ishihara, K. et al., *Soils Found.*, 30(3), 69, 1990. With permission of Japanese Geotechnical Society.)

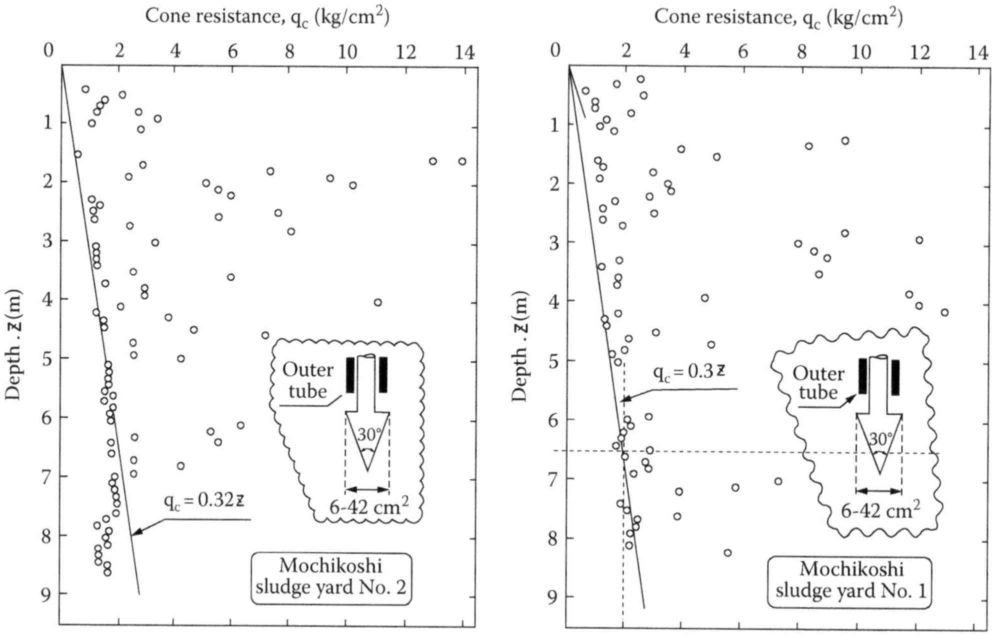

Figure F.15 Double-tube cone penetration test at Mochikoshi Tailings Dams. (From Ishihara, K. et al., *Soils Found.*, 30(3), 69, 1990. With permission of Japanese Geotechnical Society.)

of 25 kPa $< s_u <$ 38 kPa for Slide 1 and 22 kPa $< s_u <$ 32 kPa for Slide 2. Similarly, for the post-failure slopes, $s_r/\gamma_s H \approx 0.06$ giving $s_r \approx 15$ kPa for Slide 1 and $s_r \approx 21$ kPa for Slide 2. These estimates compare well with the Ishihara et al. However, for Dam No. 2, Lucia (1981) quoted $s_r \approx 10$ kPa, while Davis et al. (1988) report $s_r \approx 12$ kPa. It is difficult to understand these lower estimates of s_r, since the post-liquefaction slope stood at 1V:6H with a toe-to-crest height of 22 m.

The estimated average in-situ vertical effective stress is $\sigma'_{vo} \approx 195$ kPa for Dam No. 1 and $\sigma'_{vo} \approx 130$ kPa for Dam No. 2 according to Stark and Mesri (1992). However, there is a lack of data on the location of the failure plane, except that it can be assumed to lie below the post-failure ground surface, which is about 14 m lower than the crest in Dam No. 1 and 10 m lower than the crest in Dam No. 2. Assuming a water table about 2 m below ground, the lower bound estimates of σ'_{vo} are therefore 135 and 100 kPa, respectively, which suggest that Stark and Mesri's estimates are reasonable. Expressing the calculated strengths as ratios gives $s_r/\sigma'_{vo} \approx 15/195 \approx 0.08$ and $s_r/\sigma'_{vo} \approx 21/130 \approx 0.16$ for Dam No. 1 and Dam No. 2, respectively.

CSL parameters and the in-situ state for this case history are estimated as follows. The soils involved in the slides are quoted as being silty sand, but the CPT penetration indicates a very variable deposit. The characteristic trend lines drawn by Ishihara et al. on Figure F.15 are clearly through the low bound to the data and which experience of other tailings deposits would suggest are the slimes (i.e. silt) layers, not silty sands. Based on slimes from other mines that have been tested by Golder Associates, a typical slope to the CSL is credibly in the range of about $0.15 < \lambda_{10} < 0.25$. There are no direct data on the critical friction ratio or angle, and $M_{tc} \approx 1.25$ is adopted for the lack of other evidences.

The CPT coefficients, at an average depth of 6 m below ground surface, are estimated to be about $7 < k < 13$ and $m \approx 3$ (calculated using $H = 25$–50, $I_r = 250$, choices influenced by the calibration of *NorSand* to slimes reported by Shuttle and Cunning, 2007). Adopting $K_o = 0.7$, the estimated characteristic in-situ state range is about $+0.13 < \psi_k < +0.25$. This is extremely loose and at the looser end of experience with other tailings impoundments.

F.8 1982/3: NERLERK (CANADA)

This case history was described in outline in Chapter 1 and is revisited here to provide additional data before developing estimates of in-situ state and field-scale strengths in large displacement slides.

The case histories reviewed so far are found in the literature, and there is seemingly always a shortage of information on measured soil properties for a proper back-analysis. Nerlerk is the first case history in which we were involved, which provided us with the opportunity to ensure that various bases for a full back-analysis were covered. Although Nerlerk was constructed by CanMar under the direction of engineers from Dome Petroleum and their retained consultant EBA Ltd, its failure was of such significance to the Canadian offshore oil industry at the time that Golder Associates were retained for aspects of the subsequent investigation of what had gone wrong. This involvement included testing the various Nerlerk sands for their critical state properties, testing the underlying soft clay that triggered the failure, and evaluation of the CPT data. These triaxial tests and CPT files can be downloaded from the website. Our testing was done in the context of the back-analysis of failures. EBA had also carried out some laboratory tests previously, and these data were published by Sladen et al. (1985a). About the only thing missing from this case history is direct calibration chamber tests on the Nerlerk sands, but there are calibrations for the similar Erksak sand.

Nerlerk B-67 was a sand berm for a bottom-founded mobile drilling unit, Dome Petroleum's SSDC. The project location was towards the outer edge of the Beaufort shelf in a water depth of about 45 m, the depth slightly increasing in every direction away from the chosen berm location at about an average seabed slope of 2%. The target founding level for the drilling unit was 9 m below mean sea level, giving a berm height of nominally 36 m. The berm was nominally 200 m long by 100 m wide in plan at the crest elevation and with the long axis aligned about 20 degrees off an east–west direction. The designed slopes of the berm were a nominal 1V:5H, which is actually quite steep for totally hydraulically placed soils and to this height. Foundation conditions consisted of a 1–2 m thick veneer of soft Holocene clay underlain by dense sand. Nerlerk was constructed only during the summer open water season (approximately July to October), and Nerlerk was so large that two open water seasons were needed.

Berm construction started in 1982, using dredged sand fill from the distant Ukalerk borrow source brought to site in hopper dredges and bottom dumped in the central area of the berm. This was the usual method of berm construction that had provided stable berms elsewhere. However, soft soils had been removed at other berm locations but were left in place at Nerlerk – apparently, the hope at Nerlerk was that bottom dumping would produce a mud wave and displace the soft soils out of the foundation. This did not happen, and layer of soft clay 1–2 m thick is apparent on CPT soundings that extend through the berm into the foundation (see Figure 1.6). Because of the large fill volumes required for an island of this height, the local seabed sand was also exploited by dredging and placing of so-called Nerlerk sand through a pipeline.

A typical cross section through the Nerlerk berm at the time of the first failure in 1983 is shown in Figure F.16 (Been et al., 1987a). Figure F.17 shows a plan sketch of the failures reported by Sladen et al. (1985a), but some caution is needed here. Nerlerk berm was entirely underwater and was never seen by human eye. Knowledge of its morphology comes

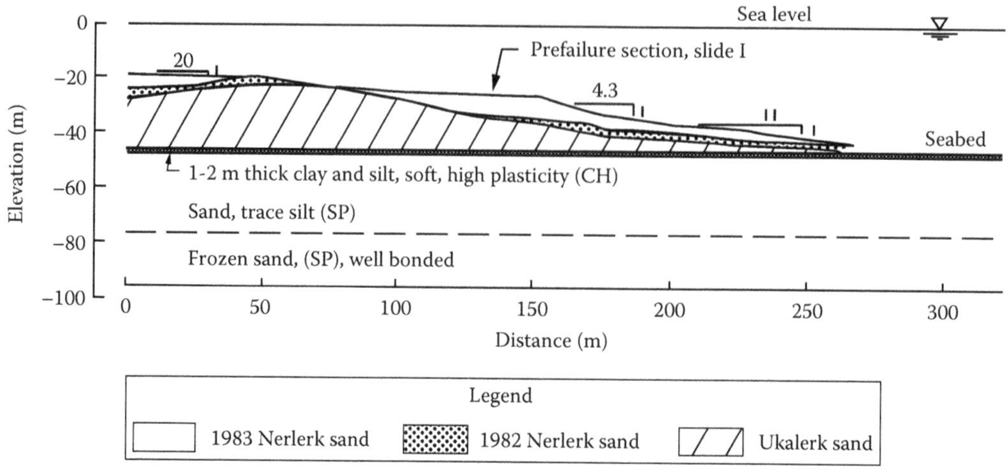

Figure F.16 Nerlerk B-67 berm and foundation cross section. (From Been, K. et al., *Can. Geotech. J.*, 24(1), 170, 1987a. With permission NRC of Canada.)

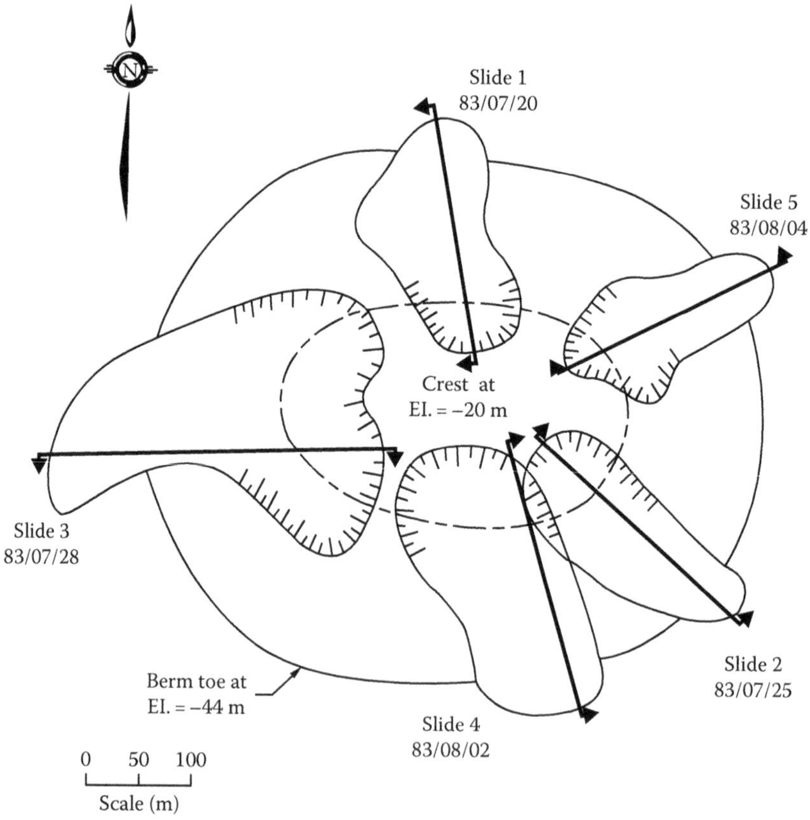

Figure F.17 Plan of failures that occurred at Nerlerk B-67 as reported by Sladen et al. (1985a). (Reproduced from NRC of Canada, Ottawa, Ontario, Canada. With permission.)

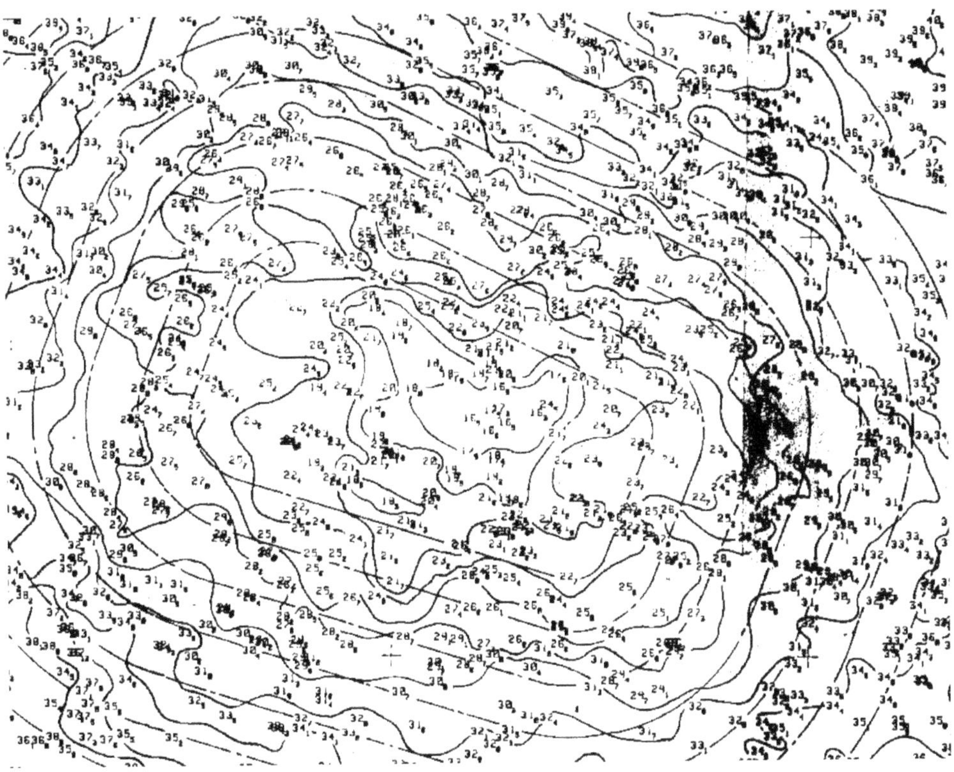

Figure F.18 Example of bathymetric survey data at Nerlerk showing interpolation of berm contours. (The rectangular area outlined in the centre is the berm top for the SSDC and is nominally 100 m × 200 m.)

from precision echo soundings from a moving survey vessel that traversed the berm site in a series of parallel tracks. The location of the survey vessel used shore-based radio beacons (the Syledis system) with a precision of about ±5 m. These time-based position and depth records were then post-processed into contour drawings. Figure F.18 shows a fragment of a survey with the nominal contours of the designed berm superimposed. Both the somewhat wide spacing of the individual survey points and the manner in which the contours have been drawn are apparent. The morphology of the slides shown in Figure F.17 necessarily involves some artistic license. That is a fact of life when dealing with underwater slides, but it also means that considerable caution is needed about the actual slopes and the extent of the slides. Further, the precision surveying did not extend much beyond the nominal berm toe, and in the case of Slide 3, it did not even go that far.

Given these limitation of the survey data, the best assessment is that the post-failure slopes of the Nerlerk slides were predominantly about 1V:16H on average and quite steep in the back-scarp zone (about 1V:7H). These slopes are markedly steeper than quoted by Sladen et al. (1985a), the survey data simply not supporting their quoted very flat final slopes (Been et al., 1987a).

As far as can be ascertained from the survey data, all slides appear to have involved mainly the local Nerlerk sand placed through a pipeline. The denser bottom-dumped Ukalerk sand in the centre of the berm was largely unaffected.

Typical grain size distributions of Nerlerk and Ukalerk sand are given in Figure 1.5. As was usual in Arctic island construction, considerable effort was put into monitoring the silt content of the fills with the dredge masters doing their best to minimize these silt contents. In the case of the local Nerlerk sand placed in the berm, the median grain size generally lay between 0.260 and 0.290 mm. The silt content was less than 2% in most cases, but did range up to as much as 5% in a few samples. Part of the higher fines content was caused by small clay balls caught up in the bulk fills, and the quantity of fines distributed through the sand fill was much less. Some of the 1983 Nerlerk fill was retrieved from the berm in 1988, and it was this material that was tested by Golder Associates. The gradation was $D_{50}=270$ μm and 1.9% fines as tested in our laboratory and is referred to as the Nerlerk 270/1 sand. We also prepared a much higher fines content sample by washing out the fines from the bulk field sample and then blending it back into a small subsample for triaxial tests. This sample had the same sand matrix but measured at 12% fines content and is referred to as the Nerlerk 270/12 sand.

X-ray diffraction was carried out to check mineralogy. This showed that the Nerlerk sand was 84% quartz and 13% feldspar plagioclase. The silt-sized and finer particles were mainly quartz as well, but with traces of illite and kaolinite. The specific gravity of the 270/1 sand was measured as 2.66.

In the case of the 270/1 sand, triaxial testing comprised seven undrained tests on loose samples to define critical state properties and a further six drained tests on denser samples to determine the dilation potential of the sand. In the case of the 270/12 Nerlerk sand, our testing was limited to four dense drained samples as the CSL was determined earlier by EBA. Figure F.19 shows the results of our critical state testing of the 270/1 sand, with Table F.2 comparing our results with those determined by EBA on slightly different gradations. These properties are not that different from the much tested Erksak sand, which came from a part of the Beaufort Shelf not that far away.

Eleven CPTs (D1–D11 inclusive) were carried out during berm construction, most being at the start of 1983 operations. Once it was realized that the berm was failing, a further 15 CPTs (D12–D26 inclusive) were carried out to try and determine what was going on. Subsequently, in 1988, a further 17 CPTs were carried out to fully characterize the Nerlerk sand fills. These CPTs can be downloaded, and a statistical summary of the measured penetration resistance data is shown in Figure F.20. The statistical processing was to divide the depth below fill surface (which varied) into 1 m zones, and within each zone, 100 'bins' were allocated at 0.25 MPa intervals. Each CPT was then scanned from top to bottom with the individual q_c values assigned to the appropriate bin. Subsequent adding up the numbers gave the cumulative probability distribution of q_c for 1 m intervals of depth, which is plotted in Figure F.20. Obviously, this method of data processing neglects any structure that may exist with loose zones parallel to the slope. However, based on the construction records (regular bathymetric surveys were undertaken during fill placement), it is possible to identify which CPTs are in Nerlerk sand only, which in mixed Nerlerk and Ukalerk sand and which in Ukalerk sand only. The data were then separated to produce the sand-specific distributions that are also shown in Figure F.20. In round numbers, the Nerlerk sand, placed by the umbrella nozzle, has about half the penetration resistance of the bottom-dumped Ukalerk sand. These differences in fill state as a result of different methods of hydraulic placement were repeatedly produced across many berms and islands even with the same sand (see Jefferies et al., 1988a). The issue is not the Nerlerk sand properties, but rather its looser initial state.

If the in-situ characteristic penetration resistance is assessed as lying between the 80th and 90th percentile intervals, as discussed in Chapter 5, then there is clearly a very nearly linear trend in this characteristic penetration resistance with depth, which gives $44 < Q_k < 52$ for the local Nerlerk sand fill.

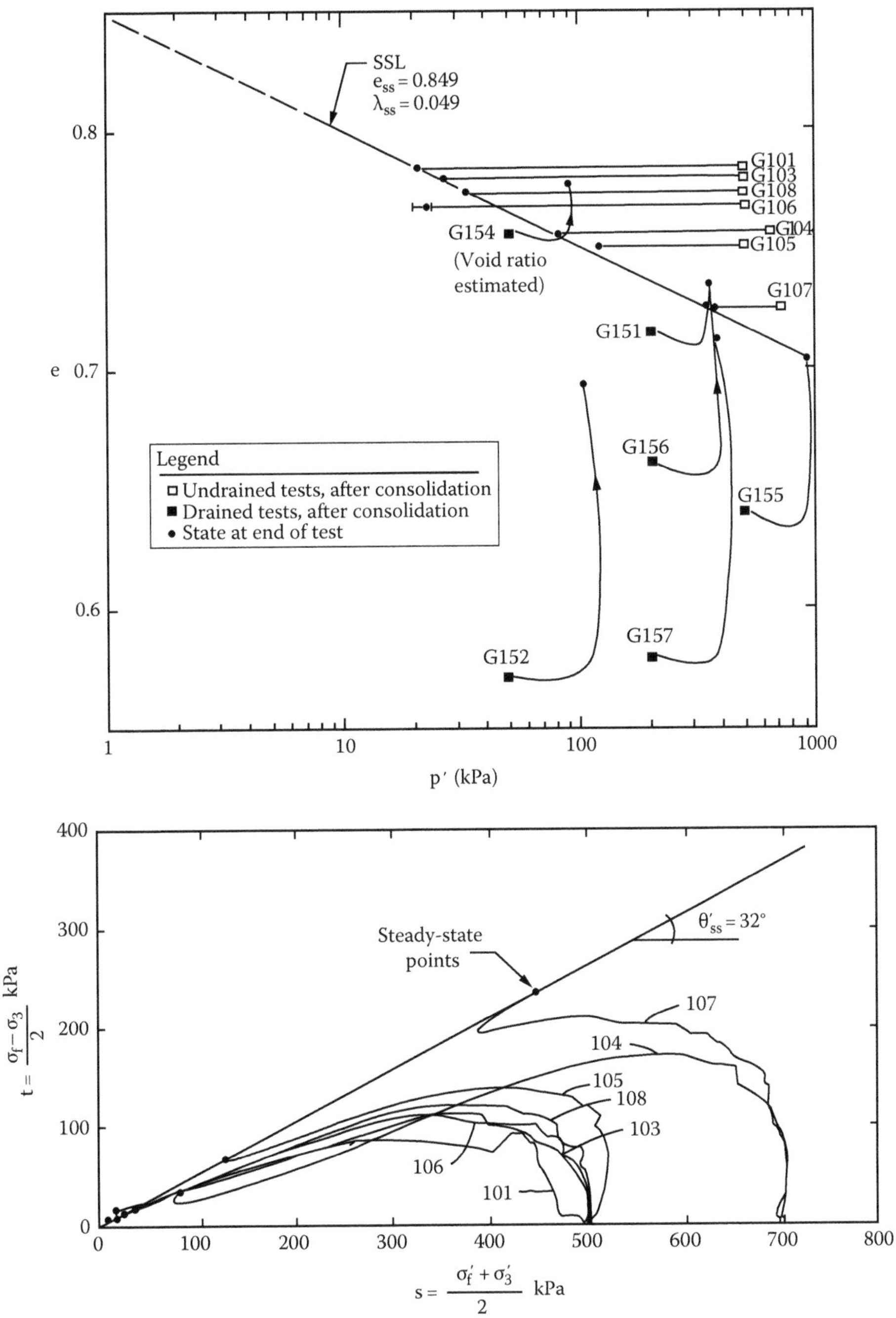

Figure F.19 Summary of state and stress paths in triaxial tests of reconstituted Nerlerk 270/1 samples.

Table F.2 Summary of index and critical state properties for Nerlerk sands

Property	*270/1*	*280/2*	*280/12*
D_{50} (mm)	0.270	0.280	0.280
% Passing #200 sieve	1.9	2	12
D_{60}/D_{10}	1.7	2.0	–
e_{min}	0.536	0.62	0.430
Γ	0.849	0.88	0.80
λ_{10}	0.049	0.04	0.07
M_{tc}	1.28	1.20	1.25

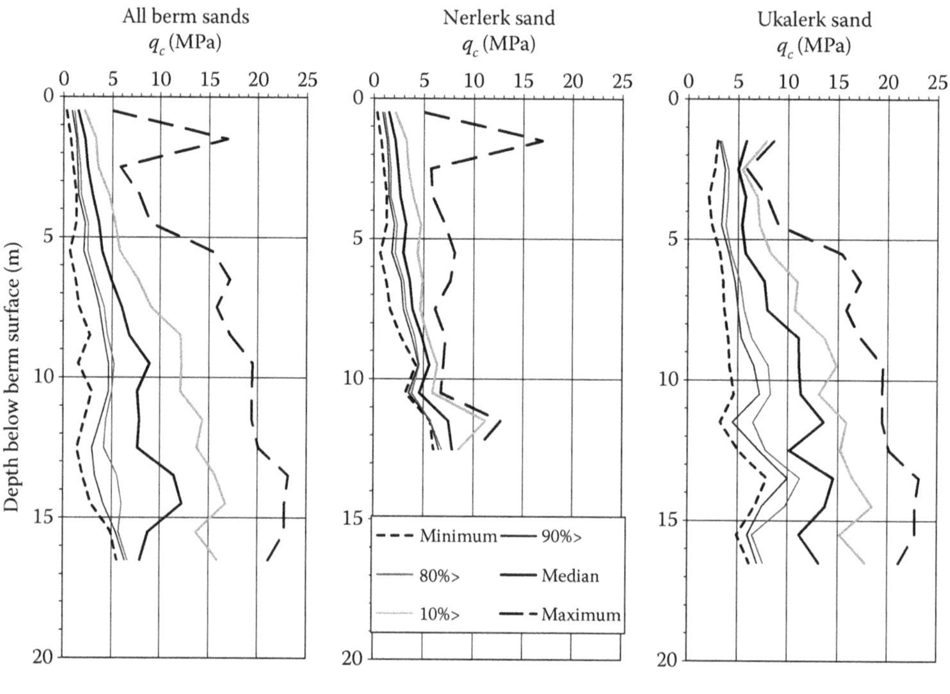

Figure F.20 Summary of CPT distributions in Nerlerk B-67 berm, in Nerlerk sand and Ukalerk sand.

The period of intense interest in Nerlerk was shortly after the CPT chamber test database had been transformed into a state parameter framework, and these Q_k trends were used to assess the characteristic state. A value $\psi_k \approx -0.03$ was obtained (Been et al., 1987a). Today, much more is appreciated about the CPT in sand (discussed in Chapter 4), in particular the effect of elastic shear rigidity on the coefficients k, m that relate Q to ψ. And the very similar Erksak sand was subsequently tested in the cone calibration chamber. However, none of this subsequent work leads to much change in the estimated characteristic in-situ state. The credible range is, if anything, a little denser than was first thought: $-0.05 < \psi_k < -0.03$.

Turning to the mobilized residual strength during the failure, there is a question about the role of the underlying soft clay that was left in place. For any chosen and credible sand state, limit equilibrium calculations give a lower factor of safety through the clay and sand fill rather than through the sand alone. However, if it is accepted, as a working hypothesis, that the role of the clay was basal straining that triggered the failure through a decreasing mean effective stress path in the overlying sand fill, then an infinite slope analysis gives a strength range of $0.09 < s_r/\sigma'_{vo} < 0.15$. This strength range can be compared with the estimate of $s_r/\sigma'_{vo} \approx 0.11$ by Stark and Mesri (1992).

Nerlerk continues to be studied, and one of the most important findings of the statistical analyses carried out 20 years later is that the looser zones in the Nerlerk fill were preferentially orientated parallel to the slope (Hicks and Onisiphorou, 2005). This type of macro-scale fabric is missed in the statistical processing we used and would bias the assessment of characteristic state to less dilatant values than quoted here.

F.9 1985: LA MARQUESA (CHILE)

Details of this case history are from De Alba et al. (1988). La Marquesa is a water retention dam, located about 60 km west of Santiago, that is some 10 m high by 220 m crest length. The dam was rebuilt in 1943 over an earlier dam that had been washed away in 1928. The dam was raised by 1.5 m in 1965. The estimated cross section through the dam is shown in Figure F.21. It is a central core earth fill dam, but without any discrete drains or filters (not unusual given the date of construction). Foundation treatment prior to fill placement was likely limited to removal of topsoil and organics.

The dam suffered extensive damage during the $M = 7.8$ central Chilean earthquake on 3 March 1985, which caused peak ground accelerations at the dam site of about 0.6 g (a very severe motion). Both slopes moved substantially, horizontal displacements were about 6.5 m at the toe of the downstream slope and 11 m at the toe of the upstream slope. The crest dropped 2 m over the middle third of the dam. The profile through the failed dam is sketched in Figure F.22.

The dam was investigated post-earthquake in two stages. Initially, two borings (B-II and B-III) and test pits at the cross-section location were used to investigate the situation with a view to dam reconstruction. A year later, in 1986, a further four borings (B-1 to B-4) and a test pit were carried out to extend the initial findings. The dam configuration in Figure F.21 is based on the results of these investigations together with information from the 1943 construction plans. The location of the borings on a plane through the central part of the failed part of the dam is also indicated. The boring used a tricone bit inside a 100 mm ID casing and with water as the drilling fluid. SPTs were carried out at frequent intervals using an energy-calibrated hammer, and the results are plotted in Figure F.22 in terms of

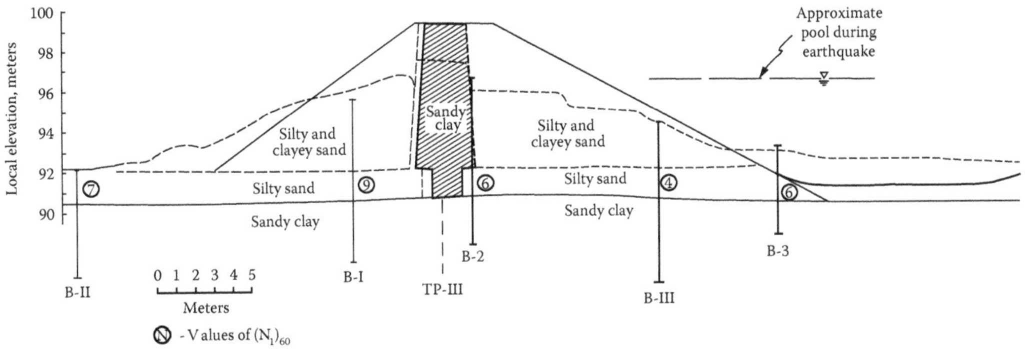

Figure F.21 Reconstructed cross section through failed portion of La Marquesa Dam. (From De Alba, P.A. et al., *J. Geotech. Eng. ASCE*, 114(12), 1414, 1988. With permission ASCE.)

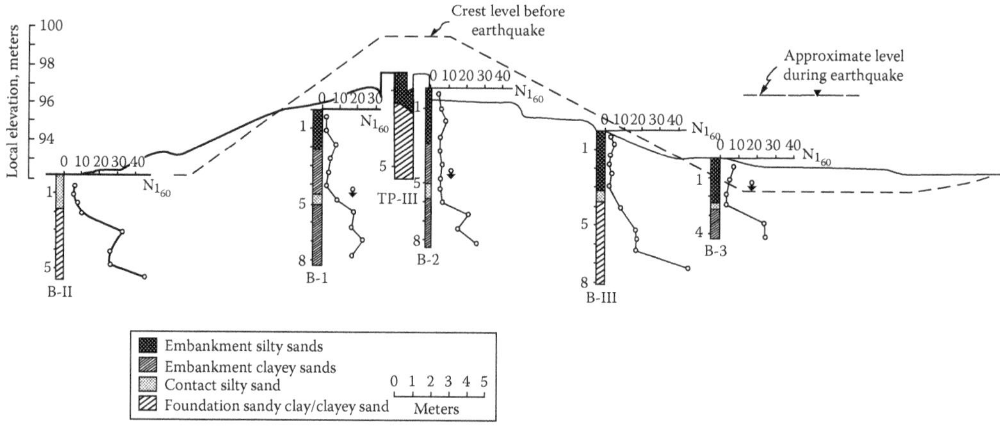

Figure F.22 Cross section through the failure zone of La Marquesa Embankment. (From De Alba, P.A. et al., *J. Geotech. Eng. ASCE*, 114(12), 1414, 1988. With permission ASCE.)

the stress-level adjusted value $(N_1)_{60}$. Fines contents were measured on each SPT sample, but there appears to have been no measurement of soil properties such as compressibility or critical friction angle. No CPTs were carried out either, so the knowledge of the dam's condition post-earthquake is constrained to these few SPTs.

De Alba et al. identified a layer of silty sand in the borings that appeared to be the contact of the fill with the foundation, what they referred to as 1–2 m thick contact silty sand. The movement of the dam shells was attributed to liquefaction of this zone. Upstream, the contact silty sand zone was estimated to have had an average pre-earthquake penetration resistance of $(N_1)_{60} \approx 4$ at a typical fines content of about 30%. Downstream, the contact silty sand zone was estimated to have had an average pre-earthquake penetration resistance of $(N_1)_{60} \approx 9$ at a typical fines content of about 20%.

Although there were only six borings to work from and they provided only about six SPT results in the contact silty sand, de Alba et al. evaluated the spatial distribution in assessing their estimates for characteristic resistance. The estimate for the upstream is certainly in accordance with boring B-3, but given the usual distribution of SPT values, it would be unlikely for a single test to give an 80% value (actually there is only a one in five chance of this happening by definition). Downstream, boring B-II would suggest a slightly lower estimate for characteristic resistance than $(N_1)_{60} \approx 9$ (mainly as a function of 80% versus average). There are really insufficient data to distinguish different characteristic values in the upstream and downstream slopes, and they are apparently geologically similar. Treating both slopes as a single material, and using an SPT–CPT conversion factor of about 3.5 (Figure 4.3), leads to a characteristic range $15 < Q_k < 25$.

The critical state line for the silty sand material (fines content 20%–30%) is estimated to have a slope $0.08 < \lambda_{10} < 0.13$ and $M_{tc} \approx 1.25$. In combination with $K_o = 0.7$ and CPT inversion coefficients $k \approx 20$–25 and $m \approx 4.5$, this leads to a best estimate of characteristic state parameter $-0.05 < \psi_k < +0.05$.

The simplest of dynamic analysis for earthquake response of the dam quickly leads to the view that (unsurprisingly) the severe ground motions liquefied the contact silty sand by cyclic loading. De Alba et al. concluded that the contact silty sand then dropped to a residual strength so allowing the large (relative to the dam height) movement of the shells. Given this block sliding mechanism, a wedge analysis is natural for back-analysis, and this was the method adopted by de Alba et al. Strengths at the onset of sliding and at the end of sliding

Table F.3 Summary of shear strengths from back-analysis of La Marquesa Dam by De Alba et al. (1988)

Slope	*Mode*	s_u	s_r
Upstream	Prefailure configuration with earthquake inertial forces	14–16 kPa	
	Postslump configuration, no inertial forces		≈4 kPa
Downstream	Prefailure configuration with earthquake inertial forces	≈28 kPa	
	Postslump configuration, no inertial force		≈13 kPa

were estimated, and while de Alba et al. refer to these as upper and lower bound estimates, they are in fact different with the former being s_u and the latter a somewhat conservative estimate of s_r. The values computed are summarized in Table F.3.

The wedge analysis offered by de Alba et al. is a little simplistic and with seemingly no effort to locate the lowest energy mode of failure. As an alternative, consider the following. Post-failure slopes are about 1V:3.8H downstream and 1V:6H upstream, with respective heights of 4.5 and 3.5 m. In the case of the downstream slope, Taylor's stability chart gives $s_r/\gamma_s H \approx 0.12$ (for the depth factor $D = 1.5$), and using a unit weight of 19 kN/m^3 for the fill, a strength of $s_r \approx 10$ kPa is obtained. For the upstream slope, Taylor's stability chart gives $s_r/\gamma_s H \approx 0.1$ (also for $D = 1.5$). However, half the slope was below the retained pool so a reduced unit weight is required in the strength estimate. Adopting 14 kN/m^3 as a reasonable average gives a mobilized strength of $s_r \approx 5$ kPa. These values are in reasonable agreement with the analysis reported by De Alba et al. (Table F.3).

If the contact silty sand was truly the controlling zone for the observed displacement pattern, then the average initial vertical effective stress in the middle of this layer is straightforward to calculate. Assuming that the water table in the downstream shell is marginally above the ground downstream of the dam, an average stress is $\sigma'_{vo} \approx 85$ kPa. Upstream the initial effective stress was less because of the retained reservoir, giving an average stress $\sigma'_{vo} \approx 50$ kPa. The overall range in mobilized post-liquefaction strength ratio is therefore about $0.08 < s_r/\sigma'_{vo} < 0.15$.

F.10 1985: LA PALMA (CHILE)

Details of this case history are also from De Alba et al. (1988), this being the second of two case histories reported in some detail by them. La Palma de Quilpue is a water retention dam, located about 50 km northwest of Santiago, that is some 10 m high by 220 m crest length. The dam was built before 1935, and a cross section through the dam is shown in Figure F.23. It is a central core earth fill dam, but without any discrete drains or filters. Foundation treatment prior to fill placement was likely limited to topsoil and organic removal.

The dam suffered extensive damage during the M = 7.8 central Chilean earthquake on March 3, 1985, which caused peak ground accelerations at the dam site of about 0.46 g. The upstream toe moved out about 5 m over about the middle third of the dam, with the failed embankment zone breaking into blocks with longitudinal cracks.

Five borings were put down through the dam in 1986, largely in the plane of the maximum height section of the dam. The location of four borings is indicated in Figure F.23. The fifth boring (B-5) was a duplicate of B-2 but located 35 m towards the right abutment from the principal investigation plane in an area that did not fail. The borings used a tricone bit inside a 100 mm ID casing and with water as the drilling fluid. SPTs were carried out at frequent intervals using an energy-calibrated hammer, and the results are plotted in Figure F.24 in terms of the stress-level adjusted value $(N_1)_{60}$. Fines contents were measured on each

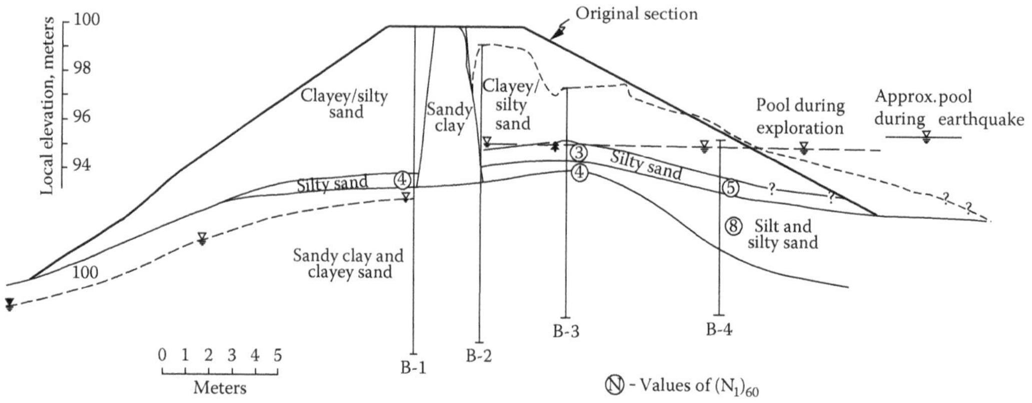

Figure F.23 Reconstructed cross section through failed portion of La Palma Embankment. (From De Alba, P.A. et al., *J. Geotech. Eng. ASCE*, 114(12), 1414, 1988. With permission ASCE.)

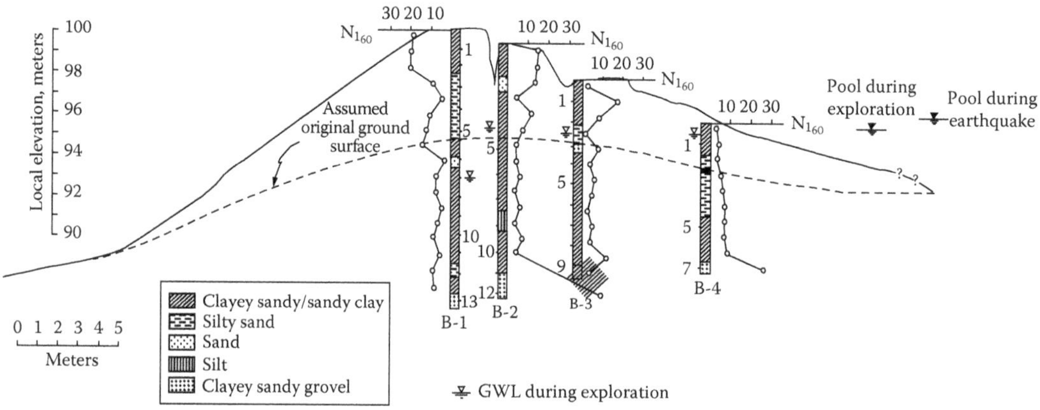

Figure F.24 Cross section through failure zone of La Palma embankment. (From De Alba, P.A. et al., *J. Geotech. Eng. ASCE*, 114(12), 1414, 1988. With permission ASCE.)

SPT sample, but there appears to have been no measurement of basic soil properties such as compressibility or critical friction angle. No CPTs were carried out either, so the knowledge of the dam's condition post-earthquake is constrained to these few SPTs.

De Alba et al. concluded that the dam failed on the inferred loose silty sand layer at the base of the embankment (i.e. at the assumed original ground surface) for which they quote a characteristic penetration resistance of $(N_1)_{60} = 3$ at a characteristic fines content of 15%. While this is an interesting hypothesis, much of the foundation does not have substantially greater penetration resistance than this supposedly liquefying loose layer, and deeper failure mechanisms are equally plausible. Also, no corrections were considered for penetration resistance increase as the cyclically induced excess pore pressures dissipated (i.e. resistance in 1985 might reasonably be somewhat less than in 1986). The characteristic value is based on essentially just two SPT values.

Mobilized strengths were calculated for the supposed liquefied zone by De Alba et al., giving a range 6 kPa $< s_r <$ 14 kPa. The upper end of the range includes inertial forces assuming that failure arose during the earthquake, while the lower end is based on the final configuration with no inertial forces.

The post-failure slope from back-scarp to toe is about 1V:3.5H, which is actually quite steep. Based on the measured SPTs, some allowance needs to be considered for a depth factor in applying Taylor's stability charts, but this probably should not be more than about 1.3. Accordingly, for the post-failure geometry and neglecting inertial forces, Taylor's charts give $0.10 < s_r/\gamma_s H < 0.12$. Noting that the reservoir pool during the earthquake was at about half the height of the sliding mass, an average unit weight of about 14 kN/m^3 would compensate for the balancing force of the water on the slope. Thus, a credible strength range is 10 kPa $< s_r <$ 12 kPa. The bounding estimates by De Alba et al. seem too conservative.

The average initial vertical effective stress in the middle of the liquefying silty sand layer is straightforward to calculate and is about an average stress $\sigma'_{vo} \approx 80$ kPa for the documented retained pool at the time of the earthquake. This leads to a credible mobilized post-liquefaction strength ratio $0.12 < s_r/\sigma'_{vo} < 0.15$.

There is now the thorny issue of trying to make sense of the measured SPTs. If the usual SPT–CPT conversion factors are applied, something in the range of $12 < Q_k < 15$ might be inferred. A constraint here is the lack of detailed gradation information for each SPT, although this is not really any worse than issues raised from estimating characteristic values from just two blowcounts. To prevent overconfidence given to little data, the adopted characteristic range is $9 < Q_k < 15$. The critical state line for the loose silty sand layer is estimated with a slope $0.06 < \lambda_{10} < 0.12$, while the CPT inversion coefficients are $k = 15$–25 and $m \approx 4.5$. The resultant characteristic state parameter is $+0.01 < \psi_k < +0.08$.

F.11 1991: SULLIVAN MINE TAILINGS SLIDE (BRITISH COLUMBIA)

This case history was described by Davies et al. (1998) but with only limited data being presented. We have kindly been given access by Klohn Crippen to their files and are particularly indebted to Howard Plewes, P.Eng. for this. The investigation and back-analysis of this case history were carried out under the direction of Bill Chin, P.Eng. What follows is based on their work. Finally, the CPT data were archived at UBC, and we appreciate being provided a copy of this archive by Dr. John Howie, P.Eng.

The Sullivan Mine is located near Kimberly in southeastern British Columbia and was established in 1905. It is a base metal mine with conventional disposal of tailings into earth fill retained impoundments, with the tailings impoundment developing over the years on an ongoing basis. Little appears known about the early stages of the tailings impoundment, whether its design or construction. However, about one million tons of iron tailings were released in 1948 during an embankment failure (Robinson, 1977). From at least the early 1970s, each raise of the impoundment was independently engineered and inspected by experienced consulting engineers. This was a modern tailings management approach, and there was no lack of care by the Mine. Nevertheless, one of the earth fill retaining dykes failed suddenly on August 23, 1991 during routine dyke raising at a height of 21 m. Failure was in the downstream direction, but there was no environmental impact from this failure because the tailings were contained by other structures and because of prompt and appropriate action by the Mine.

Construction of the failed dyke followed usual mine practice with the upstream method in which an exterior bund of mechanically placed and compacted tailings is progressively stepped upstream onto a spigotted tailings beach. The failure took place in the Active Iron Pond, which is formed by some 1500 m of containment dykes which had reached a maximum height of 21 m. About 300 m of the crest suddenly slumped during a 2.4 m raise, with the movement happening quickly during the placement of the final lift. Figure F.25 shows a cross section through the failed dyke section, together with a photograph illustrating the

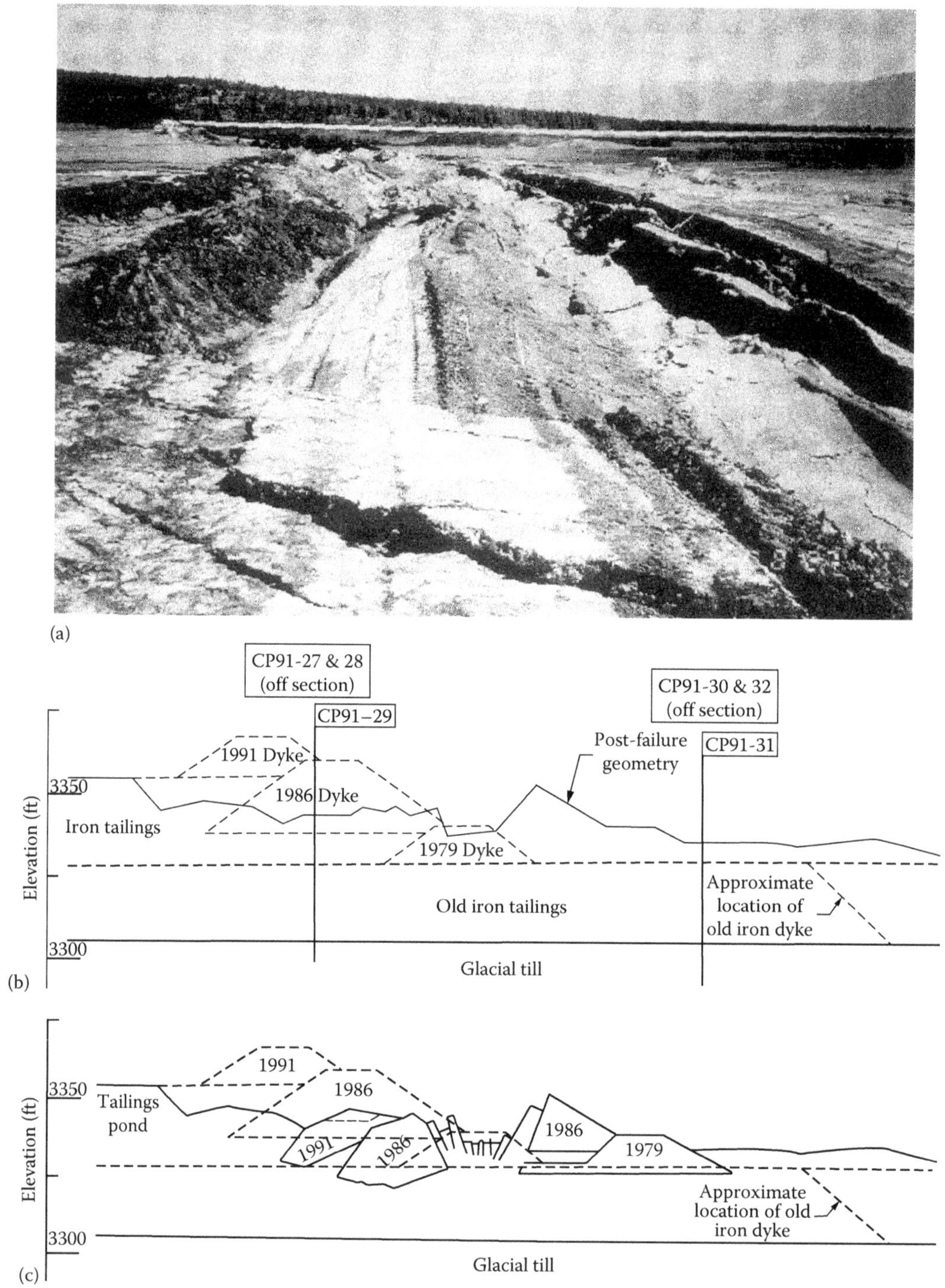

Figure F.25 Illustration and cross section through failure zone of Sullivan Mine tailings dyke slide. (a) Photograph illustrating nature of failure, (b) comparison of pre- and post-failure geometry and (c) reconstructed section. (After Davies, M.P. et al., Static liquefaction slump of mine tailings – a case history, *Proceedings 51st Canadian Geotechnical Conference*, 1998, Vol. 1, pp. 123–131. With permission from Klohn Crippen Ltd.)

nature of the failure. A full flowslide did not develop, although the toe area was displaced as much as 45 m downstream. About 75,000 m^3 of tailings were involved, with post-slump slopes in the range of 1V:10H to 1V:15H. Numerous sand boils were observed just after the failure and continued for several hours, and these together with the speed of the failure lead to the conclusion that it was a static liquefaction event.

The mechanics of the failure was investigated by taking the post failure survey, and then matching known points (such as edges of roads, crest of dyke extensions) to their pre-failure position, using geometric and volume constraints. Necessarily the process involved judgement, but lead to the failure model illustrated in Figure F.25c. It appears that the toe area comprising the 1979 dyke and part of the 1986 dyke moved first, generally horizontally. The remaining dyke sections then failed by a combination of rotations and sliding, triggered by the loss of support at the toe.

An interesting aspect of this particular failure is that engineers were aware of the importance of pore pressure, and there were numerous standpipe piezometers at the site, including in the failed area. Recorded piezometric pressures were generally within a few feet of ground surface and were above ground level at the dyke toe. But curiously, some piezometers had begun to show a declining trend a month before the failure. There are no readings between the last set of piezometric observations in mid-July and the failure on August 21.

An extensive investigation was undertaken after the slump. This included 42 CPTs of which 12 were in the vicinity of the slumped zone. The results of CPT91-29, which was through the failed mass, is presented in Figure F.26a. This CPT shows a range of soils and includes dense compacted sand which comprised the containment dykes. Interestingly, layers of very loose silts lie between 10 m depth and the underlying dense till encountered at approximately 12 m depth. Figure F.26b compares the penetration resistance profiles in six soundings in the failed mass, three along the centreline of the 1986 crest and three in the toe area, the locations of these CPTs relative to the failed mass being illustrated in Figure F.25b. The loose silts identified on the sounding shown in detail are pervasive at depth and dominate the soil profile in the toe area. In terms of characteristic penetration resistance, similar dimensionless resistances are found for the silts whether they are beneath the failed dykes or in the toe area. Bulk unit weights for the compacted fill and the iron silt tailings were estimated at 22.4 and 24.0 kN/m^3 respectively, and using these with the measured hydrostatic pore pressures at the time of the CPT soundings gives a characteristic normalized penetration resistance in the range $10 < Q_k < 14$ for the iron silt tailings.

The iron tailings involved in the failure were sandy silts, with 50% or more passing the #200 sieve. They were non-plastic. The specific gravity for the iron tailings was about 4.2, while that of the silica tailings used for dyke construction was about 3.3.

There was no determination of the CSL from laboratory tests, but a credible range is about $0.1 < \lambda_{10} < 0.2$. Similarly, there are no direct data on the critical friction ratio, and $M_{tc} \approx 1.25$ is adopted for lack of other evidence (M_{tc} should always be measured with tailings, as noted earlier). The inversion parameters are estimated as $k \approx 18$ and $m \approx 4.5$. The CPT data have been processed using these values to infer ψ. The results are shown in Figure F.27. Much of the profile is actually reasonably dense and dilatant, presumed to be the original containment dykes rather than the tailings. The siltier soils at the base are loose and average about $\psi_k \approx +0.05$, with $\psi_k \approx +0.10$ as the most contractive limit. The fluctuations in the estimated ψ are caused by the fluctuating excess pore pressure rather than fluctuations in tip resistance.

A difficulty with estimating the state parameter from the CPT soundings at the Sullivan failure is that these CPTs were not drained in the loose silts that were the triggering

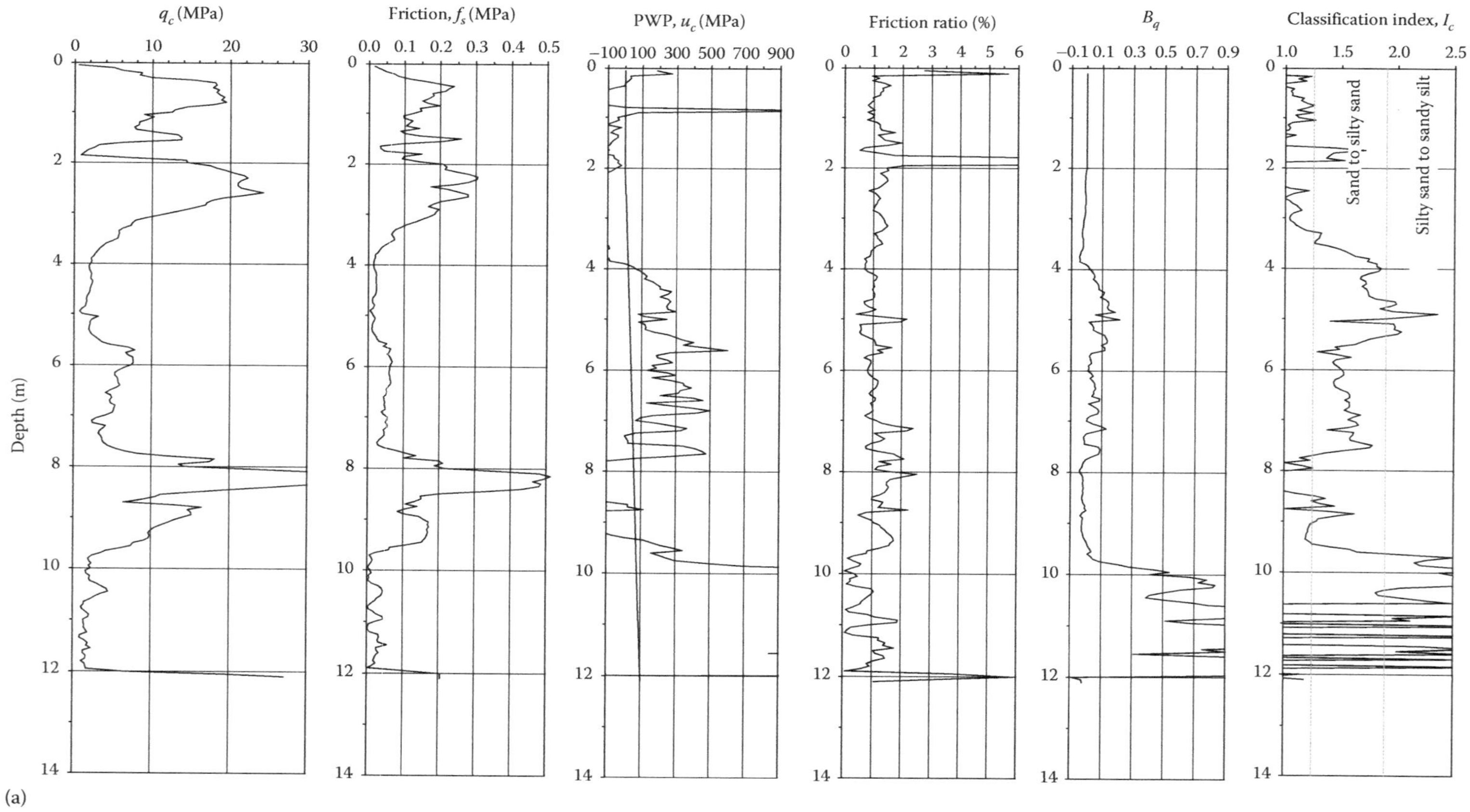

Figure F.26 CPT soundings through Sullivan Dyke failure (see Figure F.25b for location). (a) Example of measured CPT data (CP91-29 through centreline of failed mass).

(Continued)

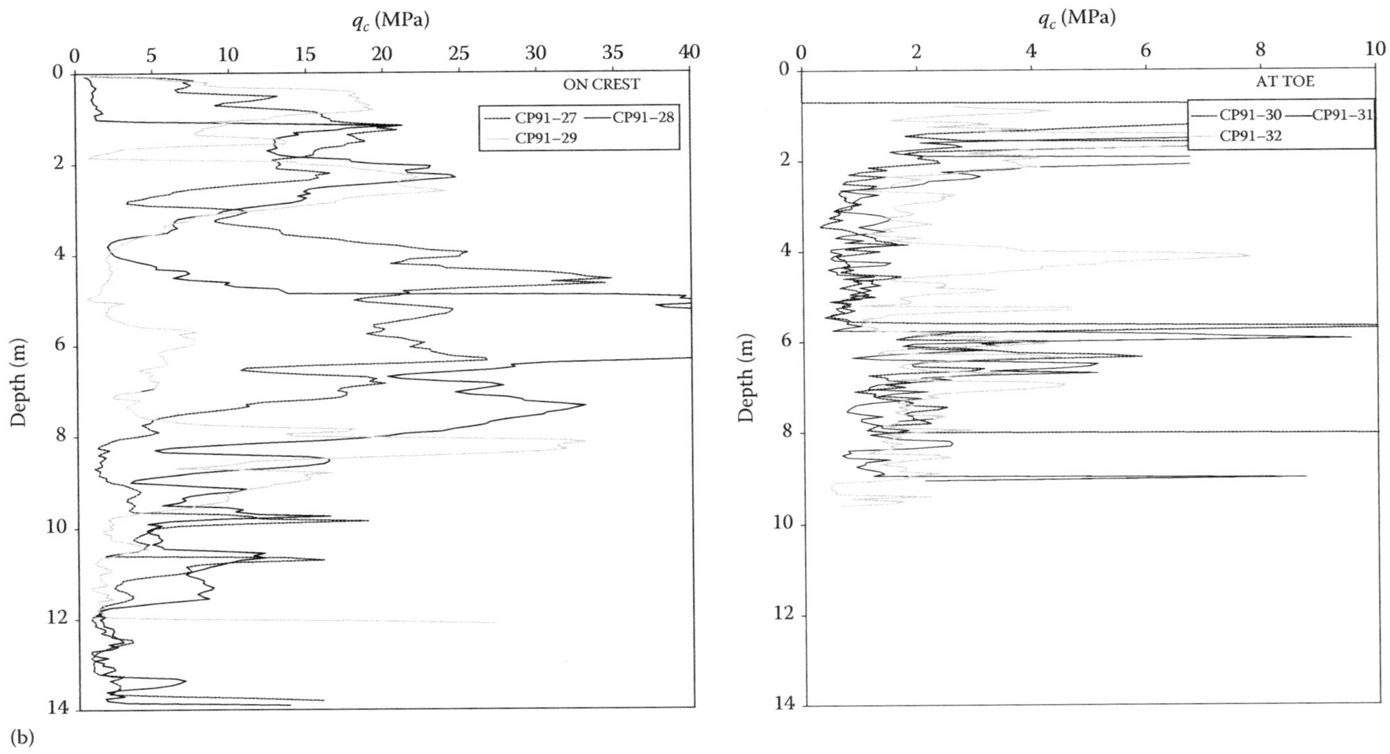

Figure F.26 (*Continued*) CPT soundings through Sullivan Dyke failure (see Figure F.25b for location). (b) Comparison of six CPT soundings from failure area (note different scales).

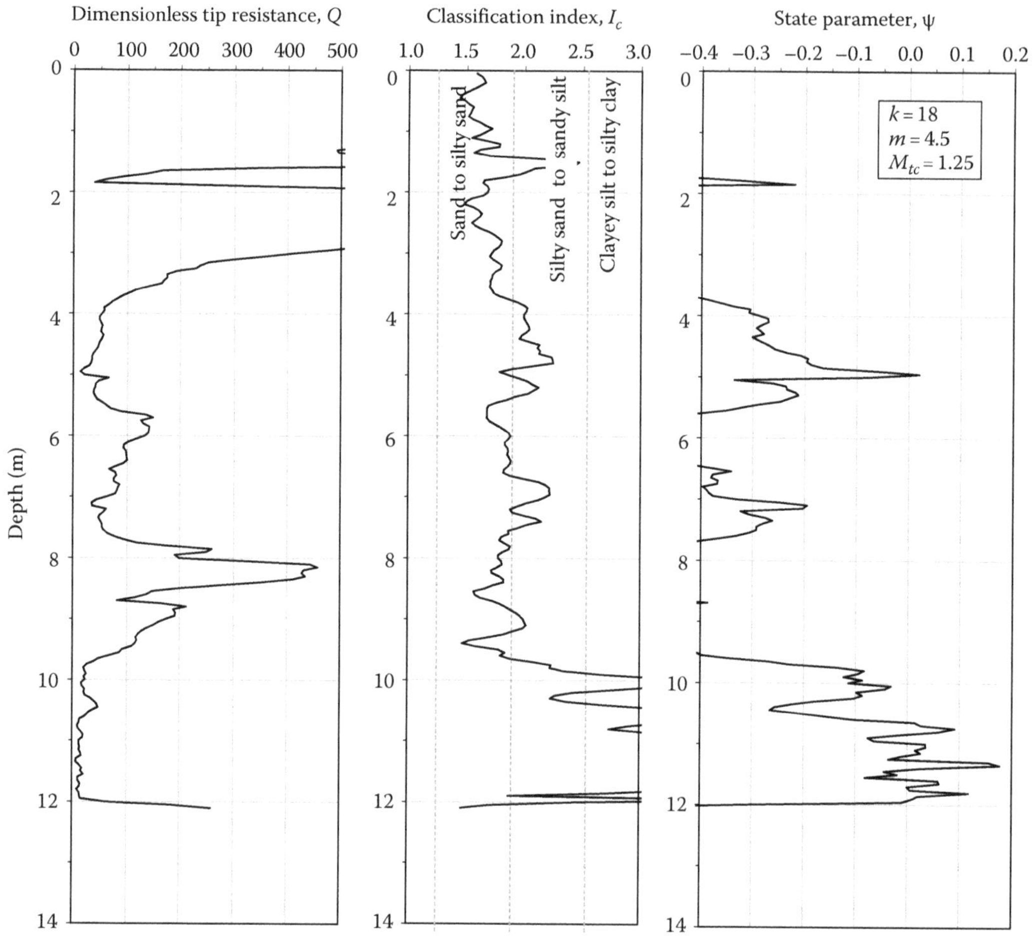

Figure F.27 Estimated in-situ state from CP91-29 data by screening method.

soils for the failure, as can be seen from the inspection of Figure F.26a. Measured excess pore pressure during cone sounding fluctuated substantially, and the magnitude of these excess pore pressures gives $B_q > 0.6$ for much of the silt. Such high B_q values confirm the presence of a loose soil (recall that soft clays may have markedly lower B_q than this).

Back-analysis of the failure was extensive and primarily used the CLARA program. This program allows non-circular failure surfaces, essential when there is a weak layer or zone which may be the controlling feature for stability. Plane strain was assumed, as with other case histories discussed in this appendix. The back-analysis was carried out for the inferred failure sequence, that is, a toe failure followed by retrogressive sliding. The failure plane was chosen to resemble the field conditions, with iteration about the estimated location to minimize the estimated residual strength. Interestingly, the same strength of $s_r \approx 10$ kPa (200 psf) was computed for both the initial toe slip mechanism and the subsequent retrogressive rotational sliding.

For the failure surfaces analysed, a reasonable average initial stress state is about $\sigma'_{vo} \approx 80$ kPa for the initial toe failure and about $\sigma'_{vo} \approx 140$ kPa for the later retrogressive failure. The credible mobilized post-liquefaction strength ratio is $0.07 < s_r/\sigma'_{vo} < 0.13$.

F.12 1994: JAMUNA (BANGABANDHU) BRIDGE (BANGLADESH)

This case history is taken from Yoshimine et al. (1999, 2001). In addition, Prof. Yoshimine kindly provided digital CPT and laboratory data from his files, which are also used here.

The Jamuna Bridge is approximately 110 km northwest of Dhaka. At 4.8 km long, it is the longest bridge in South Asia and crosses the Jamuna, the fifth largest river in the world. The bridge was built over almost 4 years between 1994 and 1998 at a cost of $900 million.

The Jamuna is a shifting braided river, consisting of numerous channels whose width and course change significantly with the seasons. Training the river to ensure it would continue to flow under the bridge corridor was one of the most difficult technical challenges of the project and the most costly of its components. The river training works comprise two guide bunds, one on each side of the river, to lead the river through the bridge corridor. More than 30 submarine flowslides occurred along the West Guide Bund.

The Guide Bund slopes were in very young sediments deposited by the Jamuna, primarily micaeous fine sands with a mean grain size of about 100–200 µm and a silt-sized fraction of 2–10%. These were normally consolidated sands. The flowslides developed on relatively gentle slopes, between about 1V:5H and 1V:3.5H, and came to rest on flatter slopes at about 1V:10H. An example of a flowslide geometry from Yoshimine et al. (1999) is given in Figure F.28. Interestingly, the slide extends above the river level, presumably with a regressive like mechanism as a noticeable scarp is evident at the river level. A plan view of the dredged area, which was about 300 m wide by 3 km long, is shown in Figure F.29 and with the various slides being indicated by arrows (from Yoshimine et al., 2001). Slides seem to be randomly distributed with the whole area being viewed as having much the same potential for flowslides.

Twenty-two CPTs were carried out along the shoulder of the slope, as shown on the West Bund plan. These CPTs supported the view that the area was geologically similar with the statistical measures of mean and standard deviations for the q_c versus depth profiles being essentially identical between south and north areas of the site. A summary of the CPT data is given in Figure F.30, for both mean and mean minus one standard deviation (approximately 83 percentile exceedance) values of penetration resistance (which is taken as characteristic). A straight line can be fitted to the mean q_c profile, but this straight line does not go through the origin of the plot. Much of the offset appears attributable to the river level which is at +7.9 m, or roughly 7 m below the top of the bund. If attention focuses on the underwater sands, and further looks at the characteristic dimensionless resistance, then the entire below water sands could be viewed as uniform with a narrow range of $14 < Q_k < 16$. Interestingly, the friction ratio for these below river sands is somewhat high for relatively

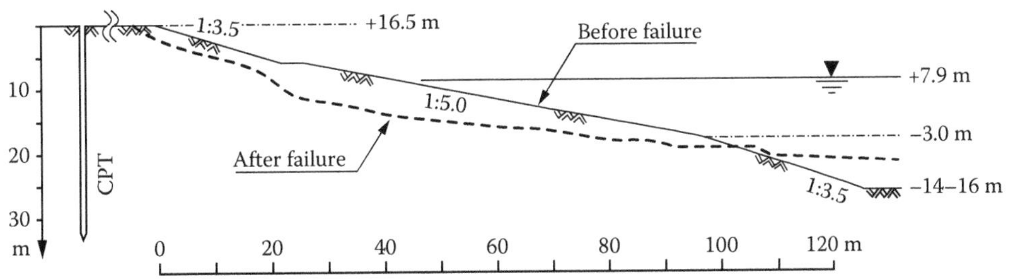

Figure F.28 Example of flowslide geometry at Jamuna. (From Yoshimine, M. et al., *Can. Geotech. J.*, 36(5), 891, 1999. With permission NRC of Canada.)

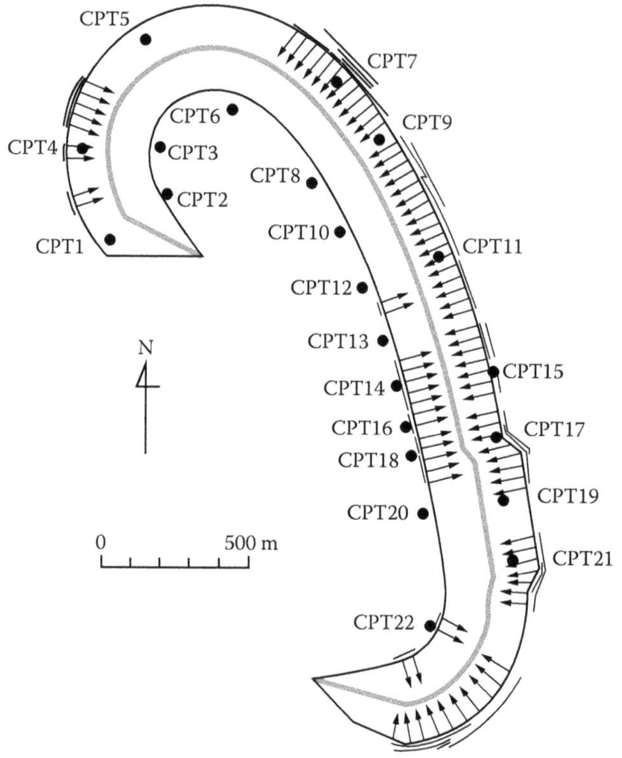

Figure F.29 Plan view of West Guide Bund of Jamuna Bridge showing CPT locations. (From Yoshimine, M. et al., *Can. Geotech. J.*, 38(3), 654, 2001. With permission NRC of Canada.)

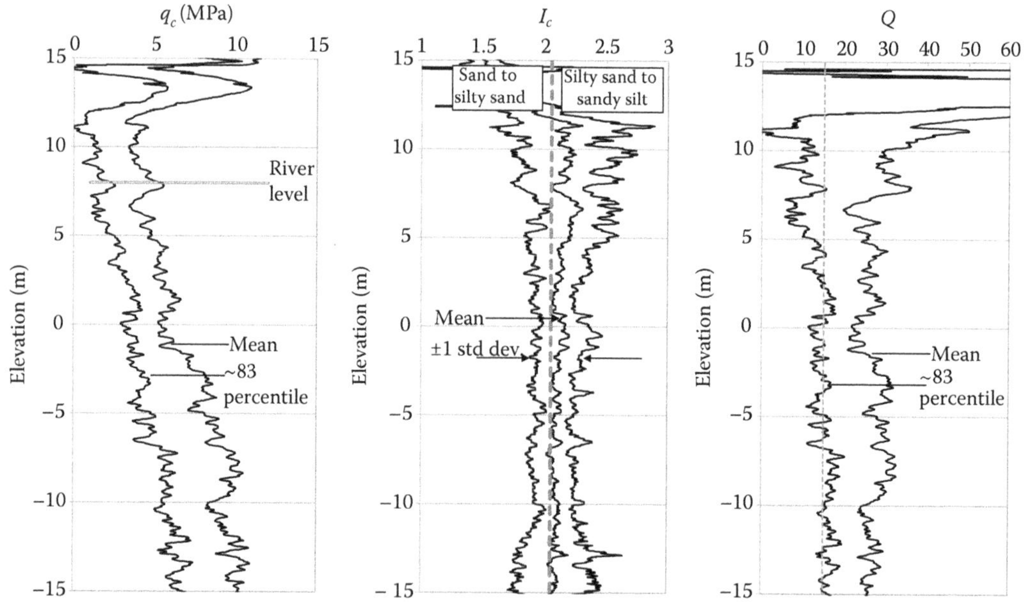

Figure F.30 Statistical summary of Jamuna West Bund CPT results. (Based on data provided by Prof. Yoshimine.)

clean sands and is $F \approx 0.8\%$ on average (Yoshimine et al., 1999), and when plotted using the soil type index I_c, these sands classify on the boundary between 'sand to silty sand' and 'silty sand to sandy silt'.

Hight, in his unpublished Rankine Lecture, referred to triaxial tests that were carried out on Jamuna sand, but we have been unable to obtain records of these tests. None of the test data appears in the public domain. This is an unfortunate missing element to this case history, since clearly the Jamuna sands were unusual – the well-constrained Q_k value at Jamuna is only about one-third of that encountered at Nerlerk. The most likely explanation is that the mica content of the sand strongly affects the sand's overall behaviour. Accepting this hypothesis regarding mica, the CSL slope is estimated as follows. The I_c value gives a measure of soil type mechanical response, and at Jamuna, this is like that of a very silty sand rather than the clean sand implied by the grain size curves. Based on the data discussed in Chapter 2, the slope of the CSL can hardly be less than $\lambda_{10} = 0.09$. On the other hand, tests on sands with mica have shown CSL slopes as great as $\lambda_{10} = 0.2$ at usual stress levels (Hird and Hassona, 1986). Therefore, a range of about $0.1 < \lambda_{10} < 0.2$ would be credible. There are no direct data on the critical friction ratio or angle, and this is a further issue as mica would be expected to lead to relatively low values. $M_{tc} \approx 1.20$ is adopted here for lack of other evidence, but it might be even less.

The confining stress of the liquefying soil is from 150 kPa $< \sigma'_{vo} <$ 300 kPa for the zone of soil that appears to have participated in the liquefaction event. For this stress range, the soil rigidity could credibly lie in the range $300 < I_r < 600$ if the mica has only limited effect on the small strain stiffness of the sand. However, mica will most certainly reduce the plastic hardening modulus, and a first estimate would be to assume that plastic hardening reduces proportionately with increase in λ. Thus something like $50 < H < 100$ would seem reasonable. This gives a range for the CPT coefficients, of about $10 < k < 12$ and $m \approx 3.5$ from Equation 4.12. Adopting $0.7 < K_o < 0.9$ (this was not a hydraulic fill, but is natural ground), the estimated characteristic in-situ state range is presently about $-0.04 < \psi_k < +0.05$. This is a rather large range for ψ, a direct consequence of the lack of available soil properties in what is clearly an unusual soil. It would serve the profession well if the triaxial tests on Jamuna sand were made public, and there may be a case for further testing as this is such an unusual case history.

Turning to the post-liquefaction strength, a difficulty with reported back-analyses of slopes is just where the final surface is drawn; this becomes especially difficult when dealing with underwater surveys and working from bathymetric surveys with a precision of usually no better than ±0.5 m (on a good day). Nevertheless, taking the Jamuna slide geometry shown earlier at face value, a characteristic initial slope would be 1V:4H. Plausibly, it might also be argued that failure was triggered by over-steepening to 1V:3.5H at the toe. The post-failure slope was 1V:10H. These slopes correspond to angles of 14.3°, 15.9° and 5.7°, respectively.

The Yoshimine et al. (1999) paper attracted discussion by Sladen (2001) on several issues, one of which was the equation used to calculate mobilized strengths in an underwater slope. For the infinite slope idealization of the failure, the failure plane is parallel to the slope surface. This configuration means that water has no stabilizing or destabilizing component on the failure mechanism since water pressure acts perpendicular to the slope. Since post-liquefaction strength is idealized as undrained, a total stress infinite slope analysis is adopted. For a sliding surface at a depth z below the slope,

$$s = \gamma_t \, z \cos\theta \sin\theta \qquad \text{(F.2 bis)}$$

The strength obtained by Equation F.2 is normalized by the initial vertical effective stress on the failure plane which is (for a submerged slope)

$$\sigma'_{vo} = \gamma' z \tag{F.3}$$

Substituting (F.3) in (F.2) immediately gives

$$\frac{s}{\sigma'_{vo}} = \frac{\gamma_t}{\gamma'} \cos\theta \sin\theta \tag{F.1 bis}$$

Equation F.1 was used correctly by Yoshimine et al.; Sladen's error was not recognizing the fact that a total stress cohesive strength was being normalized by an initial effective stress, not the current effective stress (and for which the excess pore water pressures are unknown).

Putting the initial slope, angles discussed earlier (14.3°–15.9°) in (F.1) give the range $0.50 < s_u/\sigma'_{vo} < 0.52$. Similarly, the final post-liquefaction slope of 5.7° in (F.1) gives $s_r/\sigma'_{vo} \approx 0.20$. In reality, the configuration suggested by the final slope shown in Figure F.28 suggests a degree of rotational failure, which would reduce the estimated undrained strength ratios given by the infinite slope idealization. By comparison with these strength estimates, Yoshimine et al. (2001) obtain a post-liquefaction strength ratio range of $0.12 < s_r/\sigma'_{vo} < 0.26$, entirely attributable to using some post-failure slope angles that were flatter than inferred from the cross section shown in Figure F.28.

Allowing for the wider range of post-failure angles, but which were nevertheless flatter than the steepest quoted by Yoshimine et al., together with the possible effects of non-planar failure, a credible post-liquefaction strength is about $0.12 < s_r/\sigma'_{vo} < 0.20$.

Appendix G: Seismic liquefaction case histories*

G.1 INTRODUCTION

Professor H.B. Seed largely led the development of empirical methods for seismic liquefaction assessment based on case histories, referred to in Chapter 7 as the 'Berkeley School' approach. Since his death, many other researchers and practitioners have continued the development of these empirical techniques, and the consensus method based on the NCEER workshops in 1996 and 1998 published in Youd et al. (2001) was a milestone in this development. Development has continued in a somewhat disordered fashion since and there tend to be disagreements between different individuals and groups of researchers. Some of this disagreement arises because some of the database is difficult to access (so people are working from different data sets), and some because of different interpretations of the same data.

In Chapter 7, we presented an assessment of the Class A (and to a lesser extent Class B) case histories in terms of state parameter. The data source for this assessment is the PhD thesis of Professor Robb Moss, from University of California at Berkeley and now at California Polytechnic State University. Moss delved into the Berkeley database in great detail, and what we present in this appendix is taken directly from his thesis, with his permission of course, so that the reader can see exactly what has been done and repeat the calculations. We present only the Class A case histories here, summarized in Table G.1. Class A is the terminology of Moss, indicating sites where there is good CPT data through the soil layer of interest, no requirement for a thin layer correction, and ground motion stations generally within 500 m resulting in determination of cyclic stress ratio (CSR) to a coefficient of variation (C.O.V.) less than 0.2.

The 28 Class A 'CPT' case history sites are all in California and arise from just four earthquakes. Ten of these sites are 'no-liquefaction' and eighteen are 'liquefaction' sites. Each section that follows provides a brief background to each earthquake followed by information on each site (from Moss' thesis), such as the nature of the observed liquefaction, references and applicable data, such as depth of critical layer, soil type, CPT measurements and ground motions. References are provided within each section of this appendix for ease of use by the reader.

We have added a set of 'Author's notes' that attempt to capture observations from the CPT traces and borehole logs that may cast some light on the liquefaction and uncertainties around these case histories. We asked Professor Moss to review this appendix, but the additional author's notes reflect the views of the author's alone, not Professor Moss.

* With contribution by Professor R.E.S. Moss, California Polytechnic State University at San Luis Obispo.

Table G.1 Class A seismic (CPT) liquefaction case histories

No.[a]	*Earthquake*	M_w	*Site*	*Liq (Y/N)*
30	1979 Imperial Valley	6.5	Radio Tower B1	Y
31	1979 Imperial Valley	6.5	McKim Ranch A	Y
32	1979 Imperial Valley	6.5	Kornbloom B	N
34	1979 Imperial Valley	6.5	Radio Tower B2	N
42	1981 Westmorland	5.9	Radio Tower B1	Y
44	1981 Westmorland	5.9	Radio Tower B2	N
71	1989 Loma Prieta	7	SFOBB-1	Y
72	1989 Loma Prieta	7	SFOBB-2	Y
78	1989 Loma Prieta	7	Marine Lab. C4	Y
80	1989 Loma Prieta	7	Sandholdt Rd. UC-4	Y
81	1989 Loma Prieta	7	Moss Landing State Beach 14	Y
87	1989 Loma Prieta	7	Farris Farm Site	Y
88	1989 Loma Prieta	7	Miller Farm CMF8	Y
89	1989 Loma Prieta	7	Miller Farm CMF10	Y
90	1989 Loma Prieta	7	Miller Farm CMF5	Y
91	1989 Loma Prieta	7	Miller Farm CMF3	Y
110	1989 Loma Prieta	7	Alameda Bay Farm Is.	N
111	1989 Loma Prieta	7	MBARI 3 RC-6	N
112	1989 Loma Prieta	7	MBARI 3 RC-7	N
113	1989 Loma Prieta	7	Sandholdt Rd. UC-2	N
114	1989 Loma Prieta	7	General Fish CPT-6	N
115	1989 Loma Prieta	7	MBARI 4 CPT-1	N
116	1989 Loma Prieta	7	Sandholdt Rd. UC-6	N
117	1989 Loma Prieta	7	Moss Landing State Beach 18	N
126	1994 Northridge	6.7	Balboa Blvd. Unit	Y
128	1994 Northridge	6.7	Potrero Canyon Unit C1	Y
129	1994 Northridge	6.7	Wynne Ave. Unit C1	Y
130	1994 Northridge	6.7	Rory Lane	Y

Source: Moss, R.E.S., CPT-based probabilistic assessment of seismic soil liquefaction initiation, PhD thesis, University of California, Berkeley, CA, 2003.

[a] Case history numbers correspond to Moss (2003) PhD thesis numbering.

G.2 IMPERIAL VALLEY EVENT (1979)

Strike slip movement on the Imperial Valley fault occurred on 15 October 1979 with a magnitude of $M_S = 6.6$. There was 35 km of surface rupture along the fault with a right lateral offset of up to 0.56 m. Large areas experienced liquefaction and lateral spreading.

The main references used for the Imperial Valley sites are Bennett et al. (1984), Bierschwale and Stokoe (1984) and Youd and Wieczorek (1984). However, there are other references to supplement this information.

Numerous strong motion recordings of the main shock were acquired. Cetin et al. (2000) performed a detailed site response analysis for each site. The CPT sites are adjacent to the Cetin (2000) standard penetration test (SPT) sites, therefore the Cetin (2000) peak ground acceleration (PGA) values values are used.

G.2.1 References

Bennett, M.J., McLaughlin, P.V., Sarmiento, J.S., and Youd, T.L. (1984) Geotechnical investigation of liquefaction sites, Imperial Valley, California. Open File Report 84-252, Department of the Interior, U.S. Geological Survey, Menlo Park, CA.

Bierschwale, J.G. and Stokoe, K.H. II (1984) Analytical evaluation of liquefaction potential of sands subjected to the 1981 Westmorland earthquake. Geotechnical Engineering Report GR 84-15, University of Texas, Austin, TX.

Cetin, K.O. (2000) Reliability-based assessment of seismic soil liquefaction initiation hazard. PhD dissertation, University of California, Berkeley, CA.

Cetin, K.O., Seed, R.B., Moss, R.E.S., Der Kiureghian, A., Tokimatsu, K., Harder, L.F. Jr., and Kayen, R.E. (2000) Field case histories for SPT-based in situ liquefaction potential evaluation. Geotechnical Engineering Research Report No.UCB/GT-2000/09, University of California, Berkeley, CA.

Youd, T.L. and Wieczorek, G.F. (1984) Liquefaction during the 1981 and previous earthquakes near Westmorland, CA. Open-File Report 84-680, U.S.G.S., Menlo Park, CA.

G.2.2 1979 Imperial Valley, Radio Tower B1 (30)

References:	Bennett et al. (1984) and Bierschwale and Stokoe (1984).
Nature of failure:	Liquefaction.
Comments:	Sand boils and water issued from the ground, resulting in ponding around the Radio Tower. Fissures developed around the pond and at the edge of the river flood plain.
	Point bar sand deposits.
	Vicinity of the Alamo and New Rivers.
	Located in the Salton Basin, formed by tectonic rifting.
	Correlated with site from Cetin et al. (2000) who performed site response analysis to estimate PGA.
Author's notes:	Large range in q_c (*C.O.V.* = 0.5) giving $\psi_k \approx -0.08$. Material classifies as silt, normalized friction ratio $F = 0.96$. Liquefaction likely occurred is looser material (Table G.2).

Table G.2 Summary data 1979 Imperial Valley, Radio Tower B1 site

Stress		*Strength*	
Liquefied	Y		
Data class	A	Soil class	ML
Critical layer (m)	3.0–5.5	D_{50} (mm)	0.05
Median depth (m)	4.25	% fines	75
St. dev	0.42	*PI*	na
Depth to GWT (m)	2.01		
St. dev	1.00		
σ_v (kPa)	74.72	q_c (MPa)	3.14
St. dev	8.20	St. dev	1.58
σ'_v (kPa)	52.75	f_s (kPa)	30.28
St. dev	4.53	St. dev	10.09
a_{max} (g)	0.18	Norm. exp	0.52
St. dev	0.02	C_q, C_f	1.39
r_d	0.89	C_{thin}	1.00
St. dev	0.08	f_{s1} (kPa)	42.23
M_w	6.50	St. dev	14.07
St. dev	0.13	q_{c1} (MPa)	4.38
CSReq	0.16	St. dev	2.21
St. dev	0.03	R_{f1} (%)	0.96
$C.O.V._{CSR}$	0.15	St. dev	0.58

G.2.3 1979 Imperial Valley, McKim Ranch A (31)

References:	Bennett et al. (1984) and Bierschwale and Stokoe (1984).
Nature of failure:	Liquefaction.
Comments:	Many sand boils and fissures with associated sand boils along a zone of approx. 1.8 km.
	Point bar sand deposits.
	Vicinity of the Alamo and New Rivers.
	Located in the Salton Basin, formed by tectonic rifting.
	Correlated with site from Cetin et al. (2000) who performed site response analysis to estimate PGA.
Author's notes:	High CSR resulting in liquefaction despite estimated giving $\psi_k \approx -0.15$ and high fines content (Table G.3).

Table G.3 Summary data 1979 Imperial Valley, McKim Ranch A

Stress		*Strength*	
Liquefied	Y		
Data class	A	Soil class	SM/ML
Critical layer (m)	1.5–4.0	D_{50} (mm)	0.11
Median depth (m)	2.75	% fines	31
St. dev	0.42	*PI*	na
Depth to GWT (m)	1.5		
St. dev	1.00		
σ_v (kPa)	47.75	q_c (MPa)	2.69
St. dev	8.12	St. dev	0.87
σ'_v (kPa)	35.49	f_s (kPa)	30.56
St. dev	4.38	St. dev	4.35
a_{max} (g)	0.51	Norm. exp	0.52
St. dev	0.05	C_q, C_f	1.71
r_d	0.91	C_{thin}	1.00
St. dev	0.05	f_{s1} (kPa)	52.37
M_w	6.50	St. dev	7.46
St. dev	0.13	q_{c1} (MPa)	4.61
CSReq	0.44	St. dev	1.48
St. dev	0.07	R_{f1} (%)	1.13
$C.O.V._{CSR}$	0.16	St. dev	0.40

G.2.4 1979 Imperial Valley, Kornbloom B (32)

References:	Bennett et al. (1984), Bierschwale and Stokoe (1984) and Youd and Wieczorek (1984).
Nature of failure:	No liquefaction.
Comments:	Point bar sand deposits.
	Vicinity of the Alamo and New Rivers.
	Located in the Salton Basin, formed by tectonic rifting.
	Correlated with site from Cetin et al. (2000) who performed site response analysis to estimate PGA.
Author's notes:	Low CSR. Material is silt. Large range in q_c ($C.O.V.$ = 0.68) but $\psi_k \approx -0.11$ (Table G.4).

Table G.4 Summary data 1979 Imperial Valley, Kornbloom B

Stress		*Strength*	
Liquefied	N		
Data class	A	Soil class	ML
Critical layer (m)	2.6–5.2	D_{50} (mm)	0.05
Median depth (m)	3.90	% fines	92
St. dev	0.43	*PI*	na
Depth to GWT (m)	2.74		
St. dev	1.00		
σ_v (kPa)	65.88	q_c (MPa)	2.80
St. dev	8.50	St. dev	1.90
σ'_v (kPa)	54.50	f_s (kPa)	68.56
St. dev	4.58	St. dev	24.38
a_{max} (g)	0.13	Norm. exp	0.44
St. dev	0.04	C_q, C_f	1.31
r_d	0.91	C_{thin}	1.00
St. dev	0.07	f_{s1} (kPa)	89.54
M_w	6.50	St. dev	31.84
St. dev	0.13	q_{c1} (MPa)	3.65
CSReq	0.09	St. dev	2.48
St. dev	0.01	R_{f1} (%)	2.45
$C.O.V._{CSR}$	0.11	St. dev	1.87

G.2.5 1979 Imperial Valley, Radio Tower B2 (34)

References:	Bennett et al. (1984).
Nature of failure:	No liquefaction.
Comments:	No evidence of liquefaction near this boring.
	Point bar sand deposits.
	Vicinity of the Alamo and New Rivers.
	Located in the Salton Basin, formed by tectonic rifting.
	Correlated with site from Cetin et al. (2000) who performed site response analysis to estimate PGA.
Author's notes:	Modest CSR. Shallow layer and high fines content. Large range in q_c (*C.O.V.* = 0.64) but 85th percentile $\psi_k \approx -0.18$ and mean $\psi \approx -0.28$ (Table G.5).

Table G.5 Summary data 1979 Imperial Valley, Radio Tower B2

Stress		*Strength*	
Liquefied	N		
Data class	A	Soil class	SM/ML
Critical layer (m)	2.0–3.0	D_{50} (mm)	0.10
Median depth (m)	2.5	% fines	30
St. dev	0.17	*PI*	na
Depth to GWT (m)	2.01		
St. dev	1.00		
σ_v (kPa)	41.47	q_c (MPa)	5.75
St. dev	3.65	St. dev	3.66
σ'_v (kPa)	36.66	f_s (kPa)	80.79
St. dev	3.71	St. dev	39.15
a_{max} (g)	0.16	Norm. exp	0.40
St. dev	0.02	C_q, C_f	1.49
r_d	0.95	C_{thin}	1.00
St. dev	0.05	f_{s1} (kPa)	120.68
M_w	6.50	St. dev	58.49
St. dev	0.13	q_{c1} (MPa)	8.59
CSReq	0.12	St. dev	5.47
St. dev	0.02	R_{f1} (%)	1.41
$C.O.V._{CSR}$	0.16	St. dev	1.12

G.3 WESTMORLAND EVENT

On April 26, 1981, a magnitude $M_S = 6.0$ occurred in the same vicinity as the Imperial Valley event. Liquefaction and lateral spreading was widespread.

The main references used for the Imperial Valley sites are Bennett et al. (1984), Bierschwale and Stokoe (1984) and Youd and Wieczorek (1984).

G.3.1 References

Bennett, M.J., McLaughlin, P.V., Sarmiento, J.S., and Youd, T.L. (1984) Geotechnical investigation of liquefaction sites, Imperial Valley, California. Open File Report 84-252, Department of the Interior, U.S. Geological Survey, Menlo Park, CA.

Bierschwale, J.G. and Stokoe, K.H. II (1984) Analytical evaluation of liquefaction potential of sands subjected to the 1981 Westmorland earthquake. Geotechnical Engineering Report GR 84-15, University of Texas, Austin, TX.

Cetin, K.O. (2000) Reliability-based assessment of seismic soil liquefaction initiation hazard. PhD dissertation, University of California, Berkeley, CA.

Cetin, K.O., Seed, R.B., Moss, R.E.S., Der Kiureghian, A., Tokimatsu, K. Harder, L.F. Jr., and Kayen, R.E. (2000) Field case histories for SPT-Based in situ liquefaction potential evaluation. Geotechnical Engineering Research Report No.UCB/GT-2000/09, University of California, Berkeley, CA.

Youd, T.L. and Wieczorek, G.F. (1984) Liquefaction during the 1981 and previous earthquakes near Westmorland, CA. Open-File Report 84-680, U.S.G.S., Menlo Park, CA.

G.3.2 1981 Westmoreland, Radio Tower B1 (42)

References:	Bennett et al. (1984) and Bierschwale and Stokoe (1984).
Nature of failure:	Liquefaction, sand boils, ground fissures.
Comments:	Vicinity of the Alamo and New Rivers.
	Located in the Salton Basin, formed by tectonic rifting.
	Correlated with site from Cetin et al. (2000) who performed site response analysis to estimate PGA.
Author's notes:	Same site that liquefied in 1979 Imperial Valley earthquake (see Section G.2.2). Material classifies as silt, range in q_c (*C.O.V.* = 0.43) giving $\psi_k \approx -0.1$. normalized friction ratio *F* = 0.96 (Table G.6).

Table G.6 Summary data 1981 Westmoreland, Radio Tower B1

Stress		*Strength*	
Liquefied	Y		
Data class	A	Soil class	ML
Critical layer (m)	3.0–5.5	D_{50} (mm)	0.05
Median depth (m)	4.25	% fines	75
St. dev	0.42	*PI*	na
Depth to GWT (m)	2.00		
St. dev	0.3		
σ_v (kPa)	72.50	q_c (MPa)	3.23
St. dev	7.71	St. dev	1.39
σ'_v (kPa)	50.43	f_s (kPa)	28.53
St. dev	4.92	St. dev	5.88
a_{max} (g)	0.17	Norm. exp	0.52
St. dev	0.02	C_q, C_f	1.43
r_d	0.89	C_{thin}	1.00
St. dev	0.08	f_{s1} (kPa)	40.73
Corrected M_w	5.90	St. dev	8.39
St. dev	0.15	q_{c1} (MPa)	4.61
CSReq	0.14	St. dev	1.99
St. dev	0.02	R_{f1} (%)	0.88
$C.O.V._{CSR}$	0.16	St. dev	0.42

G.3.3 1981 Westmoreland, Radio Tower B2 (44)

References:	Bennett et al. (1984) and Bierschwale and Stokoe (1984).
Nature of failure:	No liquefaction.
Comments:	Vicinity of the Alamo and New Rivers.
	Located in the Salton Basin, formed by tectonic rifting.
	Correlated with site from Cetin et al. (2000) who performed site response analysis to estimate PGA.
Author's notes:	Site also did not liquefy in 1979 Imperial Valley earthquake (see Section G.2.5). Modest CSR and high fines content. Large range in q_c (*C.O.V.* = 0.48) (Table G.7).

Table G.7 Summary data 1981 Westmoreland, Radio Tower B2

Stress		*Strength*	
Liquefied	N		
Data class	A	Soil class	SM/ML
Critical layer (m)	2.0–3.0	D_{50} (mm)	0.10
Median depth (m)	2.5	% fines	30
St. dev	0.17	*PI*	na
Depth to GWT (m)	2.01		
St. dev	0.3		
σ_v (kPa)	40.98	q_c (MPa)	5.51
St. dev	3.33	St. dev	2.64
σ'_v (kPa)	36.17	f_s (kPa)	75.17
St. dev	4.17	St. dev	28.61
a_{max} (g)	0.16	Norm. exp	0.40
St. dev	0.02	C_q, C_f	1.50
r_d	0.94	C_{thin}	1.15
St. dev	0.05	f_{s1} (kPa)	112.91
Corrected M_w	5.90	St. dev	42.96
St. dev	0.15	q_{c1} (MPa)	9.52
CSReq	0.12	St. dev	4.57
St. dev	0.02	R_{f1} (%)	1.36
$C.O.V._{CSR}$	0.17	St. dev	0.73

G.4 1989 LOMA PRIETA EVENT

The 1989 Loma Prieta earthquake occurred in northern California on October 17 at 5:04 p.m. local time, around the start of a World Series baseball game at Candlestick Park. The shock was centred on a section of the San Andreas Fault System with a moment magnitude of 6.9. Damage was heavy in Santa Cruz County and less so to the south in Monterey County, but effects extended well to the north (and further from the epicentre) into the San Francisco Bay Area, both on the San Francisco Peninsula and across the bay in Oakland.

No surface faulting occurred, though a large number of other ground failures and landslides occurred. Liquefaction was also a significant issue, especially in the heavily damaged Marina District of San Francisco. Abundant strong motion records were captured due to a large number of seismometers that were operating in the region. A readily accessible references for the Moss Landing state beach site, for which both liquefaction and non-liquefaction observations exist, is Boulanger et al. (1997).

G.4.1 References

Bennett, M.J. and Tinsley, J.C.I. (1995) Geotechnical data from surface and subsurface samples outside of and within liquefaction-related ground failures caused by the October 17, 1989, Loma Prieta earthquake, Santa Cruz and Monterey Counties, California. Open-File Report 95-663, U.S. Department of the Interior, U.S. Geological Survey, Menlo Park, CA.

Boulanger, R.W., Idriss, I.M., and Mejia, L.H. (1995). Investigation and evaluation of liquefaction related ground displacements at Moss Landing during the 1989 Loma Prieta earthquake. Report No. UCD/CGM-95/02, Center for Geotechnical Modeling, Department of Civil & Environmental Engineering, University of California, Davis, CA.

Boulanger, R.W., Mejia, L.H., and Idriss, I.M. (1997). Liquefaction at Moss Landing during Loma Prieta earthquake. *Journal of Geotechnical and Geoenvironmental Engineering*, 123(5), 453–467.

Cetin, K.O. (2000) Reliability-based assessment of seismic soil liquefaction initiation hazard. PhD dissertation, University of California, Berkeley, CA.

Cetin, K.O., Seed, R.B., Moss, R.E.S., Der Kiureghian, A., Tokimatsu, K. Harder, L.F. Jr., and Kayen, R.E. (2000): Field Case Histories for SPT-Based In Situ Liquefaction Potential Evaluation. Geotechnical Engineering Research Report No.UCB/GT-2000/09.

Holzer, T.L., Tinsley, J.C.I., Bennett, M.J., and Mueller, C.S. (1994). Observed and predicted ground deformation-Miller Farm lateral spread, Watsonville, California. In *Proceedings of Fifth U.S.-Japan Workshop on Earthquake Resistant Design of Lifeline Facilities and Countermeasures for Soil Liquefaction*, Technical Report NCEER-94-0026 79-99.

Kayen, R.E. and Mitchell, J.K. (1997). Arias Intensity Assessment of Liquefaction Test Sites on the East Side of San Francisco Bay Affected by the Loma Prieta, California, Earthquake of 17 October 1989. Natural Hazards, 16, 243–265.

Kayen, R.E., Mitchell, J.K., Seed, R.B., and Nishio, S. (1998). Soil liquefaction in the East Bay during the earthquake. Professional Paper 1551-B, U.S. Department of Interior, U.S. Geological Survey, Menlo Park, CA.

Mitchell, J.K., Lodge, A.L., Coutinho, R.Q., Kayen, R.E., Seed, R.B., Nishio, S., and Stokoe, K.H. II (1994). In situ test results from four Loma Prieta earthquake liquefaction sites: SPT, CPT, DMT, and shear wave velocity. UCB/EERC-94/04, Earthquake Engineering Research Center, College of Engineering, University of California, Berkeley, CA.

Rutherford and Chekene (1988). Geotechnical investigation: Moss Landing facility (technology building), Monterey Bay Aquarium Research Institute. Report prepared for Monterey Bay Aquarium Research Institute., San Francisco, CA.

Rutherford and Chekene (1993). Geologic hazards evaluation/geotechnical investigation: Monterey Bay Aquarium Research Institute Buildings 3 and 4. Report prepared for Monterey Bay Aquarium Research Institute, San Francisco, CA.

Woodward-Clyde Consultants (1990). *Phase I - Geotechnical Study, Marine Biology Laboratory, California State University, Moss Landing, California*. Report Prepared for California State University, San Jose, California, Oakland, California.

G.4.2 1989 Loma Prieta, San Francisco-Oakland Bay Bridge site 1 (71)

References:	Mitchell et al. (1994) and Kayen et al. (1998).
Nature of failure:	Lateral spreading, sand boils and fissures.
Comments:	PGA from strong motion instrument OHW (0.29 and 0.27g).
	Site response analyses had difficulty in achieving strong motion peaks (max 0.25g).
	Correlated with SPT from Cetin et al. (2000). Site response analysis performed.
Author's notes:	Clean sand (8% fines), medium to loose, $\psi_k \approx -0.1$. Layers within a much thicker sand deposit, but critical layer appears to have been selected based on SPT (2 low *N* values) (Table G.8).

Table G.8 Summary data 1989 Loma Prieta, San Francisco-Oakland Bay Bridge site 1

Stress		*Strength*	
Liquefied	Y		
Data class	A	Soil class	SP-SM
Critical layer (m)	6.25–7.0	D_{50} (mm)	0.28
Median depth (m)	6.75	% fines	8
St. dev	0.13	*PI*	
Depth to GWT (m)	2.99		
St. dev	0.30		
σ_v (kPa)	127.53	q_c (MPa)	5.28
St. dev	4.03	St. dev	0.68
σ'_v (kPa)	90.64	f_s (kPa)	34.83
St. dev	3.90	St. dev	4.99
a_{max} (g)	0.28	Norm. exp	0.66
St. dev	0.01	C_q, C_f	1.07
r_d	0.79	C_{thin}	1.00
St. dev	0.01	f_{s1} (kPa)	37.16
Corrected M_w	7.00	St. dev	5.32
St. dev	0.12	q_{c1} (MPa)	5.63
CSReq	0.17	St. dev	0.73
St. dev	0.01	R_{f1} (%)	0.66
$C.O.V._{CSR}$	0.06	St. dev	0.13

G.4.3 1989 Loma Prieta, San Francisco-Oakland Bay Bridge site 2 (72)

References:	Mitchell et al. (1994) and Kayen et al. (1998).
Nature of failure:	Lateral spreading, sand boils and fissures.
Comments:	PGA from strong motion instrument OHW (0.29 and 0.27g).
	Site response analyses had difficulty in achieving strong motion peaks (max 0.25g).
	Correlated with SPT from Cetin et al. (2000). Site response analysis performed.
Author's notes:	Top 2 m of a 4.5 m thick sand layer identified as critical, apparently based on SPT. $\psi_k \approx -0.12$ (Table G.9).

Table G.9 Summary data 1989 Loma Prieta, San Francisco-Oakland Bay Bridge site 2

Stress		*Strength*	
Liquefied	Y		
Data class	A	Soil class	SP-SM
Critical layer (m)	6.5–8.5	D_{50} (mm)	0.26
Median depth (m)	7.5	% fines	10
St. dev	0.34	*PI*	
Depth to GWT (m)	2.99		
St. dev	0.30		
σ_v (kPa)	141.03	q_c (MPa)	8.66
St. dev	7.74	St. dev	1.91
σ'_v (kPa)	96.79	f_s (kPa)	47.96
St. dev	4.72	St. dev	16.72
a_{max} (g)	0.28	Norm. exp	0.62
St. dev	0.01	C_q, C_f	1.02
r_d	0.76	C_{thin}	1.00
St. dev	0.02	f_{s1} (kPa)	48.94
Corrected M_w	7.00	St. dev	17.07
St. dev	0.12	q_{c1} (MPa)	8.84
CSReq	0.18	St. dev	1.95
St. dev	0.01	R_{f1} (%)	0.55
$C.O.V._{CSR}$	0.06	St. dev	0.23

G.4.4 1989 Loma Prieta, Marine Laboratory C4 (78)

References:	Boulanger et al. (1995) and Woodward-Clyde (1990).
Nature of failure:	Lateral spreading and clayey-silt boils.
Comments:	Correlates with Cetin B-2.
Author's notes:	Low tip resistance layer ($\psi_k \approx -0.02$) just above much denser material (Table G.10).

Table G.10 Summary data 1989 Loma Prieta, Marine Laboratory C4

Stress		*Strength*	
Liquefied	Y		
Data class	A	Soil class	SW
Critical layer (m)	5.2–5.8	D_{50} (mm)	0.50
Median depth (m)	5.50	% fines	3
St. dev	0.10	*PI*	
Depth to GWT (m)	2.50		
St. dev	0.30		
σ_v (kPa)	95.75	q_c (MPa)	2.12
St. dev	3.31	St. dev	0.42
σ'_v (kPa)	66.32	f_s (kPa)	10.91
St. dev	3.19	St. dev	2.53
a_{max} (g)	0.25	Norm. exp	0.78
St. dev	0.03	C_q, C_f	1.38
r_d	0.84	C_{thin}	1.00
St. dev	0.10	f_{s1} (kPa)	15.02
Corrected M_w	7.00	St. dev	3.49
St. dev	0.12	q_{c1} (MPa)	2.92
CSReq	0.20	St. dev	0.58
St. dev	0.03	R_{f1} (%)	0.51
$C.O.V._{CSR}$	0.16	St. dev	0.16

G.4.5 1989 Loma Prieta, Sandholdt Road UC-4 (80)

References:	Boulanger et al. (1995).
Nature of failure:	Liquefaction.
Comments:	Correlated with SPT UC-B10 from Cetin et al. (2000). Site response analysis performed.
	Clay layers suspect, increase the variance of measurements near inclinometer SI-2.
	Water different from what reported in the log, reflects tide at the time of earthquake.
Author's notes:	Variable q_c (*C.O.V.* = 0.62), critical layer has $\psi_k \approx -0.047$. Clean sand, but possible silt inclusion in layer based on CPT and friction ratio (Table G.11).

Table G.11 Summary data 1989 Loma Prieta, Sandholdt Road UC-4

Stress		*Strength*	
Liquefied	Y		
Data class	A	Soil class	SW
Critical layer (m)	2.4–4.6	D_{50} (mm)	0.80
Median depth (m)	3.50	% fines	2
St. dev	0.37	*PI*	
Depth to GWT (m)	2.70		
St. dev	0.30		
σ_v (kPa)	56.40	q_c (MPa)	6.60
St. dev	7.28	St. dev	4.06
σ'_v (kPa)	48.55	f_s (kPa)	33.47
St. dev	2.99	St. dev	9.71
a_{max} (g)	0.25	Norm. exp	0.60
St. dev	0.03	C_q, C_f	1.54
r_d	0.99	C_{thin}	1.00
St. dev	0.01	f_{s1} (kPa)	51.63
Corrected M_w	7.00	St. dev	14.98
St. dev	0.12	q_{c1} (MPa)	10.18
CSReq	0.23	St. dev	6.27
St. dev	0.03	R_{f1} (%)	0.51
$C.O.V._{CSR}$	0.15	St. dev	0.35

G.4.6 1989 Loma Prieta, Moss Landing State Beach UC-14 (81)

References:	Boulanger et al. (1995).
Nature of failure:	Lateral spreading, flow liquefaction, sand boils.
Comments:	Corresponds to Cetin UC-B1.
	Extensive lateral spreading caused damage to the access road. Deformations on the order of 0.3–0.6 m horizontal and 0.3 m vertical were observed near the location of the boring.
	This site is over the old Salinas River channel, over alluvial and estuarine deposits. It was low tide at the time of the earthquake.
	The critical layer consists of poorly graded sand.
	PGA from site response analysis (Cetin, 2000).
	Water different from what reported in the log, reflects tide at the time of earthquake.
Author's notes:	Critical layer appears to be a combination of CSR (higher near surface, reducing with depth) and q_c (gradually increasing with depth). $\psi_k \approx -0.142$ (Table G.12).

Table G.12 Summary data 1989 Loma Prieta, Moss Landing State Beach UC-14

Stress		*Strength*	
Liquefied	Y		
Data class	A	Soil class	SP
Critical layer (m)	2.4–4.0	D_{50} (mm)	0.28
Median depth (m)	3.20	% fines	1
St. dev	0.27	*PI*	
Depth to GWT (m)	2.40		
St. dev	0.50		
σ_v (kPa)	52.40	q_c (MPa)	4.68
St. dev	5.60	St. dev	0.68
σ'_v (kPa)	44.55	f_s (kPa)	25.76
St. dev	3.86	St. dev	3.03
a_{max} (g)	0.25	Norm. exp	0.65
St. dev	0.03	C_q, C_f	1.69
r_d	0.95	C_{thin}	1.00
St. dev	0.01	f_{s1} (kPa)	43.57
Corrected M_w	7.00	St. dev	5.13
St. dev	0.12	q_{c1} (MPa)	7.91
CSReq	0.21	St. dev	1.15
St. dev	0.03	R_{f1} (%)	0.55
$C.O.V._{CSR}$	0.13	St. dev	0.10

G.4.7 1989 Loma Prieta, Farris Farm Site (87)

References:	Holzer et al. (1994).
Nature of failure:	Lateral spreading and sand boils.
Comments:	Corresponds with SPT from Cetin et al. (2000).
	PGA based on site response analysis and calibrated attenuation relationship pinned to local strong ground motion stations.
Author's notes:	Clean sand (8% fines), moderate CSR and lower tip resistance layer, $\psi_k \approx -0.1$ (Table G.13).

Table G.13 Summary data 1989 Loma Prieta, Farris Farm Site

Stress		*Strength*	
Liquefied	Y		
Data class	A	Soil class	SP-SM
Critical layer (m)	6.0–7.0	D_{50} (mm)	0.20
Median depth (m)	6.50	% fines	8
St. dev	0.17	*PI*	
Depth to GWT (m)	4.50		
St. dev	0.30		
σ_v (kPa)	106.75	q_c (MPa)	4.05
St. dev	4.50	St. dev	0.48
σ'_v (kPa)	87.13	f_s (kPa)	28.58
St. dev	3.87	St. dev	2.28
a_{max} (g)	0.31	Norm. exp	0.67
St. dev	0.08	C_q, C_f	1.10
r_d	0.90	C_{thin}	1.00
St. dev	0.02	f_{s1} (kPa)	31.34
Corrected M_w	7.00	St. dev	2.50
St. dev	0.12	q_{c1} (MPa)	4.44
CSReq	0.28	St. dev	0.52
St. dev	0.05	R_{f1} (%)	0.71
$C.O.V._{CSR}$	0.18	St. dev	0.10

G.4.8 1989 Loma Prieta, Miller Farm CMF8 (88)

References:	Bennett and Tinsley (1995).
Nature of failure:	Lateral spreading, sliding and sand boils.
Comments:	Correlated with SPT from Cetin et al. (2000). Site response analysis performed.
	PGA from calibrated attenuation relationship pinned to local strong ground motion stations. Epicentral dist ~ 12 km.
Author's notes:	Clear lower tip resistance layer within about 7 m thick material. $\psi_k \approx 0$ (Table G.14).

Table G.14 Summary data 1989 Loma Prieta, Miller Farm CMF8

Stress		*Strength*	
Liquefied	Y		
Data class	A	Soil class	SM
Critical layer (m)	6.8–8.0	D_{50} (mm)	0.20
Median depth (m)	7.40	% fines	15
St. dev	0.20	*PI*	
Depth to GWT (m)	4.91		
St. dev	0.30		
σ_v (kPa)	123.42	q_c (MPa)	4.79
St. dev	5.29	St. dev	0.94
σ'_v (kPa)	98.99	f_s (kPa)	12.19
St. dev	4.16	St. dev	9.08
a_{max} (g)	0.30	Norm. exp	0.81
St. dev	0.07	C_q, C_f	1.01
r_d	0.73	C_{thin}	1.00
St. dev	0.01	f_{s1} (kPa)	12.29
Corrected M_w	7.00	St. dev	9.15
St. dev	0.12	q_{c1} (MPa)	4.83
CSReq	0.25	St. dev	0.94
St. dev	0.03	R_{f1} (%)	0.25
$C.O.V._{CSR}$	0.13	St. dev	0.20

G.4.9 1989 Loma Prieta, Miller Farm CMF10 (89)

References:	Bennett and Tinsley (1995).
Nature of failure:	Lateral spreading, sliding and sand boils.
Comments:	Correlated with SPT from Cetin et al. (2000). Site response analysis performed.
	PGA from calibrated attenuation relationship pinned to local strong ground motion stations. Epicentral distance ~12 km.
Author's notes:	Silty sand layer covered by 7 m of high plasticity silt or clay (MH and CL) Critical layer has $\psi_k \approx -0.12$ despite low q_c because of high friction ratio when using CPT screening method. CSR is high (Table G.15).

Table G.15 Summary data 1989 Loma Prieta, Miller Farm CMF10

Stress		*Strength*	
Liquefied	Y		
Data class	A	Soil class	SM
Critical layer (m)	7.0–9.7	D_{50} (mm)	0.15
Median depth (m)	8.65	% fines	20
St. dev	0.45	*PI*	
Depth to GWT (m)	3.00		
St. dev	0.30		
σ_v (kPa)	155.35	q_c (MPa)	4.79
St. dev	9.52	St. dev	2.41
σ'_v (kPa)	99.92	f_s (kPa)	92.40
St. dev	5.36	St. dev	9.12
a_{max} (g)	0.30	Norm. exp	0.45
St. dev	0.07	C_q, C_f	1.00
r_d	0.88	C_{thin}	1.00
St. dev	0.02	f_{s1} (kPa)	92.43
Corrected M_w	7.00	St. dev	9.12
St. dev	0.12	q_{c1} (MPa)	4.80
CSReq	0.37	St. dev	2.41
St. dev	0.06	R_{f1} (%)	1.93
$C.O.V._{CSR}$	0.15	St. dev	0.99

G.4.10 1989 Loma Prieta, Miller Farm CMF5 (90)

References:	Bennett and Tinsley (1995).
Nature of failure:	Lateral spreading, sliding and sand boils.
Comments:	Correlated with SPT from Cetin et al. (2000). Site response analysis performed.
	PGA from calibrated attenuation relationship pinned to local strong ground motion stations. Epicentral dist ~ 12 km.
Author's notes:	Identified critical layer has $\psi_k \approx -0.09$ and CSR is relatively high. Other potential loose layers exist in the profile (Table G.16).

Table G.16 Summary data 1989 Loma Prieta, Miller Farm CMF5

Stress		*Strength*	
Liquefied	Y		
Data class	A	Soil class	SM
Critical layer (m)	5.5–8.50	D_{50} (mm)	0.19
Median depth (m)	7.0	% fines	13
St. dev	0.51	*PI*	
Depth to GWT (m)	4.70		
St. dev	0.30		
σ_v (kPa)	122.40	q_c (MPa)	7.13
St. dev	10.47	St. dev	1.57
σ'_v (kPa)	99.84	f_s (kPa)	34.88
St. dev	5.18	St. dev	12.20
a_{max} (g)	0.30	Norm. exp	0.63
St. dev	0.07	C_q, C_f	1.00
r_d	0.77	C_{thin}	1.00
St. dev	0.12	f_{s1} (kPa)	34.91
Corrected M_w	7.00	St. dev	12.21
St. dev	0.12	q_{c1} (MPa)	7.13
CSReq	0.29	St. dev	1.57
St. dev	0.04	R_{f1} (%)	0.49
$C.O.V._{CSR}$	0.13	St. dev	0.20

G.4.11 1989 Loma Prieta, Miller Farm CMF3 (91)

References:	Bennett and Tinsley (1995).
Nature of failure:	Lateral spreading, sliding and sand boils.
Comments:	Correlated with SPT from Cetin et al. (2000). Site response analysis performed.
	PGA from calibrated attenuation relationship pinned to local strong ground motion stations. Epicentral dist ~ 12 km.
Author's notes:	Identified critical layer is just part of a much greater thickness of low tip resistance sands and silty sands and sandy silts. $\psi_k \approx -0.02$ in the identified critical layer (Table G.17).

Table G.17 Summary data 1989 Loma Prieta, Miller Farm CMF3

Stress		*Strength*	
Liquefied	Y		
Data class	A	Soil class	SM/ML
Critical layer (m)	5.7–7.50	D_{50} (mm)	0.12
Median depth (m)	6.50	% fines	27
St. dev	0.29	*PI*	
Depth to GWT (m)	5.70		
St. dev	0.30		
σ_v (kPa)	103.55	q_c (MPa)	3.17
St. dev	6.74	St. dev	1.40
σ'_v (kPa)	95.70	f_s (kPa)	22.66
St. dev	4.46	St. dev	9.52
a_{max} (g)	0.30	Norm. exp	0.71
St. dev	0.07	C_q, C_f	1.03
r_d	0.83	C_{thin}	1.00
St. dev	0.02	f_{s1} (kPa)	23.38
Corrected M_w	7.00	St. dev	9.82
St. dev	0.12	q_{c1} (MPa)	3.27
CSReq	0.26	St. dev	1.44
St. dev	0.04	R_{f1} (%)	0.72
$C.O.V._{CSR}$	0.16	St. dev	0.44

G.4.12 1989 Loma Prieta, Alameda Bay Farm Island (Dike location) (110)

References:	Mitchell et al. (1994) and Kayen and Mitchell (1997).
Nature of failure:	No failure, DDC (deep dynamic compaction) improved site.
Comments:	Western portion consists of sandy hydraulic fill, underlain by bay mud and deeper stiffer soil.
	Liquefaction occurred along the western and northern sections of the island.
	Deep dynamic compaction was performed in the western perimeter dike to prevent liquefaction.
	PGA was recorded at 0.27 and 0.21 at the Alameda Naval Air Station.
	Correlated with SPT from Cetin et al. (2000), corrected water table from Kayen and Mitchell (1997)
Author's notes:	Identified 'critical layer' is lower penetration resistance layer, likely just below effective depth of DDC, but still dense. $\psi_k \approx -0.24$ (Table G.18).

Table G.18 Summary data 1989 Loma Prieta, Alameda Bay Farm Island (Dike location)

Stress		*Strength*	
Liquefied	N		
Data class	A	Soil class	SP-SM
Critical layer (m)	5–6	D_{50} (mm)	0.28
Median depth (m)	5.50	% fines	7
St. dev	0.17	*PI*	
Depth to GWT (m)	2.50		
St. dev	0.30		
σ_v (kPa)	103.75	q_c (MPa)	7.10
St. dev	4.23	St. dev	2.70
σ'_v (kPa)	74.32	f_s (kPa)	152.37
St. dev	3.56	St. dev	25.35
a_{max} (g)	0.24	Norm. exp	0.34
St. dev	0.02	C_q, C_f	1.11
r_d	0.95	C_{thin}	1.00
St. dev	0.09	f_{s1} (kPa)	168.54
Corrected M_w	7.00	St. dev	28.04
St. dev	0.12	q_{c1} (MPa)	7.85
CSReq	0.16	St. dev	2.98
St. dev	0.03	R_{f1} (%)	2.15
$C.O.V._{CSR}$	0.15	St. dev	0.89

G.4.13 1989 Loma Prieta, Monterey Bay Aquarium Research Institute 3 RC-6 (111)

References:	Boulanger et al. (1995) and Rutherford and Chekene (1988).
Nature of failure:	No liquefaction.
Comments:	Corresponds to Cetin EB-1.
Author's notes:	Relatively dense, clean coarse sand and moderate CSR ($\psi_k \approx -0.15$) (Table G.19).

Table G.19 Summary data 1989 Loma Prieta, Monterey Bay Aquarium Research Institute 3 RC-6

Stress		*Strength*	
Liquefied	N		
Data class	A	Soil class	SW
Critical layer (m)	3.0–4.5	D_{50} (mm)	0.60
Median depth (m)	3.75	% fines	1
St. dev	0.25	*PI*	
Depth to GWT (m)	2.60		
St. dev	0.30		
σ_v (kPa)	64.03	q_c (MPa)	13.38
St. dev	5.31	St. dev	0.87
σ'_v (kPa)	52.74	f_s (kPa)	27.51
St. dev	3.05	St. dev	7.88
a_{max} (g)	0.25	Norm. exp	0.74
St. dev	0.03	C_q, C_f	1.61
r_d	0.91	C_{thin}	1.00
St. dev	0.07	f_{s1} (kPa)	44.16
Corrected M_w	7.00	St. dev	12.64
St. dev	0.12	q_{c1} (MPa)	21.48
CSReq	0.18	St. dev	1.39
St. dev	0.03	R_{f1} (%)	0.21
$C.O.V._{CSR}$	0.16	St. dev	0.06

G.4.14 1989 Loma Prieta, Monterey Bay Aquarium Research Institute 3 RC-7 (112)

References:	Boulanger et al. (1995) and Rutherford and Chekene (1988).
Nature of failure:	No liquefaction.
Comments:	
Author's notes:	Similar to RC-6 site at MBARI Building 3. Relatively dense, clean coarse sand and moderate CSR ($\psi_k \approx -0.12$) (Table G.20).

Table G.20 Summary data 1989 Loma Prieta, Monterey Bay Aquarium Research Institute 3 RC-7

Stress		*Strength*	
Liquefied	N		
Data class	A	Soil class	SW
Critical layer (m)	4.0–5.0	D_{50} (mm)	0.60
Median depth (m)	4.50	% fines	1
St. dev	0.17	*PI*	
Depth to GWT (m)	3.70		
St. dev	0.30		
σ_v (kPa)	74.80	q_c (MPa)	9.33
St. dev	4.19	St. dev	0.62
σ'_v (kPa)	66.95	f_s (kPa)	27.61
St. dev	3.24	St. dev	4.96
a_{max} (g)	0.25	Norm. exp	0.70
St. dev	0.03	C_q, C_f	1.32
r_d	0.88	C_{thin}	1.00
St. dev	0.08	f_{s1} (kPa)	36.56
Corrected M_w	7.00	St. dev	6.57
St. dev	0.12	q_{c1} (MPa)	12.35
CSReq	0.16	St. dev	0.81
St. dev	0.02	R_{f1} (%)	0.30
$C.O.V._{CSR}$	0.16	St. dev	0.06

G.4.15 1989 Loma Prieta, Sandholdt Road UC-2 (113)

References:	Boulanger et al. (1995).
Nature of failure:	No liquefaction.
Comments:	Near inclinometer SI-4.
	Water different from what reported in the log, reflects tide at the time of earthquake.
Author's notes:	'Critical layer' taken as bottom third of 4.5 m surficial sand layer. Identified layer has $\psi_k \approx -0.17$ (Table G.21).

Table G.21 Summary data 1989 Loma Prieta, Sandholdt Road UC-2

Stress		*Strength*	
Liquefied	N		
Data class	A	Soil Class	SW
Critical layer (m)	3.0–4.5	D_{50} (mm)	0.70
Median depth (m)	3.75	% fines	4
St. dev	0.25	*PI*	
Depth to GWT (m)	2.70		
St. dev	0.30		
σ_v (kPa)	61.20	q_c (MPa)	16.47
St. dev	5.40	St. dev	4.91
σ'_v (kPa)	50.90	f_s (kPa)	48.62
St. dev	3.51	St. dev	8.72
a_{max} (g)	0.25	Norm. exp	0.65
St. dev	0.03	C_q, C_f	1.55
r_d	0.91	C_{thin}	1.00
St. dev	0.07	f_{s1} (kPa)	75.42
Corrected M_w	7.00	St. dev	13.53
St. dev	0.12	q_{c1} (MPa)	25.55
CSReq	0.18	St. dev	7.61
St. dev	0.03	R_{f1} (%)	0.30
$C.O.V._{CSR}$	0.17	St. dev	0.10

G.4.16 1989 Loma Prieta, General Fish CPT-6 (114)

References:	Boulanger et al. (1995) and Rutherford and Chekene (1993).
Nature of failure:	No liquefaction.
Comments:	
Author's notes:	Clean sand (4% fines), shallow layer just below ground water table (GWT), $\psi_k \approx -0.17$, therefore no liquefaction (Table G.22).

Table G.22 Summary data 1989 Loma Prieta, General Fish CPT-6

Stress		*Strength*	
Liquefied	N		
Data class	A	Soil class	SW
Critical layer (m)	2.2–3.2	D_{50} (mm)	0.60
Median depth (m)	2.70	% fines	4
St. dev	0.17	*PI*	
Depth to GWT (m)	1.70		
St. dev	0.30		
σ_v (kPa)	48.90	q_c (MPa)	9.36
St. dev	3.79	St. dev	1.44
σ'_v (kPa)	39.09	f_s (kPa)	29.57
St. dev	3.74	St. dev	2.46
a_{max} (g)	0.25	Norm. exp	0.70
St. dev	0.03	C_q, C_f	1.93
r_d	0.94	C_{thin}	1.00
St. dev	0.05	f_{s1} (kPa)	57.07
Corrected M_w	7.00	St. dev	4.75
St. dev	0.12	q_{c1} (MPa)	18.06
CSReq	0.19	St. dev	2.78
St. dev	0.03	R_{f1} (%)	0.32
$C.O.V._{CSR}$	0.17	St. dev	0.06

G.4.17 1989 Loma Prieta, Monterey Bay Aquarium Research Institute 4 CPT-1 (115)

References:	Boulanger et al. (1995) and Rutherford and Chekene (1993).
Nature of failure:	No liquefaction.
Comments:	
Author's notes:	No loose layer evident. Relatively dense, clean coarse sand ($\psi_k \approx -0.16$) (Table G.23).

Table G.23 Summary data 1989 Loma Prieta, Monterey Bay Aquarium Research Institute 4 CPT-1

Stress		*Strength*	
Liquefied	N		
Data class	A	Soil class	SW
Critical layer (m)	2.3–3.5	D_{50} (mm)	0.60
Median depth (m)	2.90	% fines	4
St. dev	0.20	*PI*	
Depth to GWT (m)	1.90		
St. dev	0.30		
σ_v (kPa)	48.08	q_c (MPa)	9.59
St. dev	4.46	St. dev	1.02
σ'_v (kPa)	38.27	f_s (kPa)	27.03
St. dev	3.28	St. dev	4.50
a_{max} (g)	0.25	Norm. exp	0.70
St. dev	0.03	C_q, C_f	1.96
r_d	0.93	C_{thin}	1.00
St. dev	0.06	f_{s1} (kPa)	52.94
Corrected M_w	7.00	St. dev	8.81
St. dev	0.12	q_{c1} (MPa)	18.79
CSReq	0.19	St. dev	1.99
St. dev	0.03	R_{f1} (%)	0.28
$C.O.V._{CSR}$	0.17	St. dev	0.06

G.4.18 1989 Loma Prieta, Sandholdt Road UC-6 (116)

References:	Boulanger et al. (1995).
Nature of failure:	No liquefaction.
Comments:	Near inclinometer SI-5
	Water different from what reported in the log, reflects tide at the time of earthquake.
Author's notes:	Slightly lower tip resistance layer within about 8 m of dense surficial sand. 'Critical' layer has $\psi_k \approx -0.17$ (Table G.24).

Table G.24 Summary data 1989 Loma Prieta, Sandholdt Road UC-6

Stress		*Strength*	
Liquefied	N		
Data class	A	Soil class	SW
Critical layer (m)	6.2–7.0	D_{50} (mm)	0.60
Median depth (m)	6.60	% fines	1
St. dev	0.13	*PI*	
Depth to GWT (m)	2.70		
St. dev	0.30		
σ_v (kPa)	123.90	q_c (MPa)	18.83
St. dev	3.87	St. dev	0.61
σ'_v (kPa)	85.64	f_s (kPa)	56.94
St. dev	4.26	St. dev	8.30
a_{max} (g)	0.25	Norm. exp	0.70
St. dev	0.03	C_q, C_f	1.11
r_d	0.80	C_{thin}	1.00
St. dev	0.12	f_{s1} (kPa)	63.47
Corrected M_w	7.00	St. dev	9.25
St. dev	0.12	q_{c1} (MPa)	20.99
CSReq	0.19	St. dev	0.68
St. dev	0.03	R_{f1} (%)	0.30
$C.O.V._{CSR}$	0.19	St. dev	0.05

G.4.19 1989 Loma Prieta, Moss Landing State Beach UC-18 (117)

References:	Boulanger et al. (1995).
Nature of failure:	No liquefaction.
Comments:	Corresponds to Cetin UC-B1.
	The critical layer is composed of beach and dune deposits, differentiating it from the other borings that encountered alluvial and estuarine deposits associated with the old Salinas River Channel.
	Water different from what reported in the log, reflects tide at the time of earthquake.
Author's notes:	Critical layer appears to be a combination of CSR (higher near surface, reducing with depth) and q_c (gradually increasing with depth) within about 10 m of surficial dune sand. $\psi_k \approx -0.16$ (Table G.25).

Table G.25 Summary data 1989 Loma Prieta, Moss Landing State Beach UC-18

Stress		*Strength*	
Liquefied	N		
Data class	A	Soil class	SP
Critical layer (m)	2.4–3.4	D_{50} (mm)	0.60
Median depth (m)	2.90	% fines	1
St. dev	0.17	*PI*	
Depth to GWT (m)	2.40		
St. dev	0.50		
σ_v (kPa)	48.40	q_c (MPa)	10.40
St. dev	4.08	St. dev	0.76
σ'_v (kPa)	43.50	f_s (kPa)	52.53
St. dev	3.32	St. dev	3.25
a_{max} (g)	0.25	Norm. exp	0.72
St. dev	0.03	C_q, C_f	1.82
r_d	0.93	C_{thin}	1.00
St. dev	0.06	f_{s1} (kPa)	95.66
Corrected M_w	7.00	St. dev	5.92
St. dev	0.12	q_{c1} (MPa)	18.94
CSReq	0.17	St. dev	1.38
St. dev	0.03	R_{f1} (%)	0.27
$C.O.V._{CSR}$	0.16	St. dev	0.05

G.5 1994 NORTHRIDGE EVENT

This Northridge, California, earthquake (M = 6.7; M_s = 6.8) occurred on January 17, 1994 and was associated with a blind reverse fault, but did not produce primary surface faulting. Ground failures included slope failures in sloping ground and cracking in alluvium filled valleys. Most failures were not accompanied by sand boils and this absence prompted speculation as to the cause of failure (Holtzer et al., 1999). Holtzer et al. concluded from the studies undertaken to examine this issue that the failures were indeed caused by liquefaction.

There were sites that correlated with Cetin (2000) sites, and the PGA and critical depth were taken from that reference. Cetin (2000) performed site response analyses for these.

G.5.1 References

Abdel-Haq, A. and Hryciw, R.D. (1998). Ground settlement in Simi Valley following the Northridge earthquake. *Journal of Geotechnical Engineering*, 124(1), 80–89.

Bennett, M.J., Ponti, D.J., Tinsley, J.C.I., Holzer, T.L., and Conaway, C.H. (1998). Subsurface geotechnical investigations near sites of ground deformation caused by the January 17, 1994, Northridge, California, earthquake. Open File Report 98-373, U.S. Department of the Interior, U.S. Geological Survey, Menlo Park, CA.

Cetin, K.O. (2000) Reliability-based assessment of seismic soil liquefaction initiation hazard. PhD dissertation, University of California, Berkeley, CA.

Cetin, K.O., Seed, R.B., Moss, R.E.S., Der Kiureghian, A., Tokimatsu, K., Harder, L.F. Jr., and Kayen, R.E. (2000) Field case histories for SPT-based in situ liquefaction potential evaluation. Geotechnical Engineering Research Report No. UCB/GT-2000/09, University of California, Berkeley, CA.

Holzer, T.L., Bennett, M.J., Ponti, D.J., and Tinsley, J.C.I. (1999). Liquefaction and soil failure during 1994 Northridge earthquake. *Journal of Geotechnical Engineering*, 125(6), 438–452.

G.5.2 1994 Northridge, Balboa Boulevard, Unit C (126)

References:	Bennett et al. (1988) and Holtzer et al. (1999).
Nature of failure:	Ground cracking, cracked foundations and ruptured buried utilities.
Comments:	Balboa Blvd. is the northern extent of the San Fernando Valley.
	Deformation occurred on the gently sloping alluvial fan surface of Valley. Upper sediments are dominated by alluvial gravels, sands and finer sediments.
	Many strong motion recordings were acquired in the direct vicinity. PGA estimates from Cetin et al. (2000) site response study.
Author's notes:	Critical layer is relatively deep, layered silty sand between plastic silts and clays. High *C.O.V.* (0.57) and friction ratio. $\psi_k \approx -0.15$. High CSR (Table G.26).

Table G.26 Summary data 1994 Northridge, Balboa Boulevard, Unit C

Stress		*Strength*	
Liquefied	Y		
Data class	A	Soil class	SM/ML
Critical layer (m)	8.3–9.8	D_{50} (mm)	0.11
Median depth (m)	9.0	% fines	43
St. dev	0.25	*PI*	11
Depth to GWT (m)	7.19		
St. dev	0.3		
σ_v (kPa)	162.74	q_c (MPa)	7.26
St. dev	6.91	St. dev	4.11
σ'_v (kPa)	144.99	f_s (kPa)	187.30
St. dev	5.59	St. dev	52.01
a_{max} (g)	0.69	Norm. exp	0.33
St. dev	0.06	C_q, C_f	0.88
r_d	0.54	C_{thin}	1.00
St. dev	0.15	f_{s1} (kPa)	165.69
Corrected M_w	6.70	St. dev	46.01
St. dev	0.13	q_{c1} (MPa)	6.43
CSReq	0.36	St. dev	3.63
St. dev	0.04	R_{f1} (%)	2.58
$C.O.V._{CSR}$	0.10	St. dev	1.62

G.5.3 1994 Northridge, Potrero Canyon, Unit C1 (128)

References:	Bennett et al. (1988) and Holtzer et al. (1999).
Nature of failure:	Ground cracking and lateral spreading.
Comments:	CPT, SPT, pocket pen, torvane and various lab tests available.
	Potrero site located in Potrero Canyon, in the Santa Susana Mtns.
	Ground cracking occurred at the interface between the valley alluvial sediments and the mountainside bedrock.
	The site lies in the region of up dip projection of the seismogenic rupture surface.
	Many strong motion recordings in the area. Cetin et al. (2000) performed a site response study for this case history.
Author's notes:	Critical layer sandwiched between low plasticity silts and clays. $\psi_k \approx -0.13$ (Table G.27).

Table G.27 Summary data 1994 Northridge, Potrero Canyon, Unit C1

Stress		*Strength*	
Liquefied	Y		
Data class	A	Soil class	SM
Critical layer (m)	6–7	D_{50} (mm)	0.10
Median depth (m)	6.50	% fines	37
St. dev	0.17	*PI*	
Depth to GWT (m)	3.30		
St. dev	0.30		
σ_v (kPa)	122.67	q_c (MPa)	6.22
St. dev	4.51	St. dev	2.40
σ'_v (kPa)	91.27	f_s (kPa)	67.31
St. dev	3.92	St. dev	15.82
a_{max} (g)	0.40	Norm. exp	0.50
St. dev	0.04	C_q, C_f	1.05
r_d	0.76	C_{thin}	1.00
St. dev	0.11	f_{s1} (kPa)	70.45
Corrected M_w	6.70	St. dev	16.56
St. dev	0.13	q_{c1} (MPa)	6.52
CSReq	0.25	St. dev	2.51
St. dev	0.04	R_{f1} (%)	1.08
$C.O.V._{CSR}$	0.16	St. dev	0.49

G.5.4 1994 Northridge, Wynne Avenue, Unit C1 (129)

References:	Bennett et al. (1988) and Holtzer et al. (1999).
Nature of failure:	Ground cracking and lateral spreading.
Comments:	CPT, SPT, pocket pen, torvane and various lab tests available.
	Wynne Ave site located within a few kilometre of the epicentre.
	Ground deformation in the form of a down dropped block and other cracking occurred.
	Site response was performed by Cetin et al. (2000).
Author's notes:	Critical layer below low permeability (CL). Very high *C.O.V.* (0.64) on q_c and high fines content. $\psi_k \approx -0.11$. High CSR (Table G.28).

Table G.28 Summary data 1994 Northridge, Wynne Avenue, Unit C1

Stress		*Strength*	
Liquefied	Y		
Data Class	A	Soil Class	SM
Critical Layer (m)	5.8–6.5	D_{50} (mm)	0.15
Median depth (m)	6.13	% fines	38
St. dev	0.13	*PI*	np
Depth to GWT (m)	4.30		
St. dev	0.30		
σ_v (kPa)	112.76	q_c (MPa)	8.77
St. dev	3.50	St. dev	5.64
σ'_v (kPa)	94.85	f_s (kPa)	98.79
St. dev	3.38	St. dev	41.27
a_{max} (g)	0.54	Norm. exp	0.42
St. dev	0.04	C_q, C_f	1.02
r_d	0.74	C_{thin}	1.00
St. dev	0.11	f_{s1} (kPa)	101.01
Corrected M_w	6.70	St. dev	42.20
St. dev	0.13	q_{c1} (MPa)	8.96
CSReq	0.35	St. dev	5.77
St. dev	0.03	R_{f1} (%)	1.13
$C.O.V._{CSR}$	0.10	St. dev	0.87

G.5.5 1994 Northridge, Rory Lane (130)

References:	Abdel-Haq and Hryciw (1998).
Nature of failure:	Ground cracking and sand boils.
Comments:	Liquefaction and ground fissures in the eastern Simi Valley.
	Strong motion station USC Station #55 is located 0.8 km from site. Max PGA 0.73g N–S, 0.81g E–W. Located 14 km northwest of epicentre.
	CPT, dilatometer test (DMT) and soil sampling was performed.
	North–South ground cracking occurred with up to 20 cm of displacement. Other fissures and sand boils observed.
	Site is flat ground. Subsurface is composed of highly stratified silty sands, sandy silts and sandy silty clays.
	A liquefaction evaluation of the site occurred in 1992. Site and seismograph station on an alluvial deposit of apparently similar geomorphologic origin.
Author's notes:	$\psi_k \approx -0.21$, but very high CSR (Table G.29).

Table G.29 Summary data 1994 Northridge, Rory Lane

Stress		*Strength*	
Liquefied	Y		
Data class	A	Soil class	SM
Critical layer (m)	3–5	D_{50} (mm)	
Median depth (m)	4.00	% fines	
St. dev	0.33	*PI*	
Depth to GWT (m)	2.70		
St. dev	0.30		
σ_v (kPa)	66.60	q_c (MPa)	3.62
St. dev	6.33	St. dev	0.45
σ'_v (kPa)	53.85	f_s (kPa)	65.07
St. dev	3.66	St. dev	31.62
a_{max} (g)	0.77	Norm. exp	0.45
St. dev	0.11	C_q, C_f	1.32
r_d	0.81	C_{thin}	1.00
St. dev	0.08	f_{s1} (kPa)	85.97
Corrected M_w	6.70	St. dev	41.77
St. dev	0.13	q_{c1} (MPa)	4.78
CSReq	0.50	St. dev	0.59
St. dev	0.10	R_{f1} (%)	1.80
$C.O.V._{CSR}$	0.21	St. dev	0.90

Appendix H: Cam Clay as a special case of *NorSand*

H.1 INTRODUCTION

The names *Cam Clay* and *NorSand* have resulted in some confusion regarding the applicability of each of these models. There is a view that 'Cam Clay is perfectly suitable for soft clays' and '*NorSand* is best-in-class for sands': 'horses for courses', as the British expression goes. But both models are theoretical constructs tracing back to the second law of thermodynamics and with some simple idealizations about particulate behaviour. Each model can be applied to sands, silts or clays. The model names derive from the academic limitation that if the model is cited as 'Bloggs et al.', nobody but Bloggs et al. will use the model – to make a framework of ideas widely accepted, at least within academic circles, requires the ideas be depersonalized. Schofield and Wroth were aware of this issue and chose to give a somewhat neutral name to their set of ideas, which then established the protocol that developers of critical state type models would combine the name of the developer's adjacent body of water with a soil type. Hence, Cam Clay was named after River Cam, Superior Sand after Lake Superior, Severn-Trent Sand after the River Severn, etc. *NorSand* ought to have been Yare Sand by this scheme, but as Norfolk is a senior county to Cambridgeshire in East Anglia, and this seniority reflects the relative sophistication of the theoretical idealizations, it started life as Norfolk Sand, shortened to *NorSand*. Regardless of name, all these models address the constitutive behaviour of particulate materials with no bonds between the particles (these particles do not have even to be soils). The actual particle size (which determines whether a soil is viewed as sand, silt or clay) is irrelevant to the physics and the mathematics (although, of course, the numerical values of the properties differ from one soil to another).

This appendix considers how *NorSand* includes Original Cam Clay (OCC; Schofield and Wroth, 1968). In essence, OCC exists within *NorSand* as a particular choice of initial conditions and soil properties. It is not a case of OCC for soft clays and *NorSand* for dense sand. *NorSand* can duplicate the OCC stress–strain behaviour. Thus, this appendix looks to the underlying physical ideas and shows how OCC makes particular choices within a more general framework.

What does make OCC especially interesting is that, while all general work hardening plasticity models need numerical integration, under some circumstances OCC has closed-form solutions (i.e., simple equations giving the stress–strain curve directly). This makes OCC a valuable model for verifying numerical methods. Looking into OCC is useful.

Despite this usefulness of OCC, many engineers find the derivations in Schofield and Wroth (1968) confusing. However, the OCC ideas can be clarified if changed from the 'state view' of the original derivations (which are expressed using the soil's specific volume) to a conventional plasticity framework – and which then leads directly to *NorSand* by what amounts to two additional numerical steps.

This appendix then (1) describes OCC as per Schofield and Wroth; (2) puts OCC into a conventional basis; (3) highlights the particular choices made in OCC and (4) shows that *NorSand* replicates OCC with those particular choices.

H.2 ORIGINAL CAM CLAY

Schofield and Wroth's (1968) book presenting what is known today as OCC ('original' to distinguish it from the subsequent 'modified' variant) is now a free-to-all publication that can be downloaded at http://www.geotechnique.info. The derivation of OCC is discussed in Section 3.4, but here we review the equations from Chapter 6 of the Schofield and Wroth book that derives the OCC model (indicated by the prefix 'S&W' to the equation number).

The first key idea in OCC is that plastic work is only dissipated into heat by plastic shear strain, with the soil fictional property M scaling the mechanism. Considering the work done on the sample during ongoing plastic deformation this equation is derived:

$$\frac{p\dot{v}}{v} + q\dot{\varepsilon} - \frac{\kappa\dot{p}}{v} = Mp\left|\dot{\varepsilon}\right| \quad (\dot{\varepsilon} \neq 0) \qquad \text{(S\&W 6.10)}$$

where v is the specific volume and ε the shear strain invariant (= ε_q in our notation). A better way of looking at things is to subtract the elastic changes in strain from the total, to express changes in element geometry as strain, and to divide through by the stress p. Doing these three things changes (S&W 6.10) into the general stress dilatancy equation:

$$D^p = M - \eta \qquad \text{(3.10 bis)}$$

The second key idealization of OCC is that plastic strains are normal to the yield surface, which is used to transform (3.10) into a yield surface by separating variables and integrating (again, go back to Section 3.4.2). The resulting equation of the yield surface is

$$\frac{|q|}{Mp} + \ln\left(\frac{p}{p_x}\right) = 1 \qquad \text{(S\&W 6.17)}$$

where p_x is an integration constant. This equation is more elegantly expressed by removing concern for triaxial extension (we generalize to 3D, which includes triaxial extension, by full consideration of strain rates as set out in Appendix C) and gives

$$\frac{\eta}{M} = 1 - \ln\left(\frac{p}{p_c}\right) \qquad \text{(3.14 bis)}$$

Notice that we have changed notation between (S&W 6.17) and (3.14) with p_x being replaced by p_c. This change of notation has a reason. The contribution of OCC was to first link specific volume ($v = 1 + e$) to mechanical behaviour. This was done by identifying that the peak deviator stress on the yield surface in a q–p view was the critical state and also the same critical state as in v–p space; Figure H.1 shows the scheme. Quite why the subscript 'x' was used to denote critical state conditions remains a mystery; we prefer the obvious (and widely used) subscript 'c' as denoting the critical state.

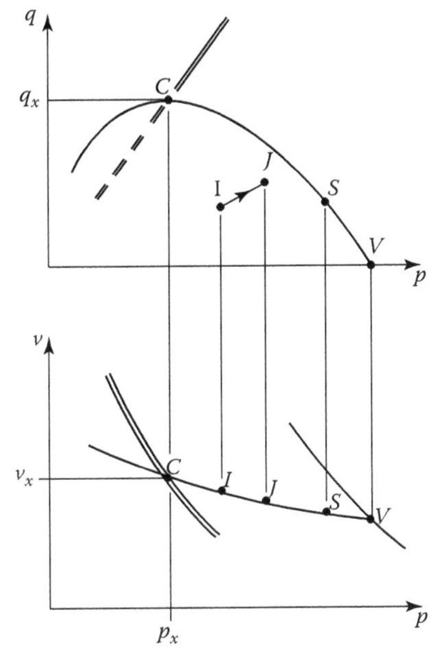

Figure H.1 Linking of stress space (q–p) with state space (v–p) in OCC. (From Schofield, A. and Wroth, C.P., *Critical State Soil Mechanics*, McGraw-Hill, London, U.K., 1968.)

Notice in Figure H.1 that a particular yield surface is associated with a range of specific volume. But, we can freely change stress states within a yield surface and with corresponding recoverable volumetric strains. Thus, a particular yield surface is not a horizontal line on a v – log(p) plot but is inclined reflecting this elastic behaviour. OCC idealized elastic stiffness as proportional to the confining stress, using an elastic compressibility κ (analogous to the familiar coefficient of rebound/recompression in an oedometer test).

A particular feature of the Cambridge approach to soil behaviour is that a 'state view' is followed, with specific volume being treated analogously with how temperature is treated when considering the compression of gas. In such an approach, the scaling coefficient for the size of the yield surface p_c (which has units of stress) is replaced by a function of specific volume, making clear the dependence of soil behaviour on the soil's current void ratio. This replacement of p_c is derived as follows. The critical state p_c, q_c on a yield surface is given by the intersection of the current elastic κ-line with the critical state locus (CSL), the respective equations for these being:

$$v + \kappa \ln(p) = v_c + \kappa \ln(p_c) \tag{H.1}$$

$$v_c = \Gamma - \lambda \ln(p_c) \tag{2.1 bis}$$

Substituting (2.1) in (H.1) allows the yield surface scaling parameter to be expressed in terms of the current specific volume and the soil properties:

$$(\lambda - \kappa)\ln(p_c) = \Gamma - v - \kappa \ln(p) \tag{H.2}$$

On substituting (H.2) in (S&W 6.17):

$$|q| = \frac{Mp}{\lambda - \kappa}(\Gamma + \lambda - \kappa - v - \lambda \ln p) \tag{S\&W 6.19}$$

This equation directly shows the effect of specific volume (or void ratio) on the stress state during plastic yielding. As specific volume evolves during plastic yielding, so does the yield surface. Schofield and Wroth note that their equation (S&W 6.19) describes a surface in v–q–p space, which they refer to as the 'state boundary surface'.

The fundamental objection to (S&W 6.19) is that emphasizing the role of specific volume has obscured the fact that OCC is a work hardening plastic model. There are no plastic strain increments in the governing equation. Also we have further lost the clarity that no plastic strain means no yielding. What happens if you do things conventionally?

H.3 PLASTICITY VIEW OF OCC

Conventional work hardening plasticity involves three ideas: (1) a flowrule, (2) a yield surface, and (3) a hardening law. OCC is simpler if viewed in this standard way.

There is nothing wrong with the Schofield and Wroth flowrule and yield surface, with the equations being simple and straightforward:

$$\text{Flowrule: } D^P = M - \eta \tag{3.10 bis}$$

$$\text{Yield surface: } \frac{\eta}{M} = 1 - \ln\left(\frac{p}{p_c}\right) \tag{3.14 bis}$$

Now, differentiate the equation for the yield surface scaling, (H.2):

$$(\lambda - \kappa)\left(\frac{\dot{p}_c}{p_c}\right) = -\dot{v} - \kappa\frac{\dot{p}}{p} \tag{H.3}$$

Recognizing the κ term as being the elastic change in specific volume (S&W 6.6), rewrite (H.3) as:

$$(\lambda - \kappa)\left(\frac{\dot{p}_c}{p_c}\right) = -\dot{v} + \dot{v}^e \tag{H.4}$$

On changing from increments of specific volume to increments of volumetric strain by diving through by specific volume, invoking the elastic–plastic strain decomposition, and then rearranging we recover a conventional hardening law for OCC:

$$\text{Hardening: } \frac{\dot{p}_c}{p_c} = \frac{v}{\lambda - \kappa}\dot{\varepsilon}_v^p \tag{H.5}$$

Notice the clarity in (H.5): (1) no plastic strain increment, no change in the yield surface and (2) the evolution of OCC is controlled by plastic volumetric strain, not specific volume per se. The parameter group $\nu/(\lambda - \kappa)$ is a conventional and dimensionless plastic hardening modulus.

With OCC now reduced to a conventional framework it is trivial to integrate the equations numerically using the Euler approach. This is set up in the downloadable spreadsheet *CamClay.xls* with the integration being provided for isotropic drained and undrained tests (in the VBA subroutines *CamClay_CID* and *CamClay_CIU*, respectively).

The conventional work hardening approach to OCC is quickly verified. Schofield and Wroth derived a closed-form solution for undrained triaxial compression, starting from a normally consolidated state under an isotropic stress p_o, using their 'state hardening' view which is given by a pair of equations:

$$\ln\frac{p}{p_u} = \Lambda \exp\left(-\frac{M\nu_o}{\kappa\Lambda}\varepsilon\right) \qquad \text{(S\&W 6.30)}$$

$$\frac{q}{Mp} = 1 - \exp\left(-\frac{M\nu_o}{\kappa\Lambda}\varepsilon\right) \qquad \text{(S\&W 6.31)}$$

where

$\Lambda = \lambda - \kappa$

p_u is a constant for a particular test that depends on the initial confining stress and the soil properties:

ε is the shear strain invariant (= ε_q in our notation)

$$\ln(p_u) = \ln(p_o) - \Lambda \qquad \text{(H.6)}$$

The *CamClay.xls* spreadsheet implements the closed-form solution on the worksheet *OCC closed-form undrained.* Figure H.2 compares a numerical integration from incremental plasticity with the closed-form solution derived by Schofield and Wroth. They are identical within numerical accuracy.

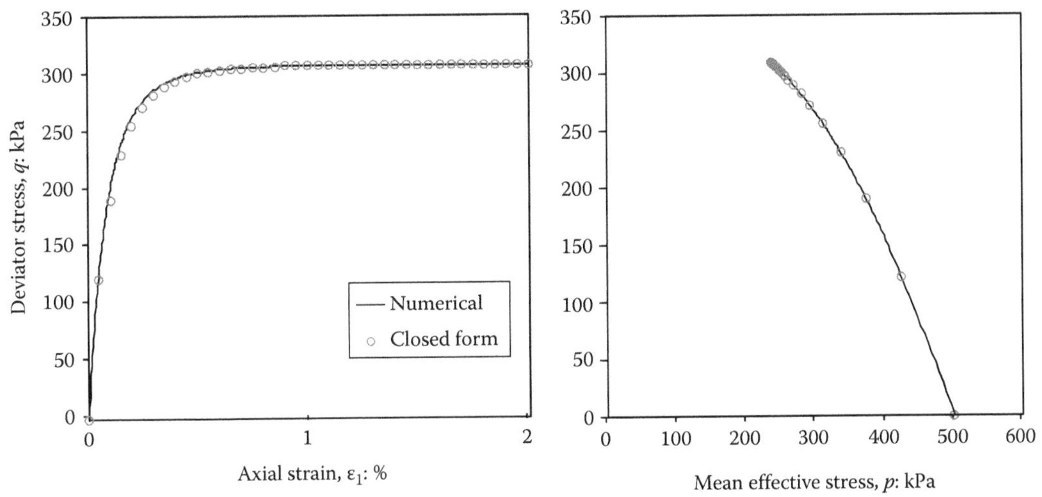

Figure H.2 Comparison of Schofield and Wroth closed-form integration of OCC for undrained triaxial compression test with numerical integration of OCC as work hardening plasticity (from CamClay.xls) for $M = 1.3$, $\lambda = 0.22$, $\kappa = 0.005$.

H.4 OCC WITHIN *NorSand*

Table H.1 compares *NorSand* with OCC, for loading in triaxial compression, in terms of flowrule, yield surface and hardening law. The similarity is evident.

Looking at the flowrule first, *NorSand* allows for the possibility of inelastic energy storage using a modified variant of Nova's proposed flowrule and results in an operating critical friction ratio that depends on the state parameter. As can be seen from Table H.1, setting the material property $N=0$ in this flowrule loses this additional detail and gets directly back to the OCC flowrule.

In terms of the yield surfaces, a kernel idealization of all variants of Cam Clay is that yield surfaces always intersect the CSL and this is captured using the scaling parameter p_c. Equally, a kernel idealization of *NorSand* is that, in general, yield surfaces do not intersect the CSL until the soil has been deformed to critical conditions – captured using a scaling parameter p_i that does not lie on the CSL. *NorSand* implements $p_i \rightarrow p_c$ in the hardening law to satisfy the axioms. However, it is perfectly acceptable to choose the initial state of the soil so that the *NorSand* yield surface does intercept the CSL as a particular case – this choice depends on the geostatic stress state and is $\psi_o = \lambda - \kappa$ for an isotropic initial stress state. Or, put another way, OCC corresponds to a particular initial choice of state parameter in *NorSand*.

However, it is not sufficient to choose the initial state parameter to capture OCC within *NorSand*. As can be seen from Table H.1 the hardening law is quite different between the two models. What is needed is also to choose the *NorSand* hardening modulus H such that the yield surface tracks the CSL as the soil deforms. This choice was derived in Appendix C when considering how H and λ are correlated, with the result that a reasonable approximation for soil states $\psi=0$ is that:

$$H = \frac{1}{\lambda - \kappa} \qquad \text{(C.42 bis)}$$

The soil property χ has no effect when the yield surface intersects the CSL, so any value can be given to that property.

Finally, there is the question of elasticity with *NorSand* having a general power law model with constant Poisson's ratio. This needs to be overridden with G_{max} set to 'a very large number' (OCC has infinite shear stiffness), the G_{max} exponent set to unity and Poisson's ratio set to matching small number (which will be near zero) to obtain the desired κ.

Table H.1 Comparison of original Cam Clay and *NorSand* as work-hardening plastic models

Model aspect	*OCC*	*NorSand*
Flowrule	$D^p = M - \eta$	$D^p = M_i - \eta$ $M_i = M - N\chi_i\lvert\psi_i\rvert$
Yield surface	$\frac{\eta}{M} = 1 - \ln\left(\frac{p}{p_c}\right)$	$\frac{\eta}{M_i} = 1 - \ln\left(\frac{p}{p_i}\right)$
Hardening law	$\frac{\dot{p}_c}{p_c} = \frac{v}{\lambda - \kappa}\dot{\varepsilon}_v^p$	$\frac{\dot{p}_i}{p_i} = H\left(\frac{p}{p_i}\right)^2\left(\frac{p_{mx}}{p} - \frac{p_i}{p}\right)\dot{\varepsilon}_q^p$ $p_{mx} = p\exp(-\chi_i\psi_i/M_i)$

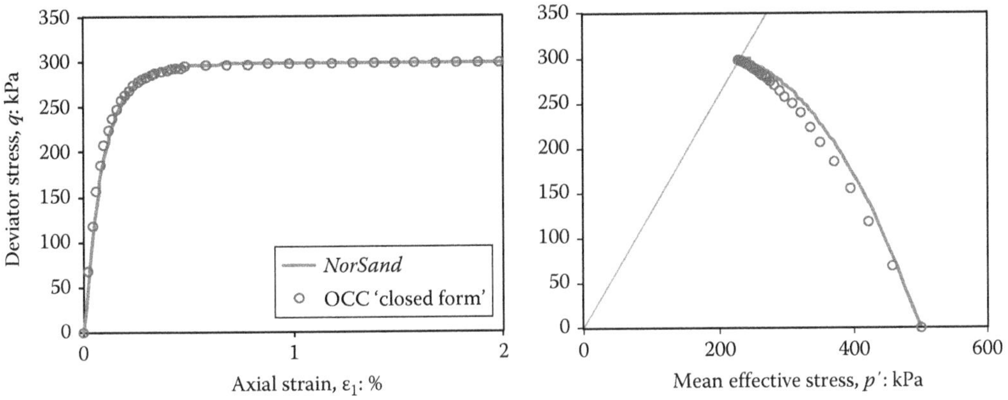

Figure H.3 Comparison of Schofield and Wroth closed-form integration of OCC for undrained triaxial compression test with *NorSand* using $\psi_o = \lambda - \kappa$, $N=0$, $H=1/(\lambda - \kappa)$.

Figure H.3 compares the results of *NorSand*, with these particular choices of properties and initial state, with the closed-form OCC solution discussed in the previous section and using the same soil properties as for Figure H.2. The fit to the stress–strain curve is a near perfect match to OCC, with slightly less perfection on the stress path. The lack of perfection in the stress path arises because OCC includes an elastic volumetric strain in its state measure that *NorSand* does not. If the state diagram is considered, to lock the yield surface to the CSL requires $\psi_o = \lambda$ but that then leads to a different final stress state than OCC. Thus, the choice $\psi_o = \lambda - \kappa$ gives the correct final state but with the effect of implying a slightly too dense yield surface for the initial loading, which is why the stress path goes a little too dilatant from the OCC path. The state parameter could be redefined to include a κ term (see (3.39) in Chapter 3), but the κ idealization is not a particularly general one for soils and further complexity is unwarranted.

References

Abdel-Haq, A. and Hryciw, R.D. (1998) Ground settlement in Simi Valley following the Northridge earthquake. *Journal of Geotechnical Engineering*, ASCE, 124(1), 80–89.

Ajalloeian, R. and Yu, H.S. (1998) Chamber studies of the effects of pressuremeter geometry on test results in sand. *Géotechnique*, 48(5), 621–636.

Alarcon-Guzman, A., Leonards, G. and Chameau, J.L. (1988) Undrained monotonic and cyclic strength of sands. *Journal of Geotechnical Engineering*, ASCE, 114(10), 1089–1109.

Ali, S.R., Pyrah, I.C. and Anderson, W.F. (1995) A novel technique for the evaluation of membrane penetration. *Géotechnique*, 45(3), 545–548.

Ambraseys, N.N. (1988) Engineering seismology. *Earthquake Engineering and Structural Dynamics*, 17, 1–105.

Andrus, R.D. and Stokoe, K.H., II (1997) Liquefaction resistance based on shear wave velocity. In *Proceedings of the NCEER Workshop on Evaluation of Liquefaction Resistance of Soils*, pp. 89–128. Buffalo, NY: National Center for Earthquake Engineering Research, State University of New York at Buffalo.

Arango, I. (1996) Magnitude scaling factors for soil liquefaction evaluations. *Journal of Geotechnical Engineering*, ASCE, 122(11), 929–936.

Arthur, J.R.F., Chua, K.S. and Dunstan, T. (1979) Dense sand weakened by continuous principal stress direction rotation. *Géotechnique*, 29(1), 91–96.

Arthur, J.R.F., Chua, K.S., Dunstan, T. and Rodriguez, J.I. (1980) Principal stress rotation: A missing parameter. *Journal of Geotechnical Engineering*, ASCE, 106(GT 4), 419–433.

Arthur, J.R.F. and Menzies, B.K. (1972) Inherent anisotropy in sand. *Géotechnique*, 22(1), 115–128.

Arulanandan, K. and Scott, R.F. (1993) Verification of numerical procedures for the analysis of soil liquefaction problems. In *Proceedings of the International Conference on the Verification of Numerical Procedures for the Analysis of Soil Liquefaction Problems*, Vols. 1 and 2. Rotterdam, the Netherlands: A.A. Balkema.

Arulmoli, K., Muraleetharan, K.K., Hosain, M.M. and Fruth, L.S. (1992) VELACS laboratory testing program. Soil Data Report, The Earth Technology Corporation, Irvine, CA, Report to the National Science Foundation, Washington, DC.

Atkinson, J.H. and Bransby, P.L. (1978) *The Mechanics of Soils: An Introduction to Critical State Soil Mechanics*. London, U.K.: McGraw-Hill.

Baki, Md. A.L., Rahman, M.M., Lo, S.R. and Gnanendran, C.T. (2012) Linkage between static and cyclic liquefaction of loose sand with a range of fines contents. *Canadian Geotechnical Journal*, 49(8), 891–906.

Baldi, G., Bellotti, R., Ghionna, V.N., Jamiolkowski, M. and Pasqualini, E. (1982) Design parameters for sand from CPT. In *Proceedings of the Second European Symposium on Penetration Testing*, Amsterdam, the Netherlands.

Baldi, G., Bellotti, R., Ghionna, V.N., Jamiolkowski, M. and Pasqualini, E. (1986) Interpretation of CPT's and CPTU's, 2nd part. In *Proceedings of the fourth International Geotechnical Seminar*, pp. 143–156. Singapore: Nanyang Technological Institute.

Baldi, G. and Nova, R. (1984) Membrane penetration tests in triaxial testing. *Journal of Geotechnical Engineering*, ASCE, 110(3), 403–420.

Baligh, M.M. (1976) Cavity expansion in sands with curved envelopes. *Journal of the Geotechnical Engineering Division*, ASCE, 102(GT11), 1131–1146.

Baxter, C.D.P., Ravi Sharma, M.S., Seher, N.V. and Jander, M. (2010) Evaluation of liquefaction resistance of non-plastic silt from mini-cone calibration chamber tests. In *Proceedings of the Second International Symposium on Cone Testing CPT 10*, Huntington Beach, CA, paper 3–38.

Becker, D.E., Crooks, J.H.A., Jefferies, M.G. and McKenzie, K. (1984) Yield behaviour and consolidation. II: Strength gain. In *Proceedings, Symposium on Sedimentation Consolidation Models*, ASCE, San Francisco, CA, pp. 382–398.

Been, K., Conlin, B.H., Crooks, J.H.A., Jefferies, M.G., Rogers, B.T., Shinde, S.B. and Williams-Fitzpatrick, S. (1987a) Back analysis of the Nerlerk berm liquefaction slides: Discussion. *Canadian Geotechnical Journal*, 24(1), 170–179.

Been, K., Crooks, J.H.A., Becker, D.E. and Jefferies, M.G. (1986) The cone penetration test in sands: Part I, state parameter interpretation. *Géotechnique*, 36(2), 239–249.

Been, K., Crooks, J.H.A. and Jefferies, M.G. (1988) Interpretation of material state from the CPT in sands and clays. In *Proceedings of the Conference on Penetration Testing in the U.K.*, Birmingham, pp. 89–92. London, U.K.: Thomas Telford.

Been, K. and Jefferies, M.G. (1985) A state parameter for sands. *Géotechnique*, 35(2), 99–112.

Been, K. and Jefferies, M.G. (1986) A state parameter for sands: Reply to Discussion. *Géotechnique*, 36(1), 123–132.

Been, K. and Jefferies, M.G. (1992) Towards systematic CPT interpretation. In *Proceedings of the Wroth Memorial Symposium*, Oxford, U.K., pp. 121–134. London, U.K.: Thomas Telford.

Been, K. and Jefferies, M.G. (1993) Determination of sand strength for limit state design. In *Proceedings of the International Symposium on Limit State Design in Geotechnical Engineering*, Copenhagen, Denmark, May 1993, Vol. 1, pp. 101–110.

Been, K. and Jefferies, M.G. (2004) Stress dilatancy in very loose sand. *Canadian Geotechnical Journal*, 41(1), 972–989.

Been, K., Jefferies, M.G., Crooks, J.H.A. and Rothenberg, L. (1987c) The cone penetration test in sands: Part II, General inference of state. *Géotechnique*, 37(3), 285–299.

Been, K., Jefferies, M.G. and Hachey, J.E. (1991) The critical state of sands. *Géotechnique*, 41(3), 365–381.

Been, K., Jefferies, M.G. and Hachey, J.E. (1992) The critical state of sands: Reply to Discussion. *Géotechnique*, 42(4), 655–663.

Been, K., Jefferies, M.G., Hachey, J.E. and Rothenburg, L. (1993) Class A prediction for model 2. In *Proceedings of the International Conference on the Verification of Numerical Procedures for the Analysis of Soil Liquefaction Problems (VELACS)* (eds. K. Arulanandan and R.F. Scott). Rotterdam, the Netherlands: A. A. Balkema.

Been, K., Lingnau, B.E., Crooks, J.H.A. and Leach, B. (1987b) Cone penetration test calibration for Erksak (Beaufort Sea); sand. *Canadian Geotechnical Journal*, 24(4), 601–610.

Been, K., Lingnau, B.E., Crooks, J.H.A. and Leach, B. (1989) Cone penetration test calibration for Erksak (Beaufort Sea); sand: Reply to discussion. *Canadian Geotechnical Journal*, 26(1), 177–182.

Been, K., Obermeyer, J., Parks, J. and Quinonez, A. (2012) Post liquefaction undrained shear strength of sandy silt and silty sand tailings. In *Proceedings of the International Conference on Tailings and Mine Waste '12*, Keystone, CO, 14–17 October, 2012, pp. 325–335.

Bellotti, R., Jamiolkowski, M., Lo Presti, D.C.F. and O'Neill, D.A. (1996) Anisotropy of small strain stiffness in Ticino sand. *Géotechnique*, 46(1), 115–131.

Bennett, M.J., McLaughlin, P.V., Sarmiento, J. and Youd, T.L. (1984) Geotechnical investigation of liquefaction sites. U.S. Geological Survey Open File Report, Imperial Valley, CA, pp. 84–252.

Bennett, M.J., Ponti, D.J., Tinsley, J.C.I., Holzer, T.L. and Conaway, C.H. (1998) Subsurface geotechnical investigations near sites of ground deformation caused by the January 17, 1994, Northridge, California, Earthquake. Open File Report 98–373, U.S. Department of the Interior, U.S. Geological Survey, Menlo Park, CA.

Bennett, M.J. and Tinsley, J.C.I. (1995) Geotechnical data from surface and subsurface samples outside of and within liquefaction-related ground failures caused by the October 17, 1989, Loma Prieta Earthquake, Santa Cruz and Monterey counties, California. Open-File Report 95–663, U.S. Department of the Interior, U.S. Geological Survey, Menlo Park, CA.

Berrill, J.B., Le Kouby, A., Canou, J. and Foray, P.Y. (2003) The effect of layering on cone resistance: Calibration chamber tests. In *Proceedings of the Ninth Australia New Zealand Conference on Geomechanics*, Auckland, New Zealand.

Bierschwale, J.G. and Stokoe, K.H., II (1984) Analytical evaluation of liquefaction potential of sands subjected to the 1981 Westmorland earthquake. Geotechnical Engineering Report GR 84–15, University of Texas, Austin, TX.

Bishop, A.W. (1950) Reply to discussion on "Measurement of shear strength of soils" by A.W. Skempton and A.W. Bishop. *Géotechnique*, 2, 90–108.

Bishop, A.W. (1966) Strength of soils as engineering materials. Sixth Rankine Lecture. *Géotechnique*, 16, 89–130.

Bishop, A.W. (1971) Shear strength parameters for undisturbed and remoulded soil specimens. In *Stress–Strain Behaviour of Soils: Proceedings of the Roscoe Memorial Symposium, Cambridge* (ed. R.H.G. Parry), pp. 3–58. London, U.K.: Foulis.

Bishop, A.W. (1973) The stability of tips and spoil heaps. *Quarterly Journal of Engineering Geology*, 6(1973), 335–376.

Bishop, A.W. and Eldin, G. (1953) The effect of stress history on the relation between φ and porosity in sand. In *Proceedings* of the *Third International Conference on Soil Mechanics and Foundation Engineering*, Zurich, Switzerland, Vol. 1, pp. 100–105.

Bishop, A.W. and Henkel, D.J. (1957) *The Measurement of Soil Properties in the Triaxial Test*. London, U.K.: Arnold.

Bishop, A.W. and Henkel, D.J. (1962) *The Measurement of Soil Properties in the Triaxial Apparatus*, 2nd ed. London, U.K.: Edward Arnold.

Bishop, A.W. and Penman, A.D.M. (1969) Informal discussion. Aberfan disaster: Technical aspects. *Proceedings Institution of Civil Engineers*, 42(2), 317–318.

Bishop, A.W., Webb, K.L. and Skinner, A.E. (1965) Triaxial tests on soil at elevated cell pressures. In *Proceedings of the Sixth International Conference on Soil Mechanics and Foundations*, Montreal, Canada, Vol. 1, pp. 170–174.

Bishop, A.W. and Wesley, L.D. (1975) A hydraulic triaxial apparatus for controlled stress path testing. *Géotechnique*, 25(4), 657–670.

Bjerrum, L. and Landva, A. (1966) Direct simple shear tests on a Norwegian quick clay. *Géotechnique*, 16(1), 1–20.

Bolton, M.D. (1986) Strength and dilatancy of sands. *Géotechnique*, 36(1), 65–78.

Boulanger, R.W. (2003) Relating K_α to relative state parameter index. *Journal of Geotechnical and Geoenvironmental Engineering*, ASCE, 129(8), 770–773.

Boulanger, R.W., Idriss, I.M. and Mejia, L.H. (1995) Investigation and evaluation of liquefaction related ground displacements at moss landing during the 1989 Loma Prieta earthquake. Report No. UCD/CGM-95/02, Center for Geotechnical Modeling, Department of Civil and Environmental Engineering, University of California, Davis, CA.

Boulanger, R.W., Mejia, L.H. and Idriss, I.M. (1997) Liquefaction at moss landing during Loma Prieta earthquake. *Journal of Geotechnical and Geoenvironmental Engineering*, ASCE, 123(5), 453–467.

Brandon, T.L., Clough, G.W. and Rajardjo, R.P. (1990) Evaluation of liquefaction potential of silty sands based on cone penetration resistance. Report to National Science Foundation, Grant # ECE-8614516, Virginia Polytechnic Institute, Blacksburg, VA.

Brandon, T.L., Clough, G.W. and Rajardjo, R.P. (1991) Fabrication of silty sand specimens for large and small scale tests. *Geotechnical Testing Journal*, 14(1), 46–55.

Burland, J.B. (1965) Deformation of soft clay. PhD Thesis, Cambridge University.

Burland, J.B. and Burbidge, M.C. (1985) Settlement of foundations on sand and gravel. *Proceedings of the Institution of Civil Engineers*, Part 1, 78, 1325–1381.

Calladine, C.R. (1969) Correspondence. *Géotechnique*, 13, 250–255.

Canadian Standards Association (1992) *Code for the Design, Construction and Installation of Fixed Offshore Structures*. Published in 5 Parts. Toronto, Canada: CSA.

Carter, J.P., Booker, J.R. and Yeung, S.K. (1986) Cavity expansion in cohesive frictional soils. *Géotechnique*, 36(3), 349–358.

Casagrande, A. (1936) Characteristics of cohesionless soils affecting the stability of earth fills. *Journal of Boston Society of Civil Engineers*, 23, 257–276.

Casagrande, A. (1950) Notes on the design of earth dams. *Journal of Boston Society of Civil Engineers*, 37, 231–255.

Casagrande, A. (1965) Role of calculated risk in earthwork and foundation engineering. *Journal of Soil Mechanics and Foundations Division*, ASCE, 91(4), 1–40.

Casagrande, A. (1975) Liquefaction and cyclic deformation of sands: A critical review. In *Proceedings of the Fifth Pan–American Conference on Soil Mechanics and Foundation Engineering*, Buenos Aires, Argentina, Vol. 5, pp. 79–133.

Castro, G. (1969) Liquefaction of sands. PhD Thesis, Harvard University, Cambridge, MA (Harvard Soil Mechanics Series 81).

Castro, G., Keller, T.O. and Boynton, S.S. (1989) Re-evaluation of the lower San Fernando Dam, GEI Consultants, Inc. Contract Report GL-89-2 Volume 1, US Army Corps of Engineers, Washington, DC.

Cetin, K.O. (2000) Reliability-based assessment of seismic soil liquefaction initiation hazard. PhD Dissertation, University of California, Berkeley, CA.

Cetin, K.O., Seed, R.B., Moss, R.E.S., Der Kiureghian, A., Tokimatsu, K., Harder, L.F. Jr. and Kayen, R.E. (2000) Field case histories for SPT-based *in situ* liquefaction potential evaluation. Geotechnical Engineering Research Report No.UCB/GT-2000/09, Berkeley.

Chadwick, P. (1959) The quasi-static expansion of a spherical cavity in metals and ideal soils. *Quarterly Journal Mechanics and Applied Mathematics*, Part 1, 12, 52–71.

Chillarige, A.V., Morgenstern, N.R., Robertson, P.K. and Christian, H.A. (1997b) Seabed instability due to flow liquefaction in the Fraser River delta. *Canadian Geotechnical Journal*, 34(4), 520–533.

Chillarige, A.V., Robertson, P.K., Morgenstern, N.R. and Christian, H.A. (1997a) Evaluation of the in situ state of Fraser River sand. *Canadian Geotechnical Journal*, 34(4), 510–519.

Chu, J. and Leong, W.K. (2002) Effect of fines on instability behaviour of loose sand. *Géotechnique*, 52(10), 751–755.

Chu, J., Leong, W.K., Loke, W.L. and Wanatowski, D. (2012) Instability of loose sand under drained conditions. *Journal of Geotechnical and Geoenvironmental Engineering*, ASCE, 138(2), 207–216.

Chu, J., Leroueil, S. and Leong, W.K. (2003) Unstable behaviour of sand and its implication for slope instability. *Canadian Geotechnical Journal*, 40(5), 873–885.

Clayton, C.R.I., Hababa, M.B. and Simons, N.E. (1985) Dynamic penetration resistance and the prediction of the compressibility of a fine grained sand – A laboratory study. *Géotechnique*, 35(1), 19–31.

Collins, I.F. and Muhunthan, B. (2003) On the relationship between stress-dilatancy, anisotropy, and plastic dissipation of granular materials. *Géotechnique*, 53, 611–618.

Coop, M.R. (1990) The mechanics of uncemented carbonate sands. *Géotechnique*, 40(4), 607–626.

Cormeau, I.C. (1975) Numerical stability in quasi-static elasto-viscoplasticity. *International Journal of Numerical Methods in Engineering*, 9(7), 109–127.

Cornforth, D.H. (1961) Plane strain failure characteristics of a saturated sand. PhD Thesis, University of London, London, U.K.

Cornforth, D.H. (1964) Some experiments on the effect of strain condition on the strength of sand. *Géotechnique*, 14, 143–167.

Cornforth, D.H. (2005) *Landslides in Practice: Investigation, Analysis and Remedial/Preventative Options in Soils*. Hoboken, NJ: Wiley.

Daouadji, A., AlGali, H., Darve, F. and Zeghloul, A. (2010) Instability of granular materials: Experimental evidence of diffuse mode of failure for loose sands. *Journal of Engineering Mechanics*, 136(5), 575–588.

Davies, M.P., Dawson, B.D. and Chin, B.G. (1998) Static liquefaction slump of mine tailings – A case history. In *Proceedings of the 51st Canadian Geotechnical Conference*, Edmonton, Canada, Vol. 1, pp. 123–131.

Davis, A.P., Castro, G. and Poulos, S.J. (1988) Strengths backfigured from liquefaction case histories. In *Proceedings of the Second International Conference on Case Histories in Geotechnical Engineering*, St Louis, MO, pp. 1693–1701.

De Alba, P.A., Seed, H.B. and Chan, C.K. (1976) Sand liquefaction in large-scale simple shear tests. *Journal of the Geotechnical Engineering Division*, ASCE, 102(9), 909–927.

De Alba, P.A., Seed, H.B., Retamal, E. and Seed, R.B. (1988) Analysis of dam failures in 1985 Chilean earthquake. *Journal of Geotechnical Engineering*, ASCE, 114(12), 1414–1434.

De Josselin de Jong, G. and Verruijt, A. (1969) Étude photo-élastique d'un empilement de disques. *Cahier Grpe.Fr. Etude Rheol.*, 2(1), 73–86.

Desrues, J., Chambon, R., Mokni, M. and Mazerolle, F. (1996) Void ratio evolution inside shear bands in triaxial sand specimens studied by computed tomography. *Géotechnique*, 46(3), 529–546.

DiPrisco, C. and Imposimato, S. (1997) Experimental analysis and theoretical interpretation of triaxial load controlled loose sand specimen collapses. *Mechanics of Cohesive Frictional Materials*, 2, 93–120.

Donaghe, R.T., Chaney, R.C. and Silver, M.L., eds. (1988) Advanced Triaxial Testing of Soil and Rock, Special Technical Publication STP 977. Baltimore, MD: ASTM.

Drucker, D.C. (1951) A more fundamental approach to stress-strain relations. In *Proceedings of the First US National Congress of Applied Mechanics*, Chicago, IL, pp. 487–491, ASME.

Drucker, D.C. (1959) A definition of stable inelastic material. *Journal of Applied Mechanics*, 26, 101–106.

Drucker, D.C., Gibson, R.E. and Henkel, D.J. (1957) Soil mechanics and work hardening theories of plasticity. *Transactions American Society of Civil Engineers*, 122, 338–346.

Egan, J.A. and Sangrey, D.A. (1978) Critical state model for cyclic load pore pressure. In *Proceedings of the ASCE Specialty Conference on Earthquake Engineering and Soil Dynamics*, Pasadena, CA, Vol. 1, pp. 410–424.

Eldin, A.K.G. (1951) Fundamental factors controlling shear properties of sands. PhD Thesis, University of London, London, U.K.

European Committee for Standardisation (1994). ENV 1997-1, Eurocode 7: Geotechnical design, Part 1: General rules. European Committee for Standardisation, Brussels.

Fear, C.E. (1996) In-situ testing for liquefaction evaluation of sandy soils. PhD Thesis, University of Alberta, Edmonton, Canada.

Fioravante, V., Giretti, D. and Jamiolkowski, M. (2013) Small strain stiffness of carbonate Kenya Sand. *Engineering Geology*, 161, 65–80.

Fioravante, V., Jamiolkowski, M., Tanizawa, F. and Tatsuoka, F. (1991) Results of CPTs in Toyoura quartz sand. In *Proceedings of the First International Symposium on Calibration Chamber Testing/ISOCCT1*, Potsdam, NY, ed. A.-B. Huang, pp. 135–146.

Fourie, A.B., Blight, G.E. and Papageorgiou, G. (2001) Soil liquefaction as a possible explanation for Merriespruit tailings dam failure. *Canadian Geotechnical Journal*, 38(4), 707–719.

Fourie, A.B. and Papageorgiou, G. (2001) Defining an appropriate steady state line for Merriespruit gold tailings. *Canadian Geotechnical Journal*, 38(4), 695–706.

Gajo, A. and Muir Wood, D. (1999) Severn-Trent sand, a kinematic hardening constitutive model: The q-p formulation. *Géotechnique*, 49, 595–614.

Gajo, A., Piffer, L. and De Polo, F. (2000) Analysis of certain factors affecting the unstable behaviour of saturated loose sand. *Mechanics of Cohesive-Frictional Materials*, 5(3), 215–237.

Garga, V.K. and McKay, L.D. (1984) Cyclic triaxial strength of mine tailings. *Journal of Geotechnical Engineering Division*, ASCE, 110(8), 1091–1105.

GDS Instruments. (2014) GDS blender element system, Hampshire, U.K. http://www.gdsinstruments.com/gds-products/gds-bender-element-system (accessed March 12, 2015.)

Ghafghazi, M. and Shuttle, D.A. (2008) Evaluation of soil state from SBP and CPT: A case history. *Canadian Geotechnical Journal*, 45(6), 824–844.

Ghionna, V. (1984) Influence of chamber size and boundary conditions on the measured cone resistance. In *Seminar on Cone Penetration Testing in the Laboratory*, University of Southampton, Southampton, U.K.

Gibbs, H.J. and Holtz, W.G. (1957) Research on determining the density of sands by spoon penetration testing. In *Proceedings of the Fourth International Conference on Soil Mechanics and Foundation Engineering*, London, U.K., Vol. 1, pp. 35–39.

Golder Associates (2015) www.golderfoundation.org (accessed March 15, 2015).

Graham, J.P. and Jefferies, M.G. (1986) Some examples of in situ lateral stress determinations in hydraulic fills using the self-boring pressuremeter. In *Proceedings of the 39th Canadian Geotechnical Conference*, Ottawa, Canada.

Green, G.E. (1971) Strength and deformation of sand measured in an independent stress control cell. In *Stress–Strain Behaviour of Soils: Proceedings of the Roscoe Memorial Symposium, Cambridge, 29–31 March 1971* (ed. R.H.G. Parry), pp. 285–323. London, U.K.: Foulis.

Green, G.E. and Bishop, A.W. (1969) A note on the drained strength of sand under generalized strain conditions. *Géotechnique*, 19(1), 144–149.

Griffiths, D.V. and Fenton, G.A. (2004) Probabilistic slope stability analysis by finite elements. *Journal of Geotechnical and Geoenvironmental Engineering*, ASCE, 130(5), 507–518.

Griffiths, D.V. and Lane, P.A. (1999) Slope stability analysis by finite elements. *Géotechnique*, 49(3), 387–403.

Harder, L.F. Jr. and Boulanger, R.W. (1997) Application of K_σ and K_α correction factors. In *Proceedings of the NCEER Workshop on Evaluation of Liquefaction Resistance of Soils*, National Center for Earthquake Engineering Research, State University of New York, Buffalo, NY, pp. 167–190.

Harman, D.E. (1976) A statistical study of static cone bearing capacity, vertical effective stress, and relative density of dry and saturated fine sands in a large triaxial testing chamber. MSc Thesis, University of Florida, Gainesville, FL.

Hazen, A. (1918) A study of the slip in the Calaveras Dam. *Engineering News Record*, 81(26), 1158–1164.

Hazen, A. (1920) Hydraulic fill dams. *Transactions of the American Society of Civil Engineers*, 83, 1713–1745.

Hazen, A. and Metcalf, L. (1918) Middle section of upstream side of Calaveras Dam slips into reservoir. *Engineering News Record*, 80(14), 679–681.

Hicks, M.A. and Boughrarou, R. (1998) Finite element analysis of the Nerlerk underwater berm failures. *Géotechnique*, 48, 169–185.

Hicks, M.A. and Onisiphorou, C. (2005) Stochastic evaluation of static liquefaction in a predominantly dilative sand. *Géotechnique*, 55(2), 123–133.

Hight, D.W., Gens, A. and Symes, M.J. (1983) The development of a new hollow cylinder apparatus for investigating the effects of principal stress rotation in soils. *Géotechnique*, 33(4), 355–383.

Hird, C.C. and Hassona, F. (1986) A state parameter for sands: Discussion. *Géotechnique*, 36(1), 123–132.

Hird, C.C. and Hassona, F. (1990) Some factors affecting the liquefaction and flow of saturated sands in laboratory tests. *Engineering Geology*, 28, 149–170.

Holzer, T.L., Bennett, M.J., Ponti, D.J. and Tinsley, J.C.I. (1999) Liquefaction and soil failure during 1994 Northridge earthquake. *Journal of Geotechnical Engineering*, ASCE, 125(6), 438–452.

Holzer, T.L., Tinsley, J.C.I., Bennett, M.J. and Mueller, C.S. (1994) Observed and predicted ground deformation – Miller farm lateral spread, Watsonville, California. In *Proceedings of the Fifth U.S. – Japan Workshop on Earthquake Resistant Design of Lifeline Facilities and Countermeasures for Soil Liquefaction*, Technical Report NCEER-94-0026 79–99.

Horsfield, D. and Been, K. (1987) Computer controlled triaxial testing. In *Proceedings of the First Canadian Symposium on Microcomputer Applications to Geotechnique*, Regina, Canada, pp. 55–62.

Houlsby, G.T. (1988) Session leader's introduction. In *Proceedings Conference on Penetration Testing in the U.K.*, Birmingham, U.K., pp. 141–146. London, U.K.: Thomas Telford.

Hryciw, R.D., Vitton, S. and Thomann, T.G. (1990) Liquefaction and flow failure during seismic exploration. *Journal of Geotechnical Engineering*, ASCE, 116(12), 1881–1899.

Hsu, H.-H. and Huang, A.-B. (1999) Calibration of cone penetration test in sand. *Proceedings Natural Science Council*, ROC(A), 23(5), 579–590.

Hughes, J.M.O., Wroth, C.P. and Windle, D. (1977) Pressuremeter tests in sands. *Géotechnique*, 27(4), 455–477.

Huntsman, S.R. (1985) Determination of in situ lateral pressure of cohesionless soils by static cone penetrometer. PhD Thesis, University of California at Berkeley, Berkeley, CA.

Huntsman, S.R., Mitchell, J.K., Klejbuk, L. and Shinde, S.B. (1986) Lateral stress measurement during cone penetration. In *Proceedings of the ASCE Specialty Conference on* in situ *Testing*, Blacksburg, VA.

Hvorslev, M.J. (1950) Triaxial tests on sands, Reid Bedford Bend, Mississipi River. Potamology Investigation Report, 5–3, Waterways Experiment Station (USCE), Vicksburg, MS.

Hynes, M.E. and Olsen, R.S. (1999) Influence of confining stress on liquefaction resistance. In *Proceedings of the International Workshop on Physics and Mechanics of Soil Liquefaction*, pp. 145–152. Rotterdam, the Netherlands: Balkema.

Idriss, I.M. and Boulanger, R.W. (2004) Semi-empirical procedures for evaluating liquefaction potential during earthquakes. In *Proceedings of the 11th International Conference on Soil Dynamics and Earthquake Engineering, and 3rd International Conference on Earthquake Geotechnical Engineering*, (eds. D. Doolin, A. Kammerer, T. Nogami, R.B. Seed and I. Towhata), Vol. 1, pp. 32–56. Stallion Press.

Idriss, I.M. and Boulanger, R.W. (2007) SPT- and CPT-based relationships for the residual shear strength of liquefied soils. In *Earthquake Geotechnical Engineering, 4th International Conference on Earthquake Geotechnical Engineering – Invited Lectures* (ed. K.D. Pitilakis), pp. 1–22. Dordrecht, Netherlands: Springer.

Idriss, I.M. and Boulanger, R.W. (2008) *Soil Liquefaction during Earthquakes*, Monograph MNO-12. Oakland, CA: Earthquake Engineering Research Institute.

Imrie, A. (1994) Overview of the liquefaction assessment and seismic stability of Duncan Dam; Specialty session, 46th Annual Canadian Geotechnical Conference, Saskatoon, Saskatchewan, September 27–29, 1993. *Canadian Geotechnical Journal*, 31(6), 918.

Ingenjörsfirman Geotech AB. (2014). http://www.geotech.eu/ (accessed March 27, 2015).

Ishihara, K. (1993) Thirty third Rankine lecture: Liquefaction and flow failure during earthquakes. *Géotechnique*, 43(3), 349–415.

Ishihara, K. and Koga, Y. (1981) Case studies of liquefaction in the 1964 Niigata earthquake. *Soils and Foundations*, 21(3), 35–52.

Ishihara, K., Tatsuoka, F. and Yasuda, S. (1975) Undrained deformation and liquefaction of sand under cyclic stresses. *Soils and Foundations*, 15(1), 29–44.

Ishihara, K. and Towhata, I. (1983) Sand response to cyclic rotation of principal stress directions as induced by wave loads. *Soils and Foundations*, 23(4), 11–26.

Ishihara, K., Yasuda, S. and Yoshida, Y. (1990) Liquefaction-induced flow failure of embankments and residual strength of silty sands. *Soils and Foundations*, 30(3), 69–80.

ISO (2010) International Standard ISO 19906, Petroleum and natural gas industries – Arctic offshore structures. Reference number ISO 19906:2010 (E).

Jefferies, M.G. (1993) NorSand: A simple critical state model for sand. *Géotechnique*, 43, 91–103.

Jefferies, M.G. (1997) Plastic work and isotropic softening in unloading. *Géotechnique*, 47, 1037–1042.

Jefferies, M.G. (1998) A critical state view of liquefaction. In *Physics and Mechanics of Soil Liquefaction* (eds. P.V. Lade and J.A. Yamamuro), pp. 221–235. Rotterdam, the Netherlands: Balkema.

Jefferies, M.G. and Been, K. (1992) Undrained Response of Norsand. In *Proceedings of the 45th Canadian Geotechnical Conference*, Toronto, October 26–28, 1992.

Jefferies, M.G. and Been, K. (2000) Implications for critical state theory from isotropic compression of sand. *Géotechnique*, 50(4), 419–429.

Jefferies, M.G., Been, K. and Hachey, J.E. (1990) The influence of scale on the constitutive behaviour of sand. In *Proceedings of the 43rd Canadian Geotechnical Conference*, Quebec City, Canada, Vol. 1, pp. 263–273.

Jefferies, M.G. and Davies, M.P. (1991) Soil classification by the cone penetration test: Discussion. *Canadian Geotechnical Journal*, 28(1), 173–468.

Jefferies, M.G. and Davies, M.P. (1993) Use of CPTu to estimate equivalent SPT N_{60}. *Geotechnical Testing Journal*, ASTM, 16(4), 458–468.

Jefferies, M.G., Jönsson, L. and Been, K. (1987) Experience with measurement of horizontal geostatic stress in sand during cone penetration test profiling. *Géotechnique*, 37(4), 483–498.

Jefferies, M.G., Rogers, B.T., Griffin, K.M. and Been, K. (1988b) Characterization of sandfills with the cone penetration test. In *Proceedings of the Conference on Penetration Testing in the U.K.*, Birmingham, U.K., pp. 73–76. London, U.K.: Thomas Telford.

Jefferies, M.G., Rogers, B.T., Stewart, H.R., Shinde, S., James, D. and Williams-Fitzpatrick, S. (1988a) Island construction in the Canadian Beaufort Sea. In *Proceedings of the ASCE Conference on Hydraulic Fill Structures*, Fort Collins, CO, pp. 816–883.

Jefferies, M.G. and Shuttle, D.A. (2002) Dilatancy in general Cambridge-type models. *Géotechnique*, 52(9), 625–637.

Jefferies, M.G. and Shuttle, D.A. (2005) NorSand: Features, calibration and use. In *Soil Constitutive Models: Evaluation, Selection, and Calibration*, ASCE Geotechnical Special Publication No. 128 (eds. J.A. Yamamuro and V.N. Kaliakin), pp. 204–236. Reston, VA: ASCE.

Jefferies, M.G. and Shuttle, D.A. (2011) On the operating critical friction ratio in general stress states. *Géotechnique*, 61(8), 709–713.

Jefferies, M.G., Stewart, H.R., Thomson, R.A.A. and Rogers, B.T. (1985) Molikpaq deployment at Tarsiut P-45. In *Proceedings of the ASCE Specialty Conference on Civil Engineering in the Arctic Offshore*, San Francisco, CA, pp. 1–27.

Johnes, M. and Maclean, I. (2008), http://www. nuff.ox.ac.uk/politics/aberfan/home2.htm (accessed March 15, 2015).

Junaideen, S.M., Tham, L.G., Law, K.T., Dai, F.C. and Lee, C.F. (2010) Behaviour of recompacted residual soils in a constant shear stress path. *Canadian Geotechnical Journal*, 47(4), 648–661.

Kayen, R.E., Mitchell, J.K., Seed, R.B. and Nishio, S. (1998) Soil liquefaction in the east bay during the earthquake. Professional Paper 1551-B, U.S. Department of Interior, U.S. Geological Survey.

Kjellman, W. (1951) Testing the shear strength of clay in Sweden. *Géotechnique*, 2(3), 225–232.

Konrad, J.-M. (1988) Interpretation of flat plate dilatometer tests in sands in terms of the state parameter. *Géotechnique*, 38(2), 263–278.

Konrad, J.-M. (1990a) Minimum undrained strength of two sands. *Journal of Geotechnical Engineering*, ASCE, 116(6), 932–947.

Konrad, J.-M. (1990b) Minimum undrained strength versus steady state strength of sands. *Journal of Geotechnical Engineering*, ASCE, 116(6), 948–963.

Konrad, J.-M. (1991) The Nerlerk berm case history: Some considerations for the design of hydraulic sand fills. *Canadian Geotechnical Journal*, 28(4), 601–612.

Konrad, J.-M. (1993) Undrained response of loosely compacted sands during monotonic and cyclic compression tests. *Géotechnique*, 43(1), 69–90.

Konrad, J.-M. (1997) In situ sand state from CPT: Evaluation of a unified approach at two CANLEX sites. *Canadian Geotechnical Journal*, 34(1), 120–130.

Konrad, J.-M., Watts, B. and Stewart, R. (1997) Assigning the ultimate strength at Duncan Dam. In *Proceedings of the 14th International Conference on Soil Mechanics and Foundation Engineering*, Hamburg, Germany, Vol. 1, pp. 143–146.

Koppejan, A.W., van Wamelen, B.M. and Weinberg, L.J.H. (1948) Coastal flow slides in the Dutch province of Zeeland. In *Proceedings of the Second International Conference on Soil Mechanics and Foundation Engineering*, Rotterdam, the Netherlands, Vol. V, pp. 89–96.

Kramer, S.L., Sivaneswaran, N. and Davis, R.O. (1990) Analysis of membrane penetration in triaxial test. *Journal of Engineering Mechanics*, ASCE, 116(4), 773–789.

Kuerbis, R.H., Negussey, D. and Vaid, Y.P. (1988) Effect of gradation and fines content on the undrained response of sand. In *ASCE Conference on Hydraulic Fill Structures*, Geotechnical Special Publication 21, pp. 330–345. Fort Collins, CO: ASCE.

Kuerbis, R.H. and Vaid, Y.P. (1988) Sand sample preparation – The slurry deposition method. *Soils and Foundations*, 28(4), 107–118.

Ladanyi, B. and Foriero, A. (1998) A numerical solution of cavity expansion problem in sand based directly on experimental stress-strain curves. *Canadian Geotechnical Journal*, 35(4), 541–559.

Ladanyi, B. and Roy, M. (1987) Point resistance of piles in sand. In *Proceedings of the Ninth Southeast Asian Geotechnical Conference*, Bangkok, Thailand.

Ladd, C.C. and Foott, R. (1974) New design procedure for stability of soft clays. *Journal of the Geotechnical Engineering Division*, ASCE, 100(GT7), 763–786.

Ladd, R.S. (1977) Specimen preparation and cyclic stability of sands. *Journal of Geotechnical Engineering Division*, ASCE, 103, 535–547.

Ladd, R.S. (1978) Preparing test specimens using undercompaction. *Geotechnical Testing Journal, GTJOAD*, ASTM, 1(1), 1–8.
Lade, P.V. (1993) Initiation of static instability in the submarine Nerlerk berm. *Canadian Geotechnical Journal*, 30, 895–904.
Lade, P.V. (1999) Instability of granular materials. In *Physics and Mechanics of Soil Liquefaction* (eds. P.V. Lade and J.A. Yamamuro), pp. 3–16, Rotterdam, the Netherlands: Balkema.
Lade, P.V. and Duncan, J.M. (1974) Elastoplastic stress–strain theory for cohesionless soil. *Journal of the Geotechnical Engineering Division*, ASCE, 101(10), 1037–1053.
Lade, P.V. and Pradel, D. (1990) Instability and plastic flow of soils. I: Experimental Observations. *Journal of Engineering Mechanics*, ASCE, 116(11), 2532–2550.
Lade, P.V. and Yamamuro, J.A. (2010) Evaluation of static liquefaction potential of silty sand slopes. *Canadian Geotechnical Journal*, 48(2), 247–264.
Lambe, T.W. and Whitman, R.V. (1968) *Soil Mechanics*. New York: John Wiley.
Lee, K.L. and Seed, H.B. (1967) Dynamic strength of anisotropically consolidated sand. *Journal of Soil Mechanics and Foundation Engineering Division*, ASCE, Vol. 93, SM5,169–190.
Lee, K.M. (2001) Influence of placement method on the cone penetration resistance of hydraulically placed sand fills. *Canadian Geotechnical Journal*, 38(3), 592–607.
Lhuer, J.-M. (1976) An experimental study of quasi-static cone penetration in saturated sands. MSc Thesis, University of Florida.
Li, X.-S. and Dafalias, Y.F. (2000) Dilatancy for cohesionless soils. *Géotechnique*, 50(4), 449–460.
Li, X.-S., Dafalias, Y.F. and Wang, Z.-L. (1999) State dependent dilatancy in critical state constitutive modelling of sand. *Canadian Geotechnical Journal*, 36, 599–611.
Li, X.-S. and Wang, Z.-L. (1998) Linear representation of steady state line for sand. *Journal of Geotechnical and Geoenvironmental Engineering*, ASCE, 124(12), 1215–1217.
Liao, S.S.C. and Whitman, R.V. (1986) Overburden correction factors for SPT in sand. *Journal of Geotechnical Engineering*, ASCE, 112(3), 373–377.
Lindenberg, J. and Koning, H.L. (1981) Critical density of sand. *Géotechnique*, 31(2), 231–245.
Lode, W. (1926) Versuche ueber den Einfluss der mitt leren Hauptspannung auf das Fliessen der Metalle Eisen Kupfer und Nickel. *Zeitschrift fuer Physik*, 36, 913–939.
Lucia, P.C. (1981) Review of experience with flow failures of tailings dams and waste impoundments. PhD Thesis, University of California at Berkeley, Berkeley, CA.
Lunne, T. (1986) Personal Communication. Results obtained at Southampton and some preliminary interpretation. In *Seminar or Cone Penetration Testing in the Laboratory*, University of Southampton, 1984.
Lunne, T., Berre, T., Andersen, K.H., Strandvik, S. and Sjursen, M. (2006) Effects of sample disturbance and consolidation procedures on measured shear strength of soft marine Norwegian clays. *Canadian Geotechnical Journal*, 43(7), 726–750.
Lyman (1938) Construction of Franklin Falls Dam. Report, US Army Corps of Engineers.
Maki, I., Boulanger, R., DeJong, J. and Jaeger, R. (2014) State-based overburden normalization of cone penetration resistance in clean sand. *Journal of Geotechnical and Geoenvironmental Engineering*, ASCE, 140(2), 04013006.
Maki, I.M. (2012) State normalization of cone penetration resistance. MSc Thesis, University of California at Davis, Davis, CA.
Manzari, M.T. and Dafalias, Y.F. (1997) A critical state two-surface plasticity model for sands. *Géotechnique*, 47, 255–272.
Marcuson, W.F. and Bieganousky, W.A. (1977) Laboratory standard penetration tests on fine sands. *Journal of Geotechnical Engineering Division*, ASCE, 103(GT6), 565–588.
Marcuson, W.F., Hynes, M.E. and Franklin, A.G. (1990) Evaluation and use of residual strength in seismic safety analysis of embankments. *Earthquake Spectra*, 6(3), 529–572.
Marcuson, W.F. and Krinitzsky, E.L. (1976) Dynamic analysis of Fort Peck Dam. Technical Report S-76–1, Soils and Pavements Laboratory, U.S. Army Engineer Waterways Experiment Station, Vicksburg, MS.
Matiotti, R., di Prisco, C. and Nova, R. (1995) Experimental observations on static liquefaction of loose sands. In *Earthquake Geotechnical Engineering* (ed. K. Ishihara), pp. 817–822. Rotterdam, the Netherlands: Balkema.

Matsuoka, H. and Nakai, T. (1974) Stress-deformation and strength characteristics of soil under three different principal stresses. *Transactions of the Japanese Society of Civil Engineer*, 6, 108–109.

MCEER. (2015). State University of New York at Buffalo, Buffalo, NY. http://mceer.buffalo.edu (accessed March 27, 2015).

Menzies, B. (1988) A computer controlled hydraulic triaxial testing system. In *Advanced Triaxial Testing of Soil and Rock*, ASTM STP 977 (eds. R.T. Donaghe, R.C. Chaney and M.L. Silver), pp. 82–94. Philadelphia, PA: American Society for Testing Material.

Meyerhof, G.G. (1984) Safety factors and limit states analysis in geotechnical engineering. *Canadian Geotechnical Journal*, 21(1), 1–7.

Middlebrooks, T.A. (1940) Fort Peck slide. *Transactions of the ASCE*, 107(Paper 2144), 723–764.

Mitchell, J.K., Lodge, A.L., Coutinho, R.Q., Kayen, R.E., Seed, R.B., Nishio, S. and Stokoe, K.H., II (1994) *In Situ Test Results from Four Loma Prieta Earthquake Liquefaction Sites: SPT, CPT, DMT, and Shear Wave Velocity*, UCB/EERC-94/04. Berkeley, CA: Earthquake Engineering Research Center, College of Engineering, University of California Berkeley.

Mohajeri, M. and Ghafghazi, M. (2012) Ground sampling and laboratory testing on a low plasticity clay. In *Proceedings of the 15th World Conference on Earthquake Engineering*, Lisbon, Portugal.

Molenkamp, F. (1981) Elasto-plastic double hardening model Monot. LGM Report, CO0218595, Delft Geotechnics, the Netherlands.

Molenkamp, F. and Luger, H.T. (1981) Modelling and minimization of membrane penetration effects in tests on granular soils. *Géotechnique*, 31(4), 471–486.

Moss, R.E.S. (2003) CPT-based probabilistic assessment of seismic soil liquefaction initiation. PhD Thesis, University of California at Berkeley, Berkeley, CA.

Moss, R.E.S. (2014) A critical state framework for seismic soil liquefaction triggering using *CPT*. In *Proceedings of the Third International Symposium on Cone Penetration Testing*, CPT14, Las Vegas, NV, May 2014, pp. 477–486.

Moss, R.E.S., Seed, R.B., Kayen, R.E., Stewart, J.P., Der Kiureghian, A. and Cetin, K.O. (2006) CPT-based probabilistic and deterministic assessment of in situ seismic soil liquefaction potential. *Journal of Geotechnical and Geoenvironmental Engineering*, ASCE 132(8), 1032–1051.

Mroz, Z. and Norris, V.A. (1982) Elastoplastic and viscoplastic constitutive models for soils with application to cyclic loading. In *Soil Mechanics – Transient and Cyclic Loads* (eds. G.N. Pande and O.C. Zienkiewicz), pp. 343–373. Chichester, U.K.: Wiley.

Muir Wood, D. (1990) *Soil Behaviour and Critical State Soil Mechanics*. London, U.K.: McGraw Hill.

Muir Wood, D., Belkheir, K. and Liu, D.F. (1994) Strain softening and state parameter for sand modelling. *Géotechnique*, 44(2), 335–339.

Muir Wood, D., Drescher, A. and Budhu, M. (1979) On the determination of stress state in the simple shear apparatus. *Geotechnical Testing Journal*, 2(4), 211–221.

Muira, S., Toki, S. and Tanizawa, F. (1984) Cone penetration characteristics and its correlation to static and cyclic deformation-strength behaviours of anisotropic sand. *Soils and Foundations*, 24(2), 58–74.

National Research Council (1985) *Liquefaction of Soils during Earthquakes*. Washington, DC: National Academy Press.

Nayak, G.C. and Zienkiewicz, O.C. (1972) Convenient form of stress invariants for plasticity. *Journal Structural Engineering*, ASCE, 98(ST4), 949–953.

Nazarian, S., Stokoe, K.H., II, and Hudson, W.R. (1983) Use of spectral analysis of surface waves method for determination of moduli and thicknesses of pavement systems. *Transportation Research Record* 930, 38–45.

Negussey, D. and Islam, M.S. (1994) Uniqueness of steady state and liquefaction potential. *Canadian Geotechnical Journal*, 31(1), 132–139.

Negussey, D., Wijewickreme, W.K.D. and Vaid, Y.P. (1988) Constant-volume friction angle of granular materials. *Canadian Geotechnical Journal*, 25(1), 50–55.

Nemat-Nasser, S. and Tobita, Y. (1982) The influence of fabric on liquefaction and densification potential of cohesionless sand. *Mechanics of Materials*, 1(1), 43–62.

Norwegian Geotechnical Institute (1977) Undrained simple shear tests on Oosterschelde sand. Contract Report 77302-3, Oslo, Norway.

Nova, R. (1982) A constitutive model under monotonic and cyclic loading. In *Soil Mechanics – Transient and Cyclic Loads* (eds. G.N. Pande and O.C. Zienkiewicz), pp. 343–373. Chichester, U.K.: Wiley.
Oda, M. (1972a) Initial fabrics and their relations to mechanical properties of granular materials. *Soils and Foundations*, 12(1), 17–36.
Oda, M. (1972b) The mechanism of fabric changes during compressional deformation of sand. *Soils and Foundations*, 12(2), 1–18.
Oda, M. and Kazama, H. (1998) Microstructure of shear bands and its relation to the mechanisms of dilatancy and failure of dense granular soils. *Géotechnique*, 48(4), 465–481.
Oda, M., Konishi, J. and Nemat-Nasser, S. (1980) Some experimentally based fundamental results on the mechanical behaviour of granular materials. *Géotechnique*, 30(4), 479–495.
Olsen, R.S. (1984) Liquefaction analysis using the cone penetrometer test (CPT). In *Proceedings of the Eighth World Conference on Earthquake Engineering*, San Francisco, CA, Vol. 3, pp. 247–254.
Olsen, R.S. (1988) Using the CPT for dynamic response characterization. In *Proceedings of the Earthquake Engineering and Soil Dynamics II Conference*, ASCE, New York, pp. 111–117.
Olsen, R.S. and Koester, J.P. (1995) Prediction of liquefaction resistance using the CPT. In *Proceedings of International Symposium on Cone Penetration Testing, CPT'95*, Linköping, Sweden, Vol. 2, pp. 251–256.
Olsen, R.S. and Malone, P.G. (1988) Soil classification and site characterization using the cone penetrometer test. In *Penetration Testing 1988*, ISOPT 1 (ed. J. De Ruiter), Vol. 2, pp. 887–893. Rotterdam, the Netherlands: Balkema.
Olson, S.M. (2001) Liquefaction analysis of level and sloping ground using field case histories and penetration resistance. PhD Thesis, University of Illinois at Urbana-Champaign, Champaign, IL.
Olson, S.M. (2006) Liquefaction analysis of Duncan Dam using strength ratios. *Canadian Geotechnical Journal*, 43(5), 484–499.
Olson, S.M. and Stark, T.D. (2001) Liquefaction analysis of Lower San Fernando Dam using strength ratios. In *Proceedings of the Fourth International Conference on Geotechnical Earthquake Engineering and Soil Dynamics*, San Diego, CA, ed. S. Prakash, Paper 4.05.
Olson, S.M. and Stark, T.D. (2002) Liquefied strength ratio from liquefaction flow failure case histories. *Canadian Geotechnical Journal*, 39, 629–647.
Olson, S.M., Stark, T.D., Walton, W.H. and Castro, G. (2000) 1907 Static liquefaction flow failure of the north dike of Wachusett Dam. *Journal of Geotechnical and Geoenvironmental Engineering*, 26(GE12), 1184–1193.
Olszak, W. and Perzyna, P. (1964) On elastic/viscoplastic soils. In *Proceedings of the IUTAM Symposium: Rheology and Soil Mechanics*, pp. 47–57. Grenoble, France: Springer.
Onisiphorou, C. (2000) Stochastic analysis of saturated soils using finite elements. PhD Thesis, University of Manchester, Manchester, U.K.
O'Shaughnessy, M.M. (1925) Letter to editor. *Engineering News Record*, 9 July 1925.
Palmer, A.C. (1967) Stress-strain relations for clays: An energy theory. *Géotechnique*, 17, 348–358.
Park, C.B., Miller, R.D. and Xia, J. (1999) Multi-channel analysis of surface waves (MASW). *Geophysics*, 64(3), 800–808.
Parkin, A., Holden, J., Aamot, K., Last, N. and Lunne, T. (1980) Laboratory investigations of CPTs in sand. NGI Report S2108–9, 9 October 1980.
Parkin, A. and Lunne, T. (1982) Boundary effects in the laboratory calibration of a cone penetrometer in sand. In *Proceedings of the Second European Symposium on Penetration Testing*, Amsterdam, the Netherlands, Vol. 2, pp. 761–768.
Parry, R.H.G. (1958) On the yielding of soils: Correspondence. *Géotechnique*, 8(4), 183–186.
Perzyna, P. (1966) Fundamental problems in viscoplasticity. *Advances in Applied Mechanics*, 9, 243–368.
Pestana, J.M. and Whittle, A.J. (1995) Compression model for cohesionless soils. *Géotechnique*, 45, 611–631.
Pillai, V.S. and Salgado, F.M. (1994) Post-liquefaction stability and deformation analysis of Duncan Dam. *Canadian Geotechnical Journal*, 31(6), 967–978.
Pillai, V.S. and Stewart, R.A. (1994) Evaluation of liquefaction potential of foundation soils at Duncan Dam. *Canadian Geotechnical Journal*, 31(6), 951–966.

Pitman, T.D., Robertson, P.K. and Sego, D.C. (1994) Influence of fines on the collapse of loose sands. *Canadian Geotechnical Journal*, 31(5), 728–739.

Plewes, H.D., Davies, M.P. and Jefferies, M.G. (1992) CPT based screening procedure for evaluating liquefaction susceptibility. In *Proceedings of the 45th Canadian Geotechnical Conference*, Toronto, Canada.

Plewes, H.D., Pillai, V.S., Morgan, M.R. and Kilpatrick, B.L. (1994) In situ sampling, density measurements, and testing of foundation soils at Duncan Dam. *Canadian Geotechnical Journal*, 31(6), 927–938.

Poorooshasb, H.B., Holubec, I. and Sherbourne, A.N. (1966) Yielding and flow of sand in triaxial compression Part I. *Canadian Geotechnical Journal*, 3, 179–190.

Poorooshasb, H.B., Holubec, I. and Sherbourne, A.N. (1967) Yielding and flow of sand in triaxial compression Parts II and III. *Canadian Geotechnical Journal*, 4, 376–397.

Popescu, R. (1995) Stochastic variability of soil properties: Data analysis, digital simulation, effects on system behaviour. PhD Thesis, Princeton University, Princeton, NJ.

Popescu, R., Prevost, J.H. and Deodatis, G. (1997) Effects of spatial variability on soil liquefaction: Some design recommendations. *Géotechnique*, 47, 1019–1036.

Potts, D.M., Dounias, G.T. and Vaughan, P.R. (1987) Finite element analysis of the direct shear box. *Géotechnique*, 37(1), 11–23.

Poulos, S.J. (1981) The steady state of deformation. *Journal of the Geotechnical Engineering Division*, ASCE, 107(5), 553–562.

Poulos, S.J. (1988) Liquefaction and related phenomena. In *Advanced Dam Engineering for Design Construction and Rehabilitation* (ed. R.B. Jansen), pp. 292–320. Reinhold, NY: Van Nostrand.

Poulos, S.J., Castro, G. and France, J.W. (1985) Liquefaction evaluation procedure. *Journal of Geotechnical Engineering Division*, ASCE, 111(GT6), 772–792.

Poulos, S.J., Castro, G. and France, J.W. (1988) Liquefaction evaluation procedure: Closure to discussion. *Journal of Geotechnical Engineering*, 114(2), 251–259.

Pournaghiazar, M., Russell, A.R. and Khalili, N. (2011) Development of a new calibration chamber for conducting cone penetration tests in unsaturated soils. *Canadian Geotechnical Journal*, 48(2), 314–321.

Pournaghiazar, M., Russell, A.R. and Khalili, N. (2012) Linking cone penetration resistances measured in calibration chambers and the field. *Géotechnique Letters*, 2, 29–35. http://dx.doi.org/10.1680/geolett.11.00040.

Rad, N.S. and Tumay, M.T. (1987) Factors affecting sand specimen preparation by raining. *Geotechnical Testing Journal*, 10(1), 31–37.

Reades, D.W. (1971) Stress-strain characteristics of sand under three dimensional loading. PhD Thesis, University of London, London, U.K.

Reid, D. (2012) Update on the Plewes method for liquefaction screening. In *Proceedings, Tailings and Mine Waste 2012*, Keystone, CO, pp. 337–345.

Reynolds, O. (1885) On the dilatancy of media composed of rigid particles in contact, with experimental illustrations. *Philosophical Magazine*, 20, 469–481.

Robertson, P.K. (1990) Soil classification using the cone penetration test. *Canadian Geotechnical Journal*, 27(1), 151–158.

Robertson, P.K. (2008) Discussion: Liquefaction potential of silts from CPTu. *Canadian Geotechnical Journal*, 45(1), 140–141.

Robertson, P.K. (2010) Evaluation of flow liquefaction and liquefied strength using the cone penetration test. *Journal of Geotechnical and Geoenvironmental Engineering*, ASCE, 136(6), 842–853.

Robertson, P.K. (2012) Evaluating flow (static) liquefaction using the CPT: An update. In *Proceedings, Tailings and Mine Waste '12*, Keystone, CO, 14–17 October 2012.

Robertson, P.K. and Campanella, R.G. (1983) Interpretation of cone penetration tests. Part I: Sand. *Canadian Geotechnical Journal*, 20(4), 718–733.

Robertson, P.K. and Campanella, R.G. (1985) Liquefaction potential of sands using the cone penetration test. *Journal of Geotechnical Engineering Division*, ASCE, 22(GT3), 298–307.

Robertson, P.K., Campanella, R.G., Gillespie, D. and Rice, A. (1986) Seismic CPT to measure in situ shear wave velocity. *Journal of Geotechnical Engineering*, ASCE, 112(GT8), 791–803.

Robertson, P.K., Campanella, R.G. and Wightman, A. (1983) SPT-CPT correlations. *Journal of Geotechnical Engineering*, ASCE, 109(GT11), 1449–1459.

Robertson, P.K., Woeller, D.J. and Addo, K.O. (1992) Standard penetration test energy measurements using a system based on the personal computer. *Canadian Geotechnical Journal*, 29(4), 551–557.

Robertson, P.K. and Wride, C.E. (1998) Evaluating cyclic liquefaction potential using the cone penetration test. *Canadian Geotechnical Journal*, 35(3), 442–459.

Robertson, P.K., Wride, C.E., List, B.R., Atukorala, U., Biggar, K.W., Byrne, P.M., Campanella, R.G. et al. (2000) The CANLEX project: Summary and conclusions. *Canadian Geotechnical Journal*, 37(3), 563–591.

Robinson, K.E. (1977) Tailings dam constructed on very loose saturated sandy silt. *Canadian Geotechnical Journal*, 14, 399–407.

Rogers, B.T., Been, K., Hardy, M.D., Johnson, G.J. and Hachey, J.E. (1990) Re-analysis of Nerlerk B-67 berm failures. In *Proceedings of the 43rd Canadian Geotechnical Conference*, Quebec City, Canada, Vol. 1, pp. 227–237.

Roscoe, K. (1953) An apparatus for the application of simple shear to soil samples. In *Proceedings of the Third International Conference on Soil Mechanics and Foundation Engineering*, London, U.K., Vol. 1, pp. 186–191.

Roscoe, K., Schofield, A.N. and Wroth, C.P (1958) On the yielding of soils. *Géotechnique*, 8(1), 22–53.

Roscoe, K.H. and Burland, J.B. (1968) On the generalized stress-strain behaviour of 'wet' clay. In *Engineering Plasticity* (eds. J. Heyman and F.A. Leckie), pp. 535–609. Cambridge, U.K.: Cambridge University Press.

Roscoe, K.H., Schofield, A.N. and Thurairajah, A. (1963) Yielding of clays in states wetter than critical. *Géotechnique*, 13, 211–240.

Rothenburg, L. and Bathurst, R.J. (1989) Analytical study of induced anisotropy in idealized granular materials. *Géotechnique*, 39(4), 601–614.

Rothenburg, L. and Bathurst, R.J. (1992) Micromechanical features of granular assemblies with planar elliptical particles. *Géotechnique*, 42(1), 79–95.

Rowe, P.W. (1962) The stress dilatancy relation for static equilibrium of an assembly of particles in contact. *Proceedings of the Royal Society of London A*, 269, 500–527.

Rowe, P.W. and Barden, L. (1964) Importance of free ends in triaxial testing. *Journal of the Soil Mechanics and Foundations Division*, ASCE, 90(1), 1–28.

Rowe, P.W. and Craig, W.H. (1976) Studies of offshore caissons founded on Oosterschelde sand. In *Design and Construction of Offshore Structures* (eds. J.P. Blanc and Mary Monro), pp. 49–55. London, U.K.: Institution of Civil Engineers.

Russell, A.R. and Khalili, N. (2004) A bounding surface plasticity model for sands exhibiting particle crushing. *Canadian Geotechnical Journal*, 41, 6, 1179–1192.

Rutherford and Chekene (1988) Geotechnical investigation: Moss landing facility (technology building), Monterey Bay Aquarium Research Institute. Report prepared for Monterey Bay Aquarium Research Institute, San Francisco, CA.

Rutherford and Chekene (1993) Geologic hazards evaluation/geotechnical investigation: Monterey Bay Aquarium Research Institute Buildings 3 and 4. Report prepared for Monterey Bay Aquarium Research Institute, San Francisco, CA.

Saada, A.S. (1987) *Proceedings International Workshop of Constitutive Equations for Granular Non-Cohesive Soils*, Case Western Reserve University, Cleveland, OH.

Salgado, R., Mitchell, J.K. and Jamiolkowski, M. (1997) Cavity expansion and penetration resistance in sand. *Journal of Geotechnical and Geoenvironmental Engineering*, 123(4), 344–354.

Salgado, R., Mitchell, J.K. and Jamiolkowski, M. (1998) Calibration chamber size effects on penetration resistance in sand. *Journal of Geotechnical and Geoenvironmental Engineering*, 124(9), 878–888.

Sangrey, D.A., Castro, G., Poulos, S.J. and France, J.W. (1978) Cyclic loading of sands, silts and clays. In *Proceedings of the ASCE Specialty Conference on Earthquake Engineering and Soil Dynamics*, Pasadena, CA, Vol. 2, pp. 836–851.

Sasitharan, S., Robertson, P.K., Sego, D.C. and Morgenstern, N.R. (1993) Collapse behaviour of sand. *Canadian Geotechnical Journal*, 30(4), 569–577.

Schanz, T. and Vermeer, P.A. (1996) Angles of friction and dilatancy of sand. *Géotechnique* 46, 145–151.

Schnaid, F. and Houlsby, G.T. (1991) An assessment of chamber size effects in the calibration of in situ tests in sand. *Géotechnique*, 41(3), 437–445.

Schofield, A. and Wroth, C.P. (1968) *Critical State Soil Mechanics*. London, U.K.: McGraw-Hill.

Scott, R.F. (1987) Twenty seventh Rankine Lecture: Failure. *Géotechnique*, 37, 423–466.

Seed, H.B. (1983) Earthquake-resistant design of earth dams. In *Proceedings Symposium on Seismic Design of Embankments and Caverns*, ASCE, New York, ed. T.R. Howard, pp. 41–64.

Seed, H.B. (1987) Design problems in soil liquefaction. *Journal of Geotechnical Engineering*, ASCE, 113(GT8), 827–845.

Seed, H.B. and De Alba, P. (1986) Use of SPT and CPT test for evaluating the liquefaction resistance of sands. In *Use of In Situ Tests in Geotechnical Engineering*, ASCE Geotechnical Special Publication (ed. S.P. Clemence), pp. 281–302.

Seed, H.B. and Idriss, I.M. (1982) *Ground Motions and Soil Liquefaction during Earthquakes*, Earthquake Engineering Research Institute Monograph. Oakland, CA: Earthquake Engineering Research Institute.

Seed, H.B., Idriss, I.M. and Arango, I. (1983) Evaluation of liquefaction potential using field performance data. *Journal of Geotechnical Engineering Division*, ASCE, 109(GT3), 458–482.

Seed, H.B. and Lee, K.L. (1966) Liquefaction of saturated sands during cyclic loading. *Journal of Soil Mechanics and Foundation Engineering*, ASCE, 92(SM6), 105–134.

Seed, H.B., Lee, K.L. and Idriss, I.M. (1969) Analysis of Sheffield Dam failure. *Journal of Soil Mechanics and Foundations*, ASCE, 95(SM6), 1453–1490.

Seed, H.B. and Peacock, W.H. (1971) Test procedure for measuring soil liquefaction characteristics. *Journal of Soil Mechanics and Foundation Engineering*, ASCE, 97(SM8), 1099–1119.

Seed, H.B., Seed, R.B., Harder, L.F. and Jong, H.-L. (1988) Re-evaluation of the slide in the lower San Fernando dam in the earthquake of February 9, 1971. Report UCB/EERC-88/04, Earthquake Engineering Research Centre, University of California at Berkeley, Berkeley, CA.

Seed, R.B. and Harder, L.F. (1990) SPT-based analysis of cyclic pore pressure generation and undrained residual strength. In *Proceedings of the H.B. Seed Memorial Symposium*, Berkeley, CA, Vol. 2, pp. 351–376.

Sego, D.C., Robertson, P.K., Sasitharan, S., Kilpatrick, B.L. and Pillai, V.S. (1994) Ground freezing and sampling of foundation soils at Duncan Dam. *Canadian Geotechnical Journal*, 31(6), 939–950.

Shen, C.K. and Lee, K.-M. (1995) A study of hydraulic fill performance in Hong Kong. GEO Report No. 40, Report to Geotechnical Engineering Office of the Hong Kong Government, by the Hong Kong University of Science and Technology, Hong Kong.

Shibata, T. and Teparaska, W. (1988) Evaluation of liquefaction potentials of soils using cone penetration tests. *Soils and Foundations*, 28(2), 49–60.

Shozen, T. (1991) Deformation under the constant stress state and its effect on stress-strain behaviour of Fraser River Sand. MASc Thesis, Department of Civil Engineering, University of British Columbia, Vancouver, British Columbia, Canada.

Shuttle, D.A. (1988) Numerical modelling of localization in soils. PhD Thesis, University of Manchester, Manchester, U.K.

Shuttle, D.A. (2006) Can the effect of sand fabric on plastic hardening be determined using a self-bored pressuremeter? *Canadian Geotechnical Journal*, 43(7), 659–673.

Shuttle, D.A. (2008) Importance of small strain response to prediction of large scale behavior in sand. *Keynote Presentation at the Fourth International Symposium on Deformation Characteristics of Geomaterials (IS Atlanta 2008)*, Atlanta, GA, 22–24 September 2008.

Shuttle, D.A. and Cunning, J. (2007) Liquefaction potential of silts from CPTu. *Canadian Geotechnical Journal*, 44(1), 1–19.

Shuttle, D.A. and Cunning, J. (2008) Reply to discussion: Liquefaction potential of silts from CPTu. *Canadian Geotechnical Journal*, 45(1), 142–145.

Shuttle, D.A. and Jefferies, M.G. (1998) Dimensionless and unbiased CPT interpretation in sand. *International Journal of Numerical and Analytical Methods in Geomechanics*, 22, 351–391.

Sigmundstad, R. (2015) http://www.fortpeckdam.com (accessed March 27, 2015).

Sills, G.C., Nyirenda, Z., May, R.E. and Henderson, T. (1988) Piezocone measurements with four pressure positions. In *Proceedings of the Conference on Penetration Testing in the U.K.*, Birmingham, U.K. London, U.K.: Thomas Telford.

Silver, M.L., Chan, C.K., Ladd, R.S., Lee, K.L., Tiedemann, D.A., Townsend, F.C., Valera, J.E. and Wilson, J.H. (1976) Cyclic triaxial strength of standard test sand. *Journal of Geotechnical Engineering Division*, ASCE, 102(GT5), 511–523.

Simpson, B. and Driscoll, R. (1998) Eurocode 7: A commentary. Building Research Establishment, Watford, U.K.

Skempton, A.W. (1954) The pore-pressure coefficients A and B. *Géotechnique*, 4(4), 143–147.

Skopek, P., Morgenstern, N.R., Robertson, P.K. and Sego, D.C. (1994) Collapse of dry sand. *Canadian Geotechnical Journal*, 31(6), 1008–1014.

Sladen, J.A. (1989a) Problems with interpretation of sand state from cone penetration test. *Géotechnique*, 39(2), 323–332.

Sladen, J.A. (1989b) Cone penetration test calibration for Erksak sand: Discussion. *Canadian Geotechnical Journal*, 26(1), 173–177.

Sladen, J.A. (2001) Undrained shear strength of clean sands to trigger flow liquefaction: Discussion. *Canadian Geotechnical Journal*, 38(3), 652–653.

Sladen, J.A., D'Hollander, R.D. and Krahn, J. (1985b) The liquefaction of sands, a collapse surface approach. *Canadian Geotechnical Journal*, 22(4), 564–578.

Sladen, J.A., D'Hollander, R.D., Krahn, J. and Mitchell, D.E. (1985a) Back analysis of the Nerlerk Berm liquefaction slides. *Canadian Geotechnical Journal*, 22(4), 579–588.

Sladen, J.A., D'Hollander, R.D., Krahn, J. and Mitchell, D.E. (1987) Back analysis of the Nerlerk Berm liquefaction slides: Reply. *Canadian Geotechnical Journal*, 24(1), 179–185.

Sladen, J.A. and Handford, G. (1987) A potential systematic error in laboratory testing of very loose sands. *Canadian Geotechnical Journal*, 24(3), 462–466.

Smith, I.M. and Griffiths, D.V. (1988, 1998) *Programming the Finite Element Method*, 2nd and 3rd edns. Chichester, U.K./New York: Wiley. ISBN: 0-471-96542-1.

Smith, I.M., Griffiths, D.V. and Margetts, L. (2013) *Programming the Finite Element Method*, 5th edn. Wiley. ISBN: 978-1-119-97334-8.

Sriskandakumar, S. (2004) Cyclic loading response of Fraser river sand for validation of numerical models simulating centrifuge tests. MASc Thesis, Department of Civil Engineering, University of British Columbia, Vancouver, British Columbia, Canada.

Stark, T.D. and Mesri, G.M. (1992) Undrained shear strength of liquefied sands for stability analysis. *Journal of Geotechnical Engineering*, ASCE, 118(GT11), 1727–1747.

Stark, T.D. and Olson, S.M. (1995) Liquefaction resistance using CPT and field case histories. *Journal of Geotechnical Engineering*, ASCE, 121(12), 856–869.

Stewart, H.R., Jefferies, M.G. and Goldby, H.M. (1983) Berm construction for the Gulf Canada Mobile Arctic Caisson. In *Proceedings of the 15th Offshore Technology Conference*, Houston, TX, Paper OTC 4552.

Stokoe, K.H., II, Wright, G.W., Bay, J.A. and Roesset, J.M. (1994) Characterization of geotechnical sites by SASW method. In *Geophysical Characterization of Sites* (ed. R.D. Woods). New Delhi: Oxford Publishers.

Suzuki, Y., Tokimatsu, K., Koyamada, K., Taya, Y. and Kubota, Y. (1995) Field correlation of soil liquefaction based on CPT data. In *Proceedings of the International Symposium on Cone Penetration Testing, CPT'95*, Lingköping, Sweden, Vol. 2, pp. 583–588.

Take, W.A. and Beddoe, R.A. (2014) Base liquefaction: A mechanism for shear-induced failure of loose granular slopes. *Canadian Geotechnical Journal*, 51(5), 496–507.

Tatsuoka, F. (1987) Strength and dilatancy of sands: Discussion. *Géotechnique*, 37(2), 219–225.

Tatsuoka, F. and Ishihara, K. (1974) Yielding of sand in triaxial compression. *Soils and Foundations*, 14, 63–76.

Tatsuoka, F., Ochi, K., Fujii, S. and Okamoto, M. (1986) Cyclic undrained triaxial and torsional shear strength of sands for different sample preparation methods. *Soils and Foundations*, 26(3), 23–41.

Taylor, D.W. (1948) *Fundamentals of Soil Mechanics*. New York: John Wiley.

Terzaghi, K. and Peck, R.B. (1948) *Soil Mechanics in Engineering Practice*. New York: Wiley and Sons.

Thevanayagam, S., Shenthan, T., Mohan, S. and Liang, J. (2002) Undrained fragility of clean sands, silty sands, and sandy silts. *Journal of Geotechnical and Geoenvironmental Engineering*, ASCE, 128(10), 849–859.

Toki, S., Tatsuoka, F., Miura, S., Yoshimi, Y., Yasuda, S. and Makihara, Y. (1986) Cyclic undrained triaxial strength of sand by a cooperative test program. *Soils and Foundations*, 26(3), 117–128.

Townsend, F.C. (1978) A review of factors affecting cyclic triaxial tests. In *Dynamic Geotechnical Testing* (eds. M.L. Silver and D. Tiedemann), ASTM STP 654, pp. 356–383. Baltimore, MD: ASTM.

Tresca, H.E. (1864) Sur l'écoulement des corps solides soumis á de fortes pressions. *Comptes Rendus de l'Académie des Sciences (Paris)*, 59, 754.

Tringale, P.T. (1983) Soil identification in situ using an acoustic cone penetrometer. PhD Thesis, University of California at Berkeley, Berkeley, CA.

U.S. Army Corp of Engineers (1939) Report on the slide of a portion of the upstream face at Fort Peck Dam. U.S. Government Printing Office, Washington, DC.

U.S. Army Corp of Engineers (1949) Report on investigation of failure of Sheffield Dam, Santa Barbara. Los Angeles, CA.

Vaid, Y.P., Chung, E.K.F. and Kuerbis, R.H. (1990) Stress path and steady state. *Canadian Geotechnical Journal*, 27(1), 1–7.

Vaid, Y.P. and Eliadorani, A. (1998) Instability and liquefaction of granular soils under undrained and partially drained states. *Canadian Geotechnical Journal*, 35(6), 1053–1062.

Vaid, Y.P. and Negussey, D. (1982) *A Critical Assessment of Membrane Penetration in the Triaxial Test*, Soil Mechanics Series No. 61. Vancouver, Canada: University of British Columbia.

Vaid, Y.P. and Sasitharan, S. (1992) The strength and dilatancy of sand. *Canadian Geotechnical Journal*, 29, 522–526.

Vaid, Y.P., Sayao, A., Hou, E. and Negussey, D. (1990b) Generalized stress-path dependent soil behaviour with a new hollow cylinder torsional apparatus. *Canadian Geotechnical Journal*, 27(5), 601–616.

Vaid, Y.P. and Sivathalayan, S. (1996a) Static and cyclic liquefaction potential of Fraser Delta sand in simple shear and triaxial tests. *Canadian Geotechnical Journal*, 33(2), 281–289.

Vaid, Y.P. and Sivathalayan, S. (1996b) Errors in estimates of void ratio of laboratory sand specimens. *Canadian Geotechnical Journal*, 33(6), 1017–1020.

Vaid, Y.P. and Sivathayalan, S. (1998) Fundamental factors affecting liquefaction susceptibility of sand. In *Proceedings of the International Workshop on the Physics and Mechanics of Soil Liquefaction*, 10–11 September 1998, John Hopkins University, Baltimore, MD, ed. P.V. Lade and J.A. Yamamuro, pp. 105–120.

Vaid, Y.P., Stedman, J.D. and Sivathayalan, S. (2001) Confining stress and static shear effects in cyclic liquefaction. *Canadian Geotechnical Journal*, 38, 580–591.

Vaid, Y.P. and Thomas, J. (1995) Liquefaction and post liquefaction behaviour of sand. *Journal of Geotechnical and Geoenvironmental Engineering*, 121(2), 163–173.

Van den Berg, P. (1994) *Analysis of Soil Penetration*. Delft, the Netherlands: Delft University Press.

Van Eekelen, H.A.M. (1977) Single-parameter models for progressive weakening of soils by cyclic loading. *Géotechnique*, 27(3), 357–368.

Van Eekelen, H.A.M. and Potts, D.M. (1978) The behaviour of Drammen Clay under cyclic loading. *Géotechnique*, 28(2), 173–196.

Vasquez-Herrera, A. and Dobry, R. (1988) *The Behavior of Undrained Contractive Sand and Its Effect on Seismic Liquefaction Flow Failures of Earth Structures*. Troy, NY: Rensselaer Polytechnic Institute.

Verdugo, R. (1992) The critical state of sands: Discussion. *Géotechnique*, 42(4), 655–663.

Vermeer, P.A. (1978) A double hardening model for sand. *Géotechnique*, 28(4), 413–433.

Vesic, A.S. (1972) Expansion of cavities in infinite soil mass. *Journal of the Soil Mechanics and Foundations*, ASCE, 98(SM3), 265–290.

Villet, W.C.B. (1981) Acoustic emissions during the static penetration of soils. PhD Thesis, University of California at Berkeley, Berkeley, CA.

Vreugdenhil, R., Davis, R. and Berrill, J.R. (1994) Interpretation of cone penetration results in multi-layered soils. *International Journal for Numerical and Analytical Methods in Geomechanics*, 18, 585–599.

Wagener, F., Craig, H.J., Blight, G.E., McPhail, G., Williams, A.B. and Strydom, J.H. (1998) The Merriespruit tailings dam failure – A review. In *Proceedings of the Conference on Tailings and Mines Waste'98*, Colorado State University, Fort Collins, CO, pp. 925–952.

Wan, R.G. and Guo, P.J. (1998) A simple constitutive model for granular soils: Modified stress-dilatancy approach. *Computers and Geotechnics*, 22, 109–133.

Wanatowski, D. and Chu, J. (2006) Stress-strain behavior of a granular fill measured by a new plane-strain apparatus. *Geotechnical Testing Journal*, 29(2), 149–157.

Wanatowski, D. and Chu, J. (2007) Static liquefaction of sand in plane strain. *Canadian Geotechnical Journal*, 44(3), 299–313.

Wanatowski, D. and Chu, J. (2012) Factors affecting pre-failure instability of sand under plane-strain conditions. *Géotechnique*, 62(2), 121–135.

Welsh Office (1969) *A Selection of Technical Reports Submitted to the Aberfan Tribunal.* London, U.K.: Her Majesty's Stationary Office.

Wesley, L.D. (2002) Interpretation of calibration chamber tests involving cone penetrometers in sands. *Géotechnique*, 52(4), 289–293.

Wijewickreme, D. and Vaid, Y.P. (1991) Stress non-uniformities in hollow cylinder torsional specimens. *Geotechnical Testing Journal*, ASTM, 14(4), 349–362.

Willis, B. (1925) A study of the Santa Barbara Earthquake of June 29, 1925. *Bulletin Siesmological Society America*, 15(4), 255–278.

Wong, R.K.S. and Arthur, J.R.F. (1986) Sand sheared by stresses with cyclic variation in direction. *Géotechnique*, 36(2), 215–226.

Wood, C.C. (1958) Shear strength and volume change characteristics of compacted soil under conditions of plane strain. PhD Thesis, University of London, London, U.K.

Wride, C.E., Hofmann, B.A., Sego, D.C., Plewes, H.D., Konrad, J.-M., Biggar, K.W., Robertson, P.K. and Monahan, P.A. (2000) Ground sampling program at CANLEX test sites. *Canadian Geotechnical Journal*, 37(3), 530–542.

Wride, C.E., McRoberts, E.C. and Robertson, P.K. (1999) Reconsideration of case histories for estimating undrained shear strength in sandy soils. *Canadian Geotechnical Journal*, 36(5), 907–933.

Wroth, C.P. (1975) In situ measurement of initial stresses and deformation characteristics. In *Proceedings, In Situ Stress Measurement of Soil Properties*, North Carolina State University, Geotechnical Engineering Division, Raleigh, NC, pp. 181–230.

Wroth, C.P. (1984) Twenty fourth Rankine Lecture: The interpretation of in situ soil tests. *Géotechnique*, 34(4), 449–489.

Wroth, C.P. (1988) Penetration testing – A more rigorous approach to interpretation. In *Proceedings of the First International Symposium on Penetration Testing*, Orlando, FL, Vol. 1, pp. 303–311.

Yamamuro, J.A. and Lade, P.V. (1997) Static liquefaction of very loose sands. *Canadian Geotechnical Journal*, 34(6), 905–917.

Yamamuro, J.A. and Lade, P.V. (1998) Steady-state concepts and static liquefaction of silty sands. *Journal of Geotechnical and Geoenvironmental Engineering*, ASCE, 124(9), 868–877.

Yang, J. (2002) Non-uniqueness of flow liquefaction line for loose sand. *Géotechnique*, 52(10), 757–760.

Yang, S., Sandven, R. and Grande, L. (2006) Steady-state lines of sand–silt mixtures. *Canadian Geotechnical Journal*, 43(11), 1213–1219.

Yoshimine, M., Robertson, P.K. and Wride, C.E. (1999) Undrained shear strength of clean sands to trigger flow liquefaction. *Canadian Geotechnical Journal*, 36(5), 891–906.

Yoshimine, M., Robertson, P.K. and Wride, C.E. (2001) Undrained shear strength of clean sands to trigger flow liquefaction: Reply to Discussion. *Canadian Geotechnical Journal*, 38(3), 654–657.

Youd, T.L. (1972) Compaction of sands by repeated shear straining. *Journal of the Soil Mechanics and Foundations*, ASCE, 98(SM7), 709–725.

Youd, T.L. and Craven, T.N. (1975) Lateral stress in sands during cyclic loading. *Journal of the Geotechnical Engineering Division*, ASCE, 101(GT2), 217–221.

Youd, T.L. and Holzer, T.L. (1994) Piezometer performance at Wildlife liquefaction site, California. *Journal of Geotechnical Engineering*, ASCE, 120(GT6), 975–995.

Youd, T.L., Idriss, I.M., Andrus, R.D., Arango, I., Castro, G., Christian, J.T., Dobry, R. et al. (2001) Liquefaction resistance of soils: Summary report from the 1996 NCEER and 1998 NCEER/NSF Workshops on evaluation of liquefaction resistance of soils. *Journal of Geotechnical and Geoenvironmental Engineering*, ASCE, 127(10), 817–833.

Youd, T.L. and Noble, S.K. (1997) Magnitude scaling factors. In *Proceedings of the NCEER Workshop on Evaluation of Liquefaction Resistance of Soils*, National Center for Earthquake Engineering Research, State University of New York at Buffalo, Buffalo, NY, pp. 149–165.

Youd, T.L. and Wieczorek, G.F. (1984) Liquefaction during the 1981 and previous earthquakes near Westmorland, CA. Open-File Report 84–680, U.S.G.S., Menlo Park, CA.

Yu, H.S. (1996) Interpretation of pressuremeter unloading tests in sand. *Géotechnique*, 46(1), 17–32.

Yu, H.S. (1998) CASM: A unified state parameter model for clay and sand. *International Journal for Numerical and Analytical Methods in Geomechanics*, 22, 621–653.

Yu, H.S. and Houlsby, G.T. (1991) Finite cavity expansion in dilatant soils: Loading analysis. *Géotechnique*, 41(2), 173–183.

Yu, H.S. and Houlsby, G.T. (1992) Finite cavity expansion in dilatant soils: Reply to Discussion. *Géotechnique*, 42(4), 649–654.

Zeghal, M. and Elgamal, A.-W. (1994) Analysis of site liquefaction using earthquake records. *Journal of Geotechnical Engineering*, ASCE, 120(6), 996–1017.

Zhu, F. and Clark, J.I. (1994) The effect of dynamic loading on lateral stress in sand. *Canadian Geotechnical Journal*, 31(2), 308–311.

Zienkiewicz, O.C. and Cormeau, I.C. (1974) Viscoplasticity, plasticity and creep in elastic solids: A unified numerical solution approach. *International Journal for Numerical Methods in Engineering*, 8(4), 821–845.

Zienkiewicz, O.C., Humpheson, C. and Lewis, R.W. (1975) Associated and non-associated viscoplasticity and plasticity in soil mechanics. *Géotechnique*, 25(4), 671–689.

Zienkiewicz, O.C. and Naylor, D.J. (1971) The adaptation of critical state soil mechanics theory for use in finite elements. In *Stress–Strain Behaviour of Soils: Proceedings of the Roscoe Memorial Symposium, Cambridge* (ed. R.H.G. Parry), pp. 537–547. London, U.K.: Foulis.

Zienkiewicz, O.C., Valliappan, S. and King, I.P. (1969) Elasto-plastic solutions of engineering problems; initial stress finite element approach. *International Journal for Numerical Methods in Engineering*, 1(1), 75–100.

Index

D

E

I

J

K

L

M

N

R

S

T

U

V

W

Y